D1340256

MULTIVARIATE DATA ANALYSIS

MULTIVARIATE DATA ANALYSIS

Fifth Edition

JOSEPH F. HAIR, JR.

Louisiana State University

ROLPH E. ANDERSON

Drexel University

RONALD L. TATHAM

Burke Marketing Research

WILLIAM C. BLACK

Louisiana State University

PRENTICE-HALL INTERNATIONAL, INC.

Acquisitions Editor: Whitney Blake
Assistant Editor: John Larkin
Editorial Assistant: Rachel Falk
Vice President/Editorial Director: James Boyd
Marketing Manager: John Chillingworth
Marketing Director: Brian Kibby
Production Editor: Aileen Mason
Production Coordinator: Carol Samet
Managing Editor: Dee Josephson
Associate Managing Editor: Linda DeLorenzo
Manufacturing Supervisor: Arnold Vila
Manufacturing Manager: Vincent Scelta
Design Manager: Pat Smythe
Interior Design: Brian Deep
Cover Design: Jayne Conte
Illustrator (Interior): ElectraGraphics, Inc.
Composition: York Graphic Service, Inc.
Cover Photo: William Whitehurst/The Stock Market

This edition may be sold only in those countries to which it is consigned by
Prentice-Hall International. It is not to be re-exported, and it is not for sale in the
U.S.A., Mexico, or Canada.

ISBN: 0-13-930587-4

PRENTICE-HALL INTERNATIONAL (UK) LIMITED, *LONDON*
PRENTICE-HALL OF AUSTRALIA PTY. LIMITED, *SYDNEY*
PRENTICE-HALL CANADA INC., *TORONTO*
PRENTICE-HALL HISPANOAMERICANA, S.A., *MEXICO*
PRENTICE-HALL OF INDIA PRIVATE LIMITED, *NEW DELHI*
PRENTICE-HALL OF JAPAN, INC., *TOKYO*
PEARSON EDUCATION ASIA PTE. LTD., *SINGAPORE*
EDITORA PRENTICE-HALL DO BRASIL, LTDA., *RIO DE JANEIRO*

Printed in the United States of America

10 9 8 7 6 5

Contents

Preface

Who would have thought when the first edition of *Multivariate Data Analysis* was published almost 20 years ago that the use of multivariate statistics would be as pervasive as it is today. During this time we have seen a dramatic change in the research environment faced by both the academic and applied researcher. First, the personal computer revolution has provided power to the desktop that was unimaginable even just a few years ago. In our time we have gone from punched cards to speech recognition, revolutionizing the way we can interact and use the personal computer. At the same time we have seen tremendous advances in both the availability of statistical programs and their ease of use, ranging from the completely integrated computer packages such as SPSS and SAS to the specialized programs for such techniques as neural networks and conjoint analysis. Today, the researcher can find almost any conceivable technique in a PC format and at a reasonable price.

On the statistical front, we have seen a continual development of new techniques, such as conjoint analysis, structural equation modeling, and neural networks. All of these technological advances, however, have been matched by an ever-increasing need for more analytical capability. The data explosion of recent years has not only taxed our resources to physically handle and analyze all of the available information, but also required a reassessment of our approach to data analysis. Finally, the complexity of the topics being addressed and theory's increased role in research design have combined to require more rigorous and sophisticated techniques to perform the necessary confirmatory analyses.

These events have all contributed to the acceptance of the last four editions of this text and the demand for this fifth edition. In approaching this revision, we have tried to embrace both the academic and applied researchers with a presentation strongly grounded in statistical techniques, but focusing on design, estimation, and interpretation. We continually strive to reduce our reliance on statistical notation and terminology and instead to identify the fundamental concepts that affect our use of these techniques and express them in simple terms: an applications-oriented introduction to multivariate analysis for the nonstatistician. We continue our commitment to providing a firm understanding of the statistical and managerial principles underlying multivariate analysis so as to develop a "comfort zone" for not only the statistical but also the practical issues involved.

What's New

The most obvious change in the fifth edition is its reorganization into four sections. This organization parallels the research process more closely, particularly in focusing on data preparation and scale development before employing dependence or interdependence techniques. Section 1, "Preparing for a Multivariate Analysis," focuses on data preparation, such as missing data analysis and assessing statistical assumptions along with data reduction, with a particular emphasis on summated scale development. Section 2, "Dependence Techniques," contains discussions on five multivariate dependence techniques: multiple regression, discriminant analysis–logistic regression, multivariate analysis of variance, conjoint analysis, and canonical correlation. Section 3, "Interdependence Techniques," provides coverage of cluster analysis and multidimensional scaling and their uses in addressing the structure among observations. The final section, section 4, "Advanced and Emerging Techniques," introduces structural equation modeling along with some emerging areas of multivariate analysis, including data mining and warehousing, neural networks, and resampling.

Two additions were made in discussing each multivariate technique. First, a simple example is provided at the beginning of each chapter to illustrate the basic principles and objectives and how a particular method would work in an actual situation. Second, each of the detailed examples using the HATCO data set are followed by a managerial overview, which provides a perspective on the interpretation of the results and how they might be employed in addressing the research question. Both of these additions should provide a more grounded context for each technique.

What's Expanded and Updated

Each chapter has been revised to incorporate advances in technology, and several chapters have undergone more extensive change. With the inclusion of chapter 3, "Factor Analysis," in section 1, emphasis has been placed on summated scale development and the application of scales or factor scores in other applications. Chapter 5, "Multiple Discriminant Analysis and Logistic Regression," now provides complete coverage of analysis of categorical dependent variables by including both discriminant analysis and logistic regression, and contains an expanded discussion of other categorical models. Chapter 7, "Conjoint Analysis," has a revised examination of issues of research design, which focuses on the development of the conjoint stimuli in a concise and straightforward manner. Finally, chapter 11, "Structural Equation Modeling," has been updated to reflect the many changes in the field in the past several years. Three topics that are discussed are second-order factor models; alternative estimations techniques, such as simulation or bootstrapping; and some of the most common operational problems, such as missing data or obtaining a not positive definite matrix during model estimation. Each of these changes, and others not mentioned, will assist in gaining a thorough understanding of both the statistical and applied issues underlying these techniques.

Looking to the Future

We have made two other additions that we think are innovative and will substantially add to your understanding of multivariate analysis. The first is a new chapter—"Emerging Techniques in Multivariate Analysis" (chapter 12)—focusing on new topics in the field of multivariate analysis. As the research environment evolves, researchers must adapt to the changing conditions. Today's researcher is faced with ever-increasing amounts of information and the need for an objective method of discovery as well as explanation. We introduce the topics of data warehousing and data mining to expose the researcher to their basic objectives and the principles involved. Using the perspectives gained in the discussion of other techniques, we contrast the more exploratory nature of data mining. We also discuss neural networks, one of the techniques associated quite strongly with data mining and used in many applications today. A third topic is resampling, also known as bootstrapping and the jackknife. Available but not widely utilized for many years, this approach is gaining wider acceptance as an alternative to parametric estimation. The discussions provide a brief overview of the topics and then a simple empirical illustration. For example, we demonstrate the ability of neural networks to perform a discriminant analysis and resampling results when applied to multiple regression.

The final development is the creation of a Web site http://www.prenhall.com/hair devoted to multivariate analysis, titled *"Great Ideas in Teaching Multivariate Statistics."* This Web site acts as a resource center for everyone interested in multivariate analysis, providing links to resources for each technique and providing a forum for identifying new topics or statistical methods. In this way we can provide more timely feedback to researchers other than just through a new edition of the book. We also plan for the Web site to be a clearinghouse for materials on teaching multivariate statistics, providing exercises, datasets, and project ideas.

Acknowledgments

A number of individuals assisted us in completing the fifth edition of this text. Barabra Ross, doctoral candidate at Louisiana State University, provided invaluable assistance in all phases of the revision. Stern Neill, also a doctoral candidate at Louisiana State University, provided assistance on the analysis of the data and the addition of new statistical techniques. We are indebted to the following reviewers for their invaluable assistance in the additions to the fifth edition:

> David Booth, *Kent State University*
> Robert Bush, *Memphis State University*
> Rabikar Chatterjee, *University of Michigan*
> Kerri Curtis, *Golden Gate University*
> Muzaffar Shaikh, *Florida Institute of Technology*

We would also like to acknowledge the assistance of the following individuals on prior editions of the text: Bruce Alford, University of Evansville; David Andrus, Kansas State University; Alvin C. Burns, Louisiana State University; Alan J. Bush, University of Memphis; Robert Bush, University of Memphis; Chaim Ehrman,

University of Illinois at Chicago; Joel Evans, Hofstra University; Thomas L. Gillpatrick, Portland State University; Dipak Jain, Northwestern University; John Lastovicka, University of Kansas; Maragret Liebman, La Salle University; Richard Netemeyer, Louisiana State University; Scott Roach, Northeast Louisiana University; Walter A. Smith, Tulsa University; Ronald D. Taylor, Mississippi State University; and Jerry L. Wall, Northeast Louisiana University.

J.F.H.
R.E.A.
R.L.T.
W.C.B.

Introduction

LEARNING OBJECTIVES

Upon completing this chapter, you should be able to do the following:

- Explain what multivariate analysis is and when its application is appropriate.
- Define and discuss the specific techniques included in multivariate analysis.
- Determine which multivariate technique is appropriate for a specific research problem.
- Discuss the nature of measurement scales and their relationship to multivariate techniques.
- Describe the conceptual and statistical issues inherent in multivariate analyses.

CHAPTER PREVIEW

Chapter 1 presents a simplified overview of multivariate analysis. It stresses that multivariate analysis methods will increasingly influence not only the analytical aspects of research but also the design and approach to data collection for decision making and problem solving. Although multivariate techniques share many characteristics with their univariate and bivariate counterparts, several key differences arise in the transition to a multivariate analysis. To illustrate this transition, this chapter presents a classification of multivariate techniques. It then provides general guidelines for the application of these techniques as well as a structured approach to the formulation, estimation, and interpretation of multivariate results. The chapter concludes with a discussion of the database utilized throughout most of the text to illustrate application of the techniques.

KEY TERMS

Before starting the chapter, review the key terms to develop an understanding of the concepts and terminology used. Throughout the chapter, the key terms appear in **boldface**. Other points of emphasis in the chapter are *italicized*. Also, cross-references within the Key Terms appear in *italics*.

Alpha (α)   See *Type I error.*

Beta (β)   See *Type II error.*

Bivariate partial correlation   Simple (two-variable) correlation between two sets of residuals (unexplained variances) that remain after the association of other independent variables is removed.

Composite measure   See *summated scale.*

Dependence technique   Classification of statistical techniques distinguished by having a variable or set of variables identified as the *dependent variable(s)* and the remaining variables as *independent*. The objective is prediction of the dependent variable(s) by the independent variable(s). An example is regression analysis.

Dependent variable   Presumed effect of, or response to, a change in the *independent variable(s)*.

Dummy variable   Nonmetrically measured variable transformed into a metric variable by assigning a 1 or a 0 to a subject, depending on whether it possesses a particular characteristic.

Effect size   Estimate of the degree to which the phenomenon being studied (e.g., correlation or difference in means) exists in the population.

Independent variable   Presumed cause of any change in the *dependent variable*.

Indicator   Single variable used in conjunction with one or more other variables to form a *composite measure*.

Interdependence technique   Classification of statistical techniques in which the variables are not divided into *dependent* and *independent sets* (e.g., factor analysis); rather, all variables are analyzed as a single set.

Measurement error   Inaccuracies of measuring the "true" variable values due to the fallibility of the measurement instrument (i.e., inappropriate response scales), data entry errors, or respondent errors.

Metric data   Also called *quantitative data, interval data,* or *ratio data,* these measurements identify or describe subjects (or objects) not only on the possession of an attribute but also by the amount or degree to which the subject may be characterized by the attribute. For example, a person's age and weight are metric data.

Multicollinearity   Extent to which a variable can be explained by the other variables in the analysis. As multicollinearity increases, it complicates the interpretation of the *variate* as it is more difficult to ascertain the effect of any single variable, owing to their interrelationships.

Multivariate analysis   Analysis of multiple variables in a single relationship or set of relationships.

Multivariate measurement   Use of two or more variables as *indicators* of a single *composite measure*. For example, a personality test may provide the answers to a series of individual questions (indicators), which are then combined to form a single score (*summated scale*) representing the personality trait.

Nonmetric data   Also called *qualitative data,* these are attributes, characteristics, or categorical properties that identify or describe a subject or object. They differ from *metric data* by indicating the presence of an attribute, but not the amount. Examples are occupation (physician, attorney, professor) or buyer status (buyer, nonbuyer). Also called *nominal data* or *ordinal data.*

Power Probability of correctly rejecting the null hypothesis when it is false, that is, correctly finding a hypothesized relationship when it exists. Determined as a function of (1) the statistical significance level (α) set by the researcher for a *Type I error*, (2) the sample size used in the analysis, and (3) the *effect size* being examined.

Practical significance Means of assessing multivariate analysis results based on their substantive findings rather than their statistical significance. Whereas statistical significance determines whether the result is attributable to chance, practical significance assesses whether the result is useful (i.e., substantial enough to warrant action).

Reliability Extent to which a variable or set of variables is consistent in what it is intended to measure. If multiple measurements are taken, the reliable measures will all be very consistent in their values. It differs from *validity* in that it relates not to what should be measured, but instead to how it is measured.

Specification error Omitting a key variable from the analysis, thus impacting the estimated effects of included variables.

Summated scales Method of combining several variables that measure the same concept into a single variable in an attempt to increase the *reliability* of the measurement through *multivariate measurement.* In most instances, the separate variables are summed and then their total or average score is used in the analysis.

Treatment Independent variable the researcher manipulates to see the effect (if any) on the dependent variable(s), such as in an experiment.

Type I error Probability of incorrectly rejecting the null hypothesis—in most cases, this means saying a difference or correlation exists when it actually does not. Also termed *alpha* (α). Typical levels are 5 or 1 percent, termed the .05 or .01 level, respectively.

Type II error Probability of incorrectly failing to reject the null hypothesis—in simple terms, the chance of not finding a correlation or mean difference when it does exist. Also termed *beta* (β), it is inversely related to *Type I error.* The value 1 minus the Type II error is defined as *power.*

Univariate analysis of variance (ANOVA) Statistical technique to determine, on the basis of one dependent measure, whether samples are from populations with equal means.

Validity Extent to which a measure or set of measures correctly represents the concept of study—the degree to which it is free from any systematic or nonrandom error. Validity is concerned with how well the concept is defined by the measure(s), whereas *reliability* relates to the consistency of the measure(s).

Variate Linear combination of variables formed in the multivariate technique by deriving empirical weights applied to a set of variables specified by the researcher.

What Is Multivariate Analysis?

The computer technology available today, almost unimaginable just two short decades ago, has made possible extraordinary advances in the analysis of psychological, sociological, and other types of behavioral data. This impact is most evident in the relative ease with which computers can analyze large quantities of complex data. Almost any problem today is easily analyzed by any number of statistical programs on microcomputers. In addition, the effects of technological progress have extended beyond the ability to manipulate data, releasing researchers from past constraints on data analysis and affording them the ability to engage in more substantive development and testing of their theoretical models. No longer are methodological limitations a critical concern to the theorist striving for empirical support. Much of this increased understanding and mastery of data analysis has come about through the study of statistics and statistical inference. Equally important, however, has been the expanded understanding and application of a group of statistical techniques known as **multivariate analysis.**

Multivariate analytical techniques are being widely applied in industry, government, and university-related research centers. Moreover, few fields of study or research have failed to integrate multivariate techniques into their analytical "toolbox." To serve this increased interest, numerous books and articles have been published on the theoretical and mathematical aspects of these tools, and introductory texts have appeared in almost every field as well. Few books, however, have been written for the researcher who is not a specialist in math or statistics. Still fewer books discuss *applications* of multivariate statistics while they provide a conceptual discussion of the statistical methods. This book was written to fill this gap.

Applications-oriented books are of crucial interest to behavioral scientists and business or government managers of all backgrounds who have to expand their knowledge of multivariate analysis to gain a better understanding of the complex phenomena in their work environment. Any researcher who examines only two-variable relationships and avoids multivariate analysis is ignoring powerful tools that can provide potentially useful information. As one researcher states, "For the purposes of . . . any . . . applied field, most of our tools are, or should be, multivariate. One is pushed to a conclusion that unless a . . . problem is treated as a multivariate problem, it is treated superficially" [7, p. 158]. According to statisticians Hardyck and Petrinovich [8, p. 7]:

> Multivariate analysis methods will predominate in the future and will result in drastic changes in the manner in which research workers think about problems and how they design their research. These methods make it possible to ask specific and precise questions of considerable complexity in natural settings. This makes it possible to conduct theoretically significant research and to evaluate the effects of naturally occurring parametric variations in the context in which they normally occur. In this way, the natural correlations among the manifold influences on behavior can be preserved and separate effects of these influences can be studied statistically without causing a typical isolation of either individuals or variables.

For example, businesspeople in most markets today are not able to follow the simplistic approach whereby consumers were considered homogeneous and characterized by a small number of demographic variables. Instead, they must develop strategies to appeal to numerous segments of customers with varied demographic and psychographic characteristics in a marketplace with multiple constraints (e.g., legal, economic, competitive, technological). It is only through multivariate techniques that multiple relationships of this type can be adequately examined to obtain a more complete, realistic understanding for decision making.

Throughout the text we use the generic term "researcher" when referring to a data analyst within either the practitioner or academic communities. We feel it inappropriate to make any distinction between these two areas, as research in either must rely on both theoretical and quantitative bases. Although the research objectives and the emphasis in interpretation may vary, a researcher within either area must address all of the issues, both conceptual and empirical, raised in the discussions of the statistical methods.

Impact of the Computer Revolution

It is almost impossible to discuss the application of multivariate techniques without a discussion of the impact of the computer. As mentioned earlier, the widespread application of computers (first mainframe and, more recently, personal or microcomputers) to process large, complex databases has dramatically spurred the use of multivariate statistical methods. The statistical theory for today's multivariate techniques was developed well before the appearance of computers, but these techniques remained almost unknown outside the field of theoretical statistics until computational power became available to perform their increasingly complex calculations. The continued technological advances in computing, particularly personal computers, have provided any interested researcher ready access to all the resources needed to address almost any size multivariate problem. In fact, many researchers call themselves *data analysts* instead of statisticians or (in the vernacular) "quantitative types." These data analysts have contributed substantially to the increase in usage and acceptance of multivariate statistics in the business and government sectors. Within the academic community, disciplines in all fields have embraced multivariate techniques, and academicians increasingly must be versed in the appropriate multivariate techniques for their empirical research. Even for people with strong quantitative training, the availability of prepackaged programs for multivariate analysis has facilitated the complex manipulation of data matrices that has long hampered the growth of multivariate techniques.

Many major universities already require entering students to purchase their own microcomputers before matriculating, and students and professors now routinely analyze multivariate data for answers to questions in fields of study from anthropology to zoology. All of the comprehensive statistical packages designed for mainframe computers (e.g., SPSS, SAS, and BMDP) are now available also for personal computers. Specialized programs for other types of multivariate analysis, including multidimensional scaling, simultaneous/structural equation modeling, and conjoint analysis, once were available only—if at all—on mainframe computers, but today they are personal computer-compatible. Expert systems are being developed to address even such issues as selecting a statistical technique [4] or designing a sampling plan to ensure desired statistical and practical objectives [3].

No longer are statistical programs first developed on mainframe systems and then migrated to micro- and personal computers; instead, programs are now initially developed for the personal computer. Perhaps the fastest growing category of statistical programs are the statistical packages designed specifically to take advantage of the flexibility of the personal computer. Multivariate techniques are so widespread that *all* the techniques illustrated in this text can be estimated with statistical packages readily available for either a mainframe, minicomputer, or personal computer. A comprehensive list of the major software programs available for multivariate analyses can be found in Appendix A. Special attention is given to the personal computer-based programs.

Multivariate Analysis Defined

Multivariate analysis is not easy to define. Broadly speaking, it refers to all statistical methods that simultaneously analyze multiple measurements on each individual or object under investigation. Any simultaneous analysis of more than two variables can be loosely considered multivariate analysis. As such, many multivariate techniques are extensions of univariate analysis (analysis of single-variable distributions) and bivariate analysis (cross-classification, correlation, analysis of variance, and simple regression used to analyze two variables). For example, simple regression (with one predictor variable) is extended in the multivariate case to include several predictor variables. Likewise, the single dependent variable found in analysis of variance is extended to include multiple dependent variables in multivariate analysis of variance. In many instances, multivariate techniques are a means of performing in a single analysis what once took multiple analyses using univariate techniques. Other multivariate techniques, however, are uniquely designed to deal with multivariate issues, such as factor analysis, which identifies the structure underlying a set of variables, or discriminant analysis, which differentiates among groups based on a set of variables.

One reason for the difficulty of defining multivariate analysis is that the term *multivariate* is not used consistently in the literature. Some researchers use multivariate simply to mean examining relationships between or among more than two variables. Others use the term only for problems in which all the multiple variables are assumed to have a multivariate normal distribution. To be considered truly multivariate, however, all the variables must be random and interrelated in such ways that their different effects cannot meaningfully be interpreted separately. Some authors state that the purpose of multivariate analysis is to measure, explain, and predict the degree of relationship among variates (weighted combinations of variables). Thus the multivariate character lies in the multiple variates (multiple combinations of variables), and not only in the number of variables or observations. For the purposes of this book, we do not insist on a rigid definition of multivariate analysis. Instead, **multivariate analysis** will include both multivariable techniques and truly multivariate techniques, because we believe that knowledge of multivariable techniques is an essential first step in understanding multivariate analysis.

Some Basic Concepts of Multivariate Analysis

Although multivariate analysis has its roots in univariate and bivariate statistics, the extension to the multivariate domain introduces additional concepts and issues that have particular relevance. These concepts range from the need for a conceptual understanding of the basic building block of multivariate analysis—the variate—to specific issues dealing with the types of measurement scales used and the statistical issues of significance testing and confidence levels. Each concept plays a significant role in the successful application of any multivariate technique.

The Variate

As previously mentioned, the building block of multivariate analysis is the **variate,** a linear combination of variables with empirically determined weights. The variables are specified by the researcher, whereas the weights are determined by the multivariate technique to meet a specific objective. A variate of n weighted variables (X_1 to X_n) can be stated mathematically as:

$$\text{Variate value} = w_1X_1 + w_2X_2 + w_3X_3 + \ldots + w_nX_n$$

where X_n is the observed variable and w_n is the weight determined by the multivariate technique.

The result is a single value representing a combination of the *entire set* of variables that best achieves the objective of the specific multivariate analysis. In multiple regression, the variate is determined so as to best correlate with the variable being predicted. In discriminant analysis, the variate is formed so as to create scores for each observation that maximally differentiates among groups of observations. In factor analysis, variates are formed to best represent the underlying structure or dimensionality of the variables as represented by their intercorrelations.

In each instance, the variate captures the multivariate character of the analysis. Thus, in our discussion of each technique, the variate is the focal point of the analysis in many respects. We must understand not only its collective impact in meeting the technique's objective but also each separate variable's contribution to the overall variate effect.

Measurement Scales

Data analysis involves the partitioning, identification, and measurement of variation in a set of variables, either among themselves or between a dependent variable and one or more independent variables. The key word here is *measurement* because the researcher cannot partition or identify variation unless it can be measured. Measurement is important in accurately representing the concept of interest and is instrumental in the selection of the appropriate multivariate method of analysis. Next we discuss the concept of measurement as it relates to data analysis and particularly to the various multivariate techniques.

There are two basic kinds of data: **nonmetric** (qualitative) and **metric** (quantitative). Nonmetric data are attributes, characteristics, or categorical properties that identify or describe a subject. Nonmetric data describe differences in type or kind by indicating the presence or absence of a characteristic or property. Many properties are discrete in that by having a particular feature, all other features are excluded; for example, if one is male, one cannot be female. There is no "amount" of gender, just the state of being male or female. In contrast, metric data measurements are made so that subjects may be identified as differing in amount or degree. Metrically measured variables reflect relative quantity or degree. Metric measurements are appropriate for cases involving amount or magnitude, such as the level of satisfaction or commitment to a job.

Nonmetric Measurement Scales

Nonmetric measurements can be made with either a nominal or an ordinal scale. Measurement with a nominal scale assigns numbers used to label or identify subjects or objects. Nominal scales, also known as categorical scales, provide the number of occurrences in each class or category of the variable being studied. Therefore, the numbers or symbols assigned to the objects have no quantitative meaning beyond indicating the presence or absence of the attribute or characteristic under investigation. Examples of nominally scaled data include an individual's sex, religion, or political party. In working with these data, the researcher might assign numbers to each category, for example, 2 for females and 1 for males. These numbers only represent categories or classes and do not imply amounts of an attribute or characteristic.

Ordinal scales are the next higher level of measurement precision. Variables can be ordered or ranked with ordinal scales in relation to the amount of the attribute possessed. Every subclass can be compared with another in terms of a "greater than" or "less than" relationship. For example, different levels of an individual consumer's satisfaction with several new products can be illustrated on an ordinal scale. The following scale shows a respondent's view of three products. The respondent is more satisfied with A than B and more satisfied with B than C.

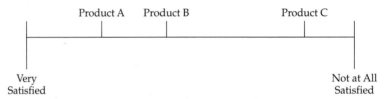

Numbers utilized in ordinal scales such as these are nonquantitative because they indicate only relative positions in an ordered series. There is no measure of how much satisfaction the consumer receives in absolute terms, nor does the researcher know the exact difference between points on the scale of satisfaction. Many scales in the behavioral sciences fall into this ordinal category.

Metric Measurement Scales

Interval scales and ratio scales (both metric) provide the highest level of measurement precision, permitting nearly all mathematical operations to be performed. These two scales have constant units of measurement, so differences between any two adjacent points on any part of the scale are equal. The only real difference between interval and ratio scales is that interval scales have an arbitrary zero point, whereas ratio scales have an absolute zero point. The most familiar interval scales are the Fahrenheit and Celsius temperature scales. Each has a different arbitrary zero point, and neither indicates a zero amount or lack of temperature, because we can register temperatures below the zero point on each scale. Therefore, it is not possible to say that any value on an interval scale is a multiple of some other point on the scale. For example, an 80°F day cannot correctly be said to be twice as hot as a 40°F day, because we know that 80°F, on a different scale, such as Celsius, is 26.7°C. Similarly, 40°F on Celsius is 4.4°C. Although 80°F is indeed twice 40°F, one cannot state that the heat of 80°F is twice the heat of 40°F because, using different scales, the heat is not twice as great; that is, $4.4°C \times 2 \neq 26.7°C$.

Ratio scales represent the highest form of measurement precision because they possess the advantages of all lower scales plus an absolute zero point. All mathematical operations are permissible with ratio-scale measurements. The bathroom scale or other common weighing machines are examples of these scales, for they have an absolute zero point and can be spoken of in terms of multiples when relating one point on the scale to another; for example, 100 pounds is twice as heavy as 50 pounds.

Understanding the different types of measurement scales is important for two reasons. First, the researcher must identify the measurement scale of each variable used, so that nonmetric data are not incorrectly used as metric data and vice versa. Second, the measurement scale is critical in determining which multivariate techniques are the most applicable to the data, with considerations made for both independent and dependent variables. In the discussion of the techniques and their classification in later sections of this chapter, the metric or nonmetric properties of independent and dependent variables are the determining factors in selecting the appropriate technique.

Measurement Error and Multivariate Measurement

The use of multiple variables and the reliance on their combination (the variate) in multivariate techniques also focuses attention on a complementary issue—measurement error. **Measurement error** is the degree to which the observed values are not representative of the "true" values. Measurement error has many sources, ranging from data entry errors to the inprecision of the measurement (e.g., imposing seven-point rating scales for attitude measurement when the researcher knows the respondents can accurately respond only to a three-point rating) to the inability of respondents to accurately provide information (e.g., responses as to household income may be reasonably accurate but rarely totally precise). Thus, all variables used in multivariate techniques must be assumed to have some degree of measurement error. The impact of measurement error is to add "noise" to the observed or measured variables. Thus, the observed value obtained represents both the "true" level and the "noise." When used to compute correlations or means, the "true" effect is partially masked by the measurement error, causing the correlations to weaken and the means to be less precise. The specific impact of measurement error and its accommodation in dependence relationships is covered in more detail in chapter 11.

The researcher's goal of reducing measurement error can follow several paths. In assessing the degree of measurement error present in any measure, the researcher must address both the **validity** and **reliability** of the measure. Validity is the degree to which a measure accurately represents what it is supposed to. For example, if we want to measure discretionary income, we should not ask about total household income. Ensuring validity starts with a thorough understanding of what is to be measured and then making the measurement as "correct" and accurate as possible. However, accuracy does not ensure validity. In our income example, the researcher could very precisely define total household income but still have an invalid measure of discretionary income because the "correct" question was not being asked.

If validity is assured, the researcher must still consider the reliability of the measurements. Reliability is the degree to which the observed variable measures the "true" value and is "error free"; thus, it is the opposite of measurement error. If the same measure is asked repeatedly, for example, more reliable measures will show greater consistency than less reliable measures. The researcher should always assess the variables being used and if valid alternative measures are available, choose the variable with the higher reliability.

The researcher may also choose to develop **multivariate measurements,** also known as **summated scales,** for which several variables are joined in a **composite measure** to represent a concept (e.g., multiple-item personality scales or summed ratings of product satisfaction). The objective is to avoid the use of only a single variable to represent a concept, and instead to use several variables as **indicators** (see Key Terms), all representing differing facets of the concept to obtain a more "well-rounded" perspective. The use of multiple indicators allows the researcher to more precisely specify the desired responses. It does not place total reliance on a single response, but instead on the "average" or "typical" response to a set of related responses. For example, in measuring satisfaction, one could ask a single question, "How satisfied are you?" and base the analysis on the single response. Or a summated scale could be developed that combined several responses of satisfaction, perhaps in different response formats and in differing areas of interest thought to comprise overall satisfaction. The guiding premise is that multiple responses reflect the "true" response more accurately than does a single response. Assessing reliability and incorporating scales in the analysis are methods the researcher should employ. For a more detailed introduction to multiple measurement models and scale construction, see further discussion in chapter 3 (Factor Analysis) and chapter 11 (Structural Equation Modeling) or additional resources [10]. In addition, compilations of scales that can provide the researcher a "ready-to-go" scale with demonstrated reliability have been published in recent years [1, 5].

The impact of measurement error and poor reliability cannot be directly seen because they are embedded in the observed variables. The researcher must therefore always work to increase reliability and validity, which in turn will result in a "truer" portrayal of the variables of interest. Poor results are not always due to measurement error, but the presence of measurement error is guaranteed to distort the observed relationships and make multivariate techniques less powerful. Reducing measurement error, although it takes effort, time, and additional resources, may improve weak or marginal results and strengthen proven results as well.

Statistical Significance versus Statistical Power

All the multivariate techniques, except for cluster analysis and multidimensional scaling, are based on the statistical inference of a population's values or relationships among variables from a randomly drawn sample of that population. If we have conducted a census of the entire population, then statistical inference is unnecessary, because any difference or relationship, however small, is "true" and does exist. But rarely, if ever, is a census conducted; therefore, the researcher is forced to draw inferences from a sample.

Interpreting statistical inferences requires that the researcher specify the acceptable levels of statistical error. The most common approach is to specify the level of **Type I error,** also known as **alpha** (α). The Type I error is the probability of rejecting the null hypothesis when actually true, or in simple terms, the chance of the test showing statistical significance when it actually is not present—the case of a "false positive." By specifying an alpha level, the researcher sets the allowable limits for error by specifying the probability of concluding that significance exists when it really does not.

When specifying the level of Type I error, the researcher also determines an associated error, termed the **Type II error** or **beta** (β). The Type II error is the probability of failing to reject the null hypothesis when it is actually false. An even more interesting probability is $1 - \beta$, termed the **power** of the statistical inference test. Power is the probability of correctly rejecting the null hypothesis when it should be rejected. Thus, power is the probability that statistical significance will be indicated if it is present. The relationship of the different error probabilities in the hypothetical setting of testing for the difference in two means is shown here:

		Reality	
		H_0: No Difference	H_a: Difference
Statistical Decision	H_0: No Difference	$1 - \alpha$	β Type II error
	H_a: Difference	α Type I error	$1 - \beta$ Power

Although specifying alpha establishes the level of acceptable statistical significance, it is the level of power that dictates the probability of "success" in finding the differences if they actually exist. Then why not set both alpha and beta at acceptable levels? Because the Type I and Type II errors are inversely related, and as the Type I error becomes more restrictive (moves closer to zero), the Type II error increases. Reducing the Type I errors therefore reduces the power of the statistical test. Thus, the researcher must strike a balance between the level of alpha and the resulting power.

Power is not solely a function of α. It is actually determined by three factors:

1. Effect size—The probability of achieving statistical significance is based not only on statistical considerations but also on the actual magnitude of the effect of interest (e.g., a difference of means between two groups or the correlation between variables) in the population, termed the **effect size** (see Key Terms). As one would expect, a larger effect is more likely to be found than a smaller effect, and thus to impact the power of the statistical test. To assess the power of any statistical test, the researcher must first understand the effect being examined. Effect sizes are defined in standardized terms for ease of comparison. Mean differences are stated in terms of standard deviations, so that an effect size of .5 indicates that the mean difference is one-half of a standard deviation. For correlations, the effect size is based on the actual correlation between the variables.

2. Alpha (α)—As already discussed, as alpha becomes more restrictive, power decreases. This means that as the researcher reduces the chance of finding an incorrect significant effect, the probability of correctly finding an effect also decreases. Conventional guidelines suggest alpha levels of .05 or .01. But the researcher must consider the impact of this decision on the power before selecting the alpha level. The relationship of these two probabilities is illustrated in later discussions.

3. Sample size—At any given alpha level, increased sample sizes always produce greater power of the statistical test. But increasing sample size can also produce "too much" power. By this we mean that by increasing sample size, smaller and smaller effects will be found to be statistically significant, until

at very large sample sizes almost any effect is significant. The researcher must always be aware that sample size can impact the statistical test by either making it insensitive (at small sample sizes) or overly sensitive (at very large sample sizes).

The relationships among alpha, sample size, effect size, and power are quite complicated, and a number of sources of guidance are available. Cohen [6] examines power for most statistical inference tests and provides guidelines for acceptable levels of power, suggesting that studies be designed to achieve alpha levels of at least .05 with power levels of 80 percent. To achieve such power levels, all three factors—alpha, sample size, and effect size—must be considered simultaneously. These interrelationships can be illustrated by two simple examples. The first example involves testing for the difference between the mean scores of two groups. Assume that the effect size is thought to range between small (.2) and moderate (.5). The researcher must now determine the necessary alpha level and sample size of each group. Table 1.1 illustrates the impact of both sample size and alpha level on power. As can be seen, power becomes acceptable at sample sizes of 100 or more in situations with a moderate effect size at both alpha levels. But when the effect size is small, the statistical tests have little power, even with expanded alpha levels or samples of 200 or more. For example, a sample of 200 in each group with an alpha of .05 still has only a 50 percent chance of significant differences being found if the effect size is small. This suggests that the researcher, if anticipating the effects to be small, must design the study with much larger sample sizes and/or less restrictive alpha levels (.05 or .10).

In the second example, Figure 1.1 graphically presents the power for significance levels of .01, .05, and .10 for sample sizes of 30 to 300 per group when the effect size (.35) falls between small and moderate. Faced with such prospects, specification of a .01 significance level requires a sample of 200 per group to achieve the desired level of 80 percent power. But if the alpha level is relaxed, 80 percent power is reached at samples of 130 for a .05 alpha level and at samples of 100 for a .10 alpha level.

Such analyses allow the researcher to make more informed decisions in study design and interpretation of the results. In planning the research, the researcher must estimate the expected effect size and then select the sample size and alpha to achieve the desired power level. In addition to its uses for planning, power

TABLE 1.1 Power Levels for the Comparison of Two Means: Variations by Sample Size, Significance Level, and Effect Size

Sample Size	alpha (α) = .05 Effect Size (ES)		alpha (α) = .01 Effect Size (ES)	
	Small (.2)	Moderate (.5)	Small (.2)	Moderate (.5)
20	.095	.338	.025	.144
40	.143	.598	.045	.349
60	.192	.775	.067	.549
80	.242	.882	.092	.709
100	.290	.940	.120	.823
150	.411	.990	.201	.959
200	.516	.998	.284	.992

Source: *Solo Power Analysis*, BMDP Statistical Software, Inc.

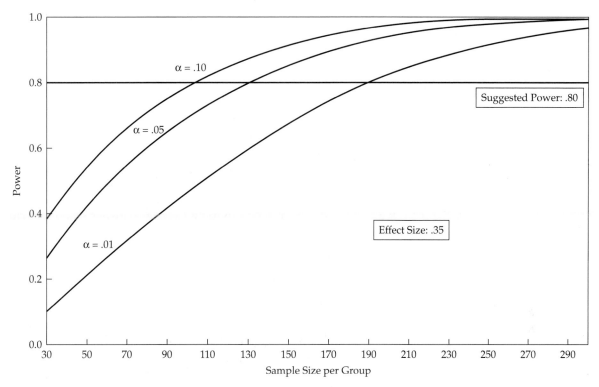

FIGURE 1.1 Impact of Sample Size on Power for Various Alpha Levels (.01, .05, .10) with Effect Size of .35

analysis is also utilized after the analysis is completed to determine the actual power achieved, so that the results can be properly interpreted. Are the results due to effect sizes, sample sizes, or significance levels? The researcher can assess each of these factors for their impact on the significance or nonsignificance of the results. The researcher today can refer to published studies detailing the specifics of power determination [6] or turn to several personal computer-based programs that assist in planning studies to achieve the desired power or calculate the power of actual results [2, 3]. Specific guidelines for multiple regression and multivariate analysis of variance—the most common applications of power analysis—are discussed in more detail in chapters 4 and 6.

Having addressed the issues in extending multivariate techniques from their univariate and bivariate origins, we now briefly introduce each multivariate method discussed in the text. Following the introductions of the techniques, we present a classification scheme to assist in the selection of the appropriate technique by specifying the research objectives (independence or dependence relationship) and the data type (metric or nonmetric).

Types of Multivariate Techniques

Multivariate analysis is an ever-expanding set of techniques for data analysis. Among the more established techniques discussed in this text are (1) principal components and common factor analysis, (2) multiple regression and multiple correlation, (3) multiple discriminant analysis, (4) multivariate analysis of variance and covariance, (5) conjoint analysis, (6) canonical correlation, (7) cluster analysis, and (8) multidimensional scaling. Among the emerging techniques also included are (9) correspondence analysis, (10) linear probability models such as logit and probit, and (11) simultaneous/structural equation modeling. Here we introduce each of the multivariate techniques, briefly defining the technique and the objective for its application.

Principal Components and Common Factor Analysis

Factor analysis, including both principal component analysis and common factor analysis, is a statistical approach that can be used to analyze interrelationships among a large number of variables and to explain these variables in terms of their common underlying dimensions (factors). The objective is to find a way of condensing the information contained in a number of original variables into a smaller set of variates (factors) with a minimum loss of information. By providing an empirical estimate of the "structure" of the variables considered, factor analysis becomes an objective basis for creating summated scales.

Multiple Regression

Multiple regression is the appropriate method of analysis when the research problem involves a single metric dependent variable presumed to be related to two or more metric independent variables. The objective of multiple regression analysis is to predict the changes in the dependent variable in response to changes in the independent variables. This objective is most often achieved through the statistical rule of least squares.

Whenever the researcher is interested in predicting the amount or magnitude of the dependent variable, multiple regression is useful. For example, monthly expenditures on dining out (dependent variable) might be predicted from information regarding a family's income, its size, and the age of the head of household (independent variables). Similarly, the researcher might attempt to predict a company's sales from information on its expenditures for advertising, the number of salespeople, and the number of stores carrying its products.

Multiple Discriminant Analysis

Multiple discriminant analysis (MDA) is the appropriate multivariate technique if the single dependent variable is dichotomous (e.g., male–female) or multichotomous (e.g., high–medium–low) and therefore nonmetric. As with multiple regression, the independent variables are assumed to be metric. Discriminant analysis is applicable in situations in which the total sample can be divided into groups based on a nonmetric dependent variable characterizing several known classes. The primary objectives of multiple discriminant analysis are to understand group differences and to predict the likelihood that an entity (individual or object) will belong to a particular class or group based on several metric independent variables. For example, discriminant analysis might be used to distinguish innovators from noninnovators according to their demographic and psychographic profiles. Other applications include distinguishing heavy product users from light users, males from females, national-brand buyers from private-label buyers, and good credit risks from poor credit risks. Even the Internal Revenue Service uses discriminant analysis to compare selected federal tax returns with a composite, hypothetical, normal taxpayer's return (at different income levels) to identify the most promising returns and areas for audit.

Multivariate Analysis of Variance and Covariance

Multivariate analysis of variance (MANOVA) is a statistical technique that can be used to simultaneously explore the relationship between several categorical independent variables (usually referred to as **treatments**) and two or more metric dependent variables. As such, it represents an extension of **univariate analysis of variance (ANOVA).** Multivariate analysis of covariance (MANCOVA) can be used in conjunction with MANOVA to remove (after the experiment) the effect of any uncontrolled metric independent variables (known as covariates) on the dependent variables. The procedure is similar to that involved in **bivariate partial correlation,** in which the effect of a third variable is removed from the correlation. MANOVA is useful when the researcher designs an experimental situation (manipulation of several nonmetric treatment variables) to test hypotheses concerning the variance in group responses on two or more metric dependent variables.

Conjoint Analysis

Conjoint analysis is an emerging dependence technique that has brought new sophistication to the evaluation of objects, such as new products, services, or ideas. The most direct application is in new product or service development, allowing for the evaluation of complex products while maintaining a realistic decision context for the respondent. The market researcher is able to assess the importance of attributes as well as the levels of each attribute while consumers evaluate only a few product profiles, which are combinations of product levels. For example, assume a product concept has three attributes (price, quality, and color), each at three possible levels (e.g., red, yellow, and blue). Instead of having to evaluate all 27 ($3 \times 3 \times 3$) possible combinations, a subset (9 or more) can be evaluated for their attractiveness to consumers, and the researcher knows not only how important each attribute is but also the importance of each level (the attractiveness of red versus yellow versus blue). Moreover, when the consumer evaluations are completed, the results of conjoint analysis can also be used in product design simulators, which show customer acceptance for any number of product formulations and aid in the design of the optimal product.

Canonical Correlation

Canonical correlation analysis can be viewed as a logical extension of multiple regression analysis. Recall that multiple regression analysis involves a single metric dependent variable and several metric independent variables. With canonical analysis the objective is to correlate simultaneously several metric dependent variables and several metric independent variables. Whereas multiple regression involves a single dependent variable, canonical correlation involves multiple dependent variables. The underlying principle is to develop a linear combination of each set of variables (both independent and dependent) to maximize the correlation between the two sets. Stated differently, the procedure involves obtaining a set of weights for the dependent and independent variables that provide the maximum simple correlation between the set of dependent variables and the set of independent variables.

Cluster Analysis

Cluster analysis an analytical technique for developing meaningful subgroups of individuals or objects. Specifically, the objective is to classify a sample of entities (individuals or objects) into a small number of mutually exclusive groups based on the similarities among the entities. In cluster analysis, unlike discriminant analysis, the groups are not predefined. Instead, the technique is used to identify the groups.

Cluster analysis usually involves at least three steps. The first is the measurement of some form of similarity or association among the entities to determine how many groups really exist in the sample. The second step is the actual clustering process, whereby entities are partitioned into groups (clusters). The final step is to profile the persons or variables to determine their composition. Many times this may be accomplished by applying discriminant analysis to the groups identified by the cluster technique.

Multidimensional Scaling

In multidimensional scaling, the objective is to transform consumer judgments of similarity or preference (e.g., preference for stores or brands) into distances represented in multidimensional space. If objects A and B are judged by respondents as being the most similar compared with all other possible pairs of objects, multidimensional scaling techniques will position objects A and B in such a way that the distance between them in multidimensional space is smaller than the distance between any other pairs of objects. The resulting perceptual maps show the relative positioning of all objects, but additional analyses are needed to describe or assess which attributes predict the position of each object.

Correspondence Analysis

Correspondence analysis is a recently developed interdependence technique that facilitates both dimensional reduction of object ratings (e.g., products, persons) on a set of attributes and the perceptual mapping of objects relative to these attributes. Researchers are constantly faced with the need to "quantify the qualitative data" found in nominal variables. Correspondence analysis differs from the interdependence techniques discussed earlier in its ability to accommodate both nonmetric data and nonlinear relationships.

In its most basic form, correspondence analysis employs a contingency table, which is the cross-tabulation of two categorical variables. It then transforms the nonmetric data to a metric level and performs dimensional reduction (similar to factor analysis) and perceptual mapping (similar to multidimensional analysis). As an example, respondents' brand preferences can be cross-tabulated on demographic variables (e.g., gender, income categories, occupation) by indicating how many people preferring each brand fall into each category of the demographic variables. Through correspondence analysis, the association, or "correspondence," of brands and the distinguishing characteristics of those preferring each brand are then shown in a two- or three-dimensional map of both brands and respondent characteristics. Brands perceived as similar are located close to one another. Likewise, the most distinguishing characteristics of respondents preferring each brand are also determined by the proximity of the demographic variable categories to the brand's position. Correspondence analysis provides a multivariate representation of interdependence for nonmetric data that is not possible with other methods.

Linear Probability Models

Linear probability models, often referred to as *logit analysis*, are a combination of multiple regression and multiple discriminant analysis. This technique is similar to multiple regression analysis in that one or more independent variables are used to predict a single dependent variable. What distinguishes a linear probability model from multiple regression is that the dependent variable is nonmetric, as in discriminant analysis. The nonmetric scale of the dependent variable requires differences in the estimation method and assumptions about the type of underlying distribution, yet in most other facets it is quite similar to multiple regression. Thus, once the dependent variable is correctly specified and the appropriate estimation technique is employed, the basic factors considered in multiple regression are used here as well. Linear probability models are distinguished from discriminant analysis primarily in that they accommodate all types of independent variables (metric and nonmetric) and do not require the assumption of multivariate normality. However, in many instances, particularly with more than two levels of the dependent variable, discriminant analysis is the more appropriate technique.

Structural Equation Modeling

Structural equation modeling, often referred to simply as LISREL (the name of one of the more popular software packages), is a technique that allows separate relationships for each of a set of dependent variables. In its simplest sense, structural equation modeling provides the appropriate and most efficient estimation technique for a series of separate multiple regression equations estimated simultaneously. It is characterized by two basic components: (1) the structural model and (2) the measurement model. The *structural model* is the "path" model, which relates independent to dependent variables. In such situations, theory, prior experience, or other guidelines allow the researcher to distinguish which independent variables predict each dependent variable. Models discussed previously that accommodate multiple dependent variables—multivariate analysis of variance and canonical correlation—are not applicable in this situation because they allow only a *single* relationship between dependent and independent variables.

The *measurement model* allows the researcher to use several variables (**indicators;** see Key Terms) for a single independent or dependent variable. For example, the dependent variable might be a concept represented by a summated scale, such as self-esteem. In the measurement model the researcher can assess the contribution of each scale item as well as incorporate how well the scale measures the concept (reliability) into the estimation of the relationships between dependent and independent variables. This procedure is similar to performing a factor analysis (discussed in a later section) of the scale items and using the factor scores in the regression.

Other Emerging Multivariate Techniques

The advent of widespread computing power ushered in the era of multivariate analysis as we know it today, with a number of specialized techniques applicable to a wide range of situations. We are now poised, however, at the beginning of an era in which multivariate analysis incorporates new approaches to identifying and representing multivariate relationships. One area of development is in multivariate systems, which involves work on data mining and neural networks. Data mining is the attempt to quantify relationships among large amounts of information with minimal prespecification of the nature of the relationships. One technique often used in conjunction with data mining is that of neural networks, a flexible analysis technique capable of performing both relationship identification (similar to multiple regression or discriminant analysis) or data reduction and structure analysis (analogous to factor or cluster analysis). Neural networks differ from the more traditional multivariate techniques discussed previously in both the formulation of the model and the more complex types of relationships that can be accommodated. Another area involves a move away from the traditional statistical theory of inferential statistics through the development of the technique of resampling or bootstrapping. This technique eliminates the need for the statistical assumptions of sampling distributions (such as normality) by actually using the computer to "resample" the original sample with replacement and generate an empirical estimate of the sampling distribution. An overview of these new techniques are provided in chapter 12.

A Classification of Multivariate Techniques

To assist you in becoming familiar with the specific multivariate techniques, we present a classification of multivariate methods in Figure 1.2 (pages 20–21). This classification is based on three judgments the researcher must make about the research objective and nature of the data: (1) Can the variables be divided into independent and dependent classifications based on some theory? (2) If they can, how many variables are treated as dependent in a single analysis? (3) How are the variables, both dependent and independent, measured? Selection of the appropriate multivariate technique depends on the answers to these three questions.

When considering the application of multivariate statistical techniques, the first question to be asked is: Can the data variables be divided into independent and dependent classifications? The answer to this question indicates whether a dependence or interdependence technique should be utilized. Note that in Figure 1.2, the dependence techniques are on the left side and the interdependence techniques are on the right. A **dependence technique** may be defined as one in which a variable or set of variables is identified as the **dependent variable** to be predicted or explained by other variables known as **independent variables.** An example of a dependence technique is multiple regression analysis. In contrast, an **interdependence technique** is one in which no single variable or group of variables is defined as being independent or dependent. Rather, the procedure involves the simultaneous analysis of all variables in the set. Factor analysis is an example of an interdependence technique. Let us focus on dependence techniques first and use the classification in Figure 1.2 to select the appropriate multivariate method.

The different dependence techniques can be categorized by two characteristics: (1) the number of dependent variables and (2) the type of measurement scale employed by the variables. First, regarding the number of dependent variables, dependence techniques can be classified as those having a single dependent variable, several dependent variables, or even several dependent/independent relationships. Second, dependence techniques can be further classified as those with either metric (quantitative/numerical) or nonmetric (qualitative/categorical) dependent variables. If the analysis involves a single dependent variable that is metric, the appropriate technique is either multiple regression analysis or conjoint analysis. Conjoint analysis is a special case. It is a dependence procedure that may treat the dependent variable as either nonmetric or metric, depending on the type of data collected. On the other hand, if the single dependent variable is nonmetric (categorical), then the appropriate techniques are multiple discriminant analysis and linear probability models. In contrast, when the research problem involves several dependent variables, four other techniques of analysis are appropriate. If the several dependent variables are metric, we must then look to the independent variables. If the independent variables are nonmetric, the technique of multivariate analysis of variance (MANOVA) should be selected. If the independent variables are metric, canonical correlation is appropriate. If the several dependent variables are nonmetric, then they can be transformed through dummy variable coding (0–1) and canonical analysis can again be used.* Finally, if a set of dependent/independent variable relationships is postulated, then structural equation modeling is appropriate.

There is a close relationship between the various dependence procedures, which can be viewed as a family of techniques. Table 1.2 defines the various multivariate dependence techniques in terms of the nature and number of dependent and independent variables. As we can see, canonical correlation can be considered to be the general model upon which many other multivariate techniques are based, because it places the least restrictions on the type and number of variables in both the dependent and independent variates. As restrictions are placed on the variates, more precise conclusions can be reached based on the specific scale of data measurement employed. Thus multivariate techniques range from the quite general method of canonical analysis to the quite specialized technique of structural equation modeling.

TABLE 1.2 The Relationship between Multivariate Dependence Methods

Canonical Correlation

$$Y_1 + Y_2 + Y_3 + \ldots + Y_n = X_1 + X_2 + X_3 + \ldots + X_n$$

(metric, nonmetric)   *(metric, nonmetric)*

Multivariate Analysis of Variance

$$Y_1 + Y_2 + Y_3 + \ldots + Y_n = X_1 + X_2 + X_3 + \ldots + X_n$$

(metric)   *(nonmetric)*

Analysis of Variance

$$Y_1 = X_1 + X_2 + X_3 + \ldots + X_n$$

(metric)   *(nonmetric)*

Multiple Discriminant Analysis

$$Y_1 = X_1 + X_2 + X_3 + \ldots + X_n$$

(nonmetric)   *(metric)*

Multiple Regression Analysis

$$Y_1 = X_1 + X_2 + X_3 + \ldots + X_n$$

(metric)   *(metric, nonmetric)*

Conjoint Analysis

$$Y_1 = X_1 + X_2 + X_3 + \ldots + X_n$$

(nonmetric, metric)   *(nonmetric)*

Structural Equation Modeling

$$Y_1 = X_{11} + X_{12} + X_{13} + \ldots + X_{1n}$$
$$Y_2 = X_{21} + X_{22} + X_{23} + \ldots + X_{2n}$$
$$Y_m = X_{m1} + X_{m2} + X_{m3} + \ldots + X_{mn}$$

(metric)   *(metric, nonmetric)*

*****Dummy variables** (see Key Terms) are discussed in greater detail later. Briefly, dummy variable coding is a means of transforming nonmetric data into metric data. It involves the creation of so-called dummy variables, in which ones and zeros are assigned to subjects, depending on whether they do or do not possess a characteristic in question. For example, if a subject is male, assign him a 0, if the subject is female, assign her a 1, or the reverse.

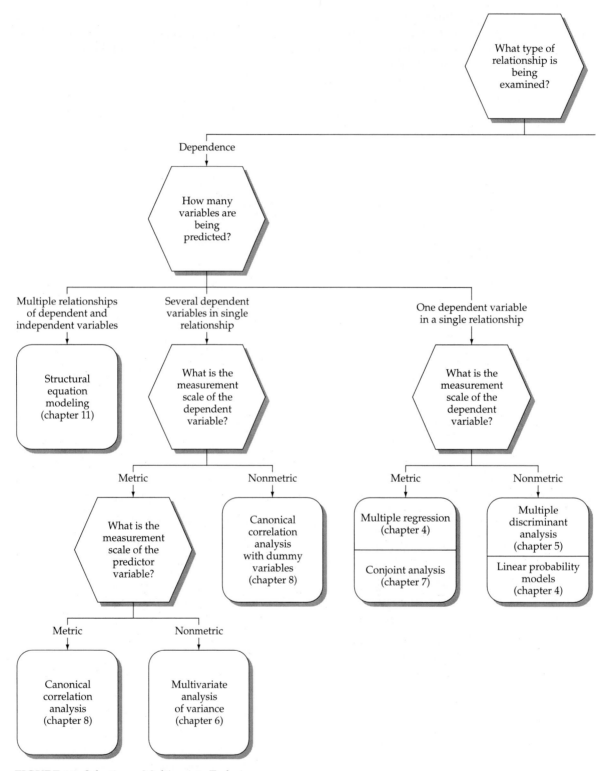

FIGURE 1.2 Selecting a Multivariate Technique

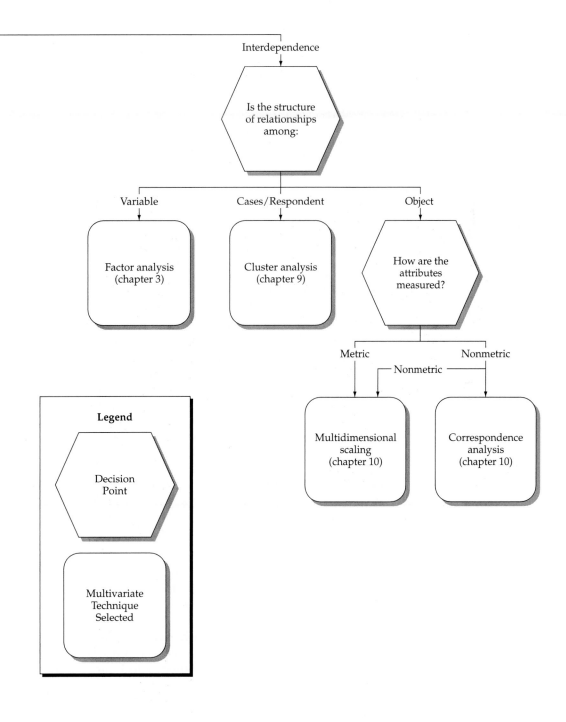

Interdependence techniques are shown on the right side of Figure 1.2. Readers will recall that with interdependence techniques the variables cannot be classified as either dependent or independent. Instead, all the variables are analyzed simultaneously in an effort to find an underlying structure to the entire set of variables or subjects. If the structure of variables is to be analyzed, then factor analysis is the appropriate technique. If cases or respondents are to be grouped to represent structure, then cluster analysis is selected. Finally, if the interest is in the structure of objects, the techniques of multidimensional scaling should be applied. As with dependence techniques, the measurement properties of the techniques should be considered. Generally, factor analysis and cluster analysis are considered to be metric interdependence techniques. However, nonmetric data may be transformed through dummy variable coding for use with special forms of factor analysis and cluster analysis. Both metric and nonmetric approaches to multidimensional scaling have been developed. If the interdependencies of objects measured by nonmetric data are to be analyzed, correspondence analysis is also an appropriate technique.

Guidelines for Multivariate Analyses and Interpretation

As just shown, multivariate analyses have a very diverse character and can be quite powerful. This power is especially tempting when the researcher is unsure of the most appropriate analysis design and relies instead on the multivariate technique as a substitute for the necessary conceptual development. Yet even when applied correctly, the strengths of accommodating multiple variables and relationships create complexity in the results and their interpretation. Therefore, we caution their use without the requisite conceptual foundation to support the selected technique on those basic concepts discussed earlier and the issues addressed in the following section.

We have already discussed several issues particularly applicable to multivariate analyses, and although no single "answer" exists, we have found that analysis and interpretation of any multivariate problem can be helped by following a set of general guidelines. By no means an exhaustive list of considerations, these guidelines represent more of a "philosophy of multivariate analysis" that has served us well. The following sections discuss these points in no particular order, and with equal emphasis on all.

Establish Practical Significance as Well as Statistical Significance

The strength of multivariate analysis is its seemingly magic means of sorting through a myriad number of possible alternatives and finding those that have statistical significance. But with this power must come caution. Many researchers become myopic in focusing solely on the achieved significance of the results without understanding their interpretations, good or bad. A researcher must instead look not only at the statistical significance of the results but also at their practical significance. **Practical significance** asks the question, "So what?" For any managerial application, the results must have a demonstrable effect that justifies action. In academic settings, research is becoming more focused on not only the statistically significant results but also their substantive and theoretical implications, which are many times drawn from their practical significance.

For example, a regression analysis is undertaken to predict repurchase intentions, measured as the probability between 0 and 100 that the customer will shop again with the firm. The study is conducted and the results come back significant at the .05 significance level. Executives rush to embrace the results and modify firm strategy accordingly. But what goes unnoticed is that even though the relationship was significant, the predictive ability was poor—so poor that the estimate of repurchase probability could vary by as much as ±20 percent at the .05 significance level. The "statistically significant" relationship could thus have a range of error of 40 percentage points! A customer predicted to have a 50 percent chance of return could really have probabilities from 30 percent to 70 percent, representing unacceptable levels upon which to take action. Researchers and managers did not probe the practical or managerial significance of the results, wherein they would have seen that the relationship still needed refinement if it were to be relied upon to guide strategy in any substantive sense.

Sample Size Affects All Results

The discussion of statistical power demonstrated the substantial impact sample size plays in achieving statistical significance, both in small and large sample sizes. For smaller samples, the sophistication and complexity of the multivariate technique may easily result in either (1) too little statistical power for the test to realistically identify significant results or (2) too easily an "overfitting" of the data such that the results are artificially good because they fit the sample very well, yet have no generalizability. A similar impact also occurs for large sample sizes, which, as discussed earlier, can make the statistical tests overly sensitive. Any time that sample sizes exceed 200 to 400 respondents, the researcher should examine all significant results to ensure that they have practical significance due to the increased statistical power from the sample size. Sample sizes also affect the results when the analyses involve groups of respondents, such as discriminant analysis or MANOVA. Unequal sample sizes among groups influence the results and require additional interpretation and/or analysis. Thus, a researcher or user of multivariate techniques should always assess the results in light of the sample used in the analysis.

Know Your Data

Multivariate techniques, by their very nature, identify complex relationships that are very difficult to represent simply. As a result, the tendency is to accept the results without the typical examination one undertakes in univariate and bivariate analyses (e.g., scatterplots of correlations and boxplots of mean comparisons). But such "shortcuts" can be a prelude to disaster. Multivariate analyses require an even *more rigorous* examination of the data because the influence of outliers, violations of assumptions, and missing data can be compounded across several variables to have quite substantial effects. To utilize the full benefits of multivariate techniques, the researcher also must "know where to look" with alternative formulations of the original model, such as nonlinear and interactive relationships. The researcher has, however, an ever-expanding set of diagnostic techniques that enable these multivariate relationships to be discovered in ways quite similar to the univariate and bivariate methods. The multivariate researcher must take the time to utilize these diagnostic measures for a greater understanding of the data and the basic relationships that exist.

Strive for Model Parsimony

Multivariate techniques are designed to accommodate multiple variables in the analysis. This feature, however, should not substitute for conceptual model development *before* the multivariate techniques are applied. Although it is always more important to avoid omitting a critical predictor variable, termed **specification error,** for several reasons the researcher must also avoid inserting variables indiscriminately and letting the multivariate technique "sort out" the relevant variables. First, irrelevant variables usually increase a technique's ability to fit the sample data, but at the expense of overfitting the data and making them less generalizable to the population. Second, irrelevant variables do not typically bias the estimates of the relevant variables, but they can mask the true effects because of multicollinearity. **Multicollinearity** represents the degree to which any variable's effect can be predicted or accounted for by the other variables in the analysis. As multicollinearity rises, the ability to define any variable's effect is diminished. Thus, including variables that are conceptually not relevant can have several potentially harmful effects, even if the additional variables do not directly bias the model results.

Look at Your Errors

Even with the statistical prowess of multivariate techniques, rarely do we achieve the best prediction in the first analysis. The researcher is then faced with the question, "Where does one go from here?" The best answer is to look at the errors in prediction, whether they are the residuals from regression analysis, the misclassification of observations in discriminant analysis, or outliers in cluster analysis. In each case, the researcher should use the errors in prediction not as a measure of failure or merely something to eliminate but as a starting point for diagnosing the validity of the obtained results and an indication of the remaining unexplained relationships.

Validate Your Results

The ability of multivariate analyses to identify complex interrelationships also means that results can be found that are specific only to the sample and not generalizable to the population. The researcher must always ensure that there are sufficient observations per estimated parameter to avoid "overfitting" the sample, as discussed earlier. But just as important are the efforts to validate the results by one of several methods, including (1) splitting the sample and using one subsample to estimate the model and the second subsample to estimate the predictive accuracy, (2) employing a bootstrapping technique [9], or (3) even gathering a separate sample to ensure that the results are appropriate for other samples. Whatever multivariate technique is employed, the researcher must strive not only to estimate a significant model but to ensure that it is representative of the population as a whole. Remember, the objective is not to find the best "fit" just to the sample data but instead to develop a model that best describes the population as a whole.

A Structured Approach to Multivariate Model Building

As we discuss the numerous multivariate techniques available to the researcher and the myriad set of issues involved in their application, it becomes apparent that the successful completion of a multivariate analysis involves more than just the selection of the correct method. Issues ranging from problem definition to a critical diagnosis of the results must be addressed. To aid the researcher or user in applying multivariate methods, a six-step approach to multivariate analysis is presented. The intent is not to provide a rigid set of procedures to follow but, instead, to provide a series of guidelines that emphasize a model-building approach. A model-building approach focuses the analysis on a well-defined research plan, starting with a conceptual model detailing the relationships to be examined. Once defined in conceptual terms, the empirical issues can be addressed, including the selection of the specific multivariate technique and the implementation issues. After significant results have been obtained, their interpretation becomes the focus, with special attention directed toward the variate. Finally, the diagnostic measures ensure that the model is not only valid for the sample data but that it is as generalizable as possible. The following discussion briefly describes each step in this approach.

This six-step model-building process provides a framework for developing, interpreting, and validating any multivariate analysis. Each researcher must develop criteria for "success" or "failure" at each stage, but the discussions of each technique provide guidelines whenever available. Emphasis on a model-building approach here, rather than just the specifics of each technique, should provide a broader base of model development, estimation, and interpretation that will improve the multivariate analyses of practitioner and academician alike.

Stage 1: Define the Research Problem, Objectives, and Multivariate Technique to Be Used

The starting point for any multivariate analysis is to define the research problem and analysis objectives in conceptual terms before specifying any variables or measures. The role of conceptual model development, or theory, cannot be overstated. No matter whether in academic or applied research, the researcher must first view the problem in conceptual terms by defining the concepts and identifying the fundamental relationships to be investigated. Developing a conceptual model is not the exclusive domain of academicians; it is just as suited to the application of real-world experience.

A conceptual model need not be complex and detailed; instead, it can be just a simple representation of the relationships to be studied. If a dependence relationship is proposed as the research objective, the researcher needs to specify the dependent and independent concepts. For an application of an interdependence technique, the dimensions of structure or similarity should be specified. Note that a concept, rather than a variable, is defined in both dependence and interdependence situations. The researcher first identifies the ideas or topics of interest rather than focusing on the specific measures to be used. This minimizes the chance that relevant concepts will be omitted in the effort to develop measures and to define the specifics of the research design. Readers interested in conceptual model development should see chapter 11.

With the objectives and conceptual model specified, the researcher has only to choose the appropriate multivariate technique. Having selected the use of either a dependence or interdependence method, the last decision is to select the specific technique based on the measurement characteristics of the dependent and independent variables. The variables may be specified prior to the study in its design or be defined after the data have been collected when specific analyses are defined.

Stage 2: Develop the Analysis Plan

With the conceptual model established and the multivariate technique selected, attention turns to the implementation issues. For each technique, the researcher must develop an analysis plan that addresses the set of issues particular to its purpose and design. The issues include general considerations such as minimum or desired sample sizes and allowable or required types of variables (metric versus nonmetric) and estimation methods, as well as specific issues such as the type of association measures used in multidimensional scaling, the estimation of aggregate or disaggregate results in conjoint analysis or the use of special variable formulations to represent nonlinear or interactive effects in regression. In each instance, these issues resolve specific details and finalize the model formulation and requirements for data collection.

Stage 3: Evaluate the Assumptions
Underlying the Multivariate Technique

With data collected, the first task is not to estimate the multivariate model but to evaluate the underlying assumptions. All multivariate techniques have underlying assumptions, both statistical and conceptual, that substantially impact their ability to represent multivariate relationships. For the techniques based on statistical inference, the assumptions of multivariate normality, linearity, independence of the error terms, and equality of variances in a dependence relationship must all be met. Assessing these assumptions is discussed in more detail in chapter 2. Each technique also has a series of conceptual assumptions dealing with such issues as model formulation and the types of relationships represented. Before any model estimation is attempted, the researcher must ensure that both statistical and conceptual assumptions are met.

Stage 4: Estimate the Multivariate Model
and Assess Overall Model Fit

With the assumptions satisfied, the analysis proceeds to the actual estimation of the multivariate model and an assessment of overall model fit. In the estimation process, the researcher may choose among options to meet specific characteristics of the data (e.g., use of covariates in MANOVA) or to maximize the fit to the data (e.g., rotation of factors or discriminant functions). After the model is estimated, the overall model fit is evaluated to ascertain whether it achieves acceptable levels on the statistical criteria (e.g., level of significance), identifies the proposed relationships, and achieves practical significance. Many times, the model will be respecified in an attempt to achieve better levels of overall fit and/or explanation. In all cases, however, an acceptable model must be obtained before proceeding.

No matter what level of overall model fit is found, the researcher must also determine if the results are unduly affected by any single or small set of observations that indicate the results may be unstable or not generalizable. These efforts ensure that the results are "robust" and stable by applying reasonably well to all observations in the sample. Ill-fitting observations may be identified as outliers, influential observations, or other disparate results (e.g., single-member clusters or seriously misclassified cases in discriminant analysis).

Stage 5: Interpret the Variate(s)

With an acceptable level of model fit, interpreting the variate(s) reveals the nature of the multivariate relationship. The interpretation of effects for individual variables is made by examining the estimated coefficients (weights) for each variable in the variate (e.g., regression weights, factor loadings, or conjoint utilities). Moreover, some techniques also estimate multiple variates that represent underlying dimensions of comparison or association (i.e., discriminant functions or principal components). The interpretation may lead to additional respecifications of the variables and/or model formulation, wherein the model is reestimated and then interpreted again. The objective is to identify empirical evidence of multivariate relationships in the sample data that can be generalized to the total population.

Stage 6: Validate the Multivariate Model

Before accepting the results, the researcher must subject them to one final set of diagnostic analyses that assess the degree of generalizability of the results by the available validation methods. The attempts to validate the model are directed toward demonstrating the generalizability of the results to the total population (see the earlier discussion of validation techniques). These diagnostic analyses add little to the interpretation of the results but can be viewed as "insurance" that the results are the most descriptive of the data and generalizable to the population.

A Decision Flowchart

For each multivariate technique, the six-step approach to multivariate model building will be portrayed in a decision flowchart partitioned into two sections. The first section (stages 1 through 3) deals with the issues addressed while preparing for actual model estimation (i.e., research objectives, research design considerations, and testing for assumptions). The second section of the decision flowchart (stages 4 through 6) deals with the issues pertaining to model estimation, interpretation, and validation. The decision flowchart provides the researcher with a simplified but systematic method of applying the structural approach to multivariate model building to any application of the multivariate technique.

Databases

To explain and illustrate each of the multivariate techniques more fully, we use hypothetical data sets throughout the book. These data sets were all obtained from the Hair, Anderson, and Tatham Company (HATCO), a large (though nonexistent) industrial supplier. Each data set was obtained from surveys of HATCO customers collected through an established marketing research firm.

Primary Database

The primary database, consisting of 100 observations on 14 separate variables, is an example of a segmentation study for a business-to-business situation, specifically a survey of existing customers of HATCO. Three types of information were collected. The first type is the perception of HATCO on seven attributes identified in past studies as the most influential in the choice of suppliers. The respondents, purchasing managers of firms buying from HATCO, rated HATCO on each attribute. The second type of information relates to actual purchase outcomes, either the evaluations of each respondent's satisfaction with HATCO or the percentage of that respondent's product purchases from HATCO. The third type of information contains general characteristics of the purchasing companies (e.g., firm size, industry type).

The data provided should give HATCO a better understanding of both the characteristics of its customers and the relationships between their perceptions of HATCO and their actions toward HATCO (purchases and satisfaction). A brief description of the database variables is provided in Table 1.3, in which the variables are classified as either independent or dependent and either metric or nonmetric. A listing of the database is provided in appendix A for those who wish to reproduce the solutions reported in this book. A definition of each variable and an explanation of its coding is given in the following sections.

Perceptions of HATCO

Each of the variables was measured on a graphic rating scale, where a 10-centimeter line was drawn between the endpoints, labeled "Poor" and "Excellent."

Poor   Excellent

Respondents indicated their perceptions by making a mark anywhere on the line. The mark was then measured and the distance from 0 (in centimeters) was recorded. The result was a scale ranging from 0 to 10, rounded to a single decimal place. The seven HATCO attributes rated by each respondent are as follows:

X_1 Delivery speed—amount of time it takes to deliver the product once an order has been confirmed

X_2 Price level—perceived level of price charged by product suppliers

X_3 Price flexibility—perceived willingness of HATCO representatives to negotiate price on all types of purchases

X_4 Manufacturer's image—overall image of the manufacturer or supplier

X_5 Overall service—overall level of service necessary for maintaining a satisfactory relationship between supplier and purchaser

X_6 Salesforce image—overall image of the manufacturer's salesforce

X_7 Product quality—perceived level of quality of a particular product (e.g., performance or yield)

Purchase Outcomes

Two specific measures were obtained that reflected the outcomes of the respondent's purchase relationships with HATCO. These measures include:

X_9 Usage level—how much of the firm's total product is purchased from HATCO, measured on a 100-point percentage scale, ranging from 0 to 100 percent

X_{10} Satisfaction level—how satisfied the purchaser is with past purchases from HATCO, measured on the same graphic rating scale as perceptions X_1 to X_7

Purchaser Characteristics

The five characteristics of the responding firms used in the study, some metric and some nonmetric, are as follows:

X_8 Size of firm—size of the firm relative to others in this market. This variable has two categories: 1 = large, 0 = small

X_{11} Specification buying—extent to which a particular purchaser evaluates each purchase separately (total value analysis) versus the use of specification buying, which details precisely the product characteristics desired. This variable has two categories: 1 = employs total value analysis approach, evaluating each purchase separately; 0 = use of specification buying

X_{12} Structure of procurement—method of procuring or purchasing products within a particular company. This variable has two categories: 1 = centralized procurement, 0 = decentralized procurement

X_{13} Type of industry—industry classification in which a product purchaser belongs. This variable has two categories: 1 = industry A, 0 = other industries

X_{14} Type of buying situation—type of situation facing the purchaser. This variable has three categories: 1 = new task, 2 = modified rebuy, 3 = straight rebuy

TABLE 1.3 Description of Database Variables

Variable Description		Variable Type
PERCEPTIONS OF HATCO		
X_1	Delivery speed	Metric
X_2	Price level	Metric
X_3	Price flexibility	Metric
X_4	Manufacturer's image	Metric
X_5	Overall service	Metric
X_6	Salesforce image	Metric
X_7	Product quality	Metric
PURCHASE OUTCOMES		
X_9	Usage level	Metric
X_{10}	Satisfaction level	Metric
PURCHASER CHARACTERISTICS		
X_8	Size of firm	Nonmetric
X_{11}	Specification buying	Nonmetric
X_{12}	Structure of procurement	Nonmetric
X_{13}	Type of industry	Nonmetric
X_{14}	Type of buying situation	Nonmetric

Other Databases

Three other specialized databases are used in the text. Chapter 2 employs a smaller database of many of these variables obtained in some pretest surveys. The purpose is to illustrate the identification of outliers, handling of missing data, and testing of statistical assumptions. Chapters 8 and 10 examine databases with the unique data needed for these techniques. In each instance, the database is described more fully in those chapters. A full listing of these databases is given in appendix A as well.

Organization of the Remaining Chapters

The remaining chapters of the text are organized into four sections, each addressing a separate stage in performing a multivariate analysis.

- *Section 1: Preparing for a Multivariate Analysis* addresses issues that must be resolved before a multivariate analysis can be performed. This section begins with chapter 2, which covers the topics of accommodating missing data, assurance of meeting the underlying statistical assumptions, and identifying outliers that might disproportionately impact the results. Chapter 3 covers factor analysis, a technique particularly suited to examining the relationships among variables and the opportunities for creating summated scales. These two chapters combine to provide the researcher not only the diagnostic tools necessary for preparing the data for analysis, but also the means for data reduction and scale construction that can be included in other multivariate techniques.
- *Section 2: Dependence Techniques* covers five dependence techniques—multiple regression, discriminant analysis, multivariate analysis of variance, conjoint analysis, and canonical correlation (chapters 4–8, respectively). Dependence techniques, as noted earlier, allow the researcher to assess the degree of relationship between the dependent and independent variables. The dependence techniques vary in the type and character of the relationship as reflected in the measurement properties of the dependent and independent variables. Each technique is examined for its unique perspective on assessing a dependence relationship and its ability to address a particular type of research objective.
- *Section 3: Interdependence Techniques* (chapters 9–10) covers the techniques of cluster analysis and multidimensional scaling. These techniques present the researcher with tools particulary suited to assessing structure by focusing on the portrayal of the relationships among and between objects, whether they are respondents (cluster analysis) or objects such as firms, products, and so forth (multidimensional scaling). It should be noted that one of the primary interdependence techniques, factor analysis and its ability to assess the relationship among variables, has already been covered in section 1.
- *Section 4: Advanced and Emerging Techniques* (chapters 11 and 12) provides the researcher an introduction to a widely used advanced multivariate technique, structural equation modeling, as well as some newly emerging techniques in the areas of data mining, neural networks, and bootstrapping. The objectives of these two chapters is not to present a complete treatment of the issues and uses of these techniques, but instead to provide an introduction that will enable the researcher to evaluate the potential use of these techniques in specific research situations.

Summary

This chapter has introduced the exciting, challenging topic of multivariate data analysis. The following chapters discuss each of the techniques in sufficient detail to enable the novice researcher to understand what a particular technique can achieve, when and how it should be applied, and how the results of its application are to be interpreted. End-of-chapter summaries of readings from the professional and academic literature further demonstrate the application and interpretation of the techniques.

Questions

1. In your own words, define multivariate analysis.
2. Name several factors that have contributed to the increased application of techniques for multivariate data analysis in recent years.
3. List and describe the multivariate data analysis techniques described in this chapter. Cite examples for which each technique is appropriate.
4. Explain why and how the various multivariate methods can be viewed as a family of techniques.
5. Why is knowledge of measurement scales important to an understanding of multivariate data analysis?
6. What are the differences between statistical and practical significance? Is one a prerequisite for the other?
7. What are the implications of low statistical power? How can the power be improved if it is deemed too low?
8. Detail the model-building approach to multivariate analysis, focusing on the major issues at each step.

References

1. Bearden, William O., Richard G. Netemeyer, and Mary F. Mobley (1993), *Handbook of Marketing Scales, Multi-Item Measures for Marketing and Consumer Behavior.* Newbury Park, Calif.: Sage.
2. BMDP Statistical Software, Inc. (1991), *SOLO Power Analysis.* Los Angeles.
3. Brent, Edward E., Edward J. Mirielli, and Alan Thompson (1993), *Ex-Sample™: An Expert System to Assist in Determining Sample Size, Version 3.0.* Columbia, Mo.: Idea Works.
4. Brent, Edward E., et al. (1991), *Statistical Navigator Professional™: An Expert System to Assist in Selecting Appropriate Statistical Analyses, Version 1.0.* Columbia, Mo.: Idea Works.
5. Brunner, Gordon C., and Paul J. Hensel (1993), *Marketing Scales Handbook, A Compilation of Multi-Item Measures.* Chicago: American Marketing Association.
6. Cohen, J. (1977), *Statistical Power Analysis for the Behavioral Sciences.* New York: Academic Press.
7. Gatty, R. (1966), "Multivariate Analysis for Marketing Research: An Evaluation." *Applied Statistics* 15 (November): 157–172.
8. Hardyck, C. D., and L. F. Petrinovich (1976), *Introduction to Statistics for the Behavioral Sciences,* 2d ed. Philadelphia: Saunders.
9. Mooney, Christopher Z., and Robert D. Duval (1993), *Bootstrapping: A Nonparametric Approach to Statistical Inference.* Beverly Hills, Calif.: Sage.
10. Sullivan, John L., and Stanley Feldman (1979), *Multiple Indicators: An Introduction.* Beverly Hills, Calif.: Sage.

PREPARING FOR A MULTIVARIATE ANALYSIS

Overview

Section 1 provides a set of tools and analyses that help to prepare the researcher for the increased complexity of a multivariate analysis. The prudent researcher appreciates the need for a higher level of understanding of the data, both in statistical and conceptual terms. Although the multivariate techniques discussed in this text present the researcher with a powerful set of analytical tools, they also pose the risk of (a) further separating the researcher from a solid understanding of the data and (b) leading to the misplaced notions that they present a "quick and easy" means of identifying relationships. As the researcher relies more heavily on these techniques to find the answer and less on a conceptual basis and understanding of the fundamental properties of the data, the risk increases for serious problems in the misapplication of techniques, violation of statistical properties, or the inappropriate inference and interpretation of the results. These risks can never be totally eliminated, but the tools and analyses discussed in this section will improve the researcher's ability to recognize many of these problems as they occur and to apply the appropriate remedy.

Chapters in Section 1

This section begins with chapter 2, Examining Your Data, which covers the topics of accommodating missing data, meeting the underlying statistical assumptions, and identifying outliers that might disproportionately impact the results. These analyses provide simple empirical assessments that detail the critical statistical properties of the data. Chapter 3, Factor Analysis, presents a discussion of an interdependence technique particularly suited to examining the relationships among variables and the creation of summated scales. The "search for structure" with factor analysis can reveal substantive interrelationships among variables and provide an objective basis for both conceptual model development and improved parsimony among the variables in a multivariate analysis. Thus, the two chapters in this section combine to provide the researcher not only the diagnostic tools necessary for preparing data for analysis, but also the means for data reduction and scale construction that can markedly improve other multivariate techniques.

CHAPTER **2**

Examining Your Data

LEARNING OBJECTIVES

Upon completing this chapter, you should be able to do the following:

- Select the appropriate graphical method to examine the characteristics of the data or relationships of interest.
- Understand the different types of missing data processes.
- Assess the type and potential impact of missing data.
- Explain the advantages and disadvantages of the approaches available for dealing with missing data.
- Identify univariate, bivariate, and multivariate outliers.
- Test your data for the assumptions underlying most multivariate techniques.
- Determine the best method of data transformation given a specific problem.
- Understand how to incorporate nonmetric variables as metric variables.

CHAPTER PREVIEW

This chapter reviews and describes the methods currently available to examine data. Data examination is a time-consuming, but necessary, step that is sometimes overlooked by researchers. Careful analysis of data leads to better prediction and more accurate assessment of dimensionality. The introductory section of this chapter offers a summary of various graphical techniques available to the researcher as a means of representing data. These techniques provide the researcher with a set of simple yet comprehensive ways to examine both the individual variables and the relationships among them. Other important concerns to the researcher when examining data are how to assess and overcome pitfalls resulting from the research design and data collection. Specifically, this chapter addresses the evaluation of missing data, identification of outliers, and testing of the assumptions underlying most multivariate techniques. Missing data are a nuisance to researchers and may result from data entry errors or from the omission of answers by respondents. Classification of missing data and the processes, or reasons, underlying their presence are discussed in this chapter. Outliers, or extreme responses, may unduly influence the outcome of any multivariate analysis. For this reason, methods to assess their impact are discussed. Finally, the assumptions underlying most multivariate analyses are reviewed. Before applying any multivariate technique, the researcher must assess the fit of the sample data with the statistical assumptions underlying that multivariate technique. For example, researchers wishing to apply regression analysis (chapter 4) would be particularly interested in assessing the assumptions of normality, homoscedasticity, independence of error, and linearity. Each of these issues should be addressed to some extent for each application of a multivariate technique.

In addition, this chapter introduces the researcher to methods of incorporating nonmetric variables in applications that require metric variables through the creation of a special type of metric variable known as dummy variables. The applicability of using dummy variables varies with each data analysis project.

KEY TERMS

Before starting the chapter, review the key terms to develop an understanding of the concepts and terminology used. Throughout the chapter the key terms appear in **boldface.** Other points of emphasis in the chapter are *italicized.* Also, cross-references within the Key Terms appear in *italics.*

All-available approach   *Imputation method* for missing data that computes values based on all available valid observations.

Boxplot   Method of representing the distribution of a variable. A box represents the major portion of the distribution, and the extensions—called whiskers—reach to the extreme points of the distribution. Very useful in making comparisons of one or more variables across groups.

Censored data   Observations that are incomplete in a systematic and known way. One example occurs in the study of causes of death in a sample in which some individuals are still living. Censored data are an example of *ignorable missing data.*

Comparison group   The category of a nonmetric variable that receives all zeros when *indicator coding* is used or all minus ones (−1s) when *effects coding* is used in creating *dummy variables.*

Complete case approach   Approach for handling missing data that computes values based on data from only complete cases, that is, cases with no *missing data.*

Data transformations   A variable may have an undesirable characteristic, such as nonnormality, that detracts from its use in a multivariate technique. A transformation, such as taking the logarithm or square root of the variable, creates a transformed variable that is more suited to portraying the relationship. Transformations may be applied to either the dependent or independent variables, or both. The need and specific type of transformation may be based on theoretical reasons (e.g., transforming a known nonlinear relationship) or empirical reasons (e.g., problems identified through graphical or statistical means).

Dummy variable   Special metric variable used to represent a single category of a nonmetric variable. To account for L levels of an nonmetric variable, $L - 1$ dummy variables are needed. For example, gender is measured as male or female and could be represented by two dummy variables (X_1 and X_2). When the respondent is male, $X_1 = 1$ and $X_2 = 0$. Likewise, when the respondent is female, $X_1 = 0$ and $X_2 = 1$. However, when $X_1 = 1$, we know that X_2 must equal 0. Thus we need only one variable, either X_1 or X_2, to represent the variable gender. If a nonmetric variable has three levels, only two dummy variables are needed. We always have one dummy variable less than the number of levels for the nonmetric variable.

Effects coding   Method for specifying the reference category for a set of *dummy variables* where the reference category receives a value of minus one (-1) across the set of dummy variables. With this type of coding, the dummy variable coefficients represent group deviations from the mean of all groups. This is in contrast to *indicator coding.*

Heteroscedasticity   See *homoscedasticity.*

Histogram   Graphical display of the distribution of a single variable. By forming frequency counts in categories, the shape of the variable's distribution can be shown. Used to make a visual comparison to the *normal distribution.*

Homoscedasticity   When the variance of the error terms (ϵ) appears constant over a range of predictor variables, the data are said to be homoscedastic. The assumption of equal variance of the population error E (where E is estimated from ϵ) is critical to the proper application of linear regression. When the error terms have increasing or modulating variance, the data are said to be *heteroscedastic.* Analysis of *residuals* best illustrates this point.

Ignorable missing data   *Missing data process* that is explicitly identifiable and/or is under the control of the researcher. Ignorable missing data do not require a remedy because the missing data are explicitly handled in the technique used.

Imputation methods   Process of estimating the *missing data* of an observation based on valid values of the other variables. The objective is to employ known relationships that can be identified in the valid values of the sample to assist in representing or even estimating the replacements for missing values.

Indicator coding   Method for specifying the reference category for a set of *dummy variables* where the reference category receives a value of zero across the set of dummy variables. The dummy variable coefficients represent the category differences from the reference category. Also see *effects coding.*

Kurtosis   Measure of the peakedness or flatness of a distribution when compared with a normal distribution. A positive value indicates a relatively peaked distribution, and a negative value indicates a relatively flat distribution.

Linearity   Used to express the concept that the model possesses the properties of additivity and homogeneity. In a simple sense, linear models predict values that fall in a straight line by having a constant unit change (slope) of the dependent variable for a constant unit change of the independent variable. In the population model $Y = b_0 + b_1X_1 + E$, the effect of a change of 1 in X_1 is to add b_1 (a constant) units of Y.

Missing at random (MAR)   Classification of *missing data* applicable when missing values of Y depend on X, but not on Y. When missing data are MAR, observed data for Y are a truly random sample for the X values in the sample, but not a random sample of all Y values due to missing values of X.

Missing completely at random (MCAR)   Classification of *missing data* applicable when missing values of Y are not dependent on X. When missing data are MCAR, observed values of Y are a truly random sample of all Y values, with no underlying process that lends bias to the observed data.

Missing data   Information not available for a subject (or case) about whom other information is available. Missing data often occur when a respondent fails to answer one or more questions in a survey.

Missing data process   Any systematic event external to the respondent (such as data entry errors or data collection problems) or any action on the part of the respondent (such as refusal to answer a question) that leads to *missing data.*

Multivariate graphical display   Method of presenting a multivariate profile of an observation on three or more variables. The methods include approaches such as glyphs, mathematical transformations, and even iconic representations (e.g., faces).

Normal distribution   Purely theoretical continuous probability distribution in which the horizontal axis represents all possible values of a variable and the vertical axis represents the probability of those values occurring. The scores on the variable are clustered around the mean in a symmetrical, unimodal pattern known as the bell-shaped, or normal, curve.

Normal probability plot   Graphical comparison of the form of the distribution to the *normal distribution*. In the normal probability plot, the normal distribution is represented by a straight line angled at 45 degrees. The actual distribution is plotted against this line, so that any differences are shown as deviations from the straight line, making identification of differences quite apparent and interpretable.

Normality   Degree to which the distribution of the sample data corresponds to a *normal distribution.*

Outlier   An observation that is substantially different from the other observations (i.e., has an extreme value). At issue is its representativeness of the population.

Residual   Portion of a dependent variable not explained by a multivariate technique. Associated with dependence methods that attempt to predict the dependent variable, the residual represents the unexplained portion of the dependent variable. Residuals can be used in diagnostic procedures to identify problems in the estimation technique or to identify unspecified relationships.

Scatterplot   Representation of the relationship between two metric variables portraying the joint values of each observation in a two-dimensional graph.

Skewness   Measure of the symmetry of a distribution; in most instances the comparison is made to a *normal distribution*. A positively skewed distribution has relatively few large values and tails off to the right, and a negatively skewed distribution has relatively few small values and tails off to the left. Skewness values falling outside the range of -1 to $+1$ indicate a substantially skewed distribution.

Stem and leaf diagram   A variant of the *histogram* which provides a visual depiction of the variable's distribution as well as an enumeration of the actual data values.

Variate   Linear combination of variables formed in the multivariate technique by deriving empirical weights applied to a set of variables specified by the researcher.

Introduction

The tasks involved in examining your data may seem mundane and inconsequential but they are an essential part of any multivariate analysis. Multivariate techniques place tremendous analytical power in the researcher's hands, but they also place a greater burden on the researcher to ensure that the statistical and theoretical underpinning on which they are based are also supported. By examining the data before the application of a multivariate technique, the researcher gains several critical insights into the characteristics of the data. First and foremost, the researcher attains a basic understanding of the data and relationships between variables. Multivariate techniques place great demands on the researcher to understand, interpret, and articulate results based on relationships that are ever increasing in complexity. Knowledge of variable interrelationships can aid immeasurably in the specification and refinement of the multivariate model as well as provide a reasoned perspective for interpretation of the results. Second, multivariate techniques demand much more from the data they are to analyze. The statistical power of the multivariate techniques requires larger data sets and more complex assumptions than encountered with univariate analyses. The analytical sophistication needed to ensure that the statistical requirements are met has forced the researcher to use a series of data examination techniques that in many instances match the complexity of the multivariate techniques. Moreover, the effects of missing data, which by definition are not directly represented in the results, can nevertheless be substantial in their impact on the nature and character of the results. The purpose of this chapter is to provide an overview of the available data examination techniques, ranging from the simple process of visual inspection of graphical displays to the multivariate statistical methods involved in handling missing data and testing the assumptions underlying all multivariate methods.

Both novice and experienced researchers may be tempted to skim or even skip this chapter to spend more time in gaining knowledge of a multivariate technique(s). Although the time, effort, and resources devoted to the data examination process may seem almost wasted because often no corrective action is warranted, the researcher should view these techniques as an "investment in multivariate insurance." Even though a technique may estimate properly and obtain results, the "hidden" problems arising from issues in the chapter can lead to potentially catastrophic problems. These problems can be avoided by following these analyses each and every time a multivariate technique is applied. These efforts will more than pay for themselves in the long run, as the occurrence of one serious and possibly fatal problem will make a convert of any researcher. We encourage you to embrace these techniques before problems raised during analysis force you to do so.

This chapter addresses four separate phases of examining your data: (1) a graphical examination of the nature of the variables in the analysis and the relationships that form the basis of multivariate analysis; (2) an evaluation process for understanding the impact missing data can have on the analysis, plus alternatives for retaining cases with missing data in the analysis; (3) the techniques best suited for identifying outliers, those cases that may distort the relationships by their uniqueness on one or more of the variables under study; and (4) the analytical methods necessary to assess the ability of the data to meet the statistical assumptions specific to many multivariate techniques. The chapter concludes by introducing a technique for incorporating nonmetric variables when metric variables are required. A set of replacement metric variables are created to represent the categories of the nonmetric variables.

Graphical Examination of the Data

As discussed earlier, the use of multivariate techniques places an increased burden on the researcher to understand, evaluate, and interpret the more complex results. One aid in these tasks is a thorough understanding of the basic characteristics of the underlying data and relationships. When univariate analyses are considered, the level of understanding is fairly simple. But as the researcher moves to more complex multivariate analyses, the need and level of understanding increase dramatically. This section reviews some of the graphical methods available to assist in gaining a basic understanding of the characteristics of the data, particularly in a multivariate sense.

The advent and widespread use of statistical programs designed for the personal computer has led to increased access to such methods. Most statistical programs have comprehensive modules of graphical techniques available for data examination that are augmented with more detailed statistical measures of data description. The following sections detail some of the more widely used techniques for examining the characteristics of the distribution, bivariate relationships, group differences, and even multivariate profiles.

The Nature of the Variable:
Examining the Shape of the Distribution

The starting point for understanding the nature of any variable is to characterize the shape of its distribution. A number of statistical measures are discussed in a later section on normality, but many times the researcher can gain an adequate perspective on the variable through a **histogram**. A histogram is a graphical representation of a single variable that represents the frequency of occurrences (data values) within data categories. The frequencies are plotted to examine the shape of the distribution of responses. If the integer values ranged from one to ten, the researcher could construct a histogram by counting the number of responses that were a one, a two, and so on. For continuous variables, categories are formed within which the frequency of data values are tabulated. For example, the responses for X_1 from the database introduced in chapter 1 are represented in Figure 2.1. Categories with midpoints of 0.0, .5, 1.0, 1.5, . . . , 6.0 are used. The height of the bars represents the frequencies of data values within each category. If examination of the distribution is to assess its normality (see section on testing assumptions for details on this issue), the normal curve can be superimposed on the distribution as well, as was done in Figure 2.1. The histogram can be used to examine any type of metric variable, from original values to residuals from a multivariate technique.

A variant of the histogram is the **stem and leaf diagram,** which presents the same graphical picture as the histogram but also provides an enumeration of the actual data values. The stem and leaf diagram in Figure 2.2 is composed of *stems* and *leaves.* The stem is the root value, to which the leaves are added. For example, in Figure 2.2, the first stem is 0.0. To this is added the leaf of 0, resulting in a

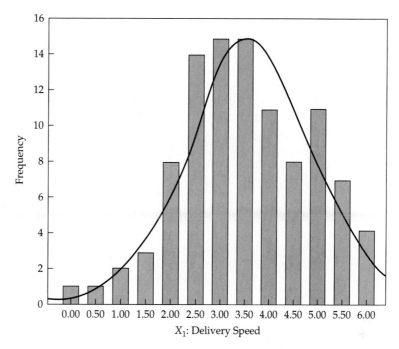

FIGURE 2.1 Graphical Representation of a Univariate Distribution: The Histogram

value of 0. In the next stem, a value of .6 is added to the stem of 0, resulting in a value of .6. If the frequencies of X_1 are compiled, 0.0 and .6 are the first two values. At the other end of the figure, the stem is 6.0. It is associated with two leaves (0 and 1), representing the values 6.0 and 6.1. These are the two highest values for X_1. The stem and leaf diagram provides a general shape of the distribution, as found with the histogram, as well as providing the actual data values.

Frequency	Stem and Leaf		
1.00	0	*	0
1.00	0	.	6
3.00	1	*	013
7.00	1	.	6688999
12.00	2	*	001333444444
10.00	2	.	5566788899
18.00	3	*	000001111233444444
10.00	3	.	5666777889
10.00	4	*	001122233
10.00	4	.	556778999
11.00	5	*	00112223344
5.00	5	.	55689
2.00	6	*	01

Stem width: 1.0
Each leaf: 1 case (s)

Valid cases: 100.0   Missing cases: .0   Percent missing: .0

FIGURE 2.2 Stem and Leaf Plot of X_1 (Delivery Speed)

Examining the Relationship between Variables

Whereas examining the distribution of a variable is essential, many times the researcher is also interested in examining relationships between two or more variables. The most popular method for examining bivariate relationships is the **scatterplot,** a graph of data points based on two variables. One variable defines the horizontal axis and the other variable defines the vertical axis. Variables may be observations, expected values, or even **residuals.** The points in the graph represent the corresponding joint values of the variables for any given case. The pattern of points represents the relationship between variables. A strong organization of points along a straight line characterizes a linear relationship or correlation. A curved set of points may denote a nonlinear relationship, which can be accommodated in many ways (see later discussion on linearity). Or there may be only a seemingly random pattern of points, indicating no relationship.

There are many types of scatterplots. One format particularly suited to multivariate techniques is the scatterplot matrix (see Figure 2.3), in which the scatterplots are represented for all combinations of variables in the lower portion of the matrix. The diagonal contains histograms of the variables. Included in the upper portion of the matrix are the corresponding correlations so that the reader can assess the correlation represented in each scatterplot. Figure 2.3 presents the scatterplots for a set of variables from the HATCO database (X_1, X_2, X_3, X_4, X_5, X_6, X_7, and X_9). For example, the scatterplot in the bottom left corner (X_1 versus X_9) represents a correlation of .676. The points are closely aligned around a straight line, indicative of a high correlation. The scatterplot in the leftmost column, third from the top (X_1 versus X_4) demonstrates the opposite, an almost total lack of relationship as evidenced by the widely dispersed pattern of points and the correlation .050. Scatterplot matrices and individual scatterplots are now available in all popular statistical programs. A variant of the scatterplot is discussed in the following section on outlier detection, where an ellipse representing a specified confidence interval for the bivariate normal distribution is superimposed to allow for outlier identification.

Examining Group Differences

The researcher is also faced with understanding the extent and character of differences between two or more groups for one or more metric variables, such as found in discriminant analysis, analysis of variance, and multivariate analysis of variance. In these cases, the researcher needs to understand how the values are distributed for each group and if there are sufficient differences between the groups to support statistical significance. Another important aspect is to identify **outliers** that may become apparent only when the data values are separated into groups. The method used for this task is the **boxplot,** a pictorial representation of the data distribution. The upper and lower boundaries of the box mark the upper and lower quartiles of the data distribution. Thus, the box length is the distance between the 25th percentile and the 75th percentile, so that the box contains the middle 50 percent of the data values. The asterisk (*) inside the box identifies the median. If the median lies near one end of the box, skewness in that direction is indicated. The larger the box, the greater the spread of the observations. The lines extending from each box (called *whiskers*) represent the distance to the smallest and the largest observations that are less than one quartile range from the box. Outliers are observations that range between 1.0 and 1.5 quartiles away from the

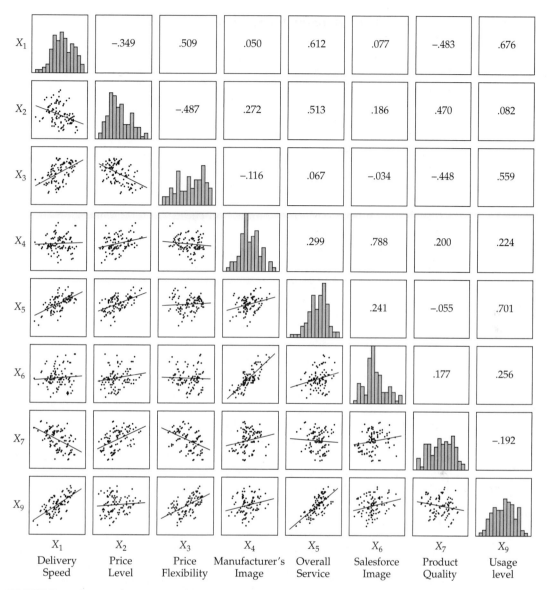

FIGURE 2.3 Scatterplot Matrix of Metric Variables
Note: Values above the diagonal are bivariate correlations, with corresponding scatterplots below the
diagonal. Diagonal portrays the distribution of each variable.

box. Extreme values are those observations greater than 1.5 quartiles away from
the end of the box.

Figure 2.4 (p. 44) shows the boxplots for X_1 (delivery speed) for each group of
X_{14} (type of buying situation) from the HATCO database. The three groups have
markedly different boxplots, indicating differences among the groups in terms of
perceptions of delivery speed. The boxplot for the first type of buying situation
also indicates that an outlier exists. The researcher should examine this observation and then consider the possible remedies. The remedies available for outliers
are discussed in more detail later.

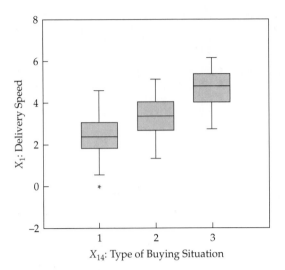

FIGURE 2.4 Box and Whiskers Plot

Multivariate Profiles

To this point the graphical methods have been restricted to univariate or bivariate portrayals. But in many instances, the researcher may desire to compare observations characterized on more than two variables. The need is for a means of presenting a multivariate profile of an observation, whether it be for descriptive purposes or as a complement to analytical procedures. To address this need, a number of multivariate graphical methods have been devised that center around one of three approaches [7]. The first are (a) glyphs, or metroglyphs, which are some form of circle with radii that correspond to a data value; or (b) a multivariate profile, which portrays a barlike profile for each observation. A second form of multivariate display involves a mathematical transformation of the original data into a mathematical relationship, which can then be portrayed graphically. The most common technique is Andrew's Fourier transformation. The final approach is the use of graphical displays with iconic representativeness, the most popular being a face [3]. The value of this type of display is the inherent processing capacity humans have for their interpretation. As noted by Chernoff [3, p. 9]:

> I believe that we learn very early to study and react to real faces. Our library of responses to faces exhausts a large part of our dictionary of emotions and ideas. We perceive the faces as a gestalt and our built-in computer is quick to pick out the relevant information and to filter out the noise when looking at a limited number of faces.

Facial representations provide a potent graphical format but also give rise to a number of considerations that impact the assignment of variables to facial features, unintended perceptions, and the quantity of information that can actually be accommodated. Discussion of these issues is beyond the scope of this text, and interested readers are encouraged to review them before attempting to use these methods [10, 11].

Figure 2.5 contains illustrations of three types of **multivariate graphical displays** produced using SYSTAT, which are also available in several other personal computer-based statistical programs. The upper portion of Figure 2.5 contains ex-

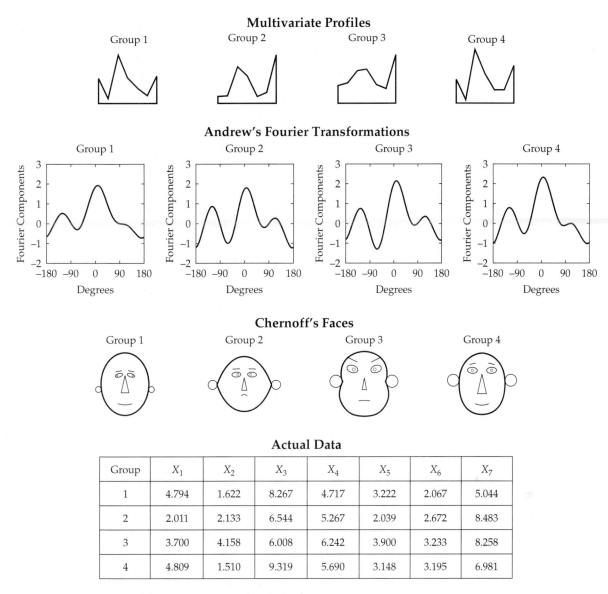

Multivariate Profiles

Group 1　　Group 2　　Group 3　　Group 4

Andrew's Fourier Transformations

Group 1　　Group 2　　Group 3　　Group 4

Chernoff's Faces

Group 1　　Group 2　　Group 3　　Group 4

Actual Data

Group	X_1	X_2	X_3	X_4	X_5	X_6	X_7
1	4.794	1.622	8.267	4.717	3.222	2.067	5.044
2	2.011	2.133	6.544	5.267	2.039	2.672	8.483
3	3.700	4.158	6.008	6.242	3.900	3.233	8.258
4	4.809	1.510	9.319	5.690	3.148	3.195	6.981

FIGURE 2.5 Examples of Multivariate Graphical Displays

amples of each type of multivariate graphical display: profiles, Fourier transformations, and iconic faces. Data values for four observations on seven variables are contained in a table at the bottom of the figure. In this instance, the data are profiles of four customer groups for the seven performance factors from the HATCO database. From the actual data values in the table, similarities and differences are difficult to distinguish, even to the extent that there may not be any differences. The objective of the multivariate profiles is to portray the data in a manner that enables each identification of differences and similarities. The first display in Figure 2.5 contains multivariate profiles, which show the leftmost portion is lowest for group two, and highest for groups one and four. This pattern corresponds to the values of X_1, and comparisons can be made between groups on a single variable or across variables for a single group. The second type of multivariate graphical display in the figure is Andrew's Fourier transformation, which represents the data values by a mathematical expression. Although comparisons on a single value are more difficult, this form of graphical display provides a single representation for generalized comparison and grouping of observations. Finally, iconic symbols (Chernoff's faces) were constructed with the seven variables assigned to various facial features. In this example, X_1 controls depiction of the mouth, X_2 is assigned to the eyebrow features, X_3 is assigned to the nose characteristics, X_4 is assigned to the eyes, X_5 controls the face shape, X_6 relates to the ear, and X_7 is assigned to the pupil position. Regarding X_1, groups one and four have smiles, and group two has a frown. This corresponds to large values for groups one and four and small values for group two. This form of graphical display combines the ability to make specific comparisons between or within groups seen in the profiles as well as the more generalized overall comparisons found in the Andrew's Fourier transformations. The researcher can employ any of these methods when examining multivariate data to provide a format that is many times more insightful than just a review of the actual data values.

Summary

The graphical displays in this section are not intended as a replacement for the statistical diagnostic measures discussed in later sections of this chapter and in other chapters. But they do provide an alternative means of developing a perspective on the character of the data and the interrelationships that exist, even if multivariate in nature. The old adage "a picture is worth a thousand words" holds true many times in the use of graphical displays for comparative or diagnostic applications.

Missing Data

Missing data are a fact of life in multivariate analysis; in fact, rarely does the researcher avoid some form of missing data problem. For this reason, the researcher's challenge is to address the issues raised by missing data that affect the generalizability of the results. To do so, the researcher's primary concern is to determine the reasons underlying the missing data, with the extent of missing data being a secondary issue in most instances. This need to focus on the reasons for missing data comes from the fact that the researcher must understand the processes leading to the missing data in order to select the appropriate course of action.

A **missing data process** is any systematic event external to the respondent (such as data entry errors or data collection problems) or action on the part of the respondent (such as refusal to answer) that leads to missing values. The effects of some missing data processes are known and directly accommodated in the research plan. But others, particularly those based on actions by the respondent, are rarely known. When the missing data processes are unknown, the researcher attempts to identify any patterns in the missing data that would characterize the missing data process. In doing so, the researcher asks such questions as: (1) Are the missing data scattered randomly throughout the observations or are distinct patterns identifiable? and (2) How prevalent are the missing data? If patterns are found and the extent of missing data is sufficient to warrant action, then it is assumed that some missing data process is in operation. Any statistical results based on these data would be biased to the extent that the variables included in the analysis are influenced by the missing data process. The concern for understanding the missing data processes is similar to the need to understand the causes of nonresponse in the data collection process. For example, are those individuals who did not respond different from those who did? If so, do these differences have any impact on the analysis, the results, or their interpretation? Concerns similar to these also arise from missing responses for individual variables.

The impact of missing data is detrimental not only through its potential "hidden" biases of the results but also in its practical impact on the sample size available for analysis. For example, if remedies for missing data are not applied, any observation with missing data on any of the variables will be excluded from the analysis. In many multivariate analyses, particularly survey research applications, missing data may eliminate so many observations that what was an adequate sample is reduced to an inadequate sample. In such situations, the researcher must either gather additional observations or find a remedy for the missing data in the original sample. Although finding a remedy for missing data is the most practical solution, few guidelines exist pertaining to the diagnosis and remedy of missing data. For this reason, the following sections discuss the different types of missing data processes, methods to identify the nature of the missing data processes, and available remedies for accommodating missing data into multivariate analyses.

A Simple Example of a Missing Data Analysis

Table 2.1 (p. 48) contains a simple example of missing data among 20 cases. As typical of many data sets, particularly in survey research, the number of missing data vary widely among both cases and variables. In this example, we can see that all of the variables (V_1 to V_5) have some missing data, with V_3 missing over one-half (55 percent) of all values. Three cases (3, 13, and 15) have more than 50 percent missing data and only five cases have complete data. Overall, 22 percent of the data values are missing. If a multivariate analysis was run that required complete data, the data would be reduced to only five cases, too few for any type of analysis. This level of reduction in available cases is not uncommon in many applications.

More sophisticated remedies for missing data will be discussed in detail in later sections, but an obvious option is the elimination of variables and/or cases. In our example, assuming that the conceptual foundations of the research are not altered substantially by the deletion of a variable, eliminating V_3 is one approach to reducing the number of missing data. By just eliminating V_3, seven additional cases, for a total of 12, now have complete information. If the three cases (3, 13, 15) with exceptionally high numbers of missing data are also eliminated, the total number of missing data is now reduced to only five instances, or 7.4 percent of all values. These five missing data, however, are all present in V_4, and we must look for any patterns among these data as well. By comparing the cases with missing data for V_4 with those having valid V_4 values, we see a pattern emerge with respect to V_2. The five cases with missing values for V_4 also have the five lowest values for V_2, indicating that missing data for V_4 are strongly associated with lower scores on V_2. This systematic association between missing and valid data

TABLE 2.1 Hypothetical Example of Missing Data

						Missing Data by Case	
Case ID	V_1	V_2	V_3	V_4	V_5	Number	Percent
1	1.3	9.9	6.7	3.0	2.6	0	0
2	4.1	5.7			2.9	2	40
3		9.9		3.0		3	60
4	.9	8.6		2.1	1.8	1	20
5	.4	8.3		1.2	1.7	1	20
6	1.5	6.7	4.8		2.5	1	20
7	.2	8.8	4.5	3.0	2.4	0	0
8	2.1	8.0	3.0	3.8	1.4	0	0
9	1.8	7.6		3.2	2.5	1	20
10	4.5	8.0		3.3	2.2	1	20
11	2.5	9.2		3.3	3.9	1	20
12	4.5	6.4	5.3	3.0	2.5	0	9
13					2.7	4	80
14	2.8	6.1	6.4		3.8	1	20
15	3.7			3.0		3	60
16	1.6	6.4	5.0		2.1	1	20
17	.5	9.2		3.3	2.8	1	20
18	2.8	5.2	5.0		2.7	1	20
19	2.2	6.7		2.6	2.9	1	20
20	1.8	9.0	5.0	2.2	3.0	0	0
MISSING DATA BY VARIABLE						TOTAL MISSING VALUES	
Number	2	2	11	6	2	Number: 23	
Percent	10	10	55	30	10	Percent: 23	

directly impacts any analysis in which V_4 and V_2 are both included. In this instance, the researcher must always scrutinize results including either V_4 and V_2 for the possible impact of this missing data process on the results.

Understanding the Reasons Leading to Missing Data

Before any missing data remedy can be implemented, the researcher must first diagnose and understand the missing data processes underlying the missing data. Sometimes these processes are under the control of the researcher and can be explicitly identified. In such instances, the missing data are termed **ignorable,** which means that specific remedies for missing data are not needed because the allowances for missing data are inherent in the technique used [9].

Ignorable Missing Data

One example of an ignorable missing data process is the "missing data" of those observations in a population that are not included when taking a sample. The purpose of multivariate techniques is to generalize from the sample observations to the entire population, which is really an attempt to overcome the missing data of observations not in the sample. The researcher makes this missing data ignorable by using probability sampling to select respondents. Probability sampling allows the researcher to specify that the missing data process leading to the omitted observations is random and that the missing data can be accounted for as sampling error in the statistical procedures. Thus, the "missing data" of the nonsampled observations is ignorable.

Another instance of ignorable missing data occurs when the data are censored. **Censored data** are observations not complete because of their stage in the missing data process. A typical example is an analysis of the causes of death. Respondents who are still living cannot provide complete information (i.e., cause or time of death) and are thus censored. Another interesting example of censored data is found in the attempt to estimate the heights of the U.S. general population based on the heights of armed services recruits (as cited in [9]). The data are censored because in certain years the armed services had height restrictions that varied in level and enforcement. Thus, the researchers are faced with the task of estimating the heights of the entire population when it is known that certain individuals (i.e., all those below the height restrictions) are not included in the sample. In both instances the researcher's knowledge of the missing data process allows for the use of specialized methods, such as event history analysis, to accommodate censored data [9].

The justification for designating missing data as ignorable is that the missing data process is operating at random (i.e., the observed values are a random sample of the total set of values, observed and missing) or explicitly accommodated in the technique used. However, in most instances the missing data process is not explicitly addressed by the techniques used. Thus, the researcher must assess the extent and impact of the missing data to determine whether they are due to a random process or, if not, whether they are amenable to one of the available remedies.

Other Types of Missing Data Processes

Missing data can occur for many reasons and in many situations. One type of missing data process that may occur in any situation is due to procedural factors, such as errors in data entry that create invalid codes, disclosure restrictions (e.g., small counts in U.S. census data), failure to complete the entire questionnaire, or even the morbidity of the respondent. In these situations, the researcher has little control over the missing data processes, but some remedies may be applicable if the missing data are found to be random. Another type of missing data process occurs when the response is inapplicable, such as questions regarding the years of marriage for adults who have never been married. Again, the analyses can be specifically formulated to accommodate these respondents.

Other missing data processes may be less easily identified and accommodated. Most often these are related directly to the respondent. One example is the refusal to respond to certain questions. This is common in questions of a sensitive nature (e.g., income or controversial issues) or when the respondent has no opinion or insufficient knowledge to answer the question. The researcher should anticipate these problems and attempt to minimize them in the research design and data collection stages of the research. However, they still may occur, and the researcher must now deal with the resulting missing data. But all is not lost. When the missing data occur in a random pattern, remedies may be available to mitigate their effect.

Examining the Patterns of Missing Data

To decide whether a remedy for missing data can be applied, the researcher must first ascertain the degree of randomness present in the missing data. Assume for the purposes of illustration that two variables (X and Y) are collected. X has no missing data, but Y does have some missing data. If a missing data process is found between X and Y where there are significant differences in the values of X

between cases for Y with valid and missing data, then the missing data are not at random. Any analysis must explicitly accommodate the missing data process between X and Y or else bias is introduced into the results.

Missing data are termed **missing at random (MAR)** if the missing values of Y depend on X, but not on Y. By this we mean that the observed Y values represent a random sample of the actual Y values for each value of X, but the observed data for Y do not necessarily represent a truly random sample of all Y values. Even though the missing data process is random in the sample, its values are not generalizable to the population. For example, assume that we know the gender of respondents (the X variable) and are asking about household income (the Y variable). We find that the missing data are random for both males and females but occur at a much higher frequency for males than females. Even though the missing data process is operating in a random manner, any remedy applied to the missing data will still reflect the missing data process because gender affects the ultimate distribution of the household income values.

A higher level of randomness is termed **missing completely at random (MCAR).** In these instances the observed values of Y are truly a random sample of all Y values, with no underlying process that lends bias to the observed data. In our earlier example, this would be shown by the fact that the missing data for household income were randomly missing in equal proportions for both males and females. If this is the form of the missing data process, any of the remedies can be applied without making allowances for the impact of any other variable or missing data process.

Diagnosing the Randomness of the Missing Data Process

As previously noted, the researcher must ascertain whether the missing data process occurs in a completely random manner. Three methods are available for this diagnosis. The first assesses the missing data process of a single variable Y by forming two groups—observations with missing data for Y and those with valid values of Y. Statistical tests are then performed to determine whether significant differences exist between the two groups on other variables of interest. Significant differences indicate the possibility of a nonrandom missing data process. Let us use our earlier example of household income and gender. We would first form two groups of respondents, those with missing data on the household income question and those who answered the question. We would then compare the percentages of gender for each group. If one gender (e.g., males) was found in greater proportion in the missing data group, we would suspect a nonrandom missing data process. If the variable we were comparing was metric (e.g., an attitude or perception) instead of categorical (gender), then t tests could be performed. The researcher should examine a number of variables to see whether any consistent pattern emerges. Remember that some differences will occur by chance, but any series of differences may indicate an underlying nonrandom pattern.

A second approach utilizes dichotomized correlations to assess the correlation of missing data for any pair of variables. For each variable, valid values are represented by the value of one, and missing data are replaced by the value of zero. These missing value indicators for each variable are then correlated. The correlations indicate the degree of association between the missing data on each variable pair. Low correlations denote randomness in the missing data for each pair of variables. Although no strict guidelines exist for identifying the level of correlation needed to indicate a nonrandom missing data process, statistical significance tests of the correlations provide a conservative estimate of the degree of randomness. If randomness is indicated for all variable pairs, then the researcher can assume that the missing data can be classified as MCAR. If significant correlations exist between some pairs of variables, then the researcher may have to assume that the data are only MAR, and these relationships must be accommodated in any remedies that are applied.

Finally, an overall test of randomness can be performed that determines whether the missing data can be classified as MCAR. This test analyzes the pattern of missing data on all variables and compares it with the pattern expected for a random missing data process. If no significant differences are found, the missing data can be classified as MCAR. If significant differences are found, however, the researcher must use the approaches described previously to identify the specific missing data processes that are nonrandom.

Approaches for Dealing with Missing Data

The approaches or remedies for dealing with missing data can be classified into one of four categories based on the randomness of the missing data process and the method used to estimate the missing data [9]. If nonrandom or MAR missing data processes are found, the researcher should apply only one remedy—the specifically designed modeling approach [9]. Application of any other method introduces bias into the results. Only if the researcher determines that the missing data process can be classified as MCAR can the approaches discussed in the following sections be used.

However, researchers often make the assessment for randomness of the missing data process before applying one of these missing data remedies. And even if the remedy is appropriate, the researcher must note the specific impact on the results associated with that remedy. Too often a remedy is applied without an assessment of the missing data processes, the appropriateness of the selected remedy, or its consequences. Thus, the researcher never realizes the effects because they are hidden in the overall results.

Use of Observations with Complete Data Only

The simplest and most direct approach for dealing with missing data is to include only those observations with complete data, also known as the **complete case approach.** This method is available in all statistical programs and is the default method in many programs. Yet the complete case approach should be used only if the missing data are MCAR, because missing data that are not MCAR have nonrandom elements that bias the results. Thus, even though only valid observations are used, the results are not generalizable to the population. Moreover, in many situations, the resulting sample size is reduced to an inappropriate size. The complete case approach is best suited for instances in which the extent of missing data is small, the sample is sufficiently large to allow for deletion of the cases with missing data, and the relationships in the data are so strong as to not be affected by any missing data process.

Delete Case(s) and/or Variable(s)

Another simple remedy for missing data is to delete the offending case(s) and/or variable(s). In this approach, the researcher determines the extent of missing data on each case and variable and then deletes the case(s) or variable(s) with excessive levels. In many cases where a nonrandom pattern of missing data is present, this may be the most efficient solution. The researcher may find that the missing data are concentrated in a small subset of cases and/or variables, with their exclusion substantially reducing the extent of the missing data. Again, no firm guidelines exist on the necessary level for exclusion, but any decision should be based on both empirical and theoretical considerations. If missing values are found for what will be a dependent variable in the proposed analysis, the case is usually excluded. This avoids any artificial increases in the explanatory power of the analysis, which can occur when the researcher first estimates the missing data for the dependent variable by one of the imputation processes described next and then uses the estimated values in the analysis of the dependence relationship. If a variable other than a dependent variable has missing values and is a candidate for deletion, the researcher should be sure that alternative variables, hopefully highly correlated, are available to represent the intent of the original variable. The researcher must always consider the gain of eliminating a source of missing data versus the deletion of a variable in the multivariate analysis.

Imputation Methods

A third category of remedies for handling missing data is through one of the many **imputation methods**. Imputation is the process of estimating the missing value based on valid values of other variables and/or cases in the sample. The objective is to employ known relationships that can be identified in the valid values of the sample to assist in estimating the missing values. However, the researcher should carefully consider the use of imputation in each instance because of its potential impact on the analysis [6]:

> The idea of imputation is both seductive and dangerous. It is seductive because it can lull the user into the pleasurable state of believing that the data are complete after all, and it is dangerous because it lumps together situations where the problem is sufficiently minor that it can be legitimately handled in this way and situations where standard estimators applied to the real and imputed data have substantial biases.

The methods discussed in this section are used primarily with metric variables for two reasons. First, estimates of the missing data for metric variables can be made with such values as a mean of all valid values. Second, nonmetric variables require an estimate of a specific value rather than an estimate on a continuous scale. It is very different to estimate a missing value for a metric variable, such as an attitude or perception—even income—than it is to estimate the respondent's gender when missing. Thus, nonmetric variables are typically not filled by the imputation process, but require either the specific modeling approach discussed in the next section or are left as missing.

Imputation methods can be defined as one of two types: (1) use of all of the available information from a subset of cases to generalize to the entire sample, or (2) methods of estimating replacement values for the missing data, which are then analyzed by standard multivariate techniques. The following discussion will describe the various options within each type and their advantages and disadvantages.

Using All Available Information as the Imputation Technique

The first type of imputation method does not actually replace the missing data, but instead imputes the distribution characteristics (e.g., means or standard deviations) or relationships (e.g., correlations) from all available valid values. Known as the **all-available approach,** this method (the PAIRWISE option in SPSS, and the CORPAIR, COVPAIR, or ALLVALUE options in BMDP) is primarily used to estimate correlations and maximize the pairwise information available in the sample. The distinguishing characteristic of this approach is that each correlation for a pair of variables is based on a potentially unique set of observations and the number of observations used in the calculations can vary for each correlation. The imputation process occurs not by replacing the missing data on the remaining cases, but instead by using the obtained correlations as representative for the entire sample. This approach can be compared to the complete-case approach discussed earlier, which uses data only from observations that have no missing data. Either approach can introduce bias if the missing data process is not MCAR.

Even though the all-available method maximizes the data utilized and overcomes the problem of missing data on a single variable eliminating a case from the entire analysis, several problems can also arise from this approach. First, correlations may be calculated that are "out of range" and inconsistent with the other correlations in the correlation matrix. Any correlation between X and Y is constrained by their correlation to a third variable Z, as shown in the following formula:

$$\text{Range of } r_{xy} = r_{xz}r_{yz} \pm \sqrt{(1 - r_{xz}^2)(1 - r_{yz}^2)}$$

The correlation between X and Y can range only from $+1$ to -1 if both X and Y have zero correlation with all other variables in the correlation matrix. Yet rarely are the correlations with other variables zero. As the correlations with other variables increase, the range of the correlation between X and Y decreases. This increases the potential for the correlation in a unique set of cases to be inconsistent with correlations derived from other sets of cases. For example, if X and Y have correlations of .6 and .4, respectively, with Z, then the possible range of correlation between X and Y is $.24 \pm .73$, or from $-.49$ to $.97$. Any value outside this range is mathematically inconsistent, yet may occur if the correlation is obtained with a differing number and set of cases for the two correlations in the all-available approach.

An associated problem is that the eigenvalues in the correlation matrix can become negative, thus altering the variance properties of the correlation matrix. Although the correlation matrix can be adjusted to eliminate this problem (e.g., the ALLVALUE option in BMDP), many procedures do not include this adjustment process. In extreme cases, the estimated variance/covariance matrix is not positive definite. All of these problems must be considered when selecting the all-available approach.

The Replacement of Missing Data

The second form of imputation involves replacing missing values with estimated values based on other information available in the sample. There are many options, varying from the direct substitution of values to estimation processes based on relationships among the variables. The following discussion focuses on the more widely used methods, although many other forms of imputation are available [9].

Case Substitution   In this method, observations with missing data are replaced by choosing another nonsampled observation. A common example is to replace a sampled household that cannot be contacted or that has extensive missing data with another household not in the sample, preferably very similar to the original observation. This method is most widely used to replace observations with complete missing data, although it can be used to replace observations with lesser amounts of missing data as well.

Mean Substitution   One of the more widely used methods, mean substitution replaces the missing values for a variable with the mean value of that variable based on all valid responses. In this manner, the valid sample responses are used to calculate the replacement value. The rationale of this approach is that the mean is the best single replacement value. This approach, although it is used extensively, has three disadvantages. First, it makes the variance estimates derived from the standard variance formulas invalid by understating the true variance in the data. Second, the actual distribution of values is distorted by substituting the mean for the missing values. Third, this method depresses the observed correlation because all missing data will have a single constant value. It does have the advantage, however, of being easily implemented and providing all cases with complete information.

Cold Deck Imputation   In this method, the researcher substitutes a constant value derived from external sources or previous research for the missing values. This is similar in nature to the mean substitution method, differing only in the source of the substitution value. Cold deck imputation has the same disadvantages as the mean substitution method, and the researcher must be sure that the replacement value from an external source is more valid than an internally generated value, such as the mean. This method can provide the researcher with the option of replacing the missing data with a value that may be deemed more valid than the mean of the sample.

Regression Imputation   In this method, regression analysis (described in chapter 4) is used to predict the missing values of a variable based on its relationship to other variables in the data set. Although it has the appeal of using relationships already existing in the sample as the basis of prediction, this method also has several disadvantages. First, it reinforces the relationships already in the data. As the use of this method increases, the resulting data become more characteristic of the sample and less generalizable. Second, unless stochastic terms are added to the estimated values, the variance of the distribution is understated. Third, this method assumes that the variable with missing data has substantial correlations with the other variables. If these correlations are not sufficient to produce a meaningful estimate, then other methods, such as mean substitution, are preferable. Finally, the regression procedure is not constrained in the estimates it makes. Thus, the predicted values may not fall in the valid ranges for variables (e.g., a value of 11 may be predicted for a 10-point scale), thereby requiring some form of additional adjustment. Even with all of these potential problems, the regression method of imputation holds promise in those instances for which moderate levels of widely scattered missing data are present and for which the relationships between variables are sufficiently established so that the researcher is confident that using this method will not impact the generalizability of the results.

Multiple Imputation   The final imputation method is actually a combination of several methods. In this approach, two or more methods of imputation are used to derive a composite estimate—usually the mean of the various estimates—for the missing value. The rationale of this approach is that the use of multiple approaches minimizes the specific concerns with any single method and the composite will be the best possible estimate. The choice of this approach is primarily based on the trade-off between the researcher's perception of the potential benefits versus the substantially higher effort required to make and combine the multiple estimates.

Model-Based Procedures

The final set of procedures explicitly incorporates the missing data into the analysis, either through a process specifically designed for missing data estimation or as an integral portion of the standard multivariate analysis. The first approach involves maximum likelihood estimation techniques that attempt to model the processes underlying the missing data and to make the most accurate and reasonable estimates possible [9]. One example is the EM approach in SPSS. It is an iterative two-stage method (the E and M stages) in which the E-stage makes the best possible estimates of the missing data and the M-stage then makes estimates of the parameters (means, standard deviations, or correlations) assuming the missing data were replaced. The process continues going through the two stages until the change in the estimated values is negligible and they replace the missing data.

The second approach involves the inclusion of missing data directly into the analysis, defining observations with missing data as a select subset of the sample. This approach is most applicable for dealing with missing values on the independent variables of a dependent relationship. Its premise has best been characterized by Cohen and Cohen [4, p. 299]:

> We thus view missing data as a pragmatic fact that must be investigated, rather than a disaster to be mitigated. Indeed, implicit in this philosophy is the idea that like all other aspects of sample data, missing data are a property of the population to which we seek to generalize.

When the missing values occur on a nonmetric variable, the researcher can easily define those observations as a separate group and then include them in any analysis, such as ANOVA, MANOVA, or even discriminant analysis. When the missing data are present on a metric independent variable in a dependence relationship, a procedure has been developed to incorporate the observations into the analysis while maintaining the relationships among the valid values [4]. This procedure is best illustrated in the context of regression analysis, although it can be used in other dependence relationships as well. The first step is to code all observations with missing data as dummy variables (where the cases with missing data receive a value of one and the other cases have a value of zero). The missing values are then imputed by the mean substitution method. Finally, the relationship is estimated by normal means. The dummy variable represents the difference for the dependent variable between those observations with missing data and those observations with valid data. The test of the dummy variable coefficient assesses the statistical significance of this difference. The coefficient of the original variable represents the relationship for all cases with nonmissing data. This method allows the researcher to retain all the observations in the analysis for purposes of maintaining the sample size. It also provides a direct test for the differences between the two groups along with the estimated relationship between the dependent and independent variables.

An Illustration of Missing Data Diagnosis

To illustrate the process of diagnosing the patterns of missing data and the application of possible remedies, a new data set is introduced (see appendix A for a complete listing of the observations). This data set was collected during the pretest of the questionnaire used to collect the data described in chapter 1. The pretest involved 70 individuals and collected responses on all 14 variables. In the course of pretesting, however, missing data occurred. The following sections detail the diagnosis of the extent of missing data in the data set and the analyses available for selecting and applying the various missing data remedies available in most statistical programs. A number of software programs are adding analyses of missing data, among them BMDP and SPSS. The analyses described next can all be replicated by data manipulation and conventional analysis. Examples are provided in appendix A.

Examining the Patterns of Missing Data

Table 2.2 contains the descriptive statistics for the observations with valid values, including the percentage of cases with missing data on each variable. Six cases were eliminated from the analysis owing to missing data on more than half of the variables of interest. The extent of missing data for the remaining 64 observations ranges from a high of 30 percent of the cases for X_1 to a low of a single case (1.6 percent) for X_6. For the variables with the higher levels of missing data (X_1, X_2, and X_3), the levels are not so excessive that they automatically dictate exclusion of the variable. Given the integral role these variables are expected to play in the various multivariate analyses, all efforts should be made to retain them in the analysis.

Summary Statistics of Pretest Data

One factor that could alleviate some of the high levels of missing data for certain variables is the elimination of cases from the analysis. To determine whether the missing data are concentrated on a select set of cases, Table 2.3 provides a graphical display of the missing data for each case that has missing data. Except for the six cases already eliminated because of extremely high levels of missing data, we see that no other cases have a disproportionate number of missing values. In fact, of the 38 cases with missing data, only four cases have more than two missing values.

TABLE 2.2 Summary Statistics of Pretest Data

	Number of Cases with Valid Data	Mean	Standard Deviation	Missing Data	
				Number	Percent
X_1 Delivery speed	45	4.0133	.9664	19	29.7
X_2 Price level	54	1.8963	.8589	10	15.6
X_3 Price flexibility	50	8.1300	1.3194	14	21.9
X_4 Manufacturer image	60	5.1467	1.1877	4	6.3
X_5 Overall service	59	2.8390	.7541	5	7.8
X_6 Salesforce image	63	2.6016	.7192	1	1.6
X_7 Product quality	60	6.7900	1.6751	4	6.3
X_9 Usage level	60	45.9667	9.4204	4	6.3
X_{10} Satisfaction level	60	4.7983	.8194	4	6.3

Note: Six of the original 70 cases had more than 50 percent missing data and were excluded from the analysis. All analyses are based on the remaining 64 cases. Twenty-six cases had no missing data.

TABLE 2.3 Graphical Display of Missing Data

		Variables								
	Number of	Missing Data								
Case	Missing Values	X_1	X_2	X_3	X_4	X_5	X_6	X_7	X_9	X_{10}
202	2	S		S						
203	2		S					S		
204	3	S		S						S
205	1			S						
207	3	S		S						S
213	2		S	S						
216	2	S				S				
218	2	S				S				
219	2							S	S	
220	1		S							
221	3	S		S				S		
222	2			S		S				
224	3	S	S						S	
225	2			S	S					
227	2		S						S	
228	2	S			S					
229	1					S				
231	1							S		
232	2	S	S							
235	2						S			S
237	1		S							
238	1	S								
240	1	S								
241	2			S		S				
244	1								S	
246	1				S					
248	2	S	S							
249	1		S							
250	2	S		S						
253	1	S								
255	2	S		S						
256	1	S								
257	2		S	S						
259	1	S								
260	1	S								
267	2			S	S					
268	1									S
269	2	S		S						

Legend: S = a missing value

Table 2.4 (p. 58) portrays the patterns of missing data. The most prevalent pattern is the missing data for X_1 found in six cases, with the second most common pattern being missing data for X_1 and X_3 in four cases. All of the remaining cases exhibit patterns that are essentially unique or shared with only a very small number of cases. As this analysis demonstrates, no patterns occur with a frequency that suggests an underlying missing data process. Thus, no case or set of cases with a missing data pattern can be eliminated that would markedly improve the missing data problem.

TABLE 2.4 Tabulated Missing Data Patterns

Number of Cases	Missing Data Patterns[a]								
	X_6	X_{10}	X_4	X_7	X_9	X_5	X_2	X_3	X_1
26									
1								X	
4								X	X
6									X
1			X						X
1			X						
2			X					X	
2						X		X	
1						X			
2						X			X
2							X		X
3							X		
2							X	X	
1				X			X		
1				X					
1				X	X				
1					X				
1					X		X		
1					X		X		X
1		X							
1	X	X							
2		X						X	X
1				X				X	X

[a]Variables are sorted on missing patterns.

Diagnosing Randomness of the Missing Data

The next step is an empirical examination of the patterns of missing data to determine whether the missing data are distributed randomly across the cases and the variables. The first test for assessing randomness is to compare the observations with and without missing data for each variable on the other variables. For example, the observations with missing data on X_1 are placed in one group and those observations with valid responses for X_1 are placed in another group. Then, these two groups are compared to identify any differences on the remaining metric variables (X_2 through X_{10}). Once comparisons have been made on all of the variables, new groups are formed based on the missing data for the next variable (X_2) and the comparisons are performed again on the remaining variables. This process continues until each variable (X_1 through X_{10}) has been examined for any differences. The objective is to identify any systematic missing data process that would be reflected in patterns of significant differences.

Table 2.5 contains the results for this analysis of the 64 remaining observations from the pretest sample. The first noticeable pattern of significant t values occurs for X_9, for which six of the nine comparisons found significant differences between the two groups. However, the impact of these differences is marginal as the number of cases with missing data on X_9 ranged from only three to five. X_7 exhibited a pattern of differences similar to X_9 with four significant differences, but small groups of cases with missing data. This analysis indicates that although

TABLE 2.5 Assessing the Randomness of Missing Data through Group Comparisons of Observations with Missing versus Valid Data

Groups Formed by Missing Data on:	X_1	X_2	X_3	X_4	X_5	X_6	X_7	X_9	X_{10}
X_1 t		−.3	1.3	2.2	2.6	1.9	−1.1	2.6	2.1
Significance		.763	.223	.033	.017	.065	.273	.017	.049
Number present	45	38	38	42	42	44	42	42	43
Number missing	0	16	12	18	17	19	18	18	17
Mean (present)	4.01	1.87	8.27	5.34	3.02	2.71	6.61	48.17	4.95
Mean (missing)		1.95	7.68	4.69	2.39	2.36	7.20	40.83	4.42
X_2 t	−.5		.7	−2.2	−4.2	−2.4	−1.2	−1.1	−1.2
Significance	.646		.528	.044	.001	.034	.260	.318	.233
Number present	38	54	42	50	49	53	51	52	50
Number missing	7	0	8	10	10	10	9	8	10
Mean (present)	3.97	1.90	8.18	4.99	2.70	2.51	6.68	45.46	4.75
Mean (missing)	4.23		7.86	5.94	3.50	3.11	7.40	49.25	5.02
X_3 t	.4	1.4		1.1	2.0	.2	.0	1.9	.9
Significance	.693	.180		.286	.066	.818	.965	.073	.399
Number present	38	42	50	48	47	49	47	46	48
Number missing	7	12	0	12	12	14	13	14	12
Mean (present)	4.03	1.98	8.13	5.24	2.95	2.61	6.80	47.02	4.84
Mean (missing)	3.90	1.60		4.79	2.42	2.56	6.77	42.50	4.62
X_4 t	−.2	2.6	−.3		.2	1.4	1.5	.2	−2.4
Significance	.882	.046	.785		.888	.249	.197	.830	.064
Number present	42	50	48	60	55	59	56	56	56
Number missing	3	4	2	0	4	4	4	4	4
Mean (present)	4.01	1.94	8.12	5.15	2.84	2.63	6.83	46.02	4.76
Mean (missing)	4.07	1.33	8.35		2.80	2.25	6.20	45.25	5.38
X_5 t	−.1	−.3	.8	.4		−.9	−.4	.5	.6
Significance	.900	.749	.502	.734		.423	.696	.669	.605
Number present	42	49	47	55	59	58	55	55	55
Number missing	3	5	3	5	0	5	5	5	5
Mean (present)	4.01	1.89	8.20	5.16	2.84	2.58	6.76	46.18	4.82
Mean (missing)	4.10	1.98	7.10	5.04		2.86	7.14	43.60	4.56
X_7 t	3.0	.9	.2	−2.1	.9	−1.5		.5	.4
Significance	.036	.440	.864	.118	.441	.193		.658	.704
Number present	42	51	47	56	55	59	60	57	56
Number missing	3	3	3	4	4	4	0	3	4
Mean (present)	4.07	1.92	8.14	5.07	2.86	2.58	6.79	46.14	4.81
Mean (missing)	3.27	1.50	8.00	6.18	2.55	2.90		42.67	4.70
X_9 t	6.1	−1.4	2.2	−1.1	−.9	−1.8	1.7		1.6
Significance	.000	.384	.101	.326	.401	.149	.128		.155
Number present	42	52	46	56	55	59	57	60	56
Number missing	3	2	4	4	4	4	3	0	4
Mean (present)	4.08	1.85	8.26	5.11	2.82	2.57	6.82	45.97	4.82
Mean (missing)	3.10	3.00	6.63	5.62	3.08	3.03	6.30		4.47
X_{10} t	1.7	.8	−2.1	2.5	2.7	1.3	.9	2.4	
Significance	.249	.463	.235	.076	.056	.302	.409	.066	
Number present	43	50	48	56	55	60	56	56	60
Number missing	2	4	2	4	4	3	4	4	0
Mean (present)	4.03	1.92	8.09	5.23	2.89	2.62	6.83	46.43	4.80
Mean (missing)	3.55	1.60	9.20	3.95	2.08	2.17	6.30	39.50	

Each cell contains six values: 1) t value for comparison of the column variable between group a (observations with valid data on the row variable) and group b (observations with missing data on the row variable); 2) significance of t value for group comparisons; 3) & 4) number of observations for group a (valid data) and group b (missing data); 5) & 6) mean of variable for group a (valid data) and group b (missing data)

Interpretation of the table:

The upper right cell indicates that the t value for the comparison of X_{10} between group a (valid data) and group b (missing data) on X_1 is 2.1, which has a significance level of .049. The sample sizes of group a and group b are 43 and 17, respectively. Finally, the mean of X_{10} for group a (valid data) is 4.95, whereas the mean for group b (missing data) is 4.42.

significant differences can be found due to the missing data on two variables (X_7 and X_9), the small number of cases involved makes this of marginal concern. If later tests of randomness indicate a nonrandom pattern of missing data, these results would then provide a starting point for possible remedies.

A second test for randomness involves the use of correlations between dichotomous variables. The dichotomous variables are formed by replacing valid values with a value of one and missing data with a value of zero. The resulting correlations between the dichotomous variables indicate the extent to which missing data are related in pairs of variables. Low correlations indicate low association between the missing data process for those two variables. Table 2.6 contains the correlations between the nine dichotomized metric variables. Review of the values indicates that only one correlation is in the moderate range (X_6 and X_{10} have a correlation of .488). This suggests that the missing data process influencing X_{10} corresponds to the missing data process affecting X_6. However, given the absence of any other correlations with even moderate values, the researcher can be assured that no single missing data process is significantly affecting a substantial number of variables.

The final test is an overall test of the missing data for being missing completely at random (MCAR). The test makes a comparison of the actual pattern of missing data with what would be expected if the missing data were totally randomly distributed. In this instance, as shown in Table 2.6, the significance level of the

TABLE 2.6 Assessing the Randomness of Missing Data through Dichotomized Variable Correlations and the Multivariate Test for Missing Completely at Random (MCAR)

	X_1 Delivery Speed	X_2 Price Level	X_3 Price Flexibility	X_4 Manufacturer Image	X_5 Overall Service	X_6 Salesforce Image	X_7 Product Quality	X_9 Usage Level	X_{10} Satisfaction Level
X_1	1.000 45								
X_2	0.003 38	1.000 54							
X_3	0.235 38	−0.020 42	1.000 50						
X_4	−0.026 42	−0.111 50	0.176 48	1.000 60					
X_5	0.066 42	−0.125 49	0.128 47	−0.075 55	1.000 59				
X_6	−0.082 44	−0.054 53	−0.067 49	−0.033 59	−0.037 58	1.000 63			
X_7	−0.026 42	0.067 51	0.020 47	−0.067 56	−0.075 55	−0.033 59	1.000 60		
X_9	−0.026 42	0.244 52	−0.137 46	−0.067 56	−0.075 55	−0.033 59	0.200 57	1.000 60	
X_{10}	0.115 43	−0.111 50	0.176 48	−0.067 56	−0.075 55	0.488* 60	−0.067 56	−0.067 56	1.000 60

Little's MCAR Test: Chi-square: 174.464
                       Degrees of freedom: 159
                       Probability: .190

Interpretation:
First value in the table represents the correlation between the dichotomized variables, where cases with a valid value receive a 1 and missing data receive a 0. The second value, below the correlation, represents the number of cases having valid data on both variables in that specific correlation pair.
*Significant at the .05 level.

MCAR tests was .190, indicating that the missing data process can be considered to be MCAR. As a result, the researcher may employ any of the remedies for missing data, because no potential biases exist in the patterns of missing data.

Remedies for Missing Data

As discussed earlier, numerous remedies are available for dealing with missing data. In this instance, several of the remedies have definite disadvantages. If the complete case approach is taken, the sample size is reduced to 26 observations, barely sufficient for even the simplest univariate analyses, much less multivariate applications. Our earlier examination of the patterns of the missing data demonstrated that there was not a small set of cases that could be deleted and thereby markedly reduce the extent of missing data. Moreover, the only viable alternative in eliminating a variable is the elimination of X_1, which has missing data on almost 30 percent of the cases. But even if X_1 were deleted, all of the cases with missing data would still have at least one other variable with missing data. Even eliminating X_1 is a relatively ineffective approach for creating more observations with complete data on all variables.

The remaining option is to employ some form of imputation to estimate replacement values for the missing values. The first possibility is to use only observations with complete data or using all available information to estimate the correlations. The advantage of the complete information approach is that it maintains consistency in the correlation matrix. However, it may also reduce the number of observations used to such a small subset of the sample (28 cases) that the resulting correlations used for imputation differ markedly from those obtained using all available information. The all-available approach maximizes the number of cases used in calculating the correlations, but may introduce inconsistencies in the calculated correlations. A third option is to use a mean substitution for all the missing data and then calculate the correlations.

Table 2.7 (p. 62) contains the correlations obtained from the all-available, complete information, and mean substitution approaches. In most instances the correlations are similar, but there are several substantial differences. First, there is a consistency between the correlations obtained with the all-available and mean substitution approaches. The differences occur among the correlations obtained with the complete information approach. Second, the notable differences are concentrated in the correlations of X_1 and X_{10} with X_4, X_5, X_6, and X_7. These differences may indicate the impact of a missing data process, but this was not detected by the earlier diagnostic measures. Although the researcher has no proof of greater validity for either approach, these results demonstrate the marked differences sometimes obtained between the approaches. Whichever approach is chosen, the researcher should examine the correlations obtained by alternative methods to understand the range of possible values.

The researcher may also select a specific estimation approach for estimating replacement values for the missing data. Table 2.8 (p. 63) contains the results of employing the mean substitution regression and EM approaches for missing value imputation. These results include the means and standard deviations obtained after missing values are replaced by the imputed data. As seen in the earlier comparisons of correlations, some differences can be detected, but no consistent pattern emerges. For variables X_1 and X_2, there are marked differences in the estimated values. In general, for the remaining variables the estimates are very similar, if not identical. Thus, the researcher does not have a definitive indication of which

TABLE 2.7 Comparison of Correlations Obtained with the All-Available (Pairwise), Complete Case (Listwise), and Mean Substitution Approaches

	X_1 Delivery Speed	X_2 Price Level	X_3 Price Flexibility	X_4 Manufacturer Image	X_5 Overall Service	X_6 Salesforce Image	X_7 Product Quality	X_9 Usage Level	X_{10} Satisfaction Level
X_1	1.000 1.000 1.000								
X_2	−.479 −.502 −.349	1.000 1.000 1.000							
X_3	.416 .429 .329	−.357 −.294 −.289	1.000 1.000 1.000						
X_4	−.099 −.245 −.086	.299 .320 .245	−.065 −.061 −.057	1.000 1.000 1.000					
X_5	.366 .566 .232	.440 .421 .382	.047 .157 .042	.432 .046 .422	1.000 1.000 1.000				
X_6	.031 −.094 .027	.260 .356 .219	−.035 −.066 −.032	.810 .804 .769	.344 .213 .323	1.000 1.000 1.000			
X_7	−.138 −.416 −.106	.348 .354 .310	−.358 −.230 −.297	.398 .382 .374	.066 −.150 .061	.402 .529 .395	1.000 1.000 1.000		
X_9	.376 .599 .265	.149 .048 .134	.601 .648 .503	.223 .191 .216	.712 .683 .656	.268 .301 .260	−.202 −.099 −.195	1.000 1.000 1.000	
X_{10}	.514 .549 .381	−.184 −.278 −.173	.702 .725 .626	.378 .170 .344	.533 .304 .477	.233 .064 .229	−.256 −.405 −.250	.669 .566 .647	1.000 1.000 1.000

Interpretation: The top value is the correlation obtained with a pairwise or all-available approach, the second value is the correlation obtained with a listwise or complete information approach, and the third value is the correlation obtained with mean substitution. Sample sizes for the all-available information approach varied; the actual sample sizes are listed in Table 2.5. A sample size of 26 was used for the complete information correlations; there were no missing data after mean substitution, so the sample size for this approach was 64.

approach is appropriate. Instead the researcher must coalesce the missing data patterns with the strengths and weaknesses of each approach and then select the most appropriate method. In the instance of differing estimates, the more conservative approach of combining the estimates into a single estimate (the multiple imputation approach) may be the most appropriate choice. Whichever approach is used, the data set with replacement values should be saved for further analysis.

A Recap of the Missing Value Analysis

Our evaluation of the issues surrounding missing data in the pretest data can be summarized in four conclusions:

1. *The missing data process is MCAR.* All of the diagnostic techniques support the conclusion that no systematic missing data process exists, making the missing data MCAR (missing completely at random). Such a finding provides two ad-

TABLE 2.8 Results of the Regression and EM Imputation Methods

					Estimated Means				
Imputation Methods	X_1 Delivery Speed	X_2 Price Level	X_3 Price Flexibility	X_4 Manufacturer Image	X_5 Overall Service	X_6 Salesforce Image	X_7 Product Quality	X_9 Usage Level	X_{10} Satisfaction Level
EM	3.71	2.03	8.11	5.15	2.82	2.60	6.84	45.85	4.77
Regression	3.84	1.96	8.10	5.15	2.81	2.59	6.88	45.77	4.77

					Estimated Standard Deviations				
Imputation Methods	X_1 Delivery Speed	X_2 Price Level	X_3 Price Flexibility	X_4 Manufacturer Image	X_5 Overall Service	X_6 Salesforce Image	X_7 Product Quality	X_9 Usage Level	X_{10} Satisfaction Level
EM	1.15	1.00	1.27	1.16	.75	.71	1.68	9.29	.82
Regression	.99	.83	1.26	1.15	.75	.72	1.69	9.18	.82

vantages to the researcher. First, there should be no "hidden" impact on the results that need to be considered when interpreting the results. Second, any of the imputation methods can be applied as remedies for the missing data. Their selection need not be based on their ability to handle nonrandom processes, but instead on the applicability of the process and its impact on the results.

2. *Imputation is the most logical course of action.* Given the minimal benefit of deleting cases and variables, the researcher is precluded from the simple solution of deleting cases or variables. Moreover, the complete case method would result in an inadequate sample size. Some form of imputation is therefore needed to maintain an adequate sample size for any multivariate analysis.

3. *Imputed correlations differ across techniques.* When estimating correlations among the variables in the presence of missing data, the researcher can choose from the three most common techniques: the complete information method, the all-available information method, and the mean substitution method. The researcher is faced in this situation, however, with differences in the results among these three methods. The all-available information and mean substitution approaches lead to generally consistent results, although the mean substitution values are generally somewhat lower in magnitude. Notable differences are found between these two approaches and the complete information approach. While the complete information approach would seem the most "safe" and conservative, in this case it is not recommended due to the small sample used (only 26 observations) and its marked differences from the other two methods. The researcher should choose, if necessary, among the two other approaches.

4. *Multiple methods for replacing the missing data are available and appropriate.* As mentioned above, mean substitution is one acceptable means of generating replacement values for the missing data. The researcher also has available the regression and EM imputation methods, each of which give reasonably consistent estimates for most variables. The presence of three acceptable methods also allows the researcher to combine the three estimates into a single composite, hopefully mitigating any effects strictly due to one of the methods.

In conclusion, the analytical tools and the diagnostic processes presented in the earlier section have provided an adequate basis for understanding and accommodating the missing data found in the pretest data. As this example demonstrates, the researcher need not fear that missing data will always preclude a multivariate analysis or always limit the generalizability of the results. Instead, the possibly "hidden" impact of missing data can be identified and actions taken to minimize the effect of missing data on the analyses performed.

Summary

The procedures available for handling missing data are varied in form, complexity, and intent. The researcher must always be prepared to assess and deal with missing data, as it is frequently encountered in multivariate analysis. The decision to use only observations with complete data may seem to be conservative and "safe," but as the preceding discussion illustrated, there are inherent limitations and biases in this and the other approaches. The researcher has no single method best suited in every situation, but instead must make a reasoned judgment of the situation, considering all of the factors described above.

Outliers

Outliers are observations with a unique combination of characteristics identifiable as distinctly different from the other observations. Outliers cannot be categorically characterized as either beneficial or problematic, but instead must be viewed within the context of the analysis and should be evaluated by the types of information they may provide. When beneficial, outliers—although different from the majority of the sample—may be indicative of characteristics of the population that would not be discovered in the normal course of analysis. In contrast, problematic outliers are not representative of the population, are counter to the objectives of the analysis, and can seriously distort statistical tests. Owing to the variability in the impact of outliers, it is imperative that the researcher examine the data for the presence of outliers to ascertain their type of influence. The reader is also referred to the discussions in chapter 4 and the appendix to that chapter, which relate to the topic of influential observations. In these discussions, outliers are placed in a framework particularly suited for assessing the influence of individual observations and determining whether this influence is helpful or harmful.

Why do outliers occur? Outliers can be classified into one of four classes. The first class arises from a procedural error, such as a data entry error or a mistake in coding. These outliers should be identified in the data cleaning stage, but if overlooked, they should be eliminated or recorded as missing values. The second class of outlier is the observation that occurs as the result of an extraordinary event, which then is an explanation for the uniqueness of the observation. The researcher must decide whether the extraordinary event should be represented in the sample. If so, the outlier should be retained in the analysis; if not, it should be deleted. The third class of outlier comprises extraordinary observations for which the researcher has no explanation. Although these are the outliers most likely to be omitted, they may be retained if the researcher feels they represent a valid segment of the population. The fourth and final class of outlier contains observations that fall within the ordinary range of values on each of the variables but are unique in their combination of values across the variables. In these situations, the researcher should retain the observation unless specific evidence is available that discounts the outlier as a valid member of the population.

The following sections detail the methods used in detecting outliers in univariate, bivariate, and multivariate situations. Once the outliers have been identified, they may be profiled to aid in placing them into one of the four classes described above. Finally, the researcher must decide on the retention or exclusion of each outlier, judging not only from the characteristics of the outlier but also from the objectives of the analysis.

Detecting Outliers

Outliers can be identified from a univariate, bivariate, or multivariate perspective. The researcher should utilize as many of these perspectives as possible, looking for a consistent pattern across methods to identify outliers. The following discussion details the processes involved in each of the three perspectives.

Univariate Detection

The univariate perspective for identifying outliers examines the distribution of observations and selects as outliers those cases falling at the outer ranges of the distribution. The primary issue is establishing the threshold for designation of an outlier. The typical approach first converts the data values to standard scores, which have a mean of 0 and a standard deviation of 1. Because the values are expressed in a standardized format, comparisons across variables can be made easily. For small samples (80 or fewer observations), the guidelines suggest identifying those cases with standard scores of 2.5 or greater as outliers. When the sample sizes are larger, the guidelines suggest that the threshold value of standard scores range from 3 to 4. If standard scores are not used, then the researcher can identify cases falling outside the ranges of 2.5 versus 3 or 4 standard deviations, depending on the sample size. In either case, the researcher must recognize that a certain number of observations may occur normally in these outer ranges of the distribution. The researcher should strive to identify only those truly distinctive observations and designate them as outliers.

Bivariate Detection

In addition to the univariate assessment, pairs of variables can be assessed jointly through a scatterplot. Cases that fall markedly outside the range of the other observations can be noted as isolated points in the scatterplot. To assist in determining the expected range of observations, an ellipse representing a specified confidence interval (varying between 50 and 90 percent of the distribution) for a bivariate normal distribution can be superimposed over the scatterplot. This provides a graphical portrayal of the confidence limits and facilitates identification of the outliers. Another variant of the scatterplot is termed the influence plot. In this type of scatterplot, each point varies in size in relation to its influence on the relationship. These methods provide some assessment of the influence of each observation to complement the designation of cases as outliers.

Multivariate Detection

The third perspective for identifying outliers involves a multivariate assessment of each observation across a set of variables. Because most multivariate analyses involve more than two variables, the researcher needs a means to objectively measure the multidimensional position of each observation relative to some common point. The Mahalanobis D^2 measure can be used for this purpose. Mahalanobis D^2 is a measure of the distance in multidimensional space of each observation from the mean center of the observations. It provides a common measure of multidimensional centrality and also has statistical properties that allow for significance testing. Given the nature of the statistical tests, it is suggested that a very conservative level, such as .001, be used as the threshold value for designation as an outlier.

Outlier Designation

When observations that are candidates for designation as an outlier have been identified by the univariate, bivariate, or multivariate methods, the researcher must then select observations that demonstrate real uniqueness in comparison with the remainder of the population. The researcher should refrain from designating too many observations as outliers and not succumb to the temptation of eliminating those cases not consistent with the remaining cases just because they are different.

Outlier Description and Profiling

Once the potential outliers have been identified, the researcher should generate profiles on each outlier observation and carefully examine the data for the variable(s) responsible for its being an outlier. In addition to this visual examination, the researcher can also employ multivariate techniques such as discriminant analysis or multiple regression to identify the differences between outliers and the other observations. The researcher should continue this analysis until satisfied with the aspects of the data that distinguish the outlier from the other observations. If possible the researcher should assign the outlier to one of the four classes described earlier.

Retention or Deletion of the Outlier

After the outliers have been identified, profiled, and categorized, the researcher must decide on the retention or deletion of each one. There are many philosophies among researchers as to how to deal with outliers. Our belief is that they should be retained unless there is demonstrable proof that they are truly aberrant and not representative of any observations in the population. But if they do represent a segment of the population, they should be retained to ensure generalizability to the entire population. As outliers are deleted, the researcher runs the risk of improving the multivariate analysis but limiting its generalizability. If outliers are problematic in a particular technique, many times they can be accommodated in the analysis in a manner in which they do not seriously distort the analysis.

An Illustrative Example of Analyzing Outliers

As an example of outlier detection, the observations of the HATCO database introduced in chapter 1 are examined here for outliers. The variables considered in the analysis are the metric variables X_1, X_2, X_3, X_4, X_5, X_6, X_7, and X_9. The outlier analysis includes univariate, bivariate, and multivariate diagnoses. If candidates for outlier designation are found, they are examined, and a decision on retention or deletion is made.

Univariate and Bivariate Detection

The first step is to examine the observations on each of the variables individually. Table 2.9 contains the observations with standardized variable values exceeding ±2.5. From this univariate perspective, a few observations exceed the threshold on a single variable, but no observation was a univariate outlier on more than one variable. For a bivariate perspective, scatterplots are formed for X_1, X_2, X_3, X_4, X_5, X_6, and X_7 versus X_9, one of the metric variables used as a dependent variable in many of the multivariate techniques. An ellipse representing the 90 percent confidence interval of a bivariate normal distribution is then superimposed on the scatterplot (see Figure 2.6, p. 68). The second part of Table 2.9 contains observations falling outside this ellipse. This is a 90 percent confidence interval; thus we would expect some observations normally to fall outside the ellipse. However, some observations (3, 5, 57, and 96) appear several times, perhaps indicating they are bivariate outliers.

Multivariate Detection

The final diagnostic method is to assess multivariate outliers with the Mahalanobis D^2 measure (see Table 2.10, p. 69). This evaluates the position of each observation compared with the center of all observations on a set of variables. In this case, all the metric variables were used for the evaluation of observations. As noted earlier, the statistical tests for significance with this measure should be very conservative (exceeding .001). With this threshold, two observations (22 and 55) are identified as significantly different. It is interesting that these observations were not seen in earlier univariate and bivariate analyses but appear only in the multivariate tests. This indicates they are not unique on any single variable but instead are unique in combination.

Retention or Deletion of the Outliers

As a result of these diagnostic tests, no observations seem to demonstrate the characteristics of outliers that should be eliminated. Each variable has some observations that are extreme, and they should be considered if that variable is used in an analysis. But no observations are extreme on a sufficient number of variables

TABLE 2.9 Identification of Univariate and Bivariate Outliers

Univariate Outliers Cases with Standardized Values (Z scores) Exceeding ± 2.5		Bivariate Outliers Cases Lying Outside the 90% Confidence Interval Ellipse	
Variable	*Cases*	X_9 *with*	*Cases*
X_1	39	X_1	1, 39, 95, 96
X_2	71	X_2	3, 49, 57, 7, 96, 97
X_3	none	X_3	11, 57, 96, 100
X_4	82	X_4	5, 22, 42, 50, 72, 82, 93, 96
X_5	96	X_5	3, 22, 39, 57, 71, 96
X_6	5, 42	X_6	5, 7, 42, 82, 96
X_7	none	X_7	57, 58, 95, 96
X_9	none		
X_{10}	none		

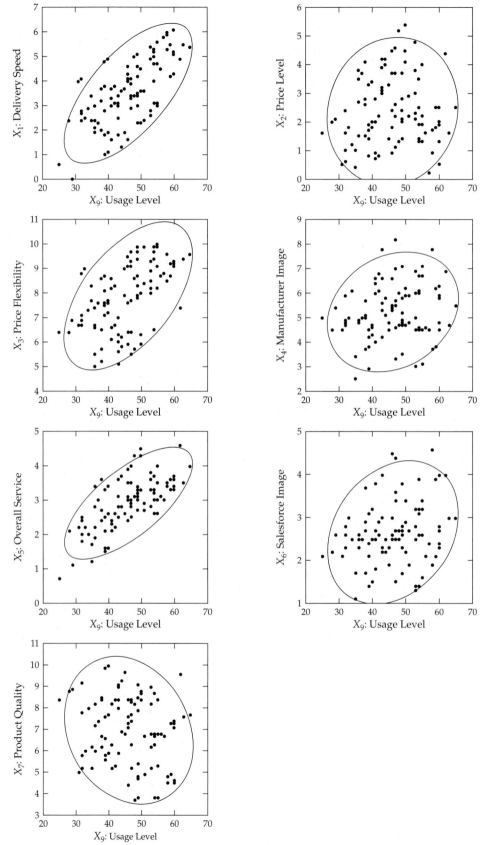

FIGURE 2.6 Graphical Identification of Bivariate Outliers

TABLE 2.10 Identification of Multivariate Outliers

Case Number	Mahalanobis D^2	D^2/df	df	Significance	Case Number	Mahalanobis D^2	D^2/df	df	Significance
1	7.031	1.004	7	0.4256	51	6.362	0.909	7	0.4982
2	6.691	0.956	7	0.4617	52	8.467	1.210	7	0.2932
3	7.567	1.081	7	0.3723	53	6.913	0.988	7	0.4380
4	7.103	1.015	7	0.4182	54	3.244	0.463	7	0.8615
5	12.870	1.839	7	0.0753	55	35.197	5.028	7	0.0000
6	.517	0.931	7	0.4809	56	3.082	0.440	7	0.8773
7	8.634	1.233	7	0.2800	57	10.488	1.498	7	0.1626
8	6.563	0.938	7	0.4758	58	5.265	0.752	7	0.6276
9	6.375	0.911	7	0.4967	59	4.348	0.621	7	0.7390
10	3.626	0.518	7	0.8217	60	7.012	1.002	7	0.4276
11	4.237	0.605	7	0.7522	61	13.001	1.857	7	0.0721
12	3.389	0.484	7	0.8468	62	5.798	0.828	7	0.5635
13	3.768	0.538	7	0.8061	63	3.322	0.475	7	0.8537
14	5.030	0.719	7	0.6563	64	6.926	0.989	7	0.4367
15	8.962	1.280	7	0.2554	65	11.683	1.669	7	0.1115
16	6.398	0.914	7	0.4942	66	2.109	0.301	7	0.9536
17	7.212	1.030	7	0.4071	67	4.382	0.626	7	0.7349
18	5.350	0.764	7	0.6173	68	5.925	0.846	7	0.5486
19	5.899	0.843	7	0.5516	69	4.878	0.697	7	0.6749
20	8.962	1.280	7	0.2554	70	5.057	0.722	7	0.6530
21	2.978	0.425	7	0.8870	71	8.294	1.185	7	0.3074
22	35.390	5.056	7	0.0000	72	10.095	1.442	7	0.1833
23	8.333	1.190	7	0.3042	73	5.887	0.841	7	0.5530
24	2.974	0.425	7	0.8874	74	5.363	0.766	7	0.6157
25	4.909	0.701	7	0.6711	75	6.471	0.924	7	0.4859
26	3.463	0.495	7	0.8391	76	4.925	0.704	7	0.6691
27	3.171	0.453	7	0.8687	77	5.847	0.835	7	0.5577
28	5.765	0.824	7	0.5674	78	7.522	1.075	7	0.3766
29	7.601	1.086	7	0.3691	79	12.279	1.754	7	0.0918
30	5.188	0.741	7	0.6370	80	2.270	0.324	7	0.9434
31	2.751	0.393	7	0.9071	81	4.943	0.706	7	0.6669
32	7.024	1.003	7	0.4264	82	14.118	2.017	7	0.0491
33	5.678	0.811	7	0.5778	83	6.837	0.977	7	0.4460
34	3.529	0.504	7	0.8321	84	2.366	0.338	7	0.9369
35	6.539	0.934	7	0.4784	85	3.016	0.431	7	0.8835
36	2.900	0.414	7	0.8941	86	3.493	0.499	7	0.8359
37	6.704	0.958	7	0.4603	87	3.354	0.479	7	0.8504
38	3.030	0.433	7	0.8823	88	2.417	0.345	7	0.9332
39	10.213	1.459	7	0.1768	89	6.011	0.859	7	0.5385
40	3.827	0.547	7	0.7995	90	4.860	0.694	7	0.6771
41	2.898	0.414	7	0.8943	91	3.763	0.538	7	0.8067
42	12.282	1.755	7	0.0917	92	5.841	0.834	7	0.5584
43	7.129	1.018	7	0.4156	93	14.328	2.047	7	0.0456
44	4.819	0.688	7	0.6821	94	5.407	0.772	7	0.6105
45	6.670	0.953	7	0.4640	95	7.391	1.056	7	0.3893
46	7.475	1.068	7	0.3811	96	16.708	2.387	7	0.0194
47	14.094	2.013	7	0.0495	97	8.195	1.171	7	0.3157
48	6.152	0.879	7	0.5221	98	4.990	0.713	7	0.6612
49	7.561	1.080	7	0.3729	99	5.587	0.798	7	0.5888
50	9.029	1.290	7	0.2506	100	4.704	0.672	7	0.6960

df = degrees of freedom

Mahalanobis D^2 value based on the following variables ($X_1, X_2, X_3, X_4, X_5, X_6,$ and X_7). The D^2/df value is approximately distributed as a t value.

to be considered unrepresentative of the population. In all instances, the observations designated as outliers, even with the multivariate tests, seem similar enough to the remaining observations to be retained in the multivariate analyses. However, the researcher should always examine the results of each specific multivariate technique to identify observations that may become outliers in that particular application.

Testing the Assumptions of Multivariate Analysis

The final step in examining the data involves testing the assumptions underlying multivariate analysis. The need to test the statistical assumptions is increased in multivariate applications because of two characteristics of multivariate analysis. First, the complexity of the relationships, owing to the typical use of a large number of variables, makes the potential distortions and biases more potent when the assumptions are violated. This is particularly true when the violations compound to become even more detrimental than if considered separately. Second, the complexity of the analyses and of the results may mask the "signs" of assumption violations apparent in the simpler univariate analyses. In almost all instances, the multivariate procedures will estimate the multivariate model and produce results even when the assumptions are severely violated. Thus, the researcher must be aware of any assumption violations and the implications they may have for the estimation process or the interpretation of the results.

Assessing Individual Variables versus the Variate

Multivariate analysis requires that the assumptions underlying the statistical techniques be tested twice: first for the separate variables, akin to the tests of assumption for univariate analyses, and second for the multivariate model **variate,** which acts collectively for the variables in the analysis and thus must meet the same assumptions as individual variables. This chapter focuses on the examination of individual variables for meeting the assumptions underlying the multivariate procedures. Discussions in each chapter address the methods used to assess the assumptions underlying the variate for each multivariate technique.

Normality

The most fundamental assumption in multivariate analysis is **normality,** referring to the shape of the data distribution for an individual metric variable and its correspondence to the **normal distribution,** the benchmark for statistical methods. If the variation from the normal distribution is sufficiently large, all resulting statistical tests are invalid, as normality is required to use the F and t statistics. Both the univariate and the multivariate statistical methods discussed in this text are based on the assumption of univariate normality, with the multivariate methods also assuming multivariate normality. Univariate normality for a single variable is easily tested, and a number of corrective measures are possible, as shown later. In a simple sense, multivariate normality (the combination of two or more variables) means that the individual variables are normal in a univariate sense and that their combinations are also normal. Thus, if a variable is multivariate normal, it is also univariate normal. However, the reverse is not necessarily true (two or more univariate normal variables are not necessarily multivariate normal). Thus a situation in which all variables exhibit univariate normality will help gain, although not guarantee, multivariate normality. Multivariate normality is more difficult to test, but some tests are available for situations in which the multivariate technique is particularly affected by a violation of this assumption. In this text, we focus on assessing and achieving univariate normality for all variables, and address multivariate normality only when it is especially critical. Even though large sample sizes tend to diminish the detrimental effects of nonnormality, the researcher should assess the normality for all variables included in the analysis.

Graphical Analyses of Normality

The simplest diagnostic test for normality is a visual check of the histogram that compares the observed data values with a distribution approximating the normal distribution (see Figure 2.1). Although appealing because of its simplicity, this method is problematic for smaller samples, where the construction of the histogram (e.g., the number of categories or the width of categories) can distort the visual portrayal to such an extent that the analysis is useless. A more reliable approach is the **normal probability plot,** which compares the cumulative distribution of actual data values with the cumulative distribution of a normal distribution. The normal distribution forms a straight diagonal line, and the plotted data values are compared with the diagonal. If a distribution is normal, the line representing the actual data distribution closely follows the diagonal.

Figure 2.7 (p. 72) shows several normal probability plots and the corresponding univariate distribution of the variable. One characteristic of the distribution's shape, the kurtosis, is reflected in the normal probability plots. **Kurtosis** refers to the "peakedness" or "flatness" of the distribution compared with the normal distribution. When the line falls below the diagonal, the distribution is flatter than expected. When it goes above the diagonal, the distribution is more peaked than the normal curve. For example, in the normal probability plot of a peaked distribution (Figure 2.7d), we see a distinct S-shaped curve. Initially the distribution is flatter, and the plotted line falls below the diagonal. Then the peaked part of the distribution rapidly moves the plotted line above the diagonal, and eventually the line shifts to below the diagonal again as the distribution flattens. A nonpeaked distribution has the opposite pattern (Figure 2.7c). Another common pattern is a simple arc, either above or below the diagonal, indicating the **skewness** of the distribution. A negative skewness (Figure 2.7e) is indicated by an arc below the diagonal, whereas an arc above the diagonal represents a positively skewed distribution (Figure 2.7f). An excellent source for interpreting normal probability plots, showing the various patterns and interpretations is Daniel and Wood [5]. These specific patterns not only identify nonnormality but also tell us the form of the original distribution and the appropriate remedy to apply.

Statistical Tests of Normality

In addition to examining the normal probability plot, one can also use statistical tests to assess normality. A simple test is a rule of thumb based on the skewness and kurtosis values (available as part of the basic descriptive statistics for a

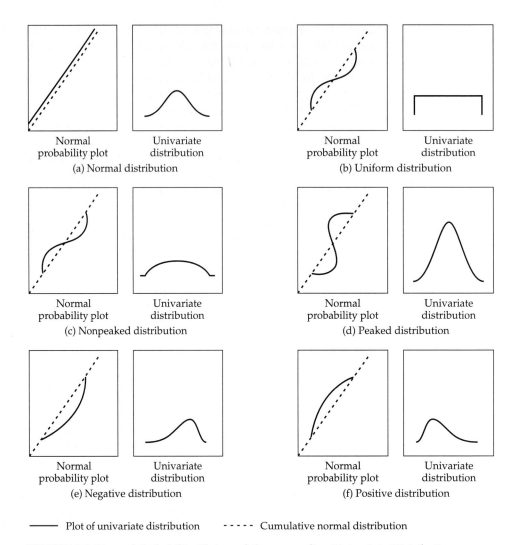

FIGURE 2.7 Normal Probability Plots and Corresponding Univariate Distributions

variable computed by all statistical programs). The statistic value (*z*) for the skewness value is calculated as:

$$z_{\text{skewness}} = \frac{\text{skewness}}{\sqrt{\dfrac{6}{N}}}$$

where *N* is the sample size. A *z* value can also be calculated for the kurtosis value using the following formula:

$$z_{\text{kurtosis}} = \frac{\text{kurtosis}}{\sqrt{\dfrac{24}{N}}}$$

If the calculated *z* value exceeds a critical value, then the distribution is nonnormal in terms of that characteristic. The critical value is from a *z* distribution, based on the significance level we desire. For example, a calculated value exceeding

± 2.58 indicates we can reject the assumption about the normality of the distribution at the .01 probability level. Another commonly used critical value is ±1.96, which corresponds to a .05 error level.

Specific statistical tests are also available in SPSS, SAS, BMDP, and most other programs. The two most common are the Shapiro-Wilks test and a modification of the Kolmogorov-Smirnov test. Each calculates the level of significance for the differences from a normal distribution. The researcher should always remember that tests of significance are less useful in small samples (fewer than 30) and quite sensitive in large samples (exceeding 1,000 observations). Thus, the researcher should always use both the graphical plots and any statistical tests to assess the actual degree of departure from normality.

Remedies for Nonnormality

A number of data transformations available to accommodate nonnormal distributions are discussed later in the chapter. This chapter confines the discussion to univariate normality tests and transformations. However, when we examine other multivariate methods, such as multivariate regression or multivariate analysis of variance, we discuss tests for multivariate normality as well. Moreover, many times when nonnormality is indicated, it is actually the result of other assumption violations; therefore, remedying the other violations eliminates the nonnormality problem. For this reason, the researcher should perform normality tests after or concurrently with analyses and remedies for other violations. (For those interested in multivariate normality, see references [8, 11].)

Homoscedasticity

Homoscedasticity is an assumption related primarily to dependence relationships between variables. It refers to the assumption that dependent variable(s) exhibit equal levels of variance across the range of predictor variable(s). Homoscedasticity is desirable because the variance of the dependent variable being explained in the dependence relationship should not be concentrated in only a limited range of the independent values. Although the dependent variables must be metric, this concept of an equal spread of variance across independent variables can be applied when the independent variables are either metric or nonmetric. With metric independent variables, the concept of homoscedasticity is based on the spread of dependent variable variance across the range of independent variable values, which is encountered in techniques such as multiple regression. The same concept also applies when the independent variables are nonmetric. In these instances, such as is found in ANOVA and MANOVA, the focus now becomes the equality of the variance (single dependent variable) or the variance/covariance matrices (multiple independent variables) across the groups formed by the nonmetric independent variables. The equality of variance/covariance matrices is also seen in discriminant analysis, but in this technique the emphasis is on the spread of the independent variables across the groups formed by the nonmetric dependent measure. In each of these instances, the purpose is the same: to ensure that the variance used in explanation and prediction is distributed across the range of values, thus allowing for a "fair test" of the relationship across all values of the nonmetric variables.

In most situations, we have many different values of the dependent variable at each value of the independent variable. For this relationship to be fully captured, the dispersion (variance) of the dependent variable values must be equal at each value of the predictor variable. Most problems with unequal variances stem from one of two sources. The first source is the type of variables included in the model. For example, as a variable increases in value (e.g., units ranging from near zero to millions), there is a naturally wider range of possible answers for the larger values. The second source results from a skewed distribution that creates heteroscedasticity. In Figure 2.8a, the scatterplots of data points for two variables (V_1 and V_2) with normal distributions exhibit equal dispersion across all data values (i.e., homoscedasticity). However, in Figure 2.8b, we see unequal dispersion **(heteroscedasticity)** caused by skewness of one of the variables (V_3). For the different values of V_3, there are different patterns of dispersion for V_1. This will cause the predictions to be better at some levels of the independent variable than at others. Violating this assumption often makes hypothesis tests either too conservative or too sensitive.

The effect of heteroscedasticity is also often related to sample size, especially when examining the variance dispersion across groups. For example, in ANOVA or MANOVA, the impact of heteroscedasticity on the statistical test depends on the sample sizes associated with the groups of smaller and larger variances. In multiple regression analysis, similar effects would occur in highly skewed distributions where there were disproportionate numbers of respondents in certain ranges of the independent variable.

Graphical Tests of Equal Variance Dispersion

The test of homoscedasticity for two metric variables is best examined graphically. The most common application of this form of assessment occurs in multiple regression, which is concerned with the dispersion of the dependent variable across the values of the metric independent variables. Because the focus of regression analysis is on the regression variate, the graphical plot of residuals is used to reveal the presence of homoscedasticity (or its opposite, heteroscedasticity). The discussion of residual analysis in chapter 4 details these procedures. Boxplots work well to represent the degree of variation between groups formed by a categorical variable. The length of the box and the whiskers each portray the variation of data within that group.

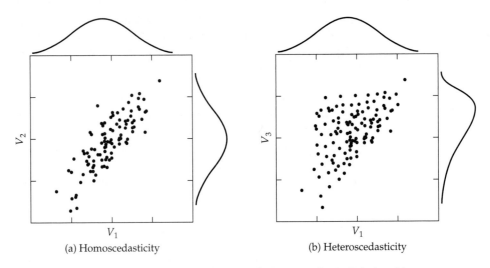

(a) Homoscedasticity   (b) Heteroscedasticity

FIGURE 2.8 Scatterplots of Homoscedastic and Heteroscedastic Relationships

Statistical Tests for Homoscedasticity

The statistical tests for equal variance dispersion relate to the variances within groups formed by nonmetric variables. The most common test, the Levene test, can be used to assess whether the variances of a single metric variable are equal across any number of groups. If more than one metric variable is being tested, so that the comparison involves the equality of variance/covariance matrices, the Box's M test is applicable. The Box's M test is available in both multivariate analysis of variance and discriminant analysis and is discussed in more detail in later chapters pertaining to these techniques.

Remedies for Heteroscedasticity

Heteroscedastic variables can be remedied through data transformations similar to those used to achieve normality. As mentioned earlier, many times heteroscedasticity is the result of nonnormality of one of the variables, and correction of the nonnormality also remedies the unequal dispersion of variance. A later section discusses data transformations of the variables to make all values have a potentially equal effect in prediction.

Linearity

An implicit assumption of all multivariate techniques based on correlational measures of association, including multiple regression, logistic regression, factor analysis, and structural equation modeling, is **linearity**. Because correlations represent only the linear association between variables, nonlinear effects will not be represented in the correlation value. This results in an underestimation of the actual strength of the relationship. It is always prudent to examine all relationships to identify any departures from linearity that may impact the correlation.

Identifying Nonlinear Relationships

The most common way to assess linearity is to examine scatterplots of the variables and to identify any nonlinear patterns in the data. An alternative approach is to run a simple regression analysis (the specifics of this technique are covered in chapter 4) and to examine the residuals. The residuals reflect the unexplained portion of the dependent variable; thus, any nonlinear portion of the relationship will show up in the residuals. The examination of residuals can also be applied to multiple regression, where the researcher can detect any nonlinear effects not represented in the regression variate. A more detailed discussion of residual analysis is included in chapter 4.

Remedies for Nonlinearity

If a nonlinear relationship is detected, the most direct approach is to transform one or both variables to achieve linearity. A number of available transformations are discussed later in this chapter. An alternative to data transformation is the creation of new variables to represent the nonlinear portion of the relationship. The process of creating and interpreting these additional variables, which can be used in all linear relationships, is discussed in chapter 4.

Absence of Correlated Errors

Predictions in any of the dependence techniques are not perfect, and we will rarely find a situation in which they are. However, we do attempt to ensure that any prediction errors are uncorrelated with each other. For example, if we found a pattern that suggests every other error is positive while the alternative error terms are negative, we would know that some unexplained systematic relationship exists in the dependent variable. If such a situation exists, we cannot be confident that our prediction errors are independent of the levels at which we are trying to predict. Some other factor is affecting the results, but is not included in the analysis.

Identifying Correlated Errors

The most common violations of the assumption that errors are uncorrelated are due to the data collection process. Similar factors that affect one group may not affect the other. If the groups are analyzed separately, the effects are constant within each group and do not impact the estimation of the relationship. But if the observations from both groups are combined, then the final estimated relationship must be a "compromise" between the two actual relationships. This causes the results to be biased because an unspecified cause is impacting the estimation of the relationship.

To identify correlated errors, the researcher must first identify possible causes. In our earlier example, this would be the two separate groups in data collection. Once the potential cause is identified, the researcher could see if differences did exist between the groups. Finding differences in the prediction errors in the two groups would then be the basis for determining that an unspecified effect was "causing" the correlated errors.

Remedies for Correlated Errors

Correlated errors must be corrected by including the omitted causal factor into the multivariate analysis. In our earlier example, the researcher would add a variable indicating in which class the respondents were. The most common remedy is the addition of a variable(s) to the analysis that represents the omitted factor. The key task facing the researcher is not the actual remedy, but rather the identification of the unspecified effect and a means of representing it in the analysis.

Data Transformations

Data transformations provide a means of modifying variables for one of two reasons: (1) to correct violations of the statistical assumptions underlying the multivariate techniques, or (2) to improve the relationship (correlation) between variables. Data transformations may be based on reasons that are either "theoretical" (transformations whose appropriateness is based on the nature of the data) or "data derived" (where the transformations are suggested strictly by an examination of the data). Yet in either case the researcher must proceed many times by trial and error, monitoring the improvement versus the need for additional transformations.

All the transformations described here are easily carried out by simple commands in the popular statistical packages. We focus on transformations that can be computed in this manner, although more sophisticated and complicated methods of data transformation are available (e.g., see Box and Cox [2]).

Transformations to Achieve Normality and Homoscedasticity

Data transformations provide the principal means of correcting nonnormality and heteroscedasticity. In both instances, patterns of the variables suggest specific transformations. For nonnormal distributions, the two most common patterns are "flat" distributions and skewed distributions. For the flat distribution, the most common transformation is the inverse (e.g., $1/Y$ or $1/X$). Skewed distributions can be transformed by taking the square root, logarithms, or even the inverse of the variable. Usually negatively skewed distributions are best transformed by employing a square root transformation, whereas the logarithm typically works best on positive skewness. In any instance, the researcher should apply all of the possible transformations and then select the most appropriate transformed variable.

Heteroscedasticity is an associated problem, and in many instances "curing" this problem will deal with normality problems as well. Heteroscedasticity is also due to the distribution of the variable(s). When examining the residuals of regression analysis for heteroscedasticity, we note that an indication of unequal variance is a cone-shaped distribution of the residuals (see chapter 4 for more specific details of the graphical analysis of residuals). If the cone opens to the right, take the inverse; if the cone opens to the left, take the square root. Some transformations can be associated with certain types of data. For example, frequency counts suggest a square root transformation; proportions are best transformed by the arcsin transformation ($X_{new} = 2$ arcsin $\sqrt{X_{old}}$); and proportional change is best handled by taking the logarithm of the variable. In all instances, once the transformations have been performed, the transformed data should be tested to see whether the desired remedy was achieved.

Transformations to Achieve Linearity

There are numerous procedures for achieving linearity between two variables, but most simple nonlinear relationships can be placed in one of four categories (see Figure 2.9). In each quadrant, the potential transformations for both dependent

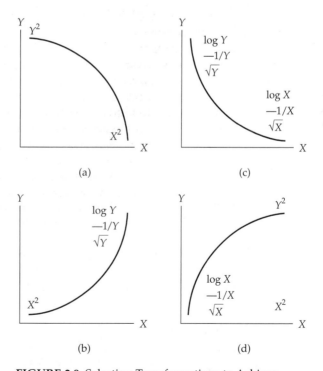

FIGURE 2.9 Selecting Transformations to Achieve
    Linearity
Source: F. Mosteller and J. W. Tukey, *Data Analysis and Regression.*
    Reading, Mass.: Addison-Wesley, 1977.

and independent variables are shown. For example, if the relationship looks like that in Figure 2.9a, then either variable can be squared to achieve linearity. When multiple transformation possibilities are shown, start with the top method in each quadrant and move downward until linearity is achieved. An alternative approach is to use additional variables, termed polynomials, to represent the nonlinear components. This method is discussed in more detail in chapter 4.

General Guidelines for Transformations

There are several points to remember when performing data transformations:

1. For a noticeable effect from transformations, the ratio of a variable's mean to its standard deviation should be less than 4.0.
2. When the transformation can be performed on either of two variables, select the variable with the smallest ratio from item 1.
3. Transformations should be applied to the independent variables except in the case of heteroscedasticity.
4. Heteroscedasticity can be remedied only by transformation of the dependent variable in a dependence relationship. If a heteroscedastic relationship is also nonlinear, the dependent variable, and perhaps the independent variables, must be transformed.
5. Transformations may change the interpretation of the variables. For example, transforming variables by taking their logarithm translates the relationship into a measure of proportional change (elasticity). Always be sure to explore thoroughly the possible interpretations of the transformed variables.

An Illustration of Testing the Assumptions Underlying Multivariate Analysis

To illustrate the techniques involved in testing the data for meeting the assumptions underlying multivariate analysis and to provide a foundation for use of the data in the subsequent chapters, the data set introduced in chapter 1 will be examined. In the course of the analysis, the assumptions of normality, homoscedasticity, and linearity will be covered. The fourth basic assumption, the absence of correlated errors, can be addressed only in the context of a specific multivariate model; this will be covered in later chapters for each multivariate technique. Emphasis will be placed on examining the metric variables, although the nonmetric variables will be assessed where appropriate.

Normality

The first analysis to be conducted in assessing the normality of the metric variables is the derivation of normal probability plots. Figure 2.10 contains the plots for each of the nine variables. In our examination of the graphs, we see some departures from the diagonal, indicative of a departure from normality. Referring to the patterns seen in Figure 2.7, we see that X_2 seems positively skewed, X_3 approximates a uniform distribution, and X_5 seems negatively skewed.

We can complement this visual analysis with statistics reflecting the shape of the distribution (skewness and kurtosis) as well as a statistical test for normality (the modified Kolmogorov-Smirnov test). Table 2.11 (p. 80) shows these values for all the metric variables. Four variables (X_2, X_3, X_4, and X_6) exhibit a statistically significant departure from normality. Table 2.11 also suggests the appropriate remedy. Two variables (X_2 and X_6) were transformed by taking the square root. In each case, the transformed variable demonstrated normality (see Table 2.11).

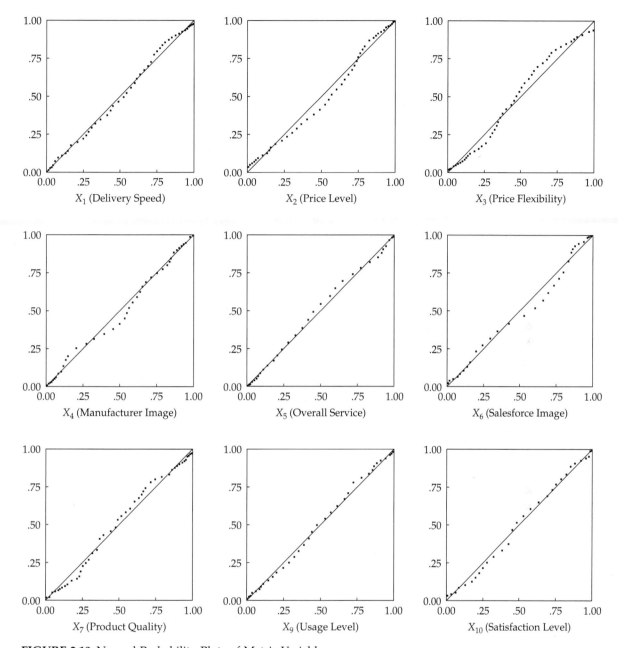

FIGURE 2.10 Normal Probability Plots of Metric Variables

Figure 2.11 (p. 81) demonstrates the effect of the transformation on X_2 in achieving normality. The transformed X_2 appears markedly more normal in both graphical portrayals, and the statistical descriptors are also improved. The researcher should always examine the transformed variables as rigorously as the original variables in terms of their normality and distribution shape.

In the case of the two remaining variables (X_3 and X_4), none of the transformations could improve the normality. These variables will have to be used in their original form. In situations where the normality of the variables is critical, the

TABLE 2.11 Distributional Characteristics, Testing for Normality, and Possible Remedies

Variable	Shape Descriptors[a]				Test of Normality		Description of the Distribution	Possible Remedies	
	Skewness		Kurtosis						
	Statistic	z value	Statistic	z value	Statistic	Significance		Transformation	Significance After Remedy
X_1 Delivery speed	-.085	-.35	-.511	-1.07	.063	>.200	Normal distribution	None	
X_2 Price level	.469	1.95*	-.509	1.06	.095	.028	Positive skewness	Square root	>.200
X_3 Price flexibility	-.289	1.19	-1.073	2.24*	.095	.027	Approaching uniform distribution	None	
X_4 Manufacturer image	.218	.91	.085	.18	.107	.007	Slight positive skewness	No improvement possible	
X_5 Overall service	-.373	1.55	.141	.29	.085	.069	Normal distribution	None	
X_6 Salesforce image	.493	2.04*	.107	.22	.122	.001	Heavy tails with positive skewness	Square root	.032
X_7 Product quality	-.229	.95	-.850	1.77	.091	.041	Slightly flat	None	
X_9 Usage level	-.069	.26	-.725	1.52	.079	.131	Normal distribution	None	
X_{10} Satisfaction level	.089	.37	-.763	1.60	.078	.142	Normal distribution	None	

[a]The z values are derived by dividing the statistics by the appropriate standard errors of .241 (skewness) and .478 (kurtosis). The equations for calculating the standard errors are given in the text.
*Significant at the .05 level.

Original Variable

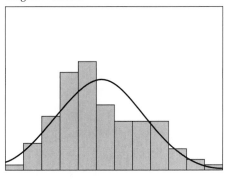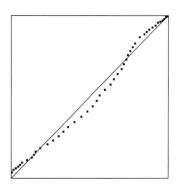

Transformed Variable

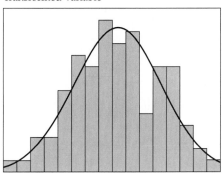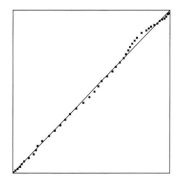

Distribution Characteristics Before and After Transformation

| | Shape Descriptors[a] | | | | Test of Normality | |
| | Skewness | | Kurtosis | | | |
Variable Form	Statistic	z Value	Statistic	z Value	Statistic	Significance
Original X_2	.469	1.95	−.509	1.06	.095	.028
Transformed X_2	−.106	.44	−.465	.97	.062	> .200

[a]The z values are derived by dividing the statistics by the appropriate standard errors of 0.241 (skewness) and 0.478 (kurtosis). The equations for calculating the standard errors are given in the text.

FIGURE 2.11 Transformation of X_2 (Price Level) to Achieve Normality

transformed variables can be used with the assurance that they meet the assumptions of normality. But the departures from normality were not so extreme in any of the original variables that they should never be used in any analysis in their original form. If the technique has a robustness to departures from normality, then the original variables may be preferred for the comparability in the interpretation phase.

Homoscedasticity

All statistical packages have tests to assess homoscedasticity on a univariate basis (e.g., the Levene test in SPSS) where the variance of a metric variable is compared across levels of a nonmetric variable. For our purposes, we examine each of the metric variables across the five nonmetric variables in the data set. These are appropriate analyses in preparation for analysis of variance or multivariate analysis of variance, in which the nonmetric variables are the independent variables, or for discriminant analysis, in which the nonmetric variables are the dependent measures.

Table 2.12 contains the results of the Levene test for each of the nonmetric variables. The nonmetric variables X_8 and X_{11} both show significant heteroscedasticity on the same performance factors (X_4, X_5, X_6, and X_7), whereas X_{12} and X_{14} have fewer occurrences among the entire set of variables. The implications of these instances of heteroscedasticity must be examined whenever group differences are examined using these nonmetric variables as independent variables and these metric variables as dependent variables. If the assumption violations are found, variable transformations are available to help remedy the variance dispersion.

The tests for homoscedasticity of two metric variables, encountered in methods such as multiple regression, are best accomplished through graphical analysis, particularly an analysis of the residuals. The interested reader is referred to chapter 4 for a complete discussion of residual analysis and the patterns of residuals indicative of heteroscedasticity.

Linearity

The final assumption to be examined is the linearity of the relationships. In the case of individual variables, this relates to the patterns of association between each pair of variables and the ability of the correlation coefficient to adequately

TABLE 2.12 Testing For Homoscedasticity

| | Nonmetric/Categorical Variable | | | | | | | | | |
| Metric Variable | X_8 Size of Firm | | X_{11} Specification Buying | | X_{12} Structure of Procurement | | X_{13} Type of Industry | | X_{14} Type of Buying Situation | |
	Levene Statistic	Sig.	Levene Statistic	Sig.	Levene Statistic	Sig.	Levene Statistic	Sig.	Levene Statistic	Sig.
X_1 Delivery speed	.934	.336	.934	.336	.382	.538	.377	.540	.114	.892
X_2 Price level	1.582	.211	1.582	.211	13.761	.000	1.345	.249	8.081	.001
X_3 Price flexibility	1.194	.277	1.194	.277	4.765	.031	.192	.662	14.383	.000
X_4 Manufacturer image	6.549	.012	6.549	.012	.281	.597	.040	.842	2.030	.137
X_5 Overall service	7.819	.006	7.819	.006	5.141	.026	.003	.957	2.888	.060
X_6 Salesforce image	5.279	.024	5.279	.024	1.626	.205	.264	.609	1.735	.182
X_7 Product quality	8.748	.004	8.748	.004	4.129	.045	2.532	.115	2.051	.134
X_9 Usage level	1.377	.243	1.377	.243	1.575	.212	.091	.763	.056	.945
X_{10} Satisfaction level	.323	.571	.323	.571	.000	.986	.054	.817	3.302	.041

Note: Values represent the value and statistical significance (Sig.) of the Levene test assessing the variance dispersion of each metric variable across the levels of the nonmetric/categorical variables.

represent the relationship. If nonlinear relationships are indicated, then the researcher can either transform one or both of the variables to achieve linearity or create additional variables to represent the nonlinear components. For our purposes, we rely on the visual inspection of the relationships to determine whether nonlinear relationships are present. The reader can refer to Figure 2.3, the scatterplot matrix containing the scatterplot for all the metric variables in the data set. Examination of the scatterplots does not reveal any apparent nonlinear relationships. Thus, transformations are not deemed necessary. The assumption of linearity will also be checked for the entire multivariate model, as is done in the examination of residuals in multiple regression.

Summary

The series of graphical and statistical tests directed toward assessing the assumptions underlying the multivariate techniques revealed relatively little in terms of violations of the assumptions. Where violations were indicated, they were relatively minor and should not present any serious problems in the course of the data analysis. The researcher is encouraged always to perform these simple, yet revealing, examinations of the data to ensure that potential problems can be identified and resolved before the analysis begins.

Incorporating Nonmetric Data with Dummy Variables

A critical factor in choosing and applying the correct multivariate technique is the measurement properties of the dependent and independent variables. Some of the techniques, such as discriminant analysis or multivariate analysis of variance, specifically require nonmetric data as dependent or independent variables. But in many instances, metric variables must be used as independent variables, such as in regression analysis, discriminant analysis, and canonical correlation. Moreover, the interdependence techniques of factor and cluster analysis generally require metric variables. To this point, all discussions have assumed metric measurement for variables. But what can we do when the variables are nonmetric, with two or more categories? Are nonmetric variables, such as gender, marital status, or occupation, precluded from use in many multivariate techniques? The answer is no, and we now discuss how to incorporate nonmetric variables into many of these situations that require metric variables.

The researcher has available a method for using dichotomous variables, known as **dummy variables,** which act as replacement variables. A dummy variable is a dichotomous variable that represents one category of a nonmetric independent variable. Any nonmetric variable with k categories can be represented as $k - 1$ dummy variables. The following example will help clarify this concept.

First, assume we wish to include gender, which has two categories, female and male. We also have measured household income level by three categories (see Table 2.13, p. 84). To represent the nonmetric variable gender, we would create two new dummy variables (X_1 and X_2), as shown in Table 2.13. X_1 would represent those individuals who are female with a value of 1, and would give all males a value of 0. Likewise, X_2 would represent all males with a value of 1 and give females a value of 0. Both variables (X_1 and X_2) are not necessary, however,

TABLE 2.13 Representing Nonmetric Variables with Dummy Variables

Nonmetric Variable with Two Categories (Gender)		Nonmetric Variable with Three Categories (Household Income Level)	
Gender	Dummy Variables	Household Income Level	Dummy Variables
Female	$X_1 = 1$, else $X_1 = 0$	if <\$15,000	$X_3 = 1$, else $X_3 = 0$
Male	$X_2 = 1$, else $X_2 = 0$	if $\geq$\$15,000 & $\leq$\$25,000	$X_4 = 1$, else $X_4 = 0$
		if >\$25,000	$X_5 = 1$, else $X_5 = 0$

because when $X_1 = 0$, gender must be female by definition. Thus we need include only one of the variables (X_1 or X_2) to test the effect of gender.

Correspondingly, if we had also measured household income with three levels, as shown in Table 2.13, we would first define three dummy variables (X_3, X_4, and X_5). But as in the case of gender, we would not need the entire set of dummy variables, and instead use $k - 1$ dummy variables, where k is the number of categories. Thus, we would use two of the dummy variables to represent the effects of household income.

There are three ways to represent the household income levels with two dummy variables, as shown in Table 2.14. This form of dummy-variable coding is known as **indicator coding.** An important consideration in this form of dummy-variable coding is to remember the category that is omitted, known as the **comparison group.** This is the category that received all zeros for the dummy variables. For example, in regression analysis, the regression coefficients for the dummy variables represent deviations from the comparison group on the criterion variable. The deviations represent the differences between means for each group of respondents formed by a dummy variable and the comparison group. This form is most appropriate when there is a logical comparison group, such as in an experiment. In an experiment with a control group acting as the comparison group, the coefficients are the mean differences on the dependent variable for each treatment group from the control group. Any time dummy-variable coding is used, we must be aware of the comparison group and remember the impacts it has in our interpretation of the remaining variables.

An alternative method of dummy-variable coding is termed **effects coding.** It is the same as indicator coding except that the comparison group (the group that got all zeros in indicator coding) is now given the value of -1 instead of 0 for the dummy variables. Now the coefficients represent differences for any group from the mean of all groups rather than from the omitted group. Both forms of dummy-variable coding will give the same results; the only differences will be in the interpretation of the dummy-variable coefficients.

TABLE 2.14 Alternative Dummy Variable Coding Patterns for a Three-Category Nonmetric Variable

Household Income Level	Pattern 1		Pattern 2		Pattern 3	
	X_3	X_4	X_3	X_4	X_3	X_4
If <\$15,000	1	0	1	0	0	0
If $\geq$\$15,000 & $\leq$\$25,000	0	1	0	0	1	0
If >\$25,000	0	0	0	1	0	1

Dummy variables are used most often in regression and discriminant analysis, where the coefficients have direct interpretation. Their use in other multivariate techniques is more limited, especially for those that rely on correlational patterns, such as factor analysis, because the correlation of a binary variable is not well represented by the traditional Pearson correlation coefficient. However, special considerations can be made in these instances, as discussed in the appropriate chapters.

Summary

This chapter has provided the researcher with the necessary tools to examine and explore the nature of the data and the relationships among variables before the application of any of the multivariate techniques. Considerable time and effort can be expended in these activities, but the prudent researcher wisely invests the necessary resources to thoroughly examine the data to ensure that the multivariate methods are applied in appropriate situations and to assist in a more thorough and insightful interpretation of the results.

Questions

1. List potential underlying causes of outliers. Be sure to include attributions to both the respondent and the researcher.
2. Discuss why outliers might be classified as beneficial and as problematic.
3. Distinguish between data that are missing at random (MAR) and missing completely at random (MCAR). Explain how each type impacts the analysis of missing data.
4. Describe the conditions under which a researcher would delete a case with missing data versus the conditions under which a researcher would use an imputation method.
5. Evaluate the following statement: In order to run most multivariate analyses, it is not necessary to meet all the assumptions of normality, linearity, homoscedasticity, and independence.
6. Discuss the following statement: Multivariate analyses can be run on any data set, as long as the sample size is adequate.

References

1. Anderson, Edgar (1969), "A Semigraphical Method for the Analysis of Complex Problems." *Technometrics* 2 (August): 387–91.
2. Box, G. E. P., and D. R. Cox (1964), "An Analysis of Transformations." *Journal of the Royal Statistical Society* B (26): 211–43.
3. Chernoff, Herman. "Graphical Representation as a Discipline," in *Graphical Representation of Multivariate Data*, Peter C. C. Wang, ed. New York: Academic Press, pp. 1–11.
4. Cohen, Jacob, and Patricia Cohen (1983), *Applied Multiple Regression/Correlation Analysis for the Behavioral Sciences*, 2d ed. Hillsdale, N.J.: Lawrence Erlbaum Associates.
5. Daniel, C., and F. S. Wood (1980), *Fitting Equations to Data*, 2d ed. New York: Wiley-Interscience.
6. Dempster, A. P., and D. B. Rubin (1983), "Overview," in *Incomplete Data in Sample Surveys: Theory and Annotated Bibliography*, vol. 2. Madow, Olkin, and Rubin, eds. New York: Academic Press.

7. Feinberg, Stephen (1979), "Graphical Methods in Statistics." *American Statistician* 33 (November): 165–78.

8. Johnson, R. A., and D. W. Wichern (1982), *Applied Multivariate Statistical Analysis.* Upper Saddle River, N.J.: Prentice-Hall.

9. Little, Roderick J. A., and Donald B. Rubin (1987), *Statistical Analysis with Missing Data.* New York: Wiley.

10. Wang, Peter C. C., ed. (1978), *Graphical Representation of Multivariate Data.* New York: Academic Press.

11. Weisberg, S. (1985), *Applied Linear Regression.* New York: Wiley.

12. Wilkinson, L. (1982), "An Experimental Evaluation of Multivariate Graphical Point Representations." In *Human Factors in Computer Systems: Proceedings,* New York: ACM Press, pp. 202–9.

CHAPTER 3

Factor Analysis

LEARNING OBJECTIVES

Upon completing this chapter, you should be able to do the following:

- Differentiate factor analytic techniques from other multivariate techniques.
- State the major purposes of factor analytic techniques.
- Distinguish between exploratory and confirmatory uses of factor analytic techniques.
- Identify the differences between component analysis and common factor analysis models.
- Tell when component analysis and common factor analysis should be used.
- Identify the difference between R and Q factor analysis.
- Explain the concept of rotation of factors.
- Tell how to determine the number of factors to extract.
- Explain how to name a factor.
- Understand the advantages and disadvantages associated with summated scales.
- Explain the purpose of factor scores and how to use them.
- Explain how to select surrogate variables for subsequent analysis.
- State the major limitations of factor analytic techniques.

CHAPTER PREVIEW

The multivariate statistical technique of factor analysis has found increased use during the past decade in all fields of business-related research. As the number of variables to be considered in multivariate techniques increases, there is a commensurate need for increased knowledge of the structure and interrelationships of the variables. This chapter describes factor analysis, a technique particularly suitable for analyzing the patterns of complex, multidimensional relationships encountered by researchers. It defines and explains in broad, conceptual terms the fundamental aspects of factor analytic techniques. Factor analysis can be utilized to examine the underlying patterns or relationships for a large number of variables and to determine whether the information can be condensed or summarized in a smaller set of factors or components. To further clarify the methodological concepts, basic guidelines for presenting and interpreting the results of these techniques are also included.

KEY TERMS

Before starting the chapter, review the key terms to develop an understanding of the concepts and terminology used. Throughout the chapter the key terms appear in **boldface.** Other points of emphasis in the chapter are *italicized.* Also, cross-references within the Key Terms appear in *italics.*

Anti-image correlation matrix   Matrix of the partial correlations among variables after factor analysis, representing the degree to which the factors "explain" each other in the results. The diagonal contains the *measures of sampling adequacy* for each variable, and the off-diagonal values are partial correlations among variables.

Bartlett test of sphericity   Statistical test for the overall significance of all correlations within a correlation matrix.

Cluster analysis   Multivariate technique with the objective of grouping respondents or cases with similar profiles on a defined set of characteristics. Similar to *Q factor analysis.*

Common factor analysis   Factor model in which the factors are based on a reduced correlation matrix. That is, *communalities* are inserted in the diagonal of the *correlation* matrix, and the extracted factors are based only on the *common variance*, with *specific* and *error variances* excluded.

Common variance   Variance shared with other variables in the factor analysis.

Communality   Total amount of variance an original variable shares with all other variables included in the analysis.

Component analysis   Factor model in which the factors are based on the total variance. With component analysis, unities (1s) are used in the diagonal of the *correlation matrix;* this procedure computationally implies that all the variance is *common* or shared.

Composite measure   See *summated scale.*

Conceptual definition   Specification of the theoretical basis for a concept that is represented by a factor.

Content validity   Assessment of the degree of correspondence between the items selected to constitute a *summated scale* and its *conceptual definition.*

Correlation matrix   Table showing the intercorrelations among all variables.

Cronbach's alpha   Measure of *reliability* that ranges from 0 to 1, with values of .60 to .70 deemed the lower limit of acceptability.

Dummy variable   Binary metric variable used to represent a single category of a nonmetric variable.

Eigenvalue   Column sum of squared loadings for a factor; also referred to as the *latent root*. It represents the amount of variance accounted for by a factor.

Error variance   Variance of a variable due to errors in data collection or measurement.

Face validity   See *content validity*.

Factor   Linear combination (variate) of the original variables. Factors also represent the underlying dimensions (constructs) that summarize or account for the original set of observed variables.

Factor indeterminancy   Characteristic of *common factor analysis* such that several different factor scores can be calculated for a respondent, each fitting the estimated factor model. This means the factor scores are not unique for each individual.

Factor loadings   Correlation between the original variables and the factors, and the key to understanding the nature of a particular factor. Squared factor loadings indicate what percentage of the variance in an original variable is explained by a factor.

Factor matrix   Table displaying the *factor loadings* of all variables on each factor.

Factor pattern matrix   One of two *factor matrices* found in an *oblique rotation* that is most comparable to the factor matrix in an *orthogonal rotation*.

Factor rotation   Process of manipulation or adjusting the factor axes to achieve a simpler and pragmatically more meaningful factor solution.

Factor score   Composite measure created for each observation on each factor extracted in the factor analysis. The factor weights are used in conjunction with the original variable values to calculate each observation's score. The factor score then can be used to represent the factor(s) in subsequent analyses. Factor scores are standardized to have a mean of 0 and a standard deviation of 1.

Factor structure matrix   A *factor matrix* found in an *oblique rotation* that represents the simple correlations between variables and factors, incorporating the unique variance and the correlations between factors. Most researchers prefer to use the *factor pattern matrix* when interpreting an oblique solution.

Indicator   Single variable used in conjunction with one or more other variables to form a *composite measure*.

Latent root   See *eigenvalue*.

Measure of sampling adequacy (MSA)   Measure calculated both for the entire correlation matrix and each individual variable evaluating the appropriateness of applying factor analysis. Values above .50 for either the entire matrix or an individual variable indicate appropriateness.

Measurement error   Inaccuracies in measuring the "true" variable values due to the fallibility of the measurement instrument (i.e., inappropriate response scales), data entry errors, or respondent errors.

Multicollinearity   Extent to which a variable can be explained by the other variables in the analysis.

Oblique factor rotation   *Factor rotation* computed so that the extracted factors are correlated. Rather than arbitrarily constraining the factor rotation to an *orthogonal* solution, the oblique rotation identifies the extent to which each of the factors are correlated.

Orthogonal   Mathematical independence (no correlation) of factor axes to each other (i.e., at right angles, or 90 degrees).

Orthogonal factor rotation   Factor rotation in which the factors are extracted so that their axes are maintained at 90 degrees. Each factor is independent of, or *orthogonal* to, all other factors. The correlation between the factors is determined to be 0.

Q factor analysis   Forms groups of respondents or cases based on their similarity on a set of characteristics (also see the discussion of cluster analysis in chapter 9).

R factor analysis   Analyzes relationships among variables to identify groups of variables forming latent dimensions (*factors*).

Reliability   Extent to which a variable or set of variables is consistent in what it is intended to measure. If multiple measurements are taken, reliable measures will all be very consistent in their values. It differs from *validity* in that it does not relate to what should be measured, but instead to how it is measured.

Reverse scoring   Process of reversing the scores of a variable, while retaining the distributional characteristics, to change the relationships (correlations) between two variables. Used in *summated scale* construction to avoid a "canceling out" between variables with positive and negative *factor loadings* on the same factor.

Specific variance   Variance of each variable unique to that variable and not explained or associated with other variables in the factor analysis.

Summated scales   Method of combining several variables that measure the same concept into a single variable in an attempt to increase the *reliability* of the measurement. In most instances, the separate variables are summed and then their total or average score is used in the analysis.

Surrogate variable   Selection of a single variable with the highest *factor loading* to represent a factor in the data reduction stage instead of using a *summated scale* or *factor score.*

Trace   Represents the total amount of variance on which the factor solution is based. The trace is equal to the number of variables, based on the assumption that the variance in each variable is equal to 1.

Validity   Extent to which a measure or set of measures correctly represents the concept of study—the degree to which it is free from any systematic or non-random error. Validity is concerned with how well the concept is defined by the measure(s), whereas *reliability* relates to the consistency of the measure(s).

VARIMAX   One of the most popular orthogonal factor rotation methods.

What Is Factor Analysis?

Factor analysis is a generic name given to a class of multivariate statistical methods whose primary purpose is to define the underlying structure in a data matrix. Broadly speaking, it addresses the problem of analyzing the structure of the interrelationships (correlations) among a large number of variables (e.g., test scores, test items, questionnaire responses) by defining a set of common underlying dimensions, known as **factors.** With factor analysis, the researcher can first identify the separate dimensions of the structure and then determine the extent to which each variable is explained by each dimension. Once these dimensions and the explanation of each variable are determined, the two primary uses for factor analysis—summarization and data reduction—can be achieved. In summarizing the data, factor analysis derives underlying dimensions that, when interpreted and understood, describe the data in a much smaller number of concepts than the original individual variables. Data reduction can be achieved by calculating scores for each underlying dimension and substituting them for the original variables.

We introduce factor analysis as our first multivariate technique because it can play a unique role in the application of other multivariate techniques. As already discussed, the primary advantage of multivariate techniques is their ability to accommodate multiple variables in an attempt to understand the complex relationships not possible with univariate and bivariate methods. Increasing the number of variables also increases the possibility that the variables are not all uncorrelated and representative of distinct concepts. Instead, groups of variables may be interrelated to the extent that they are all representative of a more general concept. This may be by design, such as the attempt to measure the many facets of personality or store image, or may arise just from the addition of new variables. In either case, the researcher must know how the variables are interrelated to better interpret the results. Finally, if the number of variables is too large or there is a need to better represent a smaller number of concepts rather than the many facets, factor analysis can assist in selecting a representative subset of variables or even creating new variables as replacements for the original variables while still retaining their original character.

Factor analysis differs from the dependence techniques discussed in the next section (i.e., multiple regression, discriminant analysis, multivariate analysis of variance, or canonical correlation), in which one or more variables are explicitly considered the criterion or dependent variables and all others are the predictor or independent variables. Factor analysis is an interdependence technique in which all variables are simultaneously considered, each related to all others, and still employing the concept of the variate, the linear composite of variables. In factor analysis, the variates (factors) are formed to maximize their explanation of the entire variable set, not to predict a dependent variable(s). If we were to draw an analogy to dependence techniques, it would be that each of the observed (original) variables is a dependent variable that is a function of some underlying and latent set of factors (dimensions) that are themselves made up of all other variables. Thus, each variable is predicted by all others. Conversely, one can look at each factor (variate) as a dependent variable that is a function of the entire set of observed variables. Either analogy illustrates the differences in purpose between dependence (prediction) and interdependence (identification of structure) techniques.

Factor analytic techniques can achieve their purposes from either an exploratory or confirmatory perspective. There is continued debate concerning the appropriate role for factor analysis. Many researchers consider it only exploratory, useful in searching for structure among a set of variables or as a data reduction method. In this perspective, factor analytic techniques "take what the data give you" and do not set any a priori constraints on the estimation of components or the number of components to be extracted. For many—if not most—applications, this use of factor analysis is appropriate. However, in other situations, the researcher has preconceived thoughts on the actual structure of the data, based on theoretical support or prior research. The researcher may wish to test hypotheses involving issues such as which variables should be grouped together on a factor or the precise number of factors. In these instances, the researcher requires that factor analysis take a confirmatory approach—that is, assess the degree to which the data meet the expected structure. The methods we discuss in this chapter do not directly provide the necessary structure for formalized hypothesis testing. We explicitly address the confirmatory perspective of factor analysis in chapter 11. In this chapter, however, we view factor analytic techniques principally from an exploratory or nonconfirmatory viewpoint.

A Hypothetical Example of Factor Analysis

Assume that through qualitative research a retail firm has identified 80 different characteristics of retail stores and their service that consumers have mentioned as affecting their patronage choice among stores. The retailer wants to understand how consumers make decisions but feels that it cannot evaluate 80 separate characteristics or develop action plans for this many variables, because they are too specific. Instead, it would like to know if consumers think in more general evaluative dimensions rather than in just the specific items. To identify these dimensions, the retailer could commission a survey asking for consumer evaluations on each of these specific items. Factor analysis would then be used to identify the underlying evaluative dimensions. Specific items that correlate highly are assumed to be a "member" of that broader dimension. These dimensions become composites of specific variables, which in turn allow the dimensions to be interpreted and described. In our example, the factor analysis might identify such dimensions as product assortment, product quality, prices, store personnel, service, and store atmosphere as the evaluative dimensions used by the respondents. Each of these dimensions contains specific items that are a facet of the broader evaluative dimension. From these findings, the retailer may then use the dimensions (factors) to define broad areas for planning and action.

An illustrative example of a simple application of factor analysis is shown in Figure 3.1, which represents the correlation matrix for nine store image elements. Included in this set are measures of the product offering, store personnel, price levels, and in-store service and experiences. The question a researcher may wish to address is: Are all of these elements separate in their evaluative properties or do they "group" into some more general areas of evaluation? For example, do all of the product elements group together? Where does price level fit, or is it separate? How do the in-store features (e.g., store personnel, service, and atmosphere) relate to one another? Visual inspection of the original correlation matrix (Figure 3.1, part 1) does not easily reveal any specific pattern. There are scattered high correlations, but variable groupings are not apparent. The application of factor analysis results in the grouping of variables as reflected in part 2 of Figure 3.1. Here some interesting patterns emerge. First, four variables all relating to the in-store experience of shoppers are grouped together. Then, three variables describing the product assortment and availability are grouped together. Finally, product quality and price levels are grouped. Each group represents a set of highly interrelated variables that may reflect a more general evaluative dimension. In this case, we might label the three variable groupings by the labels in-store experience, product offerings, and value. This would give store management a smaller set of concepts to consider in any strategic or tactical marketing plans, while still providing insight into what constitutes each general area.

PART 1: ORIGINAL CORRELATION MATRIX

	V_1	V_2	V_3	V_4	V_5	V_6	V_7	V_8	V_9
V_1 Price level	1.000								
V_2 Store personnel	.427	1.000							
V_3 Return policy	.302	.771	1.000						
V_4 Product availability	.470	.497	.427	1.000					
V_5 Product quality	.765	.406	.307	.472	1.000				
V_6 Assortment depth	.281	.445	.423	.713	.325	1.000			
V_7 Assortment width	.354	.490	.471	.719	.378	.724	1.000		
V_8 In-store service	.242	.719	.733	.428	.240	.311	.435	1.000	
V_9 Store atmosphere	.372	.737	.774	.479	.326	.429	.466	.710	1.000

PART 2: CORRELATION MATRIX OF VARIABLES AFTER GROUPING ACCORDING TO FACTOR ANALYSIS

	V_3	V_8	V_9	V_2	V_6	V_7	V_4	V_1	V_5
V_3 Return policy	1.000								
V_8 In-store service	.733	1.000							
V_9 Store atmosphere	.774	.710	1.000						
V_2 Store personnel	.741	.719	.787	1.000					
V_6 Assortment depth	.423	.311	.429	.445	1.000				
V_7 Assortment width	.471	.435	.468	.490	.724	1.000			
V_4 Product availability	.427	.428	.479	.497	.713	.719	1.000		
V_1 Price level	.302	.242	.372	.427	.281	.354	.470	1.000	
V_5 Product quality	.307	.240	.326	.406	.325	.378	.472	.765	1.000

Shaded areas represent variables grouped together by factor analysis

FIGURE 3.1 Illustrative Example of the Use of Factor Analysis to Identify Structure within a Group of Variables

Factor Analysis Decision Process

We center the discussion of factor analysis on the six-stage model-building paradigm introduced in chapter 1. Figure 3.2 (p. 94) shows the first three stages of the structured approach to multivariate model-building, and Figure 3.4 on p. 101 details the final three stages, plus an additional stage (stage 7) beyond the estimation,

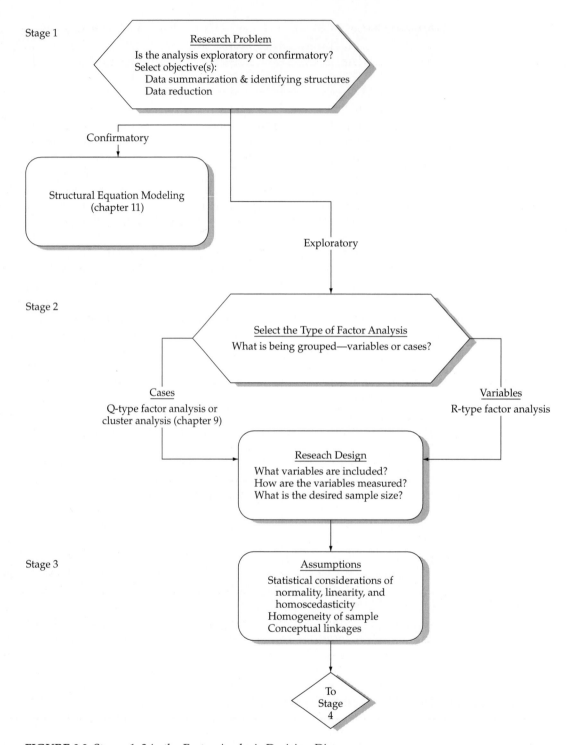

FIGURE 3.2 Stages 1–3 in the Factor Analysis Decision Diagram

interpretation, and validation of the factor models, which aids in selecting surrogate variables, computing factor scores, or creating summated scales for use in other multivariate techniques. A discussion of each stage follows.

Stage 1: Objectives of Factor Analysis

The starting point in factor analysis, as with other statistical techniques, is the research problem. The general purpose of factor analytic techniques is to find a way to condense (summarize) the information contained in a number of original variables into a smaller set of new, composite dimensions or variates (factors) with a minimum loss of information—that is, to search for and define the fundamental constructs or dimensions assumed to underlie the original variables [17, 32]. More specifically, factor analysis techniques can satisfy either of two objectives: (1) identifying structure through data summarization or (2) data reduction.

Identifying Structure through Data Summarization

Factor analysis can identify the structure of relationships among either variables or respondents by examining either the correlations between the variables or the correlations between the respondents. For example, suppose we have data on 100 respondents in terms of 10 characteristics. If the objective of the research were to summarize the characteristics, the factor analysis would be applied to a correlation matrix of the variables. This is the most common type of factor analysis, and is referred to as R factor analysis. **R factor analysis** analyzes a set of variables to identify the dimensions that are latent (not easily observed). Factor analysis also may be applied to a correlation matrix of the individual respondents based on their characteristics. This is referred to as **Q factor analysis,** a method of combining or condensing large numbers of people into distinctly different groups within a larger population. The Q factor analysis approach is not utilized very frequently because of computational difficulties. Instead, most researchers utilize some type of **cluster analysis** (see chapter 9) to group individual respondents. Also see Stewart [37] for other possible combinations of groups and variable types.

Data Reduction

Factor analysis can also (1) identify representative variables from a much larger set of variables for use in subsequent multivariate analyses or (2) create an entirely new set of variables, much smaller in number, to partially or completely replace the original set of variables for inclusion in subsequent techniques. In both instances, the purpose is to retain the nature and character of the original variables, but reduce their number to simplify the subsequent multivariate analysis. Even though the multivariate techniques were developed to accommodate multiple variables, the researcher is always looking for the most parsimonious set of variables to include in the analysis. As discussed in chapter 1, both conceptual and empirical issues support the creation of composite measures. Factor analysis provides the empirical basis for assessing the structure of variables and the potential for creating these composite measures or selecting a subset of representative variables for further analysis.

Data summarization makes the identification of the underlying dimensions or factors ends in themselves; the estimates of the factors and the contributions of each variable to the factors (termed loadings) are all that is required for the analysis. Data reduction relies on the factor loadings as well, but uses them as the basis for either identifying variables for subsequent analysis with other techniques or making estimates of the factors themselves (factor scores or summated scales), which then replace the original variables in subsequent analyses. The method of calculating and interpreting factor loadings is discussed later.

Using Factor Analysis with Other Multivariate Techniques

Factor analysis provides direct insight into the interrelationships among variables or respondents and empirical support for addressing conceptual issues relating to the underlying structure of the data. It also plays an important complementary role with other multivariate techniques through both data summarization and data reduction. From the data summarization perspective, factor analysis provides the researcher with a clear understanding of which variables may act in concert together and how many variables may actually be expected to have impacts in the analysis. For example, variables determined to be highly correlated and members of the same factor would be expected to have similar profiles of differences across groups in multivariate analysis of variance or in discriminant analysis. Examples that highlight the impact of correlated variables are the stepwise procedures of multiple regression and discriminant analysis. These techniques sequentially enter variables based on their additional predictive power over variables in the model. As one variable from a factor is entered, it becomes less likely that additional variables from that same factor would also be included because they are highly correlated and potentially have less additional predictive power than variables not in that factor. This does not mean that the other variables of the factor are less important or have less impact, but instead that their effect is already represented by the included variable from the factor. The researcher would better understand the reasoning behind the entry of variables in this technique with a knowledge of the structure of the variables.

The insight provided by data summarization can be directly incorporated into other multivariate techniques through any of the data reduction techniques. Factor analysis provides the basis for creating a new set of variables that incorporate the character and nature of the original variables in a much smaller number of new variables, whether using representative variables, factor scores, or summated scales. In this manner, problems associated with large numbers of variables or high intercorrelations among variables can be substantially reduced by substitution of the new variables. The researcher can benefit from both the empirical estimation of relationships and the insight into the conceptual foundation and interpretation of the results.

Variable Selection

Data reduction and summarization can be performed either with preexisting sets of variables or with variables created by new research. When using an existing set of variables, the researcher should still consider the conceptual underpinnings of the variables and use judgment as to the appropriateness of the variables for factor analysis. The use of factor analysis for data reduction becomes particularly critical when comparability over time or in multiple settings is required. When used in a new research effort, factor analysis can also determine structure and/or create new composite scores from the original variables. For example, one of the first steps in constructing a summated scale (see chapter 1) is to assess its dimensionality and the appropriateness of the selected variables through factor analysis. Thus, even though not truly confirmatory, exploratory factor analysis is used to evaluate the proposed dimensionality.

Once the purpose of factor analysis is specified, the researcher must then define the set of variables to be examined. In either R-type or Q-type factor analysis, the researcher implicitly specifies the potential dimensions that can be identified through the character and nature of the variables submitted to factor analysis. For example, in assessing the dimensions of store image, if no questions on store personnel were included, factor analysis would not be able to identify this dimension. The researcher must also remember that factor analysis will always produce factors. Thus, factor analysis is always a potential candidate for the "garbage in, garbage out" phenomenon. If the researcher indiscriminately includes a large number of variables and hopes that factor analysis will "figure it out," then the possibility of poor results is high. The quality and meaning of the derived factors reflects the conceptual underpinnings of the variables included in the analysis. The use of factor analysis as a data summarization technique does not exclude the need for a conceptual basis for any variables analyzed. Even if used solely for data reduction, factor analysis is most efficient when conceptually defined dimensions can be represented by the derived factors.

Stage 2: Designing a Factor Analysis

The design of a factor analysis involves three basic decisions: (1) calculation of the input data (a correlation matrix) to meet the specified objectives of grouping variables or respondents; (2) the design of the study in terms of number of variables, measurement properties of variables, and the types of allowable variables; and (3) the sample size necessary, both in absolute terms and as a function of the number of variables in the analysis.

Correlations among Variables or Respondents

The first decision in the design of a factor analysis focuses on the approach used in calculating the correlation matrix for either R-type or Q-type factor analysis. The researcher could derive the input data matrix from the computation of correlations between the variables. This would be an R-type factor analysis. The researcher could also elect to derive the correlation matrix from the correlations between the individual respondents. In this Q-type factor analysis, the results would be a factor matrix that would identify similar individuals. For example, if the individual respondents are identified by number, the resulting factor pattern might tell us that individuals 1, 5, 6, and 7 are similar. Similarly, respondents 2, 3, 4, and 8 would perhaps load together on another factor, and we would label these individuals as similar. From the results of a Q factor analysis, we could identify groups or clusters of individuals that demonstrate a similar pattern on the variables included in the analysis.

A logical question at this point would be, How does Q-type factor analysis differ from cluster analysis, as both approaches compare the pattern of responses across a number of variables and place the respondents in groups? The answer is that Q-type factor analysis is based on the intercorrelations between the respondents, whereas cluster analysis forms groupings based on a distance-based similarity measure between the respondents' scores on the variables being analyzed. To illustrate this difference, consider Figure 3.3 (p. 98), which contains the scores of four respondents over three different variables. A Q-type factor analysis of these four respondents would yield two groups with similar covariance structures,

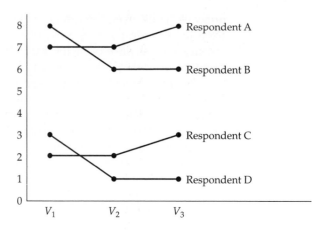

Respondent	*Variables*		
	V_1	V_2	V_3
A	7	7	8
B	8	6	6
C	2	2	3
D	3	1	1

FIGURE 3.3 Comparisons of Score Profiles for Q-type Factor Analysis and Cluster Analysis

consisting of respondents A and C versus B and D. In contrast, the clustering approach would be sensitive to the actual distances among the respondents' scores and would lead to a grouping of the closest pairs. Thus, with a cluster analysis approach, respondents A and B would be placed in one group and C and D in the other group. If the researcher decides to employ Q-type factor analysis, these distinct differences from traditional cluster analysis techniques should be noted. With the availability of other grouping techniques and the widespread use of factor analysis for data reduction and summarization, the remaining discussion in this chapter focuses on R-type factor analysis, the grouping of variables rather than respondents.

Variable Selection and Measurement Issues

Two specific questions must be answered at this point: (1) How are the variables measured? and (2) How many variables should be included? Variables for factor analysis are generally assumed to be of metric measurement. In some cases, **dummy variables** (coded 0–1), although considered nonmetric, can be used. If all the variables are dummy variables, then specialized forms of factor analysis, such as Boolean factor analysis, are more appropriate [5]. The researcher should also attempt to minimize the number of variables included but still maintain a reasonable number of variables per factor. If a study is being designed to assess a proposed structure, the researcher should be sure to include several variables (five or more) that may represent each proposed factor. The strength of factor analysis lies in finding patterns among groups of variables, and it is of little use in identifying factors composed of only a single variable. Finally, when designing a study to be factor analyzed, the researcher should, if possible, identify several key variables (sometimes referred to as key indicants or marker variables) that closely reflect the hypothesized underlying factors. This will aid in validating the derived factors and assessing whether the results have practical significance.

Sample Size

Regarding the sample size question, the researcher generally would not factor analyze a sample of fewer than 50 observations, and preferably the sample size should be 100 or larger. As a general rule, the minimum is to have at least five times as many observations as there are variables to be analyzed, and the more acceptable size would have a ten-to-one ratio. Some researchers even propose a minimum of 20 cases for each variable. One must remember, however, that with 30 variables, for example, there are 435 correlations in the factor analysis. At a .05 significance level, perhaps even 20 of those correlations would be deemed significant and appear in the factor analysis just by chance. The researcher should always try to obtain the highest cases-per-variable ratio to minimize the chances of "overfitting" the data, (i.e., deriving factors that are sample specific with little generalizability). The researcher may do this by employing the most parsimonious set of variables, guided by conceptual and practical considerations, and then obtaining an adequate sample size for the number of variables examined. When dealing with smaller sample sizes and/or a lower cases-to-variable ratio, the researcher should always interpret any findings cautiously. The issue of sample size will also be addressed in a later section on interpreting factor loadings.

Stage 3: Assumptions in Factor Analysis

The critical assumptions underlying factor analysis are more conceptual than statistical. From a statistical standpoint, the departures from normality, homoscedasticity, and linearity apply only to the extent that they diminish the observed correlations. Only normality is necessary if a statistical test is applied to the significance of the factors, but these tests are rarely used. In fact, some degree of **multicollinearity** is desirable, because the objective is to identify interrelated sets of variables.

In addition to the statistical bases for the correlations of the data matrix, the researcher must also ensure that the data matrix has sufficient correlations to justify the application of factor analysis. If visual inspection reveals no substantial number of correlations greater than .30, then factor analysis is probably inappropriate. The correlations among variables can also be analyzed by computing the partial correlations among variables, that is, the correlations between variables when the effects of other variables are taken into account. If "true" factors exist in the data, the partial correlation should be small, because the variable can be explained by the factors (variates with loadings for each variable). If the partial correlations are high, then there are no "true" underlying factors, and factor analysis is inappropriate. SPSS [36] and SAS [33] provide the **anti-image correlation matrix,** which is just the negative value of the partial correlation, whereas BMDP [5] directly provides the partial correlations. In each case, larger partial or anti-image correlations are indicative of a data matrix perhaps not suited to factor analysis.

Another mode of determining the appropriateness of factor analysis examines the entire correlation matrix. The **Bartlett test of sphericity,** a statistical test for the presence of correlations among the variables, is one such measure. It provides the statistical probability that the correlation matrix has significant correlations among at least some of the variables. The researcher should note, however, that increasing the sample size causes the Bartlett test to become more sensitive to detecting correlations among the variables. Another measure to quantify the degree of intercorrelations among the variables and the appropriateness of factor analysis is the **measure of sampling adequacy (MSA).** This index ranges from 0 to 1, reaching 1 when each variable is perfectly predicted without error by the other variables. The measure can be interpreted with the following guidelines: .80 or above, meritorious; .70 or above, middling; .60 or above, mediocre; .50 or above, miserable; and below .50, unacceptable [21, 22]. The MSA increases as (1) the sample size increases, (2) the average correlations increase, (3) the number of variables increases, or (4) the number of factors decreases [22]. The same MSA guidelines can be extended to individual variables as well. The researcher should first examine the MSA values for each variable and exclude those falling in the unacceptable range. Once the individual variables achieve an acceptable level, then the overall MSA can be evaluated and a decision made on continuance of the factor analysis.

The conceptual assumptions underlying factor analysis relate to the set of variables selected and the sample chosen. A basic assumption of factor analysis is that some underlying structure does exist in the set of selected variables. It is the responsibility of the researcher to ensure that the observed patterns are conceptually valid and appropriate to study with factor analysis, because the technique has no means of determining appropriateness other than the correlations among variables. For example, mixing dependent and independent variables in a single factor analysis and then using the derived factors to support dependence relationships is inappropriate. The researcher must also ensure that the sample is homogeneous with respect to the underlying factor structure. It is inappropriate to apply factor analysis to a sample of males and females for a set of items known to differ because of gender. When the two subsamples (males and females) are combined, the resulting correlations and factor structure will be a poor representation of the unique structure of each group. Thus, whenever differing groups are expected in the sample, separate factor analyses should be performed, and the results should be compared to identify differences not reflected in the results of the combined sample.

Stage 4: Deriving Factors and Assessing Overall Fit

Once the variables are specified and the correlation matrix is prepared, the researcher is ready to apply factor analysis to identify the underlying structure of relationships (see Fig. 3.4). In doing so, decisions must be made concerning (1) the method of extracting the factors (common factor analysis versus components analysis) and (2) the number of factors selected to represent the underlying structure in the data. Selection of the extraction method depends upon the researcher's objective. Component analysis is used when the objective is to summarize most of the original information (variance) in a minimum number of factors for prediction purposes. In contrast, common factor analysis is used primarily to identify underlying factors or dimensions that reflect what the variables share in common. For either method, the researcher must also determine the number of factors to represent the set of original variables. Both conceptual and empirical issues affect this decision.

Common Factor Analysis versus Component Analysis

The researcher can utilize two basic models to obtain factor solutions. They are known as **common factor analysis** and **component analysis**. To select the appropriate model, the researcher must first understand the differences between types of variance. For the purposes of factor analysis, three types of total variance exist: (1) **common,** (2) **specific** (also known as unique), and (3) **error.** These types of variance and their relationship to the factor model selection process are illustrated in Figure 3.5 (p. 102). Common variance is defined as that variance in a variable that is shared with all other variables in the analysis. Specific variance is

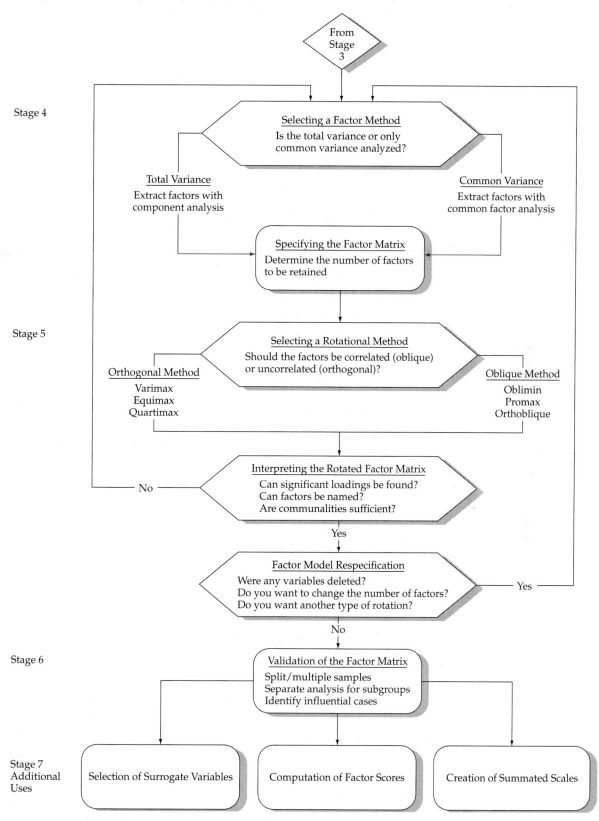

FIGURE 3.4 Stages 4–7 in the Factor Analysis Decision Diagram

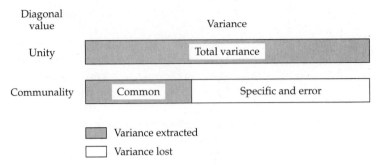

FIGURE 3.5 Types of Variance Carried into the Factor Matrix

that variance associated with only a specific variable. Error variance is that variance due to unreliability in the data-gathering process, measurement error, or a random component in the measured phenomenon. Component analysis, also known as principal components analysis, considers the total variance and derives factors that contain small proportions of unique variance and, in some instances, error variance. However, the first few factors do not contain enough unique or error variance to distort the overall factor structure. Specifically, with component analysis, unities are inserted in the diagonal of the correlation matrix, so that the full variance is brought into the factor matrix, as shown in Figure 3.5. Conversely, with common factor analysis, communalities are inserted in the diagonal. **Communalities** are estimates of the shared, or common, variance among the variables. Factors resulting from common factor analysis are based only on the common variance.

The common factor and component analysis models are both widely used. The selection of one model over the other is based on two criteria: (1) the objectives of the factor analysis and (2) the amount of prior knowledge about the variance in the variables. The component factor model is appropriate when the primary concern is about prediction or the minimum number of factors needed to account for the maximum portion of the variance represented in the original set of variables, and when prior knowledge suggests that specific and error variance represent a relatively small proportion of the total variance. In contrast, when the primary objective is to identify the latent dimensions or constructs represented in the original variables, and the researcher has little knowledge about the amount of specific and error variance and therefore wishes to eliminate this variance, the common factor model is most appropriate. Common factor analysis, with its more restrictive assumptions and use of only the latent dimensions (shared variance), is often viewed as more theoretically based. Although theoretically sound, however, common factor analysis has several problems. First, common factor analysis suffers from **factor indeterminancy**, which means that for any individual respondent, several different factor scores can be calculated from the factor model results [25]. There is no single unique solution, as found in component analysis, but in most instances the differences are not substantial. The second issue involves the calculation of the estimated communalities used to represent the shared variance. For larger-sized problems, the computations can take substantial computer time and resources. Also, the communalities are not always estimable or may be invalid (e.g., values greater than 1 or less than 0), requiring the deletion of the variable from the analysis (see the empirical example of common factor analysis later in this chapter).

The complications of common factor analysis have contributed to the widespread use of component analysis. Although there remains considerable debate over which factor model is the more appropriate [6, 18, 24, 35], empirical research has demonstrated similar results in many instances [38]. In most applications, both component analysis and common factor analysis arrive at essentially identical results if the number of variables exceeds 30 [17], or the communalities exceed .60 for most variables. If the researcher is concerned with the assumptions of components analysis, then common factor analysis should also be applied to assess its representation of structure.

When a decision has been made on the factor model, the researcher is ready to extract the initial unrotated factors. By examining the unrotated factor matrix, the researcher can explore the data reduction possibilities for a set of variables and obtain a preliminary estimate of the number of factors to extract. Final determination of the number of factors must wait, however, until the results are rotated and the factors are interpreted.

Criteria for the Number of Factors to Extract

How do we decide on the number of factors to extract? When a large set of variables is factored, the method first extracts the combinations of variables explaining the greatest amount of variance and then proceeds to combinations that account for smaller and smaller amounts of variance. In deciding when to stop factoring (that is, how many factors to extract), the researcher generally begins with some predetermined criterion, such as the percentage of variance or latent root criterion, to arrive at a specific number of factors to extract (these two techniques are discussed in more detail later). After the initial solution has been derived, the researcher computes several additional trial solutions—usually one less factor than the initial number and two or three more factors than were initially derived. Then, based on the information obtained from the trial analyses, the factor matrices are examined, and the best representation of the data is used to assist in determining the number of factors to extract. By analogy, choosing the number of factors to be interpreted is something like focusing a microscope. Too high or too low an adjustment will obscure a structure that is obvious when the adjustment is just right. Therefore, by examining a number of different factor structures derived from several trial solutions, the researcher can compare and contrast to arrive at the best representation of the data. An exact quantitative basis for deciding the number of factors to extract has not been developed. However, the following stopping criteria for the number of factors to extract are currently being utilized.

Latent Root Criterion   The most commonly used technique is the latent root criterion. This technique is simple to apply to either components analysis or common factor analysis. The rationale for the latent root criterion is that any individual factor should account for the variance of at least a single variable if it is to be retained for interpretation. Each variable contributes a value of 1 to the total eigenvalue. Thus, only the factors having **latent roots** or **eigenvalues** greater than 1 are considered significant; all factors with latent roots less than 1 are considered insignificant and are disregarded. Using the eigenvalue for establishing a cutoff is most reliable when the number of variables is between 20 and 50. If the number of variables is less than 20, there is a tendency for this method to extract a conservative number of factors (too few); whereas if more than 50 variables are involved, it is not uncommon for too many factors to be extracted.

A Priori Criterion   The a priori criterion is a simple yet reasonable criterion under certain circumstances. When applying it, the researcher already knows how many factors to extract before undertaking the factor analysis. The researcher simply instructs the computer to stop the analysis when the desired number of factors has been extracted. This approach is useful when testing a theory or hypothesis about the number of factors to be extracted. It also can be justified in attempting to replicate another researcher's work and extract the same number of factors that was previously found.

Percentage of Variance Criterion   The percentage of variance criterion is an approach based on achieving a specified cumulative percentage of total variance extracted by successive factors. The purpose is to ensure practical significance for the derived factors by ensuring that they explain at least a specified amount of variance. No absolute threshold has been adopted for all applications. However, in the natural sciences the factoring procedure usually should not be stopped until the extracted factors account for at least 95 percent of the variance or until the last factor accounts for only a small portion (less than 5 percent). In contrast, in the social sciences, where information is often less precise, it is not uncommon to consider a solution that accounts for 60 percent of the total variance (and in some instances even less) as satisfactory.

A variant of this criterion involves selecting enough factors to achieve a prespecified communality for each of the variables. If theoretical or practical reasons require a certain communality for each variable, then the research will include as many factors as necessary to adequately represent each of the original variables. This differs from focusing on just the total amount of variance explained, which neglects the degree of explanation for the individual variables.

Scree Test Criterion   Recall that with the component analysis factor model, the later factors extracted contain both common and unique variance. Although all factors contain at least some unique variance, the proportion of unique variance is substantially higher in later than in earlier factors. The scree test is used to identify the optimum number of factors that can be extracted before the amount of unique variance begins to dominate the common variance structure [9]. The scree test is derived by plotting the latent roots against the number of factors in their order of extraction, and the shape of the resulting curve is used to evaluate the cutoff point. Figure 3.6 plots the first 18 factors extracted in a study by all of the authors. Starting with the first factor, the plot slopes steeply downward initially and then slowly becomes an approximately horizontal line. The point at which the curve first begins to straighten out is considered to indicate the maximum number of factors to extract. In the present case, the first 10 factors would qualify. Beyond 10, too large a proportion of unique variance would be included; thus these factors would not be acceptable. Note that in using the latent root criterion, only eight factors would have been considered. In contrast, using the scree test provides us with two more factors. As a general rule, the scree test results in at least one and sometimes two or three *more* factors being considered for inclusion than does the latent root criterion [9].

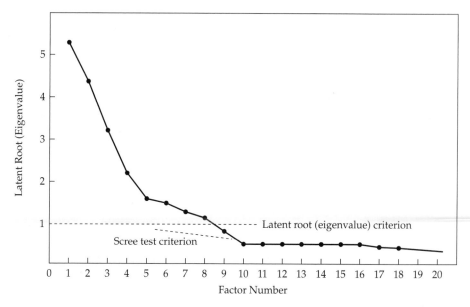

FIGURE 3.6 Eigenvalue Plot for Scree Test Criterion

Heterogeneity of the Respondents   Shared variance among variables is the basis for both common and component factor models. An underlying assumption is that shared variance extends across the entire sample. If the sample is heterogeneous with regard to at least one subset of the variables, then the first factors will represent those variables that are more homogeneous across the entire sample. Variables that are better discriminators between the subgroups of the sample will load on later factors, many times those not selected by the criteria discussed above [16]. When the objective is to identify factors that discriminate among the subgroups of a sample, the researcher should extract additional factors beyond those indicated by the methods above and examine the additional factors' ability to discriminate among the groups. If they prove less beneficial in discrimination, the solution can be run again and these later factors eliminated.

Summary of Factor Selection Criteria   In practice, most researchers seldom use a single criterion in determining how many factors to extract. Instead, they initially use a criterion such as the latent root as a guideline for the first attempt at interpretation. After the factors have been interpreted, as discussed in the following sections, the practicality of the factors is assessed. Factors identified by other criteria are also interpreted. Selecting the number of factors is interrelated with an assessment of structure, which is revealed in the interpretation phase. Thus, several factor solutions with differing numbers of factors are examined before the structure is well defined.

One word of caution in selecting the final set of factors: There are negative consequences for selecting either too many or too few factors to represent the data. If too few factors are used, then the correct structure is not revealed, and important dimensions may be omitted. If too many factors are retained, then the interpretation becomes more difficult when the results are rotated (as discussed in the next section). Although the factors are independent, you can just as easily have too many factors as having too few. As with other aspects of multivariate models, parsimony is important. The notable exception is when factor analysis is used strictly for data reduction and a set level of variance to be extracted is specified. The researcher should always strive to have the most representative and parsimonious set of factors possible.

Stage 5: *Interpreting the Factors*

Three steps are involved in the interpretation of the factors and the selection of the final factor solution. First, the initial unrotated **factor matrix** is computed to assist in obtaining a preliminary indication of the number of factors to extract. The factor matrix contains factor loadings (see discussion in the next paragraph) for each variable on each factor. In computing the unrotated factor matrix, the researcher is simply interested in the best linear combination of variables—best in the sense that the particular combination of original variables accounts for more of the variance in the data as a whole than any other linear combination of variables. Therefore, the first factor may be viewed as the single best summary of linear relationships exhibited in the data. The second factor is defined as the second-best linear combination of the variables, subject to the constraint that it is orthogonal to the first factor. To be **orthogonal** to the first factor, the second factor must be derived from the variance remaining after the first factor has been extracted. Thus the second factor may be defined as the linear combination of variables that accounts for the most residual variance after the effect of the first factor has been removed from the data. Subsequent factors are defined similarly, until all the variance in the data is exhausted.

Unrotated factor solutions achieve the objective of data reduction, but the researcher must ask if the unrotated factor solution (which fulfills desirable mathematical requirements) will provide information that offers the most adequate interpretation of the variables under examination. In most instances the answer to this question is no. Factor loading is the means of interpreting the role each variable plays in defining each factor. **Factor loadings** are the correlation of each variable and the factor. Loadings indicate the degree of correspondence between the variable and the factor, with higher loadings making the variable representative of the factor. The unrotated factor solution may not provide a meaningful pattern of variable loadings. If the unrotated factors are expected to be meaningful, the user may specify that no rotation be performed. Generally, rotation will be desirable because it simplifies the factor structure, and it is usually difficult to determine whether unrotated factors will be meaningful. Therefore, the second step employs a rotational method to achieve simpler and theoretically more meaningful factor solutions. In most cases rotation of the factors improves the interpretation by reducing some of the ambiguities that often accompany initial unrotated factor solutions.

In the third step, the researcher assesses the need to respecify the factor model owing to (1) the deletion of a variable(s) from the analysis, (2) the desire to employ a different rotational method for interpretation, (3) the need to extract a different number of factors, or (4) the desire to change from one extraction method to another. Respecification of a factor model is accomplished by returning to the extraction stage, extracting factors, and interpreting them again.

Rotation of Factors

An important tool in interpreting factors is **factor rotation.** The term rotation means exactly what it implies. Specifically, the reference axes of the factors are turned about the origin until some other position has been reached. As indicated earlier, unrotated factor solutions extract factors in the order of their importance. The first factor tends to be a general factor with almost every variable loading significantly, and it accounts for the largest amount of variance. The second and subsequent factors are then based on the residual amount of variance. Each accounts for successively smaller portions of variance. The ultimate effect of rotating the factor matrix is to redistribute the variance from earlier factors to later ones to achieve a simpler, theoretically more meaningful factor pattern.

The simplest case of rotation is an **orthogonal rotation,** in which the axes are maintained at 90 degrees. It is also possible to rotate the axes and not retain the 90-degree angle between the reference axes. When not constrained to being orthogonal, the rotational procedure is called an **oblique rotation.** Orthogonal and oblique factor rotations are demonstrated in Figures 3.7 and 3.8 (p. 108), respectively.

An Illustration of Factor Rotation   Figure 3.7, in which five variables are depicted in a two-dimensional factor diagram, illustrates factor rotation. The vertical axis represents unrotated factor II, and the horizontal axis represents unrotated factor I. The axes are labeled with 0 at the origin and extend outward to +1.0 or a −1.0. The numbers on the axes represent the factor loadings. The five variables are labeled V_1, V_2, V_3, V_4, and V_5. The factor loading for variable 2 (V_2) on unrotated factor II is determined by drawing a dashed line horizontally from the data point to the vertical axis for factor II. Similarly, a vertical line is drawn from variable 2 to the horizontal axis of unrotated factor I to determine the loading of variable 2 on factor I. A similar procedure followed for the remaining variables determines the factor loadings for the unrotated and rotated solutions, as displayed in Table 3.1 (p. 108) for comparison purposes. On the unrotated first factor, all the variables load fairly high. On the unrotated second factor, variables 1 and 2 are very high in the positive

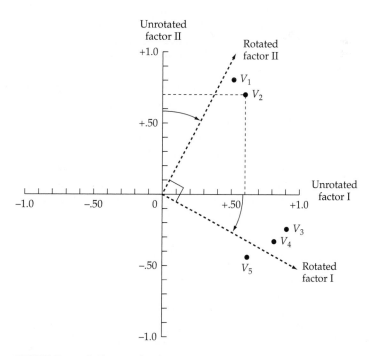

FIGURE 3.7 Orthogonal Factor Rotation

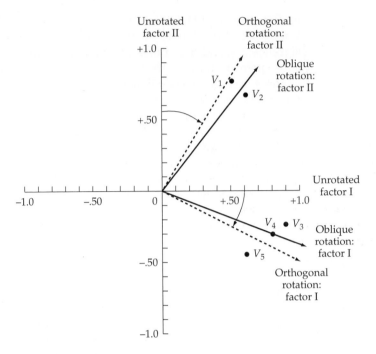

FIGURE 3.8 Oblique Factor Rotation

direction. Variable 5 is moderately high in the negative direction, and variables 3 and 4 have considerably lower loadings in the negative direction.

From visual inspection of Figure 3.7, it is obvious that there are two clusters of variables. Variables 1 and 2 go together, as do variables 3, 4, and 5. However, such patterning of variables is not so obvious from the unrotated factor loadings. By rotating the original axes clockwise, as indicated in Figure 3.7, we obtain a completely different factor loading pattern. Note that in rotating the factors, the axes are maintained at 90 degrees. This procedure signifies that the factors are mathematically independent and that the rotation has been orthogonal. After rotating the factor axes, variables 3, 4, and 5 load very high on factor I, and variables 1 and 2 load very high on factor II. Thus the clustering or patterning of these variables into two groups is more obvious after the rotation than before, even though the relative position or configuration of the variables remains unchanged.

TABLE 3.1 Comparison Between Rotated and Unrotated Factor Loadings

Variables	Unrotated Factor Loadings		Rotated Factor Loadings	
	I	*II*	*I*	*II*
V_1	.50	.80	.03	.94
V_2	.60	.70	.16	.90
V_3	.90	−.25	.95	.24
V_4	.80	−.30	.84	.15
V_5	.60	−.50	.76	−.13

The same general principles of orthogonal rotations pertain to oblique rotations. The oblique rotational method is more flexible because the factor axes need not be orthogonal. It is also more realistic because the theoretically important underlying dimensions are not assumed to be uncorrelated with each other. In Figure 3.8 the two rotational methods are compared. Note that the oblique factor rotation represents the clustering of variables more accurately. This accuracy is a result of the fact that each rotated factor axis is now closer to the respective group of variables. Also, the oblique solution provides information about the extent to which the factors are actually correlated with each other.

Most researchers agree that most direct unrotated solutions are not sufficient; that is, in most cases rotation will improve the interpretation by reducing some of the ambiguities that often accompany the preliminary analysis. The major option available is to choose an orthogonal or oblique rotation method. The ultimate goal of any rotation is to obtain some theoretically meaningful factors and, if possible, the simplest factor structure. Orthogonal rotational approaches are more widely used because all computer packages with factor analysis contain orthogonal rotation options, whereas the oblique methods are not as widespread. Orthogonal rotations are also utilized more frequently because the analytical procedures for performing oblique rotations are not as well developed and are still subject to considerable controversy. Several different approaches are available for performing either orthogonal or oblique rotations. However, only a limited number of oblique rotational procedures are available in most statistical packages; thus, the researcher will probably have to accept the one that is provided.

Orthogonal Rotation Methods   In practice, the objective of all methods of rotation is to simplify the rows and columns of the factor matrix to facilitate interpretation. In a factor matrix, columns represent factors, with each row corresponding to a variable's loading across the factors. By simplifying the rows, we mean making as many values in each row as close to zero as possible (i.e., maximizing a variable's loading on a single factor). By simplifying the columns, we mean making as many values in each column as close to zero as possible (i.e., making the number of "high" loadings as few as possible). Three major orthogonal approaches have been developed:

QUARTIMAX
The ultimate goal of a QUARTIMAX rotation is to simplify the rows of a factor matrix; that is, QUARTIMAX focuses on rotating the initial factor so that a variable loads high on one factor and as low as possible on all other factors. In these rotations, many variables can load high or near on the same factor because the technique centers on simplifying the rows. The QUARTIMAX method has not proved very successful in producing simpler structures. Its difficulty is that it tends to produce a general factor as the first factor on which most, if not all, of the variables have high loadings. Regardless of one's concept of a "simpler" structure, inevitably it involves dealing with clusters of variables; a method that tends to create a large general factor (i.e., QUARTIMAX) is not in line with the goals of rotation.

VARIMAX
In contrast to QUARTIMAX, the VARIMAX criterion centers on simplifying the columns of the factor matrix. With the VARIMAX rotational approach, the maximum possible simplification is reached if there are only 1s and 0s in a column.

That is, the VARIMAX method maximizes the sum of variances of required loadings of the factor matrix. Recall that in QUARTIMAX approaches, many variables can load high or near high on the same factor because the technique centers on simplifying the rows. With the VARIMAX rotational approach, there tend to be some high loadings (i.e., close to -1 or $+1$) and some loadings near 0 in each column of the matrix. The logic is that interpretation is easiest when the variable-factor correlations are (1) close to either $+1$ or -1, thus indicating a clear positive or negative association between the variable and the factor; or (2) close to 0, indicating a clear lack of association. This structure is fundamentally simple. Although the QUARTIMAX solution is analytically simpler than the VARIMAX solution, VARIMAX seems to give a clearer separation of the factors. In general, Kaiser's experiment [21, 22] indicates that the factor pattern obtained by VARIMAX rotation tends to be more invariant than that obtained by the QUARTIMAX method when different subsets of variables are analyzed. The VARIMAX method has proved very successful as an analytic approach to obtaining an orthogonal rotation of factors.

EQUIMAX
The EQUIMAX approach is a compromise between the QUARTIMAX and VARIMAX approaches. Rather than concentrating either on simplification of the rows or on simplification of the columns, it tries to accomplish some of each. EQUIMAX has not gained widespread acceptance and is used infrequently.

Oblique Rotation Methods   Oblique rotations are similar to orthogonal rotations, except that oblique rotations allow correlated factors instead of maintaining independence between the rotated factors. Where there were several choices among orthogonal approaches, however, there are typically only limited choices in most statistical packages for oblique rotations. For example, SPSS provides OBLIMIN; SAS has PROMAX and ORTHOBLIQUE; and BMDP provides DQUART, DOBLIMIN, and ORTHOBLIQUE. The objectives of simplification are comparable to the orthogonal methods, with the added feature of correlated factors. With the possibility of correlated factors, the factor researcher must take additional care to validate obliquely rotated factors, as they have an additional way (nonorthogonality) of becoming specific to the sample and not generalizable, particularly with small samples or a low cases-to-variable ratio.

Selecting among Rotational Methods   No specific rules have been developed to guide the researcher in selecting a particular orthogonal or oblique rotational technique. In most instances, the researcher simply utilizes the rotational technique provided by the computer program. Most programs have the default rotation of VARIMAX, but all the major rotational methods are widely available. However, there is no compelling analytical reason to favor one rotational method over another. The choice of an orthogonal or oblique rotation should be made on the basis of the particular needs of a given research problem. If the goal of the research is to reduce the number of original variables, regardless of how meaningful the resulting factors may be, the appropriate solution would be an orthogonal one. Also, if the researcher wants to reduce a larger number of variables to a smaller set of uncorrelated variables for subsequent use in regression or other prediction techniques, an orthogonal solution is the best. However, if the ultimate goal of the factor analysis is to obtain several theoretically meaningful factors or constructs, an oblique solution is appropriate. This conclusion is reached because, realistically, very few factors are uncorrelated, as in an orthogonal rotation.

Criteria for the Significance of Factor Loadings

In interpreting factors, a decision must be made regarding which factor loadings are worth considering. The following discussion details issues regarding practical and statistical significance, as well as the number of variables, that affect the interpretation of factor loadings.

Ensuring Practical Significance   The first suggestion is not based on any mathematical proposition but relates more to practical significance. It is a rule of thumb used frequently as a means of making a preliminary examination of the factor matrix. In short, factor loadings greater than ±.30 are considered to meet the minimal level; loadings of ±.40 are considered more important; and if the loadings are ±.50 or greater, they are considered practically significant. Thus the larger the absolute size of the factor loading, the more important the loading in interpreting the factor matrix. Because factor loading is the correlation of the variable and the factor, the squared loading is the amount of the variable's total variance accounted for by the factor. Thus, a .30 loading translates to approximately 10 percent explanation, and a .50 loading denotes that 25 percent of the variance is accounted for by the factor. The loading must exceed .70 for the factor to account for 50 percent of the variance. The researcher should realize that extremely high loadings (.80 and above) are not typical and that the practical significance of the loadings is an important criterion. These guidelines are applicable when the sample size is 100 or larger. The emphasis in this approach is practical, not statistical, significance.

Assessing Statistical Significance   As previously noted, a factor loading represents the correlation between an original variable and its factor. In determining a significance level for the interpretation of loadings, an approach similar to determining the statistical significance of correlation coefficients could be used. However, research [13] has demonstrated that factor loadings have substantially larger standard errors than typical correlations; thus, factor loadings should be evaluated at considerably stricter levels. The researcher can employ the concept of statistical power discussed in chapter 1 to specify factor loadings considered significant for differing sample sizes. With the stated objective of obtaining a power level of 80 percent, the use of a .05 significance level, and the proposed inflation of the standard errors of factor loadings, Table 3.2 (p. 112) contains the sample sizes necessary for each factor loading value to be considered significant. For example, in a sample of 100 respondents, factor loadings of .55 and above are significant. However, in a sample of 50, a factor loading of .75 is required for significance. In comparison with the prior rule of thumb, which denoted all loadings of .30 as having practical significance, this approach would consider loadings of .30 significant only for sample sizes of 350 or greater. These are quite conservative guidelines when compared with the guidelines of the previous section or even the statistical levels associated with conventional correlation coefficients. Thus, these guidelines should be used as a starting point in factor loading interpretation, with lower loadings considered significant and added to the interpretation based on other considerations. The next section details the interpretation process and the role that other considerations can play.

TABLE 3.2 Guidelines for Identifying Significant
Factor Loadings Based on Sample Size

Factor Loading	Sample Size Needed for Significance[a]
.30	350
.35	250
.40	200
.45	150
.50	120
.55	100
.60	85
.65	70
.70	60
.75	50

[a]Significance is based on a .05 significance level (α), a power
   level of 80 percent, and standard errors assumed to be twice
   those of conventional correlation coefficients.
Source: Computations made with SOLO *Power Analysis*, BMDP
   Statistical Software, Inc., 1993.

Adjustments Based on the Number of Variables   A disadvantage of both of the
prior approaches is that the number of variables being analyzed and the specific
factor being examined are not considered. It has been shown that as the researcher
moves from the first factor to later factors, the acceptable level for a loading to be
judged significant should increase. The fact that unique variance and error variance begin to appear in later factors means that some upward adjustment in the
level of significance should be included [21]. The number of variables being analyzed is also important in deciding which loadings are significant. As the number of variables being analyzed increases, the acceptable level for considering a
loading significant decreases. Adjustment for the number of variables is increasingly important as one moves from the first factor extracted to later factors.

To summarize the criteria for the significance of factor loadings, the following
guidelines can be stated: (1) the larger the sample size, the smaller the loading to
be considered significant; (2) the larger the number of variables being analyzed,
the smaller the loading to be considered significant; (3) the larger the number of
factors, the larger the size of the loading on later factors to be considered significant for interpretation.

Interpreting a Factor Matrix
Interpreting the complex interrelationships represented in a factor matrix is no
simple matter. By following the procedure outlined in the following paragraphs,
however, one can considerably simplify the factor interpretation procedure.

Examine the Factor Matrix of Loadings   Each column of numbers in the factor
matrix represents a separate factor. The columns of numbers are the factor loadings for each variable on each factor. For identification purposes, the computer
printout usually identifies the factors from left to right by the numbers 1, 2, 3, 4,
and so forth. It also identifies the variables by number from top to bottom. To further facilitate interpretation, the researcher should write the name of each variable in the left margin beside the variable numbers.

If an oblique rotation has been used, two factor matrices of loadings are provided. The first is the **factor pattern matrix,** which has loadings that represent the unique contribution of each variable to the factor. The second is the **factor structure matrix,** which has simple correlations between variables and factors, but these loadings contain both the unique variance between variables and factors and the correlation among factors. As the correlation among factors becomes greater, it becomes more difficult to distinguish which variables load uniquely on each factor in the factor structure matrix. Most researchers report the results of the factor pattern matrix.

Identify the Highest Loading for Each Variable   The interpretation should start with the first variable on the first factor and move horizontally from left to right, looking for the highest loading for that variable on any factor. When the highest loading (largest absolute factor loading) is identified, it should be underlined if significant. Attention then focuses on the second variable and, again moving from left to right horizontally, looking for the highest loading for that variable on any factor and underlining it. This procedure should be continued for each variable until all variables have been underlined once for their highest loading on a factor. Recall that for sample sizes of less than 100, the lowest factor loading to be considered significant would in most instances be $\pm.30$.

The process of underlining only the single highest loading as significant for each variable is an ideal that should be sought but seldom can be achieved. When each variable has only one loading on one factor that is considered significant, the interpretation of the meaning of each factor is simplified considerably. In practice, however, many variables may have several moderate-size loadings, all of which are significant, and the job of interpreting the factors is much more difficult. The difficulty arises because a variable with several significant loadings must be considered in interpreting (labeling) all the factors on which it has a significant loading. Most factor solutions do not result in a simple structure solution (a single high loading for each variable on only one factor). Thus, the researcher will, after underlining the highest loading for a variable, continue to evaluate the factor matrix by underlining all significant loadings for a variable on all the factors. Ultimately, the objective is to minimize the number of significant loadings on each row of the factor matrix (that is, make each variable associate with only one factor). A variable with several high loadings is a candidate for deletion.

Assess Communalities of the Variables   Once all the variables have been underlined on their respective factors, the researcher should examine the factor matrix to identify variables that have not been underlined and therefore do not load on any factor. The communalities for each variable are provided, representing the amount of variance accounted for by the factor solution for each variable. The researcher should view each variable's communality to assess whether it meets acceptable levels of explanation. For example, a researcher may specify that at least one-half of the variance of each variable must be taken into account. Using this guideline, the researcher would identify all variables with communalities less than .50 as not having sufficient explanation.

If there are variables that do not load on any factor or whose communalities are deemed too low, two options are available: (1) interpret the solution as it is and simply ignore those variables, or (2) evaluate each of those variables for possible deletion. Ignoring the variables may be appropriate if the objective is solely data reduction, but the researcher must still note that the variables in question are poorly represented in the factor solution. Consideration for deletion should depend on the variable's overall contribution to the research as well as its communality index. If the variable is of minor importance to the study's objective or has an unacceptable communality value, it may be eliminated and then the factor model respecified by deriving a new factor solution with those variables eliminated.

Label the Factors   When a factor solution has been obtained in which all variables have a significant loading on a factor, the researcher attempts to assign some meaning to the pattern of factor loadings. Variables with higher loadings are considered more important and have greater influence on the name or label selected to represent a factor. Thus the researcher will examine all the underlined variables for a particular factor and, placing greater emphasis on those variables with higher loadings, will attempt to assign a name or label to a factor that accurately reflects the variables loading on that factor. The signs are interpreted just as with any other correlation coefficients. On each factor, like signs mean the variables are positively related, and opposite signs mean the variables are negatively related. In orthogonal solutions the factors are independent of one another. Therefore, the signs for factor loading relate only to the factor on which they appear, not to other factors in the solution.

This label is not derived or assigned by the factor analysis computer program; rather, the label is intuitively developed by the researcher based on its appropriateness for representing the underlying dimensions of a particular factor. This procedure is followed for each extracted factor. The final result will be a name or label that represents each of the derived factors as accurately as possible.

In some instances, it is not possible to assign a name to each of the factors. When such a situation is encountered, the researcher may wish to label a particular factor or factors derived by that solution as "undefined." In such cases the researcher interprets only those factors that are meaningful and disregards undefined or less meaningful ones. In describing the factor solution, however, the researcher indicates that these factors were derived but were undefinable and that only those factors representing meaningful relationships were interpreted.

As discussed earlier, the selection of a specific number of factors and the rotation method are interrelated. Several additional trial rotations may be undertaken, and by considering the initial criterion and comparing the factor interpretations for several different trial rotations, the researcher can select the number of factors to extract. In short, the ability to assign some meaning to the factors, or to interpret the nature of the variables, becomes an extremely important consideration in determining the number of factors to extract.

Stage 6: Validation of Factor Analysis

The sixth stage involves assessing the degree of generalizability of the results to the population and the potential influence of individual cases or respondents on the overall results. The issue of generalizability is critical for each of the multivariate methods, but it is especially relevant for the interdependence methods because they describe a data structure that should be representative of the population as well. The most direct method of validating the results is to move to a confirmatory perspective and assess the replicability of the results, either with a split sample in the original data set or with a separate sample. The comparison of two or more factor model results has always been problematic. However, several options exist for making an objective comparison. The emergence of confirmatory factor analysis (CFA) through structural equation modeling has provided one option, but it is generally more complicated and requires additional software packages, such as LISREL or EQS [4, 20]. Chapter 11 discusses confirmatory factor analysis in greater detail. Apart from CFA, several other methods have been proposed, ranging from a simple matching index [10] to programs (FMATCH) designed specifically to assess the correspondence between factor matrices [34]. These methods have had sporadic use, owing in part to (1) their perceived lack of sophistication and (2) the unavailability of software or analytical programs to automate the comparisons. Thus, when CFA is not appropriate, these methods provide some objective basis for comparison.

Another aspect of generalizability is the stability of the factor model results. Factor stability is primarily dependent on the sample size and on the number of cases per variable. The researcher is always encouraged to obtain the largest sample possible and develop parsimonious models to increase the cases-to-variables ratio. If sample size permits, the researcher may wish to randomly split the sample into two subsets and estimate factor models for each subset. Comparison of the two resulting factor matrices will provide an assessment of the robustness of the solution across the sample.

In addition to generalizability, another issue of importance to the validation of factor analysis is the detection of influential observations. Discussions in chapter 2 on the identification of outliers and in chapter 4 on the influential observations in regression both have applicability in factor analysis. The researcher is encouraged to estimate the model with and without observations identified as outliers to assess their impact on the results. If omission of the outliers is justified, the results should have greater generalizability. Also, as discussed in chapter 4, several measures of influence that reflect one observation's position relative to all others (e.g., covariance ratio) are applicable to factor analysis as well. Finally, methods have been proposed for identifying influential observations specific to factor analysis [11], but complexity has limited application of these methods.

Stage 7: Additional Uses of Factor Analysis Results

Depending upon the objectives for applying factor analysis, the researcher may stop with factor interpretation or further engage in one of the methods for data reduction. If the objective is simply to identify logical combinations of variables and better understand the interrelationships among variables, then factor interpretation will suffice. This provides an empirical basis for judging the structure of the variables and the impact of this structure when interpreting the results from other multivariate techniques. If the objective, however, is to identify appropriate variables for subsequent application to other statistical techniques, then some form of data reduction will be employed. The options include (1) examining the factor matrix and selecting the variable with the highest factor loading as a surrogate representative for a particular factor dimension, or (2) replacing the original set of variables with an entirely new, smaller set of variables created either from summated scales or factor scores. Either option will provide new variables for use, for example, as the independent variables in a regression or discriminant analysis, as dependent variables in multivariate analysis of variance, or even as the clustering variables in cluster analysis. We discuss each of these options for data reduction in the following sections.

Selecting Surrogate Variables for Subsequent Analysis

If the researcher's objective is simply to identify appropriate variables for subsequent application with other statistical techniques, the researcher has the option of examining the factor matrix and selecting the variable with the highest factor loading on each factor to act as a **surrogate variable** that is representative of that factor. This is a simple and direct approach only when one variable has a factor loading that is substantially higher than all other factor loadings. In many instances, however, the selection process is more difficult because two or more variables have loadings that are significant and fairly close to each other. Such cases require a critical examination of the factor loadings of approximately the same size and selection of only one as representative of a particular dimension. This decision should be based on the researcher's a priori knowledge of theory that may suggest that one variable more than the others would logically be representative of the dimension. Also, the researcher may have knowledge suggesting that a variable with a loading slightly lower is in fact more reliable than the highest-loading variable. In such cases, the researcher may choose the variable that is loading slightly lower as the best variable to represent a particular factor.

The approach of selecting a single surrogate variable as representative of the factor—although simple and maintaining the original variable—has several potential disadvantages. First, it does not address the issue of measurement error encountered when using single measures (see the following section for a more detailed discussion) and it also runs the risk of potentially misleading results by selecting only a single variable to represent a perhaps more complex result. For example, assume that variables representing price competitiveness, product quality, and value were all found to load highly on a single factor. The selection of any one of these separate variables would create substantially different interpretations in any subsequent analysis, yet all three may be so closely related as to really preclude such an action. Second, in instances where several high loadings complicate the selection of a single variable, the researcher may have no choice but to employ factor analysis as the basis for calculating a summed scale or factor scores for use as the surrogate variable. The objective, just as in the case of selecting a single variable, is to best represent the basic nature of the factor or component.

Creating Summated Scales

Chapter 1 introduced the concept of a **summated scale,** which is formed by combining several individual variables into a single **composite measure.** In simple terms, all of the variables loading highly on a factor are combined, and the total—or more commonly the average score of the variables—is used as a replacement variable. A summated scale provides two specific benefits. First, it provides a means of overcoming to some extent the measurement error inherent in all measured variables. **Measurement error** is the degree to which the observed values are not representative of the "true" values due to any number of reasons, ranging from actual errors (e.g., data entry errors) to the inability of individuals to accurately provide information. The impact of measurement error is to partially mask any relationships (e.g., correlations or comparison of group means) and make the estimation of multivariate models more difficult. The summated scale reduces measurement error by using multiple **indicators** (variables) to reduce the reliance on a single response. By using the "average" or "typical" response to a set of related variables, the measurement error that might occur in a single question will be reduced.

A second benefit of the summated scale is its ability to represent the multiple aspects of a concept in a single measure. Many times we employ more variables in our multivariate models in an attempt to represent the many "facets" of a concept that we know is quite complex. But in doing so, we complicate the interpretation of the results because of the redundancy in the items associated with the concept. Thus, we would like not only to accommodate the "richer" descriptions of concepts by using multiple variables, but also to maintain parsimony in the number of variables in our multivariate models. The summated scale, when properly constructed, does combine the multiple indicators into a single measure representing what is held in common across the set of measures.

The process of scale construction has theoretical and empirical foundations in a number of disciplines, including psychometric theory, sociology, and marketing. Although a complete treatment of the techniques and issues involved are beyond the scope of this text, there exists a number of excellent sources for further reading on this subject [2, 12, 19, 29, 30]. Additionally, there are a series of compilations of existing scales that may be applied in a number of situations [3, 7, 31]. We discuss here, however, four issues basic to the construction of any summated scale: conceptual definition, dimensionality, reliability, and validity.

Conceptual Definition   The starting point for creating any summated scale is its **conceptual definition.** The conceptual definition specifies the theoretical basis for the summated scale by defining the concept being represented in terms applicable to the research context. In academic research, theoretical definitions are based on prior research that defines the character and nature of a concept. In a managerial setting, specific concepts may be defined that relate to proposed objectives, such as image, value, or satisfaction. In either instance, creating a summated scale is always guided by the conceptual definition specifying the type and character of the items that are candidates for inclusion in the scale.

Content validity is the assessment of the correspondence of the variables to be included in a summated scale and its conceptual definition. This form of validity, also known as **face validity,** subjectively assesses the correspondence between the individual items and the concept through ratings by expert judges, pretests with multiple subpopulations, or other means. The objective is to ensure that the selection of scale items extends past just empirical issues to also include theoretical and practical considerations [12, 30].

Dimensionality   An underlying assumption and essential requirement for creating a summated scale is that the items are unidimensional, meaning that they are strongly associated with each other and represent a single concept [19, 23]. Factor analysis plays a pivotal role in making an empirical assessment of the dimensionality of a set of items by determining the number of factors and the loadings of each variable on the factor(s). The test of unidimensionality is that each summated scale should consist of items loading highly on a single factor [1, 19, 23, 27]. If a summated scale is proposed to have multiple dimensions, each dimension should be reflected by a separate factor. The researcher can assess unidimensionality with either exploratory factor analysis, as discussed in this chapter, or confirmatory factor analysis, as described in chapter 11.

Reliability   **Reliability** is an assessment of the degree of consistency between multiple measurements of a variable. One form of reliability is test-retest, by which consistency is measured between the responses for an individual in two points in time. The objective is to ensure that responses are not too varied across time periods so that a measurement taken at any point in time is reliable. A second and more commonly used measure of reliability is internal consistency, which applies to the consistency among the variables in a summated scale. The rationale for internal consistency is that the individual items or indicators of the scale should all be measuring the same construct and thus be highly intercorrelated [12, 27].

Because no single item is a perfect measure of a concept, we must rely on a series of diagnostic measures to assess internal consistency. First, there are several measures relating to each separate item, including the item-to-total correlation (the correlation of the item to the summated scale score) or the inter-item correlation (the correlation among items). Rules of thumb suggest that the item-to-total correlations exceed .50 and that the inter-item correlations exceed .30 [30]. The second type of diagnostic measure is the reliability coefficient that assesses the consistency of the entire scale, with **Cronbach's alpha** [27, 28] being the most widely used measure. The generally agreed upon lower limit for Cronbach's alpha is .70 [30, 31], although it may decrease to .60 in exploratory research [30]. One issue in assessing Cronbach's alpha is its positive relationship to the number of items in the scale. Because increasing the number of items, even with the same degree of intercorrelation, will increase the reliability value, researchers must place more stringent requirements for scales with large numbers of items. Also available are reliability measures derived from confirmatory factor analysis. Included in these measures are the composite reliability and the average variance extracted, both discussed in greater detail in chapter 11.

Each of the major statistical programs now have reliability assessment modules or programs, such that the researcher is provided with a complete analysis of both item-specific and overall reliability measures. Any summated scale should be analyzed for reliability to ensure its appropriateness before proceeding to an assessment of its validity.

Validity   Having ensured that a scale (1) conforms to its conceptual definition, (2) is unidimensional, and (3) meets the necessary levels of reliability, the researcher must make one final assessment: scale validity. **Validity** is the extent to which a scale or set of measures accurately represents the concept of interest. We have already discussed one form of validity—content or face validity—in the discussion of conceptual definitions. Other forms of validity are measured empirically by the correlation between theoretically defined sets of variables. The three most widely accepted forms of validity are convergent, discriminant, and nomological validity [8, 29]. Convergent validity assesses the degree to which two measures of the same concept are correlated. Here the researcher may look for alternative measures of a concept and then correlate them with the summated scale. High correlations here indicate that the scale is measuring its intended concept. Discriminant validity is the degree to which two conceptually similar concepts are distinct. The empirical test is again the correlation among measures, but this time the summated scale is correlated with a similar, but conceptually distinct measure. Now the correlation should be low, demonstrating that the summated scale is sufficiently different from the other similar concept. Finally, nomological validity refers to the degree that the summated scale makes accurate predictions of other concepts in a theoretically based model. The researcher must identify theoretically supported relationships from prior research or accepted principles and then assess whether the scale has corresponding relationships. In summary, convergent validity confirms that the scale is correlated with other known measures of the concept, discriminant validity ensures that the scale is sufficiently different from other similar concepts to be distinct, and nomological validity determines if the scale demonstrates the relationships shown to exist based on theory and/or prior research.

A number of differing methods are available for assessing validity, ranging from the multitrait, multimethod (MTMM) matrices to structural equation–based approaches. Although beyond the scope of this text, a number of sources are available addressing both the range of methods available and the issues involved in the specific techniques [8, 20, 29].

Summary Summated scales, one of the recent developments in academic research, are having increased application in applied and managerial research as well. The ability of the summated scale to portray complex concepts in a single measure while reducing measurement error makes it a valuable addition in any multivariate analysis. Factor analysis provides the researcher with an empirical assessment of the interrelationships among variables, essential in forming the conceptual and empirical foundation of a summated scale through assessment content validity and scale dimensionality.

Computing Factor Scores

The third option for creating a smaller set of variables to replace the original set is the computation of factor scores. **Factor scores** are also composite measures of each factor computed for each subject. Conceptually the factor score represents the degree to which each individual scores high on the group of items that have high loadings on a factor. Thus, higher values on the variables with high loadings on a factor will result in a higher factor score. The one key characteristic that differentiates a factor score from a summated scale is that the factor score is computed based on the factor loadings of all variables on the factor, whereas the summated scale is calculated by combining only selected variables. Therefore, although the researcher is able to characterize a factor by the variables with the highest loadings, consideration must also be given to the loadings of other variables, albeit lower, and their influence on the factor score.

Most statistical programs can easily compute factor scores for each respondent. By selecting the factor score option, these scores are saved for use in subsequent analyses. The one disadvantage of factor scores is that they are not easily replicated across studies because they are based on the factor matrix, which is derived separately in each study. Replication of the same factor matrix across studies requires substantial computational programming.

Selecting among the Three Methods

To select among the three data reduction options, the researcher must make a series of decisions. The first choice is between selecting a single surrogate variable for each factor or computing a composite measure. The single surrogate variable has the advantages of being simple to administer and interpret, but has the disadvantages of not representing all of the "facets" of a factor and being prone to measurement error. If the researcher desires to employ some form of composite measure, then a choice must be made between factor scores and summated scales. Both have advantages and disadvantages, and no clear-cut choice is available for all situations. Factor scores have the advantage of representing a composite of all variables loading on the factor, although this is also a potential disadvantage in that all variables have some degree of influence in computing the factor scores and make interpretation more difficult. The summated scale is a compromise between the surrogate variable and factor score options. It is a composite measure, like factor scores, thus reducing measurement error and representing multiple facets of a concept. Yet it is similar to the surrogate variable approach because it includes only the variables that load highly on the factor and excludes those having little impact. Also, its ease of replication between samples is similar to the surrogate variable approach. Finally, like surrogate variables, summated scales are not necessarily orthogonal, whereas rotated factors can be orthogonal or uncorrelated, if needed to avoid complications in their use in other multivariate techniques. The decision rule, therefore, would be that if data are used only in the original sample or orthogonality must be maintained, factor scores are suitable. If generalizability or transferability is desired, then summated scales or surrogate variables are more appropriate. If the summated scale is a well-constructed, valid, and reliable instrument, then it is probably the best alternative. But if the summated scale is untested and exploratory, with little or no evidence of reliability or validity, surrogate variables should be strongly considered if additional analysis is not possible to improve the summated scale.

An Illustrative Example

In the preceding sections, the major questions concerning the application of factor analysis have been discussed within the model-building framework introduced in chapter 1. To clarify these topics further, we use an illustrative example of the application of factor analysis based on data from the database presented in chapter 1. Our discussion of the empirical example also follows the six-stage model-building process. The first three stages, common to either component or common factor analysis, are discussed first. Then, stages four through six for component analysis will be discussed, along with examples of the additional use of factor results. We conclude with an examination of the differences for common factor analysis in stages four and five.

Stage 1: Objectives of Factor Analysis

Factor analysis can identify the structure of a set of variables as well as provide a process for data reduction. In our example, the perceptions of HATCO on seven attributes (X_1 to X_7) are examined to (1) understand if these perceptions can be "grouped" and (2) reduce the seven variables to a smaller number. Even the relatively small number of perceptions examined here presents a complex picture of 21 separate correlations. By grouping the perceptions, HATCO will be able to see the "big picture" in terms of understanding their customers and what the customers think about HATCO. If the seven variables can be represented in a smaller number of composite variables, then the other multivariate techniques can be made more parsimonious. Of course, this approach assumes that a certain degree of underlying order exists in the data being analyzed.

Stage 2: Designing a Factor Analysis

Understanding the structure of the perceptions of variables requires R-type factor analysis and a correlation matrix between variables, not respondents. All the variables are metric and constitute a homogeneous set of perceptions appropriate for factor analysis. Regarding the adequacy of the sample size, in this example there is a 14-to-1 ratio of observations to variables, which falls within acceptable limits. Also, the sample size of 100 provides an adequate basis for the calculation of the correlations between variables.

Stage 3: Assumptions in Factor Analysis

The underlying statistical assumptions impact factor analysis to the extent that they affect the derived correlations. Departures from normality, homoscedasticity, and linearity can diminish correlations between variables. These assumptions are examined in chapter 2, and the reader is encouraged to review the findings.

The researcher must also assess the factorability of the correlation matrix. The first step is a visual examination of the correlations, identifying those that are statistically significant. Table 3.3 shows the correlation matrix for the seven perceptions of HATCO. Inspection of the correlation matrix reveals that 11 of the 21 correlations (57 percent) are significant at the .01 level. This provides an adequate

TABLE 3.3 Assessing the Appropriateness of Factor Analysis: Correlations, Measures of Sampling Adequacy, and Partial Correlations among Variables

	Correlations among Variables						
Variable	X_1	X_2	X_3	X_4	X_5	X_6	X_7
X_1 Delivery speed	1.00	−.35*	.51*	.05	.61*	.08	−.48*
X_2 Price level		1.00	−.49*	.27*	.51*	.19	.47*
X_3 Price flexibility			1.00	−.12	.07	−.03	−.45*
X_4 Manufacturer image				1.00	.30*	.79*	.20
X_5 Service					1.00	.24*	−.06
X_6 Salesforce image						1.00	.18
X_7 Product quality							1.00

*Indicates correlations significant at the .01 level.

Overall Measure of Sampling Adequacy: .446
Bartlett Test of Sphericity: 567.5   Significance: .0000

	Measures of Sampling Adequacy and Partial Correlations*						
Variable	X_1	X_2	X_3	X_4	X_5	X_6	X_7
X_1 Delivery speed	.344						
X_2 Price level	.957	.330					
X_3 Price flexibility	.018	.155	.913				
X_4 Manufacturer image	.149	.134	.095	.558			
X_5 Service	−.978	−.975	−.091	−.173	.288		
X_6 Salesforce image	−.060	−.045	−.085	−.766	.052	.552	
X_7 Product quality	−.016	−.141	.140	−.039	.088	−.092	.927

*Diagonal values are measures of sampling adequacy for individual variables; off-diagonal values are anti-image correlations (negative partial correlations).

basis for proceeding to the next level, the empirical examination of adequacy for factor analysis on both an overall basis and for each variable.

The next step is to assess the overall significance of the correlation matrix with the Bartlett test. In this example, the correlations, when taken overall, are significant at the .0001 level (see Table 3.3). But this tests only for the presence of nonzero correlations, not the pattern of these correlations. The other overall test is the measure of sampling adequacy (MSA), which in this case falls in the unacceptable range (under .50) with a value of .446. Examination of the values for each variable identifies three variables (X_1, X_2, and X_5) that also have values under .50. Because X_5 has the lowest MSA value, it will be omitted in the attempt to obtain a set of variables that can exceed the minimum acceptable MSA levels.

Table 3.4 contains the correlation matrix for the revised set of variables (X_1, X_2, X_3, X_4, X_6, and X_7) along with the measures of sampling adequacy and the Bartlett test value. In the reduced correlation matrix, 7 of the 15 correlations are statistically significant. As with the full set of variables, the Bartlett test shows that nonzero correlations exist at the significance level of .0001. The reduced set of variables collectively meets the necessary threshold of sampling adequacy with an MSA value of .665. Each of the variables also exceeds the threshold value, indicating that the reduced set of variables meets the fundamental requirements for factor analysis. Finally, with the exception of one partial correlation (X_4 and X_6)

TABLE 3.4 Assessing the Appropriateness of Factor Analysis for the Revised Set of Variables: Correlations, Measures of Sampling Adequacy, and Partial Correlations among Variables

Variable	Correlations among Variables					
	X_1	X_2	X_3	X_4	X_6	X_7
X_1 Delivery speed	1.00	−.35*	.51*	.05	.08	−.48*
X_2 Price level		1.00	−.49*	.27*	.19	.47*
X_3 Price flexibility			1.00	−.12	−.03	.45*
X_4 Manufacturer image				1.00	.79*	.20
X_6 Salesforce image					1.00	.18
X_7 Product quality						1.00

*Indicates correlations significant at the .01 level.

Overall Measure of Sampling Adequacy: .665
Bartlett Test of Sphericity: 205.965   Significance: .0000

Variable	Measures of Sampling Adequacy and Partial Correlations*					
	X_1	X_2	X_3	X_4	X_6	X_7
X_1 Delivery speed	.721					
X_2 Price level	.074	.787				
X_3 Price flexibility	−.338	.301	.748			
X_4 Manufacturer image	−.098	−.160	.081	.542		
X_6 Salesforce image	−.045	.026	−.081	−.769	.532	
X_7 Product quality	.331	−.253	.149	−.024	−.097	.779

*Diagonal values are measures of sampling adequacy for individual variables, off-diagonal values are anti-image correlations (negative partial correlations).

they are all fairly low, which is another indicator of the strength of the interrelationships among the variables in the reduced set. These measures all indicate that the reduced set of variables is appropriate for factor analysis, and the analysis can proceed to the next stages.

Component Factor Analysis: Stages 4 through 7

As noted earlier, factor analysis procedures are based on the initial computation of a complete table of intercorrelations among the variables (correlation matrix). This correlation matrix is then transformed through estimation of a factor model to obtain a factor matrix. The loadings of each variable on the factors are then interpreted to identify the underlying structure of the variables, in this case perceptions of HATCO. These steps of factor analysis, contained in stages four through seven, are examined first for component analysis. Then, a common factor analysis is performed and comparisons made between the two factor models.

Stage 4: Deriving Factors and Assessing Overall Fit

The first step is to select the number of components to be retained for further analysis. Table 3.5 contains the information regarding the six possible factors and their relative explanatory power as expressed by their eigenvalues. In addition to assessing the importance of each component, we can also use the eigenvalues to assist in selecting the number of factors. If we apply the latent root criterion, two components will be retained. The scree test (Figure 3.9, p. 124), however, indicates that three factors may be appropriate. In viewing the eigenvalue for the third factor, it was determined that its low value (.597) relative to the latent root criterion value of 1.0 precluded its inclusion. If its eigenvalue had been quite close to 1, then it might be considered for inclusion as well. These results illustrate the need for multiple decision criteria in deciding the number of components to be retained. The two factors retained represent 71 percent of the variance of the six variables.

Stage 5: Interpreting the Factors

The result of stage four is shown in Table 3.6 (p. 124), the unrotated component analysis factor matrix. To begin the analysis, let us explain the numbers included in the table. Three columns of numbers are shown. The first two are the results for the two factors that are extracted (i.e., factor loadings of each variable on each of the factors). The third column provides summary statistics detailing how well each variable is "explained" by the two components, which are discussed in the next section. The first row of numbers at the bottom of each column is the column

TABLE 3.5 Results for the Extraction of Component Factors

Factor	Eigenvalue	Percent of Variance	Cumulative Percent of Variance
1	2.51349	41.9	41.9
2	1.73952	29.0	70.9
3	.59749	10.0	80.8
4	.52956	8.8	89.7
5	.41573	6.9	96.6
6	.20422	3.4	100.0

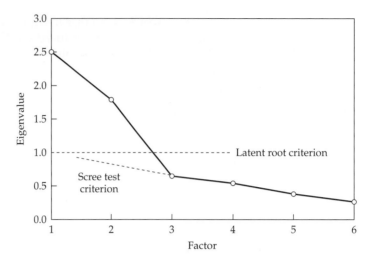

FIGURE 3.9 Scree Test for Component Analysis

sum of squared factor loadings (eigenvalues) and indicates the relative importance of each factor in accounting for the variance associated with the set of variables being analyzed. Note that the sums of squares for the two factors are 2.51 and 1.74, respectively. As expected, the factor solution has extracted the factors in the order of their importance, with factor 1 accounting for the most variance and factor 2 slightly less. At the far right-hand side of the row is the number 4.25, which represents the total explained sum of squares (2.51 + 1.74). The total sum of squared factors represents the total amount of variance extracted by the factor solution.

The total amount of variance explained by the factor solution (4.25) can be compared to the total variation in the set of variables as represented by the trace of the factor matrix. The trace is the total variance to be explained and is equal to the sum of the eigenvalues of the variable set. In components analysis, the trace is equal to the number of variables, as each variable has a possible eigenvalue of

TABLE 3.6 Unrotated Component Analysis Factor Matrix

	Factors		
Variables	1	2	Communality
X_1 Delivery speed	−.627	.514	.66
X_2 Price level	.759	−.068	.58
X_3 Price flexibility	−.730	.337	.65
X_4 Manufacturer image	.494	.798	.88
X_6 Salesforce image	.425	.832	.87
X_7 Product quality	.767	−.168	.62
			Total
Sum of squares (eigenvalue)	2.51	1.74	4.25
Percentage of trace*	41.9	29.0	70.9

*Trace = 6.0 (sum of eigenvalues).

1.0. The percentages of trace explained by each of the two factors (41.9 percent and 29.0 percent, respectively) is shown as the last row of values of Table 3.6. The percentage of trace is obtained by dividing each factor's sum of squares by the trace for the set of variables being analyzed. For example, dividing the sum of squares of 2.51 for factor 1 by the trace of 6.0 results in the percentage of trace of 41.9 percent for factor 1. By adding the percentages of trace for each of the two factors, we obtain the total percentage of trace extracted for the factor solution, which can be used as an index to determine how well a particular factor solution accounts for what all the variables together represent. If the variables are all very different from one another, this index will be low. If the variables fall into one or more highly redundant or related groups, and if the extracted factors account for all the groups, the index will approach 100 percent. The index for the present solution shows that 70.9 percent of the total variance is represented by the information contained in the factor matrix of the two-factor solution. Therefore, the index for this solution is high, and the variables are in fact highly related to one another.

The row sum of squared factor loadings is shown at the far right side of Table 3.6. These figures, referred to in the table as communalities, show the amount of variance in a variable that is accounted for by the two factors taken together. The size of the communality is a useful index for assessing how much variance in a particular variable is accounted for by the factor solution. Large communalities indicate that a large amount of the variance in a variable has been extracted by the factor solution. Small communalities show that a substantial portion of the variance in a variable is unaccounted for by the factors. For instance, the communality figure of .65 for variable X_3 indicates that it has less in common with the other variables included in the analysis than does variable X_4, which has a communality of .88. Both variables, however, still "share" over one-half of their variance with the two factors.

Having defined the various elements of the unrotated factor matrix, let us examine the factor loading patterns. As anticipated, the first factor accounts for the largest amount of variance and is a general factor, with every variable having a high loading. The loadings on the second factor show three variables (X_1, X_4, and X_6) also having high loadings. Based on this factor-loading pattern, interpretation would be extremely difficult and theoretically less meaningful. Therefore, the researcher should proceed to rotate the factor matrix to redistribute the variance from the earlier factors to the later factors. Rotation should result in a simpler and theoretically more meaningful factor pattern.

Applying an Orthogonal (VARIMAX) Rotation The VARIMAX rotated component analysis factor matrix is shown in Table 3.7 (p. 126). Note that the total amount of variance extracted is the same in the rotated solution as it was in the unrotated one, 70.9 percent. Still, two differences are apparent. First, the variance has been redistributed so that the factor-loading pattern and the percentage of variance for each of the factors is different. Specifically, in the VARIMAX rotated factor solution, the first factor accounts for 39.5 percent of the variance, compared to 41.9 percent in the unrotated solution. Likewise, the second factor accounts for 31.4 percent versus 29.0 percent in the unrotated solution. Thus the explanatory power has shifted slightly to a more even distribution because of the rotation. Second, the interpretation of the factor matrix has been simplified. Recall that in the unrotated factor solution all variables loaded significantly on

TABLE 3.7 VARIMAX-Rotated Component Analysis Factor Matrix

Variables	VARIMAX-Rotated Loadings		Communality
	Factor 1	Factor 2	
X_1 Delivery speed	−.787	.194	.66
X_2 Price level	.714	.266	.58
X_3 Price flexibility	−.804	−.011	.65
X_4 Manufacturer image	.102	.933	.88
X_6 Salesforce image	.025	.934	.87
X_7 Product quality	.764	.179	.62
			Total
Sum of squares (eigenvalue)	2.37	1.88	4.25
Percentage of trace*	39.5	31.4	70.9

*Trace = 6.0 (sum of eigenvalues).

the first factor. In the rotated factor solution, however, variables X_1, X_2, X_3, and X_7 load significantly on factor 1, and variables X_4 and X_6 load significantly on factor 2. No variable loads significantly on more than one factor. It should be apparent that factor interpretation has been simplified considerably by rotating the factor matrix.

Naming the Factors   When a satisfactory factor solution has been derived, the researcher usually attempts to assign some meaning to it. The process involves substantive interpretation of the pattern of factor loadings for the variables, including their signs, in an effort to name each of the factors. Before interpretation, a minimum acceptable level of significance for factor loadings must be selected. All significant factor loadings typically are used in the interpretation process. But variables with higher loadings influence to a greater extent the name or label selected to represent a factor.

Let us look at the results in Table 3.7 to illustrate this procedure. The factor solution was derived from component analysis with a VARIMAX rotation of the six supplier perceptions of HATCO. The cutoff point for interpretation purposes in this example is all loadings ±.55 or above (see Table 3.2). This is a conservatively high cutoff and may be adjusted if needed. But in our example, all the loadings fall substantially above or below this threshold, making interpretation quite straightforward.

Substantive interpretation is based on the significant higher loadings. Factor 1 has four significant loadings and factor 2 has two. For factor 1, we see two groups of variables. The first are price level (X_2) and product quality (X_7), both of which have positive signs. The two other variables, delivery speed (X_1) and price flexibility (X_3), have negative signs. Thus product quality and price level vary together, as do delivery speed and price flexibility. However, the two groups move in directions opposite to each other. In our example, this would indicate that as product quality and price increase, for example, delivery speed and price flexibility decrease, or vice versa. These are the four tangible characteristics of HATCO in the variable set, and are all grouped together on a single factor. This factor, perhaps named *basic value*, represents a trade-off between perceptions of price or quality of the product and perceptions of delivery speed and price flexibility. Turning to factor 2, we note that variables X_4 (manufacturer image) and X_6 (sales-force image) both relate to image components, indicating perhaps a label of *HATCO image* for the second factor. Both variables are of the same sign, suggesting that these perceptions are quite similar among respondents and do not act in differing directions, as seen in the first factor.

We should note that overall service (X_5) was not included in the factor analysis. When the factor-loading interpretations are presented, it must be noted that this variable was not included. If the results are used in other multivariate analyses, X_5 could be included as a separate variable, although it would not be assured to be orthogonal to the factor scores.

The process of naming factors has been demonstrated. You will note that it is based primarily on the subjective opinion of the researcher. Different researchers in many instances will no doubt assign different names to the same results because of differences in their backgrounds and training. For this reason, the process of labeling factors is subject to considerable criticism. But if a logical name can be assigned that represents the underlying nature of the factors, it usually facilitates the presentation and understanding of the factor solution and therefore is a justifiable procedure.

Applying an Oblique Rotation The VARIMAX rotation is orthogonal, meaning that the factors remain uncorrelated throughout the rotation process. But in many situations, the factors need not be uncorrelated and may even be conceptually linked, which requires correlation between the factors. In our example, it is quite reasonable to expect that perceptual dimensions would be correlated; thus the application of an oblique rotation is justified. Table 3.8 contains the pattern and

TABLE 3.8 Oblique Rotation of Component Analysis Factor Matrix

Variables	Oblique-Rotation Loadings		Communality[a]
	Factor 1	*Factor 2*	
Pattern Matrix			
X_1 Delivery speed	−.803	.248	.66
X_2 Price level	.704	.219	.58
X_3 Price flexibility	−.808	.043	.65
X_4 Manufacturer image	.051	.931	.88
X_6 Salesforce image	−.026	.937	.87
X_7 Product quality	.759	.129	.62
Structure Matrix			
X_1 Delivery speed	−.773	.151	
X_2 Price level	.730	.304	
X_3 Price flexibility	−.802	−.054	
X_4 Manufacturer image	.164	.938	
X_6 Salesforce image	.088	.934	
X_7 Product quality	.774	.220	
Factor Correlation Matrix			
	Factor 1	Factor 2	
Factor 1	1.00		
Factor 2	.121	1.00	

[a]Communality values are not equal to the sum of the squared loadings owing to the correlation of the factors.

structure matrices with the factor loadings for each variable on each factor. As discussed earlier, the pattern matrix is typically used for interpretation purposes, especially if the factors have a substantial correlation between them. In this case, the correlation between the factors is only .12, so that the pattern and structure matrices have quite comparable loadings. By examining the variables loading highly on each factor, we note that the interpretation is exactly the same as found with the VARIMAX rotation.

Stage 6: Validation of Factor Analysis

Validation of any factor analysis results is essential, particularly when attempting to define underlying structure among the variables. Optimally, we would always follow our use of factor analysis with some form of confirmatory factor analysis, such as structural equation modeling (see chapter 11), but this is often not feasible. We must look to other means, such as split sample analysis or application to entirely new samples.

In this example, we split the sample into two equal samples of 50 respondents and reestimated the factor models to test for comparability. Table 3.9 contains the VARIMAX rotations for the two factor models, along with the communalities. As can be seen, the two VARIMAX rotations are quite comparable in terms of both loadings and communalities for all six perceptions. One notable occurrence is the reversal of signs on factor 1 in split-sample 1 versus split-sample 2. The interpretations of the relationships among the variables (e.g., as delivery speed gets higher, price level perceptions decrease) do not change because they are relative among the loadings in each factor.

With these results we can be more assured that the results are stable within our sample. If possible, we would always like to perform additional work by gathering additional respondents and ensuring that the results generalize across the population.

TABLE 3.9 Validation of Components Factor Analysis by Split-Sample Estimation with VARIMAX Rotation

Variables	VARIMAX-Rotated Loadings		Communality
	Factor 1	Factor 2	
Split-Sample 1			
X_1 Delivery speed	−.695	.397	.64
X_2 Price level	.772	.142	.62
X_3 Price flexibility	−.822	−.098	.69
X_4 Manufacturer image	.045	.944	.89
X_6 Salesforce image	.056	.916	.84
X_7 Product quality	.811	.043	.66
Split-Sample 2			
X_1 Delivery speed	.842	−.002	.71
X_2 Price level	−.625	.396	.55
X_3 Price flexibility	.829	.107	.70
X_4 Manufacturer image	−.167	.915	.87
X_6 Salesforce image	.008	.945	.89
X_7 Product quality	−.681	.315	.56

Stage 7: Additional Uses of the Factor Analysis Results

The researcher has the option of using factor analysis not only as a data summarization tool, as seen in the prior discussion, but also as a data reduction tool. In this context, factor analysis would assist in reducing the number of variables, either through selection of a set of surrogate variables, one per factor, or by creating new composite variables for each factor. The following sections detail the issues in data reduction for this example.

Selecting Surrogate Variables for Subsequent Analysis   Let us examine the data in Table 3.7 to clarify the procedure for selecting surrogate variables. First, we recall that surrogate variables would be selected only when the rotation is orthogonal, because when we are interested in using surrogate variables in subsequent analyses, it is best, to the extent possible, that the independent variables be uncorrelated with each other. Thus an orthogonal solution would be selected instead of an oblique one.

Assuming we want to select only a single variable for further use, rather than constructing a summated scale or using factor scores (see following sections), we would examine the magnitude of the factor loadings. Focusing on the factor loadings for factor 2, we see that the loading for variable X_4 is .933 and for variable X_6, it is .934. The selection of a surrogate is difficult in cases such as this because the sizes of the loadings are essentially identical. However, if we have no a priori evidence to suggest that the reliability or validity for one of the variables is better than for the other, and if neither would be theoretically more meaningful for the factor interpretation, we would select variable X_6 as the surrogate variable, knowing that it represents both image elements to a high degree. Given the high loadings for both variables, selection of only one would be sufficient because of the high degree of intercorrelation between them (shown by their extremely high loadings on the same factor or component). Likewise, the loadings for factor 1 are .714 for variable X_2 and .764 for X_7, with comparable negative loadings for X_1 ($-.787$) and X_3 ($-.804$). For both factors, no single variable "represents" the component best; thus factor scores or a summated scale would be more appropriate.

Creating Summated Scales   A summated scale is a composite value for a set of variables calculated by such simple procedures as taking the average of the variables in the scale. This is much like the variates in other multivariate techniques, except that the weights for each variable are assumed to be equal in the averaging procedure. Factor analysis assists in the construction of the summated scale by identifying the dimensionality of the variables, which can then be related to the conceptual definition. In this example, the two-factor solution suggests that two summated scales should be constructed. The two factors, discussed earlier, correspond to dimensions that can be named and related to concepts with adequate content validity. The dimensionality of each scale is supported by the "clean" interpretation of each factor, with high factor loadings of each variable on only one factor. The reliability of the summated scales is best represented by Cronbach's alpha, which in this case is .77 for scale 1 and .85 for scale 2. Both of these reliability values exceed the recommended level of .70. Although no direct test is available to assess the validity of the summated scale, comparisons can be made to analyses made with the original variables and factor scores. Table 3.10 illustrates the use of summated scales along with factor scores as replacements for the original variables. We selected the example of identifying differences between respondents from small versus large firms (X_8). The summated scales show the same patterns of differences between small versus large firms as either the individual variables or factor scores. Thus, they do demonstrate some level of convergent validity with these other measures.

The differing signs of the loadings in factor 1 highlight an important consideration in constructing summated scales. Whenever variables have both positive and negative loadings within the same factor, either the variables with the positive or the negative loadings must have their data values reversed. Typically, the variables with the negative loadings are reverse scored so that the correlations, and the loadings, are now all positive within the factor. **Reverse scoring** is the process by which the data values for a variable are reversed so that its correlations with other variables are reversed (i.e., go from negative to positive). For ex-

TABLE 3.10 Evaluating the Replacement of the Original Variables by Factor Scores or Summated Scales

| | Mean Difference between Groups of Respondents Based on X_8, Firm Size | | | |
| | Mean Scores | | F-Test | |
Measure	*Group 1: Small Firms*	*Group 2: Large Firms*	*F Ratio*	*Significance*
Original variables				
X_1 Delivery speed[a]	4.19	2.50	64.7	.000
X_2 Price level	1.95	2.99	22.0	.000
X_3 Price flexibility[a]	8.62	6.80	70.2	.000
X_4 Manufacturer image	5.21	5.30	0.1	.709
X_6 Salesforce image	2.69	2.63	0.2	.674
X_7 Product quality	6.09	8.29	86.2	.000
Factor scores				
Factor score 1	−.640	.959	159.8	.000
Factor score 2	.052	−.078	0.41	.525
Summated scales				
Scale 1	3.81	5.49	156.8	.000
Scale 2	3.95	3.96	0.00	.957

| | Correlations between Factor Scores and Summated Scales | | | |
| | Factor Scores | | Summated Scales[b] | |
	1	*2*	*1*	*2*
Factor score 1	1.000	.000	.995	.075
Factor score 2	.000	1.00	.085	.985
Summated scale 1[b]	.995	.085	1.000	.154
Summated scale 2[b]	.075	.985	.154	1.000

[a]Have negative factor loadings.
[b]Summated scales calculated as average score across items. For example, scale 1 is average of X_1, X_2, X_3, and X_7.
Note: X_1 and X_3 are reverse-scaled, owing to their negative factor loadings.

ample, on our scale of 0 to 10, we would reverse score a variable by subtracting the original value from 10 (i.e., reverse score = 10 − original value). In this way, original scores of 10 and 0 now have the reversed scores of 0 and 10. All of the distributional characteristics are retained; only the distribution is reversed.

The purpose of reverse scoring is to prevent a "canceling out" of variables with positive and negative loadings. Let us use as an example two variables with a negative correlation, V_1 and V_2, with V_1 having a positive loading and V_2 a negative loading. This means that if 10 is the top score on V_1, the top score on V_2 would be 0. Now assume two cases. In case 1, V_1 has a value of 10 and V_2 has a value of 0 (the best case). In the second case, V_1 has a value of 0 and V_2 has a value of 10 (the worst case). If V_2 is not reverse scored, then the scale score calculated by adding the two variables for both cases 1 and 2 is 10, showing no difference, whereas we know that case 1 is the best and case 2 is the worst. If we reverse score V_2, however, the situation changes. Now case 1 has values of 10 and 10 on V_1 and V_2, respectively, and case 2 has values of 0 and 0. The summated scale scores are now 20 for case 1 and 0 for case 2, which distinguishes them as the best and worst situations.

Use of Factor Scores   Instead of calculating summated scales, we could calculate factor scores for each of the two factors in our component analysis. In this way, each respondent would have two new variables (factor scores for factors 1 and 2) that could be substituted for the original six variables in other multivariate techniques. In the test of mean differences between the two groups of respondents (Table 3.9), we see that all the variables loading highly on factor 1 (X_1, X_2, X_3, and X_7) are significantly different between the respondents from small and large firms, whereas the variables loading highly on factor 2 (X_4 and X_6) do not have significant differences. The factor scores and summated scales should show similar patterns if they are truly representative of the variables. As seen in Table 3.10, the factor scores do differ in accordance with this pattern. Factor score 1 shows significant differences, whereas factor score 2 does not. Similar differences between the two groups are seen for the summated scales. Also, the summated scales correlate very highly with the factor scores. Thus, in this instance, both the factor scores and the summated scales accurately portray the concepts they represent.

Selecting the Data Reduction Method   If the original variables are to be replaced by factor scores or summated scales, a decision must be made on which to use. This decision is based on the need for replication in other studies (which favors use of summated scales) versus the desire for orthogonality of the measures (which favors factor scores). Table 3.10 also contains the correlation matrix of factor scores and summated scales. Because we employed an orthogonal rotation, the correlation between factor scores is .000. But the summated scales can be correlated and in this case the correlation is .1545. The researcher must ascertain the need for orthogonality versus replicability in selecting factor scores versus summated scales.

Common Factor Analysis: Stages 4 and 5

Common factor analysis is the second major factor analytic model that we discuss. The primary distinction between component analysis and common factor analysis is that the latter considers only the common variance associated with a set of variables. This aim is accomplished by factoring a "reduced" correlation matrix with estimated initial communalities in the diagonal instead of unities. The differences between component analysis and common factor analysis occur only at the factor estimation and interpretation stages (stages 4 and 5). Once the communalities are substituted on the diagonal, the common factor model extracts factors in a manner similar to component analysis. The researcher uses the same criteria for factor selection and interpretation. To illustrate the differences that can occur between common factor and component analysis, the following sections detail the extraction and interpretation of a common factor analysis of the six HATCO perceptions used in the component analysis.

Stage 4: Deriving Factors and Assessing Overall Fit

The "reduced" correlation matrix with communalities on the diagonal was used in the common factor analysis. Note that X_5 was omitted from the component analysis due to an unacceptable MSA value. If included in the common factor analysis, however, the communality could not have been estimated in the original extraction of factors, an example of another problem associated with common factor analysis. Thus, the common factor analysis still would have been performed with six variables even if X_5 had not been eliminated because of the low MSA value.

The first step is to determine the number of factors to retain for examination and possible rotation. Table 3.11 shows the extraction statistics. If we were to employ the latent root criterion with a cutoff value of 1.0 for the eigenvalue, two factors would be retained. However, the scree analysis indicates that three factors be retained (see Figure 3.10). In combining these two criteria, we will retain two factors for further analysis because of the very low eigenvalue for the third factor and to maintain comparability with the component analysis. Again, as with the component analysis examined earlier, the researcher should employ a combination of criteria in determining the number of factors to retain and may even wish to examine the three-factor solution as an alternative.

The unrotated factor matrix (Table 3.12) shows that the common factor solution accounted for 58.6 percent of the total variance. Because the final common factor model sometimes differs from the initial extraction estimates (see Table 3.11), the researcher should be sure to evaluate the extraction statistics for the final common factor model. If the researcher was dissatisfied with the total variance explained, a common factor model extracting three factors could also be estimated. We note that the communalities of each variable are lower than found in component analysis. This is due primarily to the lower overall variance explained, not the performance of any one variable. Again, exploration of a three-factor model

TABLE 3.11 Results for the Extraction of Common Factors

Factor	Eigenvalue	Percent of Variance	Cumulative Percent of Variance
1	2.51349	41.9	41.9
2	1.73952	29.0	70.9
3	.59749	10.0	80.8
4	.52956	8.8	89.7
5	.41573	6.9	96.6
6	.20422	3.4	100.0

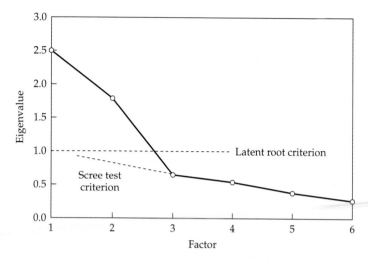

FIGURE 3.10 Scree Test for Common Factor Solution

could be made in an attempt to increase the communalities, as well as the overall variance explained. For our purposes here, we interpret the two-factor solution.

Stage 5: Interpreting the Factors

By examining the unrotated loadings, we note the need for a factor matrix rotation. Turning then to the VARIMAX rotated common factor analysis factor matrix (Table 3.13, p. 134), let us examine how it compares with the component analysis rotated factor matrix. The information that is provided in the common factor solution is similar to that provided in the component analysis solution. Sums of squares, percentages of variance, communalities, total sums of squares, and total variances extracted are all provided, just as with the component analysis solution.

Comparison of the information provided in the rotated common factor analysis factor matrix and the rotated component analysis factor matrix shows remarkable similarity. The primary differences between the component analysis and

TABLE 3.12 Unrotated Common Factor Matrix

	Factors		
Variables	*1*	*2*	*Communality*
X_1 Delivery speed	.485	.512	.50
X_2 Price level	.629	.187	.43
X_3 Price flexibility	.602	.401	.52
X_4 Manufacturer image	.625	.683	.87
X_6 Salesforce image	.526	.670	.72
X_7 Product quality	.641	.269	.48
			Total
Sum of squares (eigenvalue)	2.07	1.45	3.52
Percentage of trace*	34.5	24.1	58.6

*Trace = 6.0 (sum of eigenvalues).

TABLE 3.13 VARIMAX-Rotated Common Factor Matrix

| | VARIMAX-Rotated Loadings | | |
Variables	Factor 1	Factor 2	Communality
X_1 Delivery speed	−.693	.133	.50
X_2 Price level	.620	.215	.43
X_3 Price flexibility	−.722	−.026	.52
X_4 Manufacturer image	.109	.925	.87
X_6 Salesforce image	.037	.846	.72
X_7 Product quality	.677	.155	.48
			Total
Sum of squares (eigenvalue)	1.86	1.66	3.52
Percentage of trace*	31.0	27.6	58.6

*Trace = 6.0 (sum of eigenvalues).

common factor analysis are the generally lower loadings in the common factor analysis, owing primarily to the lower communalities of the variables used in common factor analysis. Another comparison that may be useful to the researcher is the percentage of total variance explained by each factor. In component analysis (Table 3.7), the two rotated factors differed by 8 percent (39.5 percent versus 31.4 percent, respectively). In the common factor results (Table 3.13), the rotation "spreads" the variance so that the two factors are almost equal in variance explained (31.0 percent for factor 1 and 27.6 percent for factor 2). But even with these slight differences in the variance explained, the patterns of loadings and basic interpretations are identical between the component analysis and the common factor analysis.

A Managerial Overview of the Results

Both the components and common factor analyses provide the researcher with several key insights into the structure of the variables and options for data reduction. First, concerning the structure of the variables, there are clearly two separate and distinct dimensions of evaluation used by the HATCO customers. One dimension, termed *basic value*, relates to the tangible aspects of HATCO and its products. Within this dimension there is the tradeoff between product price and quality versus the characteristics of delivery speed and price flexibility. The second dimension, *HATCO image*, pertains to the image perceptions of the manufacturer and salesforce. Business planners within HATCO can now discuss plans revolving around these two areas instead of having to deal with the separate variables.

Factor analysis also provides the basis for data reduction through either summated scales or factor scores. The researcher now has a method for combining the variables within each factor into a single score that can replace the original set of variables with two new composite variables. When looking for differences, such as between large and small firms, these new composite variables can be used so that only differences for two values, *basic value* and *image*, are analyzed.

Summary

The multivariate statistical technique of factor analysis has been presented in broad conceptual terms. Basic guidelines for interpreting the results were included to clarify further the methodological concepts. An example of the application of factor analysis was presented based on the database in chapter 1.

Factor analysis can be a highly useful and powerful multivariate statistical technique for effectively extracting information from large databases and making sense of large bodies of interrelated data. When it works well, it points to interesting relationships that might not have been obvious from examination of the raw data alone or even a correlation matrix. Factor analysis has the ability to identify sets of related variables and even develop a single composite measure to represent the entire set of related variables. This offers the researcher a powerful tool in achieving a better understanding of the structure of the data and a way to simplify other analyses of a large set of variables by using the replacement composite variables. Potential applications of factor analytic techniques to problem solving and decision making in business research are numerous. The use of these techniques will continue to grow as increased familiarity with the benefits of data summarization and data reduction is gained by researchers.

Factor analysis is a much more complex and involved subject than might be indicated by this brief exposition. Three of the most frequently cited limitations are as follows: first, there are many techniques for performing factor analyses, and controversy exists over which technique is the best. Second, the subjective aspects of factor analysis (i.e., deciding how many factors to extract, which technique should be used to rotate the factor axes, which factor loadings are significant) are all subject to many differences in opinion. Third, the problem of reliability is real. Like any other statistical procedure, a factor analysis starts with a set of imperfect data. When the data change because of changes in the sample, the data-gathering process, or the numerous kinds of measurement errors, the results of the analysis also change. The results of any single analysis are therefore less than perfectly dependable. This problem is especially critical because the results of a single-factor analytic solution frequently look plausible. It is important to emphasize that plausibility is no guarantee of validity or even stability.

Questions

1. What are the differences between the objectives of data summarization and data reduction?
2. How can factor analysis help the researcher improve the results of other multivariate techniques?
3. What guidelines can you use to determine the number of factors to extract? Explain each briefly.
4. How do you use the factor-loading matrix to interpret the meaning of factors?
5. How and when should you use factor scores in conjunction with other multivariate statistical techniques?
6. What are the differences between factor scores and summated scales? When are each most appropriate?

7. What is the difference between Q-type factor analysis and cluster analysis?

8. When would the researcher use an oblique rotation instead of an orthogonal rotation? What are the basic differences between them?

References

1. Anderson, J. C., D. W. Gerbing, and J. E. Hunter (1987), "On the Assessment of Unidimensional Measurement: Internal and External Consistency and Overall Consistency Criteria." *Journal of Marketing Research* 24 (November): 432–37.

2. American Psychological Association (1985), *Standards for Educational and Psychological Tests*. Washington, D.C.: APA.

3. Bearden, W. O., R. G. Netemeyer, and M. Moble (1993), *Handbook of Marketing Scales: Multi-Item Measures for Marketing and Consumer Behavior*. Newbury Park, Calif.: Sage.

4. Bentler, Peter M. (1992), *EQS Structural Equations Program Manual*. Los Angeles: BMDP Statistical Software.

5. BMDP Statistical Software, Inc. (1992), *BMDP Statistical Software Manual, Release 7*, vols. 1 and 2, Los Angeles: BMDP Statistical Software.

6. Borgatta, E. F., K. Kercher, and D. E. Stull (1986), "A Cautionary Note on the Use of Principal Components Analysis." *Sociological Methods and Research* 15: 160–68.

7. Bruner, G. C., and P. J. Hensel (1993), *Marketing Scales Handbook, A Compilation of Multi-Item Measures*. Chicago: American Marketing Association.

8. Campbell, D. T., and D. W. Fiske (1959), "Convergent and Discriminant Validity by the Multitrait-Multimethod Matrix." *Psychological Bulletin* 56 (March): 81–105.

9. Cattell, R. B. (1966), "The Scree Test for the Number of Factors." *Multivariate Behavioral Research* 1 (April): 245–76.

10. Cattell, R. B., K. R. Balcar, J. L. Horn, and J. R. Nesselroade (1969), "Factor Matching Procedures: An Improvement of the s index; with tables." *Educational and Psychological Measurement* 29: 781–92.

11. Chatterjee, S., L. Jamieson, and F. Wiseman (1991), "Identifying Most Influential Observations in Factor Analysis." *Marketing Science* 10 (Spring): 145–60.

12. Churchill, G. A. (1979), "A Paradigm for Developing Better Measures of Marketing Constructs." *Journal of Marketing Research* 16 (February): 64–73.

13. Cliff, N., and C. D. Hamburger (1967), "The Study of Sampling Errors in Factor Analysis by Means of Artificial Experiments." *Psychological Bulletin* 68: 430–45.

14. Cronbach, L. J. (1951), "Coefficient Alpha and the Internal Structure of Tests." *Psychometrika* 31: 93–96.

15. Dillon, W. R. and M. Goldstein (1984), *Multivariate Analysis: Methods and Applications*. New York: Wiley.

16. Dillon, W. R., N. Mulani, and D. G. Frederick (1989), "On the Use of Component Scores in the Presence of Group Structure." *Journal of Consumer Research* 16: 106–12.

17. Gorsuch, R. L. (1983), *Factor Analysis*. Hillsdale, N.J.: Lawrence Erlbaum Associates.

18. Gorsuch, R. L. (1990), "Common Factor Analysis versus Component Analysis: Some Well and Little Known Facts." *Multivariate Behavioral Research* 25: 33–39.

19. Hattie, J. (1985), "Methodology Review: Assessing Unidimensionality of Tests and Items." *Applied Psychological Measurement* 9: 139–64.

20. Joreskog, K. G., and D. Sorbo (1993), *LISREL 8: Structural Equation Modeling with the SIMPLIS Command Language*. Mooresville, Ind.: Scientific Software International.

21. Kaiser, H. F. (1970), "A Second-Generation Little Jiffy." *Psychometrika* 35: 401–15.

22. Kaiser, H. F. (1974), "Little Jiffy, Mark IV." *Educational and Psychology Measurement* 34: 111–17.

23. McDonald, R. P. (1981), "The Dimensionality of Tests and Items." *British Journal of Mathematical and Social Psychology* 34: 100–117.

24. Mulaik, S. A. (1990), "Blurring the Distinction Between Component Analysis and Common Factor Analysis." *Multivariate Behavioral Research* 25: 53–59.

25. Mulaik, S. A., and R. P. McDonald (1978), "The Effect of Additional Variables on Factor Indeterminancy in Models with a Single Common Factor." *Psychometrika* 43: 177–92.

26. Nunnally, J. L. (1978), *Psychometric Theory*, 2d ed. New York: McGraw-Hill.

27. Nunnally, J. (1979), *Psychometric Theory*. New York: McGraw-Hill.

28. Peter, J. P. (1979), "Reliability: A Review of Psychometric Basics and Recent Marketing Practices." *Journal of Marketing Research* 16 (February): 6–17.

29. Peter, J. P. (1981), "Construct Validity: A Review of Basic Issues and Marketing Practices." *Journal of Marketing Research* 18 (May): 133–45.

30. Robinson, J. P., P. R. Shaver, and L. S. Wrightsman (1991), "Criteria for Scale Selection and Evaluation," in *Measures of Personality and Social Psychological Attitudes*, J. P. Robinson, P. R. Shanver, and L. S. Wrightsman (eds.). San Diego, Calif.: Academic Press.

31. Robinson, J. P., and P. R. Shaver (1973), *Measures of Psychological Attitudes*. Ann Arbor, MI: Survey Research Center Institute for Social Research, University of Michigan.

32. Rummel, R. J. (1970), *Applied Factor Analysis.* Evanston, Ill.: Northwestern University Press.

33. SAS Institute, Inc. (1990), *SAS User's Guide: Statistics, Version 6.* Cary, N.C.: SAS Institute.

34. Smith, Scott M. (1989), *PC-MDS: A Multidimensional Statistics Package.* Provo, Utah: Brigham Young University Press.

35. Snook, S. C., and R. L. Gorsuch (1989), "Principal Component Analysis versus Common Factor Analysis: A Monte Carlo Study." *Psychological Bulletin* 106: 148–54.

36. SPSS, Inc. (1990), *SPSS Advanced Statistics Guide.* 4th ed. Chicago: SPSS.

37. Stewart, D. W. (1981), "The Application and Misapplication of Factor Analysis in Marketing Research." *Journal of Marketing Research* 18 (February): 51–62.

38. Velicer, W. F., and D. N. Jackson (1990), "Component Analysis versus Common Factor Analysis: Some Issues in Selecting an Appropriate Procedure." *Multivariate Behavioral Research* 25: 1–28.

Annotated Readings

The following readings have been selected for their illustration of the application of factor analysis to specific research problems. The annotations of each article are provided to give the reader a sense of the issues involved in each instance and the types of results achieved with this multivariate technique. Interested readers are encouraged to review the original articles in their complete form to gain a deeper appreciation for "real world" applications of factor analysis.

Richins, Marsha L., and Scott Dawson (1992), "A Consumer Values Orientation for Materialism and Its Measurement: Scale Development and Validation," *Journal of Consumer Research* 19 (December), 303–16.

This article illustrates the use of factor analysis not only to reveal the underlying structure of the data, but also as an integral component in the construction of summated scales, used in this situation to represent materialism. To better understand consumer consumption and resource allocation, this study examines the consumer value of materialism and develops a direct means of measuring individual differences in materialism. Materialism is conceptualized as containing three dimensions: acquisition centrality, acquisition as the pursuit of happiness, and possession-defined success.

To empirically capture the complex, multidimensional nature of materialism, the authors generated a large number of questions (30 items) reflecting some aspect of the three dimensions. Principal components factor analysis was then employed (1) as a further method of reducing the number of items (down to 18) and (2) to empirically reveal and demonstrate the hypothesized, underlying structure. First, the 30 items were factor analyzed and three factors emerged. An oblique rotation was then undertaken to assist in the interpretation of the factors. The result was the identification of 18 items loading highly (all above .40 and most in the .55 to .65 range) on one of the three factors. The three factors were interpreted as relating to the concepts of success, centrality, and happiness. Although the authors did not provide the overall percentage of variance extracted, the three factors were deemed sufficient and conceptually valid in their correspondence to existing theory. The results of the factor analysis were then confirmed by the use of confirmatory factor analysis (see chapter 11, this volume). The items loading highly on the factors were then summed to create three subscales of materialism, with the three subscales also summated to create an overall measure of materialism. Reliability tests for each of the summated scales all exceeded the threshold of .70 for acceptance.

Having established the multidimensional nature of the 18-item materialism scale (three subscales and an overall measure), the authors then sought to validate the results. The construct's validity was established by demonstrating that individuals who score high on the materialism scale place greater value on acquisitions, are self-centered, seek material possessions, and tend to be dissatisfied with their circumstances. Through the use of statistical methods, this study offers a measure of materialism that is both reliable and valid. Factor analysis played key roles in both revealing and measuring the underlying dimensions of materialism and providing the basis for the creation of summated scales as a more rigorous measure of materialism. Researchers now have an acceptable measure of materialism that is easily quantifiable and replicable, allowing for a better understanding of this consumer value from both its conceptual and empirical perspectives.

Deshpande, Rohit (1982), "The Organizational Context of Market Research Use," *Journal of Marketing* 46 (Fall), 91–101.

Academicians and business analysts have noted that despite the innovations in methods for the use of marketing research information, relatively few firms utilize these new techniques. This reluctance to utilize a valuable source of external market information can have substantive impacts on a firm's ability to recognize and react to market changes. One possible explanation for this lack of utilization is the influence of organizational structure. Organizational structure is conceptualized to have two dimensions: formalization and centralization. Each dimension also contains a number of subdimensions: three for formalization (job codification, rule observation, and job specificity) and two for centralization (participation in decision making and hierarchy of authority). A crucial task facing the researcher in determining the impact of organizational structure on the use of external information is to first develop a means of quantifying and measuring organizational structure that is applicable across a wide range of situations. To do so, the researcher must accommodate the myriad facets of each dimension and subdimension to ensure a complete and valid portrayal or a complex concept. Yet the researcher must also strive for parsimony in the number and character of the measures while also considering the reliability of any measures to be developed.

Factor analysis was implemented to attempt to reduce 23 scale items representing these individual facets down to a smaller set of dimensions that (1) would correspond to the conceptual model of organizational structure, while also (2) providing an adequate representation of the individual items. Factor analysis extracted five factors accounting for 70 percent of the variance in the original 23 questions. Examination of the factor matrix called for only three items to be deleted because of low or incorrect factor loadings. The criteria for inclusion on a particular factor were that items exhibited high loading for their own specific facets while loading at low levels on the other scale facets. Moreover, interpretation of the factor loadings supported the proposed structure, with the five factors directly corresponding to the hypothesized dimensions. The factor loadings then guided the creation of summated scales for each of the five dimensions. The usefulness of the summated scales was also demonstrated through their use as independent variables in a multiple regression analysis (see chapter 4, this volume), which provided acceptable levels of predictive accuracy and explanation of the relationship in accordance with the research question.

In summary, factor analysis provided for the development of a comprehensive model examining the impact of organizational structure on the usage of marketing research information while still maintaining model parsimony. The factors extracted in factor analysis provided the means for creating objective and replicable measures of the five dimensions of organizational structure that confirmed the conceptual model of organizational structure. These summated scales then acted as replacements for the individual items in the multiple regression analysis. The empirical analysis allowed the researchers to conclude that marketing managers are more likely to use research information when they work in a less structured, decentralized organizational environment in which there are few formal rules or procedures to follow.

DEPENDENCE TECHNIQUES

Overview

Whereas section 1 focused on the preparation of data for multivariate analysis, section 2 deals with what many would term the essence of multivariate analysis: dependence techniques. As noted in chapter 1, dependence techniques are based on the use of a set of independent variables to predict and explain one or more dependent variables. The researcher, whether faced with dependent variables of a metric or nonmetric nature, has a variety of dependence methods available to assist in the process of relating independent variables to dependent variables. Given the multivariate nature of these methods, all of the dependence techniques accommodate multiple independent variables while also allowing multiple dependent variables in certain situations. Thus, the researcher has a set of techniques that should allow for the analysis of almost any type of research question involving a dependence relationship. They also provide the opportunity not only for increased prediction capability, but also for enhanced explanation of the dependent variable's relationship to the independent variables. Explanation becomes increasingly important as the research questions begin to address issues concerning how the relationship between independent and dependent variables operates.

Chapters in Section 2

Section 2 covers six dependence techniques: multiple regression, discriminant analysis, logistic regression multivariate analysis of variance, conjoint analysis, and canonical correlation, in chapters 4 through 8, respectively. Dependence techniques, as noted earlier, allow the researcher to assess the degree of relationship between the dependent and independent variables. Dependence techniques vary in the type and character of the relationship, as reflected in the measurement properties of the dependent and independent variables discussed in chapter 1. For example, multiple regression, discriminant analysis, and canonical correlation all accommodate multiple metric independent variables but vary in the type of dependent variable (regression analysis—single metric; discriminant analysis—single nonmetric; and canonical correlation—multiple metric). Chapter 4, Multiple Regression Analysis, focuses on the what is perhaps the most fundamental of all multivariate techniques and a "building block" for our discussion of the other dependence methods. Whether assessing the conformity to underlying statistical assumptions, measuring predictive accuracy, or interpreting the variate of independent variables, the issues discussed in chapter 4 will be seen as crucial in many of the other techniques as well. Chapter 5, Multiple Discriminant Analysis and Logistic Regression, investigates a unique form of dependence relationship—a dependent variable that is not metric but rather is nonmetric. In this situation, the researcher is attempting to classify observations into groups. This classification into groups can be accomplished through either discriminant analysis or logistic regression, a variant of regression designed to specifically deal with nonmetric dependent variables. In chapter 6, Multivariate Analysis of Variance, the discussion differs in several ways from the prior techniques, as it is suited to the analysis of multiple metric dependent variables and nonmetric independent variables. Although this technique is a direct extension of simple analysis of variance, the multiple metric dependent variables make both prediction and explanation more difficult. Chapter 7, Conjoint Analysis, presents us with a technique unlike any of the other multivariate methods in that the researcher determines the values of the independent nonmetric variables in a quasi-experimental fashion. Once designed, the respondent provides information regarding only the dependent variable. Although it places more responsibility on the researcher, conjoint analysis provides a powerful tool for understanding complex decision processes. Finally, in chapter 8, Canonical Correlation, the researcher is exposed to the most generalized form of multivariate analysis, which accommodates multiple dependent and independent variables. In situations in which variates exist for both the dependent and independent variables, canonical correlation provides a flexible method for both prediction and explanation.

This section provides the researcher with exposure to the wide range of dependence techniques available, each suited to a specific task and relationship. When you complete this section, the issues regarding selecting from these methods should be apparent, and you should feel comfortable in selecting from these techniques and analyzing their results.

CHAPTER

Multiple Regression Analysis

LEARNING OBJECTIVES

Upon completing this chapter, you should be able to do the following:

- Determine when regression analysis is the appropriate statistical tool in analyzing a problem.
- Understand how regression helps us make predictions using the least squares concept.
- Be aware of the important assumptions underlying regression analysis and be prepared to provide remedies when violations occur.
- Interpret the results of regression from both a statistical and a managerial viewpoint.
- Apply the diagnostic procedures necessary to assess "influential" observations.
- Explain the difference between stepwise and simultaneous regression.
- Use dummy variables with an understanding of their interpretation.

CHAPTER PREVIEW

This chapter describes multiple regression analysis as it is used to solve important research problems, particularly in business. Regression analysis is by far the most widely used and versatile dependence technique, applicable in every facet of business decision making. Its uses range from the most general problems to the most specific, in each instance relating a factor (or factors) to a specific outcome. For example, regression analysis is the foundation for business forecasting models, ranging from the econometric models that predict the national economy based on certain inputs (income levels, business investment, and so forth) to models of a firm's performance in a market if a specific marketing strategy is followed. Regression models are also used to study how consumers make decisions or form impressions and attitudes. Other applications include evaluating the determinants of effectiveness for a program (e.g., what factors aid in maintaining quality) and determining the feasibility of a new product or the expected return for a new stock issue. Even though these examples illustrate only a small subset of all applications, they demonstrate that regression analysis is a powerful analytical tool designed to explore all types of dependence relationships.

Multiple regression analysis is a general statistical technique used to analyze the relationship between a single dependent variable and several independent variables. As noted in chapter 1, its basic formulation is

$$Y_1 = X_1 + X_2 + \ldots + X_n$$
$$\text{(metric)} \qquad \text{(metric)}$$

This chapter presents guidelines for judging the appropriateness of multiple regression for various types of problems. Suggestions are provided for interpreting the results of its application from a managerial as well as a statistical viewpoint. Possible transformations of the data to remedy violations of various model assumptions are examined, along with a series of diagnostic procedures that identify observations with particular influence on the results. Readers who are already knowledgeable about multiple regression procedures can skim the early portions of the chapter, but for those who are less familiar with the subject, this chapter provides a valuable background for the study of multivariate data analysis.

KEY TERMS

Before beginning this chapter, review the key terms to develop an understanding of the concepts and terminology used. Throughout the chapter the key terms appear in **boldface**. Other points of emphasis in the chapter are *italicized*. Also, cross-references within the Key Terms appear in *italics*.

Adjusted coefficient of determination (adjusted R^2) Modified measure of the *coefficient of determination* that takes into account the number of independent variables included in the regression equation and the sample size. Although the addition of independent variables will always cause the coefficient of determination to rise, the adjusted coefficient of determination may fall if the added independent variables have little explanatory power and/or if the *degrees of freedom* become too small. This statistic is quite useful for comparison between equations with different numbers of independent variables, differing sample sizes, or both.

All-possible-subsets regression Method of selecting the variables for inclusion in the regression model that considers all possible combinations of the independent variables. For example, if the researcher has specified four potential independent variables, this technique would estimate all possible regression models with one, two, three, and four variables. The technique would then identify the model(s) with the best predictive accuracy.

Backward elimination Method of selecting variables for inclusion in the regression model that starts by including all independent variables in the model and then eliminating those variables not making a significant contribution to prediction.

Beta coefficient   Standardized regression coefficient (see *standardization*) that allows for a direct comparison between coefficients as to their relative explanatory power of the dependent variable. Whereas *regression coefficients* are expressed in terms of the units of the associated variable, thereby making comparisons inappropriate, beta coefficients use standardized data and can be directly compared.

Coefficient of determination (R^2)   Measure of the proportion of the variance of the dependent variable about its mean that is explained by the independent, or predictor, variables. The coefficient can vary between 0 and 1. If the regression model is properly applied and estimated, the researcher can assume that the higher the value of R^2, the greater the explanatory power of the regression equation, and therefore the better the prediction of the dependent variable.

Collinearity   Expression of the relationship between two (collinearity) or more (multicollinearity) independent variables. Two independent variables are said to exhibit complete collinearity if their correlation coefficient is 1, and complete lack of collinearity if their correlation coefficient is 0. *Multicollinearity* occurs when any single independent variable is highly correlated with a set of other independent variables. An extreme case of collinearity/multicollinearity is *singularity*, in which an independent variable is perfectly predicted (i.e., correlation of 1.0) by another independent variable (or more than one).

Correlation coefficient (r)   Coefficient that indicates the strength of the association between any two metric variables. The sign (+ or −) indicates the direction of the relationship. The value can range from −1 to +1, with +1 indicating a perfect positive relationship, 0 indicating no relationship, and −1 indicating a perfect negative or reverse relationship (as one variable grows larger, the other variable grows smaller).

Criterion variable (Y)   See *dependent variable*.

Degrees of freedom (df)   Value calculated from the total number of observations minus the number of estimated *parameters*. These parameter estimates are restrictions on the data because, once made, they define the population from which the data are assumed to have been drawn. For example, in estimating a regression model with a single independent variable, we estimate two parameters, the intercept (b_0) and a *regression coefficient* for the independent variable (b_1). In estimating the random error, defined as the sum of the prediction errors (actual minus predicted dependent values) for all cases, we would find ($n - 2$) degrees of freedom. Degrees of freedom provide a measure of how restricted the data are to reach a certain level of prediction. If the number of degrees of freedom is small, the resulting prediction may be less generalizable because all but a few observations were incorporated in the prediction. Conversely, a large degrees-of-freedom value indicates that the prediction is fairly "robust" with regard to being representative of the overall sample of respondents.

Dependent variable (Y)   Variable being predicted or explained by the set of independent variables.

Dummy variable   Independent variable used to account for the effect that different levels of a nonmetric variable have in predicting the dependent variable. To account for L levels of a nonmetric independent variable, $L - 1$ dummy variables are needed. For example, gender is measured as male or female and could be represented by two dummy variables, X_1 and X_2. When the respondent is male, $X_1 = 1$ and $X_2 = 0$. Likewise, when the respondent is female, $X_1 = 0$ and $X_2 = 1$. However, when $X_1 = 1$, we know that X_2 must equal 0. Thus we need only one variable, either X_1 or X_2, to represent gender. We need not include both variables because one is perfectly predicted by the other (a *singularity*) and the regression coefficients cannot be estimated. If a variable has three levels, only two dummy variables are needed. Thus the number of dummy variables is one less than the number of levels of the nonmetric variable.

Effects coding   Method for specifying the reference category for a set of *dummy variables* in which the reference category receives a value of -1 across the set of dummy variables. With this type of coding, the coefficients for the dummy variables become group deviations from the means of all groups. This is in contrast to *indicator coding*, in which the reference category is given the value of zero across all dummy variables and the coefficients represent group deviations from the reference group.

Forward addition   Method of selecting variables for inclusion in the regression model by starting with no variables in the model and then adding variables based on their contribution to prediction.

Heteroscedasticity   See *homoscedasticity*.

Homoscedasticity   Description of data for which the variance of the error terms (e) appears constant over the range of values of an independent variable. The assumption of equal variance of the population error ε (where ε is estimated from e) is critical to the proper application of linear regression. When the error terms have increasing or modulating variance, the data are said to be *heteroscedastic*. The discussion of *residuals* in this chapter further illustrates this point.

Independent variable   Variable(s) selected as predictors and potential explanatory variables of the dependent variable.

Indicator coding   Method for specifying the *reference category* for a set of *dummy variables* for which the reference category receives a value of zero across the set of dummy variables. The coefficients represent the group differences from the reference category. Also see *effects coding*.

Influential observation   An observation that has a disproportionate influence on one or more aspects of the regression estimates. This influence may be based on extreme values of the independent or dependent variables, or both. Influential observations can either be "good," by reinforcing the pattern of the remaining data, or "bad," when a single or small set of cases unduly affects the regression estimates. It is not necessary for the observation to be an *outlier*, although many times outliers can be classified as influential observations as well.

Intercept (b_0)   Value on the Y axis (dependent variable axis) where the line defined by the regression equation $Y = b_0 + b_1 X_1$ crosses the axis. It is described by the constant term b_0 in the regression equation. In addition to its role in prediction, the intercept may have a managerial interpretation. If the complete absence of the independent variable has meaning, then the intercept represents that amount. For example, when estimating sales from past advertising expenditures, the intercept represents the level of sales expected if advertising is eliminated. But in many instances the constant has only predictive value because there is no situation in which all independent variables are absent. An example is predicting product preference based on consumer attitudes. All individuals have some level of attitude, so the intercept has no managerial use, but it still aids in prediction.

Least squares   Estimation procedure used in simple and multiple regression whereby the regression coefficients are estimated so as to minimize the total sum of the squared *residuals*.

Leverage points   Type of *influential observation* defined by one aspect of influence termed *leverage*. These observations are substantially different on one or more independent variables, so that they affect the estimation of one or more *regression coefficients*.

Linearity   Term used to express the concept that the model possesses the properties of additivity and homogeneity. In a simple sense, linear models predict values that fall in a straight line by having a constant unit change (slope) of the dependent variable for a constant unit change of the independent variable. In the population model $Y = b_0 + b_1X_1 + \varepsilon$, the effect of a change of 1 in X_1 is to add b_1 (a constant) units of Y.

Measurement error   Degree to which the data values do not truly measure the characteristic being represented by the variable. For example, when asking about total family income, there are many sources of measurement error (e.g., reluctance to answer full amount, error in estimating total income) that make the data values imprecise.

Moderator effect   Effect in which a third independent variable (the moderator variable) causes the relationship between a dependent/independent variable pair to change, depending on the value of the moderator variable. It is also known as an interactive effect and similar to the interaction effect seen in analysis of variance methods.

Multicollinearity   See *collinearity*.

Multiple regression   Regression model with two or more independent variables.

Normal probability plot   Graphical comparison of the shape of the sample distribution to the normal distribution. In the graph, the normal distribution is represented by a straight line angled at 45 degrees. The actual distribution is plotted against this line, so any differences are shown as deviations from the straight line, making identification of differences quite simple.

Null plot   Plot of residuals versus the predicted values that exhibits a random pattern. A null plot is indicative of no identifiable violations of the assumptions underlying regression analysis.

Outlier   In strict terms, an observation that has a substantial difference between the actual value for the dependent variable and the predicted value. Cases that are substantially "different," with regard to either the dependent or independent variables, are often termed outliers as well. In all instances, the objective is to identify observations that are inappropriate representations of the population from which the sample is drawn, so that they may be discounted or even eliminated from the analysis as unrepresentative.

Parameter   Quantity (measure) characteristic of the population. For example, μ and σ^2 are the symbols used for the population parameters mean (μ) and variance (σ^2). These are typically estimated from sample data in which the arithmetic average of the sample is used as a measure of the population average and the variance of the sample is used to estimate the variance of the population.

Part correlation   Value that measures the strength of the relationship between a dependent and a single independent variable when the predictive effects of the other independent variables in the regression model are removed. The objective is to portray the *unique* predictive effect due to a single independent variable among a set of independent variables. Differs from the *partial correlation coefficient*, which is concerned with incremental predictive effect.

Partial correlation coefficient   Value that measures the strength of the relationship between the criterion or dependent variable and a single independent variable when the effects of the other independent variables in the model are held constant. For example, r_{Y,X_2,X_1} measures the variation in Y associated with X_2 when the effect of X_1 on both X_2 and Y is held constant. This value is used in sequential variable selection methods of regression model estimation to identify the independent variable with the greatest incremental predictive power beyond the independent variables already in the regression model.

Partial F (or t) values   The partial F-test is simply a statistical test for the additional contribution to prediction accuracy of a variable above that of the variables already in the equation. When a variable (X_a) is added to a regression equation after other variables are already in the equation, its contribution may be very small even though it has a high correlation with the dependent variable. The reason is that X_a is highly correlated with the variables already in the equation. The partial F value is calculated for all variables by simply pretending that each, in turn, is the last to enter the equation. It gives the additional contribution of each variable above all others in the equation. A low or insignificant partial F value for a variable not in the equation indicates its low or insignificant contribution to the model as already specified. A t value may be calculated instead of F values in all instances, with the t value being approximately the square root of the F value.

Partial regression plot   Graphical representation of the relationship between the dependent variable and a single independent variable. The scatterplot of points depicts the partial correlation between the two variables, with the effects of other independent variables held constant (see *partial correlation coefficient*). This portrayal is particularly helpful in assessing the form of the relationship (linear versus nonlinear) and the identification of *influential observations*.

Polynomial   Transformation of an independent variable to represent a curvilinear relationship with the dependent variable. By including a squared term (X^2), a single inflection point is estimated. A cubic term estimates a second inflection point. Additional terms of a higher power can also be estimated.

Power   Probability that a significant relationship will be found if it actually exists. Complements the more widely used significance level *alpha* (α).

Prediction error   Difference between the actual and predicted values of the dependent variable for each observation in the sample (see *residual*).

Predictor variable (X_n)   See *independent variable*.

PRESS statistic   Validation measure obtained by eliminating each observation one at a time and predicting this dependent value with the regression model estimated from the remaining observations.

Reference category   The omitted level of a nonmetric variable when a *dummy variable* is formed from the nonmetric variable.

Regression coefficient (b_n)   Numerical value of the **parameter** estimate directly associated with an independent variable; for example, in the model $Y = b_0 + b_1X_1$, the value b_1 is the regression coefficient for the variable X_1. The regression coefficient represents the amount of change in the dependent variable for a one-unit change in the independent variable. In the multiple predictor model (e.g., $Y = b_0 + b_1X_1 + b_2X_2$), the regression coefficients are partial coefficients because each takes into account not only the relationships between Y and X_1 and between Y and X_2, but also between X_1 and X_2. The coefficient is not limited in range, as it is based on both the degree of association and the scale units of the independent variable. For instance, two variables with the same association to Y would have different coefficients if one independent variable was measured on a 7-point scale and another was based on a 100-point scale.

Regression variate   Linear combination of weighted independent variables used collectively to predict the dependent variable.

Residual (*e* or *ε*)   Error in predicting our sample data. Seldom will our predictions be perfect. We assume that random error will occur, but we assume that this error is an estimate of the true random error in the population ($ε$), not just the error in prediction for our sample (*e*). We assume that the error in the population we are estimating is distributed with a mean of 0 and a constant (*homoscedastic*) variance.

Semipartial correlation   See *part correlation coefficient.*

Simple regression   Regression model with a single independent variable.

Singularity   The extreme case of *collinearity* or *multicollinearity* in which an independent variable is perfectly predicted (a correlation of ± 1.0) by one or more independent variables. Regression models cannot be estimated when a singularity exists. The researcher must omit one or more of the independent variables involved to remove the singularity.

Specification error   Error in predicting the dependent variable caused by excluding one or more relevant independent variables. This omission can bias the estimated coefficients of the included variables as well as decrease the overall predictive power of the regression model.

Standardization   Process whereby raw data are transformed into new measurement variables with a mean of 0 and a standard deviation of 1. When data are transformed in this manner, the b_0 term (the intercept) assumes a value of 0. When using standardized data, the regression coefficients are known as *beta coefficients*, which allow the researcher to compare directly the relative effect of each independent variable on the dependent variable.

Standard error   Expected distribution of an estimated *regression coefficient*. The standard error is similar to the standard deviation of original data values. It denotes the expected range of the coefficient across multiple samples of the data. This is useful in statistical tests of significance that test to see if the coefficient is significantly different from zero (i.e., whether the expected range of the coefficient contains the value of zero at a given level of confidence). The *t* value of a *regression coefficient* is the coefficient divided by its standard error.

Standard error of the estimate (SEE)   Measure of the variation in the predicted values that can be used to develop confidence intervals around any predicted value. It is similar to the standard deviation of a variable around its mean.

Statistical relationship   Relationship based on the correlation of one or more independent variables with the dependent variable. Measures of association, typically correlations, represent the degree of relationship because there is more than one value of the dependent variable for each value of the independent variable.

Stepwise estimation   Method of selecting variables for inclusion in the regression model that starts by selecting the best predictor of the dependent variable. Additional independent variables are selected in terms of the incremental explanatory power they can add to the regression model. Independent variables are added as long as their *partial correlation coefficients* are statistically significant. Independent variables may also be dropped if their predictive power drops to a nonsignificant level when another independent variable is added to the model.

Studentized residual   The most commonly used form of standardized *residual*. It differs from other methods in how it calculates the standard deviation used in *standardization*. To minimize the effect of a single outlier, the residual standard deviation for observation i is computed from regression estimates omitting the ith observation in the calculation of the regression estimates.

Sum of squared errors (SSE)   Sum of the squared *prediction errors (residuals)* across all observations. It is used to denote the variance in the dependent variables not yet accounted for by the regression model. If no independent variables are used for prediction, this becomes the squared errors using the mean as the predicted value and thus equals the *total sum of squares.*

Sum of squares regression (SSR)   Sum of the squared differences between the mean and predicted values of the dependent variable for all observations. This represents the amount of improvement in explanation of the dependent variable attributable to the independent variable(s).

Tolerance   Commonly used measure of collinearity and multicollinearity. The tolerance of variable i (TOL_i) is $1 - R^{2*}_i$, where R^{2*}_i is the coefficient of determination for the prediction of variable i by the other independent variables. As the tolerance value grows smaller, the variable is more highly predicted by the other independent variables (collinearity).

Total sum of squares (TSS)   Total amount of variation that exists to be explained by the independent variables. This "baseline" is calculated by summing the squared differences between the mean and actual values for the dependent variable across all observations.

Transformation   A variable may have an undesirable characteristic, such as non-normality, that detracts from the ability of the correlation coefficient to represent the relationship between it and another variable. A transformation, such as taking the logarithm or square root of the variable, creates a new variable and eliminates the undesirable characteristic, allowing for a better measure of the relationship. Transformations may be applied to either the dependent or independent variables, or both. The need and specific type of transformation may be based on theoretical reasons (such as transforming a known nonlinear relationship) or empirical reasons (identified through graphical or statistical means).

Variance inflation factor (VIF)   Indicator of the effect that the other independent variables have on the standard error of a regression coefficient. The variance inflation factor is directly related to the *tolerance* value ($VIF_i = 1/TOL_i$). Large VIF values also indicate a high degree of *collinearity* or *multicollinearity* among the independent variables.

What Is Multiple Regression Analysis?

Multiple regression analysis is a statistical technique that can be used to analyze the relationship between a single **dependent (criterion) variable** and several **independent (predictor) variables.** The objective of multiple regression analysis is to use the independent variables whose values are known to predict the single dependent value selected by the researcher. Each independent variable is weighted by the regression analysis procedure to ensure maximal prediction from the set of independent variables. The weights denote the relative contribution of the independent variables to the overall prediction and facilitate interpretation as to the influence of each variable in making the prediction, although correlation among the independent variables complicates the interpretative process. The set of weighted independent variables forms the **regression variate,** a linear combination of the independent variables that best predicts the dependent variable (chapter 1 contains a more detailed explanation of the variate). The regression variate, also referred to as the regression equation or regression model, is the most widely known example of a variate among the multivariate techniques.

As noted in chapter 1, multiple regression analysis is a dependence technique. Thus, to use it you must be able to divide the variables into dependent and independent variables. Regression analysis is also a statistical tool that should be used only when both the dependent and independent variables are metric. However, under certain circumstances it is possible to include nonmetric data either as independent variables (by transforming either ordinal or nominal data with dummy-variable coding) or the dependent variable (by the use of a binary measure in the specialized technique of logistic regression, see chapter 5). In summary, to apply multiple regression analysis: (1) the data must be metric or appropriately transformed, and (2) before deriving the regression equation, the researcher must decide which variable is to be dependent and which remaining variables will be independent.

An Example of Simple and Multiple Regression

The objective of regression analysis is to predict a single dependent variable from the knowledge of one or more independent variables. When the problem involves a single independent variable, the statistical technique is called **simple regression.** When the problem involves two or more independent variables, it is termed **multiple regression.** To illustrate the basic principles involved, results from a small study of eight families regarding their credit card usage is provided. The purpose of the study was to determine which factors affected the number of credit cards used. Three potential factors were identified (family size, family income, and number of automobiles owned), and data were collected from each of the eight families (see Table 4.1). In the terminology of regression analysis, the dependent variable (Y) is the number of credit cards used and the three independent

TABLE 4.1 Credit Card Usage Survey Results

Family ID	Number of Credit Cards Used Y	Family Size V_1	Family Income ($000) V_2	Number of Automobiles Owned V_3
1	4	2	14	1
2	6	2	16	2
3	6	4	14	2
4	7	4	17	1
5	8	5	18	3
6	7	5	21	2
7	8	6	17	1
8	10	6	25	2

variables (V_1, V_2, and V_3) are family size, family income, and number of automobiles owned, respectively.

The discussion of this example is divided into three parts to show how regression estimates the relationship between independent and dependent variables. The topics covered are (1) setting a baseline prediction without an independent variable, using only the average; (2) prediction using a single independent variable—simple regression; and (3) prediction using several independent variables—multiple regression.

Setting a Baseline: Prediction without an Independent Variable

Before estimating the first regression equation, let us start by calculating the baseline against which we will compare the predictive ability of our regression models. The baseline should represent our best prediction without the use of any independent variables. We could use any number of options (e.g., perfect prediction, a prespecified value, or one of the measures of central tendency, such as mean, median, or mode), but the baseline predictor used in regression is the simple mean of the dependent variable, which has several desirable properties. In our example, the arithmetic average (the mean) of the number of credit cards used is seven. Our baseline prediction can then be stated as "The predicted number of credit cards used by a family is seven." We can also write this as a regression equation as follows:

Predicted number of credit cards = Average number of credit cards

or

$$\hat{Y} = \bar{y}$$

But the researcher must still answer one question: How accurate is the prediction? Because the mean will not perfectly predict each value of the dependent variable, we must create some way to assess predictive accuracy that can be used with both the baseline prediction and the regression models we create. The customary way to assess the accuracy of any prediction is to examine the errors in predicting the dependent variable. For example, with our baseline prediction of each family using seven credit cards, we overestimate the number of credit cards used by family 1 by three (see Table 4.2). Thus the error is +3. If this procedure were followed for each family, some estimates would be too high, others would be too low, and still others might be exactly correct. Although we might expect to obtain a useful measure of prediction accuracy by simply adding the errors, this would not be useful because the errors from using the mean value always sum to zero. Therefore, the simple sum of errors would never change, no matter how well or poorly we predicted the dependent variable when using the mean. To overcome this problem, we square each error and then add the results together. This total, referred to as the **sum of squared errors (SSE),** provides a measure of prediction accuracy that will vary according to the amount of **prediction errors.** The objective is to obtain the smallest possible sum of squared errors as our measure of prediction accuracy.

We choose the arithmetic average (mean) because it will always produce a smaller sum of squared errors than any other measure of central tendency, including the median, mode, any other single data value, or any other more sophisticated statistical measure. (Interested readers are encouraged to try to find a

TABLE 4.2 Baseline Prediction Using the Mean of the Dependent Variable

Regression Variate: $Y = y$
Prediction Equation: $Y = 7$

Family ID	Number of Credit Cards Used	Baseline Prediction[a]	Prediction Error[b]	Prediction Error Squared
1	4	7	−3	9
2	6	7	−1	1
3	6	7	−1	1
4	7	7	0	0
5	8	7	+1	1
6	7	7	0	0
7	8	7	+1	1
8	10	7	+3	9
Total	56		0	22

[a]Average number of credit cards used = 56 ÷ 8 = 7.
[b]Prediction error refers to the actual value of the dependent variable minus the predicted value.

better predicting value than the mean.) Therefore, for our survey of eight families, using the average as our baseline prediction gives us the best single predictor of the number of credit cards, with a sum of squared errors of 22 (see Table 4.2). In our discussion of simple and multiple regression, we use prediction by the mean as a baseline for comparison because it represents the best possible prediction without using any independent variables.

Prediction Using a Single Independent Variable—Simple Regression

As researchers, we are always interested in improving our predictions. In the preceding section, we learned that the average is the best predictor if we do not use any independent variables. But in our survey of eight families we also collected information on measures that could act as independent variables. Let us determine whether knowledge of one of these independent variables will help our predictions in what is referred to as *simple regression*.

Simple regression is a procedure for predicting data (just as the average predicts data) that uses the same rule—minimizing the sum of squared errors of prediction. We know that without using family size we can best describe the number of credit cards used as the mean value, seven. The researcher's objective for simple regression is to find an independent variable that will improve on the baseline prediction.

The Role of the Correlation Coefficient

Using our survey information, we can try to improve our predictions by reducing the prediction errors. To do so, the prediction errors in the number of credit cards used must be associated (correlated) with one of the potential independent variables (V_1, V_2, or V_3). The concept of association, represented by the **correlation coefficient (r),** is fundamental to regression analysis by describing the relationship between two variables. Two variables are said to be correlated if changes in one variable are associated with changes in the other variable. In this way, as one variable changes, we know how the other variable is changing. If V_i was correlated with credit card usage, we can use this relationship to predict the number of credit cards as follows:

Predicted number = Change in number of × Value of V_i
of credit cards   credit cards used
                           associated with unit
                           change in V_i

or

$$\hat{Y} = b_1 V_i$$

An illustration of the procedure is shown in Table 4.3 for some hypothetical data with a single independent variable X_1. If we find that as X_1 increases by one unit, the dependent variable increases (on the average) by two, we could then make predictions for each value of the independent variable. For example, when X_1 had a value of 4, we would predict a value of 8 (see part A, Table 4.3). Thus, the predicted value is always two times the value of X_1 ($2X_1$). However, we often find that the prediction is improved by adding a constant value. In part A of Table 4.3 we can see that the simple prediction of two times X_1 is wrong by two in every case. Therefore, changing our description to add a constant of two to each prediction gives us perfect predictions in all cases (see part B, Table 4.3). We will see that when estimating a regression equation, it is usually beneficial to include a constant, which is termed the **intercept.**

TABLE 4.3 Improving Prediction Accuracy by Adding an Intercept in a Regression Equation

PART A: PREDICTION WITHOUT THE INTERCEPT
Prediction Equation: $Y = 2X_1$

Value of X_1	Dependent Variable	Prediction	Prediction Error
1	4	2	2
2	6	4	2
3	8	6	2
4	10	8	2
5	12	10	2

PART B: PREDICTION WITH AN INTERCEPT OF 2.0
Prediction Equation: $Y = 2.0 + 2X_1$

Value of X_1	Dependent Variable	Prediction	Prediction Error
1	4	2	0
2	6	6	0
3	8	8	0
4	10	10	0
5	12	12	0

Specifying the Simple Regression Equation

We can select the "best" independent variable in our study of credit card usage based on the correlation coefficients because the higher the correlation coefficient, the stronger the relationship and hence the greater the predictive accuracy. Table 4.4 contains a matrix of correlations between the dependent (Y) variable and independent (V_1, V_2, or V_3) variables. Looking down the first column, we can see that V_1, family size, has the highest correlation with the dependent variable and is thus the best candidate for our first simple regression. The correlation matrix also contains the correlations among the independent variables, which we will see is very important in multiple regression (two or more independent variables).

We can now estimate our first simple regression model for the sample of eight families and see how well the description fits our data. The regression model can be stated as follows:

Predicted number = Intercept + Change in number of × Family size
of credit cards used credit cards used
associated with a unit
change in family size

or

$$\hat{Y} = b_0 + b_1V_1$$

In the regression equation, we represent the intercept as b_0, and the term b_1 is called a **regression coefficient,** denoting the estimated change in the dependent variable for a unit change of the independent variable. The **prediction error,** the difference between the actual and predicted values of the dependent variable, is termed the **residual** (e). Regression analysis also allows for the statistical testing of the intercept and regression coefficient(s) to determine if they are significantly different from zero (i.e., that they do have an impact different from zero).

Using a mathematical procedure known as **least squares** [8, 11, 15], we can estimate the values of b_0 and b_1 such that the sum of the squared errors of prediction is minimized. For this example, the appropriate values are a constant (b_0) of 2.87 and a regression coefficient (b_1) of .97 for family size. The equation indicates that for each additional family member, the credit card holdings are higher on average by .97. The constant 2.87 can be interpreted only within the range of values for the independent variable. In this case, a family size of zero is not possible, so the intercept alone does not have practical meaning. However, this does not invalidate its use, as it aids in the prediction of credit card usage for each possible family size (in our example from 1 to 5). In instances for which the independent variables can take on zero values, the intercept has a direct interpretation. For some special situations where the specific relationship is known to pass through the origin, the intercept term may be suppressed (termed "regression through the

TABLE 4.4 Correlation Matrix for the Credit Card Usage Study

Variable	Y	V_1	V_2	V_3
Y Number of credit cards used	1.000			
V_1 Family size	.866	1.000		
V_2 Family income	.829	.673	1.000	
V_3 Number of automobiles	.342	.192	.301	1.000

TABLE 4.5 Simple Regression Results Using Family Size as the Independent Variable

Regression Variate: $Y = b_0 + b_1 V_1$
Prediction Equation: $Y = 2.87 + .97 V_1$

Family ID	Number of Credit Cards Used	Family Size (V_1)	Simple Regression Prediction	Prediction Error	Prediction Error Squared
1	4	2	4.81	−.81	.66
2	6	2	4.81	1.19	1.42
3	6	4	6.75	−.75	.56
4	7	4	6.75	.25	.06
5	8	5	7.72	.28	.08
6	7	5	7.72	−.72	.52
7	8	6	8.69	−.69	.48
8	10	6	8.69	1.31	1.72
Total					5.50

origin"). In these cases, the interpretation of the residuals and the regression coefficients changes slightly. The simple regression equation and the resulting predictions and residuals for each of the eight families are shown in Table 4.5.

Because we have used the same criterion (minimizing the sum of squared errors or **least squares**), we can determine whether our knowledge of family size has helped us better predict credit card holdings by comparing the simple regression prediction with the baseline prediction. The sum of squared errors using the average (the baseline) was 22; using our new procedure with a single independent variable, the sum of squared errors decreases to 5.50 (see Table 4.5). By using the least squares procedure and a single independent variable, we see that our new approach, simple regression, is markedly better at predicting than using just the average.

Establishing a Confidence Interval for the Prediction

Because we did not achieve perfect prediction of the dependent variable, we would also like to estimate the range of predicted values that we might expect, rather than relying just on the single (point) estimate. The point estimate is our best estimate of the dependent variable and can be shown to be the average prediction for any given value of the independent variable. From this point estimate, we can also calculate the range of predicted values based on a measure of the prediction errors we expect to make. Known as the **standard error of the estimate (SEE),** this measure can be defined simply as the standard deviation of the prediction errors. We can construct a confidence interval for a variable about its mean value by adding (plus and minus) a certain number of standard deviations. For example, adding ±1.96 standard deviations to the mean defines a range for large samples that includes 95 percent of the values of a variable.

We can follow a similar method for the predictions from a regression model. Using the point estimate, we can add (plus and minus) a certain number of standard errors of the estimate (depending on the confidence level desired and sample size) to establish the upper and lower bounds for our predictions made with any independent variable(s). The standard error of the estimate (SEE) is calculated by

$$SEE = \sqrt{\frac{\text{Sum of squared errors}}{\text{Sample size} - 2}}$$

The number of *SEE*s to use in deriving the confidence interval is determined by the level of significance (α) and the sample size (N), which give a t value. The confidence interval is then calculated with the lower limit being equal to the predicted value minus (*SEE* $\times$ t value) and the upper limit calculated as the predicted value plus (*SEE* $\times$ t value). For our simple regression model, *SEE* = .957 (the square root of the value of 5.50 divided by 6). The confidence interval for the predictions is constructed by selecting the number of standard errors to add (plus and minus) by looking in a table for the t distribution and selecting the value for a given confidence level and sample size. In our example, the t value for a 95 percent confidence level with 6 degrees of freedom (sample size minus the number of coefficients, or 8 − 2 = 6) is 2.447. The amount added (plus and minus) to the predicted value is then (.957 × 2.447), or 2.34. If we substitute the average family size (4.25) into the regression equation, the predicted value is 6.99 (it differs from the average of seven only because of rounding). The expected range of credit cards becomes 4.65 (6.99 − 2.34) to 9.33 (6.99 + 2.34). For a more detailed discussion of these confidence intervals, see Neter et al. [11].

Assessing Prediction Accuracy

If the sum of squared errors (SSE) represents a measure of our prediction errors, we should also be able to determine a measure of our prediction success, which we can term the **sum of squares regression (SSR).** Together, these two measures should equal the **total sum of squares (TSS),** the same value as our baseline prediction. As the researcher adds independent variables, the total sum of squares can now be divided into (1) the sum of squares predicted by the independent variable(s), which is the sum of squares regression, and (2) the sum of squared errors:

$$\sum_{i=1}^{n}(y_i - \bar{y})^2 \quad = \quad \sum_{i=1}^{n}(y_i - \hat{y}_i)^2 \quad + \quad \sum_{i=1}^{n}(\hat{y}_i - \bar{y})^2$$

$$TSS \quad = \quad SSE \quad + \quad SSR$$

Total sum of squares = Sum of squared errors + Sum of squares regression

where

$\bar{y}$ = average of all observations

y_i = value of individual observation i

$\hat{y}$ = predicted value of observation i

We can use this division of the total sum of squares to approximate how well the regression variate describes family holdings of credit cards. The average number of credit cards held by our sampled families is our best estimate of the number held by any family. We know that this is not an extremely accurate estimate, but it is the best prediction available without using any other variables. The baseline prediction using the mean was measured by calculating the squared sum of errors in prediction (sum of squares = 22). Now that we have fitted a regression model using family size, does it explain the variation better than the average? We know it is somewhat better because the sum of squared errors is now 5.50. We can look at how well our model predicts by examining this improvement.

Sum of squared errors (baseline prediction)	SS_{Total} or SST	22.0
− Sum of squares error (simple regression)	SS_{Error} or SSE	−5.5
Sum of squares explained (simple regression)	$SS_{Regression}$ or SSR	16.5

Therefore, we explained 16.5 squared errors by changing from the average to the simple regression model using family size. This is an improvement of 75 percent ($16.5 \div 22 = .75$) over the baseline. Another way to express this level of prediction accuracy is the **coefficient of determination (R^2),** the ratio of the sum of squares regression to the total sum of squares, as shown in the following equation:

$$\text{Coefficient of determination } (R^2) = \frac{\text{Sum of squares regression}}{\text{Total sum of squares}}$$

If the regression model using family size perfectly predicted all families holding credit cards, $R^2 = 1.0$. If using family size gave no better predictions than using the average, $R^2 = 0$. When the regression equation contains more than one independent variable, the R^2 value represents the combined effect of the entire variate in prediction. The R^2 value is simply the squared correlation of the actual and predicted values.

When the coefficient of correlation (r) is used to assess the relationship between dependent and independent variables, the sign of the correlation coefficient ($+r$, $−r$) denotes the slope of the regression line. However, the "strength" of the relationship is represented by R^2, which is, of course, always positive. In our example, $R^2 = .75$, indicating that 75 percent of the variation in the dependent variable is explained by the independent variable. When discussions mention the variation of the dependent variable, they are referring to this total sum of squares that the regression analysis attempts to predict with one or more independent variables.

Prediction Using Several Independent Variables—Multiple Regression

We previously demonstrated how simple regression helped improve our prediction of credit card usage. By using data on family size, we predicted the number of credit cards a family would use much more accurately than we could by simply using the arithmetic average. This result raises the question of whether we could improve our prediction even further by using additional data obtained from the families. Would our prediction be improved if we used not only data on family size, but data on another variable, perhaps family income or number of automobiles owned by the family?

The Impact of Multicollinearity

The ability of an additional independent variable to improve the prediction of the dependent variable is related not only to its correlation to the dependent variable, but also to the correlation(s) of the additional independent variable to the independent variable(s) already in the regression equation. **Collinearity** is the association, measured as the correlation, between two independent variables. **Multicollinearity** refers to the correlation among three or more independent variables (evidenced when one is regressed against the others). Although there is a precise distinction in statistical terms, it is rather common practice to use the terms interchangeably.

The impact of multicollinearity is to reduce any single independent variable's predictive power by the extent to which it is associated with the other independent variables. As collinearity increases, the unique variance explained by each independent variable decreases and the shared prediction percentage rises. Because this shared prediction can count only once, the overall prediction increases much more slowly as independent variables with high multicollinearity are added. To maximize the prediction from a given number of independent variables, the researcher should look for independent variables that have low multicollinearity with the other independent variables but also have high correlations with the dependent variable. We revisit the issues of collinearity and multicollinearity in later sections to discuss their implications for the selection of independent variables and the interpretation of the regression variate.

The Multiple Regression Equation

To improve further our prediction of credit card holdings, let us use additional data obtained from our eight families. The second independent variable to include in the regression model is family income (V_2), which has the next highest correlation with the dependent variable. Although V_2 does have a fair degree of correlation with V_1 already in the equation, it is still the next best variable to enter because V_3 has a much lower correlation with the dependent variable. We simply expand our simple regression model to include two independent variables as follows:

$$\text{Predicted number of credit cards used} = b_0 + b_1V_1 + b_2V_2 + e$$

where

b_0 = constant number of credit cards independent of family size and income

b_1 = change in credit card usage associated with unit change in family size

b_2 = change in credit card usage associated with unit change in family income

V_1 = family size

V_2 = family income

The multiple regression model with two independent variables, when estimated with the least squares procedure, has a constant of .482 with regression coefficients of .63 and .216 for V_1 and V_2, respectively. We can again find our residual by predicting Y and subtracting the prediction from the actual value. We then square the resulting prediction error, as in Table 4.6 (p. 158). The sum of squared errors for the multiple regression model with family size and family income is 3.04. This can be compared to the simple regression model value of 5.50 (Table 4.5), which uses only family size for prediction.

When family income is added to the regression analysis, R^2 also increases to .86.

$$R^2_{(\text{family size + family income})} = \frac{22.0 - 3.04}{22.0} = \frac{18.96}{22.0} = .86$$

This means that the inclusion of family income in the regression analysis increases the prediction by 11 percent (.86 − .75), all due to the unique incremental predictive power of family income.

Adding a Third Independent Variable

We have seen an increase in prediction accuracy gained in moving from the simple to multiple regression equation, but we must also note that at some point the addition of independent variables will become less advantageous and even in

TABLE 4.6 Multiple Regression Results Using Family Size and Family Income as Independent Variables

Regression Variate: $Y = b_0 + b_1V_1 + b_2V_2$
Prediction Equation: $Y = .482 + .63V_1 + .216V_2$

Family ID	Number of Credit Cards Used	Family Size (V_1)	Family Income (V_2)	Multiple Regression Prediction	Prediction Error	Prediction Error Squared
1	4	2	14	4.76	−.76	.58
2	6	2	16	5.20	.80	.64
3	6	4	14	6.03	−.03	.00
4	7	4	17	6.68	.32	.10
5	8	5	18	7.53	.47	.22
6	7	5	21	8.18	−1.18	1.39
7	8	6	17	7.95	.05	.00
8	10	6	25	9.67	.33	.11
Total						3.04

some instances counterproductive. In our survey of credit card usage, we have one more possible addition to the multiple regression equation, the number of automobiles owned (V_3). If we now specify the regression equation to include all three independent variables, we can see some improvement in the regression equation, but not nearly of the magnitude seen earlier. The R^2 value increases to .87, only a .01 increase over the previous multiple regression model. Moreover, as we discuss in a later section, the regression coefficient for V_3 is not statistically significant. Therefore, in this instance, the researcher is best served by employing the multiple regression model with two independent variables (family size and income) and not employing the third independent variable (number of automobiles owned) in making predictions.

Summary

Regression analysis is a simple and straightforward dependence technique that can provide both prediction and explanation to the researcher. The prior example has illustrated the basic concepts and procedures underlying regression analysis in an attempt to develop an understanding of the rationale and issues of this procedure in its most basic form. The following sections detail these issues in much more detail and provide a decision process for applying regression analysis to any appropriate research problem.

A Decision Process for Multiple Regression Analysis

In the previous sections we discussed examples of simple regression and multiple regression. In those discussions, many factors affected our ability to find the best regression model. To this point, however, we have examined these issues only in simple terms, with little regard to how they combine in an overall approach to multiple regression analysis. In the following sections, the six-stage model-building process introduced in chapter 1 will be used as a framework for discussing the factors that impact the creation, estimation, interpretation, and validation of a regression analysis. The process begins with specifying the objectives of the regression analysis, including the selection of the dependent and independent variables. The researcher then proceeds to design the regression analysis, considering such factors as sample size and the need for variable transformations. With the regression model formulated, the assumptions underlying regression analysis are first tested for the individual variables. If all assumptions are met, then the model is estimated. Once results are obtained, diagnostic analyses are performed to ensure that the overall model meets the regression assumptions and that no observations have undue influence on the results. The next stage is the interpretation of the regression variate; it examines the role played by each independent variable in the prediction of the dependent measure. Finally, the results are validated to ensure generalizability to the population. Figures 4.1 (p. 160) and 4.6 (p. 177) represent stages 1–3 and 4–6 respectively in providing a graphical representation of the model-building process for multiple regression, and the following sections discuss each step in detail.

Stage 1: Objectives of Multiple Regression

Multiple regression analysis, a form of general linear modeling, is a multivariate statistical technique used to examine the relationship between a single dependent variable and a set of independent variables. The necessary starting point in multiple regression, as with all multivariate statistical techniques, is the research problem. The flexibility and adaptability of multiple regression allows for its use with almost any dependence relationship. In selecting suitable applications of multiple regression, the researcher must consider three primary issues: (1) the appropriateness of the research problem, (2) specification of a statistical relationship, and (3) selection of the dependent and independent variables.

Research Problems Appropriate for Multiple Regression

Multiple regression is by far the most widely used multivariate technique of those examined in this text. With its broad applicability, multiple regression has been used for many purposes. The ever-widening applications of multiple regression fall into two broad classes of research problems: prediction and explanation. These research problems are not mutually exclusive, and an application of multiple regression analysis can address either or both types of research problem.

Prediction with Multiple Regression

One fundamental purpose of multiple regression is to predict the dependent variable with a set of independent variables. In doing so, multiple regression fulfills one of two objectives. The first objective is to maximize the overall predictive power of the independent variables as represented in the variate. This linear combination of independent variables is formed to be the optimal predictor of the dependent measure. Multiple regression provides an objective means of assessing the predictive power of a set of independent variables. In applications focused on

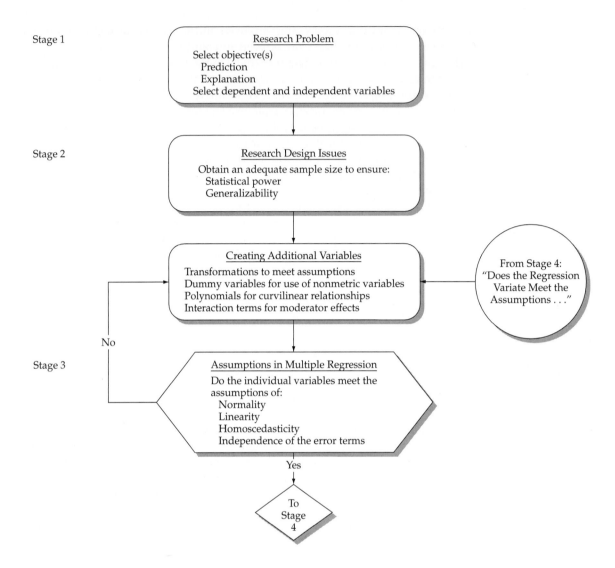

FIGURE 4.1 Stages 1–3 in the Multiple Regression Decision Diagram

this objective, the researcher is primarily interested in achieving maximum prediction. Multiple regression provides many options in both the form and the specification of the independent variables that may modify the variate to increase its predictive power. Prediction often is maximized at the expense of interpretation. One example is a variant of regression, time series analysis, in which the sole purpose is prediction and the interpretation of results is useful only as a means of increasing predictive accuracy. In other situations, predictive accuracy is crucial to ensuring the validity of the set of independent variables, thus allowing for the subsequent interpretation of the variate. Measures of predictive accuracy are formed and statistical tests regarding the significance of the predictive power can be made. In all instances, whether or not prediction is the primary focus, the regression analysis must achieve acceptable levels of predictive accuracy to justify its application. The researcher must ensure that both statistical and practical significance are considered (see the discussion of stage 4).

Multiple regression can also meet a second objective of comparing two or more sets of independent variables to ascertain the predictive power of each variate. Illustrative of a confirmatory approach to modeling, this use of multiple regression is concerned with the comparison of results across two or more alternative or competing models. The primary focus of this type of analysis is the relative predictive power among models, although in any situation the prediction of the selected model must demonstrate both statistical and practical significance.

Explanation with Multiple Regression

Multiple regression also provides a means of objectively assessing the degree and character of the relationship between dependent and independent variables by forming the variate of independent variables. The independent variables, in addition to their collective prediction of the dependent variable, may also be considered for their individual contribution to the variate and its predictions. Interpretation of the variate may rely on any of three perspectives: the importance of the independent variables, the types of relationships found, or the interrelationships among the independent variables.

The most direct interpretation of the regression variate is a determination of the relative importance of each independent variable in the prediction of the dependent measure. In all applications, the selection of independent variables should be based on their theoretical relationships to the dependent variable. Regression analysis then provides a means of objectively assessing the magnitude and direction (positive or negative) of each independent variable's relationship. The character of multiple regression that differentiates it from its univariate counterparts is the simultaneous assessment of relationships between each independent variable and the dependent measure. In making this simultaneous assessment, the relative importance of each independent variable is determined.

In addition to assessing the importance of each variable, multiple regression also affords the researcher a means of assessing the nature of the relationships between the independent variables and the dependent variable. The assumed relationship is a linear association based on the correlations among the independent variables and the dependent measure. Transformations or additional variables are also available to assess whether other types of relationships exist, particularly curvilinear relationships. This flexibility ensures that the researcher may examine the true nature of the relationship beyond the assumed linear relationship.

Finally, multiple regression provides insight into the relationships among independent variables in their prediction of the dependent measure. These interrelationships are important for two reasons. First, correlation among the independent variables may make some variables redundant in the predictive effort. As such, they are not needed to produce the optimal prediction. This does not reflect their individual relationships with the dependent variable but instead indicates that in a multivariate context, they are not needed if another set of independent variables explaining this variance is employed. The researcher must guard against determining the importance of independent variables based solely on the derived variate, because relationships among the independent variables may "mask" relationships that are not needed for predictive purposes but represent substantive findings nonetheless. The interrelationships among variables can extend not only to their predictive power but also to interrelationships among their estimated effects. This is best seen when the effect of one independent variable is contingent on another independent variable. Multiple regression provides diagnostic analyses that can determine whether such effects exist based on empirical or theoretical rationale. Indications of a high degree of interrelationships (multicollinearity) among the independent variables may suggest the use of summated scales, as discussed in chapter 2.

Specifying a Statistical Relationship

Multiple regression is appropriate when the researcher is interested in a statistical, not a functional, relationship. For example, let us examine the following relationship:

Total cost = Variable cost + Fixed cost

If the variable cost is $2 per unit, the fixed cost is $500, and we produce 100 units, we assume that the total cost will be exactly $700 and that any deviation from $700 is caused by our inability to measure cost since the relationship between costs is fixed. This is called a *functional relationship* because we expect there will be no error in our prediction.

But in our earlier example dealing with sample data representing human behavior, we assumed that our description of credit card usage was only approximate and not a perfect prediction. It was defined as a **statistical relationship** because there will always be some random component to the relationship being examined. In a statistical relationship, more than one value of the dependent value will usually be observed for any value of an independent variable. The dependent variable is assumed to be a random variable, and for a given independent variable we can only hope to estimate the average value of the dependent variable associated with it. For example, in our simple regression example, we found two families with two members, two with four members, and so on, who had different numbers of credit cards. The two families with four members held an average of 6.5 credit cards, and our prediction was 6.75. Our prediction is not as accurate as we would like, but it is better than just using the average of 7 credit cards. The error is assumed to be the result of random behavior among credit card holders.

In summary, a functional relationship calculates an exact value, whereas a statistical relationship estimates an average value. Throughout this book, we are concerned with statistical relationships. Both of these relationships are displayed in Figure 4.2.

Selection of Dependent and Independent Variables

The ultimate "success" of any multivariate technique, including multiple regression, starts with the selection of the variables to be used in the analysis. Because multiple regression is a dependence technique, the researcher must specify which variable is the dependent variable and which variables are the independent variables. The selection of both types of variables should be based principally on conceptual or theoretical grounds. Chapters 1 and 11 discuss the role of theory in multivariate analysis, and those issues strongly apply to multiple regression. The researcher must perform the fundamental decisions of variable selection, even though many options and program features are available to assist in model esti-

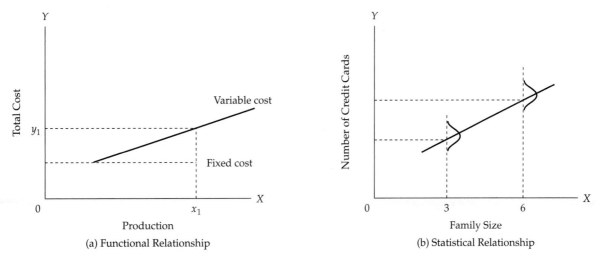

FIGURE 4.2 Comparison of Functional and Statistical Relationships

mation. If the researcher does not exert judgment during variable selection, but instead (1) selects variables indiscriminately or (2) allows for the selection of an independent variable to be based solely on empirical bases, several of the basic tenets of model development will be violated.

The selection of a dependent variable is many times dictated by the research problem. But in all instances, the researcher must be aware of the **measurement error,** especially in the dependent variable. Measurement error refers to the degree that the variable is an accurate and consistent measure of the concept being studied. If the variable used as the dependent measure has substantial measurement error, then even the best independent variables may be unable to achieve acceptable levels of predictive accuracy. Measurement error can come from several sources (see chapter 1 for a more detailed discussion). Measurement error that is problematic may be addressed through the use of summated scales, as discussed in chapters 1 and 3. The researcher must always be concerned with obtaining the best dependent and independent measures, based on both conceptual and empirical factors.

The most problematic issue in independent variable selection is **specification error,** which concerns the inclusion of irrelevant variables or the omission of relevant variables from the set of independent variables. Although the inclusion of irrelevant variables does not bias the results for the other independent variables, it does have some impact on them. First, it reduces model parsimony, which may be critical in the interpretation of the results. Second, the additional variables may mask or replace the effects of more useful variables, especially if some sequential form of model estimation is used (see the discussion of stage 4 for more detail). Finally, the additional variables may make the testing of statistical significance of the independent variables less precise and reduce the statistical and practical significance of the analysis.

Given the problems associated with adding irrelevant variables, should the researcher be concerned with excluding relevant variables? The answer is definitely yes, because the exclusion of relevant variables can seriously bias the results and negatively affect any interpretation of them. In the simplest case, the omitted variables are uncorrelated with the included variables, and the only effect is to reduce the overall predictive accuracy of the analysis. But when correlation exists between the included and omitted variables, the effects of the included variables become biased to the extent that they are correlated with the omitted variables. The greater the correlation, the greater the bias. The estimated effects for the included variables now represent not only their actual effects but also the effects that the included variables share with the omitted variables. This can lead to serious problems in model interpretation and the assessment of statistical and managerial significance.

The researcher must be careful in the selection of the variables to avoid both types of specification error. Perhaps most troublesome is the omission of relevant variables, as the variables' effect cannot be assessed without their inclusion. This heightens the need for theoretical and practical support for all variables included or excluded in a multiple regression analysis.

Measurement error also affects the independent variables by reducing their predictive power as measurement error increases. Multiple regression has no direct means of correcting for known levels of measurement error for the dependent or independent variables. If the researcher suspects that measurement error may be problematic, structural equation modeling (chapter 11) should be examined as a means of accommodating measurement error in estimating the effects of the independent variables.

Stage 2: Research Design of a Multiple Regression Analysis

In the design of a multiple regression analysis, the researcher must consider issues such as sample size, the nature of the independent variables, and the possible creation of new variables to represent special relationships between the dependent and independent variables. In doing so, the criteria of statistical and practical significance must always be maintained. The ability of multiple regression to address many types of research questions is greatly impacted by the research design issues to be discussed.

Sample Size

The sample size used in multiple regression is perhaps the most influential single element under the control of the researcher in designing the analysis. The effects of sample size are seen most directly in the statistical power of the significance testing and the generalizability of the result. Both issues are addressed in the following sections.

Statistical Power and Sample Size

The size of the sample has a direct impact on the appropriateness and the statistical power of multiple regression. Small samples, usually characterized as having fewer than 20 observations, are appropriate for analysis only by simple regression with a single independent variable. Even in these situations, only very strong relationships can be detected with any degree of certainty. Likewise, very large samples of 1,000 observations or more make the statistical significance tests overly sensitive, often indicating that almost any relationship is statistically significant. With very large samples the researcher must ensure that the criterion of practical significance is met along with statistical significance.

Power in multiple regression refers to the probability of detecting as statistically significant a specific level of R^2 or a regression coefficient at a specified significance level for a specific sample size (see chapter 1 for a more detailed discussion). Sample size has a direct and sizable impact on power. Table 4.7 illustrates the interplay among the sample size, the significance level (α) chosen, and the number of independent variables in detecting a significant R^2. The table values are the minimum R^2 that the specified sample size will detect as statistically significant at the specified α level with a probability (power) of .80. For example, if the researcher employs five independent variables, specifies a .05 significance level, and is satisfied with detecting the R^2 80 percent of the time it occurs (corresponding to a power of .80), a sample of 50 respondents will detect R^2 values of 23 percent and greater. If the sample is increased to 100 respondents, then R^2 values of 12 percent and above will be detected. But if the 50 respondents are all that are available, and the researcher wants a .01 significance level, the analysis will detect R^2 values only in excess of 29 percent. The researcher should always consider the role of sample size in significance testing before the actual collection of the data. If weaker relationships are expected, the researcher can make informed judgments as to the necessary sample size to reasonably detect the relationships, if they exist. For example, Table 4.7 demonstrates that sample sizes of 100 will detect fairly small R^2 values (10 percent to 15 percent) with up to 10 independent variables and a significance level of .05. However, if the sample size falls to 50 observations in these situations, the minimum R^2 that can be detected doubles. The researcher must always be aware of the anticipated power of any proposed multiple regression analysis and understand the elements of the research design that can be changed to meet the requirements for an acceptable analysis [9].

The researcher can also determine the sample size needed to detect effects for individual independent variables given the expected effect size (correlation), the α level, and the power desired. The possible computations are too numerous for presentation in this discussion, and the interested reader is referred to texts dealing with power analyses [5] or to a computer program to calculate sample size or power for a given situation [3].

TABLE 4.7 Minimum R^2 that Can Be Found Statistically Significant with a Power of .80 for Varying Numbers of Independent Variables and Sample Sizes

Sample Size	Significance Level (α) = .01 No. of Independent Variables				Significance Level (α) = .05 No. of Independent Variables			
	2	5	10	20	2	5	10	20
20	45	56	71	NA	39	48	64	NA
50	23	29	36	49	19	23	29	42
100	13	16	20	26	10	12	15	21
250	5	7	8	11	4	5	6	8
500	3	3	4	6	3	4	5	9
1,000	1	2	2	3	1	1	2	2

NA = not applicable.

Generalizability and Sample Size

In addition to its role in determining statistical power, sample size also affects the generalizability of the results by the ratio of observations to independent variables. A general rule is that the ratio should never fall below 5 to 1, meaning that there should be five observations for each independent variable in the variate. As this ratio falls below 5 to 1, the researcher encounters the risk of "overfitting" the variate to the sample, making the results too specific to the sample and thus lacking generalizability. Although the minimum ratio is 5 to 1, the desired level is between 15 to 20 observations for each independent variable. When this level is reached, the results should be generalizable if the sample is representative. However, if a stepwise procedure is employed, the recommended level increases to 50 to 1 [16]. In cases for which the available sample does not meet these criteria, the researcher should be certain to validate the generalizability of the results.

Fixed versus Random Effects Predictors

The examples of regression models we have discussed to this point have assumed that the levels of the independent variables are fixed. For example, if we wish to know the impact on preference of three levels of sweetener in a cola drink, we make up three different batches of cola and have a number of people sample each. We then predict the preference rating of each cola, using level of sweetener as the independent variable. We have fixed the level of sweetener and are interested in its effect at these levels. We do not assume the three levels to be a random sample for a large number of possible levels of sweetener. A random independent variable is one in which the levels are selected at random. When using a random independent variable, the interest is not just in the levels examined but rather in the larger population of possible independent variable levels from which we selected a sample.

Most regression models based on survey data are random effects models. As an illustration, a survey was conducted to help assess the relationship between age of the respondent and frequency of visits to physicians. The independent variable "age of respondent" was randomly selected from the population, and the inference regarding the population is of concern, not just knowledge of the individuals in the sample.

The estimation procedures for models using both types of independent variables are the same except for the error terms. In the random effects models, a portion of the random error comes from the sampling of the independent variables. However, the statistical procedures based on the fixed model are quite robust, so using the statistical analysis as if you were dealing with a fixed model (as most analysis packages assume) may still be appropriate as a reasonable approximation.

Creating Additional Variables

The basic relationship represented in multiple regression is the *linear* association between *metric* dependent and independent variables based on the product-moment correlation. One problem often faced by researchers is the desire to incorporate nonmetric data, such as gender or occupation, into a regression equation. Yet, as we have already discussed, regression is limited to metric data. Moreover, regression's inability to directly model nonlinear relationships may constrain the researcher when faced with situations in which a nonlinear relationship (e.g., ∪-shaped) is suggested by theory or detected when examining the data.

In these situations, new variables must be created by **transformations,** as multiple regression is totally reliant on creating new variables in the model to incorporate nonmetric variables or represent any effects other than linear relationships. Transforming the data provides the researcher with a means of modifying either the dependent or independent variables for one of two reasons: to improve or modify the relationship between independent and dependent variables or to allow the use of nonmetric variables in the regression variate. Data transformations may be based on reasons that are either "theoretical" (transformations whose appropriateness is based on the nature of the data) or "data derived" (transformations that are suggested strictly by an examination of the data). In either case the researcher must proceed many times by trial and error, constantly assessing the improvement versus the need for additional transformations. We explore these issues with discussions of data transformations that allow the regression analysis to best represent the actual data and a discussion of the creation of variables to supplement the original variables.

All the transformations we describe are easily carried out by simple commands in all the popular statistical packages. Although we focus on transformations that can be computed in this manner, other more sophisticated and complicated methods of data transformation are available (e.g., see Box and Cox [4]).

Incorporating Nonmetric Data with Dummy Variables

One common situation faced by researchers is the presence of nonmetric independent variables. Yet, to this point, all our illustrations have assumed metric measurement for both independent and dependent variables. When the dependent variable is measured as a dichotomous (0, 1) variable, either discriminant analysis or a specialized form of regression (logistic regression), both discussed in chapter 5, is appropriate. But what can we do when the independent variables are nonmetric and have two or more categories? Chapter 2 introduced the concept of dichotomous variables, known as **dummy variables,** which can act as replacement independent variables. Each dummy variable represents one category of a nonmetric independent variable, and any nonmetric variable with k categories can be represented as $k - 1$ dummy variables.

There are two forms of dummy variable coding, the most common being **indicator coding,** in which the category is represented by either 1 or 0. The regression coefficients for the dummy variables represent differences between means for each group of respondents formed by a dummy variable from the **reference category** (i.e., the omitted group that received all zeros) on the dependent variable. These group differences can be assessed directly, as the coefficients are in the same units as the dependent variable. This form of dummy variable coding can also be depicted as differing intercepts for the various groups (see Figure 4.3, p. 168). In this example, a three-category nonmetric variable is represented by two dummy variables (D_1 and D_2) representing groups 1 and 2, with group 3 the **reference category**. The regression coefficients are 2.0 for D_1 and -3.0 for D_2. These coefficients translate into three parallel lines. The reference group (in this case group 3) is defined by the regression equation with both dummy variables equaling zero. Group 1's line is two units above the line for the reference group. Group 2's line is three units below the line for reference group 3. The parallel lines indicate that dummy

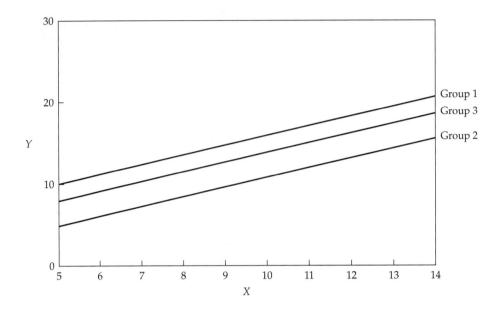

FIGURE 4.3 Incorporating Nonmetric Variables through Dummy Variables

variables do not change the nature of the relationship, but only provide for differing intercepts among the groups. This form of coding is most appropriate when there is a logical reference group, such as in an experiment. Any time dummy-variable coding is used, we must be aware of the comparison group and remember that the coefficients represent the differences in group means from this group.

An alternative method of dummy-variable coding is termed **effects coding.** It is the same as indicator coding except that the comparison or omitted group (the group that got all zeros) is now given the value of -1 instead of 0 for the dummy variables. Now the coefficients represent differences for any group from the mean of all groups rather than from the omitted group. Both forms of dummy-variable coding will give the same predictive results, coefficient of determination, and regression coefficients for the continuous variables. The only differences will be in the interpretation of the dummy-variable coefficients.

Representing Curvilinear Effects with Polynomials

Several types of data transformations are appropriate for linearizing a curvilinear relationship. Direct approaches, discussed in chapter 2, involve modifying the values through some arithmetic transformation (e.g., taking the square root or logarithm of the variable). However, such transformations have several limitations. First, they are helpful only in a simple curvilinear relationship (a relationship with only one turning or inflection point). Second, they do not provide any statistical means for assessing whether the curvilinear or linear model is more appropriate. Finally, they accommodate only univariate relationships and not the interaction between variables when more than one independent variable is involved. We now discuss a means of creating new variables to explicitly model the curvilinear components of the relationship and address each of the limitations inherent in data transformations.

Polynomials are power transformations of an independent variable that add a nonlinear component for each additional power of the independent variable. The power of 1 (X^1) represents the linear component and is the form we have discussed so far in this chapter. The power of 2, the variable squared (X^2), represents the quadratic component. In graphical terms, X^2 represents the first inflection point. A cubic component, represented by the variable cubed (X^3), adds a second inflection point. With these variables, and even higher powers, we can accommodate more complex relationships than are possible with only transformations. For example, in a simple regression model, a curvilinear model with one turning point can be modeled with the equation

$$Y = b_0 + b_1X_1 + b_2X_1^2$$

where

$$b_0 = \text{intercept}$$
$$b_1X_1 = \text{linear effect of } X_1$$
$$b_2X_1^2 = \text{curvilinear effect of } X_1$$

Although any number of nonlinear components may be added, the cubic term is usually the highest power used. As each new variable is entered into the regression equation, we can also perform a direct statistical test of the nonlinear components, which we cannot do with data transformations. Three (two nonlinear and one linear) relationships are shown in Figure 4.4 (p. 170). For interpretation purposes, the positive quadratic term indicates a ∪-shaped curve, whereas a negative coefficient indicates a ∩-shaped curve.

Multivariate polynomials are created when the regression equation contains two or more independent variables. We follow the same procedure for creating the polynomial terms as before, but must also create an additional term, the interaction term (X_1X_2), which is needed for each variable combination to represent fully the multivariate effects. In graphical terms, a two-variable multivariate polynomial is portrayed by a surface with one peak or valley. For higher-order polynomials, the best form of interpretation is obtained by plotting the surface from the predicted values.

How many terms should be added? Common practice is to start with the linear component and then sequentially add higher-order polynomials until nonsignificance is achieved. The use of polynomials, however, also has potential problems. First, each additional term requires a degree of freedom, and this may be particularly restrictive with small sample sizes. This limitation does not occur

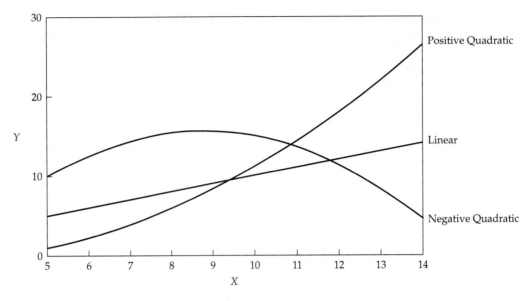

FIGURE 4.4 Representing Nonlinear Relationships with Polynomials

with data transformations. Also, multicollinearity is introduced by the additional terms and makes statistical significance testing of the polynomial terms inappropriate. Instead, the researcher must compare the R^2 values from the equation model with linear terms to the R^2 for the equation with the polynomial terms. Testing for the statistical significance of the incremental R^2 is the appropriate manner of assessing the impact of the polynomials.

Representing Interaction or Moderator Effects

The nonlinear relationships discussed above require the creation of an additional variable (for example, the squared term) to represent the changing slope of the relationship over the range of the independent variable. This focuses on the relationship between a single independent variable and the dependent variable. But what if an independent–dependent variable relationship is affected by another independent variable? This situation is termed a **moderator effect,** which occurs when the moderator variable, a second independent variable, changes the *form* of the relationship between another independent variable and the dependent variable. This is also known as an *interaction effect* and is similar to the interaction term found in analysis of variance and multivariate analysis of variance (see chapter 6 for more detail on interaction terms).

The most common moderator effect employed in multiple regression is the *quasi* or *bilinear moderator*, in which the slope of the relationship of one independent variable (X_1) changes across values of the moderator variable (X_2) [7, 14]. In our earlier example of credit card usage, assume that family income (X_2) was found to be a positive moderator of the relationship between family size (X_1) and credit card usage (Y). This would mean that the expected change in credit card usage based on family size (b_1, the regression coefficient for X_1) might be lower for families with low incomes and higher for families with higher incomes. Without the moderator effect, we assumed that family size had a "constant" effect on the number of credit cards used. But the interaction term tells us that this relationship changes, depending on family income level. Note that this does not necessarily mean the effects of family size or family income by themselves are unimportant, but instead the interaction term complements their explanation of credit card usage.

The moderator effect is represented in multiple regression by a term quite similar to the polynomials described earlier to represent nonlinear effects. The moderator term is a compound variable formed by multiplying X_1 by the moderator X_2, which is entered into the regression equation. In fact, the nonlinear term can be viewed as a form of interaction, where the independent variable "moderates" itself, thus the squared term (X_iX_i). The moderated relationship is represented as

$$Y = b_0 + b_1X_1 + b_2X_2 + b_3X_1X_2$$

where

$$b_0 = \text{intercept}$$
$$b_1X_1 = \text{linear effect of } X_1$$
$$b_2X_2 = \text{linear effect of } X_2$$
$$b_3X_1X_2 = \text{moderator effect of } X_2 \text{ on } X_1$$

Because of the multicollinearity among the old and new variables, an approach similar to testing for the significance of polynomial (nonlinear) effects is employed. To determine whether the moderator effect is significant, the researcher first estimates the original (unmoderated) equation and then estimates the moderated relationship. If the change in R^2 is statistically significant, then a significant moderator effect is present. Thus, only the incremental effect is assessed, not the individual variables.

The interpretation of the regression coefficients changes slightly in moderated relationships. The b_3 coefficient, the moderator effect, indicates the unit change in the effect of X_1 as X_2 changes. The b_1 and b_2 coefficients now represent the effects of X_1 and X_2, respectively, when the other independent variable is zero. In the unmoderated relationship, the b_1 coefficient represents the effect of X_1 across all levels of X_2, and similarly for b_2. Thus, in unmoderated regression, the regression coefficients b_1 and b_2 are "averaged" across levels of the other independent variables, whereas in a moderated relationship they are separate from the other independent variables. To determine the total effect of an independent variable, the separate and moderated effects must be combined. The overall effect of X_1 for any value of X_2 can be found by substituting the X_2 value into the following equation:

$$b_{\text{total}} = b_1 + b_3X_2$$

For example, assume a moderated regression resulted in the following coefficients: $b_1 = 2.0$ and $b_3 = .5$. If the value of X_2 ranges from 1 to 7, the researcher can calculate the total effect of X_1 at any value of X_2. When X_2 equals 3, the total effect of X_1 is 3.5 [2.0 + .5(3)]. When X_2 increases to 7, the total effect of X_1 is now 5.5 [2.0 + .5(7)]. We can see the moderator effect at work, making the relationship of X_1 and the dependent variable change, given the level of X_2. Excellent discussions of moderated relationships in multiple regression are available in a number of sources [5, 7, 14].

Summary

The creation of new variables provides the researcher with great flexibility in representing a wide range of relationships within regression models. Yet too often the desire for better model fit leads to the inclusion of these special relationships without theoretical support. In those instances the researcher is running a much greater risk of finding results with little or no generalizability. Instead, in using these additional variables, the researcher must be guided by theory that is supported by empirical analysis. In this manner, both practical and statistical significance can hopefully be met.

Stage 3: Assumptions in Multiple Regression Analysis

We have shown how improvements in prediction of the dependent variable are possible by adding independent variables and even transforming them to represent aspects of the relationship that are not linear. But to do so we must make several assumptions about the relationships between the dependent and independent variables that affect the statistical procedure (least squares) used for multiple regression. In the following sections we discuss testing for the assumptions and corrective actions to take if violations occur.

Assessing Individual Variables versus the Variate

The assumptions underlying multiple regression analysis apply both to the individual variables (dependent and independent) and to the relationship as a whole. Chapter 2 examined the available methods for assessing the assumptions for individual variables. But in multiple regression, once the variate has been derived, it acts collectively in predicting the dependent variable. This necessitates assessing the assumptions not only for individual variables but also for the variate itself. This section focuses on examining the variate and its relationship with the dependent variable for meeting the assumptions of multiple regression. These analyses actually must be performed *after* the regression model has been estimated in stage four. Thus, testing for assumptions must occur not only in the initial phases of the regression, but also after the model has been estimated.

The basic issue is whether, in the course of calculating the regression coefficients and predicting the dependent variable, the assumptions of regression analysis have been met. Are the errors in prediction a result of an actual absence of a relationship among the variables, or are they caused by some characteristics of the data not accommodated by the regression model? The assumptions to be examined are as follows:

- Linearity of the phenomenon measured
- Constant variance of the error terms
- Independence of the error terms
- Normality of the error term distribution

The principal measure of prediction error for the variate is the **residual**—the difference between the observed and predicted values for the dependent variable. Plotting the residuals versus the independent or predicted variables is a basic method of identifying assumption violations for the overall relationship. When examining residuals, some form of standardization is recommended, as it makes the residuals directly comparable. (In their original form, larger predicted values naturally have larger residuals.) The most widely used is the **studentized resid-**

ual, whose values correspond to t values. This correspondence makes it quite easy to assess the statistical significance of particularly large residuals. In the following sections, we examine a series of statistical tests that can complement the visual examination of the residual plots.

The most common residual plot involves the residuals (r_i) versus the predicted dependent values (Y_i). For a simple regression model, the residuals may be plotted against either the dependent or independent variables, as they are directly related. In multiple regression, however, only the predicted dependent values represent the total effect of the regression variate. Thus, unless the residual analysis intends to concentrate on only a single variable, the predicted dependent variables are used. Violations of each assumption can be identified by specific patterns of the residuals. Figure 4.5 (p. 174) contains a number of residual plots that address the basic assumptions discussed in the following sections. One plot of special interest is the **null plot** (Figure 4.5a), the plot of residuals when all assumptions are met. The null plot shows the residuals falling randomly, with relatively equal dispersion about zero and no strong tendency to be either greater or less than zero. Likewise, no pattern is found for large versus small values of the independent variable. The remaining residual plots will be used to illustrate methods of examining for violations of the assumptions underlying regression analysis.

Linearity of the Phenomenon

The **linearity** of the relationship between dependent and independent variables represents the degree to which the change in the dependent variable is associated with the independent variable. The regression coefficient is constant across the range of values for the independent variable. The concept of correlation is based on a linear relationship, thus making it a critical issue in regression analysis. Linearity is easily examined through residual plots. Figure 4.5b shows a typical pattern of residuals indicating the existence of a nonlinear relationship not represented in the current model. Any consistent curvilinear pattern in the residuals indicates that corrective action will increase both the predictive accuracy of the model and the validity of the estimated coefficients. Remedies through transforming the data to achieve linearity are discussed in chapter 2 [10]. The researcher may instead wish to directly include the nonlinear relationships in the regression model. Data transformations, such as the creation of polynomial terms as discussed in stage two, or specialized methods such as nonlinear regression, can accommodate the curvilinear effects of independent variables or more complex nonlinear relationships.

In multiple regression with more than one independent variable, an examination of the residuals shows the combined effects of all independent variables, but we cannot examine any independent variable separately in a residual plot. To do so, we use what are called **partial regression plots,** which show the relationship of a single independent variable to the dependent variable. They differ from the residual plots just discussed because the line running through the center of the points, which was horizontal in the earlier plots (refer to Figure 4.5), will now slope up or down depending on whether the regression coefficient for that independent variable is positive or negative. Examining the residuals around this line is done exactly as before.

In partial regression plots, the curvilinear pattern of residuals indicates a nonlinear relationship between a specific independent variable and the dependent variable. This is a more useful method when there are several independent variables,

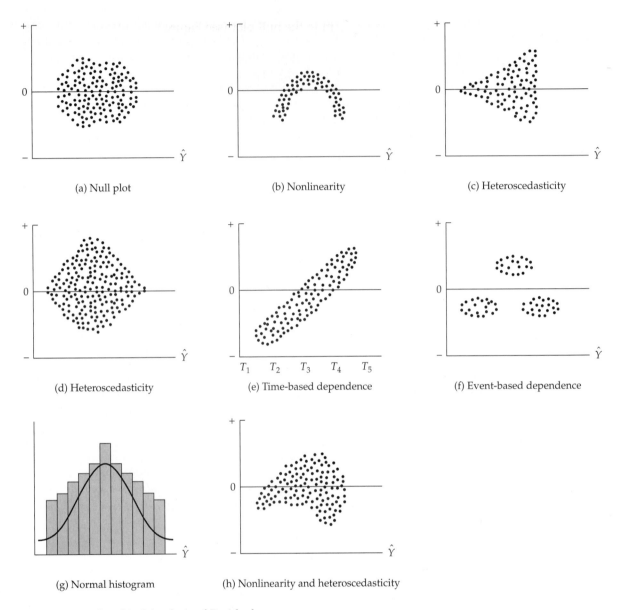

FIGURE 4.5 Graphical Analysis of Residuals

as we can tell which specific variables violate the assumption of linearity and apply the needed remedies only to them. Also, the identification of outliers or influential observations is facilitated on the basis of one independent variable at a time.

Constant Variance of the Error Term

The presence of unequal variances **(heteroscedasticity)** is one of the most common assumption violations. Diagnosis is made with residual plots or simple statistical tests. Plotting the residuals (studentized) against the predicted dependent values and comparing them to the null plot (see Figure 4.5a) shows a consistent pattern if the variance is not constant. Perhaps the most common pattern is triangle-shaped in either direction (Figure 4.5c). A diamond-shaped pattern (Figure 4.5d) can be expected in the case of percentages where more variation is expected in the midrange than at the tails. Many times, a number of violations occur simultaneously, such as in nonlinearity and heteroscedasticity (Figure 4.5h). Remedies for one of the violations often corrects problems in other areas as well.

Each statistical computer program has statistical tests for heteroscedasticity. For example, SPSS provides the Levene test for homogeneity of variance, which measures the equality of variances for a single pair of variables. Its use is particularly recommended because it is less affected by departures from normality, another frequently occurring problem in regression.

If heteroscedasticity is present, two remedies are available. If the violation can be attributed to a single independent variable, the procedure of weighted least squares can be employed. More direct and easier, however, are a number of variance-stabilizing transformations discussed in chapter 2 that allow the transformed variables to be used directly in our regression model.

Independence of the Error Terms

We assume in regression that each predicted value is independent. By this we mean that the predicted value is not related to any other prediction; that is, they are not sequenced by any variable. We can best identify such an occurrence by plotting the residuals against any possible sequencing variable. If the residuals are independent, the pattern should appear random and similar to the null plot of residuals. Violations will be identified by a consistent pattern in the residuals. Figure 4.5e displays a residual plot that exhibits an association between the residuals and time, a common sequencing variable. Another frequent pattern is shown in Figure 4.5f. This pattern occurs when basic model conditions change but are not included in the model. For example, swimsuit sales are measured monthly for 12 months, with two winter seasons versus a single summer season, yet no seasonal indicator is estimated. The residual pattern will show negative residuals for the winter months versus positive residuals for the summer months. Data transformations, such as first differences in a time series model, inclusion of indicator variables, or specially formulated regression models, can address this violation if it occurs.

Normality of the Error Term Distribution

Perhaps the most frequently encountered assumption violation is nonnormality of the independent or dependent variables or both [13]. The simplest diagnostic for the set of independent variables in the equation is a histogram of residuals, with a visual check for a distribution approximating the normal distribution (see Figure 4.5g). Although attractive because of its simplicity, this method is particularly difficult in smaller samples, where the distribution is ill-formed. A better method is the use of **normal probability plots.** They differ from residual plots in that the standardized residuals are compared with the normal distribution. The normal distribution makes a straight diagonal line, and the plotted residuals are compared with the diagonal. If a distribution is normal, the residual line closely follows the diagonal. The same procedure can compare the dependent or independent variables separately to the normal distribution [6]. Chapter 2 provides a more detailed discussion of the interpretation of normal probability plots.

Summary

Analysis of residuals, whether with the residual plots or statistical tests, provides a simple yet powerful set of analytical tools for examining the appropriateness of our regression model. Too often, however, these analyses are not made, and the violations of assumptions are left intact. Thus, users of the results are unaware of the potential inaccuracies that may be present. These range from inappropriate tests of the significance of coefficients (either showing significance when it is not present or vice versa) to the biased and inaccurate predictions of the dependent variable. We strongly recommend that these methods be applied for each set of data and regression model. Application of the remedies, particularly transformations of the data, will increase confidence in the interpretations and predictions from multiple regression.

Stage 4: Estimating the Regression Model and Assessing Overall Model Fit

Having specified the objectives of the regression analysis, selected the independent and dependent variables, addressed the issues of research design, and assessed the variables for meeting the assumptions of regression, the researcher is now ready to estimate the regression model and assess the overall predictive accuracy of the independent variables (see Figure 4.6). In this stage, the researcher must accomplish three basic tasks: (1) select a method for specifying the regression model to be estimated, (2) assess the statistical significance of the overall model in predicting the dependent variable, and (3) determine whether any of the observations exert an undue influence on the results.

General Approaches to Variable Selection

In most instances of multiple regression, the research has a number of possible independent variables from which to choose for inclusion in the regression equation. Sometimes the set of independent variables may be closely specified and the regression model is essentially used in a confirmatory approach. In other instances, the researcher may wish to pick and choose among the set of independent variables. There are several approaches (sequential search methods and combinatorial processes) to assist the researcher in finding the "best" regression model. Each of these approaches to specifying the regression model is discussed next.

Confirmatory Specification

The simplest, yet perhaps most demanding, approach for specifying the regression model is to employ a confirmatory perspective wherein the researcher completely specifies the set of independent variables to be included. As compared with the specific approaches to be discussed next, the researcher has total control over the variable selection. Although the confirmatory specification is simple in concept, the researcher must be assured that the set of variables achieves the maximum prediction while maintaining a parsimonious model. Guidelines for model development are discussed in chapters 1 and 11.

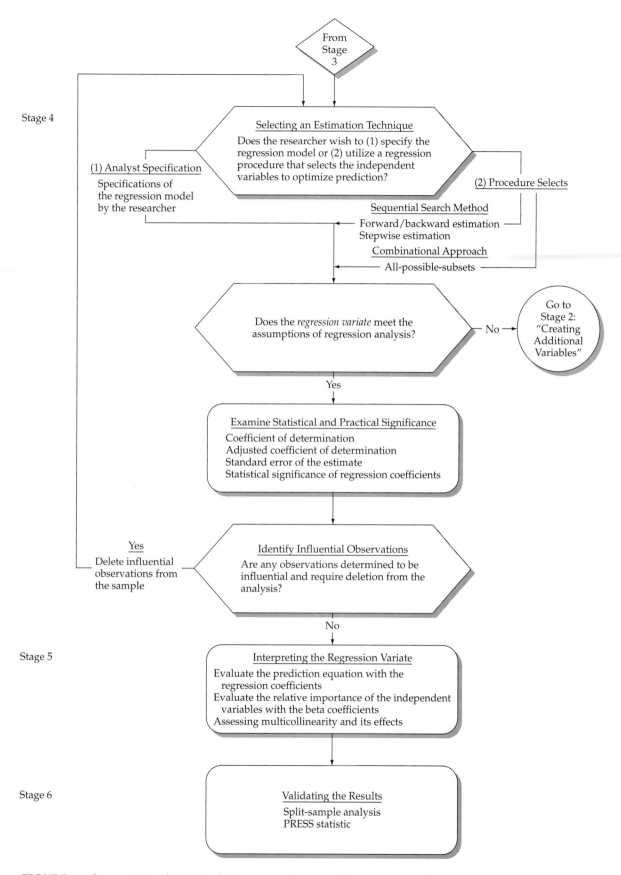

FIGURE 4.6 Stages 4–6 in the Multiple Regression Decision Diagram

Sequential Search Methods

Sequential search methods have in common the general approach of estimating the regression equation with a set of variables and then selectively adding or deleting variables until some overall criterion measure is achieved. This approach provides an objective method for selecting variables that maximizes the prediction with the smallest number of variables employed. There are two types of sequential search approaches: (1) stepwise estimation, and (2) forward addition and backward elimination. In each approach, variables are individually assessed for their contribution to prediction of the dependent variable and added to or deleted from the regression model based on their relative contribution. The stepwise procedure is discussed and then contrasted with the forward addition and backward elimination procedures.

Stepwise Estimation   **Stepwise estimation** is perhaps the most popular sequential approach to variable selection. This approach allows the researcher to examine the contribution of each independent variable to the regression model. Each variable is considered for inclusion prior to developing the equation. The independent variable with the greatest contribution is added first. Independent variables are then selected for inclusion based on their incremental contribution over the variable(s) already in the equation. The stepwise procedure is illustrated in Figure 4.7. The specific issues at each stage are as follows:

1. Start with the simple regression model in which only the one independent variable that is the most highly correlated with the dependent variable is used. The equation would be $Y = b_0 + b_1X_1$.
2. Examine the **partial correlation coefficients** to find an additional independent variable that explains the largest statistically significant portion of the error remaining from the first regression equation.
3. Recompute the regression equation using the two independent variables, and examine the **partial F value** for the original variable in the model to see whether it still makes a significant contribution, given the presence of the new independent variable. If it does not, eliminate the variable. This ability to eliminate variables already in the model distinguishes the stepwise model from the forward addition/backward elimination models. If the original variable still makes a significant contribution, the equation would be $Y = b_0 + b_1X_1 + b_2X_2$.
4. Continue this procedure by examining all independent variables not in the model to determine whether one should be included in the equation. If a new independent variable is included, examine all independent variables previously in the model to judge whether they should be kept. A potential bias in the stepwise procedure results from considering only one variable for selection at a time. Suppose variables X_3 and X_4 together would explain a significant portion of the variance (each given the presence of the other), but neither is significant by itself. In this situation, neither would be considered for the final model.

Forward Addition and Backward Elimination   **Forward addition** and **backward elimination** procedures are largely trial-and-error processes for finding the best regression estimates. The forward addition model is similar to the stepwise procedure described above, whereas the backward elimination procedure computes a regression equation with all the independent variables, and then deletes independent variables that do not contribute significantly. The primary distinction of

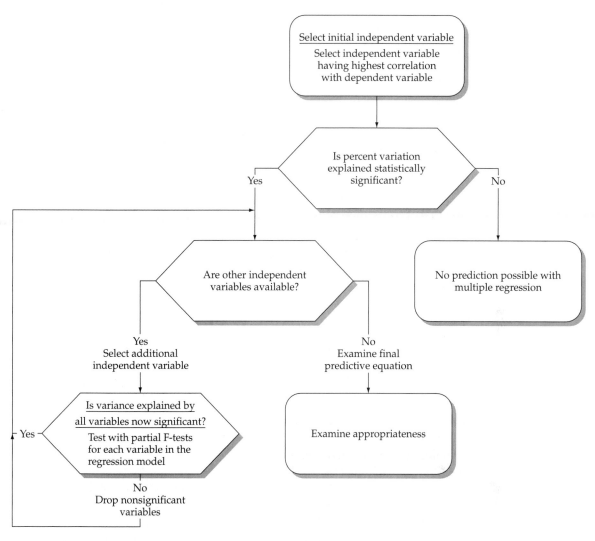

FIGURE 4.7 Flowchart of the Stepwise Estimation Method

the stepwise approach from the forward addition and backward elimination procedures is its ability to add or delete variables at each stage. Once a variable is added or deleted in the forward addition or backward elimination schemes, there is no chance of reversing the action at a later stage.

Caveats to Sequential Search Methods The researcher must be aware of two caveats in the use of any sequential search procedures. First, the multicollinearity among independent variables can have a substantial impact on the final model specification. Let us examine the situation with two highly correlated independent variables that have almost equal correlations with the dependent variable. The criterion for inclusion or deletion in these approaches is to maximize the incremental predictive power of the additional variable. If one of these variables enters the regression model, it is highly unlikely that the other variable will also enter because these variables are highly correlated and there is little unique variance for each variable separately (see the later discussion on multicollinearity). For this reason, the researcher must assess the effects of multicollinearity in model interpretation and examine the direct correlations of all potential independent variables. This will help the researcher to avoid concluding that the independent variables that do not enter the model are inconsequential when in fact they are highly related to the dependent variable, but also correlated with variables already in the model. Although the sequential search approaches will maximize the predictive ability of the regression model, the researcher must be quite careful in model interpretation. A second caveat pertains primarily to the stepwise procedure. In this approach, multiple significance tests are performed in the model estimation process. To ensure that the overall error rate across all significance tests is reasonable, the researcher should employ more conservative thresholds (e.g., .01) in adding or deleting variables.

Combinatorial Approach

The combinatorial approach is primarily a generalized search process across all possible combinations of independent variables. The best-known procedure is **all-possible-subsets regression,** which is exactly as the name suggests. All possible combinations of the independent variables are examined, and the best-fitting set of variables is identified. For example, for a model with 10 independent variables, there exist 1,024 possible regressions (one equation with only the constant, 10 equations with a single independent variable, 45 equations with all combinations of two variables, and so on). With computerized estimation procedures, this process can be managed today for even rather large problems, identifying the best overall regression equation for any number of measures of predictive fit. The researcher must remember that issues such as multicollinearity, the identification of outliers and influentials, and the interpretability of the results are not addressed in selecting the final model. When these issues are considered, the "best" equation may have serious problems that affect its appropriateness, and another model may ultimately be selected.

Overview of the Model Selection Approaches

Whether a confirmatory, sequential search, or combinatorial method is chosen, the most important criterion is the researcher's substantive knowledge of the research context that allows for an objective and informed perspective as to the variables to be included as well as the expected signs and magnitude of their coefficients. Without this knowledge, the regression results can have high predictive accuracy without any managerial or theoretical relevance. The researcher should never be totally guided by any one of these approaches, but instead should use them after careful consideration of all of them, and then accept the results only after careful scrutiny.

Testing the Regression Variate for Meeting the Regression Assumptions

With the independent variables selected and the regression coefficients estimated, the researcher must now assess the estimated model for meeting the assumptions underlying multiple regression. As discussed in stage 3, the individual variables must meet the assumptions of linearity, constant variance, independence, and normality. In addition to the individual variables, the regression variate must also meet these assumptions. The diagnostic tests discussed in stage 3 can be applied to assessing the collective effect of the variate through examination of the residuals. If substantial violations are found, the researcher must take corrective actions and then reestimate the regression model.

Examining the Statistical Significance of Our Model

If we were to take repeated samples of eight families and ask them how many family members and credit cards they have, we would seldom get exactly the same values for $Y = b_0 + b_1X_1$ from all the samples. We would expect chance variation to cause differences among many samples. Usually, we take only one sample and base our predictive model on it. With only one sample and regression model, we need to test the hypothesis that our predictive model can represent the population of all families having credit cards rather than just our one sample of eight families. These tests may take one or two basic forms: a test of the variation explained (coefficient of determination) and a test of coefficients.

Significance of the Overall Model: The Coefficient of Determination

To test the hypothesis that the amount of variation explained by the regression model is more than the variation explained by the average (i.e., that R^2 is greater than zero), the F ratio is used. The test statistic F ratio is calculated as:

$$F \text{ ratio} = \frac{\dfrac{\text{Sum of squared errors}_{\text{regression}}}{\text{Degrees of freedom}_{\text{regression}}}}{\dfrac{\text{Sum of squared errors}_{\text{total}}}{\text{Degrees of freedom}_{\text{residual}}}} = \frac{SSE_{\text{regression}}/df_{\text{regression}}}{SSE_{\text{total}}/df_{\text{residual}}}$$

where

$df_{\text{regression}}$ = number of estimated coefficients (including intercept) − 1

df_{residual} = sample size − number of estimated coefficients (including intercept)

Two important features of this ratio should be noted:

1. Each sum of squares divided by its appropriate **degrees of freedom** (*df*) is simply the variance of the prediction errors.
2. Intuitively, if the ratio of the explained variance to the baseline variance (about the mean) is high, the regression variate must be of significant value in explaining the dependent variable.

In our example of credit card usage, the F ratio for the simple regression model is $(16.5 \div 1)/(5.50 \div 6) = 18.0$. The tabled F statistic of 1 with 6 degrees of freedom at a significance level of .05 yields 5.99. Because the F ratio is greater than the table value, we reject the hypothesis that the reduction in error we obtained by using family size to predict credit card usage was a chance occurrence. This outcome means that, considering the sample used for estimation, we can explain 18 times more variation than when using the average, and that this is not very likely to happen by chance (less than 5 percent of the time). Likewise, the F ratio for the multiple regression model with two independent variables is $(18.96 \div 2)/(3.04 \div 5) = 15.59$. The multiple regression model is also statistically significant, indicating that the additional independent variable was substantial in adding to the regression model's predictive ability.

We also know that R^2 is influenced by the number of independent variables relative to the sample size. Several rules of thumb have been proposed, ranging from 10 to 15 observations per independent variable to an absolute minimum of 4 observations per independent variable. As we approach or fall below these limits, we need to adjust for the inflation in R^2 from "overfitting" the data. As part of all regression programs, an **adjusted coefficient of determination (adjusted R^2)** is given along with the coefficient of determination. Interpreted the same as the unadjusted coefficient of determination, the adjusted R^2 becomes smaller as we have fewer observations per independent variable. The adjusted R^2 value is particularly useful in comparing across regression equations involving different numbers of independent variables or different sample sizes because it makes allowances for the specific number of independent variables and the sample size upon which each model is based. In our example of credit card usage, R^2 for the simple regression model is .751, and the adjusted R^2 is .709. As we add the second independent variable, R^2 increases to .861, but the adjusted R^2 increases to only .806. In both instances, the adjusted R^2 reflects the decreasing ratio of estimated coefficients to the sample size and compensates for "overfitting" of the data.

Significance Tests of Regression Coefficients

Statistical significance testing for the estimated coefficients in regression analysis is appropriate and necessary when the analysis is based on a sample of the population rather than a census. When using a sample for estimating the regression model, the researcher is not interested in the regression estimate for just that sample, but is really interested in how generalizable the results are to the population. Each sample drawn from a population will yield a different value for the coefficient. For small sample sizes, the estimated coefficients will most likely vary widely from sample to sample. But as the size of the sample increases, the samples become more representative of the population, and the variation in the estimated coefficients for these large samples will become smaller. This is true until the analysis is estimated using the population. Then there is no need for significance testing because the "sample" is equal to, and thus perfectly representative of, the population. The expected variation of the estimated coefficients (both the constant and the regression coefficients) is termed the **standard error** of the coefficients.

Significance testing of regression coefficients provides a statistically based probability estimate of whether the estimated coefficients across a large number of samples of a certain size will indeed be different than zero. If the sample size is small, the sampling error may be too great to say with a needed degree of certainty (what we refer to as the *significance level*) that the coefficient is not equal to zero. However, if the sample size is larger, the test has greater precision because the variation in the coefficients becomes less. Larger samples do not guarantee that the coefficients will not equal zero, but instead make the test more precise.

An Example of Sampling Variation in a Regression Coefficient   To illustrate this point, 20 random samples for four sample sizes (10, 25, 50, and 100 respondents) were drawn from a large database. A simple regression was performed for each sample, and the estimated regression coefficients are recorded in Table 4.8. As we can see, the variation in the estimated coefficients is greatest for samples of 10 respondents, varying from a low coefficient of 2.20 to a high of 6.06. As the sample sizes increase to 25 and 50 respondents, the sampling error decreases considerably. Finally, the samples of 100 respondents have a range of almost one-half that

TABLE 4.8 Sampling Variation for Estimated Regression Coefficients

	Sample Size			
Sample	*10*	*25*	*50*	*100*
1	2.58	2.52	2.97	3.60
2	2.45	2.81	2.91	3.70
3	2.20	3.73	3.58	3.88
4	6.06	5.64	5.00	4.20
5	2.59	4.00	4.08	3.16
6	5.06	3.08	3.89	3.68
7	4.68	2.66	3.07	2.80
8	6.00	4.12	3.65	4.58
9	3.91	4.05	4.62	3.34
10	3.04	3.04	3.68	3.32
11	3.74	3.45	4.04	3.48
12	5.20	4.19	4.43	3.23
13	5.82	4.68	5.20	3.68
14	2.23	3.77	3.99	4.30
15	5.17	4.88	4.76	4.90
16	3.69	3.09	4.02	3.75
17	3.17	3.14	2.91	3.17
18	2.63	3.55	3.72	3.44
19	3.49	5.02	5.85	4.31
20	4.57	3.61	5.12	4.21
Minimum	2.20	2.52	2.91	2.80
Maximum	6.06	5.64	5.85	4.90
Range	3.86	3.12	2.94	2.10
Standard deviation	1.28	.85	.83	.54

for the samples of 10 respondents (2.10 versus 3.86). From this we can see that the ability of the statistical test to determine whether the coefficient is actually greater than zero is made more precise with the larger sample sizes.

Significance Testing in the Simple Regression Example When we discussed the simple regression model for the credit card usage example, we said that the regression equation for the number of credit cards is $Y = b_0 + b_1 V_1 = 2.87 + .971$(family size). We would test two hypotheses for the regression coefficients (2.87 and .971) in this regression model.

> *Hypothesis 1. The intercept (constant term) value of 2.87 is due to sampling error, and the real constant term appropriate to the population is zero.*

With this hypothesis, we would simply be testing whether the constant term has an impact different from zero and should be included in the predictive model. If found not to differ significantly from zero, we would assume that the constant term should not be used for predictive purposes. The appropriate test is the *t* test, which is commonly available on computerized regression analysis programs. The *t* value of a coefficient is the coefficient divided by the standard error. For example, a coefficient of 2.5 with a standard error of .5 would have a *t* value of 5.0. To determine if the coefficient is significantly different from zero, the computed *t* value is compared to the table value for the sample size and confidence level selected. If our value is greater than the table value, we can be confident (at our selected confidence level) that the coefficient does have a statistically significant effect in the regression variate.

From a practical point of view, this test is seldom necessary for the intercept terms. If the data used to develop the model did not include some observations with all the independent variables measured at zero, the constant term is "outside" the data and acts only to position the model. In this case, it is not necessary to test the constant term.

> *Hypothesis 2. The coefficient .971 indicates that an increase of one unit in family size is associated with an increase in the average number of credit cards held by .971 and that this coefficient also differs significantly from zero.*

If the coefficient occurred because of sampling error, we would conclude that family size has no impact on the number of credit cards held. Note that this is not a test of any exact value of the coefficient but rather of whether it should be used at all. Again, the appropriate test is the t test. The researcher should remember that the statistical test of the regression coefficient is to ensure that, across all the possible samples that could be drawn, the regression coefficient should be different than zero. In our example, the standard error of family size in the simple regression model is .229. The calculated t value is 4.25(.971 ÷ .229), which has a probability of .005. This means that we can be sure with a high degree of certainty (99.5 percent) that the coefficient should be included in the regression equation.

Summary   Significance testing of regression coefficients provides the researcher with an empirical assessment of their "true" impact. Although this is not a test of validity, it does determine whether the impacts represented by the coefficients are generalizable to other samples from this population. One note concerning the variation in regression coefficients is that many times researchers forget that the estimated coefficients in their regression analysis are specific to the sample used in estimation. They are the best estimates for that sample of observations, but as the results above show, the coefficients can vary quite markedly from sample to sample. This points to the need for concerted efforts to validate any regression analysis on a different sample(s). In doing so, the researcher must expect the coefficients to vary, but the attempt is to demonstrate that the relationship generally holds in other samples so that the results can be assumed to be generalizable to any sample drawn from the population.

Identifying Influential Observations

Up to now, we have focused on identifying general patterns within the entire set of observations. Here we shift our attention to individual observations, with the objective of finding the observations that lie outside the general patterns of the data set or that strongly influence the regression results. We should remember that these observations are not necessarily "bad" in the sense that they must be omitted. In many instances they represent the distinctive elements of the data set. However, we must first identify them and assess their impact before we can proceed. This section introduces the concept of influential observations and their potential impact on the regression results, and appendix 4A contains a more detailed discussion of the procedures for identifying influential observations.

Influential observations are of three basic types: outliers, leverage points, and influentials. **Outliers** are observations that have large residual values and can be identified only with respect to a specific regression model. Outliers have traditionally been the only form of influential observation considered in regression models, and specialized regression methods (e.g., robust regression) have even been developed to deal specifically with outliers' impact on the regression results [1, 12]. Chapter 2 provides additional procedures for identifying outliers. **Leverage points** are observations that are distinct from the remaining observations based on their independent variable values. Their impact is particularly noticeable in the estimated coefficients for one or more independent variables. Finally, **influential observations** are the broadest category, including all observations that have a disproportionate effect on the regression results. Influential observations potentially include outliers and leverage points but may include other observations as well. Also, not all outliers and leverage points are necessarily influential observations.

Influential observations can exhibit many patterns. Figure 4.8 illustrates several forms of influential observations and their correspondence to residuals. In each instance, the residual for the influential points (the perpendicular distance from the point of the estimated regression line) would not be expected to be so large as to be classified as an outlier. Thus focusing only on large residuals would generally ignore these additional influential observations. In Figure 4.8a, the influential point is a "good" one, reinforcing the general pattern of the data and lowering

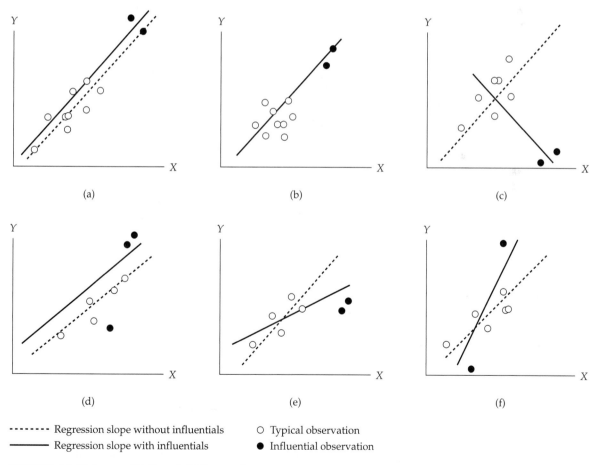

FIGURE 4.8 Patterns of Influential Observations
Adapted from Belsley et al. and Mason and Perreault [2, 9].

the standard error of the prediction and coefficients. It is a leverage point but has a small or zero residual value, as it is predicted well by the regression model. However, influential points can have an effect that is *contrary* to the general pattern of the remaining data but still have small residuals (see Figures 4.8b and 4.8c). In Figure 4.8b, two influential observations almost totally account for the observed relationship, because without them no real pattern emerges from the other data points. They also would not be identified if only large residuals were considered, because their residual value would be small. In Figure 4.8c, an even more profound effect is seen in which the influential observations counteract the general pattern of all the remaining data. In this case, the "real" data would have larger residuals than the "bad" influential points. The influential observations may affect only a portion of the results, as in Figure 4.8d, where the slope remains constant but the intercept is shifted. Finally, multiple influential points may work toward the same result. In Figure 4.8e, two influential points have the same relative position, making detection somewhat harder. In Figure 4.8f, influentials have quite different positions but a similar effect on the results. These examples illustrate that we must develop a bigger tool kit of methods for identifying these influential cases.

Procedures for identifying all types of influential observations are quite numerous and still less well defined than many other aspects of regression analysis. All computer programs provide an analysis of residuals from which those with large values (particularly standardized residuals greater than 2.0) can be easily identified. Moreover, most computer programs now provide at least some of the diagnostic measures for identifying leverage points and other influential observations.

The need for additional study of leverage points and influentials is highlighted when we see the substantial extent to which the generalizability of the results and the substantive conclusions (the importance of variables, level of fit, and so forth) can be changed by only a small number of observations. Whether "good" (accentuating the results) or "bad" (substantially changing the results), these observations must be identified to assess their impact. Influentials, outliers, and leverage points are based on one of four conditions:

1. An error in observations or data entry
2. A valid but exceptional observation that is explainable by an extraordinary situation
3. An exceptional observation with no likely explanation
4. An ordinary observation in its individual characteristics but exceptional in its combination of characteristics

Courses of action can be recommended for dealing with influentials from each condition. For an error in observation, correct the data or delete the case. With the valid but exceptional observation (condition 2), deletion of the case is warranted unless variables reflecting the extraordinary situation are included in the regression equation. The unexplained observation (condition 3) presents a special problem because there is no reason for deleting the case, but its inclusion cannot be justified either. Finally, the observation that is ordinary on each variable separately yet exceptional in its combination of characteristics (condition 4) indicates modifications to the conceptual basis of the regression model and should be retained.

In all situations, the researcher is encouraged to delete truly exceptional observations but still guard against deleting observations that, while different, are representative of the population. Remember that the objective is to ensure the most representative model for the sample data so that it will best reflect the population from which it was drawn. This extends beyond achieving the highest predictive fit, because some outliers may be valid cases that the model should attempt to predict, even if poorly. The researcher should also be aware of instances in which the results would be changed substantially by deleting just a single observation or a very small number of observations.

Stage 5: Interpreting the Regression Variate

The researcher's next task is to interpret the regression variate by evaluating the estimated regression coefficients for their explanation of the dependent variable. The researcher must evaluate not only the regression model that was estimated but also the potential independent variables that were omitted if a sequential search or combinatorial approach was employed. In those approaches, multicollinearity may substantially affect the variables ultimately included in the regression variate. Thus, in addition to assessing the estimated coefficients, the researcher must also evaluate the potential impact of omitted variables to ensure that the managerial significance is evaluated along with statistical significance.

Using the Regression Coefficients

The estimated regression coefficients are used to calculate the predicted values for each observation and to express the expected change in the dependent variable for each unit change in the independent variable(s). In addition to making the prediction, many times we also wish to engage in explanation—assessing the impact of each independent variable in predicting the dependent variable. In the multiple regression example discussed earlier, we would like to know which variable—family size or family income—has the larger effect in predicting the number of credit cards used by a family.

Unfortunately, the regression coefficients (b_0, b_1, and b_2) do not give us this information in many instances. To illustrate why, we can use a rather obvious case. Suppose we wanted to predict teenagers' monthly expenditures on CDs (Y), using two independent variables: X_1 is parents' income in thousands of dollars and X_2 is the teenager's monthly allowance measured in dollars. Suppose we found the following model by a least squares procedure:

$$Y = -.01 + X_1 + .001X_2$$

We might assume that X_1 is more important because its coefficient is 1,000 times larger than the coefficient for X_2. This assumption is not true, however. A $10 increase in the parents' income produces a $.01 change ($1 \times \$10 \div \$1,000$) in average CD purchases (we divide $10 by 1000 because the X_1 value is measured in thousands of dollars). A change of $10 in the teenager's monthly allowance also produces a $.01 change ($.001 \times \$10$) in average CD expenditures (because the teenager's allowance was measured in dollars). A $10 change in the parents' income produced the same effect as a $10 change in the teenager's allowance. Both variables are equally important, but the regression coefficients do not directly reveal this fact. We can resolve this problem in explanation by using a modified regression coefficient called the *beta coefficient*.

Standardizing the Regression Coefficients: Beta Coefficients

If each of our independent variables had been **standardized** before we estimated the regression equation, we would have found different regression coefficients. The coefficients resulting from standardized data are called **beta coefficients.** Their advantage is that they eliminate the problem of dealing with different units of measurement (as illustrated previously), thus reflecting the relative impact on the dependent variable of a change in one standard deviation in either variable. Now that we have a common unit of measurement, we can determine which variable is the most impactful.

Three cautions must be observed when using beta coefficients. First, they should be used as a guide to the relative importance of individual independent variables only when collinearity is minimal. Second, the beta values can be interpreted only in the context of the other variables in the equation. For example, a beta value for family size reflects its importance only in relation to family income, not in any absolute sense. If another independent variable were added to the equation, the beta coefficient for family size would probably change, because there would likely be some relationship between family size and the new independent variable. The third caution is that the levels (e.g., families of size 5, 6, and 7) affect the beta value. Had we found families of size 8, 9, and 10, the value of beta would likely change. In summary, beta coefficients should be used only as a guide to the relative importance of the independent variables included in the equation, and only over the range of values for which sample data actually exists.

Assessing Multicollinearity

A key issue in interpreting the regression variate is the correlation among the independent variables. This is a data problem, not a problem of model specification. The ideal situation for a researcher would be to have a number of independent variables highly correlated with the dependent variable, but with little correlation among themselves. Yet in most situations, particularly situations involving consumer response data, there will be some degree of multicollinearity. In some other occasions, such as using dummy variables to represent nonmetric variables or polynomial terms for nonlinear effects, the researcher is creating situations of high multicollinearity. The researcher's task is to (1) assess the degree of multicollinearity and (2) determine its impact on the results and the necessary remedies if needed. In the following sections we discuss the effects of multicollinearity and then detail some useful diagnostic procedures and possible remedies.

The Effects of Multicollinearity

The effects of multicollinearity can be categorized in terms of *explanation* and *estimation.* The effects on explanation primarily concern the ability of the regression procedure and the researcher to represent and understand the effects of each independent variable in the regression variate. As multicollinearity occurs (even at the relatively low levels of .30 or so), the process for separating the effects of individuals becomes more difficult. First, it limits the size of the coefficient of determination and makes it increasingly more difficult to add unique explanatory prediction from additional variables. Second, and just as important, it makes determining the contribution of each independent variable difficult because the effects of the independent variables are "mixed" or confounded. Multicollinearity results in larger portions of shared variance and lower levels of unique variance from which the effects of the individual independent variables can be determined. For example, assume that one independent variable (X_1) has a correlation of .60 with the dependent variable, and a second independent variable (X_2) has a correlation of .50. Then X_1 would explain 36 percent (obtained by squaring the correlation of .60) of the variance of the dependent variable, and X_2 would explain 25 percent (correlation of .50 squared). If the two independent variables are not correlated with each other at all, there is no "overlap," or sharing, of their predictive power. The total explanation would be their sum, or 61 percent. But as collinearity increases, there is some "sharing" of predictive power, and the collective predictive power of the independent variables decreases. Exhibit 4.1 (pp. 190–191) provides further detail on the calculation of unique and shared variance predictions among correlated independent variables.

Figure 4.9 (p. 192) portrays the proportions of shared and unique variance for our example of two independent variables in varying instances of collinearity. If the collinearity of these variables is zero, then the individual variables predict 36 and 25 percent of the variance in the dependent variable, for an overall prediction (R^2) of 61 percent. But as multicollinearity increases, the total variance explained decreases. Moreover, the amount of unique variance for the independent variables is reduced to levels that make estimation of their individual effects quite problematic.

In addition to the effects on explanation, multicollinearity can have substantive effects on the estimation of the regression coefficients and their statistical significance tests. First, the extreme case of multicollinearity in which two or more variables are perfectly correlated, termed **singularity,** prevents the estimation of any coefficients. In this instance, the singularity must be removed before the estimation of coefficients can proceed. Even if the multicollinearity is not perfect, high degrees of multicollinearity can result in regression coefficients being incorrectly estimated and even having the wrong signs. The following example (see Table 4.9, p. 192) illustrates this point. It is clear in examining the correlation matrix and the simple regressions that the relationship between Y and V_1 is positive, whereas the relationship between Y and V_2 is negative. The multiple regression equation, however, does not maintain the relationships from the simple regressions. It would appear to the casual observer examining only the multiple regression coefficients that both relationships (Y and V_1, Y, and V_2) are negative, when we know that this is not the case for Y and V_1. The sign of V_1's regression coefficient is wrong in an intuitive sense, but the strong negative correlation between V_1 and V_2 results in the reversal of signs for V_2. Even though these effects on the estimation procedure occur primarily at relatively high levels of multicollinearity (above .80), the possibility of counterintuitive and misleading results necessitates a careful scrutiny of each regression variate for possible multicollinearity.

Identifying Multicollinearity

We have seen that the effects of multicollinearity can be substantial. In any regression analysis, the assessment of multicollinearity should be undertaken in two steps: (1) identification of the extent of collinearity and (2) assessment of the degree to which the estimated coefficients are affected. If corrective action

EXHIBIT 4.1

Calculating Unique and Shared Variance Among Independent Variables

The basis for estimating all regression relationships is the correlation, which measures the association between two variables. In regression analysis, the correlations between the independent variables and the dependent variable provide the basis for forming the regression variate by estimating regression coefficients (weights) for each independent variable that maximize the prediction (explained variance) of the dependent variable. When the variate contains only a single independent variable, the calculation of regression coefficients is straightforward and based on the direct or univariate correlation between the independent and dependent variable. The percentage of explained variance of the dependent variable is simply the square of the direct correlation.

But as independent variables are added to the variate, the calculations must also consider the intercorrelations among independent variables. If the independent variables are correlated, then they "share" some of their predictive power. Because we use only the prediction of the overall variate, the shared variance must not be "doubled counted" by using the direct correlations. Thus, we calculate some additional forms of the correlation to represent these shared effects. The first is the **partial correlation coefficient,** which is the correlation of an independent (X_i) and dependent (Y) variable when the effects of the other independent variable(s) have been removed from both X_i and Y. A second form of correlation is the **part** or **semipartial correlation,** which reflects the correlation between an independent and dependent variable while controlling for the predictive effects of all other independent variables on X_i. The two forms of correlation differ in that the partial correlation removes the effects of other independent variables from X_i and Y, whereas the part correlation removes the effects only from X_i. The partial correlation represents the incremental predictive effect of one independent variable from the collective effect of all others and is used to identify independent variables that have the greatest incremental predictive power when a set of independent variables is already in the regression variate. The part correlation represents the unique relationship predicted by an independent variable after the predictions shared with all other independent variables is taken out. Thus, the part correlation is used in apportioning variance among the independent variables. Squaring the part correlation gives the unique variance explained by the independent variable.

The accompanying diagram portrays the shared and unique variance among two correlated independent variables.

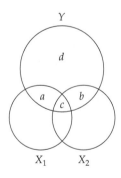

a = variance of Y uniquely explained by X_1
b = variance of Y uniquely explained by X_2
c = variance of Y explained jointly by X_1 and X_2
d = variance of Y not explained by X_1 or X_2

The variance associated with the partial correlation of X_2 controlling for X_1 can be represented as $b \div (d + b)$, where $d + b$ represents the unexplained variance after accounting for X_1. The part correlation of X_2 controlling for X_1 is $b \div (a + b + c + d)$, where $a + b + c + d$ represents the total variance of Y and b is the amount uniquely explained by X_2.

The analyst can also determine the shared and unique variance for independent variables

EXHIBIT 4.1 *(Continued)*

through simple calculations. The part correlation between the dependent variable (Y) and an independent variable (X_1) while controlling for a second independent variable (X_2) is calculated by the following equation:

Part correlation of Y, X_1, given X_2

$$= \frac{\text{Corr of } Y, X_1 - (\text{Corr of } Y, X_2 * \text{Corr of } X_1X_2)}{\sqrt{1.0 - (\text{Corr of } X_1X_2)^2}}$$

A simple example with two independent variables (X_1 and X_2) illustrates the calculation of both shared and unique variance of the dependent variable (Y). The direct correlations and the correlation between X_1 and X_2 are shown in the following correlation matrix:

	Y	X_1	X_2
Y	1.0		
X_1	.6	1.0	
X_2	.5	.7	1.0

The direct correlations of .60 and .50 represent fairly strong relationships with Y, but the correlation of .70 between X_1 and X_2 means that a substantial portion of this predictive power may be shared. The part correlation of X_1 and Y controlling for X_2 ($r_{Y, X_1(X_2)}$) and the unique variance predicted by X_1 can be calculated as

$$r_{Y, X_1(X_2)} = .60 - \frac{(.50)(.70)}{\sqrt{1.0 - .70^2}} = .35$$

unique variance predicted by $X_1 = .35^2 = .1225$

Because the direct correlation of X_1 and Y is .60, we also know that the total variance predicted by X_1 is $.60^2$, or .36. If the unique variance is .1225, then the shared variance must be .2375 (.36 − .1225).

We can calculate the unique variance explained by X_2 and confirm the amount of shared variance by the following:

$$r_{Y, X_1(X_2)} = \frac{.50 - (.60)(.70)}{\sqrt{1.0 - .70^2}} = .11$$

unique variance predicted by $X_1 = .11^2 = .0125$

With the total variance explained by X_2 being $.50^2$, or .25, the shared variance is calculated as .2375 (.25 − .0125). This confirms the amount found in the calculations for X_1.

Thus, the total variance explained (R^2) by the two independent variables is

Unique variance explained by X_1	.1225
Unique variance explained by X_2	.0125
Shared variance explained by X_1 and X_2	.2375
Total variance explained by X_1 and X_2	.3725

These calculations can be extended to more than two variables, but as the number of independent variables increases, it is easier to allow the statistical programs to perform the calculations.

The calculation of shared and unique variance illustrates the effects of multicollinearity on the ability of the independent variables to predict the dependent variable. Figure 4.9 shows these effects when faced with high to low levels of multicollinearity.

is dictated, several options exist. Here we discuss the identification and assessment procedures, and then we examine some possible remedies.

The simplest and most obvious means of identifying collinearity is an examination of the correlation matrix for the independent variables. The presence of high correlations (generally .90 and above) is the first indication of substantial collinearity. Lack of any high correlation values, however, does not ensure a lack of collinearity. Collinearity may be due to the combined effect of two or more other independent variables.

Two of the more common measures for assessing both pairwise and multiple variable collinearity are (1) the **tolerance** value and (2) its inverse—the **variance**

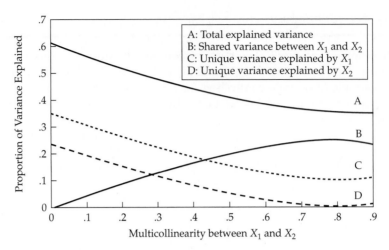

Correlation between dependent and independent variables:
X_1 and dependent (.60), X_2 and dependent (.50)

FIGURE 4.9 Proportions of Unique and Shared Variance by Levels of
Multicollinearity

TABLE 4.9 Regression Estimates with Multicollinear Data

A. DATA

| | *Variables in the Regression Analysis* | | |
| | *Dependent* | *Independent Variables* | |
Respondent	Y	V_1	V_2
1	5	6	13
2	3	8	13
3	9	8	11
4	9	10	11
5	13	10	9
6	11	12	9
7	17	12	7
8	15	14	7

B. CORRELATION MATRIX

	Y	V_1	V_2
Y	1.000		
V_1	.823	1.000	
V_2	−.977	−.913	1.000

C. REGRESSION ESTIMATES

Simple Regression (V_1):	$Y = -4.75 + 1.5V_1$
Simple Regression (V_2):	$Y = 29.75 + -1.95V_2$
Multiple Regression (V_1 and V_2):	$Y = 44.75 + -.75V_1 + -2.7V_2$

inflation factor (VIF). These measures tell us the degree to which each independent variable is explained by the other independent variables. In simple terms, each independent variable becomes a dependent variable and is regressed against the remaining independent variables. Tolerance is the amount of variability of the selected independent variable not explained by the other independent variables. Thus very small tolerance values (and thus large VIF values because $VIF = 1/tolerance$) denote high collinearity. A common cutoff threshold is a tolerance value of .10, which corresponds to a VIF value above 10. Each researcher must determine the degree of collinearity that is acceptable, because most defaults or recommended thresholds still allow for substantial collinearity. For example, the suggested cutoff for the tolerance value of .10 corresponds to a multiple correlation of .95. Moreover, a multiple correlation of .9 between one independent variable and all others (similar to the rule we applied in the pairwise correlation matrix) would result in a tolerance value of .19. Thus any variables with tolerance values below .19 (or above a VIF of 5.3) would have a correlation of more than .90.

We strongly suggest that the researcher always specify the tolerance values in regression programs, because the default values for excluding collinear variables allow for an extremely high degree of collinearity. For example, the default tolerance value in SPSS for excluding a variable is .0001, which means that until more than 99.99 percent of variance is predicted by the other independent variables, the variable could be included in the regression equation. Estimates of the actual effects of high collinearity on the estimated coefficients are possible but beyond the scope of this text (see Neter et al. [11]).

Even with diagnoses using VIF or tolerance values, we still do not necessarily know which variables are intercorrelated. A procedure development by Belsley et al. [2] allows for the intercorrelated variables to be identified, even if we have correlation among several variables. It provides the researcher greater diagnostic power in assessing the extent and impact of multicollinearity and is discussed in the appendix to this chapter.

Remedies for Multicollinearity

The remedies for multicollinearity range from modification of the regression variate to the use of specialized estimation procedures. Once the degree of collinearity has been determined, the researcher has a number of options:

- Omit one or more highly correlated independent variables and identify other independent variables to help the prediction. The researcher should be careful when following this option, however, to avoid creating specification error when deleting one or more independent variables.
- Use the model with the highly correlated independent variables for prediction only (i.e., make no attempt to interpret the regression coefficients).
- Use the simple correlations between each independent variable and the dependent variable to understand the independent–dependent variable relationship.
- Use a more sophisticated method of analysis such as Bayesian regression (or a special case—ridge regression) or regression on principal components to obtain a model that more clearly reflects the simple effects of the independent variables. These procedures are discussed in more detail in several texts [2, 11].

Each of these options requires that the researcher make a judgment on the variables included in the regression variate, which should always be guided by the theoretical background of the study.

Stage 6: Validation of the Results

After identifying the best regression model, the final step is to ensure that it represents the general population (generalizability) and is appropriate for the situations in which it will be used (transferability). The best guideline is the extent to which the regression model matches an existing theoretical model or set of previously validated results on the same topic. In many instances, however, prior results or theory are not available. Thus we also discuss empirical approaches to model validation.

Additional or Split Samples

The most appropriate empirical validation approach is to test the regression model on a new sample drawn from the general population. A new sample will ensure representativeness and can be used in several ways. First, the original model can predict values in the new sample, and predictive fit can be calculated. Second, a separate model can be estimated with the new sample and then compared with the original equation on characteristics such as the significant variables included; sign, size, and relative importance of variables; and predictive accuracy. In both instances, the researcher determines the validity of the original model by comparing it to regression models estimated with the new sample.

Many times the ability to collect new data is limited or precluded by such factors as cost, time pressures, or availability of respondents. When this is the case, the researcher may then divide the sample into two parts: an estimation subsample for creating the regression model and the holdout or validation subsample used to "test" the equation. Many procedures, both random and systematic, are available for splitting the data, each drawing two independent samples from the single data set. All the popular statistical packages have specific options to allow for estimation and validation on separate subsamples. Chapter 5 provides a discussion of the use of estimation and validation subsamples in discriminant analysis.

Whether a new sample is drawn or not, it is likely that differences will occur between the original model and other validation efforts. The researcher's role now shifts to being a mediator among the varying results, looking for the best model across all samples. The need for continued validation efforts and model refinements reminds us that no regression model, unless estimated from the entire population, is the final and absolute model.

Calculating the PRESS Statistic

An alternative approach to obtaining additional samples for validation purposes is to employ the original sample in a specialized manner by calculating the **PRESS statistic,** a measure similar to R^2 used to assess the predictive accuracy of the estimated regression model. It differs from the prior approaches in that not one, but $n - 1$ regression models are estimated. The procedure, similar to bootstrapping techniques discussed in chapter 12, omits one observation in the estimation of the regression model and then predicts the omitted observation with the estimated model. Thus, the observation cannot affect the coefficients of the model used to calculate its predicted value. The procedure is applied again, omitting another observation, estimating a new model, and making the prediction. The residuals for the observations can then be summed to provide an overall measure of predictive fit.

Comparing Regression Models

When comparing regression models, the most common standard used is overall predictive fit. We discussed earlier that R^2 provides us with this information, but it has one drawback: as more variables are added, R^2 will always increase. Thus, by including all independent variables, we will never find a higher R^2, but we may find that a smaller number of independent variables results in an almost identical value. Therefore, to compare between models with different numbers of independent variables, we use the adjusted R^2. The adjusted R^2 is also useful in comparing models between different data sets, as it will compensate for the different sample sizes.

Predicting with the Model

Model predictions can always be made by applying the estimated model to a new set of independent variable values and calculating the dependent variable values. However, in doing so, we must consider several factors that can have a serious impact on the quality of the new predictions:

1. When applying the model to a new sample, we must remember that the predictions now have not only the sampling variations from the original sample but also those of the newly drawn sample. Thus we should always calculate the confidence intervals of our predictions in addition to the point estimate to see the expected range of dependent variable values.

2. We must make sure that the conditions and relationships measured at the time the original sample was taken have not changed materially. For instance, in our credit card example, if most companies started charging higher fees for their cards, actual credit card holdings might change substantially, yet this information would not be included in the model.

3. Finally, we must not use the model to estimate beyond the range of independent variables found in the sample. For instance, in our credit card example, if the largest family had 6 members, it might be unwise to predict credit card holdings for families with 10 members. One cannot assume that the relationships are the same for values of the independent variables substantially greater or less than those in the original estimation sample.

Illustration of a Regression Analysis

The issues concerning the application and interpretation of regression analysis have been discussed in the preceding sections by following the six-stage model-building framework introduced in chapter 1 and discussed earlier in this chapter. To provide an illustration of the important questions at each stage, an illustrative example is presented here detailing the application of multiple regression to a research problem specified by HATCO. Chapter 1 introduced a research setting in which HATCO had obtained a number of measures in a survey of customers. To demonstrate the use of multiple regression, we show the procedures used by researchers to attempt to predict the product usage levels of the individuals in the sample with a set of seven independent variables.

Stage 1: Objectives of Multiple Regression

HATCO management has long been interested in more accurately predicting the level of business obtained from its customers in the attempt to provide a better basis for production controls and marketing efforts. To this end, researchers at HATCO proposed that a multiple regression analysis should be attempted to predict the product usage levels of the customers based on their perceptions of HATCO's performance. In addition to finding a way to accurately predict usage levels, the researchers were also interested in identifying the factors that led to increased product usage for application in differentiated marketing campaigns.

To apply the regression procedure, researchers selected product usage level (X_9) as the dependent variable (Y) to be predicted by independent variables representing perceptions of HATCO's performance. The following seven variables were included as independent variables:

X_1 Delivery speed
X_2 Price level
X_3 Price flexibility
X_4 Manufacturer image
X_5 Overall service
X_6 Salesforce image
X_7 Product quality

The relationship among the seven independent variables and product usage was assumed to be statistical, not functional, because it involved perceptions of performance and may have had levels of measurement error.

Stage 2: Research Design of a Multiple Regression Analysis

The HATCO survey obtained 100 respondents from their customer base. All 100 respondents provided complete responses, resulting in 100 observations available for analysis. The first question to be answered concerning sample size is the level of relationship (R^2) that can be detected reliably with the proposed regression analysis. Table 4.7 indicates that the sample of 100, with seven potential independent variables, is able to detect relationships with R^2 values of approximately 18 percent at a power of .80 with the significance level set at .01. If the significance level is relaxed to .05, then the analysis will identify relationships explaining about 13 percent of the variance. The proposed regression analysis was deemed sufficient to identify not only statistically significant relationships but also relationships that had managerial significance.

The sample of 100 observations also meets the proposed guideline for the ratio of observations to independent variables with a ratio of 15 to 1. Although the researchers can be assured that they will not be in danger of overfitting the sample, they must still validate the results to ensure the generalizability of the findings to the entire customer base.

Stage 3: Assumptions in Multiple Regression Analysis

Meeting the assumptions of regression analysis is essential to ensure that the results obtained were truly representative of the sample and that we have obtained the best results possible. Any serious violations of the assumptions must be detected and corrected if at all possible. Analysis to ensure that the research is meeting the basic assumptions of regression analysis involves two steps: (1) testing the individual dependent and independent variables, and (2) testing the overall relationship after model estimation. This section addresses the assessment of individual variables, and the overall relationship will be examined after the model has been estimated.

The three assumptions to be addressed for the individual variables are linearity, constant variance, and normality. For purposes of the regression analysis, we summarize the results found in chapter 2 detailing the examination of the dependent and independent variables. First, scatterplots of the individual variables did not indicate any nonlinear relationships between the dependent variable and the independent variables. Tests for heteroscedasticity found that only one of the variables (X_2) violated this assumption. Finally, in the tests of normality, three of the variables (X_2, X_4, and X_6) were found to violate the statistical tests. In each case, transformations by taking logarithms were indicated. The series of tests for these three assumptions underlying regression analysis indicated that the concerns should enter on the normality of three independent variables. Although regression analysis has been shown to be quite robust even when the normality assumption is violated, researchers should estimate the regression analysis with both the original and transformed variables to assess the consequences of nonnormality of the independent variables on the interpretation of the results. We estimate the regression equation with both original and transformed variables in the following section.

Stage 4: Estimating the Regression Model and Assessing Overall Model Fit

With the regression analysis specified in terms of dependent and independent variables, the sample deemed adequate for the objectives of the study, and the assumptions assessed for the individual variables, the model-building process now proceeds to estimation of the regression model and assessing the overall model fit. For purposes of illustration, the stepwise procedure is employed to select variables for inclusion in the regression variate. After the regression model has been estimated, the variate will be assessed for meeting the assumptions of regression analysis. Finally, the observations will be examined to determine whether any observation should be deemed influential. Each of these issues are discussed in the following sections.

Stepwise Estimation: Selecting the First Variable

Table 4.10 (p. 198) displays all the correlations among the seven independent variables and their correlations with the dependent variable (Y). Examination of the correlation matrix indicates that overall service (X_5) is most closely correlated with the dependent variable (.70). Our first step is to build a regression equation using this best independent variable. Note that the correlation of delivery speed (X_1) with the dependent variable is .68. However, X_1 is correlated (.61) with X_5. This is our first clue that use of both independent variables X_1 and X_5 might not be appropriate because they are as highly correlated with each other as they are with the dependent variable. The results of this first step appear as shown in Table 4.11. The concepts from Table 4.11 (p. 198) follow.

TABLE 4.10 Correlation Matrix: HATCO Data

				Predictors			
Variables	X_1	X_2	X_3	X_4	X_5	X_6	X_7
Predictors							
X_1 Delivery speed	1.00						
X_2 Price level	−.35	1.00					
X_3 Price flexibility	.51	−.49	1.00				
X_4 Manufacturer image	.05	.27	−.12	1.00			
X_5 Overall service	.61	.51	.07	.30	1.00		
X_6 Salesforce image	.08	.19	−.03	.79	.24	1.00	
X_7 Product quality	−.48	.47	−.45	.20	−.06	.18	1.00
Dependent							
$Y(X_9)$ Usage level	.68	.08	.56	.22	.70	.26	−.19

TABLE 4.11 Example Output: Step 1 of HATCO Multiple Regression Example

Variable entered: X_5 Overall service

Multiple R	.701
Multiple R^2	.491
Adjusted R^2	.486
Standard error of estimate	6.446

	Analysis of Variance			
	Sum of Squares	df	Mean Square	F Ratio
Regression	3,927.31	1	3,927.31	94.53
Residual	4,071.69	98	41.55	

	Variables in Equation				Variables Not in Equation	
Variables	Coefficient	Standard Error of Coefficient	Standardized Regression Coefficient (beta)	Partial t Value	Partial Correlation	t Value
Y-intercept	21.653	2.596		8.341		
X_5 Overall service	8.384	.862	.701	9.722		
X_1 Delivery speed					.439	4.812
X_2 Price level					−.453	−5.007
X_3 Price flexibility					.720	10.210
X_4 Manufacturer image					.022	.216
X_6 Salesforce image					.126	1.252
X_7 Product quality					−.216	−2.178

Multiple R   Multiple R is the correlation coefficient (at this step) for the simple regression of X_5 and the dependent variable. It has no plus or minus sign because in multiple regression the signs of the individual variables may vary, so this co-efficient reflects only the degree of association.

R Square   R square (R^2) is the correlation coefficient squared ($.701^2 = .491$), also referred to as the coefficient of determination. This value indicates the percentage of total variation of Y explained by X_5. The total sum of squares (3,927.31 + 4,071.69 = 7,999.0) is the squared error that would occur if we used only the mean of Y to predict the dependent variable. Using the values of X_5 reduces this error by 49.1 percent (3,927.31 ÷ 7,999.0 = .491).

Standard Error of the Estimate   The standard error of the estimate is another measure of the accuracy of our predictions. It is the square root of the sum of the squared errors divided by the degrees of freedom ($\sqrt{4.071} ÷ 98$). It represents an estimate of the standard deviation of the actual dependent values around the regression line; that is, it is a measure of variation around the regression line. The standard error of the estimate can also be viewed as the standard deviation of the prediction errors and thus becomes a measure to assess the absolute size of the prediction error. It is also used in estimating the size of the confidence interval for the predictions. See Neter et al. [11] for details regarding this procedure.

Variables in the Equation (Step 1)   In step 1, a single independent variable (X_5) is used to calculate the regression equation for predicting the dependent variable. For each variable in the equation, several measures need to be defined: the regression coefficient, the standard error of the coefficient, and the partial t value of variables in the equation.

Regression Coefficient
The value 8.38 is the regression coefficient (b_5) of the independent variable (X_5). Thus, the predicted value for each observation is the intercept (21.65) plus the regression coefficient (8.38) times its value of the independent variable ($Y = 21.65 + 8.38X_5$). The standardized regression coefficient, or beta value, of .70 is the value calculated from standardized data. With only one independent variable, the squared beta coefficient equals the coefficient of determination. The beta value allows you to compare the effect of X_5 on Y to the effect on Y of other independent variables at each stage, because this value reduces the regression coefficient to a comparable unit, the number of standard deviations. (Note that at this time we have no other variables available for comparison.)

Standard Error of the Coefficient
The standard error of the coefficient is the standard error of the estimate of b_5. The value of b_5 divided by the standard error (8.38 ÷ .86 = 9.74) is the calculated partial t value for a t-test of the hypothesis $b_5 = 0$. A smaller standard error implies more reliable prediction. Thus we would like to have small standard errors and therefore smaller confidence intervals. This standard error is also referred to as the standard error of the regression coefficient. It is an estimate of how much the regression coefficient will vary between samples of the same size taken from the same population. If one were to take multiple samples of the same size from the same population and use them to calculate the regression

equation, the standard error is an estimate of how much the regression coefficient would vary from sample to sample.

Partial t Value of Variables in the Equation

The partial t value of variables in the equation, as just calculated, measures the significance of the partial correlation of the variable reflected in the regression coefficient. It is particularly useful in Figure 4.7 in helping to determine whether any variable should be dropped from the equation once a variable has been added. Also given is the level of significance, which is compared to the threshold level set by the researcher for dropping the variable. In our example, we have set a .10 level for dropping variables from the equation. The critical value for a significance level of .10 with 98 degrees of freedom is 1.658. Therefore, X_5 meets our requirements for inclusion in the regression equation. F values may be given at this stage rather than t values. They are directly comparable because the t value is approximately the square root of the F value.

Variables Not in the Equation

Although X_5 has been included in the regression equation, six other potential independent variables remain for inclusion to improve the prediction of the dependent variable. For those values, two measures are available to assess their potential contribution: partial correlations and t values.

Partial Correlation

The partial correlation is a measure of the variation in Y not accounted for by the variables in the equation (only X_5 in step 1) that can be accounted for by each of these additional variables. For example, the value .720 represents the partial correlation of X_3 given that X_5 is in the equation. The partial correlation, however, can be misinterpreted. It does not mean that we explain 72.0 percent of the previously unexplained variance. It means that 51.8 percent ($.720^2 = .518$, the partial coefficient of determination) of the unexplained (not the total) variance can now be accounted for by X_3. Because 49.1 percent was already explained by X_5, 26.4 percent $[(1 - .491) \times .518]$ of the total variance could be explained by adding variable X_3. A Venn diagram illustrates this concept.

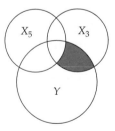

The shaded area of X_3 as a proportion of the shaded area of Y represents the partial correlation of X_3 with Y given X_5. This represents the variance in X_3 (after removal of the effects of X_5 on X_3) in common with the remaining variance in Y (after removing the effects of X_5 on Y). The calculation of the unique variance associated with adding X_3 can also be determined through the part correlation, as described in Exhibit 4.1.

t Values of Variables Not in the Equation

The column of *t* values measures the significance of the partial correlations for variables not in the equation. These are calculated as a ratio of the additional sum of squares explained by including a particular variable and the sum of squares left after adding that same variable. If this *t* value does not exceed a specified significance level, the variable will not be allowed to enter the equation. The tabled *t* value for a significance level of .05 with 97 degrees of freedom is 1.98. Looking at the column of *t* values in Table 4.12, we note that four variables (X_1, X_2, X_3, and X_7) exceed this value and are candidates for inclusion.

Table 4.11 shows that the simple correlation of X_1 with the dependent variable is .68, but the correlation is only .56 for X_3. Therefore, it may appear that variable X_1 would be included in the model next. But in deciding which additional variables to include in the equation, we first select the independent variable that exhibits the highest partial correlation with the dependent variable (not the highest correlation with Y). The partial correlation of X_3 is the largest (.720), and therefore X_3 (and even X_2) is considered for addition to the model before X_1.

We now know that a significant portion of the variance in the dependent variable is explained by X_5. We can also see that X_3 has the highest partial correlation coefficient with the dependent variable, and that the *t* value is significant at the .05 level. (It is significant at the .01 level as well.) We can now look at the new model using both X_5 and X_3.

Stepwise Estimation: Adding X_3

The multiple R and R squared values have both increased with the addition of X_3 (see Table 4.12, p. 202). The R^2 has increased by 26.4 percent, the amount we predicted when we examined the partial correlation coefficient from X_3 of .720. The increase in R^2 of 26.4 percent is derived by multiplying the 50.9 percent of variation that was not explained after step 1 by the partial correlation squared: $50.9 \times (.720)^2 = 26.4$; that is, of the 50.9 percent unexplained with X_5, $(.720)^2$ of this variance was explained by adding X_3, yielding a total variance explained of .755—that is, .491 + $[.509 \times (.720)^2]$.

The value of b_5 has changed very little. This is a further indication that variables X_5 and X_3 are relatively independent (the simple correlation between the two variables is .07). If the effect of X_3 on Y were totally independent of the effect of X_5, the b_5 coefficient would not change at all. The partial *t* values indicate that both X_5 and X_3 are statistically significant predictors of Y. The *t* value for X_5 is now 13.221, whereas it was 9.722 in step 1. The *t* value for X_3 examines the contribution of this variable given that X_5 is already in the equation. Note that the *t* value for X_3 (10.210) is the same value shown for X_3 in step 1 under the heading "Variables Not in the Equation" (see Table 4.11).

Because X_3 and X_5 both make significant contributions, neither will be dropped in the stepwise estimation procedure. We can now ask "Are other predictors available?" Looking at the partial correlations for the variables not in the equation in Table 4.12, we see that X_6 has the highest partial correlation (.236). This variable would explain 5.6 percent of the heretofore unexplained variance ($.236^2 = .056$), or 1.4 percent of the total variance [$(1 - .755) \times .056 = 0.014$]. This is a very modest contribution of explanatory power of our prediction, even though the partial correlation is significant at the .05 significance level. (The tabled *t* value for 96 degrees of freedom at a .05 level is 1.98, and the *t* value for X_6 is 2.378.)

TABLE 4.12 Example Output: Step 2 of HATCO Multiple Regression Example

Variable entered: X_3 Price flexibility

Multiple R	.869
Multiple R^2	.755
Adjusted R^2	.750
Standard error of estimate	4.498

Analysis of Variance

	Sum of Squares	df	Mean Square	F Ratio
Regression	6,036.5	2	3,018.26	149.18
Residual	1,962.5	97	20.23	

	Variables in Equation				Variables Not in Equation	
Variables	Coefficient	Standard Error of Coefficient	Standardized Regression Coefficient (beta)	Partial t Value	Partial Correlation	t Value
Y-intercept	−3.489	3.057				
X_3 Price flexibility	3.336	.327	.515	10.210		
X_5 Overall service	7.974	.603	.666	13.221		
X_1 Delivery speed					.021	.205
X_2 Price level					−.027	−.267
X_4 Manufacturer image					.181	1.808
X_6 Salesforce image					.236	2.378
X_7 Product quality					.169	1.683

Stepwise Estimation: A Third Variable Is Added—X_6

With X_6 entered into the regression equation, the results are shown in Table 4.13. As we predicted, the value of R^2 increases by 1.4 percent. In addition, examination of the partial correlations for X_1, X_2, X_4, and X_7 indicates that no additional value will be gained by adding them to the predictive equation. These partial correlations are all very small and have t values associated with them that would not be statistically significant at the level (.05) chosen for this model.

Evaluating the Variate for the Assumptions of Regression Analysis

In evaluating the estimated equation, we have considered statistical significance. We must also address two other basic issues: (1) meeting the assumptions underlying regression, and (2) identifying the influential data points. We consider each of these issues in the following sections.

The assumptions to examine are linearity, homoscedasticity, independence of the residuals, and normality. The principal measure used in evaluating the regression variate is the residual—the difference between the actual dependent vari-

TABLE 4.13 Example Output: Step 3 of HATCO Multiple Regression Example

Variable entered: X_6 Saleforce image

Multiple R	.877
Multiple R^2	.768
Adjusted R^2	.761
Standard error of estimate	4.394

	Analysis of Variance			
	Sum of Squares	*df*	*Mean Square*	*F Ratio*
Regression	6,145.7	3	2,048.6	106.12
Residual	1,853.3	96	19.3	

	Variables in Equation				*Variables Not in Equation*	
Variables	*Coefficient*	*Standard Error of Coefficient*	*Standardized Regression Coefficient (beta)*	*Partial t Value*	*Partial Correlation*	*t Value*
Y-intercept	−6.520					
X_3 Price flexibility	3.376	.320	.521	10.562		
X_5 Overall service	7.621	.607	.637	12.547		
X_6 Salesforce image	1.406	.591	.121	2.378		
X_1 Delivery speed					.040	.389
X_2 Price level					−.041	−.405
X_4 Manufacturer image					−.002	−.021
X_7 Product quality					.130	1.273

able value and its predicted value. For comparison, we use the studentized residuals. The most basic type of residual plot is shown in Figure 4.10 (p. 204), the studentized residuals versus the predicted values. As we can see, the residuals fall within a generally random pattern, very similar to the null plot in Figure 4.5a. However, we must make specific tests for each assumption to check for violations.

Linearity The first assumption, linearity, will be assessed through an analysis of residuals and partial regression plots. Figure 4.10 does not exhibit any nonlinear pattern to the residuals, thus ensuring that the overall equation is linear. But we must also be certain, when using more than one independent variable, that each independent variable's relationship is also linear to ensure its best representation in the equation. To do so, we use the partial regression plot for each independent variable in the equation. In Figure 4.11 (p. 204) we see that the relationships for X_3 and X_5 are quite well defined; thus they have strong and significant effects in the regression equation. The variable X_6 is less well defined, both in slope and scatter of the points, thus explaining its lesser effect in the equation (evidenced by the smaller coefficient, beta value, and significance level). For all three

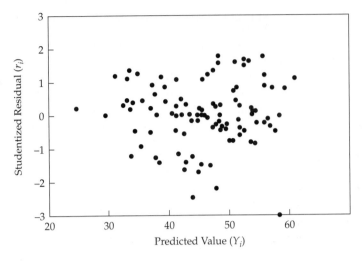

FIGURE 4.10 Analysis of Studentized Residuals

variables, no nonlinear pattern is shown, thus meeting the assumption of linearity for each independent variable.

Homoscedasticity   The next assumption deals with the constancy of the residuals across values of the independent variables. Our analysis is again through examination of the residuals (Figure 4.10), which shows no pattern of increasing or decreasing residuals. This finding indicates homoscedasticity in the multivariate (the set of independent variables) case.

Independence of the Residuals   The third assumption deals with the effect of carryover from one observation to another, thus making the residual not independent. When carryover is found in such instances as time-series data, the researcher must identify the potential sequencing variables (such as time in a time-series problem) and plot the residuals by this variable. For example, assume that the identification number represents the order in which we collect our responses. We could plot the residuals and see whether a pattern emerges. In our example, several variables, including the identification number and each independent vari-

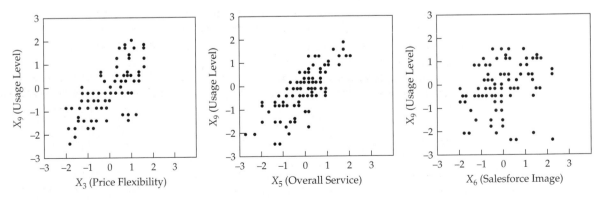

FIGURE 4.11 Standardized Partial Regression Plots

able, were tried and no consistent pattern was found. We must use the residuals in this analysis, not the original dependent variable values, because the focus is on the prediction errors, not the relationship captured in the regression equation.

Normality   The final assumption we will check is normality of the error term of the variate with a visual examination of the normal probability plots of the residuals. As shown in Figure 4.12, the values fall along the diagonal with no substantial or systematic departures; thus, the residuals are considered to represent a normal distribution. The regression variate is found to meet the assumption of normality.

Applying Remedies for Assumption Violations   After testing for violations of the four basic assumptions of multivariate regression for both individual variables and the regression variate, the researcher should assess the impact of any remedies on the results. In the examination of individual variables in chapter 2, the only remedies needed were the transformations of X_2, X_4, and X_6 to achieve normality. If we substitute these variables for their original values and reestimate the regression equation, we achieve almost identical results (see Table 4.14, p. 206). The same variables enter the equation, with the only substantive difference being a slightly stronger coefficient for the transformed X_6 variable and a slight improvement in the R^2 value. (.771 versus .768). The independent variables not in the equation still show nonsignificant levels for entry—even those that were transformed. Thus, in this case, the remedies for violating the assumptions improved the prediction slightly but did not alter the substantive findings.

Identifying Outliers as Influential Observations

For our final analysis, we attempt to identify any observations that are influential (having a disproportionate impact on the regression results) and determine whether they should be excluded from the analysis. Although more detailed procedures are

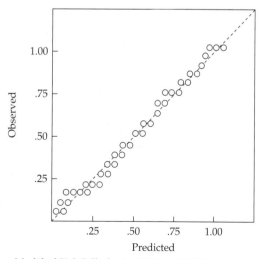

Modified K-S (Lilliefors): .0688 ($p > .2000$)

FIGURE 4.12 Normal Probability Plot:
Standardized Residuals

TABLE 4.14 Example Output Multiple Regression Results after Remedies for Violation of Assumptions

Multiple R	.878
Multiple R^2	.771
Adjusted R^2	.764
Standard error of estimate	4.368

Analysis of Variance

	Sum of Squares	df	Mean Square	F Ratio
Regression	6,167.1	3	2,055.71	107.73
Residual	1,831.9	96	19.08	

	Variables in Equation				Variables Not in Equation	
Variables	Coefficient	Standard Error of Coefficient	Standardized Regression Coefficient (beta)	Partial t Value	Partial Correlation	t Value
Y-intercept	−6.792	3.226		−2.11		
X_3 Price flexibility	3.409	.319	.526	10.70		
X_5 Overall service	7.640	.599	.639	12.75		
log X_6 Salesforce image	3.953	1.511	.131	2.62		
X_2 Delivery speed					.048	.469
log X_2 Price level					−.075	−.737
log X_4 Manufacturer image					−.047	−.463
X_7 Product quality					.118	1.163

available for identifying outliers as influential observations, we address the use of residuals in identifying outliers in the following section.

The most basic diagnostic tool involves the residuals and identification of any outliers—that is, observations not predicted well by the regression equation that have large residuals. Figure 4.13 shows the studentized residuals for each observation. Because the values correspond to t values, upper and lower limits can be set once the desired confidence interval has been established. Perhaps the most widely used level is 95 percent confidence ($\alpha = .05$). The corresponding t value is 1.96, thus identifying statistically significant residuals as those with residuals greater than this value. Four observations (7, 11, 14, 100) have significant residuals and can be classified as outliers. Outliers are important because they are observations not represented by the regression equation for one or more reasons, any one of which may be an influential effect on the equation that requires a remedy.

Examination of the residuals also can be done through the partial regression plots (see Figure 4.11). These plots help to identify influential observations for each independent–dependent variable relationship. Examining Figure 4.11, a set of separate and distinct points (observations 7, 11, 14, 100) can be identified for variables X_3 and X_6. These points are not well represented by the relationship and

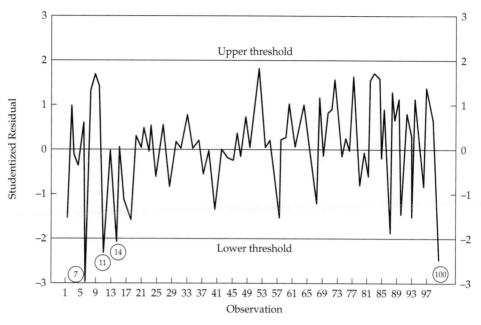

FIGURE 4.13 Plot of Studentized Residuals

thus could affect the partial correlation as well. More detailed analyses to ascertain whether any of the observations can be classified as influential observations, as well as what may be the possible remedies, are discussed in appendix 4A.

Stage 5: Interpreting the Regression Variate

With the model estimation completed, the regression variate specified, and the diagnostic tests that confirm the appropriateness of the results administered, we can now examine our predictive equation, which includes X_3, X_5, and X_6. The section of Table 4.13 headed "Variables in Equation" yields the prediction equation from the column labeled "Coefficient." From this column, we read the constant term (-6.520) and the coefficients (3.376, 7.621, and 1.406) for X_3, X_5, and X_6, respectively. The predictive equation would be written

$$Y = -6.520 + 3.376X_3 + 7.621X_5 + 1.406X_6$$

With this equation, the expected usage level for any customer could be calculated if their evaluations of HATCO are known. For illustration, let us assume that a customer rated HATCO with a value of 4.0 for each of these three measures. The predicted product usage level for that customer would be

$$\text{Predicted level of product usage} = -6.520 + 3.376(4.0) + 7.621(4.0) + 1.406(4.0)$$
$$= -6.520 + 13.504 + 30.484 + 5.624$$
$$= 43.902$$

In addition to providing a basis for predicting product usage levels, the regression coefficients also provide a means of assessing the relative importance of the individual variables in the overall prediction of product usage. In this situation, all the variables are expressed on the same scale, and thus direct comparisons can be made. But in most instances the beta coefficients are used for comparison between independent variables. In Table 4.13, the beta coefficients are listed in the column headed "Standardized Regression Coefficient." The researcher can make direct comparisons among the variables to ascertain their relative importance in the regression variate. For our example, X_5 (overall service) was the most important, followed closely by X_3 (price flexibility). The third independent variable, X_6 (salesforce image) was notably lower in importance. This supports its lower incremental amount of variance explained and the lower univariate correlation with product usage. Although significant, X_6 does not merit the attention that should be accorded to the other two independent variables.

Measuring the Degree and Impact of Multicollinearity

In any interpretation of the regression variate, the researcher must be aware of the impact of multicollinearity. As discussed earlier, highly collinear variables can distort the results substantially or make them quite unstable and thus not generalizable. Two measures are available for testing the impact of collinearity: (1) calculating the tolerance and VIF values, and (2) using the condition indices and decomposing the regression coefficient variance. The tolerance value is one minus the proportion of the variable's variance explained by the other independent variables. Thus a high tolerance value indicates little collinearity, and tolerance values approaching zero indicate that the variable is almost totally accounted for by the other variables. The variance inflation factor is the reciprocal of the tolerance value; thus we look for small VIF values as indicative of low intercorrelation among variables. In our example, tolerance values all exceed .93, indicating very low levels of collinearity (see Table 4.15). Likewise, the VIF values are all quite close to 1.0. These results indicate that interpretation of the regression variate coefficients should not be affected adversely by multicollinearity. A second approach to identifying multicollinearity and its effects is through the decomposition of the coefficient variance. Researchers are encouraged to explore this technique and the additional insights it offers into the interpretation of the regression equation. Details of this method are discussed in the appendix to this chapter.

Although multicollinearity does not have a substantial impact on the estimated regression variate, it does have an impact on the composition of the variate. After X_5 (the first variable added to the regression variate), the second-highest correlation with the dependent variable is X_1. Yet X_1 also has a fairly high level of collinearity (.61) with X_5. Because X_5 entered the regression variate first in the stepwise procedure, there is not enough unique variance in X_1 to justify its inclusion. Therefore, only X_5 entered the regression variate. However, it would be substantively incorrect to interpret from these results that X_1 has no impact on product usage, when in fact it was the independent variable with the second highest bivariate correlation with the dependent variable. The correct interpretation

TABLE 4.15 Testing for Multicollinearity: Assessing Tolerance and VIF Values

Variable	Tolerance	VIF
X_3 Price flexibility	0.99287009	1.00718111
X_5 Overall service	0.93639766	1.06792236
X_6 Salesforce image	0.93946418	1.06443654

would be that X_5 or X_1 demonstrates high impact, but that the similarity of their effect on product usage (high collinearity) dictates that only one of them is needed in the prediction process. The researcher must never allow an estimation procedure to dictate the interpretation of the results, but instead must understand the issues of interpretation accompanying each estimation procedure. For example, if all seven independent variables are entered into the regression variate, the researcher still has to contend with the effects of collinearity on the interpretation of the coefficients for X_5 and X_1, but in a different manner than if stepwise were used.

Stage 6: Validating the Results

The final task facing the researcher involves the validation process of the regression model. The primary concern of this process is to ensure that the results are generalizable to the population and not specific to the sample used in estimation. The most direct approach to validation is to obtain another sample from the population and assess the correspondence of the results from the two samples. In the absence of an additional sample, the researcher can assess the validity of the results in several approaches. The first involves examination of the adjusted R^2 value. In this situation, the adjusted R^2 value is .761 (as compared with an R^2 value of .768; see Table 4.14), which indicates that the estimated model is not overfitted to the sample and maintains an adequate ratio of observations to variables in the variate.

A second approach is to divide the sample into two subsamples, estimate the regression model for each subsample, and compare the results. Table 4.16 contains the overall stepwise results plus the results from stepwise models estimated for two subsamples of 50 observations each. Comparison of the overall model fit demonstrates a high level of similarity of the results in terms of R^2, adjusted R^2, and the standard error of the estimate. But in comparing the individual coefficients, one difference does appear. In sample 1, X_6 did not enter in the stepwise

TABLE 4.16 Split-Sample Validation of the Stepwise Estimation

Model Component	Overall (n = 100)	Sample 1 (n = 50)	Sample 2 (n = 50)
Independent Variables			
X_3 Price flexibility			
Regression coefficient	3.376	3.108	3.632
Beta coefficient	.521	.506	.529
t value	10.562	6.803	8.439
X_5 Overall service			
Regression coefficient	7.621	8.278	7.037
Beta coefficient	.637	.710	.574
t value	12.547	9.555	8.954
X_6 Salesforce image			
Regression coefficient	1.406	Not Entered	2.447
Beta coefficient	.121		.200
t value	2.378		3.166
Model Fit			
R^2	.768	.741	.826
Adjusted R^2	.761	.730	.814
Standard error of the estimate	4.394	4.764	3.816

results as it did in sample 2 and the overall sample. The omission of X_6 in one of the subsamples confirms that it was a marginal predictor, as indicated by the low beta and t values in the overall model.

Evaluating Alternative Regression Models

The stepwise regression model examined in the previous discussion provided a solid evaluation of the research problem as formulated, but the researcher is always well served in evaluating alternative regression models in the search for additional explanatory power and confirmation of earlier results. In this section we examine two additional regression models: one model including all seven independent variables in a confirmatory approach, and a second model adding a nonmetric variable (X_8) through the use of a dummy variable.

Confirmatory Regression Model

A primary alternative to the stepwise regression estimation method is the confirmatory approach, whereby the researcher specifies the independent variable to be included in the regression equation. In this manner, the researcher retains complete control over the regression variate in terms of both prediction and explanation. This approach is especially appropriate in situations of replication of prior research efforts or for validation purposes.

In this situation, the confirmatory perspective involves the inclusion of all seven perceptual measures as independent variables. These are the same variables considered in the stepwise estimation process, but in this instance all are directly entered into the regression equation at one time. Here the researcher can judge the potential impacts of multicollinearity on the selection of independent variables and the effect on overall model fit from including all seven variables.

The results in Table 4.17 are similar to the final results achieved through stepwise estimation (see Table 4.13), with two notable exceptions. First, even though more independent variables are included, the overall model fit decreases. Whereas the coefficient of determination increases (.768 to .775) because of the additional independent variables, the adjusted R^2 decreases (.761 to .758). This indicates the inclusion of several independent variables that were nonsignificant in the regression equation. Although they contribute to the overall R^2 value, they detract from the adjusted R^2. This illustrates the role of the adjusted R^2 in comparing regression variates with differing numbers of independent variables. Another indication of the overall poorer fit of the confirmatory model is the increase in the SEE (from 4.394 to 4.424). This illustrates that overall R^2 should not be the sole criterion for predictive accuracy because it can be influenced by many factors, one being the number of independent variables.

The other difference is in the regression variate, where multicollinearity affects the number and strength of the significant variables. First, only two variables (X_3 and X_5) are statistically significant, whereas the stepwise model contains a third variable (X_6). In the stepwise model, X_6 was the least significant variable, with a t value of 2.378. When the confirmatory approach is used, the multicollinearity of X_6 with X_4, which is now also included in the regression equation, decreases the unique impact for X_6 and results in a nonsignificant coefficient. Second, the strength and significance of all of the other variables is also decreased in the confirmatory model. The t values and statistical significance decrease for both X_3 and X_5 even though they remain significant. This is a result of their multicollinearity with the nonsignificant variables. Finally, the impact of multicollinearity between

TABLE 4.17 Example Output: Multiple Regression Results Using a Confirmatory Approach with All Seven Independent Variables

Multiple R	.880
Multiple R^2	.775
Adjusted R^2	.758
Standard error of estimate	4.424

	Analysis of Variance			
	Sum of Squares	df	Mean Square	F ratio
Regression	6,198.68	7	885.53	45.25
Residual	1,800.32	92	19.57	

	Variables in Equation				
Variables	Regression Coefficient	Standard Error of Coefficient	Standardized Regression Coefficient (beta)	t value	Statistical Significance
Intercept	−10.187	4.977		−2.047	.044
X_1 Delivery speed	−.058	2.013	−.008	−.029	.977
X_2 Price level	−.697	2.090	−.093	−.333	.740
X_3 Price flexibility	3.368	.411	.520	8.191	.000
X_4 Manufacturer image	−.042	.667	−.005	−.063	.950
X_5 Overall service	8.369	3.918	.699	2.136	.035
X_6 Salesforce image	1.281	.947	.110	1.352	.180
X_7 Product quality	.567	.355	.100	1.595	.114

X_1 and X_5 prevents the inclusion of X_1 in the stepwise model even though it has the second highest correlation with the dependent variable. This multicollinearity also is reflected in the confirmatory model in the nonsignificant coefficient for X_1 due to its multicollinearity with X_3 and X_5 in the equation.

The confirmatory approach provides the researcher with control over the regression variate, but at the possible cost of a regression equation with poorer prediction and explanation if the researcher does not closely examine the results. The confirmatory and sequential approaches both have strengths and weaknesses that should be considered in their use, but the prudent researcher will employ both approaches in order to address the strengths of each.

Including a Nonmetric Independent Variable

The prior discussion focused on the confirmatory estimation method as an alternative for possibly increasing prediction and explanation, but the researcher also should consider the possible improvement from the addition of nonmetric independent variables. As discussed in an earlier section and in chapter 2, nonmetric variables cannot be directly included in the regression equation, but must instead be represented by a series of newly created variables, termed dummy variables, which represent the separate categories of the nonmetric variable. In this example, the variable of firm size (X_8), which has the two categories of large and small firms, will be added to the regression variate. The variable is already coded in the appropriate form, as large firms are coded as 1 and small firms as 0. The variable can be directly included in the regression equation and will represent the difference in the dependent variable between large and small firms, given the other variables in the regression equation. Specifically, because large firms have the value 1, small firms act as the reference category. The regression coefficient is interpreted as the value for large firms compared to small firms. A positive coefficient indicates that large firms have greater product usage than small firms, whereas a negative coefficient indicates that small firms have greater product usage.

Table 4.18 contains the results of the addition of X_8 to the final stepwise results. Examination of the overall fit statistics indicates substantial improvement, with all of the measures (R^2, adjusted R^2, and SEE) increasing over the stepwise model (see Table 4.13). This is supported by the statistical significance of the regression coefficient for X_8 at a level well exceeding .05. The positive value of the coefficient (3.852) indicates that large firms, given their characteristics on the other three independent variables in the equation, still have a product usage level that is 3.8% higher, and that this is a statistically significant difference. The addition of X_8 added to both the prediction and explanation of the research question. This illustrates the manner in which the researcher can add nonmetric variables to the metric variables in the regression variate and improve both prediction and explanation.

TABLE 4.18 Example Output: Multiple Regression Results Adding X_8, Firm Size, as an Independent Variable by Using a Dummy Variable

Multiple R	.890
Multiple R^2	.793
Adjusted R^2	.784
Standard error of estimate	4.177

Analysis of Variance

	Sum of Squares	df	Mean Square	F ratio
Regression	6341.826	4	1585.457	90.889
Residual	1657.174	95	17.444	

Variables in Equation

Variables	Regression Coefficient	Standard Error of Coefficient	Standardized Regression Coefficient (beta)	t value	Statistical Significance
Intercept	−16.335	4.254		−3.840	.000
X_3 Price flexibility	4.245	.399	.655	10.630	.000
X_5 Overall service	8.055	.592	.673	13.613	.000
X_6 Salesforce image	1.462	.562	.125	2.602	.011
X_8 Firm size (large)	3.852	1.149	.211	3.353	.001

A Managerial Overview of the Results

The regression results, including the complementary evaluation of the confirmatory model and the addition of the nonmetric variable, all assist in addressing the basic research question: What affects product usage? In formulating a response, the researcher must consider two aspects: prediction and explanation. In terms of prediction, the regression models all achieve high levels of predictive accuracy. The amount of variance explained exceeds 75 percent and the expected error rate for any prediction is approximately ±9 percent. In this type of research setting, these levels, augmented by the results supporting model validity, provide the highest levels of assurance as to the quality and accuracy of the regression models as the basis for developing business strategies.

In terms of explanation, all of the estimated models arrived at essentially the same results: two strong influences (price flexibility and overall service) and a somewhat lesser influence (salesforce image) on product usage. Increases in any of these three variables will result in corresponding increases in product usage. For example, an increase of one point in the customer's perception of overall service will result in an average increase of at least 8 percent in product usage. Similar results are seen for the other two variables. Moreover, at least one firm characteristic, firm size, demonstrated a significant effect on product usage. Larger firms have levels of usage almost 4 percent higher than the smaller firms. These results provide management with a framework for developing strategies for improving product usage levels. Actions directed toward increasing the perceptions of HATCO can be justified in light of the corresponding increases in product usage.

Before developing any conclusions or business plans from these results, the researcher should note, however, that two of the influences (price flexibility and salesforce image) are contained in the two perceptual dimensions identified in chapter 3. These dimensions, which represent broad measures of customers' perceptions of HATCO, should thus also be considered in any conclusions. To state that *only* these three specific variables are influences on product usage would be a serious misstatement of the more complex patterns of collinearity among variables. Thus, these variables are better viewed as representatives of the perceptual dimensions, with the other variables in each dimension also considered in any conclusions drawn from these results. Management now has an objective analysis that confirms not only the specific influences of key variables, but also the perceptual dimensions that must be considered in any form of business planning regarding strategies aimed at impacting product usage.

Summary

This chapter presents a simplified introduction to the rationale and fundamental concepts underlying multiple regression analysis. It emphasizes that multiple regression analysis can describe and predict the relationships among two or more intervally scaled variables. Also, multiple regression analysis, which can be used to examine the incremental and total explanatory power of many variables, is a great improvement over the sequential analysis approach necessary with univariate techniques. Both stepwise and simultaneous techniques can be used to estimate a multiple regression equation, and under certain circumstances nonmetric dummy-coded variables can be included in the regression equation. Finally, numerous diagnostic techniques exist for testing both the assumptions underlying multiple regression analysis and the existence of cases exerting an undue influence on the resulting equation or predictions. This chapter provides a fundamental presentation of how regression works and what it can achieve. Familiarity with the concepts presented in this chapter will help the researcher better understand the more complex and detailed technical presentations in other textbooks while also providing a foundation for regression analyses the researcher might undertake.

Questions

1. How would you explain the relative importance of the independent variables used in a regression equation?
2. Why is it important to examine the assumption of linearity when using regression?
3. How can nonlinearity be corrected or accounted for in the regression equation?
4. Could you find a regression equation that would be acceptable as statistically significant and yet offer no acceptable interpretational value to management?
5. What is the difference in interpretation between regression coefficients associated with interval-scale independent variables and dummy-coded (0, 1) independent variables?
6. What are the differences between interactive and correlated independent variables? Do any of these differences affect your interpretation of the regression equation?
7. Are influential cases always to be omitted? Give examples of occasions when they should or should not be omitted.

References

1. Barnett, V., and T. Lewis (1984), *Outliers in Statistical Data*, 2d ed. New York: Wiley.
2. Belsley, D. A., E. Kuh, and R. E. Welsch (1980), *Regression Diagnostics: Identifying Influential Data and Sources of Collinearity*. New York: Wiley.
3. BMDP Statistical Software, Inc. (1991), *SOLO Power Analysis*. Los Angeles: BMDP.
4. Box, G. E. P., and D. R. Cox (1964), "An Analysis of Transformations." *Journal of the Royal Statistical Society B* 26: 211–43.
5. Cohen, J., and P. Cohen (1983), *Applied Multiple Regression/Correlation Analysis for the Behavioral Sciences*, 2d ed. Hillsdale, N.J.: Lawrence Erlbaum Associates.
6. Daniel, C., and F. S. Wood (1980), *Fitting Equations to Data*, 2d ed. New York: Wiley-Interscience.
7. Jaccard, J., R. Turrisi, and C. K. Wan (1990), *Interaction Effects in Multiple Regression*. Beverly Hills, Calif.: Sage Publications.
8. Johnson, R. A., and D. W. Wichern (1982), *Applied Multivariate Statistical Analysis*. Upper Saddle River, N.J., Prentice Hall.
9. Mason, C. H., and W. D. Perreault, Jr. (1991), "Collinearity, Power, and Interpretation of Multiple Regression Analysis." *Journal of Marketing Research* 28 (August): 268–80.
10. Mosteller, F., and J. W. Tukey (1977), *Data Analysis and Regression*. Reading, Mass.: Addison-Wesley.
11. Neter, J., W. Wassermann, and M. H. Kutner (1989), *Applied Linear Regression Models*. Homewood, Ill.: Irwin.
12. Rousseeuw, P. J., and A. M. Leroy Robust (1987), *Regression and Outlier Detection*. New York: Wiley.
13. Seer, G. A. F. (1984), *Multivariate Observations*. New York: Wiley.
14. Sharma, S., R. M. Durand, and O. Gur-Arie (1981), "Identification and Analysis of Moderator Variables." *Journal of Marketing Research* 18 (August): 291–300.
15. Weisberg, S. (1979), *Applied Linear Regression*. New York: Wiley, 1985.
16. Wilkinson, L. (1975), "Tests of Significance in Stepwise Regression." *Psychological Bulletin* 86: 168–74.

Annotated Articles

The following annotated articles are provided as illustrations of the application of multiple regression to substantive research questions of both a conceptual and managerial nature. The reader is encouraged to review the complete articles for greater detail on any of the specific issues regarding methodology or findings.

Hise, Richard T., Myron Gable, J. Patrick Kelly, and James B. McDonald (1983), "Factors Affecting the Performance of Individual Chain Store Units: An Empirical Analysis." *Journal of Retailing* 59(2), 22–39.

Multiple regression analysis is selected to evaluate the relative importance of eighteen variables suggested to have an impact on store performance. The independent variables fall into one of four categories: store manager, store characteristics, competition, and location. Moreover, each variable can be characterized as to its management controllability, time duration, and irreversibility. Separate regression equations are derived with three commonly used retail performance measures as the dependent variables: sales volume, contribution income, and return on assets. Although no hypotheses are provided for testing, results obtained from the study provide information on the types of variables that may be used in management's decision making in order to increase firm performance.

Given access to a sample of 132 outlets from a chain of mall retailers, the authors use a combination of forward, backward, and stepwise entry in the development of the three equations. In interpreting their findings, the authors examine standardized beta coefficients as an assessment of the relative impact of each variable on performance. The variables included in the study explain about half of the variation in the three performance measures as indicated by R^2 values of .60, .51, and .43 for sales volume, contribution income, and return on assets, respectively. Across the three regression equations, the variables affecting most strongly were inventory levels and fixed assets, manager's experience and years in that position, and number of employees. When examining the directions of the relationships, the authors are also able to generalize that these variables are more impactful when they are short-run in scope, controllable, and reversible. The application of multiple regression provides researchers with an objective means of relating a large set of possible factors to actual firm performance. From these findings managers are now able to assess the relative importance of these factors for use in developing and modifying managerial practices and strategies.

Clawson, C. Joseph (1974), "Fitting Branch Locations, Performance Standards, and Marketing Strategies to Local Conditions." *Journal of Marketing* 38 (January), 8–14.

This is one of the early attempts to apply multiple regression analysis as a more rigorous means—as opposed to more subjective techniques—for establishing goals and reevaluating marketing strategies. Specifically, the author uses multiple regression to select new locations, evaluate current site performance, and allocate marketing support among several geographically dispersed sites of savings and loan branches. The dependent variable of interest is "net savings gain," representing the increase in net deposits for each branch. Consultation with management leads to the selection of 24 independent variables representing different aspects of the population, competition, and characteristics of specific branches that might affect branch performance in attracting new deposits. The author estimates the regression equation from a sample of 26 savings and loan branches for which these data were available. The analysis, performed with the stepwise estimation method, identifies ten predictor variables explaining 91.5% of the variation in interbranch performance. The impact of multicollinearity is seen in that a number of the excluded predictor variables have high bivariate correlations with the dependent measure, but also are correlated with variables included in the equation. Positive contributors to branch performance include branch attractiveness, high percentages of persons in the age bracket of 45 to 64, savings gains by other competing branches in the area, and local promotions. Factors detracting from branch performance include the presence of competitor's main offices, concentrations of renter dwellings, or location on approaches to shopping centers. The implications for evaluating existing branches and identifying strengths and weaknesses are also discussed, along with the potential for evaluating prospective branch sites. In all, this technique is shown to be an important tool for managers in evaluating performance. As a statistical technique, multiple regression allows for the selection and assessment of independent variables in terms of their relative strengths toward explaining the dependent variable.

Alpert, Mark I., and Jon F. Bibb (1974), " 'Fitting Branch Locations, Performance Standards, and Marketing Strategies': A Clarification." *Journal of Marketing* 38 (April), 72–74.

The authors, while supporting the use of multiple regression, question the validity of the findings in Clawson's article. In their critique, they indicate that the model's lack of validity is due to one major methodological flaw—the use of too many independent variables (n) relative to the sample size (N). They demonstrate that even while the adjusted R^2 of the Clawson study remains high (86.7 percent), the use of stepwise regression when the number of independent variables falls below the recommended ratio of five cases per variable substantially increases the risk of finding results that are due solely to chance and lack generalizability. Although stepwise regression can "reduce" the number of variables to a smaller number in such a situation, it still faces the strong possibility of "overfitting" the data. The authors offer three recommendations for modifying the application to increase the value of the results: (1) increase the sample size to at least 5 to 10 times the number of independent variables (i.e., 120–240 branch locations); (2) reduce the number of independent variables through factor analysis (*see* chapter 3); or (3) cross-validate the model either through split-sample or on additional data.

Advanced Diagnostics for Multiple Regression Analysis

LEARNING OBJECTIVES

After reading our discussion of these techniques, you should be able to do the following:

- Understand how the condition index and regression coefficient variance–decomposition matrix isolate the effects, if any, of multicollinearity on the estimated regression coefficients.
- Identify those variables with unacceptable levels of collinearity or multicollinearity.
- Identify the observations with a disproportionate impact on the multiple regression results.
- Isolate influential observations and assess the relationships when the influential observations are deleted.

PREVIEW

Multiple regression is perhaps the most widely used statistical technique, and it has led the movement toward increased usage of other multivariate techniques. In moving from simple to multiple regression, the increased analytical power of the multivariate model requires additional diagnostics to deal with the correlations between variables and those observations with substantial impact on the results. This appendix describes advanced diagnostic techniques for assessing (1) the impact of multicollinearity and (2) the identity of influential observations and their impact on multiple regression analysis. Chapter 4 dealt with the basic diagnoses for these issues; here we discuss more sensitive procedures that have recently been proposed specifically for multivariate situations. These procedures are not refinements to the estimation procedures but instead address questions in interpreting the results that occur in the presence of multicollinearity and influential observations.

KEY TERMS

Before reading this appendix, review the key terms to develop an understanding of the concepts and terminology used. Throughout the appendix the key terms appear in **boldface.** Other points of emphasis in the appendix are *italicized.* Also, cross-references in the Key Terms appear in *italics.*

Collinearity   Relationship between two (collinearity) or more (*multicollinearity*) variables. Variables exhibit complete collinearity if their correlation coefficient is 1 and a complete lack of collinearity if their correlation coefficient is 0.

Condition index   Measure of the relative amount of variance associated with an *eigenvalue* so that a large condition index indicates a high degree of *collinearity.*

Cook's distance (D_i)   Summary measure of the influence of a single case (observation) based on the total changes in all other residuals when the case is deleted from the estimation process. Large values (usually greater than 1) indicate substantial influence by the case in affecting the estimated regression coefficients.

COVRATIO   Measure of the influence of a single observation on the entire set of estimated regression coefficients. A value close to 1 indicates little influence. If the COVRATIO value minus 1 is greater than $\pm 3p/n$ (where p is the number of independent variables + 1, and n is the sample size), the observation is deemed to be influential based on this measure.

Deleted residual   Process of calculating *residuals* in which the influence of each observation is removed when calculating its residual. This is accomplished by omitting the ith observation from the regression equation used to calculate its predicted value.

DFBETA   Measure of the change in a regression coefficient when an observation is omitted from the regression analysis. The value of DFBETA is in terms of the coefficient itself; a standardized form (SDFBETA) is also available. No threshold limit can be established for DFBETA, although the researcher can look for values substantially different from the remaining observations to assess potential influence. The SDFBETA values are scaled by their standard errors, thus supporting the rationale for cutoffs of 1 or 2, corresponding to confidence levels of .10 or .05, respectively.

DFFIT   Measure of an observation's impact on the overall model fit, which also has a standardized version (SDFFIT). The best rule of thumb is to classify as influential any standardized values (SDFFIT) that exceed $2/\sqrt{p/n}$, where p is the number of independent variables + 1, and n is the sample size. There is no threshold value for the DFFIT measure.

Eigenvalue   Measure of the amount of variance contained in the correlation matrix so that the sum of the eigenvalues is equal to the number of variables. Also known as the latent root or characteristic root.

Hat matrix Matrix that contains values for each observation on the diagonal, known as *hat values*, which represent the impact of the observed dependent variable on its predicted value. If all cases have equal influence, each would have a value of p/n, where p equals the number of independent variables $+ 1$, and n is the number of cases. If a case has no influence, its value would be $-1 \div n$, whereas total domination by a single case would result in a value of $(n - 1)/n$. Values exceeding $2p/n$ for larger samples, or $3p/n$ for smaller samples ($n \leq 30$), are candidates for classification as *influential observations*.

Hat value See *hat matrix*.

Influential observation Observation with a disproportionate influence on one or more aspects of the regression estimates. This influence may have as its basis (1) substantial differences from other cases on the set of independent variables, (2) extreme (either high or low) observed values for the criterion variables, or (3) a combination of these effects. Influential observations can either be "good," by reinforcing the pattern of the remaining data, or "bad," when a single or small set of cases unduly affects (biases) the regression estimates.

Leverage point An observation that has substantial impact on the regression results due to its differences from other observations on one or more of the independent variables. The most common measure of a leverage point is the *hat value*, contained in the *hat matrix*.

Mahalanobis distance (D^2) Measure of the uniqueness of a single observation based on differences between the observation's values and the mean values for all other cases across all independent variables. The source of influence on regression results is for the case to be quite different on one or more predictor variables, thus causing a shift of the entire regression equation.

Multicollinearity See *collinearity*.

Outlier In strict terms, an observation that has a substantial difference between its actual and predicted values of the dependent variable (a large *residual*) or between its independent variable values and those of other observations. The objective of denoting outliers is to identify observations that are inappropriate representations of the population from which the sample is drawn, so that they may be discounted or even eliminated from the analysis as unrepresentative.

Regression coefficient variance–decomposition matrix Method of determining the relative contribution of each *eigenvalue* to each estimated coefficient. If two or more coefficients are highly associated with a single eigenvalue (*condition index*), an unacceptable level of *multicollinearity* is indicated.

Residual Measure of the predictive fit for a single observation, calculated as the difference between the actual and predicted values of the dependent variable. Residuals are assumed to have a mean of zero and a constant variance. They not only play a key role in determining if the underlying assumptions of regression have been met, but also serve as a diagnostic tool in identifying *outliers* and *influential observations*.

SDFBETA See *DFBETA*.

SDFFIT See *DFFIT*.

Standardized residual Rescaling of the *residual* to a common basis by dividing each residual by the standard deviation of the residuals. Thus, standardized residuals have a mean of 0 and standard deviation of 1. Each standardized residual value can now be viewed in terms of standard errors in middle to large sample sizes. This provides a direct means of identifying outliers as those with values above 1 or 2 for confidence levels of .10 and .05, respectively.

Studentized residual   Most commonly used form of *standardized residual*. It differs from other standardization methods in calculating the standard deviation employed. To minimize the effect of a single *outlier*, the standard deviation of residuals used to standardize the ith residual is computed from regression estimates omitting the ith observation. This is done repeatedly for each observation, each time omitting that observation from the calculations. This approach is similar to the *deleted residual*, although in this situation the observation is omitted from the calculation of the standard deviation.

Tolerance   Commonly used measure of *collinearity* and *multicollinearity*. The tolerance of variable i (TOL_i) is $1 - R_i^{*2}$, where R_i^{*2} is the coefficient of determination for the prediction of variable i by the other predictor variables. Tolerance values approaching zero indicate that the variable is highly predicted (collinear) with the other predictor variables.

Variance inflation factor (VIF)   Measure of the effect of other predictor variables on a regression coefficient. VIF is inversely related to the *tolerance* value ($VIF_i = 1 \div TOL_i$). Large VIF values (a usual threshold is 10.0, which corresponds to a tolerance of .10) indicate a high degree of *collinearity* or *multicollinearity* among the independent variables.

Assessing Multicollinearity

As discussed in chapter 4, **collinearity** and **multicollinearity** can have several harmful effects on multiple regression, both in the interpretation of the results and in how they are obtained, such as stepwise regression. The use of several variables as predictors makes the assessment of multiple correlation between the independent variables necessary to identify multicollinearity. But this is not possible by examining only the correlation matrix (which shows only simple correlations between two variables). We now discuss a method developed specifically to diagnose the amount of multicollinearity present and the variables exhibiting the high multicollinearity. All major statistical programs have analyses providing these collinearity diagnostics.

A Two-Part Process

The method has two components. First is the **condition index,** which represents the collinearity of combinations of variables in the data set (actually the relative size of the **eigenvalues** of the matrix). The second is the **regression coefficient variance–decomposition matrix,** which shows the proportion of variance for each regression coefficient (and its associated variable) attributable to each condition index (eigenvalue). We combine these in a two-step procedure:

1. Identify all condition indices above a threshold value. The threshold value usually is in a range of 15 to 30, with 30 the most commonly used value.
2. For all condition indices exceeding the threshold, identify variables with variance proportions above 90 percent. A collinearity problem is indicated when a condition index identified in step 1 as above the threshold value accounts for a substantial proportion of variance (.90 or above) for *two or more* coefficients.

The example shown in Table 4A.1 illustrates the basic procedure and shows both the condition indices and variance decomposition values. First, a threshold of 30

TABLE 4A.1 Hypothetical Coefficient Variance–Decomposition Analysis with Condition Indices

Condition Index (u_i)	b_1	b_2	b_3	b_4	b_5	b_6
			Proportion of variance of coefficient:			
1.0 u_1	.003	.001	.000	.003	.000	.000
4.0 u_2	.000	.021	.005	.003	.000	.000
16.5 u_3	.000	.012	.003	.010	.000	.001
45.0 u_4	.001	.963	.003	.972	.983	.000
87.0 u_5	.003	.002	.000	.009	.015	.988
122.0 u_6	.991	.001	.987	.003	.002	.011

for the condition index selects three condition indices (u_4, u_5, and u_6). Second, coefficients exceeding the .90 threshold for these three condition indices are b_1 and b_3 with u_6; b_2, b_4, and b_5 with u_4; and b_6 with u_5 (see the underlined values in Table 4A.1). However, u_5 has only a single value (b_6) associated with it; thus no collinearity is shown for this coefficient. As a result, we would attempt to remedy the significant correlations among two sets of variables: (1) V_1, V_3 and (2) V_2, V_4, V_5.

An Illustration of Assessing Multicollinearity

In chapter 4, we discussed the use of multiple regression in predicting the usage level (X_9) for HATCO customers. The stepwise procedure identified three statistically significant predictors: X_3, X_5, and X_6. However, before we accept these regression results as valid, we must examine the degree of multicollinearity and its effect on the results. To do so, we employ the two-part process (condition indices and the decomposition of the coefficient variance) and make comparisons with the conclusions drawn from the **variance inflation factor (VIF)** and **tolerance** values.

As discussed in chapter 4 and also presented in Table 4A.2 (p. 222), the VIF and tolerance values indicate inconsequential collinearity. No VIF value exceeds 10.0, and the tolerance values show that collinearity does not explain more than 10 percent of any independent variable's variance. This conclusion is supported when we employ the two-step procedure. First, examine the condition indices. We fail to pass the first step, as no condition index is greater than 30.0. Even if we were to proceed to the second step by using a threshold value of 15 for the condition index, we would select only a single condition index (u_4), where only one coefficient (the intercept) loads highly. Thus, we can find no support for the existence of multicollinearity in these regression results, just as indicated by the tolerance and VIF measures.

Identifying Influential Observations

In chapter 4, we examined only one approach to identifying **influential observations,** that being the use of studentized residuals to identify outliers. As noted then, however, observations may be classified as influential even though they are not recognized as outliers. In fact, many times an influential observation will not be identified as an outlier because it has influenced the regression estimation to such a degree as to make its residual negligible. Thus, we need to examine more specific procedures to measure an observation's influence in several aspects of multiple regression [2]. In the following discussion, we discuss a four-step process of identifying outliers, leverage points, and influential observations. As noted before, an observation may fall into one or more of these classes, and the course of action to be taken depends on the judgment of the researcher, based on the best available evidence.

Step 1: Examining Residuals

Residuals are instrumental in detecting violations of model assumptions, and they also play a role in identifying observations that are **outliers** on the dependent variable. We employ two methods of detection: the analysis of residuals and partial regression plots.

Analysis of Residuals

The **residual** is the primary means of classifying an observation as an outlier. The residual for the ith observation is calculated as the actual minus predicted values of the dependent variable, or:

$$\text{Residual}_i = Y_i - \hat{Y}_i$$

The residual and its many forms, however, are actually based on two procedures: the cases used for calculating the predicted value, and the use (or nonuse) of some form of standardization. We have already seen how we calculate the residual, but a second form, the **deleted residual,** differs from the normal residual in that the ith observation is omitted when estimating the regression equation used to calculate the predicted value for that observation. Thus, each observation has no impact on its own predicted value in the deleted residual. The deleted residual is less commonly used, although it has the benefit of reducing the influence of the observation on its calculation.

TABLE 4A.2 Testing for Multicollinearity in Multiple Regression

PART A. ASSESSING TOLERANCE AND VIF VALUES

Variable	Tolerance	Variance Inflation Factor (VIF)
X_3	0.993	1.007
X_5	0.936	1.068
X_6	0.939	1.064

PART B. USING THE CONDITION INDICES AND THE DECOMPOSITION OF COEFFICIENT VARIANCE MATRIX

			Proportion of Coefficient Variance			
Number	Eigenvalue	Condition Index	Intercept	X_3	X_5	X_6
1	3.882	1.000	.001	.002	.004	.005
2	.060	8.046	.014	.110	.021	.850
3	.045	9.246	.020	.136	.909	.042
4	.012	17.719	.965	.753	.066	.103

*Underlined values exceed the .90 level

The second procedure in defining a residual involves whether to standardize the residuals. Residuals that are not standardized are in the scale of the dependent variable, which is useful in interpretation but gives no insight as to what is too large or small enough not to consider. **Standardized residuals** are the result of a process of creating a common scale by dividing each residual by the standard deviation of residuals. After standardization, the residuals have a mean of 0 and a standard deviation of 1. With a fairly large sample size (50 or above), standardized residuals approximately follow the t distribution, such that residuals exceeding a threshold such as 1.96 (the critical t value at the .05 confidence level) can be deemed statistically significant. Observations falling outside the threshold are statistically significant in their difference from 0 and can be considered outliers. This means that the predicted value is also significantly different from the actual value at the .05 level. A stricter test of significance has also been proposed, which accounts for multiple comparisons being made across various sample sizes [4].

A special form of standardized residual is the **studentized residual.** It is similar in concept to the deleted residual, but in this case the ith observation is eliminated when deriving the standard deviation used to standardize the ith residual. The rationale is that if an observation is extremely influential, it may not be identified by the normal standardized residuals because of its impact on the estimated regression model. The studentized residual eliminates the case's impact on the standardization process and offers a "less influenced" residual measure. It can be evaluated by the same criteria as the standardized residual.

The five types of residuals typically calculated by combining the options for calculation and standardization are (1) the normal residual, (2) the deleted residual, (3) the standardized residual, (4) the studentized residual, and (5) the studentized deleted residual. Each type of residual offers unique perspectives on both the predictive accuracy of the regression equation by its designation of outliers and the possible influences of the observation on the overall results.

Partial Regression Plots

To graphically portray the impact of individual cases, the partial regression plot is most effective. Because the slope of the regression line of the partial regression plot is equal to the variable's coefficient in the regression equation, an outlying case's impact on the regression slope (and the corresponding regression equation coefficient) can be readily seen. The effects of outlying cases on individual regression coefficients are portrayed visually. Again, most computer packages have the option of plotting the partial regression plot, so the researcher need look only for outlying cases separated from the main body of observations. A visual comparison of the partial regression plots with and without the observation(s) deemed influential can illustrate their impact.

Step 2: Identifying Leverage Points from the Predictors

Our next step is finding those observations that are substantially different from the remaining observations on one or more independent variables. These cases are termed **leverage points** in that they may "lever" the relationship in their direction because of their difference from the other observations (see chapter 4 for a general description of leverage points).

Hat Matrix

When only two predictor variables are involved, plotting each variable on an axis of a two-dimensional plot will show those observations substantially different from the others. Yet, when a larger number of predictor variables are included in the regression equation, the task quickly becomes impossible through univariate methods. However, we are able to use a special matrix, the **hat matrix,** which contains values **(hat values)** for each observation that indicate leverage. The hat values represent the combined effects of all independent variables for each case.

Hat values (found on the diagonal of the hat matrix) measure two aspects of influence. First, for each observation, the hat value is a measure of the distance of the observation from the mean center of all other observations on the independent variables (similar to the Mahalanobis distance discussed next). Second, large diagonal values also indicate that the observation carries a disproportionate weight in determining its predicted dependent variable value, thus minimizing its residual. This is an indication of influence, because the regression line must be closer to this observation (i.e., strongly influenced) for the small residual to occur. This is not necessarily "bad," as illustrated in chapter 4, when the influential observations fall in the general pattern of the remaining observations.

What is a large hat value? The range of possible values are between 0 and 1, and the average value is p/n, where p is the number of predictors (the number of coefficients plus one for the constant) and n is the sample size. The rule of thumb for situations in which p is greater than 10 and the sample size exceeds 50 is to select observations with a leverage value greater than twice the average ($2p/n$). When the number of predictors is less than 10 or the sample size is less than 50, use of three times the average ($3p/n$) is suggested. The more widely used computer programs all have options for calculating and printing the leverage values for each observation. The analyst must then select the appropriate threshold value ($2p/n$ or $3p/n$) and identify observations with values larger than the threshold.

Mahalanobis Distance

A measure comparable to the hat value is the **Mahalanobis distance (D^2),** which considers only the distance of an observation from the mean values of the independent variables and not the impact on the predicted value. The Mahalanobis distance (discussed in chapter 2) is another means of identifying outliers. It is limited in this purpose because threshold values depend on a number of factors, and a rule of thumb threshold value is not possible. It is possible, however, to determine statistical significance of the Mahalanobis distance from published tables [1]. Yet even without the published tables, the researcher can look at the values and identify any observations with substantially higher values than the remaining observations. For example, a small set of observations with the highest Mahalanobis values that are two to three times the next highest value would constitute a substantial break in the distribution and another indication of possible leverage.

Step 3: Single-Case Diagnostics Identifying Influential Observations

Up to now we have found outlying points on the predictor and criterion variables but have not formally estimated the influence of a single observation on the results. In this third step, all the methods rely on a common proposition: the most direct measure of influence involves deleting one or more observations and observing the changes in the regression results in terms of the residuals, individual coefficients, or overall model fit. The researcher then needs only to examine the values and select those observations that exceed the specified value. We have already discussed one such measure, the studentized deleted residual, but will now explore several other measures appropriate for diagnosing individual cases.

Influences on Individual Coefficients

The impact of deleting a single observation on each regression coefficient is shown by the **DFBETA** and its standardized version the **SDFBETA.** Calculated as the change in the coefficient when the observation is deleted, DFBETA is the relative effect of an observation on each coefficient. Guidelines for identifying particularly high values of SDFBETA suggest that a threshold of ± 1.0 or ± 2.0 be applied to small sample sizes, whereas $\pm 2\sqrt{n}$ should be used for medium and larger data sets.

Overall Influence Measures

Cook's distance (D_i) is considered the single most representative measure of influence on overall fit. It captures the impact of an observation from two sources: the size of changes in the predicted values when the case is omitted (outlying studentized residuals) as well as the observation's distance from the other observations (leverage). A rule of thumb is to identify observations with a Cook's distance of 1.0 or greater, although the threshold of $4/(n - k - 1)$, where k is the number of independent variables, is suggested as a more conservative measure in small samples or for use with larger data sets. Even if no observations exceed this threshold, however, additional attention is dictated if a small set of observations has substantially higher values than the remaining observations.

A similar measure is the **COVRATIO,** which estimates the effect of the observation on the efficiency of the estimation process. Specifically, COVRATIO represents the degree to which an observation impacts the standard errors of the regression coefficients. It differs from the DFBETA and SDFBETA in that it considers all coefficients collectively rather than each coefficient individually. A threshold can be established at $1 \pm 3p/n$. Values above the threshold of $1 + 3p/n$ make the estimation process more efficient, whereas those less than $1 - 3p/n$ detract from the estimation efficiency. This allows the COVRATIO to act as another indicator of observations that have a substantial influence both positively and negatively on the set of coefficients.

A third measure is **SDFFIT,** the degree to which the fitted values change when the case is deleted. A cutoff value of $2\sqrt{(k + 1)/(n - k - 1)}$ has been suggested to detect substantial influence. Even though both Cook's distance and SDFFIT are measures of overall fit, they must be complemented by the measures of steps 1 and 2 to enable us to determine whether influence arises from the residuals, leverage, or both. A unstandardized version **(DFFIT)** is also available.

Step 4: Selecting and Accommodating Influential Observations

The identification of influential observations is more a process of convergence by multiple methods than a reliance on a single measure. Because no single measure totally represents all dimensions of influence, it is a matter of interpretation, although these measures typically identify a small set of observations. In selecting the observations with large values on the diagnostic measures, the researcher should first identify all observations exceeding the threshold values, and then examine the range of values in the data set being analyzed and look for large gaps between the highest values and the remaining data. Some additional observations will be detected that should be classified as influential.

After identification, several courses of action are possible. First, if the number of observations is small and a justifiable argument can be made for their exclusion, the cases should be deleted and the new regression equation estimated. If, however, deletion cannot be justified, several more "robust" estimation techniques are available, among them robust regression [3]. Whatever action is taken, it should meet our original objective of making the data set most representative of the actual population, to ensure validity and generalizability.

Example from the HATCO Database

To illustrate the diagnostic procedures for influential observations, we examine the regression example from chapter 4. In identifying any observations that are "influential" (having a disproportionate impact of the regression results), we will also determine whether they should be excluded from the analysis. The four-step process described above is used in the following steps.

Step 1: Examining the Residuals

Our first diagnosis involves examination of the residuals and identification of any outliers (i.e., observations with large residuals that are not predicted well by the regression equation). Table 4A.3 shows the five types of residuals for the 100 observations. As can be seen, there is a close agreement among the related values (e.g., residual versus deleted residual, or standardized versus studentized residual). One noticeable characteristic, however, is that as the residual values get larger, the difference between the related values grows larger as well. This indicates that the process of eliminating an observation from affecting its own residual does make a difference. Studentized residuals are the most common form of residuals used to denote outliers. Figure 4A.1 (p. 229) shows the studentized residuals for each observation. Because the standardized and studentized residuals correspond to t values, upper and lower limits have been set at the 95 percent confidence interval (t value $= \pm 1.96$). Statistically significant residuals are those falling outside these limits. Four observations (7, 11, 14, 100) have significant standardized and studentized residuals and can be classified as outliers. An outlier is not necessarily an influential point, nor do all influential points have to be outliers. But outliers are important to note as they represent observations not represented by the regression equation because of one or more reasons, one of which may have an influential effect on the equation and thus may require a remedy.

Examination of the residuals can also be done though the partial regression plots (see Figure 4A.2, p. 230). These plots help identify influential observations for each predictor–criterion variable relationship. As noted in Figure 4A.2, a set of separate and distinct points (observations 7, 11, 14, 100) can be identified as possible influential observations for all three independent variables (X_3, X_5, and X_6). These points are not well represented by any of the relationships and thus affect to some degree the estimated regression coefficients.

TABLE 4A.3 Comparison of Five Basic Types of Residuals

Observation	Actual Value	Predicted Value	Residual	Deleted Residual	Standardized Residual	Studentized Residual	Studentized Deleted Residual
1	32.0	38.299	−6.299	−6.431	−1.434	−1.449	−1.457
2	43.0	39.425	3.575	3.807	.814	.840	.838
3	48.0	49.290	−1.290	−1.397	−.294	−.306	−.304
4	32.0	34.401	−2.401	−2.487	−.546	−.556	−.554
5	58.0	58.268	−.268	−.295	−.061	−.064	−.064
6	45.0	42.637	2.363	2.412	.538	.543	.541
7	46.0	58.552	−12.552	−13.687	−2.857	−2.983	−3.115
8	44.0	38.843	5.157	5.313	1.174	1.191	1.194
9	63.0	56.106	6.894	7.092	1.569	1.591	1.604
10	54.0	48.121	5.879	6.088	1.338	1.362	1.368
11	32.0	42.367	−10.367	−10.724	−2.360	−2.400	−2.462
12	47.0	50.580	−3.580	−3.646	−.815	−.822	−.821
13	39.0	38.798	.202	.211	.046	.047	.047
14	38.0	48.292	−10.292	−10.675	−2.342	−2.386	−2.447
15	54.0	53.421	.579	.598	.132	.134	.133
16	49.0	49.194	−.194	−.203	−.044	−.045	−.045
17	38.0	44.236	−6.236	−6.536	−1.419	−1.453	−1.462
18	40.0	47.496	−7.496	−7.847	−1.706	−1.746	−1.765
19	54.0	56.860	−2.860	−3.025	−.651	−.669	−.667
20	55.0	53.421	1.579	1.630	.359	.365	.363
21	41.0	41.049	−.049	−.050	−.011	−.011	−.011
22	35.0	33.036	1.964	2.111	.447	.463	.462
23	55.0	55.655	−.655	−.676	−.149	−.151	−.151
24	36.0	34.094	1.906	1.974	.434	.442	.440
25	49.0	51.656	−2.656	−2.700	−.605	−.609	−.607
26	49.0	49.999	−.999	−1.013	−.227	−.229	−.228
27	36.0	33.756	2.244	2.327	.511	.520	.518
28	54.0	56.097	−2.097	−2.219	−.477	−.491	−.489
29	49.0	52.918	−3.918	−4.115	−.892	−.914	−.913
30	46.0	47.339	−1.339	−1.410	−.305	−.313	−.311
31	43.0	41.579	1.421	1.466	.323	.329	.327
32	53.0	53.175	−.175	−.179	−.040	−.040	−.040
33	60.0	56.449	3.551	3.673	.808	.822	.820
34	47.0	46.540	.460	.476	.105	.106	.106
35	35.0	33.978	1.022	1.094	.233	.241	.240
36	39.0	41.227	−2.227	−2.259	−.507	−.510	−.508
37	44.0	44.774	−.774	−.814	−.176	−.181	−.180
38	46.0	46.147	−.147	−.151	−.034	−.034	−.034
39	29.0	28.812	.188	.203	.043	.044	.044
40	28.0	34.184	−6.184	−6.400	−1.407	−1.432	−1.440
41	40.0	42.467	−2.467	−2.498	−.562	−.565	−.563
42	58.0	58.268	−.268	−.295	−.061	−.064	−.064
43	53.0	54.141	−1.141	−1.225	−.260	−.269	−.268
44	48.0	49.654	−1.654	−1.677	−.376	−.379	−.377
45	38.0	36.105	1.895	1.970	.431	.440	.438
46	54.0	55.381	−1.381	−1.482	−.314	−.326	−.324
47	55.0	51.553	3.447	3.606	.784	.802	.801
48	43.0	43.055	−.055	−.059	−.013	−.013	−.013
49	57.0	49.426	7.574	7.694	1.724	1.737	1.756
50	53.0	51.417	1.583	1.691	.360	.372	.371
51	41.0	40.682	.318	.325	.072	.073	.073

(Continued)

TABLE 4A.3 (*Continued*)

Observation	Actual Value	Predicted Value	Residual	Deleted Residual	Standardized Residual	Studentized Residual	Studentized Deleted Residual
52	53.0	52.114	.886	.904	.202	.204	.203
53	50.0	52.300	−2.300	−2.454	−.523	−.541	−.539
54	32.0	38.807	−6.807	−6.946	−1.549	−1.565	−1.577
55	39.0	37.997	1.003	1.052	.228	.234	.233
56	47.0	45.875	1.125	1.156	.256	.260	.258
57	62.0	59.143	2.857	3.098	.650	.677	.675
58	65.0	60.592	4.408	4.612	1.003	1.026	1.027
59	46.0	46.278	−.278	−.281	−.063	−.064	−.063
60	50.0	48.731	1.269	1.293	.289	.292	.290
61	54.0	51.075	2.925	3.046	.666	.679	.677
62	60.0	55.209	4.791	4.943	1.090	1.108	1.109
63	47.0	47.725	−.725	−.744	−.165	−.167	−.166
64	36.0	41.615	−5.615	−5.893	−1.278	−1.309	−1.314
65	40.0	34.479	5.521	5.845	1.256	1.293	1.297
66	45.0	46.912	−1.912	−1.951	−.435	−.440	−.438
67	59.0	55.172	3.828	4.013	.871	.892	.891
68	46.0	41.465	4.535	4.642	1.032	1.044	1.045
69	58.0	51.521	6.479	6.693	1.474	1.499	1.509
70	49.0	47.598	1.402	1.421	.319	.321	.320
71	50.0	51.771	−1.771	−1.927	−.403	−.420	−.419
72	55.0	53.898	1.102	1.176	.251	.259	.258
73	51.0	51.675	−.675	−.703	−.154	−.157	−.156
74	60.0	53.147	6.853	6.991	1.560	1.575	1.588
75	41.0	37.848	3.152	3.226	.717	.726	.724
76	49.0	53.398	−4.398	−4.501	−1.001	−1.013	−1.013
77	42.0	41.922	.078	.079	.018	.018	.018
78	47.0	50.438	−3.438	−3.617	−.782	−.803	−.801
79	39.0	33.239	5.761	6.119	1.311	1.351	1.357
80	56.0	49.429	6.571	6.706	1.495	1.511	1.521
81	59.0	53.240	5.760	5.940	1.311	1.331	1.337
82	47.0	48.718	−1.718	−1.909	−.391	−.412	−.410
83	41.0	36.944	4.056	4.316	.923	.952	.952
84	37.0	45.516	−8.516	−8.613	−1.938	−1.949	<u>−1.979</u>
85	53.0	47.452	5.548	5.621	1.263	1.271	1.275
86	43.0	41.544	1.456	1.483	.331	.334	.333
87	51.0	46.211	4.789	4.850	1.090	1.097	1.098
88	36.0	43.035	−7.035	−7.133	−1.601	−1.612	−1.626
89	34.0	36.882	−2.882	−2.959	−.656	−.665	−.663
90	60.0	56.412	3.588	3.786	.817	.839	.838
91	49.0	48.018	.982	1.015	.223	.227	.226
92	39.0	46.256	−7.256	−7.615	−1.651	−1.692	−1.709
93	43.0	43.757	−.757	−.826	−.172	−.180	−.179
94	36.0	30.641	5.359	5.706	1.220	1.259	1.262
95	31.0	35.818	−4.818	−4.961	−1.097	−1.113	−1.114
96	25.0	23.373	1.627	1.820	.370	.392	.390
97	60.0	53.625	6.375	6.512	1.451	1.466	1.475
98	38.0	33.122	4.878	5.143	1.110	1.140	1.142
99	42.0	39.991	2.009	2.051	.457	.462	.460
100	33.0	44.848	−11.848	−12.241	<u>−2.696</u>	<u>−2.741</u>	<u>−2.840</u>
Minimum	25.0	23.373	−12.552	−13.687	−2.857	−2.983	−3.115
Maximum	65.0	60.592	7.754	7.694	1.724	1.737	1.756

Note: Statistically significant residuals are underlined.

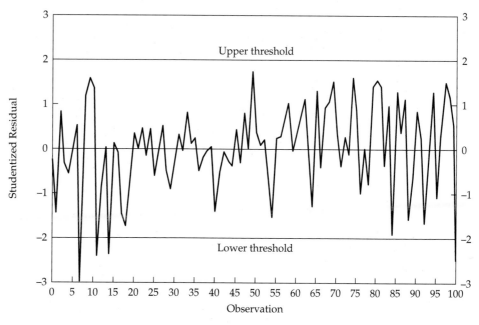

FIGURE 4A.1 Studentized Residuals

Step 2: Identifying Leverage Points

Although the residual analysis identifies outliers, we are unable to see the magnitude of each observation's impact on the predictions. To do so, we use the hat values as the measure of leverage (see Table 4A.4, pp. 231–232, and Figure 4A.3, p. 233). We use the threshold limits of $2p/n$ because the sample size exceeds 50. The calculated threshold value is $2 \times 4/100 = .08$. Using this threshold limit, we identify four cases (5, 42, 82, and 96) as leverage points. These cases do have an influential effect on the results, but as discussed earlier, all influential points are not "bad" (i.e., they do not distort or detract from the relationship). We must examine the observations further to assess what part(s) of the results they affect.

Step 3: Single-Case Diagnostics

We have identified the potentially influential cases, and we now complement this analysis with measures reflecting their impact on specific portions of the regression results. In each instance, we delete a case from the regression estimation and observe the changes in the results. Our first measure is the studentized deleted residual (see Table 4A.3). As with the residual analysis in step 1, four cases are again significant but are joined by an additional case (case 84). In each instance, these are outliers from the regression equation estimated without their influence in either the calculation of the predicted values or the standardized process.

Our next step is to identify the impact on the estimated regression coefficients by use of the SDFBETA values (see Table 4A.4 and Figure 4A.4, p. 233). Given their standardized format and the sample size, the threshold will be established at $\pm 2/\sqrt{n}$, or .08. Whereas a number of observations are identified for each coefficient, several observations are identified as affecting multiple coefficients (7, 40, 94, 98, and 100). Also, observation 7 has the greatest impact on the intercept

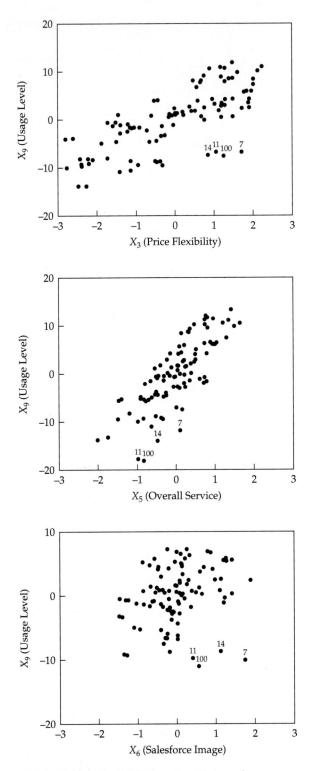

FIGURE 4A.2 Partial Regression Plots and
Influential Observations

TABLE 4A.4 Diagnostic Measures for Identifying Influential Observations

Observation	Mahalanobis Distance	Cook's Distance	Leverage (Hat Values)	COVRATIO	SDFFIT	SDFBETA Intercept	X_3	X_5	X_6
1	1.049	.011	.011	.975	−.211	−.168	.102	.079	.052
2	5.033	.011	.051	1.078	.213	.045	−.088	−.080	.162
3	<u>6.564</u>	.002	.066	<u>1.125</u>	−.087	−.020	.055	−.064	.016
4	2.453	.003	.025	1.066	−.105	−.075	.027	.078	.008
5	<u>8.040</u>	.000	<u>.081</u>	<u>1.147</u>	−.020	.014	−.009	.000	−.017
6	1.020	.002	.010	1.051	.078	.038	−.001	−.010	−.050
7	<u>7.219</u>	<u>.201</u>	.073	<u>.771</u>	<u>−.937</u>	<u>.657</u>	<u>−.402</u>	−.038	<u>−.759</u>
8	1.917	.011	.019	1.012	.208	.167	−.152	.011	−.079
9	1.775	.018	.018	.964	.272	−.191	.172	.100	.051
10	2.410	.016	.024	.999	.258	.031	−.148	.137	.056
11	2.301	<u>.050</u>	.023	<u>.842</u>	<u>−.457</u>	−.007	−.192	<u>.341</u>	−.131
12	.801	.003	.008	1.032	−.111	.040	−.071	−.008	.017
13	3.222	.000	.033	1.089	.010	.005	.001	−.003	−.007
14	2.566	<u>.053</u>	.026	<u>.847</u>	<u>−.472</u>	.172	−.152	.169	<u>−.372</u>
15	2.095	.000	.021	1.076	.024	−.013	.020	.000	−.001
16	3.163	.000	.032	1.088	−.009	.001	−.006	.000	.005
17	3.541	.025	.036	1.000	−.320	−.131	<u>.247</u>	−.148	.000
18	3.438	.036	.035	.959	−.382	−.111	.050	−.192	<u>.314</u>
19	4.412	.006	.045	1.083	−.160	.117	−.093	−.002	−.110
20	2.095	.001	.021	1.070	.065	−.036	.053	.001	−.001
21	2.255	.000	.023	1.078	−.002	−.001	−.001	.001	.001
22	5.908	.004	.060	1.111	.126	.062	.020	−.099	−.034
23	1.971	.000	.020	1.074	−.026	.019	−.013	−.008	−.013
24	2.432	.002	.025	1.071	.083	.063	−.035	−.057	.003
25	.596	.002	.006	1.043	−.078	.029	−.033	−.031	.011
26	.417	.000	.004	1.055	−.027	.004	.001	−.014	.000
27	2.548	.003	.026	1.069	.100	.077	−.045	−.067	.004
28	4.425	.003	.045	1.092	−.118	.083	−.069	.006	−.082
29	3.748	.010	.038	1.058	−.205	.052	−.128	−.043	.120
30	4.055	.001	.041	1.094	−.072	−.018	.049	−.044	−.001
31	2.079	.001	.021	1.071	.058	.030	−.046	.008	.011
32	1.132	.000	.011	1.066	−.006	.004	−.003	−.001	−.003
33	2.304	.006	.023	1.049	.152	−.075	.077	.091	−.047
34	2.421	.000	.024	1.079	.020	.003	−.012	.007	.008
35	5.491	.001	.055	1.113	.063	.042	−.003	−.029	−.042
36	.421	.001	.004	1.046	−.061	−.036	.013	.024	.012
37	3.912	.000	.040	1.096	−.041	−.018	.030	−.023	.010
38	1.331	.000	.013	1.068	−.005	.001	−.003	.003	.000
39	<u>6.388</u>	.000	.065	<u>1.127</u>	.013	.008	−.002	−.011	.002
40	2.359	.018	.024	.990	−.269	<u>−.238</u>	<u>.151</u>	.131	.060
41	.227	.001	.002	1.042	−.063	−.031	.011	.021	.007
42	<u>8.040</u>	.000	<u>.081</u>	<u>1.147</u>	−.020	.014	−.009	.000	−.017
43	5.755	.001	.058	1.116	−.072	.007	−.024	−.036	.056
44	.370	.001	.004	1.051	−.045	.010	−.018	−.010	.010
45	2.768	.002	.028	1.075	.087	.075	−.046	−.014	−.053
46	5.735	.002	.058	1.114	−.088	.014	−.031	−.047	.064
47	3.393	.007	.034	1.062	.172	−.094	.131	−.060	.074
48	4.431	.000	.045	1.103	−.003	−.002	.002	−.002	.001
49	.553	.012	.006	.932	.221	−.048	.112	.027	−.062
50	5.327	.002	.054	1.107	.097	.011	−.004	.062	−.076
51	1.270	.000	.013	1.067	.011	.003	.002	−.008	.000

(Continued)

TABLE 4A.4 (*Continued*)

Observation	Mahalanobis Distance	Cook's Distance	Leverage (Hat Values)	COVRATIO	SDFFIT	SDFBETA Intercept	X_3	X_5	X_6
52	1.047	.000	.011	1.063	.029	−.006	.003	.020	−.009
53	5.231	.005	.053	1.099	−.139	.004	.070	−.098	−.026
54	.989	.012	.010	.960	−.225	−.166	.132	.078	−.001
55	3.639	.001	.037	1.091	.052	.016	.016	−.040	−.007
56	1.687	.000	.017	1.069	.043	−.005	−.002	−.012	.034
57	6.730	.010	.068	1.110	.196	−.081	−.031	.135	.085
58	3.397	.012	.034	1.044	.221	−.161	.121	.136	.016
59	.199	.000	.002	1.055	−.007	−.002	.002	−.002	.002
60	.821	.000	.008	1.058	.040	.004	.000	.020	−.022
61	2.945	.005	.030	1.065	.138	−.072	.105	−.048	.053
62	2.056	.010	.021	1.022	.198	−.084	.096	.112	−.076
63	1.473	.000	.015	1.068	−.027	.008	−.018	.011	−.004
64	3.676	.021	.037	1.018	−.292	−.168	.242	−.111	.046
65	4.496	.025	.045	1.029	.314	.121	−.044	−.261	.152
66	.978	.001	.010	1.055	−.062	.005	−.036	.017	.015
67	3.565	.010	.036	1.057	.196	−.131	.091	.005	.144
68	1.288	.006	.013	1.020	.160	.106	−.117	.027	−.032
69	2.186	.019	.022	.980	.275	−.014	.050	.157	−.184
70	.346	.000	.003	1.052	.037	.002	.008	.007	−.016
71	7.034	.004	.071	1.126	−.124	−.014	.070	−.097	.013
72	5.257	.001	.053	1.110	.067	−.001	.001	.049	−.048
73	2.996	.000	.030	1.085	−.032	.009	−.022	−.002	.015
74	.964	.012	.010	.958	.225	−.119	.135	.072	−.006
75	1.292	.003	.013	1.044	.111	.080	−.065	−.048	.013
76	1.274	.006	.013	1.022	−.155	.088	−.109	−.030	−.001
77	1.064	.000	.011	1.065	.003	.001	.001	−.002	.000
78	3.916	.008	.040	1.068	−.183	.020	−.101	−.021	.123
79	4.808	.028	.049	1.026	.339	.149	−.055	−.288	.144
80	1.001	.012	.010	.967	.218	−.016	.081	.061	−.122
81	2.014	.014	.020	.998	.236	−.049	.064	.147	−.130
82	8.911	.005	.090	1.150	−.137	.000	.075	−.046	−.081
83	4.964	.015	.050	1.068	.241	.083	−.114	−.110	.164
84	.129	.011	.001	.897	−.212	−.068	.047	−.037	.052
85	.284	.005	.003	.987	.146	−.027	.059	−.031	.031
86	.809	.001	.008	1.057	.045	.031	−.023	.001	−.021
87	.250	.004	.003	1.004	.124	−.006	.043	−.037	.015
88	.363	.009	.004	.947	−.191	−.108	.061	.002	.077
89	1.612	.003	.016	1.051	−.109	−.068	.023	.079	−.004
90	4.176	.010	.042	1.068	.197	−.137	.092	.013	.145
91	2.230	.000	.023	1.076	.041	−.010	.000	−.006	.034
92	3.682	.035	.037	.970	−.380	−.135	.060	−.168	.322
93	7.266	.001	.073	1.136	−.054	−.012	.038	−.012	−.027
94	5.030	.026	.051	1.039	.321	.285	−.266	−.101	−.012
95	1.862	.009	.019	1.019	−.192	−.165	.094	.083	.064
96	9.485	.005	.096	1.159	.134	.098	−.037	−.115	−.003
97	1.099	.012	.011	.973	.217	−.127	.137	.062	.015
98	4.101	.018	.041	1.041	.266	.219	−.223	−.067	.014
99	1.015	.001	.010	1.055	.066	.039	−.038	−.021	.018
100	2.186	.062	.022	.778	−.517	.084	−.261	.335	−.212
Minimum	.129	.000	.001	.771	−.937	−.238	−.402	−.288	−.759
Maximum	9.485	.201	.096	1.159	.339	.657	.247	.341	.322

Note: Values exceeding thresholds are underlined.

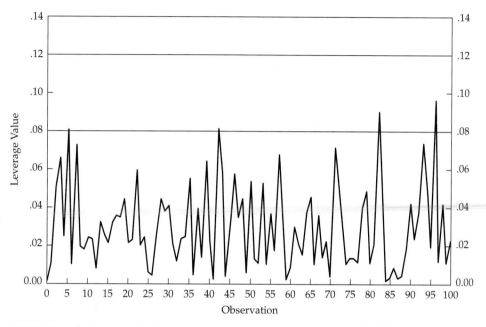

FIGURE 4A.3 Leverage Values

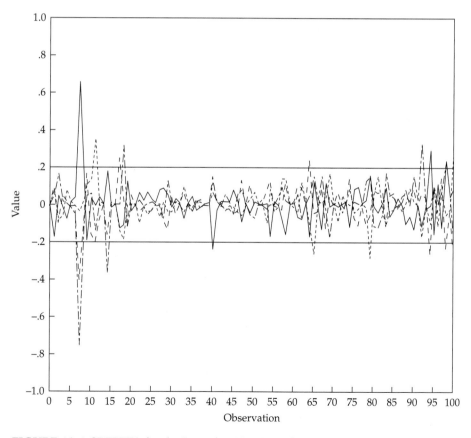

FIGURE 4A.4 SDFBETA for the Intercept, X_3, X_5, and X_6

and the coefficients for X_3 and X_5, and observation 11 has the greatest impact on the coefficient for X_5. We can combine these results with other measures to identify the final set of influential measures.

We have three measures for considering all coefficients simultaneously. The first measure is Cook's distance, which simultaneously captures both the leverage effects and the change in residuals (see Table 4A.4 and Figure 4A.5). The calculated threshold of .042 $[4/(100 - 3 - 1)]$ identifies again the four observations of 7, 11, 14, and 100. The COVRATIO (see Table 4A.4 and Figure 4A.6) identifies eight observations that contribute positively to the estimation process (3, 5, 39, 42, 71, 82, 93, and 96), whereas only four observations detract from the process (7, 11, 14, and 100). Finally, the SDFFIT measure (see Table 4A.4 and Figure 4A.7), with a threshold of .408, also selects four observations (7, 11, 14, and 100) as possible influential observations.

Table 4A.5 (p. 236) summarizes the results of the diagnostic measures for identifying influential observations. Across all of the measures, four observations (7, 11, 14, and 100) have emerged as potentially negative influential points owing to their substantial influence and differences from the remaining observations. Observation 7 is substantially higher than the other observations on almost all of the diagnostic measures, but the other three observations should also be considered as influential due to their consistent identification across almost all of the measures.

Step 4: Selecting and Accommodating Influential Cases

Although there is no single procedure for identifying influential cases and then deciding on the course of action, the basic premise is quite simple. In the absence of data entry error or other correctable reasons, influential cases that are substantially different from the remaining data on one or more variables should be

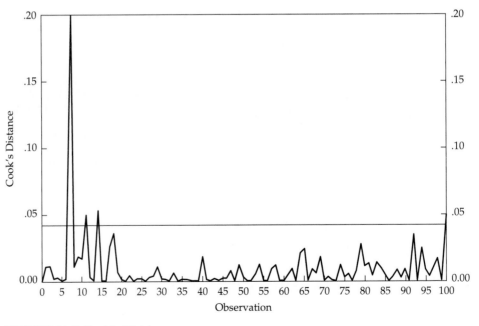

FIGURE 4A.5 Cook's Distance

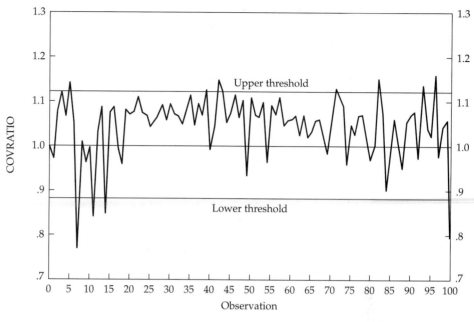

FIGURE 4A.6 COVRATIO

closely examined. If it is ascertained that a case is unrepresentative of the general population, it should be eliminated. Our objective is to estimate the regression equation on a representative sample to obtain generalizable results. If the sample contains one or more unrepresentative observations, we are hindered in achieving this objective.

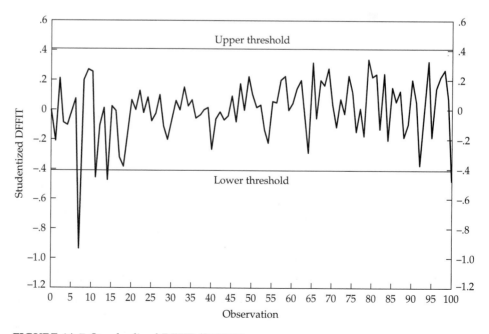

FIGURE 4A.7 Standardized DFFIT (SDFFIT)

For purposes of illustration, we select four cases (7, 11, 14, and 100) for elimination. These cases were consistently identified by the diagnostic analyses and are deemed to be the cases with the most impact on improving the regression equation. Table 4A.6 shows the final regression model with these four cases eliminated. Comparing these results with those in chapter 4, we see substantial improvement in every area. Overall prediction is improved, with the R^2 changing from .768 to .833, more than the effect we obtained by adding the third variable (X_6) to the equation. Also, the standard error decreased from 4.39 to 3.69, a 16 percent improvement. Moreover, each coefficient improved in statistical significance, indicating a strengthening of the relationships by removing these influential outliers.

Overview

The identification of influential cases is an essential step in interpreting the results of regression analysis. The analyst must be careful, however, to use discretion in the elimination of cases identified as influential. There are always outliers in any population, and the researcher must be careful not to trim the data set so

TABLE 4A.5 Summary of Diagnostic Tests for Influential Observations

Diagnostic Measure	Threshold Value Specification	Calculated Threshold Value	Observations Exceeding Threshold[a]
Residuals			
Standardized	Critical t value at specified confidence level	±1.96	**7,** 11, 14, 100
Studentized	Critical t value at specified confidence level	±1.96	**7,** 11, 14, 100
Studentized deleted	Critical t value at specified confidence level	±1.96	**7,** 11, 14, 84, 100
Leverage			
Hat values	Small sample: $3(k+1)/n$	.12	None
	Medium/large sample: $2(k+1)/n$	.08	5, 42, 82, **96**
Mahalanobis distance	Evaluate the distribution of values	None	Top ten observations: **96,** 82, 5, 42, 93, 7, 71, 57, 3, 39
Single-case measures			
SDFBETA	Small sample: Critical t value at specified confidence level	±1.96	None
	Medium/large sample: $2/\sqrt{n}$	.02	Intercept: **7,** 40, 94, 98 X_3: **7,** 17, 40, 64, 94, 98, 100 X_5: **11,** 65, 79, 100 X_6: **7,** 14, 18, 92, 100
Cook's distance	$\dfrac{4}{n-k-1}$	.042	**7,** 11, 14, 100
COVRATIO	$1 \pm \dfrac{3(k+1)}{n}$	Upper: 1.12	Upper: 3, 5, 39, 42, 71, 82, 93, **96**
		Lower: .88	Lower: **7,** 11, 14, 100
SDFFIT	$2\sqrt{\dfrac{k+1}{n-k-1}}$	.408	**7,** 11, 14, 100

[a]Observations with maximum values appear in **boldface.**
n = sample size
k = number of independent variables

TABLE 4A.6 Multiple Regression Results after Eliminating Four Influential Observations

Overall Regression Model Results

Multiple R	.913
Multiple R^2	.833
Adjusted R^2	.828
Standard error of estimate	3.698

Analysis of Variance

	Sum of Squares	Degrees of Freedom	Mean Square	F ratio
Regression	6291.8	3	2097.3	153.37
Residual	1258.1	92	13.7	

Variables in Equation

Variables	Coefficient	Standard Error of Coefficient	Standardized Regression Coefficient	Partial t value
Intercept	−9.645	2.803		−3.440
X_3 Price flexibility	3.719	.274	.582	13.566
X_5 Overall service	7.094	.521	.601	13.605
X_8 Salesforce image	2.337	.521	.198	4.482

Variables Not in Equation

	Partial Correlation	t value
X_1 Delivery speed	−.004	−.036
X_2 Price level	−.019	−.177
X_4 Manufacturer image	−.116	−1.117
X_7 Product quality	.136	1.311

that good results are almost guaranteed. Yet, one must also attempt to best represent the relationships in the sample, and the influence of just a few cases may distort or completely inhibit achieving this objective. Thus, we recommend that one use these techniques wisely and with care, as they represent both potential benefit and harm.

Summary

As the applications of multiple regression analysis increase in both scope and complexity, it becomes essential to explore the issues addressed in this appendix. Both multicollinearity and influential observations can have substantial impacts on the results and their interpretation. However, recent advances in diagnostic techniques, such as those described earlier, now provide the researcher with a simplified method of performing analyses that will identify problems in each of these areas. Whenever the regression analysis encounters either multi-collinearity or influential observations, the researcher is encouraged to investigate the issues raised here and to take the appropriate remedies if needed.

Questions

1. Describe the reasons for not relying solely on the univariate correlation matrix for diagnosing multicollinearity.
2. In what instances does the detection of outliers potentially miss other influential observations?
3. Describe the differences in using residuals (including the studentized residual) versus the single-case diagnostics of DFBETA and DFFIT.
4. What criteria would you suggest for determining whether an observation was to be deleted from the analysis?

References

1. Barnett, V., and T. Lewis (1984), *Outliers in Statistical Data*, 2d ed. New York: Wiley.
2. Belsley, D. A., E. Kuh, and R. E. Welsch (1980), *Regression Diagnostics: Identifying Influential Data and Sources of Collinearity*. New York: Wiley.
3. Rousseeuw, P. J., and A. M. Leroy (1987), *Robust Regression and Outlier Detection*. New York: Wiley.
4. Weisberg, S. (1985), *Applied Linear Regression*. New York: Wiley.

Multiple Discriminant Analysis and Logistic Regression

LEARNING OBJECTIVES

Upon completing this chapter, you should be able to do the following:

- State the circumstances under which a linear discriminant function rather than multiple regression should be used.
- Understand the assumptions underlying discriminant analysis in assessing the appropriateness of its use for a particular problem.
- Identify the major issues in the application of discriminant analysis.
- Describe the two computation approaches for discriminant analysis and state when each should be used.
- Tell how to interpret the nature of the linear discriminant function, that is, to identify independent variables with significant discriminatory power.
- Explain the usefulness of the classification matrix methodology and tell how to develop a classification matrix.
- Describe the various approaches to evaluating the classificatory power of the discriminant function, including the distinction between the hit ratio and multiple regressions R^2.
- Justify the use of a split-sample approach in validating the discriminant function.
- Understand the strengths and weaknesses of discriminant analysis compared to logistic regression.
- Intercept the results of a logistic regression analysis, with comparisons to multiple regression and discriminant analysis.

CHAPTER PREVIEW

Multiple regression is undoubtedly the most widely used multivariate dependence technique. The primary basis for the popularity of regression has been its ability to predict and explain metric variables. But what about nonmetric variables? Multiple regression is not suitable for this question. This chapter introduces two techniques—discriminant analysis and logistic regression—that address the situation when a dependent variable is nonmetric. In this situation, the researcher is interested in the prediction and explanation of the relationships that impact the category in which an object is located, such as why a person is or is not a customer, or if a firm will succeed or fail. This chapter avoids the jargon and mathematical formulas without glossing over important concepts. The two major objectives of this chapter are: (1) to introduce the underlying nature, philosophy, and conditions of multiple discriminant analysis and logistic regression; and (2) to demonstrate the application and interpretation of these techniques with an illustrative example.

Chapter 1 stated that the basic purpose of discriminant analysis is to estimate the relationship between a single nonmetric (categorical) dependent variable and a set of metric independent variables, in this general form:

$$Y_1 \qquad = X_1 + X_2 + X_3 + \ldots + X_n$$

(nonmetric)   (metric)

Multiple discriminant analysis and logistic regression have widespread application in situations in which the primary objective is to identify the group to which an object (e.g., person, firm, or product) belongs. Potential applications include predicting the success or failure of a new product, deciding whether a student should be admitted to graduate school, classifying students as to vocational interests, determining the category of credit risk for a person, or predicting whether a firm will be successful. In each instance, the objects fall into groups, and it is desired that the group membership for each object can be predicted or explained by a set of independent variables selected by the researcher with either of the two techniques discussed in this chapter.

KEY TERMS

Before starting the chapter, review the key terms to develop an understanding of the concepts and terminology to be used. Throughout the chapter the key terms appear in **boldface**. Other points of emphasis in the chapter are *italicized*. Also, cross-references within the Key Terms appear in *italics*.

Analysis sample Group used in estimating the discriminant function(s) or the logistic regression model. When constructing *classification matrices*, the original sample is divided randomly into two groups, one for model estimation (the analysis sample) and the other for validation (the *holdout sample*).

Box's M Statistical test for the equality of the covariance matrices of the independent variables across the groups of the dependent variable. If the statistical significance is greater than the critical level (e.g., .01), then the equality of the covariance matrices is supported. If the test shows statistical significance, then the groups are deemed different and the assumption is violated.

Categorical variable Variable that uses values that serve merely as a label or means of identification. It is also referred to as a nonmetric, nominal, binary, qualitative, or taxonomic variable. The number on a football jersey is an example. A more complete discussion of its characteristics and its differences from a *metric variable* is found in chapter 1.

Centroid   Mean value for the discriminant *Z* scores of all objects within a particular category or group. For example, a two-group discriminant analysis has two centroids, one for the objects in each of the two groups.

Classification function   Method of classification in which a linear function is defined for each group. Classification is performed by calculating a score for each observation on each group's classification function and then assigning the observation to the group with the highest score. This differs from the calculation of the *discriminant Z score*, which is calculated for each *discriminant function.*

Classification matrix   Matrix assessing the predictive ability of the discriminant function(s) or logistic regression. It is also called a confusion, assignment, or prediction matrix. Created by crosstabulating actual group membership with predicted group membership, this matrix consists of numbers on the diagonal representing correct classifications, and off-diagonal numbers representing incorrect classifications.

Cross-validation   Procedure of dividing the sample into two parts: the *analysis sample* used in estimation of the discriminant function(s) or logistic regression model, and the *holdout sample* used to validate the results. Cross-validation avoids the "overfitting" of the discriminant function or logistic regression by allowing its validation on a totally separate sample.

Cutting score   Criterion (score) against which each individual's discriminant *Z* score is compared to determine predicted group membership. When the analysis involves two groups, group prediction is determined by computing a single cutting score. Entities with *Z* scores below this score are assigned to one group, whereas those with scores above it are classified in the other group. For three or more groups, multiple discriminant functions are used, with a different cutting score for each function.

Discriminant coefficient   See *discriminant weight.*

Discriminant function   A variate of the independent variables selected for their discriminatory power used in the prediction of group membership. The predicted value of the discriminant function is the *discriminant Z score*, which is calculated for each object (person, firm, or product) in the analysis. It takes the form of the linear equation

$$Z_{jk} = a + W_1 X_{1k} + W_2 X_{2k} + \ldots + W_n X_{nk}$$

where

Z_{jk} = discriminant *Z* score of discriminant function *j* for object *k*

a = intercept

W_i = discriminant weight for independent variable *i*

X_{ik} = independent variable *i* for object *k*

Discriminant loadings   Measurement of the simple linear correlation between each independent variable and the discriminant *Z* score for each discriminant function; also called *structure correlations.* Discriminant loadings are calculated whether or not an independent variable is included in the discriminant function.

Discriminant Z score   Score defined by the *discriminant function* for each object in the analysis and usually stated in standardized terms. Also referred to as the

Z score, it is calculated for each object on each discriminant function, and used in conjunction with the *cutting score* to determine predicted group membership. This is different from the z score terminology used for standardized variables.

Discriminant weight   Weight whose size is determined by the variance structure of the original variables across the groups of the dependent variable. Independent variables with large discriminatory power usually have large weights, and those with little discriminatory power usually have small weights; however, multicollinearity among the independent variables will cause exceptions to this rule. Also called the *discriminant coefficient.*

Fisher's linear discriminant function   See *classification function.*

Hit ratio   Percentage of objects (individuals, respondents, firms, etc.) correctly classified by the discriminant function. It is calculated as the number of objects in the diagonal of the *classification matrix* divided by the total number of objects.

Holdout sample   Group of objects not used when the discriminant function(s) or logistic regression model is computed. This group is then used to validate the discriminant function or logistic regression model on a separate sample of respondents. Also called the *validation sample.*

Likelihood value   Measure used in *logistic regression* and *logit analysis* to represent the lack of predictive fit. Even though these methods do not use the least squares procedure in model estimation, as is done in multiple regression, the likelihood value is similar to the sum of squared error in regression analysis.

Logistic coefficient   Coefficient that acts as the weighting factor for the independent variables in relation to their discriminatory power. Similar to a regression weight or *discriminant coefficient.*

Logistic curve   An S-shaped curve formed by the *logit transformation* that represents the probability of an event. The S-shaped form is nonlinear because the probability of an event must approach 0 and 1, but never be greater. Thus, although there is a linear component in the midrange, the probabilities as they approach the lower and upper bounds of probability (0 and 1) must "flatten out" and become asymptotic to these bounds.

Logistic regression   Special form of regression in which the dependent variable is a nonmetric, dichotomous (binary) variable. Although some differences exist, the general manner of interpretation is quite similar to linear regression.

Logit analysis   See *logistic regression.*

Logit transformation   Transformation of the values of the discrete binary dependent variable of *logistic regression* into an S-shaped curve (*logistic curve*) representing the probability of an event. This probability is then used to form the *odds ratio,* which acts as the dependent variable in logistic regression.

Maximum chance criterion   Measure of predictive accuracy in the *classification matrix* that compares the percentage correctly classified (also known as the *hit ratio*) with the percentage of respondents in the largest group. The rationale is that the best uninformed choice is to classify every observation into the largest group.

Metric variable   Variable with a constant unit of measurement. If a variable is scaled from 1 to 9, the difference between 1 and 2 is the same as that between 8 and 9. A more complete discussion of its characteristics and differences from a *categorical variable* is found in chapter 1.

Odds ratio   The comparison of the probability of an event to the probability of the event not happening, which is used as the dependent variable in *logistic regression.*

Optimum cutting score   *Discriminant Z score* value that best separates the groups on each discriminant function for classification purposes.

Percentage correctly classified   See *hit ratio.*

Polar extremes approach   Method of constructing a categorical dependent variable from a metric variable. First, the metric variable is divided into three categories. Then the extreme categories are used in the discriminant analysis or logistic regression, and the middle category is not included in the analysis.

Potency index   Composite measure of the discriminatory power of an independent variable when more than one *discriminant function* is estimated. Based on *discriminant loadings*, it is a relative measure used for comparing the overall discrimination provided by each independent variable across all significant discriminant functions.

Press's *Q* statistic   Measure of the classificatory power of the *discriminant function* when compared with the results expected from a chance model. The calculated value is compared to a critical value based on the chi-square distribution. If the calculated value exceeds the critical value, the classification results are significantly better than would be expected by chance.

Proportional chance criterion   Another criterion for assessing the *hit ratio*, in which the "average" probability of classification is calculated considering all group sizes.

Pseudo R^2   A value of overall model fit that can be calculated for *logistic regression*; comparable to the R^2 measure used in multiple regression.

Simultaneous estimation   Estimation of the *discriminant function(s)* or the *logistic regression* model in a single step, where weights for all independent variables are calculated simultaneously. This is in contrast to *stepwise estimation*, where independent variables are entered sequentially according to discriminating power.

Split-sample validation   See *cross-validation.*

Stepwise estimation   Process of estimating the *discriminant function(s)* or *logistic regression* model whereby independent variables are entered sequentially according to the discriminatory power they add to the group membership prediction.

Stretching the vectors   Scaled *vector* in which the original vector is scaled to represent the corresponding *F* ratio. Used to graphically represent the *discriminant function loadings* in a combined manner with the group centroids.

Structure correlations   See *discriminant loadings.*

Territorial map   Graphical portrayal of the cutting scores on a two-dimensional graph. When combined with the plots of individual cases, the dispersion of each group can be viewed and the misclassifications of individual cases identified directly from the map.

Tolerance   Proportion of the variation in the independent variables not explained by the variables already in the model (function). It can be used to protect against multicollinearity. Calculated as $1 - R_i^{2*}$, where R_i^{2*} is the amount of variance of independent variable *i* explained by all of the other independent variables. A tolerance of 0 means that the independent variable under consideration is a perfect linear combination of independent variables already in the model (equation). A tolerance of 1 means that an independent variable is totally independent of other variables already in the model.

Validation sample   See *holdout sample.*

Variate   Linear combination that represents the weighted sum of two or more independent variables that comprise the *discriminant function*. Also called linear combination or linear compound.

Vector   Representation of the direction and magnitude of a variable's role as portrayed in a graphical interpretation of discriminant analysis results.

Wald statistic   Test used in logistic regression for the significance of the *logistic coefficient*. Its interpretation is like the *F* or *t* values used for the significance testing of regression coefficients.

Z score   See *discriminant Z score*.

What Are Discriminant Analysis and Logistic Regression?

In attempting to choose an appropriate analytical technique, we sometimes encounter a problem that involves a categorical dependent variable and several metric independent variables. For example, we may wish to distinguish good from bad credit risks. If we had a metric measure of credit risk, then we could use multivariate regression. But we may be able to ascertain only if someone is in the good or bad risk category, and this is not the metric type measure required by multivariate regression analysis.

Discriminant analysis and logistic regression are the appropriate statistical techniques when the dependent variable is **categorical** (nominal or nonmetric) and the independent variables are metric. In many cases, the dependent variable consists of two groups or classifications, for example, male versus female or high versus low. In other instances, more than two groups are involved, such as low, medium, and high classifications. Discriminant analysis is capable of handling either two groups or multiple (three or more) groups. When two classifications are involved, the technique is referred to as two-group discriminant analysis. When three or more classifications are identified, the technique is referred to as *multiple discriminant analysis (MDA).* **Logistic regression,** also known as **logit analysis,** is limited in its basic form to two groups, although alternative formulations can handle more than two groups.

Discriminant analysis involves deriving a **variate,** the linear combination of the two (or more) independent variables that will discriminate best between a priori defined groups. Discrimination is achieved by setting the variate's weights for each variable to maximize the between-group variance relative to the within-group variance. The linear combination for a discriminant analysis, also known as the **discriminant function,** is derived from an equation that takes the following form:

$$Z_{jk} = a + W_1X_{1k} + W_2X_{2k} + \ldots + W_nX_{nk}$$

where

Z_{jk} = discriminant Z score of discriminant function j for object k

a = intercept

W_i = discriminant weight for independent variable i

X_{ik} = independent variable i for object k

Discriminant analysis is the appropriate statistical technique for testing the hypothesis that the group means of a set of independent variables for two or more groups are equal. To do so, discriminant analysis multiplies each independent variable by its corresponding weight and adds these products together. The result is a single composite **discriminant Z score** for each individual in the analysis. By averaging the discriminant scores for all the individuals within a particular group, we arrive at the group mean. This group mean is referred to as a **centroid.** When the analysis involves two groups, there are two centroids; with three groups, there are three centroids; and so forth. The centroids indicate the most typical location of any individual from a particular group, and a comparison of the group centroids shows how far apart the groups are along the dimension being tested.

The test for the statistical significance of the discriminant function is a generalized measure of the distance between the group centroids. It is computed by comparing the distributions of the discriminant scores for the groups. If the overlap in the distributions is small, the discriminant function separates the groups well. If the overlap is large, the function is a poor discriminator between the groups. Two distributions of discriminant scores shown in Figure 5.1 further illustrate this concept. The top diagram represents the distributions of discriminant scores for a function that separates the groups well, whereas the lower diagram shows the distributions of discriminant scores on a function that is a relatively poor discriminator between group A and B. The shaded areas represent probabilities of misclassifying objects from group A into group B.

Multiple discriminant analysis is unique in one characteristic among the dependence relationships of interest here: if there are more than two groups in the dependent variable, discriminant analysis will calculate more than one discriminant function. As a matter of fact, it will calculate $NG - 1$ functions, where NG is

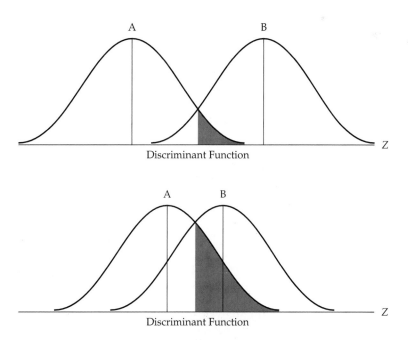

FIGURE 5.1 Univariate Representation of Discriminant Z Scores

the number of groups. Each discriminant function will calculate a discriminant Z score. In the case of a three-group dependent variable, each object will have a score for discriminant functions one and two, allowing the objects to be plotted in two dimensions, with each dimension representing a discriminant function. Thus, discriminant analysis is not limited to a single variate, as is multiple regression, but creates multiple variates representing dimensions of discrimination among the groups.

Logistic regression is a specialized form of regression that is formulated to predict and explain a binary (two-group) categorical variable rather than a metric dependent measure. The form of the logistic regression variate is similar to the variate in multiple regression. The variate represents a single multivariate relationship with regression-like coefficients that indicate the relative impact of each predictor variable. The differences between logistic regression and discriminant analysis will become more apparent in our discussion of logistic regression's unique characteristics later in this chapter. Yet many similarities also exist between the two methods. When the basic assumptions of both methods are met, they each give comparable predictive and classificatory results and employ similar diagnostic measures. Logistic regression, however, has the advantage of being less affected than discriminant analysis when the basic assumptions, particularly normality of the variables, are not met. It also can accommodate nonmetric variables through dummy-variable coding, just as regression can. Logistic regression is limited, however, to prediction of only a two-group dependent measure. Thus, in cases for which three or more groups form the dependent measure, discriminant analysis is better suited.

Analogy with Regression and MANOVA

The application and interpretation of discriminant analysis is much the same as in regression analysis; that is, the discriminant function is a linear combination (variate) of metric measurements for two or more independent variables and is used to describe or predict a single dependent variable. The key difference is that discriminant analysis is appropriate for research problems in which the dependent variable is categorical (nominal or nonmetric), whereas regression is utilized when the dependent variable is metric. As discussed earlier, logistic regression is a variant of regression, thus having many similarities except for the type of dependent variable. Discriminant analysis is also comparable to "reversing" multivariate analysis of variance (MANOVA), which we discuss in chapter 6. In discriminant analysis, the single dependent variable is categorical, and the independent variables are metric. The opposite is true of MANOVA, which involves metric dependent variables and categorical independent variable(s).

Hypothetical Example of Discriminant Analysis

Discriminant analysis is applicable to any research question with the objective of understanding group membership, whether the groups comprise individuals (e.g., customers versus noncustomers), firms (e.g., profitable versus unprofitable), products (e.g., successful versus unsuccessful), or any other object that can be evaluated on a series of independent variables. To illustrate the basic premises of discriminant analysis, we examine two research settings, one involving two groups (purchasers versus nonpurchasers) and the other three groups (levels of switching behavior). Logistic regression operates in a manner quite comparable to discriminant analysis for two groups. Therefore, we do not specifically illustrate logistic regression here, deferring our discussion until a separate discussion of logistic regression later in this chapter.

A Two-Group Discriminant Analysis: Purchasers versus Nonpurchasers

Suppose HATCO wants to determine whether one of its new products—a new and improved food mixer—will be commercially successful. In carrying out the investigation, HATCO is primarily interested in identifying (if possible) those consumers who would purchase the new product and those who would not. In statistical terminology, HATCO would like to minimize the number of errors it would make in predicting which consumers would buy the new food mixer and which would not. To assist in identifying potential purchasers, HATCO has devised rating scales on three characteristics—durability, performance, and style—to be used by consumers to evaluate the new product. Rather than relying on each scale as a separate measure, HATCO hopes that a weighted combination of all three would better predict whether a consumer is likely to purchase the new product.

Discriminant analysis can obtain a weighted combination of the three scales to be used in predicting the likelihood that a consumer will purchase the product. In addition to determining whether consumers who are likely to purchase the new product can be distinguished from those who are not, HATCO would also like to know which characteristics of its new product are useful in differentiating purchasers from nonpurchasers; that is, evaluations on which of the three characteristics of the new product best separates purchasers from nonpurchasers? For example, if the response "would purchase" is always associated with a high durability rating and the response "would not purchase" is always associated with a low durability rating, HATCO could conclude that the characteristic of durability distinguishes purchasers from nonpurchasers. In contrast, if HATCO found that about as many persons with a high rating on style said they would purchase the food mixer as those who said they would not, then style is a characteristic that discriminates poorly between purchasers and nonpurchasers.

Table 5.1 (p. 248) lists the ratings on these three characteristics of the new mixer with a specified price by a panel of 10 potential purchasers. In rating the food mixer, each panel member would be implicitly comparing it with products already on the market. After the product was evaluated, the evaluators were asked to state their buying intentions ("would purchase" or "would not purchase"). Five stated that they would purchase the new mixer and five said they would not.

Table 5.1 potentially identifies several discriminating variables. First, there is a substantial difference between the mean ratings for the "would purchase" and "would not purchase" groups on X_1, the characteristic of durability ($7.4 - 3.2 = 4.2$). Durability appears to discriminate well between the "would purchase" and "would not purchase" groups, and is likely to be an important characteristic to potential purchasers. On the other hand, the characteristic of style (X_3) has a much smaller difference of 0.2 between mean ratings ($4.0 - 3.8 = 0.2$) for the "would

TABLE 5.1 HATCO Survey Results for the Evaluation of a New Consumer Product

Groups Based on Purchase Intention	Evaluation of New Product*		
	X_1 Durability	X_2 Performance	X_3 Style
Group 1: Would purchase			
Subject 1	8	9	6
Subject 2	6	7	5
Subject 3	10	6	3
Subject 4	9	4	4
Subject 5	4	8	2
Group mean	7.4	6.8	4.0
Group 2: Would not purchase			
Subject 6	5	4	7
Subject 7	3	7	2
Subject 8	4	5	5
Subject 9	2	4	3
Subject 10	2	2	2
Group mean	3.2	4.4	3.8
Difference between group means	4.2	2.4	0.2

*Evaluations are made on a 10-point scale (1 = very poor to 10 = excellent).

purchase" and "would not purchase" groups. Therefore, we would expect this characteristic to be less discriminating in terms of a decision to purchase or not to purchase. However, before we can make such statements conclusively, we must examine the distribution of scores for each group. Large standard deviations within one or both groups might make the difference between means nonsignificant and inconsequential in discriminating between the groups.

Because we have only 10 respondents in two groups and three independent variables, we can also look at the data graphically to determine what discriminant analysis is trying to accomplish. Figure 5.2 shows the ten respondents on each of the three variables. The "would purchase" group is represented by circles and the "would not purchase" group by the squares. Respondent identification numbers are inside the shapes. Looking first at X_1, durability, which had a substantial difference in mean scores, we see that we could almost perfectly discriminate between the groups using only this variable. If we established the value of 5.5 as our cutoff point to discriminate between the two groups, then we would "misclassify" only respondent 5, one of the "would purchase" group members. This is indicative of the large difference in the means for the two groups on durability and the lack of overlap between the distributions of the two groups (see Figure 5.1). Examining X_2, performance, we see there is a less clear-cut distinction between the two groups. However, this variable does provide high discrimination for respondent 5, who was misclassified if we used only X_1. In addition, the respondents who would be misclassified using X_2 are well separated on X_1. Thus X_1 and X_2 might be used quite effectively *in combination* to predict group membership. Finally, X_3 style shows little differentiation between the groups. Thus, by forming a variate of only X_1 and X_2, and omitting X_3, a discriminant function may be formed that maximizes the separation of the groups on the discriminant score.

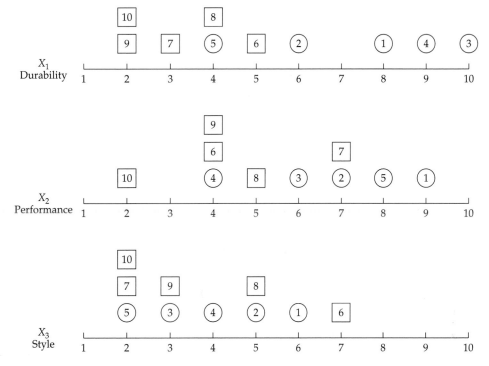

FIGURE 5.2 Graphical Representation of 10 Potential Purchasers on Three Possible Discriminating Variables

Table 5.2 (p. 250) contains the results for three different discriminant functions. The first discriminant function contains just X_1, equating the value of X_1 to the discriminant Z score. As discussed earlier, the use of only X_1, the best discriminator, results in the misclassification of subject 5. As shown in the **classification matrix** in Table 5.2, four out of five subjects in group 1 (all but subject 5) and five of five subjects in group 2 are correctly classified (i.e., lie on the diagonal of the classification matrix). The **percentage correctly classified** is thus 90 percent (9 out of 10 subjects). Because X_2 provides discrimination for subject 5, we can form a second discriminant function by equally combining X_1 and X_2 to utilize each variable's unique discriminatory powers. As we see in Table 5.2, using a cutting score of 11 with this new discriminant function achieves a perfect classification of the two groups (100 percent correctly classified). Thus, X_1 and X_2 in combination are able to make better predictions of group membership than either variable separately.

Discriminant analysis follows a procedure very similar to the above example when empirically estimating the discriminant function. It identifies the variables with the greatest differences between the groups and derives a discriminant coefficient that weights each variable to reflect these differences. The third discriminant function in Table 5.2 represents the actual estimated discriminant function ($Z = -4.53 + .476X_1 + .359X_2$). Using a cutting score of 0, this third function also achieves a 100 percent correct classification rate with the maximum separation possible between the groups.

A Geometric Representation of the Two-Group Discriminant Function

A graphic illustration of another two-group analysis will help to further explain the nature of discriminant analysis [6]. Figure 5.3 demonstrates what happens when a two-group discriminant function is computed. Assume we have two groups, A and B, and two measurements, V_1 and V_2, on each member of the two groups. We can plot in a scatter diagram the association of variable V_1 with variable V_2 for each member of the two groups. In Figure 5.3 the small dots represent the variable measurements for the members of group B and the large dots those for group A. The ellipses drawn around the large and small dots would enclose some prespecified proportion of the points, usually 95 percent or more in each group. If we draw a straight line through the two points at which the ellipses intersect and then project the line to a new Z axis, we can say that the overlap between the univariate distributions A' and B' (represented by the shaded area) is smaller than would be obtained by any other line drawn through the ellipses formed by the scatterplots [6].

The important thing to note about Figure 5.3 is that the Z axis expresses the two-variable profiles of groups A and B as single numbers (discriminant scores). By finding a linear combination of the original variables V_1 and V_2, we can project the results as a discriminant function. For example, if the large and small dots are projected onto the new Z axis as discriminant Z scores, the result condenses

TABLE 5.2 Creating Discriminant Functions to Predict Purchasers versus Nonpurchasers

	Calculated Discriminant Z Scores		
Group	*Function 1:* $Z = X_1$	*Function 2:* $Z = X_1 + X_2$	*Function 3:* $Z = -4.53 + .476X_1 + .359X_2$
Group 1: Would purchase			
Subject 1	8	17	2.51
Subject 2	6	13	.84
Subject 3	10	16	2.38
Subject 4	9	13	1.19
Subject 5	4	12	.25
Group 2: Would not purchase			
Subject 6	5	9	−.71
Subject 7	3	10	−.59
Subject 8	4	9	−.83
Subject 9	2	6	−2.14
Subject 10	2	4	−2.86
Cutting score	5.5	11	0.0

CLASSIFICATION ACCURACY:

Actual Group	*Predicted Group* 1	2	*Predicted Group* 1	2	*Predicted Group* 1	2
1: Would purchase	4	1	5	0	5	0
2: Would not purchase	0	5	0	5	0	5

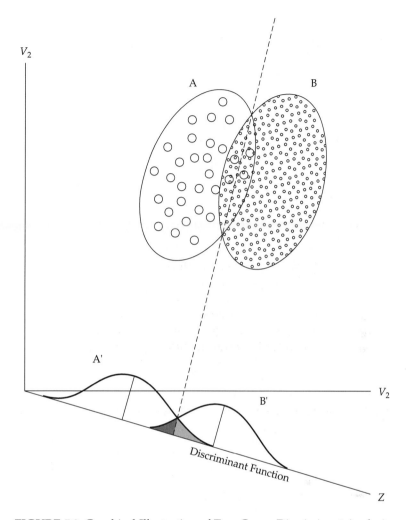

FIGURE 5.3 Graphical Illustration of Two-Group Discriminant Analysis

the information about group differences (shown in the V_1V_2 plot) into a set of points (Z scores) on a single axis, shown by distributions A′ and B′.

To summarize, for a given discriminant analysis problem, a linear combination of the independent variables is derived, resulting in a series of discriminant scores for each object in each group. The discriminant scores are computed according to the statistical rule of maximizing the variance between the groups and minimizing the variance within them. If the variance between the groups is large relative to the variance within the groups, we say that the discriminant function separates the groups well.

A Three-Group Example of Discriminant Analysis: Switching Intentions

The two-group example just examined demonstrates the rationale and benefit of combining independent variables into a variate for purposes of discriminating between groups. But discriminant analysis also has another means of discrimination: the estimation and use of multiple variates in instances where there are three or more groups. These discriminant functions now become dimensions of discrimination, each dimension separate and distinct from the other. Thus, in addition to improving the explanation of group membership, these additional discriminant functions add insight into the various combinations of independent variables that discriminate between groups.

As an illustration of a three-group application of discriminant analysis, we examine research conducted by HATCO concerning the possibility of a competitor's customers switching suppliers. A small-scale pretest involved interviews of 15 customers of a major competitor. In the course of the interviews, the customers were asked their probability of switching suppliers on a three-category scale. The three possible responses were "definitely switch," "undecided," and "definitely not switch." Customers were assigned to groups 1, 2, or 3, respectively, according to their responses. The customers also rated the competitor on the two characteristics of price competitiveness and service level. The research issue is now to determine whether the customers' ratings of their current supplier can predict their probability of leaving that supplier. Because the dependent variable of probability of switching was measured as a categorical (nonmetric) variable and the ratings of price and service are metric, discriminant analysis is appropriate.

With three categories of the dependent variable, discriminant analysis can estimate two discriminant functions, each representing a different dimension of discrimination. Table 5.3 contains the survey results for the 15 customers, 5

TABLE 5.3 HATCO Survey Results of Switching Intentions by Potential Customers

Groups Based on Switching Intention	Evaluation of Current Supplier*	
	X_1 Price Competitiveness	X_2 Service Level
Group 1: Definitely switch		
Subject 1	2	2
Subject 2	1	2
Subject 3	3	2
Subject 4	2	1
Subject 5	2	3
Group mean	2.0	2.0
Group 2: Undecided		
Subject 6	4	2
Subject 7	4	3
Subject 8	5	1
Subject 9	5	2
Subject 10	5	3
Group mean	4.6	2.2
Group 3: Definitely not switch		
Subject 11	2	6
Subject 12	3	6
Subject 13	4	6
Subject 14	5	6
Subject 15	5	7
Group mean	3.8	6.2

*Evaluations are made on a 10-point scale (1 = very poor to 10 = excellent).

in each category of the dependent variable. As we did in the two-group example, we can look at the mean scores for each group to see if one of the variables discriminates well among all the groups. For X_1, price competitiveness, we see a rather large mean difference between groups 1 and groups 2 or 3 (2.0 versus 4.6 or 3.8). X_1 may discriminate well between group 1 and groups 2 or 3, but is much less effective in discriminating between groups 2 and 3. For X_2, service level, we see that the difference between groups 1 and 2 is very small (2.0 versus 2.2), whereas a large difference exists between group 3 and groups 1 or 2 (6.2 versus 2.0 or 2.2). Thus, X_1 distinguishes group 1 from groups 2 and 3, and X_2 distinguishes group 3 from groups 1 and 2. As a result, we see that X_1 and X_2 provide different "dimensions" of discrimination between the groups.

To illustrate this graphically, Figure 5.4 (p. 254) portrays the three groups on each of the independent variables separately. Viewing the group members on any one variable, we can see that no variable discriminates well among all the groups. But if we now construct two simple discriminant functions, the results become much clearer. For illustration purposes, we calculate two discriminant functions with weights of 0.0 or 1.0 for the variables. Discriminant function 1 gives X_1 a weight of 1.0, and X_2 a weight of 0.0. Likewise, discriminant function 2 gives X_2 a weight of 1.0, and X_1 a weight of 0.0. The functions can be stated mathematically as

$$\text{Discriminant function 1} = 1.0(X_1) + 0.0(X_2)$$
$$\text{Discriminant function 2} = 0.0(X_1) + 1.0(X_2)$$

This shows in simple terms how the discriminant analysis procedure estimates weights to maximize discrimination.

With the two functions, we can now calculate two discriminant scores for each respondent. Figure 5.4 also contains a plot of each respondent in a two-dimensional representation. The separation between groups now becomes quite apparent, and each group can be easily distinguished. We can establish values on each dimension that will define regions containing each group (e.g., all members of group 1 are in the region less than 3.5 on dimension 1 and less than 4.5 on dimension 2). Each of the other groups can be similarly defined in terms of the ranges of their discriminant function scores.

Moreover, the two discriminant functions provide the dimensions of discrimination. The first discriminant function, price competitiveness, distinguishes between undecided customers and those customers who have decided to switch. But price competitiveness does not distinguish those who have decided not to switch. Instead, the perception of service level, defining the second discriminant function, predicts whether a customer will decide not to switch versus whether a customer is undecided or determined to switch suppliers. The researcher can present to management the separate impacts of both price competitiveness and service level in making this decision.

The estimation of more than one discriminant function, when possible, provides the researcher with both improved discrimination and additional perspectives on the features and the combinations that best discriminate among the groups. The following sections detail the necessary steps for performing a discriminant analysis, for assessing its level of predictive fit, and then interpreting the influence of independent variables in making that prediction.

A. Individual Variables

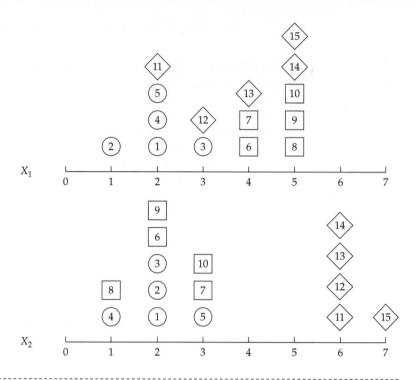

B. Two-Dimensional
Representation of
Discriminant Functions

Discriminant
Function 1 = $1.0 * X_1 + 0 * X_2$

Discriminant
Function 2 = $0 * X_1 + 1.0 * X_2$

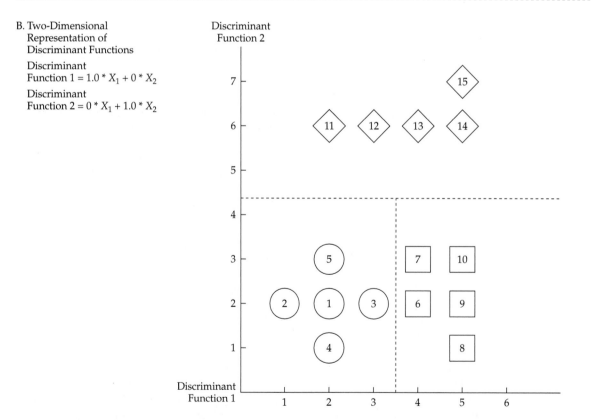

FIGURE 5.4 Graphical Representation of Potential Discriminating Variables for a Three-
Group Discriminant Analysis

The Decision Process for Discriminant Analysis

The application of discriminant analysis can be viewed from the six-stage model-building perspective introduced in chapter 1 and portrayed in Figure 5.5 (stages 1–3) and Figure 5.6 (stages 4–6; p. 261). As with all multivariate applications, setting the objectives is the first step in the analysis. Then the researcher must address specific design issues and make sure the underlying assumptions are met. The analysis proceeds with the derivation of the discriminant function and the determination of whether a statistically significant function can be derived to separate the two (or more) groups. The discriminant results are then assessed for predictive accuracy by developing a classification matrix. Next, interpretation of the discriminant function determines which of the independent variables contributes the most to discriminating between the groups. Finally, the discriminant function should be validated with a holdout sample. Each of these stages is discussed in the following sections. We discuss logistic regression in a separate section after examining the decision process for discriminant analysis. In this way, the similarities and differences between these two techniques can be highlighted.

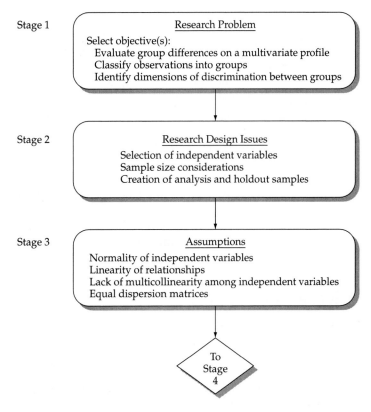

FIGURE 5.5 Stages 1–3 in the Discriminant Analysis Decision Diagram

Stage 1: Objectives of Discriminant Analysis

A review of the objectives for applying discriminant analysis should further clarify its nature. Discriminant analysis can address any of the following research objectives:

1. Determining whether statistically significant differences exist between the average score profiles on a set of variables for two (or more) a priori defined groups.
2. Determining which of the independent variables account the most for the differences in the average score profiles of the two or more groups
3. Establishing procedures for classifying objects (individuals, firms, products, and so on) into groups on the basis of their scores on a set of independent variables.
4. Establishing the number and composition of the dimensions of discrimination between groups formed from the set of independent variables.

As can be noted from these objectives, discriminant analysis is useful when the researcher is interested either in understanding group differences or in correctly classifying objects into groups or classes. Discriminant analysis, therefore, can be considered either a type of profile analysis or an analytical predictive technique. In either case, the technique is most appropriate where there is a single categorical dependent variable and several metrically scaled independent variables. As a profile analysis, discriminant analysis provides an objective assessment of differences between groups on a set of independent variables. In this situation, discriminant analysis is quite similar to multivariate analysis of variance (see chapter 6 for a more detailed discussion of multivariate analysis of variance). For understanding group differences, discriminant analysis lends insight into the role of individual variables as well as defining combinations of these variables that represent dimensions of discrimination between groups. These dimensions are the collective effects of several variables that work jointly to distinguish between the groups. The use of sequential estimation methods also allows for identifying subsets of variables with the greatest discriminatory power. Finally, for classification purposes, discriminant analysis provides a basis for classifying not only the sample used to estimate the discriminant function but also any other observations that can have values for all the independent variables. In this way, the discriminant analysis can be used to classify other observations into the defined groups.

Stage 2: Research Design for Discriminant Analysis

The successful application of discriminant analysis requires consideration of several issues. These issues include the selection of both the dependent and the independent variables, the sample size needed for estimation of the discriminant functions, and the division of the sample for validation purposes.

Selection of Dependent and Independent Variables

To apply discriminant analysis, the researcher must first specify which variables are to be independent and which variable is to be dependent. Recall that the dependent variable is categorical and the independent variables are metric.

The researcher should focus on the dependent variable first. The number of dependent variable groups (categories) can be two or more, but these groups must be mutually exclusive and exhaustive. By this we mean that each observation can be placed into only one group. In some cases, the dependent variable may involve two groups (dichotomous), such as good versus bad. In other cases, the dependent variable may involve several groups (multichotomous), such as the occupations of physician, attorney, or professor.

The preceding examples of categorical variables were true dichotomies (or multichotomies). There are some situations, however, where discriminant analysis is appropriate even if the dependent variable is not a true categorical variable. We may have a dependent variable that is of ordinal or interval measurement that we wish to use as a categorical dependent variable. In such cases, we would have to create a categorical variable. For example, if we had a variable that measured the average number of cola drinks consumed per day, and the individuals responded on a scale from zero to eight or more per day, we could create an artificial trichotomy (three groups) by simply designating those individuals who consumed no, one, or two cola drinks per day as light users, those who consumed three, four, or five per day as medium users, and those who consumed six, seven, eight, or more as heavy users. Such a procedure would create a three-group categorical variable in which the objective would be to discriminate among light, medium, and heavy users of colas.

Any number of artificial categorical groups can be developed. Most frequently, the approach would involve creating two, three, or four categories. But a larger number of categories could be established if the need arose. When three or more categories are created, the possibility arises of examining only the extreme groups in a two-group discriminant analysis. This procedure is called the polar extremes approach.

The **polar extremes approach** involves comparing only the extreme two groups and excluding the middle group from the discriminant analysis. For example, the researcher could examine the light and heavy users of cola drinks and exclude the medium users. This approach can be used any time the researcher wishes to examine only the extreme groups. However, the researcher may also want to try this approach when the results of a regression analysis are not as good as anticipated. Such a procedure may be helpful because it is possible that group differences may appear even though regression results are poor; that is, the polar extremes approach with discriminant analysis can reveal differences that are not as prominent in a regression analysis of the full data set [6]. Such manipulation of the data naturally would necessitate caution in interpreting one's findings.

After a decision has been made on the dependent variable, the researcher must decide which independent variables to include in the analysis. Independent variables usually are selected in two ways. The first approach involves identifying variables either from previous research or from the theoretical model that is the underlying basis of the research question. The second approach is intuition—utilizing the researcher's knowledge and intuitively selecting variables for which no previous research or theory exists but that logically might be related to predicting the groups for the dependent variable.

Sample Size

Discriminant analysis is quite sensitive to the ratio of sample size to the number of predictor variables. Many studies suggest a ratio of 20 observations for each predictor variable. Although this ratio may be difficult to maintain in practice, the researcher must note that the results become unstable as the sample size decreases relative to the number of independent variables. The minimum size recommended is five observations per independent variable. Note that this ratio applies to all variables considered in the analysis, even if all of the variables considered are not entered into the discriminant function (such as in stepwise estimation).

In addition to the overall sample size, the researcher must also consider the sample size of each group. At a minimum, the smallest group size must exceed the number of independent variables. As a practical guideline, each group should have at least 20 observations. But even if all groups exceed 20 observations, the researcher must also consider the relative sizes of the groups. If the groups vary widely in size, this may impact the estimation of the discriminant function and the classification of observations. In the classification stage, larger groups have a disproportionately higher chance of classification. If the group sizes do vary markedly, the researcher may wish to randomly sample from the larger group(s), thereby reducing their size to a level comparable to the smaller group(s).

Division of the Sample

One final note about the impact of sample size in discriminant analysis. As will be discussed later, many times the sample is divided into two subsamples, one used for estimation of the discriminant function and another for validation purposes. It is essential that each subsample be of adequate size to support conclusions from the results.

A number of procedures have been suggested for dividing the sample; the most popular one involves developing the discriminant function on one group and then testing it on a second group. The usual procedure is to divide the total sample of respondents randomly into two groups. One of these groups, the **analysis sample,** is used to develop the discriminant function. The second group, the **holdout sample,** is used to test the discriminant function. This method of validating the function is referred to as the **split-sample** or **cross-validation** approach [4, 8, 15].

No definite guidelines have been established for dividing the sample into analysis and holdout (or validation) groups. The most popular procedure is to divide the total group so that one-half of the respondents are placed in the analysis sample and the other half are placed in the holdout sample. However, no hard-and-fast rule has been established, and some researchers prefer a 60–40 or 75–25 split between the analysis and the holdout groups.

When selecting the individuals for the analysis and holdout groups, one usually follows a proportionately stratified sampling procedure. If the categorical groups for the discriminant analysis are equally represented in the total sample, an equal number of individuals is selected. If the categorical groups are unequal, the sizes of the groups selected for the holdout sample should be proportionate to the total sample distribution. For instance, if a sample consists of 50 males and 50 females, the holdout sample would have 25 males and 25 females. If the sample contained 70 females and 30 males, then the holdout samples would consist of 35 females and 15 males.

Several additional comments need to be made regarding the division of the total sample into analysis and holdout groups. If the researcher is going to divide the sample into analysis and holdout groups, the sample must be sufficiently large to do so. Again, no hard-and-fast rules have been established, but it seems logical that the researcher would want at least 100 in the total sample to justify dividing it into the two groups. One compromise procedure the researcher can select if the sample size is too small to justify a division into analysis and holdout groups is to develop the function on the entire sample and then use the function to classify the same group used to develop the function. This procedure results in an upward bias in the predictive accuracy of the function, but is certainly better than not testing the function at all.

Stage 3: Assumptions of Discriminant Analysis

It is desirable to meet certain conditions for proper application of discriminant analysis. The key assumptions for deriving the discriminant function are multivariate normality of the independent variables and unknown (but equal) dispersion and covariance structures (matrices) for the groups as defined by the dependent variable [7, 9]. Mixed evidence exists concerning the sensitivity of discriminant analysis to violations of these assumptions. The researcher should examine the data and if assumptions are violated, the researcher should identify the alternative methods available and the impacts on the results that can be expected. Data not meeting the multivariate normality assumption can cause problems in the estimation of the discriminant function. Therefore, it is suggested that logistic regression be used as an alternative technique, if possible.

Unequal covariance matrices can negatively affect the classification process. If the sample sizes are small and the covariance matrices are unequal, then the statistical significance of the estimation process is adversely affected. The more likely case is that of unequal covariances among groups of adequate sample size, whereby observations are "overclassified" into the groups with larger covariance matrices. This effect can be minimized by increasing the sample size and also by using the group-specific covariance matrices for classification purposes, but this approach mandates cross-validation of the discriminant results. Finally, quadratic classification techniques are available in many of the statistical programs if large differences exist between the covariance matrices of the groups and the remedies do not minimize the effect [5, 11, 13].

Another characteristic of the data that can affect the results is multicollinearity among the independent variables. Multicollinearity denotes that two or more independent variables are highly correlated, so that one variable can be highly explained or predicted by the other variable(s) and thus it adds little to the explanatory power of the entire set. This consideration becomes especially critical when stepwise procedures are employed. The researcher, in interpreting the discriminant function, must be aware of the level of multicollinearity and its impact on determining which variables enter the stepwise solution. For a more detailed discussion of multicollinearity and its impact on stepwise solutions, see chapter 4. The procedures for detecting the presence of multicollinearity are addressed in chapter 4.

As with any of the multivariate techniques employing a variate, an implicit assumption is that all relationships are linear. Nonlinear relationships are not reflected in the discriminant function unless specific variable transformations are made to represent nonlinear effects. Finally, outliers can have a substantial impact on the classification accuracy of any discriminant analysis results. The researcher is encouraged to examine all results for the presence of outliers and to eliminate true outliers if needed. For a discussion of some of the techniques for assessing the violations in the basic statistical assumptions or outlier detection, see chapter 2.

Stage 4: Estimation of the Discriminant Model and Assessing Overall Fit

To derive the discriminant function, the researcher must decide on the method of estimation and then determine the number of functions to be retained (see Figure 5.6). With the functions estimated, overall model fit can be assessed in several ways. First, **discriminant Z scores,** also known as the **Z scores,** can be calculated for each object. Comparison of the group means on the Z scores provides one measure of discrimination between groups. Predictive accuracy is measured as the number of observations classified into the correct groups. A number of criteria are available to assess whether the classification process achieves practical and/or statistical significance. Finally, casewise diagnostics can identify the classification accuracy of each case and its relative impact on the overall model estimation.

Computational Method

Two computational methods can be utilized in deriving a discriminant function: the simultaneous (direct) method and the stepwise method. **Simultaneous estimation** involves computing the discriminant function so that all of the independent variables are considered concurrently. Thus the discriminant function is computed based upon the entire set of independent variables, regardless of the discriminating power of each independent variable. The simultaneous method is appropriate when, for theoretical reasons, the researcher wants to include all the independent variables in the analysis and is not interested in seeing intermediate results based only on the most discriminating variables.

Stepwise estimation is an alternative to the simultaneous approach. It involves entering the independent variables into the discriminant function one at a time on the basis of their discriminating power. The stepwise approach begins by choosing the single best discriminating variable. The initial variable is then paired with each of the other independent variables one at a time, and the variable that is best able to improve the discriminating power of the function in combination with the first variable is chosen. The third and any subsequent variables are selected in a similar manner. As additional variables are included, some previously selected variables may be removed if the information they contain about group differences is available in some combination of the other variables included at later stages. Eventually, either all independent variables will have been included in the function or the excluded variables will have been judged as not contributing significantly to further discrimination.

The stepwise method is useful when the researcher wants to consider a relatively large number of independent variables for inclusion in the function. By sequentially selecting the next best discriminating variable at each step, variables that are not useful in discriminating between the groups are eliminated and a re-

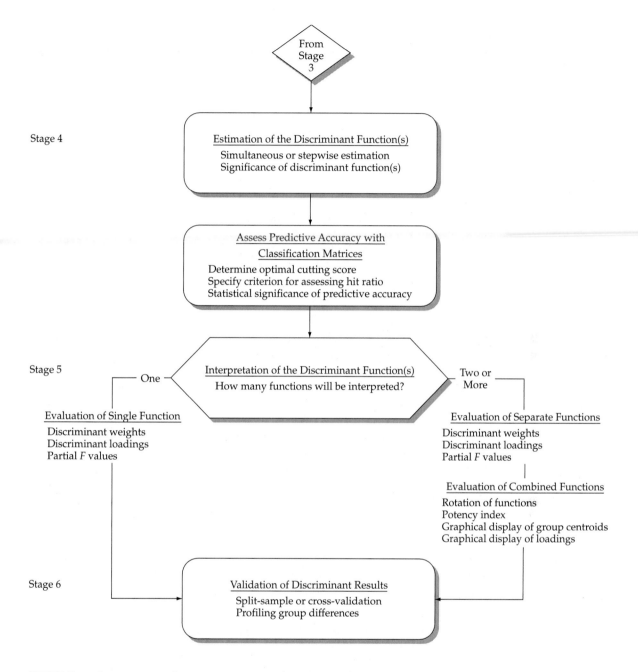

FIGURE 5.6 Stages 4–6 in the Discriminant Analysis Decision Diagram

duced set of variables is identified. The reduced set typically is almost as good as—and sometimes better than—the complete set of variables. But the researcher should note that stepwise estimation becomes less stable and generalizable as the ratio of sample size to independent variable declines below the recommended level of 20 observations per independent variable. It is particularly important in these instances to validate the results in as many ways as possible.

Statistical Significance

After the discriminant function has been computed, the researcher must assess its level of significance. A number of different statistical criteria are available. The measures of Wilks' lambda, Hotelling's trace, and Pillai's criterion all evaluate the statistical significance of the discriminatory power of the discriminant function(s). Roy's greatest characteristic root evaluates only the first discriminant function. For a more detailed discussion of the advantages and disadvantages of each criterion, see the discussion of significance testing in multivariate analysis of variance in chapter 6.

If a stepwise method is used to estimate the discriminant function, the Mahalanobis D^2 and Rao's V measures are most appropriate. Both are measures of generalized distance. The Mahalanobis D^2 procedure is based on generalized squared Euclidean distance that adjusts for unequal variances. The major advantage of this procedure is that it is computed in the original space of the predictor variables rather than as a collapsed version used in other measures. The Mahalanobis D^2 procedure becomes particularly critical as the number of predictor variables increases because it does not result in any reduction in dimensionality. A loss in dimensionality would cause a loss of information because it decreases variability of the independent variables. In general, Mahalanobis D^2 is the preferred procedure when the researcher is interested in the maximal use of available information. The Mahalanobis D^2 procedure performs a stepwise discriminant analysis similar to a stepwise regression analysis, designed to develop the best one-variable model, followed by the best two-variable model, and so forth, until no other variables meet the desired selection rule. The selection rule in this procedure is to maximize Mahalanobis D^2 between groups. Both stepwise and simultaneous methods are available in the major statistical programs.

The conventional significance criterion of .05 or beyond is often used. Many researchers believe that if the function is not significant at or beyond the .05 level, there is little justification for going further. Some researchers, however, disagree. Their decision rule for continuing to a higher significance level (e.g., .10 or more) is the cost versus the value of the information. If the higher levels of risk for including nonsignificant results (e.g., significance levels > .05) are acceptable, discriminant functions may be retained that are significant at the .2 or even the .3 level.

If the number of groups is three or more, then the researcher must decide not only if the discrimination between groups overall is statistically significant but also if each of the estimated discriminant functions is statistically significant. As discussed earlier, discriminant analysis estimates one less discriminant function than there are groups. If three groups are analyzed, then two discriminant functions will be estimated; for four groups, three functions will be estimated; and so on. The computer programs all provide the researcher the information necessary to ascertain the number of functions needed to obtain statistical significance, without including discriminant functions that do not increase the discriminatory power significantly. If one or more functions are deemed not statistically significant, the discriminant model should be reestimated with the number of functions to be derived limited to the number of significant functions. In this manner, the assessment of predictive accuracy and the interpretation of the discriminant functions will be based only on significant functions.

Assessing Overall Fit

Once the significant discriminant functions have been identified, attention shifts to ascertaining the overall fit of the retained discriminant function(s). This assessment involves three tasks: calculating discriminant Z scores for each observation, evaluating group differences on the discriminant Z scores, and assessing group membership prediction accuracy.

Calculating Discriminant Z Scores

With the retained discriminant functions defined, the basis for calculating the discriminant Z scores has been established. As discussed earlier, the discriminant Z score of any discriminant function can be calculated for each observation by the following formula:

$$Z_{jk} = a + W_1 X_{1k} + W_2 X_{2k} + \ldots + W_n X_{nk}$$

where

Z_{jk} = discriminant Z score of discriminant function j for object k

a = intercept

W_i = discriminant coefficient for independent variable i

X_{ik} = independent variable i for object k

This score, a metric variable, provides a direct means of comparing observations on each function. Observations with similar Z scores are assumed more alike on the variables constituting this function than those with disparate scores. There are versions of the discriminant function that use either standardized or unstandardized weights and values. The standardized version is more useful for interpretation purposes, but the unstandardized version is easier to use in calculating the discriminant Z score.

We should note that the discriminant function differs from the **classification function,** also known as **Fisher's linear discriminant function.** The classification functions, one for each group, can be used in classifying observations. In this method of classification, an observation's values for the independent variables are inserted in the classification functions and a classification score for each group is calculated for that observation. The observation is then classified into the group with the highest classification score. We use the discriminant function as the means of classification because it provides a concise and simple representation of each discriminant function, simplifying the interpretation process and the assessment of the contribution of independent variables.

Evaluating Group Differences

One means of assessing overall model fit is to determine the magnitude of differences between the members of each group in terms of the discriminant Z scores. A summary measure of the group differences is a comparison of the group **centroids,** the average discriminant Z score for all group members. A measure of success of discriminant analysis is its ability to define discriminant function(s) that result in significantly different group centroids. The differences between centroids is measured in terms of Mahalanobis D^2 measure, for which tests are available to determine if the differences are statistically significant. The researcher should ensure that even with significant discriminant functions, there are significant differences between each of the groups.

Group centroids on each discriminant function can also be plotted to demonstrate the results from a global perspective. Plots are usually prepared for the first two or three discriminant functions (assuming they are statistically significant and valid predictive functions). The values for each group show its position in reduced discriminant space (so called because not all of the functions and thus not all of the variance is plotted). The researcher can see the differences between the groups on each function; however, visual inspection does not totally explain what these differences are. Circles can be drawn enclosing the distribution of observations around their respective centroids to clarify group differences further, but this procedure is beyond the scope of this text (see Dillon and Goldstein [3]).

Assessing Group Membership Prediction Accuracy

Given that the dependent variable is nonmetric, it is not possible to use a measure such as R^2, as is done in multiple regression, to assess predictive accuracy. Rather, each observation must be assessed as to whether it was correctly classified. In doing so, several major considerations must be addressed: the statistical and practical rationale for developing classification matrices, the cutting score determination, construction of the classification matrices, and the standards for assessing classification accuracy.

Why Classification Matrices Are Developed   The statistical tests for assessing the significance of the discriminant functions do not tell how well the function predicts. For example, suppose the two groups are significantly different beyond the .01 level. With sufficiently large sample sizes, the group means (centroids) could be virtually identical, and we still would have statistical significance. In short, these statistics suffer the same drawbacks as the classical tests of hypotheses. Thus the level of significance can be a very poor indication of the function's ability to discriminate between the two groups. To determine the predictive ability of a discriminant function, the researcher must construct classification matrices.

To clarify further the usefulness of the classification matrix procedure, we shall relate it to the concept of R^2 in regression analysis. Most of us have probably read articles in which the author has found statistically significant relationships and yet has explained only 10 percent (or less) of the variance (i.e., $R^2 = 0.10$). Usually, this R^2 is significantly different from zero simply because the sample size is large. With multiple discriminant analysis, the **hit ratio** (percentage correctly classified) is analogous to regression's R^2. The hit ratio reveals how well the discriminant function classified the objects; the R^2 indicates how much variance the regression equation explained. The F-test for statistical significance of the R^2 is, therefore, analogous to the chi-square (or D^2) test of significance in discriminant analysis. Clearly, with a sufficiently large sample size in discriminant analysis, we could have a statistically significant difference between the two (or more) groups and yet correctly classify only 53 percent (when chance is 50 percent, with equal group sizes) [14].

Cutting Score Determination   If the statistical test indicates that the function discriminates significantly, it is customary to develop classification matrices to provide a more accurate assessment of the discriminating power of the function. Before a classification matrix can be constructed, however, the researcher must determine the cutting score. The **cutting score** is the criterion (score) against which each object's discriminant score is compared to determine into which group the object should be classified.

In constructing classification matrices, the researcher will want to determine the **optimum cutting score** (also called a critical Z value). The optimal cutting scores will differ depending on whether the sizes of the groups are equal or unequal. If the groups are of equal size, the optimal cutting score will be halfway between the two group centroids. The cutting score for two groups of equal size is therefore defined as

$$Z_{CE} = \frac{Z_A + Z_B}{2}$$

where

Z_{CE} = critical cutting score value for equal group sizes

Z_A = centroid for group A

Z_B = centroid for group B

Specifying Probabilities of Classification for Unequal Group Sizes
To correctly calculate the cutting score when there are unequal group sizes, the researcher must also determine whether to specify if the observed group sizes reflect the actual population proportions or whether the population group sizes should be assumed to be equal. The default assumption is equal probabilities; in other words, each group is assumed to have an equal chance of occurring even if the group sizes in the sample are unequal. If the researcher is unsure if the observed proportions in the sample are representative of the population proportions, the conservative approach is to employ equal probabilities. However, if the sample is randomly drawn from the population so that the groups do estimate the population proportions in each group, then the best estimates of the prior probabilities are not equal values but, instead, the sample proportions. The impact of specifying the prior probabilities as equal to the sample proportions varies by the discrepancy of the sample proportions from the population proportions. But the researcher should always specify the probabilities in all analyses (either as equal or based on sample sizes) to ensure that the correct assumptions are underlying the classification process.

Cutting Score Determination for Unequal Size Groups
If the groups are not of equal size, but are assumed to be representative of the population proportions, a weighted average of the group centroids will provide an optimal cutting score for a discriminant function, calculated as follows:

$$Z_{CU} = \frac{N_A Z_B + N_B Z_A}{N_A + N_B}$$

where

Z_{CU} = critical cutting score value for unequal group sizes

N_A = number in group A

N_B = number in group B

Z_A = centroid for group A

Z_B = centroid for group B

Both of the formulas for calculating the optimal cutting score assume that the distributions are normal and the group dispersion structures are known.

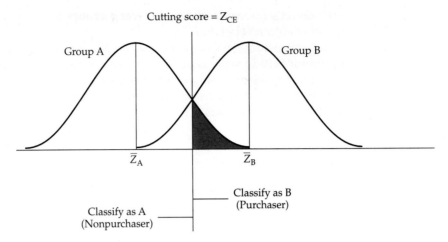

FIGURE 5.7 Optimal Cutting Score with Equal Sample Sizes

The concept of an optimal cutting score for equal and unequal groups is illustrated in Figures 5.7 and 5.8, respectively. Both the weighted and unweighted cutting scores are shown. It is apparent that if group A is much smaller than group B, the optimal cutting score will be closer to the centroid of group A than to the centroid of group B. Also, if the unweighted cutting score was used, none of the objects in group A would be misclassified, but a substantial portion of those in group B would be misclassified.

Costs of Misclassification   The optimal cutting score also must consider the cost of misclassifying an object into the wrong group. If the costs of misclassifying are approximately equal for all groups, the optimal cutting score will be the one that will misclassify the fewest number of objects across all groups. If the misclassification costs are unequal, the optimum cutting score will be the one that minimizes the costs of misclassification. More sophisticated approaches to determining cutting scores are discussed in Dillon and Goldstein [3] and Huberty [12]. These approaches are based upon a Bayesian statistical model and are appropriate when

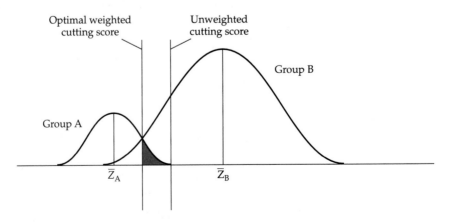

FIGURE 5.8 Optimal Cutting Score with Unequal Sample Sizes

the costs of misclassification into certain groups are very high, when the groups are of grossly different sizes, or when one wants to take advantage of a priori knowledge of group membership probabilities.

In practice, when calculating the cutting score, it is usually not necessary to insert the raw variable measurements for every individual into the discriminant function and to obtain the discriminant score for each person to use in computing the Z_A and Z_B (group A and B centroids). In many instances, the computer program will provide the discriminant scores as well as the Z_A and Z_B as regular output. When the researcher has the group centroids and sample sizes, the optimal cutting score can be obtained by merely substituting the values into the appropriate formula.

Constructing Classification Matrices To validate the discriminant function through the use of classification matrices, the sample should be randomly divided into two groups. One of the groups (the analysis sample) is used to compute the discriminant function. The other group (the holdout or validation sample) is retained for use in developing the classification matrix. The procedure involves multiplying the weights generated by the analysis sample by the raw variable measurements of the holdout sample. Then the individual discriminant scores for the holdout sample are compared with the critical cutting score value and classified as follows:

$$\text{Classify an individual into group A if } Z_n < Z_{ct}.$$

or

$$\text{Classify an individual into group B if } Z_n > Z_{ct}.$$

where

$$Z_n = \text{discriminant } Z \text{ score for the } n\text{th individual}$$

$$Z_{ct} = \text{critical cutting score value}$$

The results of the classification procedure are presented in matrix form, as shown in Table 5.4. The entries on the diagonal of the matrix represent the number of individuals correctly classified. The numbers off the diagonal represent the incorrect classifications. The entries under the column labeled "Actual Group Size" represent the number of individuals actually in each of the two groups. The entries at the bottom of the columns represent the number of individuals assigned to the groups by the discriminant function. The percentage correctly

TABLE 5.4 Classification Matrix for Two-Group Discriminant Analysis

	Predicted Group			
Actual Group	*1*	*2*	*Actual Group Size*	*Percentage Correctly Classified*
1	22	3	25	88
2	5	20	25	80
Predicted group size	27	23	50	84[a]

[a]Percent correctly classified = (Number correctly classified/Total number of observations) × 100
= [(22 + 20)/50] × 100
= 84%

classified for each group is shown at the right side of the matrix, and the over-all percentage correctly classified, also known as the hit ratio, is shown at the bottom.

In our example, the number of individuals correctly assigned to group 1 is 22, whereas 3 members of group 1 are incorrectly assigned to group 2. Similarly, the number of correct classifications to group 2 is 20, and the number of incorrect assignments to group 1 is 5. Thus the classification accuracy percentages of the discriminant function for the actual groups 1 and 2 are 88 and 80 percent, respectively. The overall classification accuracy (hit ratio) is 84 percent.

One final topic regarding classification procedures is the t test. A t test is available to determine the level of significance for the classification accuracy. The formula for a two-group analysis (equal sample size) is

$$t = \frac{p - .5}{\sqrt{\dfrac{.5(1.0 - .5)}{N}}}$$

where

$$p = \text{proportion correctly classified}$$
$$N = \text{sample size}$$

This formula can be adapted for use with more groups and unequal sample sizes.

Measuring Predictive Accuracy Relative to Chance   As noted earlier, the predictive accuracy of the discriminant function is measured by the hit ratio, which is obtained from the classification matrix. The researcher may ask, What is and what is not considered an acceptable level of predictive accuracy for a discriminant function? For example, is 60 percent an acceptable level, or should one expect to obtain 80 to 90 percent predictive accuracy? To answer this question, the researcher must first determine the percentage that could be classified correctly by chance (without the aid of the discriminant function).

Determining the Chance-Based Criteria Used
When the sample sizes of the groups are equal, the determination of the chance classification is rather simple; it is obtained by dividing 1 by the number of groups. The formula is C = 1/(number of groups). For instance, for a two-group function the chance probability would be .50; for a three-group function the chance probability would be .33; and so forth.

The determination of the chance classification for situations in which the group sizes are unequal is somewhat more involved. Let us assume that we have a sample in which 75 subjects belong to one group and 25 to the other. We could arbitrarily assign all the subjects to the larger group and achieve a 75 percent classification accuracy without the aid of a discriminant function. It could be concluded that unless the discriminant function achieves a classification accuracy higher than 75 percent, it should be disregarded because it has not helped us improve our prediction accuracy.

Determining the chance classification based on the sample size of the largest group is referred to as the **maximum chance criterion.** It is determined by computing the percentage of the total sample represented by the largest of the two (or more) groups. For example, if the group sizes are 65 and 35, the maximum chance criterion is 65 percent correct classifications. Therefore, if the hit ratio for the discriminant function does not exceed 65 percent, it has not helped us predict, based on this criterion.

The maximum chance criterion should be used when the sole objective of the discriminant analysis is to maximize the percentage correctly classified [14]. However, situations in which we are concerned only about maximizing the percentage correctly classified are rare. Usually the researcher uses discriminant analysis to correctly identify members of all groups. In cases where the sample sizes are unequal and the researcher wants to classify members of all groups, the discriminant function defies the odds by classifying a subject in the smaller group(s). But the chance criterion does not take this fact into account [14]. Therefore, another chance model—the **proportional chance criterion**—should be used in most situations.

The proportional chance criterion should be used when group sizes are unequal and the researcher wishes to correctly identify members of the two (or more) groups. The formula for this criterion is

$$C_{PRO} = p^2 + (1 - p)^2$$

where

$$p = \text{proportion of individuals in group 1}$$

$$1 - p = \text{proportion of individuals in group 2}$$

Using the group sizes from our earlier example (75 and 25), we see that the proportional chance criterion would be 62.5 percent compared with 75 percent. Therefore, in this instance, a prediction accuracy of 75 percent would be acceptable because it is above the 62.5 percent proportional chance criterion.

These chance model criteria are useful only when computed with holdout samples (split-sample approach). If the individuals used in calculating the discriminant function are the ones being classified, the result will be an upward bias in the prediction accuracy. In such cases, both of these criteria would have to be adjusted upward to account for this bias.

Comparing the Hit Ratio to Chance-Based Criteria

The question of classification accuracy is crucial. If the percentage of correct classifications is significantly larger than would be expected by chance, an attempt can be made to interpret the discriminant functions in the hope of developing group profiles. However, if the classification accuracy is no greater than can be expected by chance, whatever structural differences appear to exist merit little or no interpretation; that is, differences in score profiles would provide no meaningful information for identifying group membership.

The question, then, is how high should the classification accuracy be relative to chance? For example, if chance is 50 percent (two-group, equal sample size), does a classification (predictive) accuracy of 60 percent justify moving to the interpretation stage? No general guidelines have been developed to answer this question. Ultimately, the decision depends on the cost in relation to the value of the information. If the costs associated with a 60 percent predictive accuracy (relative to 50 percent by chance) are greater than the value to be derived from the findings, there is no justification for interpretation. If the value is high relative to the costs, 60 percent accuracy would justify moving on to interpretation.

The cost-versus-value argument offers little assistance to the neophyte data researcher, but the following criterion is suggested: The classification accuracy should be at least one-fourth greater than that achieved by chance. For example, if chance accuracy is 50 percent, the classification accuracy should be 62.5 percent ($62.5\% = 1.25 \times 50\%$). If chance accuracy is 30 percent, the classification accuracy should be 37.5 percent. This criterion provides only a rough estimate of the acceptable level of predictive accuracy. The criterion is easy to apply with groups of equal size. With groups of unequal size, an upper limit is reached when the maximum chance model is used to determine chance accuracy. This does not present too great a problem, however, because under most circumstances, the maximum chance model would not be used with unequal group sizes.

Statistically Based Measures of Classification Accuracy
Relative to Chance
A statistical test for the discriminatory power of the classification matrix when compared with a chance model is **Press's Q statistic.** This simple measure compares the number of correct classifications with the total sample size and the number of groups. The calculated value is then compared with a critical value (the chi-square value for 1 degree of freedom at the desired confidence level). If it exceeds this critical value, then the classification matrix can be deemed statistically better than chance. The Q statistic is calculated by the following formula:

$$\text{Press's } Q = \frac{[N - (nK)]^2}{N(K - 1)}$$

where

N = total sample size

n = number of observations correctly classified

K = number of groups

For example, in Table 5.4, the Q statistic would be based on a total sample of $N = 50$, $n = 42$ correctly classified observations, and $K = 2$ groups. The calculated statistic would be

$$\text{Press's } Q = \frac{[50 - (42 \times 2)]^2}{50(2 - 1)} = 23.12$$

The critical value at a significance level of .01 is 6.63. Thus we would conclude that in the example the predictions were significantly better than chance, which would have a correct classification rate of 50 percent. This simple test is sensitive to sample size; large samples are more likely to show significance than small sample sizes of the same classification rate. For example, if the sample size is increased to 100 in the example and the classification rate remains at 84 percent, the Q statistic increases to 46.24. One must be careful in drawing conclusions based solely on this statistic, however, because as the sample sizes become larger, a lower classification rate will still be deemed significant.

Casewise Diagnostics

The final means of assessing model fit is to examine the predictive results on a case-by-case basis. Similar to the analysis of residuals in multiple regression, the attempt is to understand which observations (1) have been misclassified, and (2) are not representative of the remaining group members. Although the classification matrix provides overall classification accuracy, it does not detail the individual case results. Also, even if we can denote which cases are correctly or incorrectly classified, we still need a measure of an observation's similarity to the remainder of the group.

Misclassification of Individual Cases

When analyzing residuals from a multiple regression analysis, an important decision involves setting the level of residual considered substantive and worthy of attention. In discriminant analysis, this issue is somewhat simpler because an observation is either correctly or incorrectly classified. The purpose of identifying and analyzing the misclassified observations is to identify any characteristics of these observations that could be incorporated into the discriminant analysis for improving predictive accuracy. This analysis may take the form of profiling the misclassified cases on either the independent variables or other variables not included in the model.

Examining these cases on the independent variables may identify nonlinear trends or other relationships or attributes that led to the misclassification. Screening other variables for their differences between these cases would be the first step for their possible inclusion in the discriminant analysis. Although there are no prespecified analyses, such as found in multiple regression, the researcher is encouraged to evaluate these misclassified cases from several perspectives in attempting to uncover the unique features they hold in comparison to their other group members.

The researcher can also make some assessment as to the similarity of an observation to the other group members by evaluating the Mahalanobis D^2 distance of the observation to the group centroid. Observations closer to the centroid are assumed more representative of the group than those farther away. In a graphical analysis of the observations, the researcher can identify outlying observations and make some assessment as to their impact on the results. For example, in a two group situation, a member of group A may have a large Mahalanobis D^2 distance, indicating it is less representative of the group. However, if that distance is away from the group B centroid, then it would actually increase the chance of correct classification, even though it is less representative of the group. A smaller distance that places an observation between the two centroids would probably have a lower probability of correct classification, even though it is closer to its group centroid than the earlier situation.

A graphical representation of the observations is another approach for examining the characteristics of observations, especially the misclassified observations. One common approach is to plot the observations based on their discriminant Z scores and portray the overlap among groups and the misclassified cases. If two or more functions are retained, the optimal cutting points can also be portrayed to give what is known as a **territorial map** depicting the regions corresponding to each group. Plotting the individual observations along with the group centroids, as discussed earlier, shows not only the general group characteristics depicted in the centroids, but also the variation in the group members. This is analogous to the areas defined in the three-group example at the beginning of this chapter, in which cutting scores on both functions defined areas corresponding to the classification predictions for each group.

Summary

The estimation and assessment stage has a number of similarities with the other dependence techniques, allowing for either a direct or stepwise estimation process and an analysis of overall and casewise predictive accuracy. The researcher should devote considerable attention to these issues to avoid the use of a fundamentally flawed discriminant analysis model.

Stage 5: Interpretation of the Results

If the discriminant function is statistically significant and the classification accuracy is acceptable, the researcher should focus on making substantive interpretations of the findings. This process involves examining the discriminant functions to determine the relative importance of each independent variable in discriminating between the groups. Three methods of determining the relative importance have been proposed: (1) standardized discriminant weights, (2) discriminant loadings (structure correlations), and (3) partial F values.

Discriminant Weights

The traditional approach to interpreting discriminant functions examines the sign and magnitude of the standardized **discriminant weight** (sometimes referred to as a **discriminant coefficient**) assigned to each variable in computing the discriminant functions. When the sign is ignored, each weight represents the relative contribution of its associated variable to that function. Independent variables with relatively larger weights contribute more to the discriminating power of the function than do variables with smaller weights. The sign denotes only that the variable makes either a positive or a negative contribution [3].

The interpretation of discriminant weights is analogous to the interpretation of beta weights in regression analysis and is therefore subject to the same criticisms. For example, a small weight may indicate either that its corresponding variable is irrelevant in determining a relationship or that it has been partialed out of the relationship because of a high degree of multicollinearity. Another problem with the use of discriminant weights is that they are subject to considerable instability. These problems suggest caution in using weights to interpret the results of discriminant analysis.

Discriminant Loadings

In recent years, loadings have increasingly been used as a basis for interpretation because of the deficiencies in utilizing weights. **Discriminant loadings,** referred to sometimes as **structure correlations,** measure the simple linear correlation between each independent variable and the discriminant function. The discriminant loadings reflect the variance that the independent variables share with the discriminant function and can be interpreted like factor loadings in assessing the relative contribution of each independent variable to the discriminant function. (Chapter 3 further discusses factor-loading interpretation.)

Discriminant loadings (like weights) may be subject to instability. Loadings are considered relatively more valid than weights as a means of interpreting the discriminating power of independent variables because of their correlational nature. The researcher still must be cautious when using loadings to interpret discriminant functions.

Partial **F** Values

As discussed earlier, two computational approaches—simultaneous and stepwise—can be utilized in deriving discriminant functions. When the stepwise method is selected, an additional means of interpreting the relative discriminating power of the independent variables is available through the use of partial F values. This is accomplished by examining the absolute sizes of the significant F values and ranking them. Large F values indicate greater discriminatory power.

In practice, rankings using the *F*-values approach are the same as the ranking derived from using discriminant weights, but the *F* values indicate the associated level of significance for each variable.

Interpretation of Two or More Functions

When there are two or more significant discriminant functions, we are faced with additional problems of interpretation. First, can we simplify the discriminant weights or loadings to facilitate the profiling of each function? Second, how do we represent the impact of each variable across the functions? These problems are found both in measuring the total discriminating effects across functions and in assessing the role of each variable in profiling each function separately. We address these two questions by introducing the concepts of rotation of the functions, the potency index, and stretched attribute vectors in graphical representations.

Rotation of the Discriminant Functions

After the discriminant functions have been developed, they can be "rotated" to redistribute the variance. (The concept is more fully explained in chapter 3.) Basically, rotation preserves the original structure and the reliability of the discriminant solution while making the functions easier to interpret substantively. In most instances, the VARIMAX rotation is employed as the basis for rotation.

Potency Index

Previously, we discussed using the standardized weights or discriminant loadings as measures of a variable's contribution to a discriminant function. When two or more functions are derived, however, a composite or summary measure is useful in describing the contributions of a variable across *all* significant functions. The **potency index** is a relative measure among all variables that is indicative of each variable's discriminating power [15]. It includes both the contribution of a variable to a discriminant function (its discriminant loading) and the relative contribution of the function to the overall solution (a relative measure among the eigenvalues of the functions). The composite is simply the sum of the individual potency indices across all significant discriminant functions. Interpretation of the composite measure is limited, however, by the fact that it is useful only in depicting the relative position (such as the rank order) of each variable, and the absolute value has no real meaning. The potency index is calculated by a two-step process:

> **Step 1: Calculate a potency value for each significant function.** In the first step, the discriminating power of a variable, represented by the squared value of the unrotated discriminant loading, is "weighted" by the relative contribution of the discriminant function to the overall solution. First, the relative eigenvalue measure for each significant discriminant function is calculated simply as:

$$\text{Relative eigenvalue of discriminant function } i = \frac{\text{Eigenvalue of discriminant function } i}{\text{Sum of eigenvalues across all significant functions}}$$

The potency value of each variable on a discriminant function is then

$$\text{Potency value of variable } i \text{ on function } j = (\text{Discriminant loading}_{ij})^2 \times \text{Relative eigenvalue of function } j$$

Step 2: *Calculate a composite potency index across all significant functions.* Once a potency value has been calculated for each function, the composite potency index is calculated as the sum of potency values on each significant discriminant function. The potency index now represents the total discriminating effect of the variable across all of the significant discriminant functions. It is only a relative measure, however, and its absolute value has no substantive meaning.

Graphical Display of Discriminant Loadings

To depict differences in the groups on the predictor variables, the researcher can plot the discriminant loadings. The simplest approach is to plot actual rotated or unrotated loadings on a graph. The preferred approach would be to plot the rotated loadings. An even more accurate approach, however, involves what is called **stretching the vectors.**

Before explaining the process of stretching, we must first define a vector in this context. A **vector** is merely a straight line drawn from the origin (center) of a graph to the coordinates of a particular variable's loadings. The length of each vector is indicative of the relative importance of each variable in discriminating among the groups. To stretch a vector, the researcher multiplies the discriminant loading (preferably after rotation) by its respective univariate F value.

The plotting process always involves all the variables included in the model as significant. But the researcher may also plot the other variables with significant univariate F ratios that were not significant in the discriminant function. This procedure shows the importance of collinear variables that are not included, such as in a stepwise solution. By using this procedure, we note that vectors point toward the groups having the highest mean on the respective predictor and away from the groups having the lowest mean scores. The group centroids are also stretched in this procedure by multiplying them by the approximate F value associated with each discriminant function. If the loadings are stretched, the centroids must be stretched as well to plot them accurately on the same graph. The approximate F values for each discriminant function are obtained by the following formula:

$$F \text{ value}_{\text{function}_i} = \text{Eigenvalue}_{\text{function}_i} \left(\frac{\text{Estimation sample size} - \text{No. of groups}}{\text{No. of groups} - 1} \right)$$

As an example, assume that the sample of 50 observations was divided into three groups. The multiplier of each eigenvalue would be $(50 - 3)/(3 - 1) = 23.5$. For more details on this procedure, see Dillon and Goldstein [3].

An alternative to stretching the attribute vectors and centroids is the "territorial maps" provided by most programs. They do not include the vectors, but they do plot the centroids and the boundaries for each group.

Which Interpretive Method to Use?

Several methods for interpreting the nature of discriminant functions have been discussed, both for single- and multiple-function solutions. Which methods should be used? The loadings approach is somewhat more valid than the use of weights and should be utilized whenever possible. The use of univariate and partial F values allows the researcher to use several measures and look for some consistency in evaluations of the variables. If two or more functions are estimated, then the researcher can employ several graphical techniques and the potency index, which aid in interpreting the multidimensional solution. The most basic point is that the researcher should employ all available methods to arrive at the most accurate interpretation.

Stage 6: Validation of the Results

The final stage of a discriminant analysis involves validating the discriminant results to provide assurances that the results have external as well as internal validity. With the propensity of discriminant analysis to inflate the hit ratio if evaluated only on the analysis sample, cross-validation is an essential step. Most often the cross-validation is done with the original sample, but it is possible to employ an additional sample as the holdout sample. In addition to cross-validation, the researcher should use group profiling to ensure that the group means are valid indicators of the conceptual model used in selecting the independent variables.

Split-Sample or Cross-Validation Procedures

Recall that the most frequently utilized procedure in validating the discriminant function is to divide the groups randomly into analysis and holdout samples. This involves developing a discriminant function with the analysis sample and then applying it to the holdout sample. The justification for dividing the total sample into two groups is that an upward bias will occur in the prediction accuracy of the discriminant function if the individuals used in developing the classification matrix are the same as those used in computing the function; that is, the classification accuracy will be higher than is valid for the discriminant function, if it was used to classify a separate sample. The implications of this upward bias are particularly important when the researcher is concerned with the external validity of the findings.

Other researchers have suggested, however, that greater confidence could be placed in the validity of the function by following this procedure several times [15]. Instead of randomly dividing the total sample into analysis and holdout groups once, the researcher would randomly divide the total sample into analysis and holdout samples several times, each time testing the validity of the function through the development of a classification matrix and a hit ratio. Then the several hit ratios would be averaged to obtain a single measure.

More sophisticated methods based on estimation with multiple subsets of the sample have been suggested for validating discriminant functions [2, 3]. The two most widely used approaches are the U-method and the jackknife method. Both methods are based on the "leave-one-out" principle, in which the discriminant function is fitted to repeatedly drawn samples of the original sample. The most prevalent use of this method has been to estimate $k - 1$ samples, eliminating one observation at a time from a sample of k cases. The primary difference in the two approaches is that the U-method focuses on classification accuracy, whereas the jackknife approach addresses the stability of the discriminant coefficients. Both approaches are quite sensitive to small sample sizes. Guidelines suggest that either of these two approaches be used only when the smallest group size is at least three times the number of predictor variables, and most researchers suggest a ratio of five to one [12]. In spite of these limitations, both methods provide the most valid and consistent estimate of the classification accuracy rate. The use of the *U*-method and jackknife method has been limited because many of the major computer packages provide them as a program option.

Profiling Group Differences

Another validation technique is to profile the groups on the independent variables to ensure their correspondence with the conceptual bases used in the original model formulation. When the researcher has identified the independent variables that make the greatest contribution in discriminating between the groups, the next step is to profile the characteristics of the groups based on the group means. This profile enables the researcher to understand the character of each group according to the predictor variables. For example, referring to the HATCO survey data presented in Table 5.1, we see that the mean rating on "durability" for the "would purchase" group is 7.4, whereas the comparable mean rating on "durability" for the "would not purchase" group is 3.2. Thus a profile of these two groups shows that the "would purchase" group rates the perceived durability of the new product substantially higher than the "would not purchase" group.

Another approach is to profile the groups on a separate set of variables that should mirror the observed group differences. This separate profile provides an assessment of external validity in that the groups vary on both the independent variable(s) and the set of associated variables. This is similar in character to the validation of derived clusters described in chapter 9.

Logistic Regression: Regression with a Binary Dependent Variable

As we have discussed, discriminant analysis is appropriate when the dependent variable is nonmetric. However, when the dependent variable has only two groups, logistic regression may be preferred for several reasons. First, discriminant analysis relies on strictly meeting the assumptions of multivariate normality and equal variance–covariance matrices across groups—assumptions that are not met in many situations. Logistic regression does not face these strict assumptions and is much more robust when these assumptions are not met, making its application appropriate in many more situations. Second, even if the assumptions are met, many researchers prefer logistic regression because it is similar to regression. Both have straightforward statistical tests, the ability to incorporate nonlinear effects, and a wide range of diagnostics. For these and more technical reasons, logistic regression is equivalent to two-group discriminant analysis and may be more suitable in many situations.

Our discussion of logistic regression does not cover each of the six steps of the decision process, but instead highlights the differences and similarities between logistic regression and discriminant analysis or multiple regression. Multiple regression analysis is discussed in chapter 4.

Representation of the Binary Dependent Variable

In discriminant analysis, the nonmetric character of a dichotomous dependent variable is accommodated by making predictions of group membership based on discriminant Z scores. This requires the calculation of cutting scores and the assignment of observations to groups. Logistic regression approaches this task in a manner more similar to that found in multiple regression. It differs from multiple regression, however, in that it directly predicts the probability of an event occurring. Although the probability value is a metric measure, there are fundamental differences between multiple and logistic regression. Probability values can be any value between zero and one, but the predicted value must be bounded to fall within the range of zero and one. To define a relationship bounded by zero and one, logistic regression uses an assumed relationship between the independent and dependent variables that resembles an S-shaped curve (see Figure 5.9). At very low levels of the independent variable, the probability approaches zero. As the independent variable increases, the probability increases up the curve, but then the slope starts decreasing so that at any level of the independent variable, the probability will approach one but never exceed it. As we saw in our discussions of regression in chapter 4, the linear models of regression cannot accommodate such a relationship, as it is inherently nonlinear. Moreover, such situations cannot be studied with ordinary regression, because doing so would violate several assumptions. First, the error term of a discrete variable follows the binomial distribution instead of the normal distribution, thus invalidating all statistical testing based on the assumptions of normality. Second, the variance of a dichotomous variable is not constant, creating instances of heteroscedasticity as well. Logistic regression was developed to specifically deal with these issues. Its unique relationship between dependent and independent variables requires a somewhat different approach in estimating, assessing goodness of fit, and interpreting the coefficients.

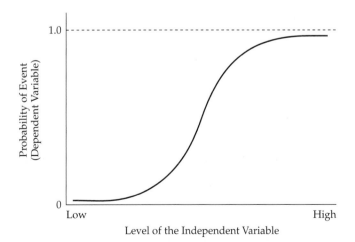

FIGURE 5.9 Form of the Logistic Relationship between Dependent and Independent Variables

Estimating the Logistic Regression Model

Logistic regression, even though it has a single variate comprised of estimated coefficients for each independent variable—as found in multiple regression—is estimated in an entirely different manner. Multiple regression employs the method of least squares, which minimizes the sum of the squared differences between the actual and predicted values of the dependent variable. The nonlinear nature of the logistic transformation requires that another procedure, the maximum likelihood procedure, be used in an iterative manner to find the "most likely" estimates for the coefficients. This results in the use of the **likelihood value** instead of the sum of squares when calculating measure of overall model fit.

The process of estimating the coefficients, however, is still quite similar in many regards to that of linear regression. As shown earlier, the logit model has the specific form of the **logistic curve.** To estimate a logistic regression model, this curve is fitted to the actual data. Figure 5.10 portrays two hypothetical examples of fitting a logistic relationship to sample data. The actual data, an event either happening or not (0 or 1), are represented as observations at either the top or bottom of the graph. These are the events that occur at each value of the independent variable (the X-axis). In part A, the logistic curve cannot fit the data well because there are a number of values of the independent variable that have both events and nonevents (i.e., high overlap of the distributions). However, in part B, there is a much more well-defined relationship, and the logistic curve fits the data quite well. This simple example, similar to a scatterplot of dependent and independent variables in regression with a line representing the "best fit" of the correlation, can be extended to include multiple independent variables just as in regression.

Interpreting the Coefficients

One of the advantages of logistic regression is that we need to know only whether an event (purchase or not, credit risk or not, firm failure or success) occurred to then use a dichotomous value as our dependent variable. From this dichotomous value, the procedure predicts its estimate of the probability that the event will or will not occur. If the predicted probability is greater than .50, then the prediction is yes, otherwise no. Logistic regression derives its name from the **logistic transformation** used with the dependent variable. When this transformation is used, however, the logistic regression and its coefficients take on a somewhat different meaning from those found in regression with a metric dependent variable.

The procedure that calculates the **logistic coefficient** compares the probability of an event occurring with the probability of its not occurring. This **odds ratio** can be expressed as

$$\frac{\text{Prob}_{(event)}}{\text{Prob}_{(no\ event)}} = e^{B_0\ +\ B_1X_1\ +\ \ldots\ +\ B_nX_n}$$

The estimated coefficients ($B_0, B_1, B_2, \ldots, B_n$) are actually measures of the changes in the ratio of the probabilities, termed the odds ratio. Moreover, they are expressed in logarithms, so they need to be transformed back (the antilog of the value has to be taken) so that their relative effect on the probabilities is assessed more easily. Computer programs perform this procedure automatically and give both the actual coefficient and the transformed coefficient. Use of this procedure does not change in any manner the way we interpret the sign of the coefficient. A positive coefficient increases the probability, whereas a negative value decreases the predicted probability.

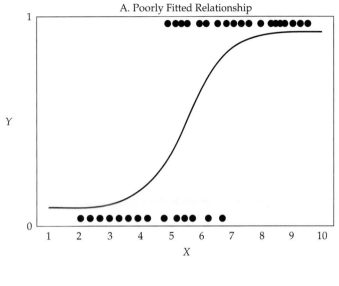

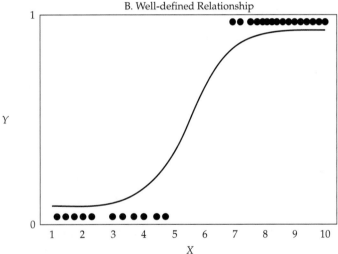

FIGURE 5.10 Examples of Fitting the Logistic Curve to Sample Data

Let us look at a simple example to see what we mean. If B_i is positive, its transformation (antilog) will be greater than 1, and the odds ratio will increase. This increase occurs when the predicted probability of the event's occurring increases and the predicted probability of its not occurring is reduced. Thus the model has a higher predicted probability of occurrence. Likewise, if B_i is negative, the antilog is less than one and the odds will be decreased. A coefficient of zero equates to a value of 1.0, resulting in no change in the odds. A more detailed discussion of interpretation of coefficients, logistic transformation, and the estimation procedures can be found in numerous texts [10].

In our earlier discussion of the assumed distribution of possible dependent variables, we described an S-shaped or logistic curve. To represent the relationship between the dependent and independent variables, the coefficients must actually represent nonlinear relationships among the dependent and independent variables. Although the transformation process of taking logarithms provides a linearization of the relationship, the researcher must remember that the coefficients actually represent different slopes in the relationship across the values of the independent variable. In this way, the S-shaped distribution can be estimated. If the researcher is interested in the slope of the relationship at various values of the independent variable, the coefficients can be calculated and the relationship assessed [5].

Assessing the Goodness-of-Fit of the Estimated Model

Logistic regression is similar to multiple regression in many of its results, but it is different in the method of estimating coefficients. Instead of minimizing the squared deviations (least squares), logistic regression maximizes the "likelihood" that an event will occur. Using this alternative estimation technique also requires that we assess model fit in different ways.

The overall measure of how well the model fits, similar to the residual or error sums of squares value for multiple regression, is given by the likelihood value. (It is actually -2 times the log of the likelihood value and is referred to as $-2LL$ or -2 log likelihood.) A well-fitting model will have a small value for $-2LL$. The minimum value for $-2LL$ is 0. (A perfect fit has a likelihood of 1, and $-2LL$ is then 0.) The likelihood value can be compared between equations as well, with the difference representing the change in predictive fit from one equation to another. Statistical programs have automatic tests for the significance of these differences.

The chi-square test for the reduction in the log likelihood value provides one measure of improvement due to the introduction of the independent variable(s). A null model, which is similar to calculating the total sum of squares using only the mean, provides the baseline for comparison. In addition to the statistical chi-square tests, several different "R^2-like" measures have been developed to represent overall model fit, as is done by the coefficient of determination in multiple regression. The researcher can also construct a "pseudo R^2" value for logistic regression similar to the R^2 value in regression analysis [5]. The R^2 for a logit model (R^2_{logit}) can be calculated as

$$R^2_{logit} = \frac{-2LL_{null} - (-2LL_{model})}{-2LL_{null}}$$

We can assess overall model fit in a manner similar to multiple regression, and we can also employ several methods that employ the nonmetric character of the dependent variable. First, we can use the method of classification matrices developed for discriminant analysis to assess predictive accuracy in terms of group membership. All of the chance-related measures used earlier are applicable here as well. Second, Hosmer and Lemeshow [10] have developed another classification test. The cases are first divided into approximately 10 equal classes. Then, the number of actual and predicted events are compared in each class with the chi-square statistic. This test provides a comprehensive measure of predictive accuracy that is based not on the likelihood value, but instead on the actual prediction of the dependent variable. The appropriate use of this test requires an adequate sample size to ensure that each group has at least five observations and never falls below one. Also, the chi-square statistic is sensitive to sample size, thus allowing this measure to find very small differences statistically significant when the sample size becomes large. The researcher should employ all of these various measures of fit to assess this technique, which has aspects of both multiple regression and discriminant analysis.

Testing for Significance of the Coefficients

Logistic regression can also test the hypothesis that a coefficient is different from zero (zero means that the odds ratio does not change and the probability is not affected), as is done in multiple regression. In multiple regression, the t value is used to assess the significance of each coefficient. Logistic regression uses a different statistic, the **Wald statistic.** It provides the statistical significance for each estimated coefficient so that hypothesis testing can occur just as it does in multiple regression.

Other Similarities to Multiple Regression   Despite the fact that a binary dependent measure is used and group membership is the predicted outcome, the format of logistic regression is much like that of multiple regression. Just as in regression, nominal and categorical data may be included as independent variables through some form of dummy-variable coding. Also, model selection procedures such as those found in multiple regression (forward and backward stepwise) are also available. Finally, to examine the results more clearly, many of the diagnostic measures, such as residuals, residual plots, and measures of influence, are also available.

The researcher faced with a dichotomous variable need not resort to methods designed to accommodate the limitations of multiple regression nor be forced to employ discriminant analysis, especially if its statistical assumptions are violated. Logistic regression addresses these problems and provides a method developed to deal directly with this situation in the most efficient manner possible.

A Two-Group Illustrative Example

To illustrate the application of two-group discriminant analysis, we use variables drawn from the HATCO database introduced in chapter 1. This example examines each of the six stages of the model-building process to a research problem particularly suited to multiple discriminant analysis.

Stage 1: Objectives of Discriminant Analysis

You will recall that one of the customer characteristics obtained by HATCO in its survey was a categorical variable indicating which purchasing approach a firm used: total value analysis versus specification buying. Firms that employ total value analysis evaluate each aspect of the purchase, including both the product and the services being purchased. Specification buying, on the other hand, defines all product and service characteristics desired, and the seller then makes a bid to fill the specifications. Both approaches have merit in certain situations, but HATCO's management team expects that firms using these two approaches would emphasize different characteristics of suppliers in their selection decision. The objective is to identify the perceptions of HATCO that differ significantly between firms using these two purchasing methods. The company would then be able to tailor sales presentations and benefits offered to best match the buyers' perceptions. To do so, discriminant analysis was selected to identify those perceptions of HATCO that best distinguish firms using each buying approach.

Stage 2: Research Design for Discriminant Analysis

The research design stage focuses on three key issues: selecting dependent and independent variables, assessing the adequacy of the sample size for the planned analysis, and dividing the sample for validation purposes.

Selection of Dependent and Independent Variables

Because the dependent variable, the purchasing approach employed by a firm, is a two-group categorical variable, discriminant analysis is the appropriate technique. The survey also collected perceptions of HATCO that can now be used to differentiate between the two groups of firms. Discriminant analysis uses as independent variables the first seven variables from the database (X_1 to X_7) to discriminate between firms applying each purchasing method (X_{11}).

Sample Size

The sample of 100 observations, when split into analysis and holdout (validation) samples, meets the suggested minimum (5-to-1) ratio for discriminant analysis by providing a 9-to-1 ratio of observations to independent variables (60 observations for 7 potential independent variables) in the analysis sample. Although this ratio would increase to 15 to 1 if the sample were not split, it was deemed more important to validate the results rather than to increase the number of observations in the analysis sample. Moreover, both groups exceed the minimum size of 20 observations per group. Finally, the two groups of firms contain 60 and 40 observations, making them comparable enough in size not to impact either the estimation or the classification processes.

Division of the Sample

Previous discussion has emphasized the need for validating the discriminant function with a split sample or holdout sample. Any time a holdout sample is used, the researcher must ensure that the resulting sample sizes are sufficient to support the number of predictors included in the analysis. The HATCO database has 100 observations; it was decided that a holdout sample of 40 observations would be sufficient for validation purposes. This split would still leave 60 observations for estimation of the discriminant function. It is important to ensure randomness in the selection of the holdout sample so that any ordering of the observations does not affect the processes of estimation and validation. The control cards necessary for both selection of the holdout sample and performance of the two-group discriminant analysis are shown in appendix A at the end of the text.

Stage 3: Assumptions of Discriminant Analysis

The principal assumptions underlying discriminant analysis involve the formation of the variate or discriminant function (normality, linearity, and multicollinearity) and the estimation of the discriminant function (equal variance and covariance matrices). How to examine the independent variables for normality, linearity, and multicollinearity is explained in chapter 2. For purposes of our illustration of discriminant analysis, these assumptions are met at acceptable levels.

Most statistical programs have one or more statistical tests for the assumption of equal covariance or dispersion matrices addressed in chapter 2. The most common test is **Box's M** (for more detail, see chapter 2). In the two-group example, the significance of differences in the covariance matrices between the two groups is .0320. Even though the significance is less than .05 (in this test the researcher looks for values above the desired significance level), the sensitivity of the test to factors other than just covariance differences (e.g., normality of the variables and increasing sample size) make this an acceptable level. However, separate variance estimates rather than the pooled estimates are used in the classification stage for illustration purposes. No additional remedies are needed before estimation of the discriminant function can be performed.

Stage 4: Estimation of the Discriminant Model and Assessing Overall Fit

Let us begin our assessment of the two-group discriminant analysis by examining Table 5.5, which shows the group means for each of the independent variables, based on the 60 observations constituting the analysis sample. Besides profiling the two groups, we can also identify the variables with the largest differences in the group means (X_1, X_3, and X_7). Table 5.5 also shows the Wilks' lambda and univariate ANOVA used to assess the significance between means of the independent variables for the two groups. These tests indicate that five of the seven independent variables show significant univariate differences between the two groups. Only X_4 (manufacturer image) and X_6 (salesforce image) are not

TABLE 5.5 Group Descriptive Statistics and Tests of Equality for the Two-Group Discriminant Analysis

Dependent Variable[a]	Group Means for the Independent Variables[b]							Sample Size
	X_1	X_2	X_3	X_4	X_5	X_6	X_7	
0: Specification buying	2.23	2.97	6.87	5.16	2.58	2.56	8.47	22
1: Total value analysis	4.26	2.08	8.57	5.44	3.18	2.83	6.01	38
Total	3.51	2.41	7.95	5.33	2.96	2.73	6.91	60

	Standard Deviations for the Independent Variables[b]							
0: Specification buying	1.05	1.19	.76	.82	.94	.58	.95	
1: Total value analysis	1.10	1.12	1.28	1.32	.50	.92	1.32	
Total	1.46	1.21	1.38	1.16	.75	.82	1.68	

	Tests for the Equality of the Group Means[c]							
Wilks' lambda	.542	.873	.645	.986	.846	.973	.499	
Univariate F ratio	48.992	8.453	31.881	.822	10.576	1.620	58.176	
Significance level	.000	.005	.000	.368	.002	.208	.000	

[a]X_{11} = specification buying
[b]X_1 = delivery speed; X_2 = price level; X_3 = price flexibility; X_4 = manufacturer image; X_5 = overall service; X_6 = salesforce image; X_7 = product quality
[c]Wilks' lambda (U statistic) and univariate F ratio with 1 and 58 degrees of freedom

significantly different. The purpose of discriminant analysis is to define the set of variables that will best discriminate between the groups. For that, we must estimate the discriminant function.

Estimation of the Discriminant Function

Because the objective of this analysis is to determine which variables are the most efficient in discriminating between firms using the two purchasing approaches, a stepwise procedure is used. If the objective were simply to determine the discriminating capabilities of the entire set of benefits, with no regard to the impact of any individual benefit sought, all variables would be entered into the model simultaneously. The Mahalanobis D^2 measure will be used in the stepwise procedure to determine the variable with the greatest power of discrimination.

The stepwise procedure begins with all of the variables excluded from the model and then selects the variable that maximizes the Mahalanobis distance between the groups. In this example, a minimum significance value of .05 is required for entry, and Mahalanobis D^2 is used to actually select the variables. The maximum Mahalanobis D^2 is associated with X_7 (see Table 5.6). After X_7 entered the model,

TABLE 5.6 Results from Step 1 of Stepwise Two-Group Discriminant Analysis Model

STEP 1: X_7 (PRODUCT QUALITY) INCLUDED IN THE ANALYSIS
SUMMARY STATISTICS

		Degrees of Freedom		Significance	Between Groups
Wilks' lambda	.499	1	1	58	
Equivalent F	58.176		1	58	.000
Minimum D^2	4.175				0 and 1
Equivalent F	58.176		1	58	.000

VARIABLES IN THE ANALYSIS AFTER STEP 1

		F to Remove	
Variables	Tolerance	Value	Significance
X_7 Product quality	1.00	58.176	.000

VARIABLES NOT IN THE ANALYSIS AFTER STEP 1

Variables	Tolerance	Minimum Tolerance	F to Enter Value	Significance	D^2	Between Groups
X_1 Delivery speed	.973	.973	16.680	.000	6.615	0 and 1
X_2 Price level	.933	.933	.454	.503	4.242	0 and 1
X_3 Price flexibility	.997	.997	18.196	.000	6.837	0 and 1
X_4 Manufacturer image	.963	.963	2.874	.095	4.596	0 and 1
X_5 Overall service	.994	.994	7.203	.010	5.229	0 and 1
X_6 Salesforce image	.962	.962	3.896	.053	4.745	0 and 1

SIGNIFICANCE TESTING OF GROUP DIFFERENCES AFTER STEP 1[a]
Group 0: Specification buying

Group 1: Total value analysis	58.176
	(.000)

[a]F statistic and significance level (in parentheses) between groups after step 1. Each F statistic has 1 and 58 degrees of freedom.

the remaining variables were evaluated on the basis of distance between their means after the variance associated with X_7 was removed. Again, variables with significance levels greater than .05 were eliminated from consideration for entry at the next step.

Three variables met the .05 significance level criteria for consideration at the next stage (X_1, X_3, and X_5). Variable X_3 is the next best candidate to enter the model because it has the highest Mahalanobis D^2 (6.837) (see Table 5.6). Given X_1's large Mahalanobis D^2 value (6.615), it is very likely that it will also enter the model at a later step if it is not highly correlated with variables previously selected. The statistical significance tests must be calculated after the effect of the variable(s) in the models are removed. For instance, high multicollinearity of X_1 with variables in the model could substantially reduce the significance level and the Mahalanobis D^2. Also, in cases where two or more variables are entered into the model, the variables already in the model are evaluated for possible removal. A variable may be removed if high multicollinearity exists between it and the other included independent variables such that its significance falls below the significance level for removal (.10).

In step 2 (see Table 5.7), X_3 enters the model, as expected. As in step 1, the overall model is significant ($F = 46.81$), as is the discriminating power of both variables included to this point (X_3 and X_7). As noted earlier, X_1 is the next candidate for inclusion, but the significance level has been reduced substantially because of X_1's multicollinearity with X_3 and X_7 already in the discriminant function. Also, the Mahalanobis D^2 has increased (from 4.175 to 6.837), indicative of a "spreading out" and separation of the groups by X_3 and X_7 already in the discriminant function. Note that X_5 is almost identical in remaining discrimination power, but X_1 will enter in the third step due to its slight advantage. Table 5.8 (p. 287) reviews the results of the third step of the stepwise process, where X_1 does enter the discriminant function. The overall results are still statistically significant and continue to improve in discrimination, as evidenced by the decrease in the Wilks' lambda value (from .378 to .331). With X_1, X_3, and X_7 included, none of the remaining four independent variables passes the entry criterion of .05. After X_1 was entered in the equation, X_6 had relatively little additional discriminatory power and did not meet the entry criterion. Thus, the estimation process stops with three variables (X_1, X_3, and X_7) constituting the discriminant function.

Table 5.9 (p. 288) provides the overall stepwise discriminant analysis results after all the significant variables have been included in the estimation of the discriminant function. This summary table describes the three variables (X_1, X_3, and X_7) that were significant discriminators based on their Wilks' lambda and minimum Mahalanobis D^2 values. The multivariate aspects of the model are reported under the heading "Canonical Discriminant Functions." Note that the discriminant function is highly significant (.000) and displays a canonical correlation of .818. We interpret this correlation by squaring it $(.818)^2 = .669$. Thus, 66.9 percent of the variance in the dependent variable (X_{11}) can be accounted for (explained) by this model, which includes only three independent variables. The standardized canonical discriminant function coefficients are provided, but are less preferred for interpretation purposes than the discriminant loadings. The unstandardized discriminant coefficients are used to calculate the discriminant Z scores that can be used in classification. The discriminant loadings are reported under the heading "Structure Matrix" and are ordered from highest to lowest by

TABLE 5.7 Results from Step 2 of Stepwise Two-Group Discriminant Analysis Model

STEP 2: X_3 (PRICE FLEXIBILITY) INCLUDED IN THE ANALYSIS

SUMMARY STATISTICS

		Degrees of Freedom			*Significance*	*Between Groups*
Wilks' lambda	.378	2	1	58		
Equivalent F	46.810		2	57	.000	
Minimum D^2	6.837					0 and 1
Equivalent F	46.810		2	57	.000	

VARIABLES IN THE ANALYSIS AFTER STEP 2

		F to Remove			*Between*
Variables	*Tolerance*	*Value*	*Significance*	D^2	*Groups*
X_3 Price flexibility	.997	18.196	.000	2.288	0 and 1
X_7 Product quality	.997	40.195	.000	4.175	0 and 1

VARIABLES NOT IN THE ANALYSIS AFTER STEP 2

		Minimum	*F to Enter*			*Between*
Variables	*Tolerance*	*Tolerance*	*Value*	*Significance*	D^2	*Groups*
X_1 Delivery speed	.932	.932	7.974	.007	8.403	0 and 1
X_2 Price level	.809	.809	.661	.419	6.967	0 and 1
X_4 Manufacturer image	.946	.946	3.884	.054	7.600	0 and 1
X_5 Overall service	.980	.980	7.770	.007	8.363	0 and 1
X_6 Salesforce image	.959	.958	3.557	.064	7.536	0 and 1

SIGNIFICANCE TESTING OF GROUP DIFFERENCES AFTER STEP 2^a
Group 0: Specification buying

Group 1: Total value analysis	46.810
	(.000)

[a]F statistic and significance level (in parentheses) between groups after step 1. Each F statistic has 2 and 57 degrees of freedom.

the size of the loading. The loadings are discussed later under the interpretation phase. The classification function coefficients, also known as Fisher's linear discriminant functions, are used in classification and are discussed later. Finally, group centroids are also reported, and they represent the mean of the individual discriminant function scores for each group.

Group centroids can be used to interpret the discriminant function results from a global or an overall perspective. Table 5.9 reveals that the group centroid for the firms using specification buying (group 0) is -1.836, whereas the group centroid for the firms using the total value analysis approach (group 1) is 1.1063. Figure 5.11 (p. 289) is a plot of the centroids showing each group's deviation from the overall mean of the two groups. To show that the overall mean is 0, multiply the number in each group by its centroid and add the result (e.g., $-1.836 \times 22 + 1.1063 \times 38 = 0.0$).

TABLE 5.8 Results from Step 3 of Stepwise Two-Group Discriminant Analysis Model

STEP 3: X_1 (DELIVERY SPEED) INCLUDED IN THE ANALYSIS

SUMMARY STATISTICS

		Degrees of Freedom		Significance	Between Groups
Wilks' lambda	.331	3	1	58	
Equivalent F	37.683		3	56	.000	
Minimum D^2	8.403					0 and 1
Equivalent F	37.683		3	56	.000	

VARIABLES IN THE ANALYSIS AFTER STEP 3

		F to Remove			
Variables	Tolerance	Value	Significance	D^2	Between Groups
X_7 Product quality	.965	.	.000	4.886	0 and 1
X_3 Price flexibility	.954	.	.004	6.615	0 and 1
X_1 Delivery speed	.932	.	.007	6.837	0 and 1

VARIABLES NOT IN THE ANALYSIS AFTER STEP 3

			F to Enter			
Variables	Tolerance	Minimum Tolerance	Value	Significance	D^2	Between Groups
X_2 Price level	.788	.788	.	.238	8.728	0 and 1
X_4 Manufacturer image	.937	.920	.	.120	8.972	0 and 1
X_5 Overall service	.570	.542	.	.248	8.716	0 and 1
X_6 Salesforce image	.957	.925	.	.109	9.010	0 and 1

SIGNIFICANCE TESTING OF GROUP DIFFERENCES AFTER STEP 3[a]

	Group 0: Specification buying
Group 1: Total value analysis	37.683
	(.000)

[a]F statistic and significance level (in parentheses) between groups after step 1. Each F statistic has 3 and 56 degrees of freedom.

Assessing Overall Fit

The second step in the estimation stage is to assess the predictive accuracy of the discriminant function. To accomplish this, we must calculate a classification matrix. Classification matrices for both the analysis and the holdout samples are developed. Although examination of the holdout sample and its predictive accuracy is actually performed in the validation stage, the results are discussed now for ease of comparison between estimation and holdout samples. To better understand the classification process, we must determine the cutting score, the criterion against which each observation's discriminant Z score is judged to determine into which group it should be classified.

In this analysis sample of 60 observations, we know that the dependent variable consists of two groups, 22 firms following the specification buying approach and the remaining 38 firms using the total value analysis method. If we are not sure that the population proportions are represented by the sample, then we should employ equal probabilities. However, because our sample of firms is

TABLE 5.9 Summary of Two-Group Discriminant Analysis Results

SUMMARY TABLE

	Action		Wilks' Lambda		Minimum D^2		Between
Steps	Entered	Removed	Value	Significance	Value	Significance	Groups
1	X_7 Product quality		.499	.000	4.175	.000	0 and 1
2	X_3 Price flexibility		.378	.000	6.837	.000	0 and 1
3	X_1 Delivery speed		.331	.000	8.403	.000	0 and 1

CANONICAL DISCRIMINANT FUNCTIONS

		Percent of Variance		Canonical	After	Wilks'			
Function	Eigenvalue	Function	Cumulative	Correlation	Function	Lambda	Chi-square	df	Significance
					0	.331	62.424	3	.000
1^a	2.019	100	100	.818					

CANONICAL DISCRIMINANT FUNCTION COEFFICIENTS

Independent Variables	Standardized	Unstandardized
X_1 Delivery speed	.447	.413
X_3 Price flexibility	.472	.421
X_7 Product quality	−.659	−.549
Constant		−1.003

STRUCTURE MATRIX[b]

Independent Variables	Discriminant Function Loadings: Function 1
X_7 Product quality	−.705
X_1 Delivery speed	.647
X_3 Price flexibility	.522
X_2 Price level	−.443
X_6 Salesforce image	−.168
X_4 Manufacturer image	.155
X_5 Overall service	−.145

CLASSIFICATION FUNCTION COEFFICIENTS

Independent Variables	Group 0: Specification Buying	Group 1: Total Value Analysis
X_1 Delivery speed	2.021	3.219
X_3 Price flexibility	4.728	5.950
X_7 Product quality	5.932	4.340
Constant	−44.606	−45.848

GROUP MEANS (CENTROIDS) OF CANONICAL DISCRIMINANT FUNCTIONS

Group	Group Centroids: Function 1
Specification buying	−1.836
Total value buying	1.063

[a]Marks the 1 canonical discriminant function remaining in the analysis.
[b]Pooled-within-groups correlations between discriminating variables and canonical discriminant functions (variables ordered by size of correlation within function).

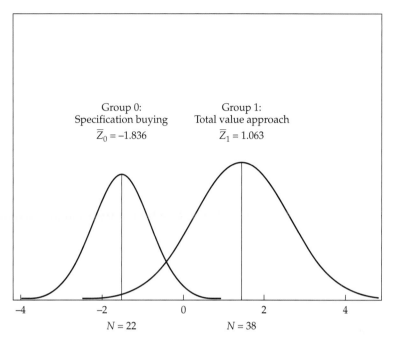

FIGURE 5.11 Plot of Group Centroids (Z).

randomly drawn, we can be reasonably sure that this sample does reflect the population proportions. Thus, this discriminant analysis uses the sample proportions to specify the prior probabilities for classification purposes.

To illustrate the importance of cutting score determination, let us focus on how the prior probabilities are used in the calculation of the cutting score. If the two groups were of equal size, the cutting score would simply be the average of the two centroids. Because the groups are of unequal size, a weighted average must be used. The weighted average is calculated as follows:

$$Z_{CU} = \frac{N_A Z_B + N_B Z_A}{N_A + N_B}$$

where

Z_{CU} = critical cutting score for unequal group sizes

N_A = number in group A

N_B = number in group B

Z_A = centroid for group A

Z_B = centroid for group B

By substitution the appropriate values in the formula, we can obtain the critical cutting score (assuming equal costs of misclassification):

$$Z_{CU} = \frac{38(-1.836) + 22(1.063)}{38 + 22} = -.773$$

The group sizes used in the preceding calculation are based on the data set used in the analysis sample and do not include the holdout sample. The procedure for classifying firms with the optimal cutting score is as follows:

1. Classify a firm as using specification buying if its discriminant score is less than $-.773$.
2. Classify a firm as using the total value analysis approach if its discriminant score is greater than $-.773$. Classification matrices for the observations in both the analysis and the holdout samples were calculated, and the results are shown in Table 5.10. The analysis sample, with 91.7 percent prediction accuracy, is slightly higher than the 85.0 percent accuracy of the holdout sample, as anticipated. Next we examine the individual observations and their classification accuracy.

The 91.7 percent classification accuracy is quite high. For illustration purposes, however, let us compare it with the two other chance measures of classifying individuals correctly without using the discriminant function. The first measure is the proportional chance criterion. We have unequal group sizes and we want to identify members of each group equally well. The first group, those using specification buying, constitute 36.7 percent of the analysis sample (22/60), with the second group, using total value analysis, forming the remaining 63.3 percent (38/60). The formula for the proportional chance criterion is

$$C_{PRO} = p^2 + (1 - p)^2$$

where

$$C_{PRO} = \text{proportional chance criterion}$$
$$p = \text{proportion of firms in group 1}$$
$$1 - p = \text{proportion of firms in group 2}$$

The calculated proportional chance value is .535 ($.367^2 + .633^2 = .535$).

The maximum chance criterion is simply the percentage correctly classified if all observations were placed in the group with the greatest probability of occurrence. Because group 1 (total value analysis) occurs 63.3 percent of the time, we would be correct 63.3 percent of the time if we assigned all observations to this group. Because the maximum chance criterion is larger than the proportional test criterion, our model should outperform the 63.3 percent level.

The classification accuracy of 91.7 percent is substantially higher than the proportional chance criterion of 53.6 percent and the maximum chance criterion of 63.3 percent. It also exceeds the suggested threshold of the value plus 25 percent, which in this case sets the threshold at 79.1 percent ($63.3 \times 1.25 = 79.1$). Note that the proportional chance criterion is compared with the percentage correctly classified in the holdout sample, reducing the upward bias seen in the classification of the analysis sample.

The final measure of classification accuracy is Press's Q. From the earlier discussion, the calculation for the analysis sample is

$$\text{Press's } Q_{\text{analysis sample}} = \frac{[60 - (55 \times 2)]^2}{60 \, (2 - 1)} = 41.7$$

And the calculation for the holdout sample is

$$\text{Press's } Q_{\text{holdout sample}} = \frac{[40 - (34 \times 2)]^2}{40 \, (2 - 1)} = 19.6$$

In both instances, the calculated values exceed the critical value of 6.63. Thus the classification accuracy for the analysis and, more important, the holdout sample exceed at a statistically significant level the classification accuracy expected by chance. The researcher must remember always to use caution in the application

TABLE 5.10 Classification Matrices for Two-Group Discriminant Analysis for Analysis and Holdout Samples

CLASSIFICATION RESULTS: ANALYSIS SAMPLE[a]

		Predicted Group Membership	
Actual Group	*Number of Cases*	*Specification Buying*	*Total Value Analysis*
Specification buying	22	21 95.5%	1 4.5%
Total value analysis	38	4 10.5%	34 89.5%
Number of Cases		25	35

CLASSIFICATION RESULTS: HOLDOUT SAMPLE[b]

		Predicted Group Membership	
Actual Group	*Number of Cases*	*Specification Buying*	*Total Value Analysis*
Specification buying	18	15 83.3%	3 16.7%
Total value analysis	22	3 13.6%	19 86.4%
Number of cases		18	22

[a]Percent of "grouped" cases correctly classified: 91.7 percent [(21 + 34)/60 = .917].
[b]Percent of "grouped" cases correctly classified: 85.0 percent [(15 + 19)/40 = .850].

of a holdout sample with small data sets. In this case the small sample size of 40 for the holdout sample was adequate, but larger sizes are always more desirable.

Casewise Diagnostics

In addition to examining the overall results, we can examine the individual observations for their predictive accuracy and identify specifically the misclassified cases. Table 5.11 (p. 292) contains the group predictions for the analysis and holdout samples. Cases 88, 84, 93, and 82 are actually in group 1 (total value analysis) but were predicted to be in group 0 (specification buying). These four misclassified cases are shown in Table 5.10. Also, case 13 is actually in group 0, but is predicted to be in group 1. This is also shown as a misclassification in Table 5.10. A similar examination can be performed for the holdout sample.

Once the misclassified cases are identified, further analysis can be performed to understand the reasons for their misclassification. In Table 5.12 (p. 293), the misclassified cases are combined from the analysis and holdout samples and then compared to the correctly classified cases. The attempt is to identify specific differences on the independent variables that might identify either new variables to be added or common characteristics that should be considered. The four cases (both analysis and holdout samples) misclassified in the specification buying group have significant differences on two of the three independent variables in the discriminant function. On both X_7 and X_3, their profile is much closer to the total value analysis group members. Also, the misclassified cases have a lower

TABLE 5.11 Group Predictions for Individual Cases in the Two-Group Discriminant Analysis

Case ID	Actual Group	Discriminant Z Score	Predicted Group	Case ID	Actual Group	Discriminant Z Score	Predicted Group
ANALYSIS SAMPLE							
79	0	−3.04	0	14	1	.47	1
65	0	−3.01	0	11	1	.51	1
39	0	−2.98	0	67	1	.60	1
96	0	−2.67	0	90	1	.63	1
48	0	−2.24	0	7	1	.73	1
2	0	−2.22	0	95	1	.73	1
6	0	−2.22	0	1	1	.74	1
45	0	−2.11	0	80	1	.79	1
54	0	−2.08	0	92	1	1.00	1
86	0	−1.96	0	58	1	1.04	1
71	0	−1.64	0	51	1	1.12	1
8	0	−1.64	0	15	1	1.37	1
31	0	−1.63	0	20	1	1.37	1
70	0	−1.55	0	49	1	1.42	1
68	0	−1.55	0	97	1	1.49	1
99	0	−1.51	0	43	1	1.53	1
52	0	−1.42	0	28	1	1.55	1
53	0	−1.32	0	59	1	1.56	1
36	0	−1.27	0	26	1	1.64	1
24	0	−1.14	0	29	1	1.65	1
89	0	−1.12	0	50	1	1.74	1
88	1	−.97	0	72	1	1.85	1
84	1	−.92	0	73	1	1.96	1
93	1	−.81	0	81	1	2.24	1
82	1	−.74	0	61	1	2.36	1
17	1	−.69	1	47	1	2.40	1
32	1	−.60	1	33	1	2.54	1
23	1	−.54	1	25	1	2.68	1
12	1	−.12	1	42	1	3.01	1
13	0	−.06	1	5	1	3.05	1
HOLDOUT SAMPLE							
94	0	−2.62	0	85	0	−.08	1
98	0	−2.60	0	76	1	.02	1
83	0	−2.28	0	38	1	.47	1
37	0	−2.18	0	21	1	.52	1
40	0	−2.15	0	100	1	.53	1
3	0	−1.70	0	63	1	.54	1
34	0	−1.52	0	66	1	.58	1
75	0	−1.49	0	74	1	.97	1
60	0	−1.46	0	18	1	1.02	1
30	0	−1.44	0	22	1	1.04	1
10	0	−1.39	0	9	1	1.05	1
41	0	−1.25	0	77	1	1.15	1
27	0	−1.19	0	55	1	1.16	1
4	0	−1.18	0	19	1	1.54	1
57	0	−1.13	0	46	1	1.56	1
91	1	−.82	0	78	1	1.59	1
56	1	−.79	0	16	1	1.85	1
64	1	−.74	0	69	1	2.13	1
35	0	−.17	1	62	1	2.51	1
87	0	−.15	1	44	1	2.65	1

Note: Cases 82 and 64 are predicted to be in Group 0 rather than Group 1 since SPSS utilizes standardized discriminant scores when separate variance estimates are used.

TABLE 5.12 Profiling Correctly Classified and Misclassified Observations in the Two-Group Discriminant Analysis

Dependent Variable Group/Profile Variables	Mean Scores			t test	
	Correctly Classified	Misclassified	Difference	t value	Significance
Specification buying ($X_{11} = 0$)	($N = 36$)	($N = 4$)			
X_1 Delivery speed[a]	2.494	2.550	−.056	−.274[b]	.786
X_2 Price level	3.089	2.075	1.014	1.681	.101
X_3 Price flexibility[a]	6.650	8.175	−1.525	−3.755	.001
X_4 Manufacturer image	5.358	4.775	.583	.838[b]	.460
X_5 Overall service	2.758	2.325	.433	.895	.376
X_6 Salesforce image	2.694	2.000	.694	2.279	.028
X_7 Product quality[a]	8.483	6.575	1.908	4.918	.000
Total value analysis ($X_{11} = 1$)	($N = 53$)	($N = 7$)			
X_1 Delivery speed[a]	4.372	2.829	1.543	7.075[b]	.000
X_2 Price level	1.730	3.600	−1.870	−5.563	.000
X_3 Price flexibility[a]	8.881	6.657	2.224	4.663[b]	.002
X_4 Manufacturer image	5.057	6.400	−1.343	−2.331	.052
X_5 Overall service	3.026	3.229	−.203	−.852	.398
X_6 Salesforce image	2.609	3.314	−.705	−2.079	.042
X_7 Product quality[a]	5.981	6.914	−.933	−1.830	.072

[a]Variables included in the discriminant function.
[b]t-test performed with separate variance estimates rather than a pooled variance estimate because the Levene test detected significant differences in the variances between the two groups.

salesforce image (X_6). This may suggest that possibly they were misclassified initially or that another variable would identify a characteristic that justifies their use of specification buying, even though they seem similar to the total value analysis group. The total value analysis group shows a similar pattern, although with different variables. In this case, the seven misclassified cases are much closer to the specification buying group on two variables (X_1 and X_3), but not on X_7. This group also demonstrates significant differences on three of the other four independent variables. Researchers should examine the patterns in both groups with the objective of understanding the characteristics common to them in an attempt at defining the reasons for misclassification.

Stage 5: Interpretation of the Results

After estimating the function, the next phase is interpretation. This stage involves examining the function to determine the relative importance of each independent variable in discriminating between the groups. Table 5.13 (p. 294) contains, among the interpretive measures, the discriminant weights, loadings for the function and the univariate F ratio. The independent variables were screened by the stepwise procedure, and three (X_1, X_3, and X_7) are significant enough to be included in the function. For interpretation purposes, we rank the independent variables in terms of both their weights and loadings—indicators of their discriminating power. Signs do not affect the rankings; they simply indicate a positive or negative relationship with the dependent variable. Because loadings are considered more valid than weights, we use loadings in our example.

TABLE 5.13 Summary of Interpretive Measures for Two-Group Discriminant Analysis

Independent Variables	Standardized Weights	Discriminant Loadings		Univariate F Ratio	
	Value	Value	Rank	Value	Rank
X_1 Delivery speed	.447	.647	2	48.992	2
X_2 Price level	NI	−.433	4	8.453	5
X_3 Price flexibility	.472	.522	3	31.881	3
X_4 Manufacturer image	NI	−.168	5	.822	7
X_5 Overall service	NI	.155	6	10.576	4
X_6 Salesforce image	NI	−.145	7	1.620	6
X_7 Product quality	−.659	−.705	1	58.176	1

NI: Not included in the stepwise solution.

When using the discriminant-loadings approach, we need to know which variables are substantive discriminators worthy of note. In simultaneous discriminant analysis, all variables are entered in the function, and generally any variables exhibiting a loading of ±.30 or higher are considered substantive. With stepwise procedures, this determination is made easier in one way because the criteria specified for the technique prevent nonsignificant variables from entering the function. However, multicollinearity and other factors may preclude a variable from entering the equation, but that does not necessarily mean that it does not have a substantial effect. This can be determined by evaluating the discriminant loadings. The loadings of the three variables entered in the discriminant function are the three highest and all exceed ±.30, along with X_2 (price level). In understanding the factors that distinguish between these two groups, the researcher should consider all four of these variables.

The researcher is interested in interpretations of the individual variables that have statistical and practical significance. Such interpretations are accomplished by identifying the variables with substantive loadings and understanding what the differing group means on each variable indicate. For example, for all the variables in this analysis, higher scores indicate more favorable perceptions of HATCO on that attribute (for more detail, see chapter 1). From Table 5.13, we can use the discriminant loadings (structure matrix information) and the univariate F values to determine the ranking of these variables in terms of their discriminating value. Both measures exhibit a high degree of correspondence. Of the three variables in the function, X_7 discriminates the most and X_3 discriminates the least. X_6 was not included in the model, even though it was very comparable in discrimination power to X_1 at the third step of the estimation process, because its collinearity with the variables already included, particularly X_1, reduced the additional discriminating power it could provide. Referring back to Table 5.5, we note that two of three variables (X_1, delivery speed, and X_3, price flexibility) have higher means for those firms employing the total value approach, meaning that they have more favorable perceptions of HATCO than do firms using specification buying. Only on product quality (X_7) is the mean for firms using specification buying higher. One can conclude that firms using the total value analysis approach employs a wider range of factors, whereas specification buying focuses on product quality.

Stage 6: Validation of the Results

The final stage addresses the internal and external validity of the discriminant function. The primary means of validation is through the use of the holdout sample and the assessment of its predictive accuracy. In this manner, validity is established if the discriminant function performs at an acceptable level in classifying observations that were not used in the estimation process. If the holdout sample is formed from the original sample, then this approach establishes internal validity. If another separate sample, perhaps from another population or segment of the population, forms the holdout sample, then this addresses the external validity of the discriminant results.

In our example, the holdout sample comes from the original sample. Thus, the acceptable levels on all measures of predictive accuracy found in the holdout sample do establish internal validity. Moreover, the analysis of the correctly classified versus misclassified cases provides additional insight into the group predictions both in the analysis and holdout samples. From the profiles of the groups as shown in Table 5.5, we know that for the variables selected in the examination of the discriminant loadings (X_1, X_2, X_3, and X_7), the specification buying group has better perceptions of HATCO regarding X_2 (price level) and X_7 (product quality), and the total value analysis group has higher perceptions on X_1 (delivery speed) and X_3 (price flexibility). This presents distinct profiles of the two groups with which to develop specialized marketing efforts. The researcher is encouraged to extend the validation process through expanded profiling of the groups and the possible use of additional samples to establish external validity.

A Managerial Overview

The discriminant analysis of HATCO customers based on their type of purchasing strategy identified a set of perceptual differences that can provide a rather succinct and powerful distinction between the two groups. A key finding is that differences are found in a subset of only four perceptions, allowing for a focus on key variables and not having to deal with the entire set. The four variables identified as discriminating between the groups (listed in order of importance) are X_7 (product quality), X_1 (delivery speed), X_3 (price flexibility), and X_2 (price level). Results also indicate that those firms following total value analysis have higher perceptions of HATCO on the delivery speed and price flexibility measures, perhaps indicative of an appreciation of the relational qualities of HATCO. The specification buyers are more favorable on the product quality and price level perceptions, perhaps indicating their focus on the more basic and objective aspects of the transactions. Thus, once a firm's purchasing strategy is identified, the key variables are known and management can employ a strategy to accentuate the positive perceptions in their dealings with these customers to further solidify their position.

The results, which are highly significant, provide the researcher the ability to correctly identify the purchasing strategy used based on these perceptions more than 90 percent of the time. This provides confidence in the development of strategies based on these results, given their high degree of consistency. Analysis of the misclassified firms revealed a small number of firms that seemed "out of place." The misclassified firms using total value analysis strongly resembled the specification buying firms on two aspects—delivery speed and price flexibility. Likewise, the misclassified total value analysis firms differed in terms of their perceptions.

These results could indicate an actual mistake in determining the type of purchasing strategy used, but more likely they indicate the possibility of another characteristic not addressed in the study that would provide some rationale for these differences.

A Three-Group Illustrative Example

To illustrate the application of a three-group discriminant analysis, we once again use the HATCO database. In the previous example, we were concerned with discriminating between only two groups, so we were able to develop a single discriminant function and a cutting score to divide the two groups. In the three-group example, it is necessary to develop two separate discriminant functions to distinguish among three groups. The first function separates one group from the other two, and the second separates the remaining two groups. As with the prior example, the six stages of the model-building process are discussed.

Stage 1: Objectives of Discriminant Analysis

HATCO's objective in this research is to determine the relationship between the firms' perceptions of HATCO and the type of purchasing situation most often faced. Firms that predominantly deal with HATCO in different purchasing situations may view and evaluate HATCO differently. The resulting discriminant model, like the two-group model, allows for a precise determination of the perceptions uniquely held by firms for each type of purchase situation. From this information, HATCO can develop targeted strategies in each purchasing situation that accentuate its perceived strengths.

Stage 2: Research Design for Discriminant Analysis

To test this relationship, a discriminant analysis is performed using X_{14} as the dependent variable and the perceptions of HATCO by these firms (X_1 to X_7) as the independent variables. Note that X_{14} differs from the dependent variable in the two-group example in that it has three categories in which to classify a firm by the purchasing situation (new task, modified rebuy, or straight rebuy) most often used with HATCO. The sample size of 100, which again will be split into analysis and holdout samples, provides an adequate case–to–independent variable ratio (9 to 1). In the analysis sample, only one group, with 15 observations, falls below the recommended level of 20 cases per group. Although the group size would exceed 20 if the entire sample were used in the analysis phase, the need for validation dictated the creation of the holdout sample. The analysis proceeds with attention paid to the classification and interpretation of this group.

Stage 3: Assumptions of Discriminant Analysis

As was the case in the two-group example, the assumptions of normality, linearity, and collinearity of the independent variables have already been discussed at length in chapter 2. The analyses performed in chapter 2 indicated that the independent variables met these assumptions at adequate levels to allow for the analysis to continue without additional remedies. The remaining assumption, the equality of the variance/covariance or dispersion matrices, is also addressed in chapter 2. Box's M test assesses the similarity of the dispersion matrices of the independent variables among the three groups (categories). The test statistic indicated differences at the .01 significance level. In many instances this significance level would necessitate remedial action, but the sensitivity of Box's M test to sample size and other characteristics of the independent variables makes it a very liberal test. Thus, the statistical test is judged to provide inadequate evidence that the dispersion matrices are sufficiently different to require corrective action, and the analysis can proceed. We employ pooled variance estimates in this analysis to allow the cross-validation procedure to be used.

Stage 4: Estimation of the Discriminant Model and Assessing Overall Fit

As in the previous example, we begin our analysis by reviewing the group means and standard deviations to see if the groups are significantly different on any single variable. Table 5.14 gives the group means, standard deviations, Wilks' lambda, and univariate F ratios (simple ANOVAs) for each independent variable. Review of the significance levels of the individual variables reveals that on a univariate basis, all of the variables except X_4 and X_6 display significant differences between the group means. Although visual inspection of the group means can provide insight into the differences between the groups (e.g., 1 versus 2 and 3, 1 and 2 versus 3, 1 versus 2 versus 3, etc.), we do not know the statistical significance of any specific comparison, only that overall significant differences exist. This will prove important in discriminant analysis for three or more groups as more than one function is created, with each function providing discrimination between sets of groups. In the simple example from the beginning of this chapter, one variable

TABLE 5.14 Group Descriptive Statistics and Tests of Equality for the Three-Group Discriminant Analysis

Dependent Variable[a]	Group Means for the Independent Variables[b]							Sample Size
	X_1	X_2	X_3	X_4	X_5	X_6	X_7	
1: New task	2.43	2.16	7.23	5.07	2.28	2.69	7.76	21
2: Modified rebuy	3.23	3.52	6.98	5.59	3.35	2.69	7.31	15
3: Straight rebuy	4.64	1.93	9.18	5.41	3.30	2.80	5.92	24
Total	3.51	2.41	7.95	5.33	2.96	2.73	6.91	60

Standard Deviations for the Independent Variables[b]								
1: New task	1.16	.92	.88	.87	.67	.73	1.37	
2: Modified rebuy	1.13	1.38	1.34	1.12	.64	.77	1.75	
3: Straight rebuy	1.02	.89	.70	1.39	.37	.94	1.41	
Total	1.46	1.21	1.38	1.16	.75	.82	1.68	

Tests for the Equality of the Group Means[c]								
Wilks' lambda	.550	.709	.461	.967	.546	.996	.754	
Univariate F ratio	23.346	11.674	33.362	.961	23.692	.126	9.293	
Significance level	.000	.000	.000	.389	.000	.882	.000	

[a]X_{14} = type of buying situation
[b]X_1 = delivery speed; X_2 = price level; X_3 = price flexibility; X_4 = manufacturer image; X_5 = overall service; X_6 = salesforce image; X_7 = product quality
[c]Wilks' lambda (U statistic) and univariate F ratio with 2 and 57 degrees of freedom

discriminated between groups 1 versus 2 and 3, whereas the other discriminated between groups 2 versus 3 and 1. This is one of the primary benefits arising from the use of discriminant analysis.

Estimation of the Discriminant Function

The stepwise procedure is performed in the same manner as in the two-group example. The data in Table 5.15 show that the first variable to enter the model is X_1 (delivery speed). As was done in the two-group analysis, variables must first have a significance level of .05 or below to be considered for inclusion. Then, from the variables meeting this criterion, the variable with the highest Mahalanobis D^2 measure is selected for inclusion in the discriminant function. Review of the significance levels reveals that of the variables not included in the model after step 1 (see Table 5.15), all but X_4, X_6, and X_7 have significance values low enough to be considered for inclusion in the model at later steps.

TABLE 5.15 Results from Step 1 of Stepwise Three-Group Discriminant Analysis Model

STEP 1: X_1 (DELIVERY SPEED) INCLUDED IN THE ANALYSIS

SUMMARY STATISTICS

		Degrees of Freedom			Significance	Between Groups
Wilks' lambda	.550	1	2	57		
Equivalent F	23.346		2	57	.000	
Minimum D^2	.526					1 and 2
Equivalent F	4.606		1	57	.036	

VARIABLES IN THE ANALYSIS AFTER STEP 1

		F to Remove	
Variables	Tolerance	Value	Significance
X_1 Delivery speed	1.00	23.346	.000

VARIABLES NOT IN THE ANALYSIS AFTER STEP 1

Variables	Tolerance	Minimum Tolerance	F to Enter		D^2	Between Groups
			Value	Significance		
X_2 Price level	.766	.766	14.609	.000	2.718	2 and 3
X_3 Price flexibility	.982	.982	17.273	.000	.660	1 and 2
X_4 Manufacturer image	.998	.998	.690	.506	.698	1 and 2
X_5 Overall service	.704	.704	13.243	.000	2.537	2 and 3
X_6 Salesforce image	.991	.991	.020	.980	.531	1 and 2
X_7 Product quality	.869	.869	1.142	.326	.529	1 and 2

SIGNIFICANCE TESTING OF GROUP DIFFERENCES AFTER STEP 1[a]

	Group 1: New task	Group 2: Modified rebuy
Group 2: Modified rebuy	4.606 (.036)	
Group 3: Straight rebuy	45.335 (.000)	15.274 (.000)

[a]F statistic and significance level (in parentheses) between pairs of groups after step 1. Each F statistic has 1 and 57 degrees of freedom.

Table 5.16 details the second step of the stepwise procedure, which added X_2 (price level) to the discriminant function. The discrimination between groups has increased, as reflected in the lower Wilks' lambda value and increase in the minimum D^2 (.526 to 2.718). Of the variables not in the equation, only X_3 (price flexibility) meets the significance level necessary for consideration. When X_3 is added to the model in the third step (see Table 5.17, p. 300), no other variables can be considered for inclusion because all of their significance levels are above .05. With no variables to add, the estimation procedure terminates.

The information provided in Table 5.18 (pp. 301–302) summarizes the steps of the three-group discriminant analysis. Only the variables X_1, X_2, and X_3 entered into the discriminant function. Discrimination increased with the addition of each

TABLE 5.16 Results from Step 2 of Stepwise Three-Group Discriminant Analysis Model

STEP 2: X_2 (PRICE LEVEL) INCLUDED IN THE ANALYSIS

SUMMARY STATISTICS

		Degrees of Freedom		Significance	Between Groups
Wilks' lambda	.361	2	2	57	
Equivalent F	18.587		4	112	.000
Minimum D^2	2.718				2 and 3
Equivalent F	12.325		2	56	.000

VARIABLES IN THE ANALYSIS AFTER STEP 2

Variables	Tolerance	F to Remove		D^2	Between Groups
		Value	Significance		
X_1 Delivery speed	.766	26.988	.000	.046	1 and 3
X_2 Price level	.766	14.609	.000	.526	1 and 2

VARIABLES NOT IN THE ANALYSIS AFTER STEP 2

Variables	Tolerance	Minimum Tolerance	F to Enter		D^2	Between Groups
			Value	Significance		
X_3 Price flexibility	.878	.684	17.162	.000	4.201	1 and 2
X_4 Manufacturer image	.963	.739	.009	.991	2.719	2 and 3
X_5 Overall service	.017	.017	.846	.435	2.975	2 and 3
X_6 Salesforce image	.952	.736	.534	.589	2.766	2 and 3
X_7 Product quality	.786	.693	2.140	.127	2.755	2 and 3

SIGNIFICANCE TESTING OF GROUP DIFFERENCES AFTER STEP 2[a]

	Group 1: New task	Group 2: Modified rebuy
Group 2: Modified rebuy	17.749 (.000)	
Group 3: Straight rebuy	26.405 (.000)	12.325 (.000)

[a]F statistic and significance level (in parentheses) between groups after step 1. Each F statistic has 2 and 56 degrees of freedom.

TABLE 5.17 Results from Step 3 of Stepwise Three-Group Discriminant Analysis Model

STEP 3: X_3 (PRICE FLEXIBILITY) INCLUDED IN THE ANALYSIS

SUMMARY STATISTICS

		Degrees of Freedom		Significance	Between Groups
Wilks' lambda	.222	3	2	57	
Equivalent F	20.540		6	110	.000
Minimum D^2	4.201				1 and 2
Equivalent F	11.824		3	55	.000

VARIABLES IN THE ANALYSIS AFTER STEP 3

Variables	Tolerance	F to Remove		D^2	Between Groups
		Value	Significance		
X_1 Delivery speed	.764	.	.000	1.757	1 and 2
X_2 Price level	.684	.	.000	.660	1 and 2
X_3 Price flexibility	.878	.	.000	2.718	1 and 3

VARIABLES NOT IN THE ANALYSIS AFTER STEP 3

Variables	Tolerance	Minimum Tolerance	F to Enter		D^2	Between Groups
			Value	Significance		
X_4 Manufacturer image	.962	.665	.	.982	4.206	1 and 2
X_5 Overall service	.017	.017	.	.585	4.376	1 and 2
X_6 Salesforce image	.944	.652	.	.535	4.432	1 and 2
X_7 Product quality	.780	.615	.	.121	4.678	1 and 2

SIGNIFICANCE TESTING OF GROUP DIFFERENCES AFTER STEP 3[a]

	Group 1: New task	Group 2: Modified rebuy
Group 2: Modified rebuy	11.824	
	(.000)	
Group 3: Straight rebuy	34.432	18.934
	(.000)	(.000)

[a]F statistic and significance level (in parentheses) between groups after step 1. Each F statistic has 3 and 55 degrees of freedom.

variable, achieving by the third step a substantial ability to discriminate between the groups. By comparing the final Wilks' lambda for the discriminant analysis (.222) with the Wilks' lambda (.461) for the best result from a single variable, X_3, we see that a marked improvement is made in using the discriminant functions rather than a single variable.

Statistical Significance

Table 5.18 also contains the overall impact of the discriminant functions. Note that the functions are statistically significant, as measured by the chi-square statistic, and that the first function accounts for 78.4 percent of the variance explained by the two functions. After the first function is extracted, the chi-square is recalculated.

TABLE 5.18 Multivariate Results for Three-Group Discriminant Analysis

SUMMARY TABLE

	Action		Wilks' Lambda		Minimum D^2		Between
Steps	Entered	Removed	Value	Significance	Value	Significance	Groups
1	X_1 Delivery speed		.550	.000	.526	.036	1 and 2
2	X_2 Price level		.361	.000	2.718	.000	2 and 3
3	X_3 Price flexibility		.222	.000	4.201	.000	1 and 2

CANONICAL DISCRIMINANT FUNCTIONS

		Percent of Variance							
Function[a]	Eigenvalue	Function	Cumulative	Canonical Correlation	After Function	Wilks' Lambda	Chi-square	df	Significance
					0	.222	84.177	6	.000
1	1.935	78.4	78.4	.812	1	.653	23.879	2	.000
2	.532	21.6	100.0	.589					

DISCRIMINANT FUNCTION COEFFICIENTS

	Standardized				Unstandardized			
	Unrotated		Rotated		Unrotated		Rotated	
Independent Variables	Function 1	Function 2	Function 1	Function 2	Function 1	Function 2	Function 1	Function 2
X_1 Delivery speed	.785	.559	.834	.088	.713	.508	.420	.768
X_2 Price level	.495	.995	.011	1.111	.476	.957	.011	1.068
X_3 Price flexibility	.788	−.285	.462	.845	.825	−.298	.872	.092
Constant					−10.207	−1.720	−8.434	−6.001

UNROTATED AND ROTATED STRUCTURE MATRIX[b]

	Unrotated Discriminant Function Loadings		Rotated Discriminant Function Loadings	
Independent Variables	Function 1	Function 2	Function 1	Function 2
X_1 Delivery speed	.650[c]	.039	.568[c]	.319
X_2 Price level	−.159	.824[c]	−.502	.672[c]
X_3 Price flexibility	.721[c]	−.566	.891[c]	−.186
X_4 Manufacturer image	.046	.188[c]	−.041	.190[c]
X_5 Overall service	.509	.817[c]	.101	.957[c]
X_6 Salesforce image	.161	.171[c]	.071	.224[c]
X_7 Product quality	−.129	.243[c]	−.222[c]	.163

UNROTATED AND ROTATED GROUP MEANS (CENTROIDS) OF CANONICAL DISCRIMINANT FUNCTIONS

	Unrotated Group Centroids		Rotated Group Centroids	
Group	Function 1	Function 2	Function 1	Function 2
New Task	−1.482	−.579	−1.081	−1.167
Modified Rebuy	−.473	1.206	−.952	.879
Straight Rebuy	1.592	−.247	1.541	.472

(Continued)

TABLE 5.18 (*Continued*)

CLASSIFICATION FUNCTION COEFFICIENTS

	X_{14} *Type of Buying Situation*		
Independent Variables	*Group 1: New Task*	*Group 2: Modified Rebuy*	*Group 3: Straight Rebuy*
X_1 Delivery speed	4.207	5.833	6.568
X_2 Price level	7.356	9.543	9.135
X_3 Price flexibility	10.060	10.360	12.498
Constant	−50.478	−63.750	−82.323

[a]Marks the one canonical discriminant function remaining in the analysis.
[b]Pooled-within-groups correlations between discriminating variables and canonical discriminant functions.
[c]Denotes largest absolute correlation between each variable and any discriminant function.

The results show that significant differences are present in the remaining variance. If more groups are used in the model (e.g., a four-group discriminant analysis), additional canonical discriminant functions would be possible, and the chi-square statistic would be continually recalculated on the residual variance to test for significant differences until the maximum number of canonical discriminant functions was extracted (maximum number of discriminant functions = number of groups − 1). The total amount of variance explained by the first function is $.812^2$, or 65.9 percent. The next function explains $.589^2$, or 34.7 percent, of the remaining variance (34.1 percent). Therefore, the total variance explained by both functions is 65.9 percent + (34.7 percent × .341), or 77.7 percent of the total variation in the dependent variable.

Assessing Overall Fit

The estimated discriminant functions are linear composites similar to a regression line (i.e., they are a linear combination of variables). Just as a regression line is an attempt to explain the maximum amount of variation in its dependent variable, these linear composites attempt to explain the variations or differences in the dependent categorical variable. The first discriminant function is developed to explain (account for) the largest amount of variation (difference) in the discriminant groups. The second discriminant function, which is orthogonal and independent of the first, explains the largest percentage of the remaining (residual) variance after the variance for the first function is removed.

Calculating Discriminant Z Scores   Table 5.18 also shows discriminant function coefficients (weights) and the structure matrix of discriminant loadings. Both unrotated and rotated values are provided. Rotation of the discriminant functions facilitates interpretation in the same way that factors were simplified for interpretation by rotation (see chapter 3 for a more detailed discussion of rotation). We examine the unrotated and rotated values more fully in step 5. The classification functions are used to make group membership predictions, and the group means section of Table 5.18 profiles each group by the mean discriminant Z scores for each function.

Because this is a three-group discriminant analysis model, two discriminant functions are calculated to discriminate among the three groups. Values for each case are entered into the discriminant procedure, and linear composites are formulated. The discriminant functions are based only on the variables included in the discriminant model (X_1, X_2, and X_3). In the examples in this chapter, the discriminant Z scores are based on the rotated coefficients.

After the linear composites are calculated, the procedure correlates all seven independent variables with the discriminant functions to develop a structure (loadings) matrix. This procedure enables us to see where the discrimination would occur if all seven variables were included in the model (that is, if none were excluded by multicollinearity or lack of statistical significance). Much of this discussion is based upon concepts presented in the chapters on canonical correlation (chapter 8) and factor analysis (chapter 3).

Evaluating Group Differences   Even though both discriminant functions are statistically significant, the researcher must always ensure that the discriminant functions provide differences among all of the groups. It is possible to have statistically significant functions, but have at least one pair of groups not be statistically different (i.e., not discriminated between). This problem becomes especially prevalent as the number of groups increases and/or a number of small groups are included in the analysis. The last section of Table 5.17 provides the significance tests for group differences between each pair of groups (e.g., group 1 versus group 2, group 1 versus group 3, etc.). All pairs of groups showed statistically significant differences, denoting that the discriminant functions created separation not only in an overall sense, but for each group as well. We also examine the group centroids graphically in a later section.

Assessing Group Membership Prediction Accuracy   The final step of assessing overall model fit is to determine the predictive accuracy level of the discriminant function(s). This determination is accomplished in the same fashion as with the two-group discriminant model, by examination of the classification matrices. Table 5.19 (p. 304) shows that the two discriminant functions in combination achieve a high degree of classification accuracy. The hit ratio for the analysis sample is 81.7 percent, whereas that for the holdout sample it is 75.0 percent. These results demonstrate the upward bias that is likely when applied only to the analysis sample and not also to a holdout sample. Although both these hit ratios are high, they must be compared with the maximum chance and the proportional chance criteria to assess their "true" effectiveness. A third classification matrix for the cross-validation procedure is discussed in step 6.

The maximum chance criterion is simply the hit ratio obtained if we assign all the observations to the group with the highest probability of occurrence. In the present sample of 100 observations, 34 were in group 1, 32 in group 2, and 34 in group 3. From this information, we can see that the highest probability would be 34 percent (groups 1 or 3). Based on the maximum chance criterion, therefore, our model is very good. The proportional chance criterion is calculated by squaring the proportions of each group, with a calculated value of 33.36 percent. Because C_{MAX} is greater than C_{PRO}, the maximum chance criterion is the measure to outperform. If we establish the threshold as 25 percent greater than the criterion value, the hit ratio must exceed 42.5 percent (34 percent $\times$ 1.25). The hit ratios of both 83.3 percent (analysis sample) and 75.0 percent (holdout sample) both exceed this

TABLE 5.19 Classification Matrices for Three-Group Discriminant Analysis

CLASSIFICATION RESULTS: ANALYSIS SAMPLE[a]

Actual Group	Number of Cases	Predicted Group Membership		
		New Task	Modified Rebuy	Straight Rebuy
New task	21	16	3	2
		76.2%	14.3%	9.5%
Modified rebuy	15	3	9	3
		20.0%	60.0%	20.0%
Straight rebuy	24	0	0	24
		0%	0%	100%
Number of cases		19	12	29

CLASSIFICATION RESULTS: HOLDOUT SAMPLE[b]

Actual Group	Number of Cases	Predicted Group Membership		
		New Task	Modified Rebuy	Straight Rebuy
New task	13	12	1	0
		92.3%	7.7%	0%
Modified rebuy	17	4	8	5
		23.5%	47.1%	29.4%
Straight rebuy	10	0	0	10
		0%	0%	100.0%
Number of cases		16	9	15

CLASSIFICATION RESULTS: CROSS-VALIDATION OF THE ANALYSIS SAMPLE[c]

Actual Group	Number of Cases	Predicted Group Membership		
		New Task	Modified Rebuy	Straight Rebuy
New task	21	16	3	2
		76.2%	14.3%	9.5%
Modified rebuy	15	3	9	3
		20.0%	60.0%	20.0%
Straight rebuy	24	1	0	23
		4.2%	0%	95.8%
Number of cases		20	12	28

[a]Percent of analysis sample cases correctly classified: 81.7 percent [(16 + 9 + 24)/60 = .817].
[b]Percent of holdout cases correctly classified: 75.0 percent [(12 + 8 + 10)/40 = .750].
[c]Percent of cross-validated cases correctly classified: 80.0 percent [(16 + 9 + 23)/60 = .800].

criterion substantially, but we should evaluate the classification rate for individual groups as well.

For both the analysis and holdout samples, groups 1 and 3 have hit ratios far exceeding the threshold of 42.5 percent. Group 2, however, has hit ratios of 69 percent and 47.5 percent in the analysis and holdout samples, respectively. Thus, although all three groups do exceed the threshold value, the predictions of group 2 should be improved if at all possible, possibly by the addition of independent variables or a review of classification of firms in this group to identify the characteristics of this group not represented in the discriminant function. When completed, the researcher can confidently conclude that the discriminant model is valid and has adequate levels of statistical and practical significance for all groups.

The final measure of classification accuracy is Press's Q, calculated for both analysis and holdout samples. It tests the statistical significance that the classification accuracy is better than chance. The calculated value for the analysis sample is

$$\text{Press's } Q_{\text{analysis sample}} = \frac{[60 - (49 \times 3)]^2}{60 (3 - 1)} = 63.1$$

The calculated value for the holdout sample is

$$\text{Press's } Q_{\text{holdout sample}} = \frac{[40 - (30 \times 3)]^2}{40 (3 - 1)} = 31.2$$

Because the critical value at a .01 significance level is 6.63, the discriminant analysis can confidently be described as predicting group membership better than chance, as indicated by this and the other classification accuracy measures.

Casewise Diagnostics

Table 5.20 (pp. 306–307) contains additional classification data for each individual case. The observation number is shown on the left side of the table, which also contains notation as to the cases included in the holdout sample. The "Actual Group" column indicates group numbers: group 1 (new task), group 2 (modified rebuy), or group 3 (straight rebuy). An asterisk beside a number indicates that particular observation was misclassified by the discriminant function. The "Highest Probability Group" column shows the most likely group assignment of an observation using the discriminant classification functions, and the "Second-Highest Group" column shows the next most likely assignment using the discriminant classification functions. The discriminant scores for each observation on each function are also provided. When there is one discriminant function (two groups), classification of cases is based on the values for the single function; multiple functions are used for three or more groups. Misclassified cases are indicated when the actual and group predictions differ. The probability values allow the researcher to assess the extent of misclassification. As an example, let us assume two situations: in the first situation, the incorrect group has a probability of .53 and the correct group has a probability of .46; in the second situation, the incorrect group has a probability of .76 and the correct group has a value of .22. In both situations there was a misclassification, but the extent or magnitude varies widely. The researcher should evaluate the extent of misclassification for each case.

Misclassification of Individual Cases   Using the information from Table 5.20, the misclassified cases for each group can be identified and then compared to the correctly classified cases of that group to identify differences. Table 5.21 (p. 308) contains an analysis of the differences for the three-group discriminant analysis, which reveals distinct patterns. The misclassified cases seem to be rather uniform in their differences, suggesting the possibility of a fourth group. First, the new task group had almost uniformly higher perceptions than the correctly classified,

TABLE 5.20 Classification Data for Three-Group Discriminant Analysis

Group Membership Prediction

Case Number	Actual Group	Highest Probability:					Second Highest Probability:			Discriminant Scores	
		Group	P(D/G)	df	P(G/D)	D^2	Group	P(G/D)	D^2	Function 1	Function 2
1	1	1	0.851	2	0.949	0.324	2	0.038	6.10	−0.686	−1.577
2	1	1	0.534	2	0.869	1.254	2	0.130	4.373	−2.150	−0.835
3[u]	2	2	0.115	2	0.993	4.330	1	0.005	15.673	−1.978	2.689
4[u]	1	1	0.582	2	0.988	1.081	2	0.010	9.54	−1.095	−2.207
5	3	3	0.649	2	0.997	0.866	2	0.003	11.899	2.471	0.451
6	2	1*	0.636	2	0.603	0.905	2	0.332	1.428	−0.709	−0.291
7	1	3*	0.852	2	0.983	0.320	2	0.016	7.645	1.812	0.968
8	2	1*	0.190	2	0.528	3.322	2	0.471	2.879	−2.435	0.053
9[u]	3	3	0.814	2	0.992	0.411	2	0.007	9.284	2.094	0.796
10[u]	2	2	0.866	2	0.908	0.288	1	0.053	6.629	−1.046	1.407
11	1	1	0.360	2	0.840	2.043	3	0.127	6.083	0.268	−1.641
12	2	3*	0.894	2	0.918	0.225	2	0.053	4.975	1.166	0.180
13	1	1	0.602	2	0.918	1.014	3	0.046	7.289	−0.177	−1.611
14	1	3*	0.309	2	0.574	2.349	1	0.342	3.117	0.639	−0.767
15	3	3	0.690	2	0.993	0.742	2	0.004	10.821	2.191	−0.093
16[u]	3	3	0.705	2	0.947	0.701	1	0.032	7.193	1.478	−0.362
17	2	2	0.475	2	0.928	1.490	1	0.069	7.373	−2.074	1.360
18[u]	2	3*	0.497	2	0.604	1.398	2	0.285	1.959	0.361	0.393
19[u]	3	3	0.767	2	0.995	0.532	2	0.004	10.555	2.269	0.456
20	3	3	0.690	2	0.993	0.742	2	0.004	10.821	2.191	−0.093
21[u]	2	1*	0.261	2	0.794	2.687	3	0.178	5.943	0.464	−1.715
22[u]	1	1	0.246	2	0.936	2.809	3	0.053	8.833	0.239	−2.200
23	3	3	0.492	2	0.803	1.420	2	0.188	3.380	0.807	1.412
24	1	1	0.660	2	0.983	0.832	2	0.017	8.320	−1.565	−1.940
25	2	3*	0.942	2	0.942	0.120	2	0.038	5.575	1.313	0.211
26	3	3	0.529	2	0.638	1.273	2	0.277	2.004	0.413	0.501
27[u]	1	1	0.626	2	0.983	0.938	2	0.016	8.486	−1.652	−1.949
28	3	3	0.775	2	0.994	0.510	2	0.004	10.468	2.226	0.273
29	3	3	0.980	2	0.978	0.040	2	0.018	7.132	1.703	0.587
30[u]	2	2	0.634	2	0.952	0.913	1	0.037	8.093	−1.525	1.643
31	1	2*	0.528	2	0.596	1.276	1	0.399	2.753	−1.905	0.273
32	3	3	0.524	2	0.702	1.294	2	0.270	2.268	0.547	1.026
33	3	3	0.827	2	0.985	0.380	2	0.013	8.058	1.885	0.984
34[u]	1	2*	0.878	2	0.890	0.260	1	0.092	5.464	−1.383	1.151
35[u]	1	1	0.444	2	0.989	1.623	2	0.008	10.698	−0.698	−2.382
36	1	1	0.748	2	0.691	0.581	2	0.287	1.669	−1.10	−0.405
37[u]	2	2	0.541	2	0.950	1.230	1	0.044	8.037	−1.819	1.570
38[u]	3	3	0.260	2	0.779	2.691	1	0.197	5.178	1.195	−1.131
39	1	1	0.062	2	0.999	5.552	2	0.001	18.10	−2.392	−3.125
40[u]	1	1	0.734	2	0.955	0.619	2	0.044	6.101	−1.821	−1.433
41[u]	1	1	0.630	2	0.593	0.925	2	0.371	1.190	−0.970	−0.212
42	3	3	0.671	2	0.996	0.799	2	0.003	11.684	2.429	0.374
43	3	3	0.814	2	0.980	0.411	2	0.019	7.404	1.762	1.074
44[u]	2	3*	0.846	2	0.916	0.334	2	0.048	5.297	1.183	0.018
45	1	1	0.715	2	0.915	0.670	2	0.084	4.769	−1.896	−1.091
46[u]	3	3	0.626	2	0.984	0.937	2	0.015	8.341	1.893	1.373
47	3	3	0.518	2	0.949	1.315	1	0.038	7.497	1.612	−0.672
48	2	2	0.487	2	0.919	1.440	1	0.078	7.039	−2.079	1.291
49	3	3	0.603	2	0.963	1.012	1	0.025	8.044	1.682	−0.524
50	3	3	0.630	2	0.826	0.923	2	0.161	3.257	0.836	1.125
51	2	1*	0.351	2	0.840	2.096	3	0.128	6.124	0.281	−1.658

TABLE 5.20 (Continued)

		Group Membership Prediction								Discriminant Scores	
		Highest Probability:					Second Highest Probability:				
Case Number	Actual Group	Group	P(D/G)	df	P(G/D)	D^2	Group	P(G/D)	D^2	Function 1	Function 2
52	2	2	0.436	2	0.738	1.659	3	0.247	4.786	−0.137	1.876
53	2	2	0.307	2	0.963	2.360	3	0.030	10.246	−1.004	2.414
54	1	1	0.844	2	0.806	0.340	2	0.187	2.592	−1.387	−0.671
55[u]	1	1	0.177	2	0.542	3.458	3	0.423	4.221	0.761	−1.429
56[u]	2	1*	0.565	2	0.565	1.143	2	0.322	1.595	−0.471	−0.289
57[u]	2	2	0.043	2	0.720	6.290	3	0.279	9.130	0.127	3.143
58	3	3	0.370	2	0.992	1.989	2	0.007	10.836	2.236	1.699
59	3	3	0.271	2	0.347	2.609	1	0.337	2.403	0.022	−0.077
60[u]	2	2	0.801	2	0.791	0.445	3	0.157	4.623	−0.441	1.307
61	3	3	0.514	2	0.937	1.331	1	0.048	7.026	1.525	−0.681
62[u]	3	3	0.929	2	0.981	0.147	2	0.017	7.333	1.755	0.791
63[u]	3	3	0.429	2	0.904	1.693	1	0.077	6.338	1.413	−0.823
64[u]	2	2	0.384	2	0.825	1.914	1	0.173	5.715	−2.335	0.868
65	1	1	0.413	2	0.993	1.767	2	0.007	10.995	−1.670	−2.359
66[u]	2	3*	0.364	2	0.753	2.021	1	0.202	4.383	0.987	−0.837
67	3	3	0.953	2	0.953	0.097	2	0.041	5.470	1.383	0.740
68	2	2	0.483	2	0.746	1.455	1	0.251	4.305	−2.131	0.622
69[u]	3	3	0.934	2	0.932	0.137	2	0.057	4.787	1.226	0.667
70	2	2	0.690	2	0.534	0.743	3	0.278	2.987	−0.188	0.480
71	2	2	0.067	2	0.994	5.408	3	0.004	17.385	−1.717	3.074
72	3	3	0.587	2	0.893	1.066	2	0.102	4.467	1.095	1.404
73	3	3	0.887	2	0.978	0.241	2	0.013	7.969	1.738	0.023
74[u]	3	3	0.955	2	0.978	0.091	2	0.015	7.485	1.703	0.218
75[u]	1	1	0.930	2	0.880	0.144	2	0.114	3.555	−1.395	−0.954
76[u]	2	3*	0.984	2	0.969	0.033	2	0.026	6.342	1.556	0.652
77[u]	2	1*	0.314	2	0.740	2.316	3	0.218	5.028	0.412	−1.465
78[u]	3	3	0.959	2	0.958	0.083	2	0.028	6.189	1.442	0.202
79	1	1	0.296	2	0.995	2.434	2	0.005	12.485	−1.800	−2.552
80	3	3	0.784	2	0.886	0.486	2	0.061	4.897	1.063	−0.036
81	3	3	0.893	2	0.962	0.227	2	0.034	5.945	1.485	0.946
82	2	2	0.249	2	0.979	2.781	1	0.019	11.381	−2.159	2.030
83[u]	1	1	0.412	2	0.938	1.772	2	0.062	6.536	−2.411	−1.221
84	1	2*	0.888	2	0.645	0.238	1	0.279	2.585	−0.798	0.416
85[u]	2	3*	0.316	2	0.406	2.301	2	0.307	1.923	0.105	−0.019
86	1	1	0.574	2	0.570	1.110	2	0.410	1.096	−1.244	−0.127
87[u]	2	1*	0.407	2	0.459	1.799	2	0.276	2.145	−0.067	−0.289
88	1	2*	0.694	2	0.486	0.731	1	0.479	1.434	−1.059	0.030
89	1	1	0.783	2	0.971	0.488	2	0.023	7.302	−0.835	−1.821
90	3	3	0.939	2	0.965	0.126	2	0.031	6.074	1.512	0.826
91[u]	2	2	0.569	2	0.423	1.129	1	0.395	1.936	−0.339	0.011
92	2	3*	0.409	2	0.510	1.790	2	0.308	1.858	0.231	0.200
93	2	2	0.217	2	0.922	3.051	1	0.077	8.687	−2.638	1.335
94[u]	1	1	0.096	2	0.942	4.687	2	0.058	9.573	−3.245	−1.198
95	1	1	0.816	2	0.970	0.406	2	0.025	7.064	−0.903	−1.779
96	1	1	0.038	2	0.999	6.565	2	0.001	19.648	−2.582	−3.244
97	3	3	0.760	2	0.993	0.550	2	0.005	10.346	2.160	0.064
98[u]	1	1	0.148	2	0.912	3.820	2	0.088	7.824	−3.028	−0.996
99	1	1	0.862	2	0.795	0.296	2	0.194	2.447	−1.263	−0.654
100[u]	1	1	0.289	2	0.657	2.484	3	0.293	4.364	0.487	−1.332

*Misclassification
[u]Unselected case used in holdout or validation sample.

TABLE 5.21 Profiling Correctly Classified and Misclassified Observations in the Three-Group Discriminant Analysis

Dependent Variable Group/Profile Variables	*Mean Scores*			*t test*	
	Correctly Classified	Misclassified	Difference	t value	Significance
New Task ($X_{14} = 1$)	(N = 28)	(N = 6)			
X_1 Delivery speed[a]	2.332	3.183	−.851	−1.989	.055
X_2 Price level[a]	1.900	3.000	−1.100	−2.831	.008
X_3 Price flexibility[a]	7.039	7.583	−.54	−1.189	.243
X_4 Manufacturer image	4.839	5.517	−.677	−1.498	.144
X_5 Overall service	2.043	3.100	−1.057	−4.897	.000
X_6 Salesforce image	2.482	3.233	−.751	−2.534	.016
X_7 Product quality	7.629	7.550	.079	.216[b]	.831
Modified Rebuy ($X_{14} = 2$)	(N = 17)	(N = 15)			
X_1 Delivery speed[a]	3.294	3.567	−.273	−.770	.448
X_2 Price level[a]	4.188	2.040	2.148	7.045[b]	.000
X_3 Price flexibility[a]	6.447	8.260	−1.813	−5.431	.000
X_4 Manufacturer image	5.953	5.127	.826	2.463	.020
X_5 Overall service	3.712	2.800	.912	5.472	.000
X_6 Salesforce image	3.012	2.373	.638	2.875	.007
X_7 Product quality	8.041	6.493	1.548	3.005	.005
Straight Rebuy ($X_{14} = 3$)	(N = 34)	(N = 0)			
X_1 Delivery speed[a]	4.365	ND	NA	NA	NA
X_2 Price level[a]	1.865	ND	NA	NA	NA
X_3 Price flexibility[a]	9.215	ND	NA	NA	NA
X_4 Manufacturer image	5.238	ND	NA	NA	NA
X_5 Overall service	3.256	ND	NA	NA	NA
X_6 Salesforce image	2.671	ND	NA	NA	NA
X_7 Product quality	6.003	ND	NA	NA	NA

Note: Contains observations from both analysis and holdout samples.
ND: No data
NA: Not Applicable due to missing data
[a]Variables included in the discriminant function.
[b]t test performed with separate variance estimates rather than a pooled variance estimate because the Levene test detected significant differences in the variances between the two groups.

with significant differences for four variables (X_1, X_2, X_5, and X_6). The almost opposite pattern occurs for the modified rebuy group—the misclassified cases are significantly lower in their perceptions on six variables (X_2 through X_7, except for X_3). Combined, these misclassified cases seem to be forming a middle group with perceptions between the two existing groups. HATCO researchers should use this finding to investigate whether an additional variable would help distinguish these cases, or if the dependent variable inadequately captures these cases in the current three-group classification.

The analysis of misclassifieds can be supplemented by the graphical examination of the individual observations. Figure 5.12 plots each observation based on its two rotated discriminant Z scores with an overlay of the territorial map representing the boundaries of the cutting scores for each function. In viewing each group's dispersion around the group centroid, we see that the straight rebuy group (group 3) is most concentrated, with little overlap with the other two groups. The compactness of group 3 is also shown by the fact that there were no misclassifications of this group. This is in contrast to group 1 (6 misclassified) and group 2

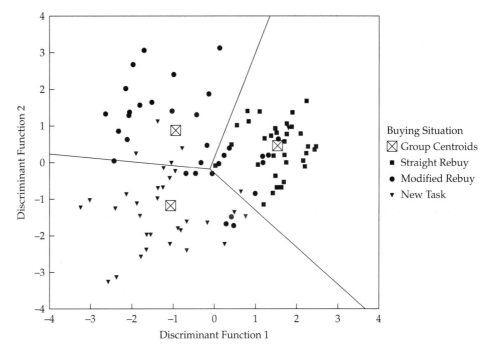

FIGURE 5.12 Territorial Map and Rotated Discriminant Z Scores

(15 misclassified). The dispersion of groups 2 and 3 are shown in the figure and the possible existence of a fourth group in the center and right of the graph can be seen. The graphical portrayal is useful not only for identifying these misclassified cases that may form a new group, but also in identifying outliers.

Stage 5: Interpretation of Three-Group Discriminant Analysis Results

The next stage of the discriminant analysis involves interpretation of the discriminant functions. The first step is to review the rotation of the functions for purposes of simplifying interpretation. The next step is to examine the contributions of the predictor variables to each function separately (i.e., discriminant loadings). The final step examines the cumulative effects of the independent variables across multiple discriminant functions with the potency index. Graphical display can also benefit the researcher in this two-dimensional solution to understand the relative position of each group and the interpretation of the relevant variables in determining this position.

Rotation

After the discriminant functions are developed, they can be "rotated" to redistribute the variance (this concept is more fully explained in chapter 3). Basically, rotation preserves the original structure and reliability of the discriminant models while making them easier to interpret substantively. In the present application we chose the most widely used procedure of VARIMAX rotation. The rotation affects the function coefficients and discriminant loadings, as well as the calculation of the discriminant Z scores and the group centroids (see Table 5.18).

Examining the rotated versus unrotated coefficients or loadings reveals a somewhat more simplified set of results (i.e., loadings tend to separate into high versus low values instead of being midrange), but in this case the rotation was not as helpful as in other situations. Yet, given a discriminant analysis with two or more estimated functions, rotation can be a powerful tool that should always be considered to increase the interpretability of the results.

Assessing the Contribution of Predictor Variables

The first task is to plot the group centroids to create an overall visual perspective of the differences in the three groups. Figure 5.12 depicts the centroids for the first two functions along with each observation. Although there was some overlap and misclassification of the cases, the three groups were well separated for most cases. This graphical approach illustrates the differences in the groups due to the discriminant functions, but does not provide a basis for explaining these differences in terms of the independent variables.

To assess the contributions of the seven predictors, the researcher has a number of measures to employ—discriminant loadings, univariate F ratios, and the potency index. The techniques involved in the use of discriminant loadings and the univariate F ratios were discussed in the two-group example. An additional interpretational technique available in a multifunction solution is the potency index. Table 5.22 illustrates the calculation of the potency index for each of the predictor variables. Loadings represent the correlation between the independent variable and the discriminant Z score. Thus, the squared loading is the variance in the independent variable associated with the discriminant function. By weighting the explained variance of each function by the relative discriminatory power of the functions and summing across functions, the potency index represents the total discriminating effect across all discriminant functions. Reviewing the potency indices for the seven independent variables reveals the interesting fact that X_5 has the second highest value, but was not included in the stepwise solution. How could this happen? First, remember that multicollinearity can affect stepwise solutions due to redundancy among highly multicollinear variables. X_5 was the second most discriminating variable in a univariate sense (see Table 5.14), but

TABLE 5.22 Calculation of the Potency Indices for the Three-Group Discriminant Analysis

	Discriminant Function 1				Discriminant Function 2				
Variables	Loading (L)	Squared Loading	Relative Eigenvalue[a]	Potency Value[b]	Loading (L)	Squared Loading	Relative Eigenvalue[a]	Potency Value[b]	Potency Index[c]
X_1 Delivery speed	0.650	0.423	0.784	0.331	0.039	0.002	0.216	0.000	0.332
X_2 Price level	−0.159	0.025	0.784	0.020	0.824	0.679	0.216	0.146	0.166
X_3 Price flexibility	0.721	0.520	0.784	0.408	−0.566	0.320	0.216	0.069	0.477
X_4 Manufacturer image	0.046	0.002	0.784	0.002	0.188	0.035	0.216	0.008	0.009
X_5 Overall service	0.509	0.259	0.784	0.203	0.817	0.667	0.216	0.144	0.347
X_6 Salesforce image	0.161	0.026	0.784	0.020	0.171	0.029	0.216	0.006	0.027
X_7 Product quality	−0.129	0.017	0.784	0.013	0.243	0.059	0.216	0.013	0.026

Note: Unrotated loadings are used to correspond to the original eigenvalues before rotation.
[a]Relative eigenvalue is the eigenvalue of the discriminant function divided by the sum of the eigenvalues for all significant functions. In our example, the relative eigenvalue for function 1 is [1.935/(1.935 + .532)] = .784.
[b]Potency value = square loading × relative eigenvalue.
[c]Potency index = potency value for discriminant function 1 + potency value for discriminant function 2.

the entry of X_3, which is correlated with X_5, into the discriminant function made X_5's discriminatory power redundant (**tolerance** of .017). But how did X_5 have a high potency value, even though it was not in the final discriminant solution? Remember that discriminant loadings are calculated for all independent variables, even if not in the discriminant solution. The intent is to provide for the interpretation in just such instances where multicollinearity is present.

Table 5.23 presents the three preferred interpretive measures for each variable. The results generally support the stepwise analysis, although X_5 has a substantial univariate F ratio and potency index, but was not included because of collinearity. The other variables not included (X_2, X_4, and X_6) all have very low loadings, nonsignificant F values, and/or low potency index values. Of particular note is the interpretation of the two dimensions of discrimination. This can be done solely through examination of the loadings, but is complemented by a graphical display of the discriminant loadings, as described in the following section.

Graphical Display of Discriminant Loadings   To depict the differences in terms of the predictor variables, the loadings and the group centroids can be plotted in reduced discriminant space. As noted earlier, the most valid representation is the use of stretched attribute vectors and group centroids. Table 5.24 (p. 312) shows the calculations for stretching both the discriminant loadings (used for attribute vectors) and the group centroids. The plotting process always involves all the variables included in the model by the stepwise procedure (in our example, X_1, X_2, and X_3). However, the researcher frequently plots the variables not included in the discriminant function if their respective univariate F ratios are significant. This procedure shows the importance of collinear variables that were not included in the final stepwise model. Our example demonstrates the significance of including these variables in the plotting process. X_7 is included in the final model because it is significant with a portion of the residual variance, and X_5 exhibits much stronger differences between the groups but was not included because of collinearity.

The plots of the stretched attribute vectors for the rotated discriminant loadings are shown in Figure 5.13 (p. 312). The vectors plotted using this procedure point to the groups having the highest mean on the respective independent variable and away from the groups having the lowest mean scores. Thus the interpretation of the plot in Figure 5.12 indicates that the first discriminant function (1) is the primary source of difference between groups 1 and 2 versus group 3. Moreover, the first function corresponds most closely to variable X_3 and somewhat with X_1.

TABLE 5.23 Summary of Interpretive Measures for Three-Group Discriminant Analysis

Independent Variables	Rotated Discriminant Function Loadings		Univariate F Ratio	Potency Index
	Function 1	*Function 2*		
X_1 Delivery speed	.568	.319	23.346	0.332
X_2 Price level	−.502	.672	11.674	0.166
X_3 Price flexibility	.891	−.186	33.362	0.477
X_4 Manufacturer image	−.041	.190	.961	0.009
X_5 Overall service	.101	.957	23.692	0.347
X_6 Salesforce image	.071	.224	.126	0.027
X_7 Product quality	−.222	.163	9.293	0.026

TABLE 5.24 Calculation of the Stretched Attribute Vectors and Group Centroids in Reduced Discriminant Space

Independent Variables	Rotated Discriminant Function Loadings		Univariate F Ratio	Reduced Space Coordinates	
	Function 1	Function 2		Function 1	Function 2
X_1 Delivery speed[a]	.568	.319	23.346	13.261	7.447
X_2 Price level[a]	−.502	.672	11.674	−5.860	7.845
X_3 Price flexibility[a]	.891	−.186	33.362	29.726	−6.205
X_4 Manufacturer image	−.041	.190	.961	−0.039[b]	0.183[b]
X_5 Overall service	.101	.957	23.692	2.393	22.673
X_6 Salesforce image	.071	.224	.126	0.009[b]	0.028[b]
X_7 Product quality	−.222	.163	9.293	−2.063	1.515

Group	Group Centroids		Approximate F Value		Reduced Space Coordinates	
	Function 1	Function 2	Function 1	Function 2	Function 1	Function 2
Group 1: New task	−1.081	−1.167	55.148	15.162	−59.614	−17.694
Group 2: Modified rebuy	−.952	.879	55.148	15.162	−52.500	13.327
Group 3: Straight rebuy	1.541	.472	55.148	15.162	84.982	7.156

[a]Variables entered in the stepwise solution.
[b]Vectors not plotted because of nonsignificant F ratio.

X_2 and X_7 have negative loadings. This is shown not only in the graph, but also from an examination of the discriminant loadings. Thus the distinguishing characteristics of firms primarily making straight rebuy decisions (group 3) are more favorable perceptions of delivery speed and price flexibility and less favorable

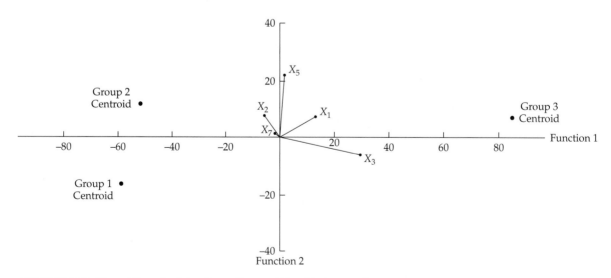

FIGURE 5.13 Plot of Stretched Attribute Vectors (Variables) in Reduced Discriminant Space

perceptions on price level and product quality. Similar profiles can be made for the two remaining groups for the first discriminant function.

The second discriminant function (2) provides the distinction between group 1 versus groups 2 and 3. The close correspondence between the X_2 and X_5 vectors and the second function signifies that the perceptions of price level and manufacturer image are most descriptive of the second discriminant function.

At this point, all this plotting may appear too complicated. What other procedure is simpler, yet effective? The easiest approach is to use the rotated correlations (loadings) provided in Table 5.18 and the group centroids. The footnoted loadings indicate on which function each variable has the highest loading (e.g., X_3 for function 1 versus X_2, and X_5 for function 2). To determine between which groups each function discriminates, simply look at the group centroids and see where differences lie. This is merely a distance assessment, not a statistical measure, but it is usually sufficient. For example, looking at function 1, we see that the centroid for group 1 is -1.081, for group 2 it is $-.952$, and for group 3 it is 1.541. From this we conclude that the primary source of differences for this function is between groups 1 and 2 versus group 3. A similar approach can be used for function 2. However, because function 1 represents substantially more variance than function 2, one must be cautious in determining the impact of variables based on their loadings on this function.

Stage 6: Validation of the Discriminant Results

The internal validity of the discriminant results are supported by the levels of predictive accuracy found in the holdout sample. For each measure of predictive accuracy, the discriminant functions applied to the holdout sample exceeded the criterion values. The validity of the discriminant results can also be validated through the use of the U-method. Here the discriminant model is estimated by leaving out one case and then predicting that case with the estimated model. This is done in turn for each observation, such that an observation never influences the discriminant model that predicts its group classification. The cross-classification results are shown in Table 5.19. The overall hit ratio is better than the holdout sample results, and the low hit ratio for group 2 improves to 60 percent, although the hit ratio for group 1 does decline somewhat to 76.2 percent. With acceptable hit ratios for both the holdout and cross-validated samples, the validity of the discriminant model is supported.

The researcher is also encouraged to extend the validation process through profiling the groups on additional sets of variables and/or applying the discriminant function to other sample(s) representative of the overall population or segments within the population. Moreover, analysis of the misclassified cases will help establish if any additional variables are needed and/or if the dependent group classifications need revision.

A Managerial Overview

The discriminant analysis aimed at understanding the perceptual differences of customers based on their typical type of buying situation with HATCO produces several major findings. First, there are two dimensions of discrimination between the buying situation types. The first dimension is typified by higher perceptions of delivery speed and price flexibility, which may be indicative of relational qualities. The second dimension is characterized best by higher perceptions on price level and service, more objective characteristics of the transaction.

The three groups can also be profiled on these two dimensions and variables associated with each dimension to understand the perceptual differences among them. The straight rebuy group, which is characterized as a routine and repetitive transaction, has much higher perceptions of HATCO than the other two groups on the first dimension, perhaps indicative of the long-standing relationship. Groups 2 and 3, lower on the first dimension, are distinguished by the second dimension, with the modified rebuy group having higher perceptions of HATCO on price level and service when compared to the new task group. These general patterns can be extended to profiling the groups on the separate independent variables and focusing on the key differentiating variables, in this case X_1 (delivery speed), X_2 (price level), X_3 (price flexibility), and X_5 (overall service). The discriminant analysis identifies the variables most impactful so that management can develop concise programs incorporating a smaller set of variables.

The results also identified a set of observations from the new task and modified rebuy groups, who potentially represent either a fourth type of buying situation or are characterized by a variable not included in the model at this time. They have fairly common perceptions, but are not typical of the other customers in those types of buying situations. In many instances, they are more like the straight rebuy group in terms of their perceptions. Thus, management is presented managerial input for strategic and tactical planning from not only the direct results of the discriminant analysis, but also from the classification errors.

An Illustrative Example of Logistic Regression

There are several reasons why logistic regression is an attractive alternative to discriminant analysis whenever the dependent variable has only two categories. First, logistic regression is less affected than discriminant analysis by the variance/covariance inequalities across the groups, a basic assumption of discriminant analysis. Second, logistic regression can handle categorical independent variables easily, whereas in discriminant analysis the use of dummy variables created problems with the variance/covariance equalities. Finally, logistic regression results parallel those of multiple regression in terms of their interpretation and the casewise diagnostic measures available for examining residuals. The following example is identical to the two-group discriminant analysis discussed earlier, with logistic regression used this time for the estimation of the model. As we will see, logistic regression as an alternative to discriminant analysis has many advantages, but the researcher must also examine the results carefully for overfitting of the data or other model estimation problems as encountered in this example.

Stages 1, 2, and 3: Research Objectives, Research Design, and Statistical Assumptions

The issues addressed in the first three stages of the decision process are identical for the two group discriminant analysis and logistic regression. The research problem is still to determine if differences in perceptions of HATCO can distinguish between customers using specification buying versus total value analysis. The sample of 100 customers is divided into an analysis sample of 60 observations, with the remaining 40 observations constituting the holdout or validation sample. Finally, given the robustness of logistic regression to violation of the assumption of equality of the variance/covariance matrices across groups, logistic regression is well suited for application in this situation. We now focus on the results stemming from the use of logistic regression to estimate and understand the differences between these two types of customers.

Stage 4: Estimation of the Logistic Regression Model and Assessing Overall Fit

If we revisit our discussion of the differences of the groups on the seven independent variables (refer to Table 5.5), we recall that X_1, X_3, and X_7 had the greatest differences between the two groups. In logistic regression, a corresponding comparison is to assess the logistic relationship between each independent variable and X_{11}, similar to the plots of correlations used in regression. Figure 5.14 (p. 316) shows the logistic relationships for all seven independent variables, along with the partial correlation. As was seen in examining the means, X_1, X_3, and X_7 had the highest correlations and thus the greatest discriminating power between the groups. In the logistic relationships in Figure 5.13, the vertical line drawn at the .50 predicted probability value is the cutting score for making group predictions. For example, in the case of X_1, observations to the left of the cutting score will be classified in the specification buying group ($X_{11} = 0$), whereas observations to the right of the cutting score will be classified in the total value analysis group ($X_{11} = 1$). Misclassified cases can then be identified for each group. For example, for the specification buying group at the bottom of the graphs, all observations to the right of the line will be misclassified into the total value analysis group. In some instances, such as X_4, the independent variable can make no distinction between the groups and all cases are classified into the larger group, in this instance group 1. In these situations, the predicted probability values never fall below .50.

Model Estimation

Logistic regression is estimated much like multiple regression in that a base model is first estimated to provide a standard for comparison (see chapter 4 for more detail). In multiple regression, the mean is used to set the base model and calculate the total sum of squares. In logistic regression, the same process is used, with the mean used in the estimated model not to set the sum of squares, but instead to set the log likelihood value. From this model, the partial correlations for each variable can be established and the most discriminating variable chosen according to the selection criteria.

Table 5.25 (p. 317) contains the base model results for the logistic regression analysis. The log likelihood value ($-2LL$) here is 78.859. The score statistics, a measure of association used in logistic regression, along with the partial correlation for each independent variable are indicators of the variable selected in the stepwise procedure. Several criteria can be used to guide entry: greatest reduction in the $-2LL$ value, greatest Wald coefficient, or highest conditional probability. In our example, we employ the reduction of the log likelihood ratio criteria.

As seen in the first step of the estimation process, X_7 was selected for entry (Table 5.26, p. 318). This corresponded to the highest score statistic and the highest partial correlation. Although the entry of X_7 into the logistic regression model obtained a reasonable model fit, examination of the remaining independent variables not in the equation indicated that several met our threshold of .05 significance for inclusion in the model and thus further model expansion was

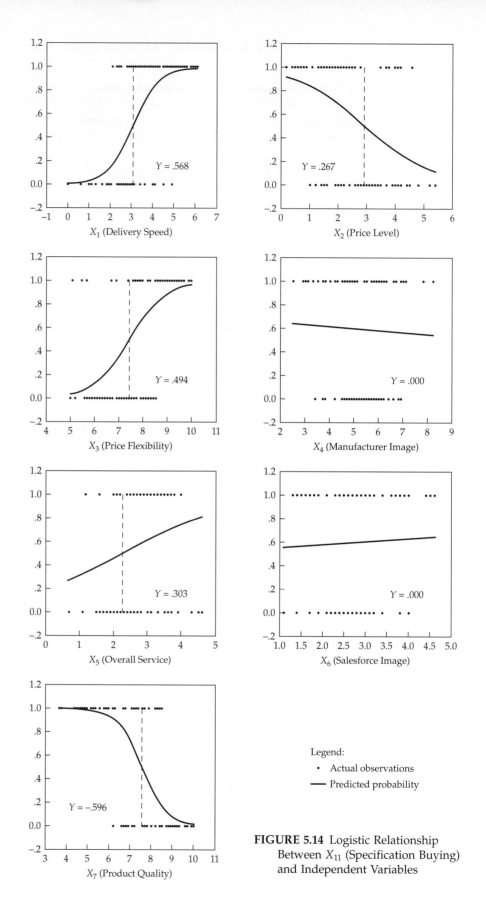

FIGURE 5.14 Logistic Relationship Between X_{11} (Specification Buying) and Independent Variables

TABLE 5.25 Logistic Regression Base Model Results

OVERALL MODEL FIT

-2 log likelihood ($-2LL$): 78.859

VARIABLES NOT IN THE EQUATION

	Score Statistic	Significance	Partial Correlation (r)
X_1 Delivery speed	27.476	.000	.568
X_2 Price level	7.631	.006	.267
X_3 Price flexibility	21.287	.000	.495
X_4 Manufacturer image	.840	.360	.000
X_5 Overall service	9.256	.002	.303
X_6 Salesforce image	1.631	.202	.000
X_7 Product quality	30.041	.000	.596

indicated. X_3, with the highest score statistic and partial correlation, was selected for entry at step 2 (Table 5.27, p. 319). There was improvement in all measures of model fit, ranging from a decrease in the $-2LL$ value to the various R^2 measures. Again, although extremely high levels of model fit were obtained with two variables, examination of the variables not in the equation indicated that other variables could enter the stepwise solution. However, when X_5, the remaining variable with the highest partial correlation and score statistic, was entered, the logistic regression model became unstable and produced inappropriate coefficients and even indicated perfect fit in some measures. Moreover, the estimated coefficients had inappropriate values and significance levels. Given the iterative nature of the estimation process, the researcher must always examine the results to see if this type of "overfitting" has occurred and select the prior model as most appropriate. In this case, we select the two-variable model and use it for further assessment and interpretation.

Statistical Significance

There are two statistical tests for the significance of the final model (see Table 5.27). First, a chi-square test for the change in the $-2LL$ value from the base model is comparable to the overall F test in multiple regression. In the two-variable model, this reduction was statistically significant at the .000 level. Also, the Hosmer and Lemeshow measure of overall fit [10] has a statistical test, which indicates that there was no statistically significant difference between the observed and predicted classifications. These two measures, in combination, provide support for acceptance of the two-variable model as a significant logistic regression model and suitable for further examination.

The estimated coefficients for the two independent variables and the constant can also be evaluated for statistical significance. The Wald statistic is used to assess significance, except in cases in which the coefficient is extremely large, when the score statistic is used. Both coefficients are statistically significant at the .01 level, although the constant is significant at only the .10 level. Thus, the individual variables are significant and should be interpreted.

TABLE 5.26 Step 1: Entry of X_7 (Product Quality) in Stepwise Logistic Regression Model

OVERALL MODEL FIT

Goodness of Fit Measures	Value	Change in $-2LL$		
			Value	Significance
-2 log likelihood $(-2LL)$	37.524			
Goodness of Fit	37.408	From base model	41.335	.000
"Pseudo" R^2	.524	From prior step	41.335	.000
Cox and Snell R^2	.498			
Nagelkerke R^2	.681			

	Chi-square	df	Significance
Hosmer and Lemeshow	2.664	8	.9535

VARIABLES IN THE EQUATION

Variable	B	S.E.	Wald	Signif.	r	Exp(B)
X_7 Product quality	-1.896	.495	14.678	.000	$-.401$	.150
Constant	14.581	3.794	14.774	.000		

VARIABLES NOT IN THE EQUATION

	Score Statistic	Significance	Partial Correlation (r)
X_1 Delivery speed	10.593	.001	.328
X_2 Price level	.214	.643	.000
X_3 Price flexibility	15.614	.000	.415
X_4 Manufacturer image	4.985	.026	.195
X_5 Overall service	6.669	.010	.243
X_6 Salesforce image	6.441	.011	.237

CLASSIFICATION MATRIX

	Predicted Group Membership[a]					
	Analysis Sample			Holdout Sample		
Actual Group	Group 0	Group 1	Total	Group 0	Group 1	Total
Group 0: Specification buying	17	5	22	14	4	18
	(77.3)	(22.7)		(77.8)	(22.2)	
Group 1: Total value analysis	4	34	38	3	19	22
	(10.5)	(89.5)		(13.6)	(86.4)	
Total	21	39	60	17	23	40

B = logistic coefficient; S.E. = standard error; Wald = Wald statistic; Signif. = significance level; r = correlation; Exp(B) = exponentiated coefficient.
[a]Values in parentheses are percent correctly classified (hit ratio).

Assessing Overall Model Fit

In assessing model fit, several measures are available. First, the $-2LL$ value is given. In the single variable model (see Table 5.26), the $-2LL$ value is reduced from the base model value of 78.859 to 37.524, a decrease of 41.335. Smaller values of the $-2LL$ measure indicate better model fit. The goodness of fit measure compares the predicted probabilities to the observed probabilities, with higher values indicating better fit. There is no upper or lower limit for this measure, and the value for the single variable model is 37.408. Next, three measures compara-

TABLE 5.27 Step 2: Entry of X_3 (Price Flexibility) in Stepwise Logistic Regression Model

OVERALL MODEL FIT

Goodness of Fit Measures	Value	Change in $-2LL$		
			Value	Significance
-2 log likelihood ($-2LL$)	20.258			
Goodness of Fit	58.967	From base model	58.601	.000
"Pseudo" R^2	.743	From prior step	17.266	.000
Cox and Snell R^2	.623			
Nagelkerke R^2	.852			
	Chi-square	df	Significance	
Hosmer and Lemeshow	10.344	8	.2417	

VARIABLES IN THE EQUATION

Variable	B	S.E.	Wald	Signif.	r	Exp(B)
X_3: Price flexibility	1.830	.717	6.517	.011	.239	6.237
X_7: Product quality	−2.912	1.135	6.581	.010	−.241	.054
Constant	8.329	5.110	2.657	.103		

VARIABLES NOT IN THE EQUATION

	Score Statistic	Significance	Partial Correlation (r)
X_1 Delivery speed	3.746	.053	.149
X_2 Price level	3.641	.056	.144
X_4 Manufacturer image	5.557	.018	.212
X_5 Overall service	8.824	.003	.294
X_6 Salesforce image	8.770	.003	.293

CLASSIFICATION MATRIX

	Predicted Group Membership[a]					
	Analysis Sample			Holdout Sample		
Actual Group	Group 0	Group 1	Total	Group 0	Group 1	Total
Group 0: Specification buying	21 (95.5)	1 (4.5)	22	15 (83.3)	3 (16.7)	18
Group 1: Total value analysis	0 (0.0)	38 (100.0)	38	2 (9.1)	20 (90.9)	22
Total	21	39	60	17	23	40

B = logistic coefficient; S.E. = standard error; Wald = Wald statistic; Signif. = significance level; r = correlation; Exp(B) = exponentiated coefficient.
[a]Values in parentheses are percent correctly classified (hit ratio).

ble to the R^2 measure in multiple regression are available. The Cox and Snell R^2 measure operates in the same manner, with higher values indicating greater model fit. However, this measure is limited in that it cannot reach the maximum value of 1, so Nagelkerke proposed a modification that had the range of 0 to 1. In our instance, the Cox and Snell value is .498 and the Negelkerke value is .681. The third measure is the "pseudo" R^2 measure based on the improvement in the $-2LL$ value. The value of .524 for the single variable model is calculated as:

$$R^2_{\text{logit}} = \frac{.2LL_{\text{null}} - (-2LL_{\text{model}})}{-2LL_{\text{null}}}$$

$$= \frac{78.859 - 37.524}{78.859}$$

$$= \frac{41.335}{78.859} = .524$$

The final measure of model fit is the Hosmer and Lemeshow value, which measures the correspondence of the actual and predicted values of the dependent variable. In this case, better model fit is indicated by a smaller difference in the observed and predicted classification. A good model fit is indicated by a nonsignificant chi-square value.

In the two-variable model, all of the measures of model fit improved. The $-2LL$ value decreased to 20.258. The R^2 values ranged from .623 to .852, all improvements from the single-variable model and indicative of good model fit when compared to the R^2 values usually found in multiple regression. The Hosmer and Lemeshow measure still showed nonsignificance, indicating no difference in the distribution of the actual and predicted dependent values.

Finally, the classification matrices, identical in nature to those used in discriminant analysis, show extremely high hit ratios of correctly classified cases for the two-variable model. The overall hit ratios are 98.3 percent and 87.5 percent for the analysis and holdout samples, respectively. Likewise, the individual group hit ratios are consistently high and do not indicate a problem in predicting either of the two groups. The two-variable model, including X_3 and X_7, demonstrates excellent model fit and statistical significance at the overall model level as well as for the variables included in the model.

Casewise Diagnostics

The analysis of the misclassification of individual observations can provide further insight into possible improvements of the model, but in this case there are a total of only six misclassified cases in both analysis and holdout samples, providing an inadequate base for making any further analysis. Casewise diagnostics such as residuals and measures of influence are available, but are of little use in this situation (for a more detailed discussion of measures such as Cook's distance and DFBETA, see the appendix to chapter 4). Therefore, given the low levels of misclassification, no further analysis of misclassification is performed.

Stage 5: Interpretation of the Results

The logistic regression model produced a variate quite similar to that of the two-group discriminant analysis, although with one less independent variable. In the case of discriminant analysis, X_1, X_3, and X_7 were included in the stepwise solution. But logistic regression included only X_3 and X_7 (see Table 5.27). The implications from both analyses were similar: price flexibility (X_3) had a positive association and product quality (X_7) had a negative association with the dependent variable. Given that the dependent variable (X_{11}) had two groups—specification buying ($X_{11} = 0$) and total value analysis ($X_{11} = 1$)—the coefficients imply that firms using total value analysis have lower perceptions of the product quality while having higher perceptions of price flexibility. Because both discriminant analysis and logistic regression predict group membership, the coefficients relate to the relative group means on the independent variables.

Stage 6: Validation of the Results

The validation of the logistic regression model is accomplished in this example through the same method used in discriminant analysis: creation of analysis and holdout samples. Although some improvement in the hit ratio was seen in the analysis sample with logistic regression, the holdout samples were almost identical for discriminant analysis and logistic regression. This leads to the conclusion that both methods have strong empirical support in their validation on separate samples at approximately the same level.

A Managerial Overview

Logistic regression presents an alternative to discriminant analysis that may be more "comfortable" to many researchers due to its similarity to multiple regression. Given its robustness in the face of data conditions that can negatively impact discriminant analysis (e.g., unequal variance/covariance matrices), logistic regression is also the preferred estimation technique in many applications. In this example, logistic regression provided a small increase in predictive accuracy with a simpler variate that had the same substantive interpretation, only with one less variable. From the logistic regression results, the researcher can focus on the tradeoff of price flexibility versus product quality when strictly concerned with prediction. However, to gain a full understanding, variables not in the analysis but still showing differences between the groups should complement these results. Collinearity among the variables can make the discriminatory power redundant among variables, but redundancy does not make variables irrelevant from a perspective of explanation.

Summary

The underlying nature, concepts, and approach to multiple discriminant analysis and logistic regression have been presented. Basic guidelines for their application and interpretation were included to clarify further the methodological concepts. Illustrative examples for both two-group and three-group solutions were presented based on the HATCO database. These applications demonstrated the major points of importance in applying discriminant analysis and logistic regression and in selecting between the two methods in certain situations.

Multiple discriminant analysis helps to understand and explain research problems that involve a single categorical dependent variable and several metric independent variables. A mixed data set (both metric and nonmetric) is also possible for the independent variables if the nonmetric variables are dummy coded (0–1). The result of a discriminant analysis and logistic regression can assist in profiling the intergroup characteristics of the subjects and in assigning them to their appropriate groups. Potential applications of these two techniques to both business and nonbusiness problems are numerous.

Some of the concepts presented in this chapter are based on material discussed in chapters 3, 4, and 6. Thus it is recommended that these three chapters be studied together.

Questions

1. How would you differentiate among multiple discriminant analysis, regression analysis, and analysis of variance?
2. When would you employ logistic regression rather than discriminant analysis? What are the advantages and disadvantages of this decision?
3. What criteria could you use in deciding whether to stop a discriminant analysis after estimating the discriminant function(s)? After the interpretation stage?
4. What procedure would you follow in dividing your sample into analysis and holdout groups? How would you change this procedure if your sample consisted of fewer than 100 individuals or objects?
5. How would you determine the optimum cutting score?
6. How would you determine whether the classification accuracy of the discriminant function is sufficiently high relative to chance classification?
7. How does a two-group discriminant analysis differ from a three-group analysis?
8. Why should a researcher stretch the loadings and centroid data in plotting a discriminant analysis solution?
9. How do logistic regression and discriminant analyses each handle the relationship of the dependent and independent variables?
10. What are the differences in estimation and interpretation between logistic regression and discriminant analysis?

References

1. Cohen, J. (1977), *Statistical Power Analysis for the Behavioral Sciences*, rev. ed. New York: Academic Press.
2. Crask, M., and W. Perreault (1977), "Validation of Discriminant Analysis in Marketing Research." *Journal of Marketing Research* 14 (February): 60–68.
3. Dillon, W. R., and M. Goldstein (1984), *Multivariate Analysis: Methods and Applications.* New York: Wiley.
4. Frank, R. E., W. E. Massey, and D. G. Morrison (1965), "Bias in Multiple Discriminant Analysis." *Journal of Marketing Research* 2(3):250–58.
5. Gessner, Guy, N. K. Maholtra, W. A. Kamakura, and M. E. Zmijewski (1988), "Estimating Models with Binary Dependent Variables: Some Theoretical and Empirical Observations." *Journal of Business Research* 16(1):49–65.
6. Green, P. E., D. Tull, and G. Albaum (1988), *Research for Marketing Decisions.* Upper Saddle River, N.J.: Prentice Hall.
7. Green, P. E. (1978), *Analyzing Multivariate Data.* Hinsdale, Ill.: Holt, Rinehart, and Winston.
8. Green, P. E., and J. D. Carroll (1978), *Mathematical Tools for Applied Multivariate Analysis.* New York: Academic Press.
9. Harris, R. J. (1975), *A Primer of Multivariate Statistics.* New York: Academic Press.
10. Hosmer, D. W., and S. Lemeshow (1989), *Applied Logistic Regression.* New York: Wiley.
11. Huberty, C. J. (1984), "Issues in the Use and Interpretation of Discriminant Analysis." *Psychological Bulletin* 95: 156–71.
12. Huberty, C. J., J. W. Wisenbaker, and J. C. Smith (1987), "Assessing Predictive Accuracy in Discriminant Analysis." *Multivariate Behavioral Research* 22 (July): 307–29.
13. Johnson, N., and D. Wichern (1982), *Applied Multivariate Statistical Analysis.* Upper Saddle River, N.J.: Prentice Hall.
14. Morrison, D. G. (1969), "On the Interpretation of Discriminant Analysis." *Journal of Marketing Research* 6(2): 156–63.
15. Perreault, W. D., D. N. Behrman, and G. M. Armstrong (1979), "Alternative Approaches for Interpretation of Multiple Discriminant Analysis in Marketing Research." *Journal of Business Research* 7: 151–73.

Annotated Articles

The following annotated articles are provided as illustrations of the application of discriminant analysis and logistic regression to substantive research questions of both a conceptual and managerial nature. The reader is encouraged to review the complete articles for greater detail on any of the specific issues regarding methodology or findings.

Dant, Rajiv P., James R. Lumpkin, and Robert P. Bush (1990), "Private Physicians or Walk-in Clinics: Do the Patients Differ?" *Journal of Health Care Marketing* 10(2): 25–35.

This article employs two methods, multivariate analysis of variance (MANOVA) and multiple discriminant analysis (MDA), to examine differences in the criteria used by patients in choosing the type of facility from which they will obtain health care. The authors seek to determine if there are differences in patronage behavior between walk-in clinics and traditional private practices in terms of (1) patients' expectations about the two delivery systems, (2) patients' performance evaluations, and (3) patients' demographic characteristics and nature of their medical needs. Information of this nature will enable the healthcare provider to better segment the market, provide expected services, and reduce the costs associated with providing unwanted services. The authors develop a list of ten characteristics or attributes, from prior literature and preliminary qualitative research, which are understood to be crucial in distinguishing between the two patient groups. MANOVA is employed to test for significant differences between the group means (see chapter 6 of this volume). MANOVA is also coupled with MDA to assist in determining the direction and strength of each criterion variable on the overall group differences.

Multiple discriminant analysis uses nonmetric dependent variables and metric independent variables. The dependent variables in this case are the two patron classes: walk-in patients and private practice patients. Independent variables used are the medical facility attributes and demographic information. Although the sample size of 602 was adequate for division into the recommended analysis and holdout samples, the authors did not follow this procedure, which may have led to an upward bias in the hit ratio used in the validation stage of the MDA procedure. The significance and contribution of each discriminant loading is assessed to understand the relative impact of each on the group separation. The authors do an excellent job of describing the hit ratio and comparing the results obtained to the maximum chance and proportional chance criteria. Overall, the results indicate that the type of treatment sought and certain demographic features of the consumer provide a better means of determining patronage than do consumer ratings of attribute importance. It is interesting to note that in this paper, MANOVA mimics the role of MDA, using attributes as independent variables in the prediction of the dependent variable, group membership.

Lussier, Robert N. (1995), "A Nonfinancial Business Success versus Failure Prediction Model for Young Firms." *Journal of Small Business Management* 33(1): 8–20.

This article employs logistical regression analysis to test a model to predict the success or failure of a young business by examining managerial factors rather than using financial ratios. The author seeks to ascertain whether successful and failed businesses start with equal resources. The dependent variable is dichotomized as either success (the business is maintaining at least industry-average profits) or failure (the business is involved in court proceedings or voluntary actions resulting in losses to creditors) of young firms (up to 10 years old). From past research, the author identifies 15 major variables contributing to success or failure. A sample of 216 respondents (108 failed and 108 successful) is used to estimate the parameters of the regression model employing the one step using the maximum-likelihood method.

The results indicate an overall significant model (chi-square < .01) with 4 of the 15 variables significant at the .05 level. The model performs better at predicting firm failure than firm success, but in either case, it performs better than would be expected from random guessing. The model accurately predicts the success or failure of a particular firm approximately 70 percent of the time. Overall, the research holds that successful and failed businesses do not have equivalent starting resources. Firms that succeed developed more specific business plans and sought more professional advice, whereas those that failed had more education and less difficulty attracting and retaining quality employees. The model may be implemented by entrepreneurs, investors, creditors, consultants, and others as a means of assessing the probability for a business's success or failure.

Roth, Kendall, and Allen J. Morrison (1992), "Business-Level Competitive Strategy: A Contingency Link to Internalization." *Journal of Management* 18(3): 473–87.

This article explores whether organizations use different strategic positioning when confronted with an international context. To support the premise, an analysis of the strategy content for businesses competing strictly domestically is compared to those engaging in both domestic and international activities. This comparison is accomplished using multiple discriminant analysis. The sample consists of 294 respondents from the pulp and paper industry. Of these, 104 are classified as international, that is, at least some of their sales are derived from international activities. The remaining 190 are classified as domestic. The discriminating variables are derived from four strategy dimensions identified in the literature. They are complex innovation, marketing differentiation, breadth, and conservative cost control. Each dimension is comprised of multiple attributes. Overall, 15 competitive attributes are identified as representing the four strategic dimensions. Due to conceptual concerns, these attributes are not aggregated. So that the authors may assess the predictive accuracy of the discriminant function, a holdout sample of 56 randomly selected observations are withheld.

All variables are entered simultaneously, which allows for the determination of the discriminating capability of all 15 competitive attributes. The findings are significant, meaning that the discriminant function effectively differentiates between domestic and international firms. Although the authors do not employ a stepwise procedure, they do attempt to assess each variable's relative discriminant ability. Based on the standardized coefficients, competitive attributes associated with breadth, marketing differentiation, and complex innovation demonstrated a high degree of

usefulness in discriminating between the two groups. These include customer service, market breadth, effective control of distribution, and developing brand identification. To assess the discriminant function's ability to predict, the authors construct classification matrices for both the analysis and holdout sample. The overall percentage correctly classified, or the hit ratio, was 81 percent for the analysis sample and 79 percent for the holdout sample, both of which exceed the proportional chance criterion of 54.5 percent. From these measures, the validity of the discriminant model is demonstrated. Based on the findings from the discriminant analysis and subsequent difference testing, the authors conclude that business-level strategy is contingent on the international context.

Montemayor, Edilberto F. (1996), "Congruence between Pay Policy and Competitive Strategy in High-Performing Firms." *Journal of Management* 22(6): 889–908.

Through the use of discriminant analysis, this article examines seven theoretical propositions concerning the matching of business strategy with pay policy. Specifically, the author examines under which business strategy does pay policy have a positive relationship with high-performing organizations. The three strategic types are identified: cost leadership, differentiation, and innovation. The author identifies ten measures representing five aspects of pay policy, which serve as the independent variables: (1) compensation philosophies, (2) external competitiveness, (3) incentive-base mix, (4) individual (merit) pay increases, and (5) pay administration. From a multi-industry pool of organizations, a random sample of 282 respondents was gathered. Data analysis consists of multivariate analysis of variance (MANOVA) followed by multiple discriminant analysis (MDA). Due to sample considerations and to extend the generalizability of the results, the author uses the averages from 26 jackknife pseudosamples to represent the discriminant coefficients and standard error estimates.

The three-group discriminant analysis (based on strategic types) results in two canonical discriminant functions used to differentiate the three groups. Only high performers were used for the discriminant analysis, which consisted of 104 of the respondents. The first discriminant function separates high-performing *cost leaders* from high-performing *innovators*, whereas the second discriminant function separates high-performing *differentiators* from high-performing *innovators*. The authors obtain a 56 percent hit rate. Based on a maximum chance criterion of 33 percent, the model is good. Although the jackknife procedure allowed for a rigorous approach and sought to account for the small sample size (20 observations for each predictor variable), the author did not provide for a holdout sample, which may have led to an upward bias in the hit ratio. The results indicate that there is a link between pay policy and business strategy and that where there is a mismatch performance suffers.

CHAPTER **6**

Multivariate Analysis of Variance

LEARNING OBJECTIVES

Upon completing this chapter, you should be able to do the following:

- Explain the difference between the univariate null hypothesis of ANOVA and the multivariate null hypothesis of MANOVA.
- Discuss the advantages of a multivariate approach to significance testing compared to the more traditional univariate approaches.
- State the assumptions for the use of MANOVA.
- Discuss the different types of test statistics that are available for significance testing in MANOVA.
- Describe the purpose of post hoc tests in ANOVA and MANOVA.
- Interpret interaction results when more than one independent variable is used in MANOVA.
- Describe the purpose of multivariate analysis of covariance (MANCOVA).

CHAPTER PREVIEW

As a theoretical construct, multivariate analysis of variance (MANOVA) was introduced several decades ago by Wilks' original formulation [22]. However, it was not until the development of appropriate test statistics with tabled distributions and the wide availability of computer programs to compute these statistics that MANOVA became a practical tool for researchers.

Multivariate analysis of variance is an extension of analysis of variance (ANOVA) to accommodate more than one dependent variable. It is a dependence technique that measures the differences for two or more metric dependent variables based on a set of categorical (nonmetric) variables acting as independent variables. ANOVA and MANOVA can be stated in the following general forms:

Analysis of Variance

$$Y_1 = X_1 + X_2 + X_3 + \cdots + X_n$$

(metric)   (nonmetric)

Multivariate Analysis of Variance

$$Y_1 + Y_2 + Y_3 + \cdots + Y_n = X_1 + X_2 + X_3 + \cdots + X_n$$

(metric)   (nonmetric)

Like ANOVA, MANOVA is concerned with differences between groups (or experimental treatments). However, ANOVA is termed a univariate procedure because we use it to assess group differences on a single metric dependent variable. MANOVA is termed a multivariate procedure because we use it to assess group differences across multiple metric dependent variables simultaneously. In MANOVA, each treatment group is observed on two or more dependent variables.

Both ANOVA and MANOVA are particularly useful when used in conjunction with **experimental designs**—that is, research designs in which the researcher directly controls or manipulates one or more independent variables to determine the effect on one (ANOVA) or more (MANOVA) dependent variables. ANOVA and MANOVA provide the tools necessary to judge the observed effects (i.e., whether an observed difference is due to a treatment effect or to random sampling variability). See chapter 1 for a discussion of how these techniques relate to other multivariate procedures.

KEY TERMS

Before starting the chapter, review the key terms to develop an understanding of the concepts and terminology to be used. Throughout the chapter the key terms appear in **boldface.** Other points of emphasis in the chapter are *italicized.* Also, cross-references within the Key Terms appear in *italics.*

Alpha (α) Significance level associated with the statistical testing of the differences between two or more groups. Typically, small values, such as .05 or .01, are specified to minimize the possibility of making a *Type I error.*

A priori test See *planned comparison.*

Analysis of variance (ANOVA) Statistical technique used to determine whether samples from two or more groups come from populations with equal means. Analysis of variance employs one dependent measure, whereas multivariate analysis of variance compares samples based on two or more dependent variables.

Beta (β) See *Type II error.*

Blocking factor Characteristic of respondents in the ANOVA or MANOVA that is used to reduce within-group variability. This characteristic becomes an additional *treatment* in the analysis. In doing so, additional groups are formed that are more homogeneous. As an example, assume that customers were asked buying intentions for a product and the independent measure used was age. Examination of the data found that substantial variation was due to gender. Then gender could be added as a further treatment so that each age category was split into male and female groups with greater within-group homogeneity.

Bonferroni inequality Approach for adjusting the selected *alpha* level to control for the overall *Type I error* rate. The procedure involves (1) computing the adjusted rate as α divided by the number of statistical tests to be performed, and then (2) using the adjusted rate as the critical value in each separate test.

Box test Statistical test for the equality of the variance/covariance matrices of the dependent variables across the groups. It is very sensitive, especially to the presence of nonnormal variables. A significance level of .01 or less is used as an adjustment for the sensitivity of the statistic.

Contrast Procedure for investigating specific group differences of interest in conjunction with ANOVA and MANOVA—for example, comparing group mean differences for a specified pair of groups.

Covariates, or **covariate analysis** Use of regression-like procedures to remove extraneous (nuisance) variation in the dependent variables due to one or more uncontrolled metric independent variables (covariates). The covariates are assumed to be linearly related to the dependent variables. After adjusting for the influence of covariates, a standard ANOVA or MANOVA is carried out. This adjustment process (known as ANCOVA or MANCOVA) usually allows for more sensitive tests of treatment effects.

Critical value Value of a test statistic (*t* test, *F* test) that denotes a specified significance level. For example, 1.96 denotes a .05 significance level for the *t* test with large sample sizes.

Discriminant function "Dimension" of difference or discrimination between the groups in the MANOVA analysis. The discriminant function is a variate of the dependent variables.

Disordinal interaction Form of *interaction effect* among independent variables that invalidates interpretation of the *main effects* of the treatments. A disordinal interaction is exhibited graphically by plotting the means for each group and having the lines intersect or cross. In this type of interaction the mean differences not only vary, given the unique combinations of independent variable levels, but the relative ordering of groups changes as well.

Effect size Standardized measure of group differences used in the calculation of statistical power. Calculated as the difference in group means divided by the standard deviation, it is then comparable across research studies as a generalized measure of effect (i.e., differences in group means).

Experimental design Research plan in which the researcher directly manipulates or controls one or more predictor variables (see *treatment*) and assesses their effect on the dependent variables. Common in the physical sciences, it is gaining in popularity in business and the social sciences. For example, respondents are shown separate advertisements that vary systematically on a characteristic, such as different appeals (emotional versus rational) or types of presentation (color versus black-and-white), and are then asked their attitudes, evaluations, or feelings toward the different advertisements.

Factor Nonmetric independent variable, also referred to as a *treatment* or experimental variable.

Factorial design Design with more than one *factor* (treatment). Factorial designs examine the effects of several factors simultaneously by forming groups based on all possible combinations of the levels (values) of the various treatment variables.

Greatest characteristic root (*gcr*) Statistic for testing the null hypothesis in MANOVA. It tests the first *discriminant function* of the dependent variables for its ability to discern group differences.

Hotelling's *T²* Test to assess the statistical significance of the difference on the means of two or more variables between two groups. It is a special case of MANOVA used with two groups or levels of a treatment variable.

Independence Critical assumption of ANOVA or MANOVA that requires that the dependent measures for each respondent be totally uncorrelated with the responses from other respondents in the sample. A lack of independence severely affects the statistical validity of the analysis unless corrective action is taken.

Interaction effect In factorial designs, the joint effects of two *treatment* variables in addition to the individual *main effects*. This means that the difference between groups on one treatment variable varies depending on the level of the second treatment variable. For example, assume that respondents were classified by income (three levels) and gender (males versus females). A significant interaction would be found when the differences between males and females on the independent variable(s) varied substantially across the three income levels.

Main effect In factorial designs, the individual effect of each *treatment* variable on the dependent variable.

Multivariate normal distribution Generalization of the univariate normal distribution to the case of *p* variables. A multivariate normal distribution of sample groups is a basic assumption required for the validity of the significance tests in MANOVA (see chapter 2 for more discussion of this topic).

Null hypothesis Hypothesis that samples come from populations with equal means for either a dependent variable (univariate test) or a set of dependent variables (multivariate test). The null hypothesis can be accepted or rejected depending on the results of a test of statistical significance.

Ordinal interaction Acceptable type of *interaction effect* in which the magnitudes of differences between groups vary but the groups' relative positions remain constant. It is graphically represented by plotting mean values and observing nonparallel lines that do not intersect.

Orthogonal Statistical independence or absence of association. Orthogonal *variates* explain unique variance, with no variance explanation shared between them. Orthogonal *contrasts* are *planned comparisons* that are statistically independent and represent unique comparisons of group means.

Planned comparison *A priori test* that tests a specific comparison of group mean differences. These tests are performed in conjunction with the tests for *main* and *interaction effects* by using a *contrast*.

Post hoc test Statistical test of mean differences performed after the statistical tests for *main effects* have been performed. Most often, post hoc tests do not use a single *contrast*, but instead test for differences among all possible combinations of groups. Even though they provide abundant diagnostic information, they do inflate the overall *Type I* error rate by performing multiple statistical tests and thus must use very strict confidence levels.

Power Probability of identifying a treatment effect when it actually exists in the sample. Power is defined as $1 - \beta$ (see *beta*). Power is determined as a function of (1) the statistical significance level (α) set by the researcher for a *Type I error*, (2) the sample size used in the analysis, and (3) the *effect size* being examined.

Repeated measures   Use of two or more responses from a single individual in an ANOVA or MANOVA analysis. The purpose of a repeated measures design is to control for individual-level differences that may affect the within-group variance. Repeated measures are a form of respondent's lack of *independence*.

Replication   Readministration of an experiment with the intent of validating the results in another sample of respondents.

Significance level   See *alpha*.

Standard error   Measure of the dispersion of the means or mean differences expected due to sampling variation. The standard error is used in the calculation of the *t statistic*.

Stepdown analysis   Test for the incremental discriminatory power of a dependent variable after the effects of other dependent variables have been taken into account. Similar to stepwise regression or discriminant analysis, this procedure, which relies on a specified order of entry, determines how much an additional dependent variable adds to the explanation of the differences between the groups in the MANOVA analysis.

t **statistic**   Test statistic that assesses the statistical significance between two groups on a single dependent variable (see *t* test).

t **test**   Test to assess the statistical significance of the difference between two sample means for a single dependent variable. The *t* test is a special case of ANOVA for two groups or levels of a treatment variable.

Treatment   Independent variable that a researcher manipulates to see the effect (if any) on the dependent variables. The treatment variable can have several levels. For example, different intensities of advertising appeals might be manipulated to see the effect on consumer believability.

Type I error   Probability of rejecting the null hypothesis when it should be accepted, that is, concluding that two means are significantly different when in fact they are the same. Small values of *alpha* (e.g., .05 or .01), also denoted as α, lead to rejection of the null hypothesis and acceptance of the alternative hypothesis that population means are not equal.

Type II error   Probability of failing to reject the null hypothesis when it should be rejected, that is, concluding that two means are not significantly different when in fact they are different. Also known as the *beta* (β) error.

U **statistic**   See *Wilks' lambda*.

Variate   Linear combination of variables. In MANOVA, the dependent variables are formed into *variates* in the discriminant function(s).

Vector   Set of real numbers (e.g., $X_1 \ldots X_n$) that can be written in either columns or rows. Column vectors are considered conventional and row vectors are considered transposed. Column vectors and row vectors are shown as follows:

$$X = \begin{bmatrix} X_1 \\ X_2 \\ . \\ . \\ . \\ X_n \end{bmatrix} \qquad X^T = [X_1 \ X_2 \ldots X_n]$$

Column vector   Row vector

The T on the row vector indicates that it is the transpose of the column vector.

Wilks' lambda   One of the four principal statistics for testing the null hypothesis in MANOVA. Also referred to as the maximum likelihood criterion or *U* statistic.

What Is Multivariate Analysis of Variance?

Multivariate analysis of variance is the multivariate extension of the univariate techniques for assessing the differences between group means. The univariate procedures include the *t* test for two-group situations and ANOVA for situations with three or more groups defined by two or more independent variables. Before proceeding with our discussion of the unique aspects of MANOVA, we review the basic principles of the univariate techniques.

Univariate Procedures for Assessing Group Differences

These procedures are classified as univariate not because of the number of independent variables, but instead because of the number of dependent variables. In multiple regression, the terms univariate and multivariate refer to the number of independent variables, but for ANOVA and MANOVA, the terminology applies to the use of single or multiple dependent variables. The following discussion addresses the two most common types of univariate procedures, the *t* test, which compares a dependent variable across two groups, and ANOVA, which is used whenever the number of groups is three or more.

The *t* Test

The *t* **test** assesses the statistical significance of the difference between two independent sample means. For example, a researcher may expose two groups of respondents to different advertisements reflecting different advertising messages—one informational and one emotional—and subsequently ask each group about the appeal of the message on a 10-point scale, with 1 being poor and 10 being excellent. The two different advertising messages represent a **treatment** with two levels (informational versus emotional). A treatment also known as a **factor,** is a nonmetric independent variable, experimentally manipulated or observed, that can be represented in various categories or levels. In our example, the treatment is the effect of emotional versus informational appeals.

To determine whether the two messages are viewed differently (meaning that the treatment has an effect), a *t* statistic is calculated. The *t* **statistic** is the ratio of the difference between the sample means ($\mu_1 - \mu_2$) to their standard error. The **standard error** is an estimate of the difference between means to be expected because of sampling error, rather than real differences between means. This can be shown in the equation

$$t \text{ statistic} = \frac{\mu_1 - \mu_2}{SE_{\mu_1 \mu_2}}$$

where

$$\mu_1 = \text{mean of group 1}$$
$$\mu_2 = \text{mean of group 2}$$
$$SE_{\mu_1 \mu_2} = \text{standard error of the difference in group means}$$

By forming the ratio of the actual difference between the means to the difference expected due to sampling error, we quantify the amount of the actual impact of the treatment that is due to random sampling error. In other words, the *t* value, or *t* statistic, represents the group difference in terms of standard errors.

If the *t* value is sufficiently large, then statistically we can say that the difference was not due to sampling variability, but represents a true difference. This is done by comparing the t statistic to the **critical value** of the t statistic (t_{crit}). If the absolute value of the t statistic is greater than the critical value, this leads to rejection of the **null hypothesis** of no difference in the appeals of the advertising messages between groups. This means that the actual difference due to the appeals is statistically larger than the difference expected from sampling error.

We determine the critical value (t_{crit}) for our t statistic and test the statistical significance of the observed differences by the following procedure:

1. Compute the t statistic as the ratio of the difference between sample means to their standard error.
2. Specify a **Type I error** level (denoted as α, or **significance level**), which indicates the probability level the researcher will accept in concluding that the group means are different when in fact they are not.
3. Determine the critical value (t_{crit}) by referring to the t distribution with $N_1 + N_2 - 2$ degrees of freedom and a specified α, where N_1 and N_2 are sample sizes.
4. If the absolute value of the computed t statistic exceeds t_{crit}, the researcher can conclude that the two advertising messages have different levels of appeal (i.e., $\mu_1 \neq \mu_2$), with a Type I error probability of α. The researcher can then examine the actual mean values to determine which group is higher on the dependent value.

Analysis of Variance

In our example for the t test, a researcher exposed two groups of respondents to different advertising messages and subsequently asked them to rate the appeal of the advertisements on a 10-point scale. Suppose we were interested in evaluating three advertising messages rather than two. Respondents would be randomly assigned to one of three groups, and we would have three sample means to compare. To analyze these data, we might be tempted to conduct separate t tests for the difference between each pair of means (i.e., group 1 versus group 2; group 1 versus group 3; and group 2 versus group 3).

However, multiple t tests inflate the overall Type I error rate (we discuss this in more detail in the next section). **ANOVA** avoids this Type I error inflation due to making multiple comparisons of treatment groups by determining in a single test whether the entire set of sample means suggests that the samples were drawn from the same general population. That is, ANOVA is used to determine the probability that differences in means across several groups are due solely to sampling error.

The logic of an ANOVA test is fairly straightforward. As the name "analysis of variance" implies, two independent estimates of the variance for the dependent variable are compared, one that reflects the general variability of respondents within the groups (MS_W) and another that represents the differences between groups attributable to the treatment effects (MS_B):

1. *Within-groups estimate of variance* (MS_W: mean square within groups): This is an estimate of the average random respondent variability on the dependent variable within a treatment group and is based on deviations of individual scores from their respective group means. MS_W is comparable to the standard error between two means calculated in the t test as it represents variability within groups. The value MS_W is sometimes referred to as the error variance.
2. *Between-groups estimate of variance* (MS_B: mean square between groups): The second estimate of variance is the variability of the treatment group means on the

dependent variable. It is based on deviations of group means from the overall grand mean of all scores. Under the null hypothesis of no treatment effects (i.e., $\mu_1 = \mu_2 = \mu_3 = \ldots = \mu_k$), this variance estimate, unlike MS_W, reflects any treatment effects that exist; that is, differences in treatment means increase the expected value of MS_B.

Given that the null hypothesis of no group differences is true, MS_W and MS_B represent independent estimates of population variance. Therefore, the ratio of MS_B to MS_W is a measure of how much variance is attributable to the different treatments versus the variance expected from random sampling. The ratio of MS_B to MS_W gives us a value for the F statistic. This is similar to the calculation of the t value, and can be shown as

$$F \text{ statistic} = \frac{MS_B}{MS_W}$$

Because group differences tend to inflate MS_B, large values of the F statistic lead to rejection of the null hypothesis of no difference in means across groups. If the analysis has several different treatments (independent variables), then estimates of MS_B are calculated for each treatment and F statistics are calculated for each treatment. This allows for the separate assessment of each treatment.

To determine if the F statistic is sufficiently large to support rejection of the null hypothesis, follow a process similar to the t test. First, determine the critical value for the F statistic (F_{crit}) by referring to the F distribution with $(k - 1)$ and $(N - k)$ degrees of freedom for a specified level of α (where $N = N_1 + \ldots + N_k$ and $k =$ number of groups). If the value of the calculated F statistic exceeds F_{crit}, conclude that the means across all groups are not all equal.

Examination of the group means then allows the researcher to assess the relative standing of each group on the dependent measure. Although the F statistic test assesses the null hypothesis of equal means, it does not address the question of which means are different. For example, in a three-group situation, all three groups may differ significantly, or two may be equal but differ from the third. To assess these differences, the researcher can employ either planned comparisons or post hoc tests. We examine each of these methods in a later section.

Multivariate Analysis of Variance

As statistical inference procedures, both the univariate techniques (t test and ANOVA) and MANOVA are used to assess the statistical significance of differences between groups. In the t test and ANOVA, the null hypothesis tested is the equality of dependent variable means across groups. In MANOVA, the null hypothesis tested is the equality of **vectors** of means on multiple dependent variables across groups. The distinction between the hypotheses tested in ANOVA and MANOVA is illustrated in Figure 6.1 (p. 334). In the univariate case, a single dependent measure is tested for equality across the groups. In the multivariate case, a **variate** is tested for equality. The concept of a variate has been instrumental in our discussions of the previous multivariate techniques and is covered in detail in chapter 1. In MANOVA, the researcher actually has two variates, one for the dependent variables and another for the independent variables. The dependent variable variate is of more interest because the metric-dependent measures can be combined in a linear combination, as we have already seen in multiple

ANOVA

$H_0 : \mu_1 = \mu_2 = \ldots \mu_k$

Null hypothesis (H_0) = all the group means are equal,
that is, they come from the same population.

MANOVA

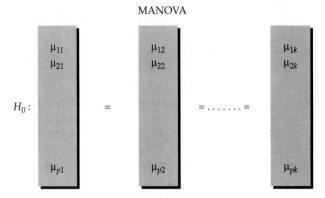

Null hypothesis (H_0) = all the group mean vectors are equal,
that is, they come from the same population.

μ_{pk} = means of variable p, group k

FIGURE 6.1 Null Hypothesis Testing of ANOVA and
MANOVA

regression and discriminant analysis. The unique aspect of MANOVA is that the
variate optimally combines the multiple dependent measures into a single value
that maximizes the differences across groups.

The Two-Group Case: Hotelling's T^2

In our earlier univariate example, researchers were interested in the appeal of two
advertising messages. But what if they also wanted to know about the purchase
intent generated by the two messages? If only univariate analyses were used, the
researchers would perform separate t tests on the ratings of both the appeal of
the messages and the purchase intent generated by the messages. Yet the two measures are interrelated; thus, what is really desired is a test of the differences between the messages on both variables collectively. This is where **Hotelling's T^2**,
a specialized form of MANOVA that is a direct extension of the univariate t test,
can be used.

Hotelling's T^2 provides a statistical test of the variate formed from the dependent variables that produces the greatest group difference. It also addresses the
problem of "inflating" the Type I error rate that arises when making a series of t
tests of group means on several dependent measures. It controls this inflation of
the Type I error rate by providing a single overall test of group differences across
all dependent variables at a specified α level.

How does Hotelling's T^2 achieve these goals? Consider the following equation
for a variate of the dependent variables:

$$C = W_1 Y_1 + W_2 Y_2 + \cdots + W_n Y_n$$

where

C = composite or variate score for a respondent

W_i = weight for dependent variable i

Y_i = dependent variable i

In our example, the ratings of message appeal are combined with the purchase intentions to form the composite. For any set of weights, we could compute composite scores for each respondent and then calculate an ordinary t statistic for the difference between groups on the composite scores. However, if we can find a set of weights that gives the maximum value for the t statistic for this set of data, these weights would be the same as the discriminant function between the two groups (as shown in chapter 5). The maximum t statistic that results from the composite scores produced by the discriminant function can be squared to produce the value of Hotelling's T^2 [10]. The computational formula for Hotelling's T^2 represents the results of mathematical derivations used to solve for a maximum t statistic (and, implicitly, the most discriminating linear combination of the dependent variables). This is equivalent to saying that if we can find a discriminant function for the two groups that produces a significant T^2, the two groups are considered different across the mean vectors.

How does Hotelling's T^2 provide a test of the hypothesis of no group difference on the vectors of mean scores? Just as the t statistic follows a known distribution under the null hypothesis of no treatment effect on a single dependent variable, Hotelling's T^2 follows a known distribution under the null hypothesis of no treatment effect on any of a set of dependent measures. This distribution turns out to be an F distribution with p and $N_1 + N_2 - 2 - 1$ degrees of freedom after adjustment (where p = the number of dependent variables). To get the critical value for Hotelling's T^2, we find the tabled value for F_{crit} at a specified α level and compute T^2_{crit} as follows:

$$T^2_{crit} = \frac{p(N_1 + N_2 - 2)}{N_1 + N_2 - p - 1} \times F_{crit}$$

The k-Group Case: MANOVA

MANOVA can be considered a simple extension of Hotelling's T^2 procedure; that is, we devise dependent variable weights to produce a variate score for each respondent, as described earlier. If we wanted to evaluate three advertising messages both for their appeal and for the purchase intentions they generate, we would use MANOVA. In MANOVA we now want to find the set of weights that maximizes the ANOVA F value computed on the variate scores for all the groups. But MANOVA can also be considered an extension of discriminant analysis (see chapter 5) in that multiple variates of the dependent measures can be formed if the number of groups is three or more. The first variate, termed a **discriminant function,** specifies a set of weights that maximize the differences between groups, thereby maximizing the F value. The maximum F value itself allows us to compute directly what is called the **greatest characteristic root (gcr)** statistic, which allows for the statistical test of the first discriminant function. The greatest characteristic root statistic can be calculated as [10]: $gcr = (k - 1) F_{max}/(N - k)$.

To obtain a single test of the hypothesis of no group differences on this first vector of mean scores, we could refer to tables of the gcr distribution. Just as the

F statistic follows a known distribution under the null hypothesis of equivalent group means on a single dependent variable, the *gcr* statistic follows a known distribution under the null hypothesis of equivalent group mean vectors (i.e., group means are equivalent on a set of dependent measures). A comparison of the observed *gcr* to *gcr*$_{crit}$ gives us a basis for rejecting the overall null hypothesis of equivalent group mean vectors.

Any subsequent discriminant functions are **orthogonal;** they maximize the differences among groups based on the remaining variance not explained by the prior function(s). Thus, in many instances, the test for differences between groups involves not just the first variate score but a set of variate scores that are evaluated simultaneously. A range of multivariate tests is available (e.g., Wilks' lambda, Pillai's criterion), each best suited to specific situations of testing these multiple variates.

Differences between MANOVA and Discriminant Analysis

In the previous section we discussed the basic elements of both the univariate and multivariate tests for assessing differences between groups on one or more dependent variables. In doing so, we noted the calculation of the discriminant function, which in the case of MANOVA is the variate of dependent variables that maximizes the difference between groups. The question may arise: What is the difference between MANOVA and discriminant analysis? In some aspects, MANOVA and discriminant analysis are "mirror images." The dependent variables in MANOVA (a set of metric variables) are the independent variables in discriminant analysis, and the single nonmetric dependent variable of discriminant analysis becomes the independent variable in MANOVA. Moreover, both use the same methods in forming the variates and assessing the statistical significance between groups.

The differences, however, center around the objectives of the analyses and the role of the nonmetric variable(s). Discriminant analysis employs a single nonmetric variable as the dependent variable. The categories of the dependent variable are assumed as given, and the independent variables are used to form variates that maximally differ between the groups formed by the dependent variable categories. In MANOVA, the set of metric variables now act as the dependent variables and the objective becomes finding groups of respondents that exhibit differences on the set of dependent variables. The groups of respondents are not prespecified; instead, the researcher uses one or more independent variables (nonmetric variables) to form groups. MANOVA, even while forming these groups, still retains the ability to assess the impact of each nonmetric variable separately.

A Hypothetical Illustration of MANOVA

A simple example can illustrate the benefits of using MANOVA. Assume that researchers identified two nonmetric variables (product type and customer status), which they thought caused differences in how people evaluated advertisements for HATCO products. Each variable has two categories: product type (product 1 versus product 2) and customer status (current customer versus excustomer). In combining these two variables, we get four distinct groups (product 1/excustomer, product 2/excustomer, product 1/customer, product 2/customer). Selected customers in each group can then be asked to evaluate HATCO advertisements with a 10-point scale as to their ability to gain attention and persuade them to purchase (see Table 6.1).

We can use MANOVA to combine the two dependent measures (recall and purchase) into a single variate. Assume for this example that the two dependent measures were equally weighted when summed into the variate value (total). This is identical to discriminant analysis and provides a single composite value with the variables weighted to achieve maximum differences among the groups. MANOVA differs from discriminant analysis in how the groups are formed and analyzed. In this case, discriminant analysis could be performed only on the set of four groups, without distinction as to a group's characteristics (product type or customer status). The researcher would be able to determine if the variate significantly differed only across the groups, but could not assess which characteristics of the groups related to these differences. With MANOVA, however, the researcher analyzes the differences in the groups while also assessing if the differences are due to product type, customer type, or both. Thus, MANOVA focuses the analysis on the composition of the groups based on their characteristics (the independent variables). This allows the researcher to propose any number of independent nonmetric variables (within limits) to form groups and then look for significant differences in the dependent variable variate associated with specific nonmetric variables.

In Table 6.1, the four groups have quite different means on the composite variable "total" (i.e., 4.25, 8.25, 11.75, and 14.0). Figure 6.2 (p. 338) portrays the four group means. The two lines connect the groups (excustomer and customer) for product 1 and product 2. Discriminant analysis would determine that there were significant differences on the composite variable and also that both dependent

TABLE 6.1 Hypothetical Example of MANOVA

		Product 1 $\bar{x}_{recall} = 3.50$ $\bar{x}_{purchase} = 4.50$ $\bar{x}_{total} = 8.00$				Product 2 $\bar{x}_{recall} = 5.50$ $\bar{x}_{purchase} = 5.625$ $\bar{x}_{total} = 11.125$		
Customer Type/Product Line	ID	Attention	Purchase	Total	ID	Attention	Purchase	Total
Excustomer	1	1	3	4	5	3	4	7
$\bar{x}_{recall} = 3.00$	2	2	1	4	6	4	3	7
$\bar{x}_{purchase} = 3.25$	3	2	3	5	7	4	5	9
$\bar{x}_{total} = 6.25$	4	3	2	5	8	5	5	10
Average		2.0	2.25	4.25		4.0	4.25	8.25
Customer	9	4	7	11	13	6	7	13
$\bar{x}_{recall} = 6.00$	10	5	6	11	14	7	8	15
$\bar{x}_{purchase} = 6.875$	11	5	7	12	15	7	7	14
$\bar{x}_{total} = 12.875$	12	6	7	13	16	8	6	14
Average		5.0	6.75	11.75		7.0	7.0	14.0

Values are responses on a 10 point scale (1 = Low, 10 = High)

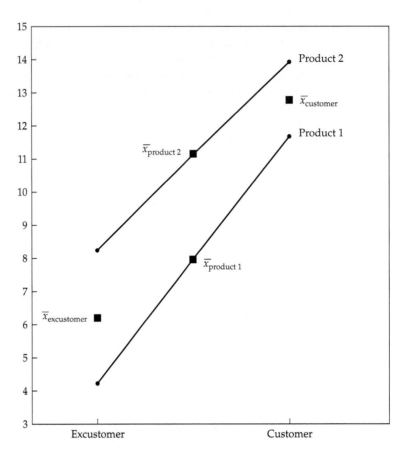

FIGURE 6.2 Graphical Display of Group Means of Variate (Total) for
Hypothetical Example

variables (recall and purchase) did contribute to the differences. MANOVA,
however, goes beyond analyzing only the differences across groups by assessing whether product type and/or customer type created groups with these differences. This is accomplished by evaluating the category means (denoted by
the symbol ■), which are also shown in Figure 6.2. If we look at product type
(ignoring distinctions as to customer type), we can see a mean value of 8.0 for
users of product 1 and a mean value of 11.125 for users of product 2. Likewise,
for customer type, excustomers had a mean value of 6.25 and customers a mean
value of 12.875. Visual examination would suggest that both of these category
means show significant differences, with the differences for customer type
(12.875 − 6.25 = 6.625) greater than that for product (11.125 − 8.00 = 3.125). By
being able to represent these independent variable category means in the analysis, the MANOVA analysis not only shows that overall differences do occur (as
was done with discriminant analysis), but also that both customer type and
product type contribute significantly to forming these differing groups.
Therefore, both characteristics "cause" significant differences, a finding not possible with discriminant analysis.

When Should We Use MANOVA?

With the ability to examine several dependent measures simultaneously, the researcher can gain in several ways from the use of MANOVA. We discuss the issues in using MANOVA from the perspectives of controlling statistical accuracy and efficiency while still providing the appropriate forum for testing multivariate questions.

Control of Experimentwide Error Rate

The use of separate univariate ANOVAs or *t* tests can create a problem when trying to control the overall, or experimentwide, error rate [11]. For example, assume that we evaluate a series of five dependent variables by separate ANOVAs, each time using .05 as the significance level. Given no real differences in the dependent variables, we would expect to observe a significant effect on any given dependent variable 5 percent of the time. However, across our five separate tests, the probability of a Type 1 error lies somewhere between 5 percent, if all dependent variables are perfectly correlated, and 23 percent $(1 - .95^5)$, if all dependent variables are uncorrelated. Thus a series of separate statistical tests leaves us without control of our effective overall or experimentwide Type I error rate. If the researcher desires to maintain control over the experimentwide error rate and there is at least some degree of intercorrelation among the dependent variables, then MANOVA is appropriate.

Differences among a Combination of Dependent Variables

A series of univariate ANOVA tests also ignores the possibility that some composite (linear combination) of the dependent variables may provide evidence of an overall group difference that may go undetected by examining each dependent variable separately. Individual tests ignore the correlations among the dependent variables and thus use less than the total information available for assessing overall group differences. In the presence of multicollinearity among the dependent variables, MANOVA will be more powerful than the separate univariate tests. In this manner, MANOVA may detect *combined* differences not found in the univariate tests. Moreover, if multiple variates are formed, then they may provide *dimensions* of differences that can distinguish among the groups better than single variables. However, in some instances of a large number of dependent variables, the statistical power of the ANOVA tests exceeds that obtained with a single MANOVA [4]. The considerations involving sample size, number of dependent variables, and statistical power are considered in a subsequent section.

A Decision Process for MANOVA

The process of performing a multivariate analysis of variance is similar to that found in many other multivariate techniques, so it can be described through the six-stage model-building process described in chapter 1. The process begins with the specification of research objectives. It then proceeds to a number of design issues facing a multivariate analysis and then an analysis of the assumptions underlying MANOVA. With these issues addressed, the process proceeds to estimation of the MANOVA model and the assessment of overall model fit. When an acceptable MANOVA model is found, then the results can be interpreted in more detail. The final step involves efforts to validate the results to ensure generalizability to the population. Figure 6.3 (stages 1–3) and Figure 6.5 (stages 4–6, p. 350) provide a graphical portrayal of the process, which is discussed in detail in the following sections.

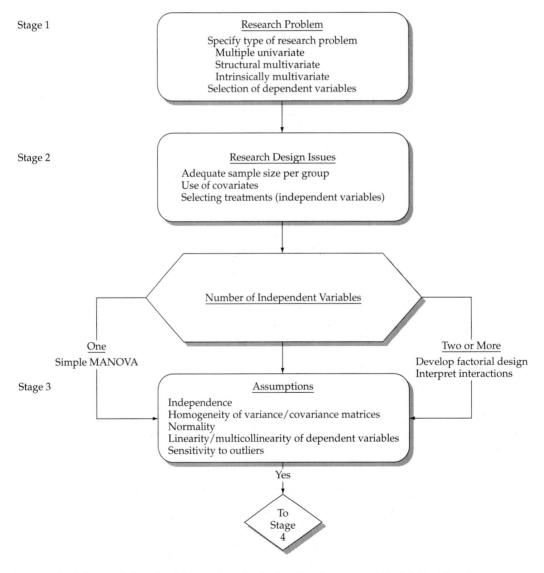

FIGURE 6.3 Stages 1–3 in the Multivariate Analysis of Variance (MANOVA) Decision Diagram

Stage 1: Objectives of MANOVA

The selection of MANOVA is based on the desire to analyze a dependence relationship represented as the differences in a set of dependent measures across a series of groups formed by one or more categorical independent measures. As such, MANOVA represents a powerful analytical tool suitable to a wide array of research questions. Whether used in actual or quasi-experimental situations (such as field settings or survey research for which the independent measures are categorical), MANOVA can provide insights into not only the nature and predictive power of the independent measures but also the interrelationships and differences seen in the set of dependent measures.

Types of Multivariate Questions Suitable for MANOVA

The advantages of MANOVA versus a series of univariate ANOVAs extend past the statistical domain discussed earlier and are also found in its ability to provide a single method of testing a wide range of differing multivariate questions. Throughout the text, we emphasize the interdependent nature of multivariate analysis. MANOVA has the flexibility to allow the researcher to select the test statistics most appropriate for the question of concern. Hand and Taylor [9] have classified multivariate problems into three categories, each of which employs different aspects of MANOVA in its resolution. These three categories are multiple univariate, structured multivariate, and intrinsically multivariate questions.

Multiple Univariate Questions

A researcher studying multiple univariate questions identifies a number of separate dependent variables (e.g., age, income, education of consumers) that are to be analyzed separately but needs some control over the experimentwide error rate. In this instance, MANOVA is used to assess whether an overall difference is found between groups, and then the separate univariate tests are employed to address the individual issues for each dependent variable.

Structured Multivariate Questions

A researcher dealing with structured multivariate questions gathers two or more dependent measures that have specific relationships among them. A common situation in this category is repeated measures, where multiple responses are gathered from each subject, perhaps over time or in a pretest–posttest exposure to some stimulus, such as an advertisement. Here MANOVA provides a structured method for specifying the comparisons of group differences on a set of dependent measures while maintaining statistical efficiency.

Intrinsically Multivariate Questions

An intrinsically multivariate question involves a set of dependent measures in which the principal concern is how they differ *as a whole* across the groups. Differences on individual dependent measures are of less interest than their collective effect. One example is the testing of multiple measures of response that should be consistent, such as attitudes, preference, and intention to purchase, all of which relate to differing advertising campaigns. The full power of MANOVA is utilized in this case by assessing not only the overall differences but also the differences among combinations of dependent measures that would not otherwise be apparent. This type of question is served well by MANOVA's ability to detect multivariate differences, even when no single univariate test shows differences.

Selecting the Dependent Measures

In identifying the questions appropriate for MANOVA, it is important also to discuss briefly the development of the research question, specifically the selection of the dependent measures. A common problem encountered with MANOVA is the tendency of researchers to misuse one of its strengths—the ability to handle multiple dependent measures—by including variables without a sound conceptual or theoretical basis. The problem occurs when the results indicate that a subset of the dependent variables has the ability to influence the overall differences among groups. If some of the dependent measures with the strong differences are not really appropriate for the research question, then "false" differences may lead the researcher to draw incorrect conclusions about the set *as a whole*. Thus, the researcher should always scrutinize the dependent measures and make sure there is a solid rationale for including them. Any ordering of the variables, such as possible sequential effects, should also be noted. MANOVA provides a special test, stepdown analysis, to assess the statistical differences in a sequential manner, much like the addition of variables to a regression analysis.

In summary, the researcher should assess all aspects of the research question carefully and ensure that MANOVA is applied in the correct and most powerful way. The following sections address many issues that have an impact on the validity and accuracy of MANOVA; however, it is ultimately the responsibility of the researcher to employ the technique properly.

Stage 2: Issues in the Research Design of MANOVA

Although MANOVA tests for assumptions in the same way as ANOVA and follows the same basic principles, several issues are unique in the application of MANOVA. These issues concern both the design and statistical testing of the MANOVA model.

Sample Size Requirements—Overall and by Group

As might be expected, MANOVA requires greater sample sizes than univariate ANOVAs, and the sample size must exceed specific thresholds in each cell (group) of the analysis a recommended minimum cell size is 20 observations, although larger cell sizes may be required for acceptable statistical power. At the minimum, the sample in each cell must be greater than the number of dependent variables included. Although this concern may seem minor, the inclusion of just a small number of dependent variables (from 5 to 10) in the analysis places a sometimes bothersome constraint on data collection. This is particularly a problem in field experimentation or survey research, where the researcher has less control over the achieved sample.

Factorial Designs—Two or More Treatments

Many times the researcher wishes to examine the effects of several independent variables or treatments rather than using only a single treatment in either the ANOVA or MANOVA tests. An analysis with more than two treatments is called a **factorial design.** In general, a design with n treatments is called an n-way factorial design.

Selecting Treatments

The most common use of factorial designs involves those research questions that relate two or more nonmetric independent variables to a set of dependent variables. In these instances, the independent variables have been specified in the design of the experiment or included in the design of the field experiment or survey questionnaire. But in some instances, treatments are added after the analysis is designed. The most common use of additional treatments is as a **blocking factor,** which is a nonmetric characteristic used post hoc to segment the respondents to obtain greater within-group homogeneity and reduce the MS_W source of variance. By doing so, the ability of the statistical tests to identify differences is enhanced. As an example, assume that in our earlier advertising example, we discovered that males reacted differently than females to the advertisements. If gender is then used as a blocking factor, the differences between the messages may become more apparent, whereas the differences were obscured when males and females were assumed to react similarly and were not separated. The effects of message type and gender are then evaluated separately, providing a more precise test of their individual effects.

A Hypothetical Example

As an example of a simple two-treatment factorial design, assume that a cereal manufacturer wishes to examine the impact of three different color possibilities (red, blue, and green) and three different shapes (stars, cubes, and balls) on the overall consumer evaluation of a new cereal. We could examine the impact of both of these independent variables simultaneously by employing a 3×3 factorial design. Respondents would be randomly assigned to evaluate one of the nine possible combinations of color and shape (using, for example, a 10-point overall evaluation scale). In analyzing this design, three different overall effects can be tested with ANOVA:

1. The **main effect** of color: Are there any differences between the mean ratings given to red (i.e., including all ratings of red stars, red cubes, and red balls), blue, and green?
2. The main effect of shape: Are there any differences between the mean ratings given to stars (i.e., including all ratings of red stars, blue stars, and green stars), cubes, and balls?
3. The **interaction effect** of color and shape: As to the overall difference between colors, is this difference the same when we examine it separately for stars, cubes, and balls? For example, if red was rated highest overall but received a very low rating when it was rated as a ball (relative to blue and green), this outcome would be evidence of an interaction effect; that is, the effect of color depends on what shape we are considering. We could pose this interaction question in an equivalent fashion by asking whether the effect of shape depends on what color we are considering.

In ANOVA factorial designs, each of these three effects would be tested with an F statistic. The MANOVA factorial design is a straightforward extension of ANOVA. For every F statistic in ANOVA that evaluates an effect on a single dependent variable, there is a corresponding multivariate statistic (e.g., *gcr* or Wilks' lambda) that evaluates the same effect on a set (vector) of dependent variable means.

Interpreting Interaction Terms

The interaction term represents the joint effect of two treatments and is the effect that must be examined first. If the interaction effect is not statistically significant, then the effects of the treatments are independent. Independence in factorial designs means that the effect of one treatment is the same for each level of the other treatment(s) and that the main effects can be interpreted directly. If the interaction term is significant, then the type of interaction must be determined. Interactions can be termed **ordinal** or **disordinal.** An ordinal interaction occurs when the effects of a treatment are not equal across all levels of another treatment, but the magnitude is always the same direction. In a disordinal interaction, the effects of one treatment are positive for some levels and negative for other levels of the other treatment.

The differences between interactions are best portrayed graphically. Figure 6.4 uses the example of cereal shapes and colors. The vertical axis represents the mean evaluations of each group of respondents across the combinations of levels. The x-axis represents the three categories for color (red, blue, and green). The lines connect the category means for each shape across the three colors. For example in the upper graph, the value for red balls is about 4.0, the value for blue balls is about 5.0, and the value increases slightly to about 5.5 for green balls.

Three forms of interaction are shown. In case a, there is no interaction. This is shown by the parallel lines representing the differences of the various shapes across the levels of color (the same effect would be seen if the differences in color were graphed across the three types of shape). In the case of no interaction, the effects of each treatment (the differences between groups) are constant at each level and the lines are roughly parallel. In case b, we see that the effects of each treatment are not constant and thus the lines are not parallel. The differences for red are large, but they decline slightly for blue cereal and even more for green cereal. Thus, the differences by color vary across the shapes. But the relative ordering among levels of shape are the same, with stars always highest, followed by the cubes and then the ball shapes. Finally, in case c, the differences in color vary not only in magnitude but also in direction. This is shown by lines that are not parallel and that cross between levels. For example, cubes have a higher evaluation than stars when the color is red, but the evaluation is lower for the colors of blue and green.

If the significant interactions are ordinal, the researcher must interpret the interaction term and ensure that its results are acceptable conceptually. If so, then the effects of each treatment can be described. But if the significant interaction is disordinal, then the main effects of the treatments cannot be interpreted and the study must be redesigned. This stems from the fact that with disordinal interactions, the effects vary not only across treatment levels but also in direction (positive or negative). Thus, the treatments do not represent a consistent effect.

MANOVA Counterparts of Other ANOVA Designs

Many types of ANOVA designs exist, and are discussed in standard experimental design texts [13, 16, 18]. For every ANOVA design, there is a multivariate counterpart; that is, any ANOVA on a single dependent variable can be extended to MANOVA designs. To illustrate this fact, we would have to discuss each ANOVA

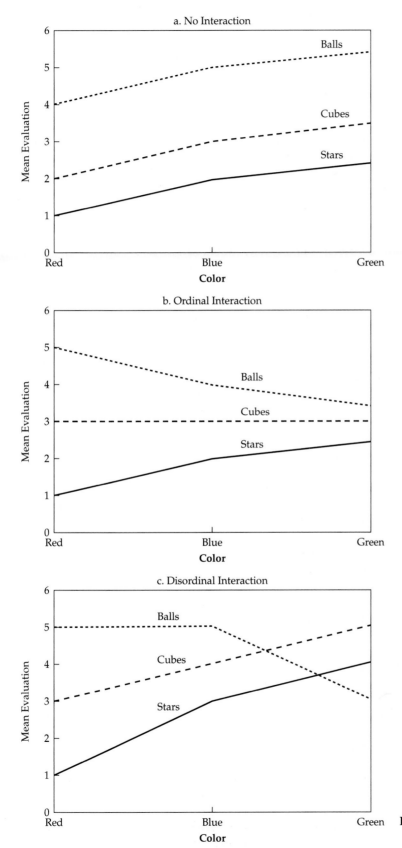

FIGURE 6.4 Interaction Effects in Factorial Designs

design in detail. Clearly, this is not possible in a single chapter because entire books are devoted to the subject of ANOVA designs. For more information, the reader is referred to more statistically oriented texts [1, 2, 5, 6, 7, 8, 10, 17, 21].

Using Covariates—ANCOVA and MANCOVA

In any univariate ANOVA design, metric independent variables, referred to as **covariates,** can be included. The design is then termed an analysis of covariance (ANCOVA) design. Metric covariates are typically included in an experimental design to remove extraneous influences from the dependent variable, thus increasing the within-group variance (MS_W). This is similar to the use of a blocking factor, only this time the variable is metric. Procedures similar to linear regression are employed to remove variation in the dependent variable associated with one or more covariates. Then a conventional ANOVA is carried out on the adjusted dependent variable. Multivariate analysis of covariance (MANCOVA) is a simple extension of the principles of ANCOVA to multivariate (multiple dependent variables) analysis; that is, MANCOVA can be viewed as MANOVA of the regression residuals, or variance in the dependent variables not explained by the covariates.

Objectives of Covariance Analysis

A **covariate analysis** is appropriate to achieve two specific purposes: (1) to eliminate some systematic error outside the control of the researcher that can bias the results, and (2) to account for differences in the responses due to unique characteristics of the respondents. A systematic bias can be eliminated by the random assignment of respondents to various treatments. However, in nonexperimental research, such controls are not possible. For example, in testing advertising, effects may differ, depending on the time of day or the composition of the audience and their reactions. The purpose of the covariate is to eliminate any effects that (1) affect only a portion of the respondents, or (2) vary among the respondents. For instance, personal differences, such as attitudes or opinions, may affect responses, but the experiment does not include them as a treatment factor. The researcher uses a covariate to take out any differences due to these factors before the effects of the experiment are calculated. This is the second role of covariance analysis.

Selecting Covariates

An effective covariate in ANCOVA is one that is highly correlated with the dependent variable but not correlated with the independent variables. Let us examine why. Variance in the dependent variable forms the basis of our error term in ANOVA. If our covariate is correlated with the dependent variable and not the independent variable(s), we can explain some of the variance with the covariate (through linear regression). We are left with only a smaller residual variance in the dependent variable that would not have been explained by the independent variable anyway (because the covariate is not correlated with the independent variable). This residual variance provides a smaller error term (MS_W) for the F statistic and thus a more efficient test of treatment effects. However, if the covariate is correlated with the independent variable(s), then the covariate will "explain" some of variance that could have been "explained" by the independent variable and reduce its effects. Because the covariate is extracted first, any variation associated with the covariate is not available for the independent variables.

A common question involves how many covariates to add to the analysis. Although the researcher wants to account for as many extraneous effects as possible, too large a number will reduce the statistical efficiency of the procedures. A rule of thumb [12] is that the number of covariates should be less than (.10 × sample size) − (number of groups − 1). For example, for a sample size of 100 respondents and 5 groups, the number of covariates should be less than 6 [.10 × 100 − (5 − 1)]. The researcher should always attempt to minimize the number of covariates, while still ensuring that effective covariates are not eliminated, because in many cases, particularly with small sample sizes, they can markedly improve the sensitivity of the statistical tests.

There are two requirements for use of an analysis of covariance: (1) the covariates must have some relationship with the dependent measures, and (2) the covariates must have a homogeneity of regression effect, meaning that the covariate(s) have equal effects on the dependent variable across the groups. In regression terms, this implies equal coefficients for all groups. Statistical tests are available to assess whether this assumption holds true for each covariate used. If either of these requirements is not met, then the use of covariates is inappropriate.

A Special Case of MANOVA: Repeated Measures

We have discussed a number of situations in which we wish to examine differences on several dependent measures. A special situation of this type occurs when the same respondent provides several measures, such as test scores over time, and we wish to examine them to see whether any trend emerges. Without special treatment, however, we would be violating the most important assumption, independence. There are special MANOVA models, termed **repeated measures,** that can account for this dependence and still ascertain whether any differences occurred across individuals for the set of dependent variables. The within-person perspective is important so that each person is placed on "equal footing." For example, assume we were assessing improvement on test scores over the semester. We must account for the earlier test scores and how they relate to later scores, as we might expect to see different trends for those with low versus high initial scores. Thus we must "match" each respondent's scores when performing the analysis. We do not address the details of repeated measures models in this text because it is a specialized form of MANOVA. The interested reader is referred to any number of excellent treatments on the subject [1, 2, 5, 6, 7, 8, 10, 17, 21].

Stage 3: Assumptions of ANOVA and MANOVA

The univariate test procedures of ANOVA described in this chapter are valid (in a formal sense) only if it is assumed that the dependent variable is normally distributed and that variances are equal for all treatment groups. There is evidence [16, 23], however, that F tests in ANOVA are robust with regard to these assumptions except in extreme cases. For the multivariate test procedures of MANOVA to be valid, three assumptions must be met: (1) the observations must be independent, (2) the variance–covariance matrices must be equal for all treatment groups, and (3) the set of p-dependent variables must follow a multivariate normal distribution (i.e., any linear combination of the dependent variables must follow a normal distribution) [10]. In addition to the strict statistical assumptions, the researcher must also consider several issues that influence the possible effects—namely, the linearity and multicollinearity of the variate of dependent variables.

Independence

The most basic, yet most serious, violation of an assumption occurs when there is a lack of **independence** among observations. There are a number of experimental as well as nonexperimental situations in which this assumption can easily be violated. For example, a time-ordered effect (serial correlation) may occur if measures are taken over time, even from different respondents. Another common problem is gathering information in group settings, so that a common experience (such as a noisy room or confusing set of instructions) would cause a subset of individuals (those with the common experience) to have answers that are somewhat correlated. Finally, extraneous and unmeasured effects can affect the results by creating dependence among the respondents. Although there are no tests with an absolute certainty of detecting all forms of dependence, the researcher should explore all possible effects and correct for them if found. If dependence is found among groups of respondents, then a possible solution is to combine those within the groups and analyze the group's average score instead of the scores of the separate respondents. Another approach is to employ a blocking factor or some form of covariate analysis to account for the dependence. In either case, or when dependence is suspected, the researcher should use a lower level of significance (.01 or even lower).

Equality of Variance–Covariance Matrices

The second assumption of MANOVA is the equivalence of covariance matrices across the groups. Here, as with the problem of heteroscedasticity addressed in multiple regression, we are concerned with substantial differences in the amount of variance of one group versus another for the same variables. In MANOVA, however, the interest is in the variance–covariance matrices of the dependent measures for each group. The requirement of equivalence is a strict test because instead of equal variances for a single variable in ANOVA, the MANOVA test examines all elements of the covariance matrix of the dependent variables. For example, for five dependent variables, the five correlations and ten covariances are all tested for equality across the groups. Fortunately, a violation of this assumption has minimal impact if the groups are of approximately equal size (i.e., if the largest group size divided by the smallest group size is less than 1.5). If the sizes differ more than this, then the researcher should test and correct for unequal variances, if possible. MANOVA programs provide the test for equality of covariance matrices—typically the **Box test**—and provide significance levels for the test statistic. If the researcher encounters a significant difference that requires a remedy, one of the many variance-stabilizing transformations available may work (see chapter 2 for a discussion of these approaches). The Box test is very sensitive to departures from normality [10, 19]. Thus one should always check for univariate normality of all dependent measures before performing this test.

If the unequal variances persist after transformation and the group sizes differ markedly, the researcher should make adjustments for their effects. First, one has to ascertain which group has the largest variance. This determination is easily made either by examining the variance–covariance matrix or by using the determinant of the variance–covariance matrix, which is provided by all statistical programs. If the larger variances are found with the larger group sizes, the alpha level is overstated, meaning that differences should actually be assessed using a somewhat lower value (e.g., use .03 instead of .05). If the larger variance is found in the smaller group sizes, then the reverse is true. The power of the test has been reduced, and the researcher should increase the significance level.

Normality

The last assumption for MANOVA to be valid concerns normality of the dependent measures. In the strictest sense, the assumption is that all the variables are **multivariate normal.** Multivariate normality assumes that the joint effect of two variables is normally distributed. Even though this assumption underlies most multivariate techniques, there is no direct test for multivariate normality. Therefore, most researchers test for univariate normality of each variable. Although univariate normality does not guarantee multivariate normality, if all variables meet this requirement, then any departures from multivariate normality are usually inconsequential. Violations of this assumption have little impact with larger sample sizes, just as is found with ANOVA. Violating this assumption primarily creates problems in applying the Box test, but transformations can correct these problems in most situations. (For a discussion of transforming variables, refer to chapter 2.) With moderate sample sizes, modest violations can be accommodated as long as the differences are due to skewness and not outliers.

Linearity and Multicollinearity among the Dependent Variables

Although MANOVA assesses the differences across combinations of dependent measures, it can construct a linear relationship only between the dependent measures (and any covariates, if included). The researcher is again encouraged first to examine the data, this time assessing the presence of any nonlinear relationships. If these exist, then the decision can be made whether they need to be incorporated into the dependent variable set, at the expense of increased complexity but greater representativeness. Chapter 2 addresses such tests.

In addition to the linearity requirement, the dependent variables should not have high multicollinearity (discussed in chapter 4) because this indicates only redundant dependent measures and decreases statistical efficiency. We discuss the impact of multicollinearity on the statistical power of the MANOVA in the next section.

Sensitivity to Outliers

In addition to the impact of heteroscedasticity discussed earlier, MANOVA (and ANOVA) are especially sensitive to outliers and their impact on the Type I error. The researcher is strongly encouraged first to examine the data for outliers and eliminate them from the analysis, if at all possible, because their impact will be disproportionate in the overall results.

Stage 4: Estimation of the MANOVA Model and Assessing Overall Fit

Once the MANOVA analysis has been formulated and the assumptions tested for compliance, the assessment of significant differences among the groups formed by the treatment(s) can proceed (see Figure 6.5). In making this assessment, the researcher must select the test statistics most appropriate for the study objectives. Moreover, in any situation, but especially as the analysis becomes more complex,

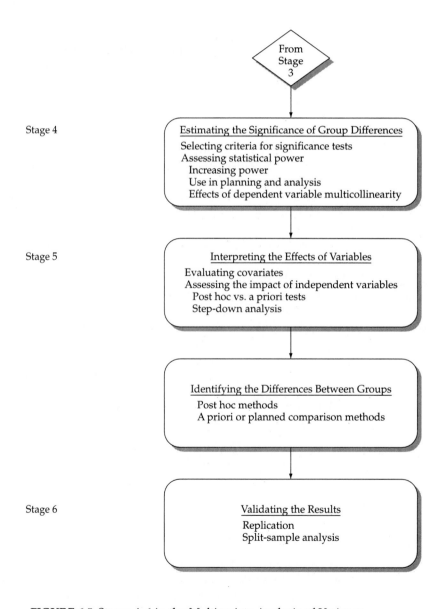

FIGURE 6.5 Stages 4–6 in the Multivariate Analysis of Variance (MANOVA) Decision Diagram

the researcher must evaluate the power of the statistical tests to provide the most informed perspective on the results obtained.

Criteria for Significance Testing

MANOVA presents the researcher with several criteria with which to assess multivariate differences across groups. The four most popular are Roy's gcr; Wilks' lambda (also known as the **U statistic**); Hotelling's trace; and Pillai's criterion. As stated in the discussions of discriminant analysis in chapter 5, these criteria assess the differences across "dimensions" of the dependent variables. Roy's greatest characteristic root, as the name implies, measures the differences on only the first canonical root (or discriminant function) among the dependent variables. This criterion provides some advantages in power and specificity of the test but makes it less useful in certain situations where all dimensions should be considered. Roy's gcr test is most appropriate when the dependent variables are strongly interrelated on a single dimension, but it is also the measure most likely to be severely affected by violations of the assumptions.

The other three measures assess all sources of difference among the groups. Readers who have some familiarity with MANOVA from other texts or statistical programs probably have encountered a more commonly used test statistic for overall significance in MANOVA called **Wilks' lambda.** We have referred to the greatest characteristic root and the first discriminant function, and these terms imply that there may be additional characteristic roots and discriminant functions. Actually, there are p or $(k - 1)$ (whichever is the smaller) characteristic roots or discriminant functions, where p is the number of dependent variables and k is the number of groups. Unlike the gcr statistic, which is based on the first (greatest) characteristic root, Wilks' lambda considers all the characteristic roots; that is, it examines whether groups are somehow different without being concerned with whether they differ on at least one linear combination of the dependent variables. As it turns out, Wilks' lambda is much easier to calculate than the gcr statistic. Its formulation is $|W|/|W + A|$, where $|W|$ is the determinant (a single number) of the within-groups multivariate dispersion matrix, and $|W + A|$ is the determinant of the sum of W and A, A being the between-groups multivariate dispersion matrix. The larger the between-groups dispersion, the smaller the value of Wilks' lambda and the greater the implied significance. Although the distribution of Wilks' lambda is complex, good approximations for significance testing are available by transforming it into an F statistic [18].

Which statistic is preferred? Researchers can choose from these two statistics plus a number of other measures. Other widely used measures include Pillai's criterion and Hotelling's trace, both of which are similar to Wilks' lambda because they consider all the characteristic roots and can be approximated by an F statistic. The measure to use is the one most immune to violations of the assumptions underlying MANOVA that yet maintains the greatest power. There is agreement that either Pillai's criterion or Wilks' lambda best meets these needs, although evidence suggests Pillai's criterion is more robust and should be used if sample size decreases, unequal cell sizes appear, or homogeneity of covariances is violated. However, if the researcher is confident that all assumptions are strictly met and the dependent measures are representative of a single dimension of effects, then Roy's gcr is the most powerful test statistic. The major statistical packages provide all these measures, and a comparison among them can be made.

Statistical Power of the Multivariate Tests

In simple terms, **power** is the probability that the statistical test will identify a treatment's effect if it actually exists. Power can be defined as one minus the probability of a **Type II error** (β). As such, power is also related to the significance, or alpha (α), level, which defines the acceptable Type I error. For a more detailed discussion of the relationships between Type I and Type II errors and power, see chapter 1.

The level of power for any of the four statistical criteria—Roy's gcr, Wilks' lambda, Hotelling's trace, or Pillai's criterion—is based on three considerations: the alpha level, the effect size of the treatment, and the sample size of the groups. Power is inversely related to the alpha level selected. As alpha increases (becomes more conservative, such as moving from .05 to .01), power decreases. Therefore, if the researcher reduces the alpha level to reduce the Type I error, such as in the case of a possible dependence among observations or an adjustment for multiple comparisons, power decreases. The researcher must always be aware of the implications of adjusting the alpha level, because the overriding objective of the analysis is not only avoiding Type I errors but also identifying the treatment effects if they do indeed exist. If the alpha level is set too stringently, then the power may be too low to identify valid results. The researcher should consider not only the alpha level but also the resulting power, and try to maintain an acceptable alpha level with power in the range of .80.

Increasing Power in ANOVA and MANOVA

How does the researcher increase power once an alpha level has been specified? The primary "tool" at the researcher's disposal is the sample size of the groups. But before we assess the role of sample size, we need to understand the impact of **effect size,** which is a standardized measure of group differences, typically expressed as the differences in group means divided by their standard deviation. The magnitude of the effect size has a direct impact on the power of the statistical test. For any given sample size, the power of the statistical test will be higher the larger the effect size. Conversely, if a treatment has a small expected effect size, it is going to take a much larger sample size to achieve the same power as a treatment with a large effect size.

With the alpha level specified and the effect size identified, the final element affecting power is the sample size. In many instances, this is the element most under the control of the researcher. As discussed before, increased sample size generally reduces sampling error and increases the sensitivity (power) of the test. In analyses with group sizes of fewer than 50 members, obtaining desired power levels can be quite problematic. Increasing sample sizes in each group has marked effects until group sizes of approximately 150 are reached, and then the increase in power slows markedly. One note of caution must be made that large sample sizes reduce the sampling error component to such a small level that any small difference is regarded as statistically significant. When the sample sizes do become large and statistical significance is indicated, the researcher must examine the power and effect sizes to ensure not only statistical significance but practical significance as well.

Using Power in Planning and Analysis

The estimation of power should be used both in planning the analysis and in assessing the results. In the planning stage, the researcher determines the sample size needed to identify the estimated effect size. In many instances, the effect size can be estimated from prior research or reasoned judgments, or even set at a minimum level of practical significance. In each case, the sample size needed to achieve a given level of power with a specified alpha level can be determined.

By assessing the power of the test criteria after the analysis has been completed, the researcher provides a context for interpreting the results, especially if significant differences were not found. The researcher must first determine whether the achieved power was sufficient (.80 or above). If not, can the analysis be reformulated to provide more power? A possibility includes some form of blocking treatment or covariate analysis that will make the test more efficient by accentuating the effect size. If the power was adequate and statistical significance was not found for a treatment effect, then most likely the effect size for the treatment was too small to be of statistical or practical significance.

Calculating Power Levels

To calculate power for ANOVA analyses, published sources [3, 20] as well as computer programs are now available. The methods of computing the power of MANOVA, however, are much more limited. Fortunately, most computer programs provide an assessment of power for the significance tests and allow the researcher to determine whether power should play a role in the interpretation of the results. In terms of published material for planning purposes, little exists for MANOVA because many elements affect the power of a MANOVA analysis. One source [15] of published tables presents power in a number of common situations for which MANOVA is applied. Table 6.2 provides an overview of the sample sizes needed for various levels of analysis complexity. A review of the table leads to several general points. First, increasing the number of dependent variables requires increased sample sizes to maintain a given level of power. The additional sample size needed is more pronounced for the smaller effect sizes. Second, if the effect sizes are expected to be small, the researcher must be prepared to engage in a substantial research effort to achieve acceptable levels of power. For example, to achieve the suggested power of .80 when assessing small effect sizes in a four-group design, 115 subjects per group are required if two dependent measures are used. The required sample size increases to 185 per group if eight dependent

TABLE 6.2 Sample Size Requirements per Group for Achieving Statistical Power of .80 in MANOVA

	Number of Groups											
	3				4				5			
	Number of Dependent Variables				Number of Dependent Variables				Number of Dependent Variables			
Effect Size	2	4	6	8	2	4	6	8	2	4	6	8
Very large	13	16	18	21	14	18	21	23	16	21	24	27
Large	26	33	38	42	29	37	44	48	34	44	52	58
Medium	44	56	66	72	50	64	74	84	60	76	90	100
Small	98	125	145	160	115	145	165	185	135	170	200	230

Source: Läuter, J. (1978), "Sample Size Requirements for the T^2 Test of MANOVA (Tables for One-Way Classification)." *Biometrical Journal* 20: 389–406.

variables are considered. The benefits of parsimony in the dependent variable set occur not only in interpretation but in the statistical tests for group differences as well.

The Effects of Dependent Variable Multicollinearity on Power

Up to this point we have discussed power from a perspective applicable to both ANOVA and MANOVA. In MANOVA, however, the researcher must also consider the effects of multicollinearity of the dependent variables on the power of the statistical tests. The researcher, whether in the planning or analysis stage, must consider the strength and direction of the correlations as well as the effect sizes of the dependent variables. If we classify variables by their effect sizes as strong or weak, then several patterns emerge [4]. First, if the correlated variable pair is made up of either strong–strong or weak–weak variables, then the greatest power is achieved when the correlation between variables is highly negative. This suggests that MANOVA is optimized by adding dependent variables that have high negative correlations. For example, rather than including two redundant measures of satisfaction, the researcher might replace them with correlated measures of satisfaction and dissatisfaction to increase power. When the correlated variable pair is a mixture (strong–weak), then power is maximized when the correlation is high, either positive or negative. One exception to this is the finding that multiple items to increase reliability results in a net gain of power, even if the items are redundant and positively correlated.

Stage 5: Interpretation of the MANOVA Results

Once the statistical significance of the treatments has been assessed, the researcher may wish to examine the results through some combination of three methods: (1) interpreting the effects of covariates, if employed; (2) assessing which dependent variable(s) exhibited differences across the groups; or (3) identifying which groups differ on a single dependent variable or the entire dependent variate. We first examine the methods by which the significant covariates and dependent variables are identified, and then we address the methods by which differences among individual groups can be measured.

Evaluating Covariates

Having met the assumptions for applying covariates, the researcher may wish to interpret the actual effect of the covariates on the dependent variate and their impact on the actual statistical tests of the treatments. Because ANCOVA and MANCOVA are applications of regression procedures within the analysis of variance method, assessing the impact of the covariates on the dependent variables is quite similar to examining regression equations. For each covariate, a regression equation that details the strength of the predictive relationship is formed. If the covariates represent theoretically based effects, then these results provide an objective basis for accepting or rejecting the proposed relationships. In a practical vein, the researcher can examine the impact of the covariates and eliminate those with little or no effect.

The researcher should also examine the overall impact of adding the covariate(s) in the statistical tests for the treatments. The most direct approach is to run the analysis with and without the covariates. Effective covariates will improve the statistical power of the tests and reduce within-group variance. If the researcher does not see any substantial improvement, then the covariates may be eliminated, because they reduce the degrees of freedom available for the tests of treatment effects. This approach also can identify those instances in which the covariate is "too powerful" and reduces the variance to such an extent that the treatments are all nonsignificant. Often this occurs when a covariate is included that is correlated with one of the independent variables and thus "removes" this variance, thereby reducing the explanatory power of the independent variable.

Assessing the Dependent Variate

The next step is an analysis of the dependent variate to assess which of the dependent variables contribute to the overall differences indicated by the statistical tests. This step is essential because there may be a subset of variables in the set of dependent variables that accentuates the differences, whereas another subset of variables may be nonsignificant or may mask the significant effects of the remainder. The procedures described in this section are termed **post hoc tests**—tests of the dependent variables that are chosen *after* examining the pattern of the data. Another approach is the use of **a priori tests**—tests planned *prior* to looking at the data from a theoretical or practical decision-making viewpoint. From a pragmatic standpoint, situations arise wherein a key dependent variable must be isolated and tested with maximum power. We recommend that an a priori test be performed in such situations. The most common approach is to perform univariate tests for the selected variables. For example, in a two-group case, an ordinary t test is an a priori test for a given dependent variable. However, researchers should be aware that as the number of these a priori tests increases, one of the major benefits of the multivariate approach to significance testing—control of the Type I error rate—is negated unless specific adjustments are made that control for the inflation of the Type I error.

For a two-group MANOVA, this adjustment involves the T^2 statistic. Given that the T^2 statistic exceeds T^2_{crit} for a specified α level, we conclude that the vectors of the mean scores are different. The discriminant function tells us what linear combination of the dependent variables produces the most reliable group difference, but other group comparisons may also be of interest. If we wish to test the group differences individually for each of the dependent variables, we could compute a standard t statistic and compare it to the square root of T^2_{crit} (i.e., T_{crit}) to judge its significance. This procedure would ensure that the probability of any Type I error across all the tests would be held to α (where α is specified in the calculation of T^2_{crit}) [10]. We could make similar tests for k-group situations by adjusting the α level by the **Bonferroni inequality,** which states that the alpha level should be adjusted for the number of tests being made. The adjusted alpha level used in any separate test is defined as the overall alpha level divided by the number of tests (adjusted alpha = [(overall alpha)/(number of tests)]).

A procedure known as **stepdown analysis** [14,19] may also be used to assess individually the differences of the dependent variables. This procedure involves computing a univariate F statistic for a dependent variable after eliminating the effects of other dependent variables preceding it in the analysis. The procedure is somewhat similar to stepwise regression, but here we examine whether a particular dependent variable contributes unique (uncorrelated) information on group differences. The stepdown results would be exactly the same as performing a covariate analysis, with the other preceding dependent variables used as the covariates. A critical assumption of stepdown analysis is that the researcher must know the order in which the dependent variables should be entered, because the interpretations can vary dramatically given different entry orders. If the ordering has theoretical support, then the stepdown test is valid. Variables indicated to be nonsignificant are "redundant" with the earlier significant variables, as they add no further information concerning differences about the groups. The order of dependent variables may be changed to test whether the effects of variables are either redundant or unique, but the process becomes rather complicated as the number of dependent variables increases.

Other procedures involve further analysis of the discriminant functions, in particular the first discriminant function, to gain additional information about which variables best differentiate between the groups. All these analyses are directed toward assisting the researcher in understanding which of the dependent variables contribute to the differences in the dependent variate across the treatment(s).

Identifying Differences between Individual Groups

Although the univariate and multivariate tests of ANOVA and MANOVA allow us to reject the null hypothesis that the groups' means are all equal, they do not pinpoint where the significant differences lie if there are more than two groups. Multiple t tests are not appropriate for testing the significance of differences between the means of paired groups because the probability of a Type I error increases with the number of intergroup comparisons made (similar to the problem of using multiple univariate ANOVAs versus MANOVA). Many procedures are available for further investigation of specific group mean differences of interest, all of which can be classified as either a priori or post hoc. These procedures use different approaches to control Type I error rates across multiple tests.

Post Hoc Methods

Among the more common post hoc procedures are (1) the Scheffé method, (2) Tukey's honestly significant difference (HSD) method, (3) Tukey's extension of the Fisher least significant difference (LSD) approach, (4) Duncan's multiple-range test, and (5) the Newman-Kuels test. Each method identifies which comparisons among groups (e.g., group 1 versus groups 2 and 3) have significant differences. These methods provide the researcher with tests of each combination of groups, thus simplifying the interpretative process.

While they simplify the identification of group differences, these methods all share the problem of having quite low levels of power. Because the post hoc tests must examine all possible combinations, the power of any individual test is rather low. These five post hoc or multiple-comparison tests of significance have been contrasted for power [19]. The conclusions are that the Scheffé method is the most conservative with respect to Type I error, and the remaining tests are ranked in this order: Tukey HSD, Tukey LSD, Newman-Kuels, and Duncan. If the effect sizes are large or the number of groups is small, the post hoc methods may identify the group differences. But the researcher must also recognize the limitations of these methods and employ other methods if more specific comparisons can be identified. A discussion of the options available with each method is beyond the scope of this chapter. Excellent discussions and explanations of these procedures can be found in other texts [12, 23].

A Priori or Planned Comparisons

The researcher can also make specific comparisons between groups by using a priori or **planned comparisons.** This method is similar to the post hoc tests but differs in that the researcher specifies which group comparisons are to be made in the planned comparisons versus testing the entire set, as done in the post hoc tests. Planned comparisons are more powerful because the number of comparisons are fewer, but more power is of little use if the researcher does not specifically test for "correct" group comparisons. Planned comparisons are most appropriate when conceptual bases can support the specific comparisons to be made. They should not be used in an exploratory manner, however, because they do not have effective controls against inflating the overall Type I error levels.

The researcher specifies the groups to be compared through a **contrast,** which is just a combination of group means that represent a specific planned comparison. Contrasts can be stated generally as

$$C = W_1G_1 + W_2G_2 + \cdots + W_kG_k$$

where

$$C = \text{contrast value}$$
$$W = \text{weights}$$
$$G = \text{group means}$$

The contrast is formulated by assigning positive and negative weights to specify the groups to be compared while ensuring that the weights sum to zero. For example, assume we have three group means. To test for a difference between G_1 and G_2, $C = (1)G_1 + (-1)G_2 + (0)G_3$. To test whether the average of G_1 and G_2 differs from G_3, the contrast is specified as $C = (.5)G_1 + (.5)G_2 + (-1)G_3$. A separate F statistic is computed for each contrast. In this manner, the researcher can create any comparisons desired and test them directly, but the probability of a Type I error for each a priori comparison is equal to α. Thus, several planned comparisons will inflate the overall Type I error level. All the statistical packages can perform either a priori or post hoc tests for single dependent variables.

If the researcher wishes to perform comparisons of the entire dependent variate, extensions of these methods are available. After concluding that the group mean vectors are not equivalent, the researcher might be interested in whether there are any group differences on the composite dependent variate. A standard ANOVA F statistic can be calculated and compared to $F_{\text{crit}} = (N - k)gcr_{\text{crit}}/(k - 1)$, where the value of gcr_{crit} is taken from the gcr distribution with appropriate degrees of freedom.

Stage 6: Validation of the Results

Analysis of variance techniques (ANOVA and MANOVA) were developed in the tradition of experimentation, with **replication** as the primary means of validation. The specificity of experimental treatments allows for a widespread use of the same experiment in multiple populations to assess the generalizability of the results. This is a principal tenet of the scientific method. In social science and business research, however, true experimentation is many times replaced with statistical tests in nonexperimental situations such as survey research. The ability to validate the results in these situations is based on the replicability of the treatments. In many instances, demographic characteristics such as age, gender, income, and the like are used as treatments. These treatments may seem to meet the requirement of comparability, but the researcher must ensure that the additional element of randomized assignment to a cell is also met; however, many times in survey research this is not true. For example, having age and gender be the independent variables is a common example of the use of ANOVA or MANOVA in survey research. But in terms of validation, the researcher must be wary of analyzing multiple populations and comparing results as the sole proof of validity. Because respondents in a simple sense "select themselves," the treatments in this case cannot be assigned by the researcher, and thus randomized assignment is impossible. So the researcher should strongly consider the use of covariates to control for other features that might be characteristic of the age or gender groups that could affect the dependent variables but are not included in the analysis.

Another issue is the claim of causation when experimental methods or techniques are employed. The principles of causation are examined in more detail in chapter 11. For our purposes here, the researcher must remember that in all research settings, including experiments, certain conceptual criteria (e.g., temporal ordering of effects and outcomes) must be established before causation may be supported. The single application of a particular technique used in an experimental setting does not ensure causation.

Summary

We have discussed the appropriate applications and important considerations of MANOVA in addressing multivariate analyses with multiple dependent measures. Although there are considerable benefits from its use, MANOVA must be carefully and appropriately applied to the question at hand. When doing so, researchers have at their disposal a technique with flexibility and statistical power. We now illustrate the applications of MANOVA (and its univariate counterpart ANOVA) in a series of examples.

Example 1: Difference between Two Independent Groups

To introduce the practical benefits of a multivariate analysis of group differences, we begin our discussion with one of the best-known experimental designs: the two-group randomized design, in which each respondent is randomly assigned to only one of the two levels (groups) of the treatment (independent variable). In the univariate case, a single metric dependent variable is measured, and the null hypothesis is that the two groups have equal means. In the multivariate case, multiple metric dependent variables are measured, and the null hypothesis is that the two groups have equal vectors of means. For a two-group univariate analysis, the appropriate test statistic is the t statistic (a special case of ANOVA); for a multivariate analysis, the appropriate test statistic is Hotelling's T^2. Because both analyses employ the same context, they do not differ in the appropriate approaches for validation, which essentially involve replication in other samples or through a split sample. Therefore, validation issues (stage 6) are not discussed for each of the separate analyses.

A Univariate Approach: The t Test

Stage 1: Objectives of the Analysis
From the HATCO survey of 100 customers, managers were interested in determining whether HATCO performed better when specification buying or total-value analysis (X_{11}) was used by the customers (chapter 1 contains a complete description of the variables and the entire study). As a measure of performance, HATCO decided to use X_9, usage level, or the percentage of the customer's business that comes from HATCO. The higher the value, the more HATCO supplies, on a percentage basis, to that firm. The null hypothesis that HATCO wishes to test is that they serve both types of firms equally well; that is, the type of buying methodology has no impact on usage level ($H_0: \mu_1 = \mu_2$). The alternative hypotheses is that the buying methodology does define groups who buy from HATCO in significantly different proportions ($H_A: \mu_1 < \mu_2$ or $H_A: \mu_1 > \mu_2$).

Stage 2: Research Design of the ANOVA
The principal consideration in the design of the two-group ANOVA is the sample size in each of the cells. As is the case in most survey research, the cells sizes are unequal. In this survey, 40 firms indicated that they used specification buying, and 60 firms indicated that they used total-value analysis. Unequal cell sizes make the statistical tests more sensitive to violations of the assumptions, especially the test for homogeneity of variance of the dependent variable. HATCO researchers did not identify any variables appropriate for inclusion as covariates. Finally, additional independent variables that would create a factorial design were deemed unsuitable at this time.

Stage 3: Assumptions in ANOVA
The independence of the respondents was ensured as much as possible by the random sampling plan. The assumption of normality and the presence of outliers for the dependent variable, X_9, were examined in chapter 2 and found to be acceptable. The assumption particularly important to ANOVA is the homogeneity of the variance of the dependent variable between groups. Several tests are available for testing this assumption (Levene, Cochran's C, and Bartlett–Box). The Levene statistic indicates no difference (significance = .2434), as do the Cochran's C (significance = .396) and the Bartlett–Box (significance = .411) tests. Thus, the unequal cell sizes should not impact the sensitivity of the statistical tests of group differences.

Stage 4: Estimation of the ANOVA Model and Assessing Overall Fit
Usage, measured in percentage terms, is shown for all firms in Table 6.3 (p. 360). The boxplots in Figure 6.6 (p. 361) portray the usage levels for respondents in two groups—customers using specification buying and customers using total value analysis. As can be seen, the customers using specification buying conducted an average of 42.1 percent of their business with HATCO, whereas those using total

TABLE 6.3 Firm Usage Percentages by Buying Method (Basic Data for Univariate *t* Test or Two-Group ANOVA)

Buying Method

Group 1: Specification Buying		Group 2: Total-Value Analysis			
Observation	X_9 (Usage Level)	Observation	X_9 (Usage Level)	Observation	X_9 (Usage Level)
2	43.0	1	32.0	51	41.0
3	48.0	5	58.0	55	39.0
4	32.0	7	46.0	56	47.0
6	45.0	9	63.0	58	65.0
8	44.0	11	32.0	59	46.0
10	54.0	12	47.0	61	54.0
13	39.0	14	38.0	62	60.0
24	36.0	15	54.0	63	47.0
27	36.0	16	49.0	64	36.0
30	46.0	17	38.0	66	45.0
31	43.0	18	40.0	67	59.0
34	47.0	19	54.0	69	58.0
35	35.0	20	55.0	72	55.0
36	39.0	21	41.0	73	51.0
37	44.0	22	35.0	74	60.0
39	29.0	23	55.0	76	49.0
40	28.0	25	49.0	77	42.0
41	40.0	26	49.0	78	47.0
45	38.0	28	54.0	80	56.0
48	43.0	29	49.0	81	59.0
52	53.0	32	53.0	82	47.0
53	50.0	33	60.0	84	37.0
54	32.0	38	46.0	88	36.0
57	62.0	42	58.0	90	60.0
60	50.0	43	53.0	91	49.0
65	40.0	44	48.0	92	39.0
68	46.0	46	54.0	93	43.0
70	49.0	47	55.0	95	31.0
71	50.0	49	57.0	97	60.0
75	41.0	50	53.0	100	33.0
79	39.0				
83	41.0				
85	53.0				
86	43.0				
87	51.0				
89	34.0				
94	36.0				
96	25.0				
98	38.0				
99	42.0				

Mean	42.100	48.767			
Variance	60.653	77.405			
Sample size	40	60			

Calculating the *t* statistic

Standard error*: $\sqrt{\dfrac{77.405}{60} + \dfrac{60.653}{40}} = 1.675$

t statistic: $\dfrac{48.767 - 42.100}{1.675} = 3.980$

*This formula for standard error is appropriate for equal cell sizes or for cases where all cells have greater than 30 observations. For situations with unequal cell sizes or small cell sizes (fewer than 30), see Stevens [20].

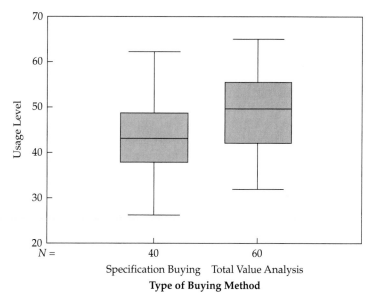

FIGURE 6.6 Boxplots of Usage Level for Two-Group ANOVA

value analysis gave HATCO 48.77 percent. The *t* test analysis examines the difference between groups and statistically tests for the equality of the two group means.

To conduct the test, we first choose the significance level .05 (the maximum allowable Type I error rate). Thus, before we conduct the study, we know that 5 times out of 100 we might conclude that the buying methodology had an impact on the firm's usage rate when in fact it did not. All statistical programs automatically calculate the significance levels of the differences; we illustrate here how those calculations are performed. To determine the value for t_{crit}, we refer to the *t* distribution with $40 + 60 - 2 = 98$ degrees of freedom and $\alpha = .05$. We find that $t_{crit} = 1.66$. Next, we compute the value of our *t* statistic. As shown at the bottom of Table 6.3, $t = 3.98$. Because this exceeds t_{crit}, we conclude that the buying methodology does affect a firm's usage rate of HATCO products. The statistical power of the test is .97, ensuring that the difference found is statistically significant.

Stage 5: Interpretation of the Results

The single dependent variable and the presence of only two groups eliminates the need to examine either the dependent variate or the differences between the groups in addition to the overall tests described in stage 4. The researcher can report statistically significant percentages of business from the HATCO customers using the two buying methods. Firms using total value analysis buy a significantly greater percentage of their products from HATCO than do firms using specification buying. The researcher must assess, however, whether the difference of approximately 6 percent has practical significance for managerial decision making.

A Multivariate Approach: Hotelling's T²

It is probably unrealistic to assume that a difference between any two experimental groups will be manifested in only a single dependent variable. For example, two advertising messages not only may produce different levels of purchase intent but also may affect a number of other (potentially correlated) aspects of the response to advertising (e.g., overall product evaluation, message credibility, interest, attention). Many researchers handle this multiple-criterion situation by repeated application of individual univariate t tests until all the dependent variables have been analyzed. This approach has serious deficiencies. As discussed earlier, consider what might happen to the Type I error rate (inflation over multiple t tests) and the inability of paired t tests to detect differences among combinations of the dependent variables that are not apparent in univariate tests.

Stage 1: Objectives of the MANOVA

In our univariate example, HATCO compared the usage level of firms (X_9 is the dependent variable) utilizing different buying methodologies (X_{11} is the independent variable). To convert this example into a multivariate example, we require at least two dependent variables. Let us assume that HATCO was also interested in the satisfaction levels of firms with the two buying approaches. We would now select X_{10} (satisfaction with HATCO) as the second dependent variable (see Table 6.4). The null hypothesis HATCO is now testing is that the vectors of the mean scores for each group are equivalent (i.e., that buying method has no effect on either usage or satisfaction).

Stage 2: Research Design of MANOVA

The primary consideration in a two-group MANOVA is still the sample size for each of the groups. As discussed in the univariate example, the group sizes are 60 and 40, exceeding both the minimum and recommended sizes. These sample sizes should be adequate to provide the recommended power of .80 for at least medium effect sizes.

Stage 3: Assumptions in MANOVA

Before calculating the test statistics for mean differences across the groups, the researcher must first determine if the dependent measures are significantly correlated. The most widely used test for this purpose is Bartlett's test for sphericity. It examines the correlations among all dependent variables and assesses whether, collectively, significant intercorrelation exists. In our example, a significant degree of intercorrelation does exist (.657) (see Table 6.5, p. 364).

The other critical assumption concerns the homogeneity of the variance–covariance matrices among the two groups. The first analysis assesses the univariate homogeneity of variance across the two groups. As shown in Table 6.5, univariate tests for both variables are nonsignificant. The next step is to assess the dependent variables collectively by testing the equality of the entire variance–covariance matrices between the groups. As shown in Table 6.5, a difference is seen in the correlation among the two dependent variables in the two groups (.823 for the customers using specification buying and .559 for those using total-value analysis). This illustrates the differing levels of covariance. The test for overall equivalence of the variance–covariance matrices is the Box's M test, which in this example has a significance level of .01. Given the sensitivity of this test, the significance level is deemed acceptable and the analysis proceeds.

TABLE 6.4 Firm Usage Percentage and Satisfaction Levels by Buying Method (Basic Data for Hotelling's T^2 or Two-Group MANOVA)

				Buying Method				
Group 1: Specification Buying			Group 2: Total-Value Analysis					
Observation	X_9 (Usage Level)	X_{10} (Satisfaction Level)	Observation	X_9 (Usage Level)	X_{10} (Satisfaction Level)	Observation	X_9 (Usage Level)	X_{10} (Satisfaction Level)
2	43.0	4.3	1	32.0	4.2	51	41.0	5.0
3	48.0	5.2	5	58.0	6.8	55	39.0	3.7
4	32.0	3.9	7	46.0	5.8	56	47.0	4.2
6	45.0	4.4	9	63.0	5.4	58	65.0	6.0
8	44.0	4.3	11	32.0	4.3	59	46.0	5.6
10	54.0	5.4	12	47.0	5.0	61	54.0	4.8
13	39.0	4.4	14	38.0	5.0	62	60.0	6.1
24	36.0	3.7	15	54.0	5.9	63	47.0	5.3
27	36.0	3.7	16	49.0	4.7	64	36.0	4.2
30	46.0	5.1	17	38.0	4.4	66	45.0	4.9
31	43.0	3.3	18	40.0	5.6	67	59.0	6.0
34	47.0	3.8	19	54.0	5.9	69	58.0	4.3
35	35.0	4.1	20	55.0	6.0	72	55.0	3.9
36	39.0	3.6	21	41.0	4.5	73	51.0	4.9
37	44.0	4.8	22	35.0	3.3	74	60.0	5.1
39	29.0	3.9	23	55.0	5.2	76	49.0	5.2
40	28.0	3.3	25	49.0	4.9	77	42.0	5.1
41	40.0	3.7	26	49.0	5.9	78	47.0	5.1
45	38.0	3.2	28	54.0	5.8	80	56.0	5.1
48	43.0	4.7	29	49.0	5.4	81	59.0	4.5
52	53.0	5.2	32	53.0	5.0	82	47.0	5.6
53	50.0	5.5	33	60.0	6.1	84	37.0	4.4
54	32.0	3.7	38	46.0	5.1	88	36.0	4.3
57	62.0	6.2	42	58.0	6.7	90	60.0	6.1
60	50.0	5.0	43	53.0	5.9	91	49.0	4.4
65	40.0	3.4	44	48.0	4.8	92	39.0	5.5
68	46.0	4.5	46	54.0	6.0	93	43.0	5.2
70	49.0	4.8	47	55.0	4.9	95	31.0	4.0
71	50.0	5.4	49	57.0	4.9	97	60.0	5.2
75	41.0	4.1	50	53.0	3.8	100	33.0	4.4
79	39.0	3.3						
83	41.0	4.1						
85	53.0	5.6						
86	43.0	3.7						
87	51.0	5.5						
89	34.0	4.0						
94	36.0	3.6						
96	25.0	3.4						
98	38.0	3.7						
99	42.0	4.3						
Mean	42.100	4.295		48.767	5.088			
Variance	60.653	.612		77.405	.569			
Sample size	40			60				

Note: Hotelling's $T^2 = 26.333$.

TABLE 6.5 Diagnostic Information for Two-Group MANOVA

TEST OF ASSUMPTIONS: HOMOGENEITY OF VARIANCE–COVARIANCE MATRICES

	Variance–Covariance Matrices (values in parentheses are correlations)			
	Group 1: Specification Buying		Group 2: Total-Value Analysis	
	X_9	X_{10}	X_9	X_{10}
X_9: Usage level	60.656		77.402	
X_{10}: Satisfaction level	5.011	.611	3.707	.569
	(.823)		(.559)	

Diagnostic Tests

	X_9: Usage Level		X_{10}: Satisfaction Level		Overall	
	Statistic	Significance	Statistic	Significance	Statistic	Significance
Univariate tests						
Cochran's C	.561	.396	.518	.803		
Bartlett–Box	.677	.411	.060	.807		
Levene test	1.377	.243	.323	.571		
Multivariate test						
Box's M					11.684	.010

TEST OF ASSUMPTION: CORRELATION OF DEPENDENT VARIABLES

	Statistic	Significance
Bartlett test of sphericity	54.474	.000
Intercorrelation: X_9 versus X_{10}	.657	

Stage 4: Estimation of the MANOVA Model and Assessing Overall Fit

The means of each dependent variable (usage level and satisfaction level) for the two groups of firms are presented in Table 6.4. To conduct the test, we again specify our significance level (.05 in this example) as the maximum allowable Type I error. To determine the value for T^2_{crit}, we refer to the F distribution with 2 and 97 degrees of freedom. With an F_{crit} of 3.09, the T^2_{crit} can be calculated as follows:

$$T^2_{crit} = \frac{p(N_1 + N_2 - 2)}{N_1 + N_2 - p - 1} \times F_{crit}$$

$$= \frac{2(60 + 40 - 2)}{60 + 40 - 2 - 1} \times 3.09$$

$$= 6.24$$

As shown in the note on Table 6.4, the computed value of Hotelling's T^2 is 26.33. Because this exceeds T^2_{crit}, we reject the null hypothesis and conclude that buying method has had some impact on the set of dependent measures. Moreover, the power for the multivariate test was almost 1.0, indicating that the sample sizes and the effect size were sufficient to ensure that the significant differences would be detected if they existed beyond the differences due to sampling error.

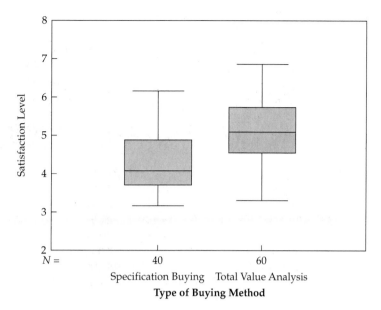

FIGURE 6.7 Boxplots of Second Dependent Variable, Satisfaction Level, for Two-Group MANOVA

Stage 5: Interpretation of the Results

Given the significance of the multivariate test indicating group differences on the dependent variate (vector of means), the researcher must examine the results to assess their logical consistency. The group using total value analysis not only has a higher usage level (refer to Figure 6.6) but also has a higher level of satisfaction (Figure 6.7). The question the researcher must now assess is whether both dependent variables are significantly different or whether the results are derived mainly from differences of only one of the two dependent variables.

One post hoc test of obvious interest is whether the buying method has an impact on usage level (X_9) or on satisfaction level (X_{10}), each considered separately. The group means and univariate tests of mean differences are as follows:

	Dependent Variables	
	X_9 *Usage Level*	X_{10} *Satisfaction Level*
Group Means		
Total-value analysis	48.77	5.09
Specification buying	42.10	4.30
Difference	+6.67	+0.79
Univariate Test of Group Differences		
t statistic	3.96	5.08
Significance level	.000	.000

TABLE 6.6 Stepdown Tests for Two-Group MANOVA

ROY-BARGMANN STEPDOWN F TESTS

Variable	Between-Groups Mean Square	Within-Groups Mean Square	Stepdown F	Degrees of Freedom		Significance of Stepdown F
				Between	Within	
X_9	1066.667	70.738	15.079	1	98	.000
X_{10}	3.246	.336	9.653	1	97	.002

The t statistic was previously computed as 3.96 for the difference in usage level, and here the t statistic for the difference in satisfaction levels is 5.08. Both of the t statistics exceed the square root of T^2_{crit} ($\sqrt{6.24} = 2.50$). Thus we can conclude that the buying method has a positive impact on usage level and satisfaction. In our use of T^2, we also are confident that the probability of a Type I error is held to .05 across both of the post hoc tests.

A second analysis of the dependent variate is the stepdown test, which examines the significance of group differences while allowing for dependent variable intercorrelation. Table 6.6 shows that both X_9 and X_{10} are significantly different, even when controlling for their intercorrelation. Thus, after examining these results and tests, the researcher can safely conclude that the two groups, both collectively and individually, differ significantly on both variables.

Example 2: Difference between k Independent Groups

The two-group randomized design (example 1) is a special case of the more general k-group randomized design. In the general case, each respondent is randomly assigned to one of k levels (groups) of the treatment (independent variable). In the univariate case, a single metric dependent variable is measured, and the null hypothesis is that all group means are equal (i.e., $\mu_1 = \mu_2 = \mu_3 = \ldots = \mu_k$). In the multivariate case, multiple metric dependent variables are measured, and the null hypothesis is that all group vectors of mean scores are equal (i.e., $\mu_1 = \mu_2 = \mu_3 = \ldots \mu_k$, where μ refers to a vector or set of mean scores). For a univariate analysis, the appropriate test statistic is the F statistic resulting from ANOVA. For a multivariate analysis, we examine two of the more widely used test statistics including the greatest characteristic root (*gcr*) statistic (also known as Roy's largest root) and Wilks' lambda (also referred to as Wilks' likelihood ratio criterion or the U statistic).

A Univariate Approach: k-*Groups ANOVA*

Stage 1: Objectives of the ANOVA

The HATCO survey also asked the customers to classify the type of purchases being made with HATCO (X_{14}) as primarily a new-task situation, a modified rebuy, or a straight rebuy (chapter 1 presents a more detailed discussion of these data). HATCO is also interested in knowing whether usage varies across the type of buying situations. The overall null hypothesis HATCO now wishes to test is that $\mu_1 = \mu_2 = \mu_3$ (i.e., all three groups are equivalent in their usage level).

Stage 2: Research Design of ANOVA

In the sample, 34 firms indicated that the new-task buying situation best characterized their relationship with HATCO, 32 firms indicated modified rebuy, and 34 firms indicated straight rebuy. These sample sizes are adequate to obtain sufficient power with medium or larger effect sizes (refer to Table 6.2). If the effect size was small or the sample sizes were to decrease owing to missing data or other factors, the power would fall below recommended levels. The researcher would then need to carefully assess the statistical tests for statistical power and the practical significance of the differences.

Stage 3: Assumptions in ANOVA

The univariate tests for homogeneity of variance for X_9, usage level, across the three groups shows no significant differences with any of the three tasks. Thus, the researcher again can eliminate, as was done in the two-group example, the varying group sizes as impacts on the statistical tests of group differences.

Stage 4: Estimation of the ANOVA Model and Assessing Overall Fit

The ANOVA model tests for the differences in group means among those HATCO customers using one of the three buying methods. Figure 6.8 (p. 369) contains a graphical depiction of the responses by group, and the usage levels of all 100 firms by group are displayed in Table 6.7. To conduct the test by hand, we specify .05 as the Type I error rate. To determine the value for F_{crit}, we refer to the F distribution with $(3 - 1) = 2$ and $(100 - 3) = 97$ degrees of freedom with $\alpha = .05$. We find that $F_{crit} = 3.09$. The calculation of the F statistic from ANOVA is usually summarized in an ANOVA table similar to that shown in Table 6.8 (p. 369). The mean square values for both between-groups and within-groups variances are calculated as the sum of squares (the sum of squared deviations) divided by the appropriate degrees of freedom. As shown in Table 6.8, the resulting F statistic = 106.66 (2749.833/25.776). Because this exceeds F_{crit} we can conclude that all group means are not equal.

Stage 5: Interpretation of the Results

As shown in Table 6.7, the group means for usage level are new task (36.91), modified rebuy (46.53), and straight rebuy (54.88). Examining these means, we note that HATCO usage increases as we proceed from new task to modified rebuy to straight rebuy. One hypothesis of interest is whether there is a significant difference between the new-task or modified-rebuy versus straight-rebuy situations [i.e., (36.91 + 46.53)/2 versus 54.88]. This type of question can be tested with one of the a priori procedures. The contrast is significant (assume $\alpha = .05$ as the criterion; formulas for calculation can be found in texts oriented more to the statistician). Thus we can conclude that the straight-rebuy situation has higher levels of usage than the other two buying situations. Another approach is to use one of the post hoc procedures that tests all group differences and identifies those differences that are statistically significant. One such statistical test is the Scheffé method, perhaps the most widely used of these post hoc methods. In this example,

TABLE 6.7 Firm Usage Percentage by Type of Buying Situation (Basic Data for Three-Group ANOVA)

			Buying Situation		
Group 1: New Task		Group 2: Modified Rebuy		Group 3: Straight Rebuy	
Observation	X_9 (Usage Level)	Observation	X_9 (Usage Level)	Observation	X_9 (Usage Level)
1	32.0	3	48.0	5	58.0
2	43.0	6	45.0	9	63.0
4	32.0	8	44.0	15	54.0
7	46.0	10	54.0	16	49.0
11	32.0	12	47.0	19	54.0
13	39.0	17	38.0	20	55.0
14	38.0	18	40.0	23	55.0
22	35.0	21	41.0	26	49.0
24	36.0	25	49.0	28	54.0
27	36.0	30	46.0	29	49.0
31	43.0	37	44.0	32	53.0
34	47.0	44	48.0	33	60.0
35	35.0	48	43.0	38	46.0
36	39.0	51	41.0	42	58.0
39	29.0	52	53.0	43	53.0
40	28.0	53	50.0	46	54.0
41	40.0	56	47.0	47	55.0
45	38.0	57	62.0	49	57.0
54	32.0	60	50.0	50	53.0
55	39.0	64	36.0	58	65.0
65	40.0	66	45.0	59	46.0
75	41.0	68	46.0	61	54.0
79	39.0	70	49.0	62	60.0
83	41.0	71	50.0	63	47.0
84	37.0	76	49.0	67	59.0
86	43.0	77	42.0	69	58.0
88	36.0	82	47.0	72	55.0
89	34.0	85	53.0	73	51.0
94	36.0	87	51.0	74	60.0
95	31.0	91	49.0	78	47.0
96	25.0	92	39.0	80	56.0
98	38.0	93	43.0	81	59.0
99	42.0			90	60.0
100	33.0			97	60.0
Mean	36.912		46.531		54.882
Variance	25.593		28.132		23.746
Sample size	34		32		34

while controlling the overall error rate so as not to exceed .05, the Scheffé method still identifies that all the groups are significantly different from each other. From this, the researcher will know that the significant differences are due to each group comparison and not specific to differences among only certain groups.

In summary, univariate ANOVA suggests that the type of buying situation leads to higher usage levels. Post hoc tests enable the researcher to identify these significant differences quite easily and to maintain statistical control on the overall significance level.

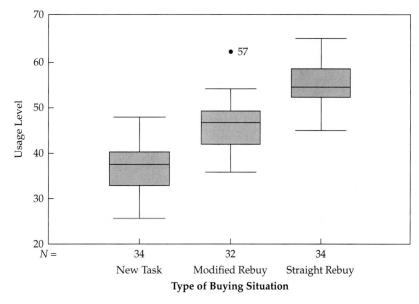

FIGURE 6.8 Boxplots of Usage Level for Three-Group ANOVA

A Multivariate Approach: k-Groups MANOVA

In *k*-group designs in which multiple dependent variables are measured, many researchers proceed with a series of individual *F* tests (ANOVAs) until all the dependent variables have been analyzed. As the reader should suspect, this approach suffers from the same deficiencies as a series of *t* tests across multiple dependent variables; that is, a series of *F* tests with ANOVA (1) results in an inflated Type I error rate, and (2) ignores the possibility that some composite of the dependent variables may provide reliable evidence of overall group differences. In addition, because individual *F* tests ignore the correlations among the independent variables, they use less than the total information available for assessing overall group differences.

MANOVA again provides a solution to these problems. MANOVA solves the Type I error rate problem by providing a single overall test of group differences at a specified α level. It solves the composite variable problem by implicitly forming and testing the linear combinations of the dependent variables that provide the strongest evidence of overall group differences.

TABLE 6.8 Three-Group ANOVA Results: Usage Level by Type of Buying Situation

Source of Variance	Sum of Squares	Mean Square	Degrees of Freedom	F Ratio
Between groups	5498.767	2749.383	2	106.666
Within-groups error	2500.233	25.776	97	

Stage 1: Objectives of the MANOVA

In our earlier univariate example, HATCO assessed its performance across firms having one of three types of buying situations (X_9) (new-task buying, modified rebuy, or straight rebuy). To convert this example to a multivariate example, we require at least two dependent variables. As in our earlier multivariate extension of a univariate example, let us assume that HATCO also wished to examine differences in satisfaction (X_{10}) with HATCO across the three groups (see Table 6.9).

TABLE 6.9 Firm Usage Percentage and Satisfaction Level by Buying Situation (Basic Data for Three-Group MANOVA)

			Buying Situation					
Group 1: New Task Scores			Group 2: Total Value Analysis			Group 3: Straight Rebuy Scores		
Observation	X_9 (Usage Level)	X_{10} (Satisfaction Level)	Observation	X_9 (Usage Level)	X_{10} (Satisfaction Level)	Observation	X_9 (Usage Level)	X_{10} (Satisfaction Level)
1	32.0	4.2	3	48.0	5.2	5	58.0	6.8
2	43.0	4.3	6	45.0	4.4	9	63.0	5.4
4	32.0	3.9	8	44.0	4.3	15	54.0	5.9
7	46.0	5.8	10	54.0	5.4	16	49.0	4.7
11	32.0	4.3	12	47.0	5.0	19	54.0	5.9
13	39.0	4.4	17	38.0	4.4	20	55.0	6.0
14	38.0	5.0	18	40.0	5.6	23	55.0	5.2
22	35.0	3.3	21	41.0	4.5	26	49.0	5.9
24	36.0	3.7	25	49.0	4.9	28	54.0	5.8
27	36.0	3.7	30	46.0	5.1	29	49.0	5.4
31	43.0	3.3	37	44.0	4.8	32	53.0	5.0
34	47.0	3.8	44	48.0	4.8	33	60.0	6.1
35	35.0	4.1	48	43.0	4.7	38	46.0	5.1
36	39.0	3.6	51	41.0	5.0	42	58.0	6.7
39	29.0	3.9	52	53.0	5.2	43	53.0	5.9
40	28.0	3.3	53	50.0	5.5	46	54.0	6.0
41	40.0	3.7	56	47.0	4.2	47	55.0	4.9
45	38.0	3.2	57	62.0	6.2	49	57.0	4.9
54	32.0	3.7	60	50.0	5.0	50	53.0	3.8
55	39.0	3.7	64	36.0	4.2	58	65.0	6.0
65	40.0	3.4	66	45.0	4.9	59	46.0	5.6
75	41.0	4.1	68	46.0	4.5	61	54.0	4.8
79	39.0	3.3	70	49.0	4.8	62	60.0	6.1
83	41.0	4.1	71	50.0	5.4	63	47.0	5.3
84	37.0	4.4	76	49.0	5.2	67	59.0	6.0
86	43.0	3.7	77	42.0	5.1	69	58.0	4.3
88	36.0	4.3	82	47.0	5.6	72	55.0	3.9
89	34.0	4.0	85	53.0	5.6	73	51.0	4.9
94	36.0	3.6	87	51.0	5.5	74	60.0	5.1
95	31.0	4.0	91	49.0	4.4	78	47.0	5.1
96	25.0	3.4	92	39.0	5.5	80	56.0	5.1
98	38.0	3.7	93	43.0	5.2	81	59.0	4.5
99	42.0	4.3				90	60.0	6.1
100	33.0	4.4				97	60.0	5.2
Mean	36.912	3.929		46.531	5.003		54.882	5.394
Variance	25.593	.282		28.132	.237		23.746	.508
Sample size	34			32			34	

The null hypothesis HATCO now wishes to test is that the three sample vectors of the mean scores are equivalent.

Stage 2: Research Design of the MANOVA
As discussed in the univariate example of the three-group analysis, the sample sizes are adequate based on the number of dependent variables. A more important consideration is the effect of the group sample sizes on the statistical power of the tests of group differences. In referring to Table 6.1, sample sizes of 30 and above will provide adequate power for large effect sizes and somewhat lower levels of power for medium effect sizes. These sample sizes, however, are not adequate to provide the recommended power of .80 for small effect sizes. The required sample size for small effect sizes in this situation would be 98 respondents per group. Thus, any nonsignificant results should be examined closely to evaluate whether the effect size has managerial significance, because the low statistical power precluded designating it as statistically significant.

Stage 3: Assumptions in MANOVA
The two univariate tests for homogeneity of variance indicate a nonsignificant difference for X_9, the usage level, but mixed results for X_{10}, the satisfaction level with HATCO (Table 6.10). In the case of satisfaction with HATCO, combining the two

TABLE 6.10 Diagnostic Information for Three-Group MANOVA

TEST OF ASSUMPTIONS: HOMOGENEITY OF VARIANCE–COVARIANCE MATRICES

Variance–Covariance Matrices (values in parentheses are correlations)

	Group 1: New Task		Group 2: Modified Rebuy		Group 3: Straight Rebuy	
	X_9	X_{10}	X_9	X_{10}	X_9	X_{10}
X_9: Usage level	25.598		28.128		23.743	
X_{10}: Satisfaction level	.648 (.241)	.282	1.366 (.529)	.237	.763 (.219)	.509

Diagnostic Tests

	X_9: Usage Level		X_{10}: Satisfaction Level		Overall	
	Statistic	Significance	Statistic	Significance	Statistic	Significance
Univariate tests						
Cochran's C	.363	.965	.495	.033		
Bartlett–Box	.114	.892	2.670	.070		
Levene test	.056	.945	3.302	.041		
Multivariate test						
Box's M					9.796	.147

TEST OF ASSUMPTION: CORRELATION OF DEPENDENT VARIABLES

	Statistic	Significance
Bartlett test of sphericity	9.480	.002
Intercorrelation: X_9 versus X_{10}	.307	

tests (.033 and .070) provides a sufficient level of nonsignificance for the test of homogeneity of variance to proceed to the multivariate test. Using Box's M test for homogeneity of the variance–covariance matrices, we find that the groups have no significant differences.

The second assumption to test is the correlation among the dependent variables. In this case Bartlett's test of sphericity has a significance level of .002, satisfying the necessary level of intercorrelation to justify MANOVA. See Table 6.10 for more details.

Stage 4: Estimation of the MANOVA Model and Assessing Overall Fit

From examination of the boxplots of responses in each group for usage level (refer to Figure 6.8) and satisfaction level (Figure 6.9), the indications are that both variables may differ across the three groups. The purpose of the multivariate test is to assess these differences collectively rather than singularly with univariate tests. Table 6.11 provides summary output from the MANOVA performed on the data in Table 6.9. Pillai's criterion has a significance level (.0000) well below our prespecified level of .05. The value of gcr is .723. Referring to the gcr distribution with appropriate degrees of freedom, and setting $\alpha = .05$, we see that $gcr_{crit} = .310$. Because .723 exceeds this value, we again conclude that the mean vectors of the three groups are not equal. As we see in Table 6.11, the statistical program estimates a significance level of .000 for this measure as well. The value of Wilks' lambda (see Table 6.11) is .264. An approximate F statistic associated with this value of Wilks' lambda is 45.4. With 4 and 192 degrees of freedom and an α level of .05, $F_{crit} = 2.41$. Because 45.4 well exceeds this value, we again reach the same conclusion that the mean vectors of the three groups are not equal. Using any of the measures of multivariate differences results in the same conclusion: The combined

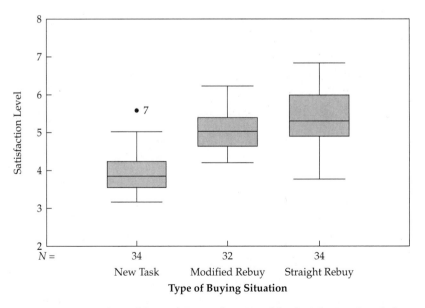

FIGURE 6.9 Boxplots of Second Dependent Variable, Satisfaction Level, for Three-Group MANOVA

TABLE 6.11 Three-Group MANOVA Summary Table

MULTIVARIATE TESTS OF SIGNIFICANCE

Test Name	Value	Approximate F	Degrees of Freedom		Significance of F Statistic
			Between Group	Within Group	
Pillai's criterion	.771	30.419	4	194	.000
Hotelling's trace	2.655	63.052	4	190	.000
Wilks' lambda	.264	45.411	4	192	.000
Roy's gcr	.723				

STATISTICAL POWER OF MANOVA TESTS

	Effect Size	Power
Pillai's criterion	.385	1.000
Hotelling's trace	.570	1.000
Wilks' lambda	.486	1.000

UNIVARIATE F TESTS

Variable	Between-Groups Sum of Squares	Within-Groups Sum of Squares	Degrees of Freedom	Between-Groups Mean Square	Within-Groups Mean Square	F Statistic	Significance
X_9	5498.767	2500.233	2 and 97	2749.383	25.776	106.666	.000
X_{10}	39.007	33.459	2 and 97	19.503	.345	56.542	.000

ROY-BARGMAN STEPDOWN F TESTS

Variable	Between-Groups Mean Square	Within-Groups Mean Square	Stepdown F	Degrees of Freedom		Significance of Stepdown F
				Between	Within	
X_9	2749.383	25.776	106.666	2	97	.000
X_{10}	2.783	.316	8.819	2	96	.000

dependent variables, usage level and satisfaction level, vary across the three buying situations.

Stage 5: Interpretation of the Results

In interpreting the results (shown in Table 6.9), the researcher must first examine the group means. We have already seen the pattern for usage level increasing across the three groups, and the group means for satisfaction level follow the same general pattern: new task (3.93), modified rebuy (5.00), and straight rebuy (5.39).

A number of post hoc tests may be of interest. For example, are group mean differences statistically significant for each dependent variable considered alone? We can examine this question with the individual F tests for the two dependent variables. Consistent with the multivariate results, both variables show significant differences across the groups. A stepdown analysis, as shown in Table 6.11, shows that both variables have unique differences across the groups; that is, they are not so highly correlated that there were no unique differences in satisfaction after the effects of usage were accounted for. This result suggests that buying situation has significant separate effects on satisfaction that are unrelated to a firm's usage level. Finally, are the differences in usage and satisfaction found between all buying situations? For example, do firms with the modified rebuy situation show a difference in satisfaction when compared with those in the straight rebuy? All such questions can be answered with the multivariate extension of contrast procedure outlined earlier. Examination of discriminant functions (which have to be obtained by using a discriminant analysis program, as described in chapter 5) are generally more useful first steps in post hoc analysis as the number of dependent variables increases.

Example 3: A Factorial Design for MANOVA with Two Independent Variables

In the prior two examples, the MANOVA analyses have been extensions of univariate two- and three-group analyses. In this example, we explore a multivariate factorial design—two independent variables used as treatments to analyze differences of two dependent variables. In the course of our discussion, we assess the interactive or joint effects between the two treatments on the dependent variables separately and collectively.

Stage 1: Objectives of the MANOVA

In the previous multivariate research questions, HATCO considered the effect of only a single-treatment variable on the dependent variables. But the possibility of joint effects among two or more independent variables must also be considered. After deliberation, one proposed analysis focused on extending the prior three-group analysis of the differences in the two dependent variables, usage level (X_9) and satisfaction level (X_{10}), by considering not only the effects of type of buying situation (X_{14}) but also industry type (X_{13}). The objective is to reduce the within-group variance of the buying situation groups by adding the second treatment—industry type. This will form a separate group for each industry type within each buying situation. The boxplots in Figure 6.10 show that the two industry types

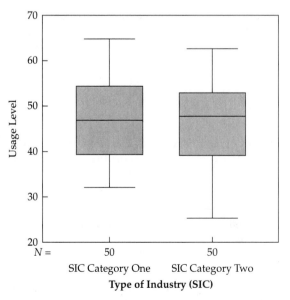
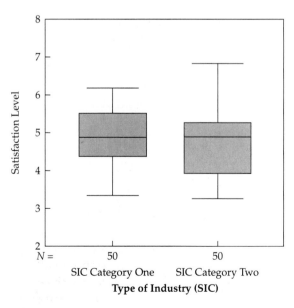

FIGURE 6.10 Boxplots of Usage Level and Satisfaction Level for Second Independent Variable, Industry Type, in 3×2 Factorial Design MANOVA

do not vary substantially on either of the dependent variables. But this does not preclude the value of including industry type as a means of reducing within-group variance. The data for each group (combination of buying situation and industry type) are shown in Table 6.12 (p. 376). In this manner, the impact of industry type can be evaluated simultaneously with buying situation along with examining for the possible occurrence of interaction effects.

Stage 2: Research Design of the MANOVA

The factorial design of two independent variables—X_{14} and X_{13}—raises the issue of adequate sample size in the various groups. Because there are three levels of X_{14} (new task, modified rebuy, and straight rebuy) and two levels of X_{13} (SIC category one and SIC category two), this is a 3×2 design with six groups. The researcher must ensure in creating the factorial design that each group has sufficient sample size to (1) meet the minimum requirements of group sizes exceeding the number of dependent variables, as well as (2) provide the statistical power to assess differences deemed practically significant. In this case, the sample sizes range from 16 to 18 respondents per group. This exceeds the number of dependent variables (two) in each group, but the statistical power is quite low. While tabled values do not deal with a factorial design, values for a six-group MANOVA indicate that this sample size will detect only moderately large or better effect sizes with a power of .70 [15]. Thus, the researcher must recognize that unless the effect sizes are substantial, the limited sample sizes in each group preclude the identification of significant differences.

Stage 3: Assumptions in MANOVA

As with the prior MANOVA analyses, the assumption of greatest importance is the homogeneity of variance–covariance matrices across the groups. In this instance, there are six groups involved in testing the assumption. The univariate tests for usage level and satisfaction are both nonsignificant, except for the Bartlett-Box test of satisfaction, which has a significance level of .038 (see Table 6.13, p. 377). With the univariate tests showing nonsignificance, the researcher can proceed to the multivariate test. Box's M test has a significance level of .09, thus allowing us to accept the null hypothesis of homogeneity of variance–covariance matrices at the .05 level. Meeting this assumption allows for direct interpretation of the results without having to consider group sizes, level of covariances in the group, and so forth.

The second assumption is the correlation of the dependent measures, which is assessed with Bartlett's test of sphericity. In this example, the significance is .004, indicative of a significant level of correlation between the two independent measures (see Table 6.13).

Stage 4: Estimation of the MANOVA Model and Assessing Overall Fit

The boxplots for each dependent variable across the six groups (Figure 6.11, p. 378) show that differences do seem to exist between them. The differences are the most pronounced across buying situations, but differences within each buying situation for the two industry types also can be seen. The MANOVA model tests not only for the main effects of both independent variables but also their interaction or joint effect on the two dependent variables. The first step is to examine the interaction effect and determine whether it is statistically significant.

TABLE 6.12 Firm Usage Percentage and Satisfaction Level by Buying Situation and Industry Type (Basic Data for 3×2 Factorial Design MANOVA)

SIC Category One

	Group 1: New Task Scores			Group 3: Total Value Analysis			Group 5: Straight Rebuy Scores	
Observation	X_9 (Usage Level)	X_{10} (Satisfaction Level)	Observation	X_9 (Usage Level)	X_{10} (Satisfaction Level)	Observation	X_9 (Usage Level)	X_{10} (Satisfaction Level)
2	43.0	4.3	8	44.0	4.3	15	54.0	5.9
11	32.0	4.3	10	54.0	5.4	16	49.0	4.7
13	39.0	4.4	18	40.0	5.6	20	55.0	6.0
22	35.0	3.3	21	41.0	4.5	23	55.0	5.2
24	36.0	3.7	25	49.0	4.9	32	53.0	5.0
27	36.0	3.7	30	46.0	5.1	33	60.0	6.1
31	43.0	3.3	44	48.0	4.8	43	53.0	5.9
34	47.0	3.8	51	41.0	5.0	46	54.0	6.0
35	35.0	4.1	53	50.0	5.5	58	65.0	6.0
55	39.0	3.7	57	62.0	6.2	62	60.0	6.1
75	41.0	4.1	66	45.0	4.9	67	59.0	6.0
83	41.0	4.1	68	46.0	4.5	69	58.0	4.3
84	37.0	4.4	77	42.0	5.1	73	51.0	4.9
88	36.0	4.3	85	53.0	5.6	80	56.0	5.1
94	36.0	3.6	87	51.0	5.5	81	59.0	4.5
98	38.0	3.7	92	39.0	5.5	90	60.0	6.1
99	42.0	4.3						
100	33.0	4.4						
Mean	38.278	3.972		46.937	5.150		56.313	5.488
Variance	15.390	.141		37.663	.252		16.761	.423
Sample size	18			16			16	

SIC Category Two

	Group 2: New Task Scores			Group 4: Total Value Analysis			Group 6: Straight Rebuy Scores	
Observation	X_9 (Usage Level)	X_{10} (Satisfaction Level)	Observation	X_9 (Usage Level)	X_{10} (Satisfaction Level)	Observation	X_9 (Usage Level)	X_{10} (Satisfaction Level)
1	32.0	4.2	3	48.0	5.2	5	58.0	6.8
4	32.0	3.9	6	45.0	4.4	9	63.0	5.4
7	46.0	5.8	12	47.0	5.0	19	54.0	5.9
14	38.0	5.0	17	38.0	4.4	26	49.0	5.9
36	39.0	3.6	37	44.0	4.8	28	54.0	5.8
39	29.0	3.9	48	43.0	4.7	29	49.0	5.4
40	28.0	3.3	52	53.0	5.2	38	46.0	5.1
41	40.0	3.7	56	47.0	4.2	42	58.0	6.7
45	38.0	3.2	60	50.0	5.0	47	55.0	4.9
54	32.0	3.7	64	36.0	4.2	49	57.0	4.9
65	40.0	3.4	70	49.0	4.8	50	53.0	3.8
79	39.0	3.3	71	50.0	5.4	59	46.0	5.6
86	43.0	3.7	76	49.0	5.2	61	54.0	4.8
89	34.0	4.0	82	47.0	5.6	63	47.0	5.3
95	31.0	4.0	91	49.0	4.4	72	55.0	3.9
96	25.0	3.4	93	43.0	5.2	74	60.0	5.1
						78	47.0	5.1
						97	60.0	5.2
Mean	35.375	3.881		46.125	4.856		53.611	5.311
Variance	34.117	.456		20.115	.192		27.668	.601
Sample size	16			16			18	

TABLE 6.13 Diagnostic Information for 3×2 Factorial Design MANOVA

TEST OF ASSUMPTIONS: HOMOGENEITY OF VARIANCE–COVARIANCE MATRICES

Variance–Covariance Matrices
(values in parentheses are correlations)

| | *SIC Category One* | | | | | |
| | *Group 1: New Task* | | *Group 3: Modified Rebuy* | | *Group 5: Straight Rebuy* | |
	X_9	X_{10}	X_9	X_{10}	X_9	X_{10}
X_9: Usage level	15.389		37.662		16.762	
X_{10}: Satisfaction level	−.162	.141	1.643	.252	.984	.423
	(−.110)		(.533)		(.370)	

| | *SIC Category Two* | | | | | |
| | *Group 2: New Task* | | *Group 4: Modified Rebuy* | | *Group 6: Straight Rebuy* | |
	X_9	X_{10}	X_9	X_{10}	X_9	X_{10}
X_9: Usage level	34.117		20.117		27.663	
X_{10}: Satisfaction level	1.461	.456	1.053	.192	.375	.600
	(.370)		(.536)		(.092)	

Diagnostic Tests

| | X_9: Usage Level | | X_{10}: Satisfaction | | Overall | |
	Statistic	*Significance*	*Statistic*	*Significance*	*Statistic*	*Significance*
Univariate tests						
Cochran's C	.248	.447	.291	.115		
Bartlett–Box	1.062	.380	2.363	.038		
Levene test	1.332	.257	1.519	.191		
Multivariate test					24.050	.090
Box's M						

TEST OF ASSUMPTIONS: CORRELATION OF DEPENDENT VARIABLES

	Statistic	*Significance*
Bartlett test of sphericity	8.225	.004
Intercorrelation: X_9 versus X_{10}	.292	

Table 6.14 (p. 379) contains the MANOVA results for testing the interaction effect. All four multivariate tests indicate that the interaction effect is not significant. This means that the differences between industry types are roughly equal across the three buying situations for both dependent variables collectively. The univariate tests confirm that this finding holds for each variable separately. Figure 6.12 (p. 380) documents the lack of interaction effect for each dependent variable. In the graphs for each dependent variable, the differences between the two industry types are relatively equal across the three buying situations. With a nonsignificant interaction effect, the main effects can be interpreted directly without adjustment.

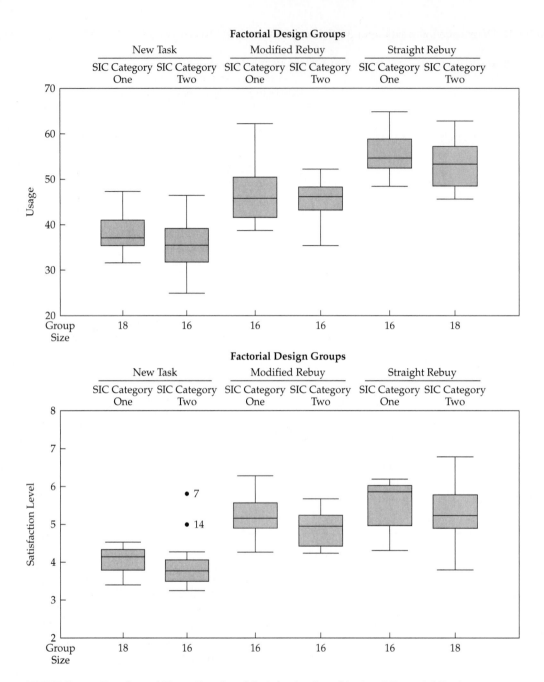

FIGURE 6.11 Boxplots of Usage Level and Satisfaction Level in 3×2 Factorial Design MANOVA.

Tables 6.15 (p. 381) and 6.16 (p. 382) contain the MANOVA results for the main effects of buying situation and industry type. Industry type (X_{13}) has a significance level of .069 for the multivariate tests, indicating a nonsignificant difference attributable to industry type. The researcher should consider, however, raising the required significance level, because the power of the multivariate tests is reduced

TABLE 6.14 3×2 Factorial Design MANOVA Summary Table: Interaction Effect

INTERACTION EFFECT: INDUSTRY TYPE (X_{13}) BY BUYING SITUATION (X_{14})

MULTIVARIATE TESTS OF SIGNIFICANCE

Test Name	Value	Approximate F	Degrees of Freedom		Significance of F Statistic
			Between Group	Within Group	
Pillai's criterion	.020	.464	4	188	.762
Hotelling's trace	.020	.458	4	184	.766
Wilks' lambda	.980	.461	4	186	.764
Roy's gcr	.019				

STATISTICAL POWER OF MANOVA TESTS

	Effect Size	Power
Pillai's criterion	.010	.16
Hotelling's trace	.010	.16
Wilks' lambda	.010	.16

UNIVARIATE F TESTS

Variable	Between-Groups Sum of Squares	Within-Groups Sum of Squares	Degrees of Freedom	Between-Groups Mean Square	Within-Groups Mean Square	F Statistic	Significance
X_9	21.682	2361.764	2 and 94	10.841	25.125	.431	.651
X_{10}	.170	32.435	2 and 94	.085	.345	.247	.782

ROY-BARGMAN STEPDOWN F TESTS

Variable	Between-Groups Mean Square	Within-Groups Mean Square	Stepdown F	Degrees of Freedom		Significance of Stepdown F
				Between Group	Within Group	
X_9	10.841	25.125	.431	2	94	.651
X_{10}	.158	.319	.495	2	93	.611

owing to the rather small sample sizes per group. If this were done, then industry type would be considered a significant effect. The second independent variable, satisfaction (X_{14}), shows highly significant effects for all the multivariate tests. In each instance, the significance level exceeds .000. Moreover, the statistical power is 1.0, indicating that the very large effect sizes ensured high levels of power even with the small sample sizes per group. The impact of the two independent variables can be compared by examining the relative effect sizes. The effect sizes for buying situation are eight to ten times larger than those associated with usage level. This comparison gives the researcher an evaluation of practical significance separate from the statistical significance tests. In this example, buying situation is the dominant effect, with industry type having a slight effect. Moreover, the interaction or joint effects between the two treatments are nonsignificant for both dependent variables.

Stage 5: Interpretation of the Results

A comparison among the six groups should not be attempted with the post hoc tests, such as the Scheffé method, which make all possible comparisons while controlling overall Type I error. In this example, the relatively small sample sizes and

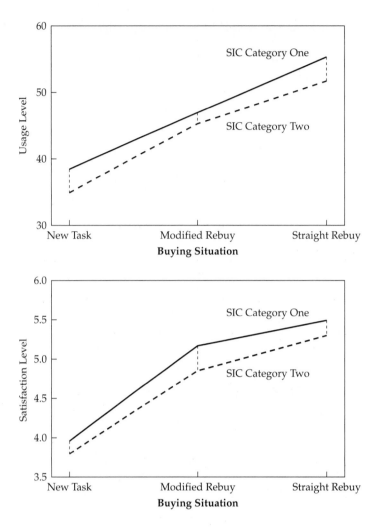

FIGURE 6.12 Plots of Interaction Effects in 3 × 2 Factorial
Design MANOVA for Usage Level and Satisfaction Level.

the large number of tests required for all comparisons of six groups result in such low levels of statistical power that only very large effect sizes can be detected reliably. The researcher thus should examine the differences for practical significance in addition to statistical significance. If specific comparisons among the groups can be formulated, then planned comparisons can be specified and tested directly in the analysis.

Even though the group comparisons are limited by the small sample sizes, examination of each dependent variable is still warranted. For example, industry type (X_{13}) is judged to be nonsignificant at the .05 alpha level when the set of dependent variables is evaluated. But if the researcher examines the univariate tests, some interesting points emerge. First, the industry type does affect the usage level (significance = .036) but not the satisfaction level (significance = .328). Thus, when evaluated collectively, the set of dependent variables is found to be nonsignifi-

TABLE 6.15 3×2 Factorial Design MANOVA Summary Table: Main Effect of Industry Type

MAIN EFFECT: INDUSTRY TYPE (X_{13})

MULTIVARIATE TESTS OF SIGNIFICANCE

Test Name	Value	Approximate F	Degrees of Freedom		Significance of F Statistic
			Between Group	Within Group	
Pillai's criterion	.056	2.752	2	93	.069
Hotelling's trace	.059	2.752	2	93	.069
Wilks' lambda	.944	2.752	2	93	.069
Roy's gcr	.056				

STATISTICAL POWER OF MANOVA TESTS

	Effect Size	Power
Pillai's criterion	.056	.53
Hotelling's trace	.056	.53
Wilks' lambda	.056	.53

UNIVARIATE F TESTS

Variable	Between-Groups Sum of Squares	Within-Groups Sum of Squares	Degrees of Freedom	Between-Groups Mean Square	Within-Groups Mean Square	F Statistic	Significance
X_9	114.019	2351.764	1 and 94	114.019	25.125	4.538	.036
X_{10}	.872	32.435	1 and 94	.872	.345	2.527	.115

ROY-BARGMAN STEPDOWN F TESTS

Variable	Between-Groups Mean Square	Within-Groups Mean Square	Stepdown F	Degrees of Freedom		Significance of Stepdown F
				Between Group	Within Group	
X_9	114.019	25.125	4.538	1	94	.036
X_{10}	.309	.319	.967	1	93	.328

cant. The practical significance of demonstrating an effect on the usage level may lead to use of the univariate results in addition to the multivariate results. This can be contrasted to the case for the second independent variable—buying situation (X_{14})—in which univariate tests for both dependent variables support its multivariate effect. This also holds in the stepdown analysis, which can be interpreted that buying situation affects not only the set of dependent variables but also these dependent variables separately, even after the impact of other dependent variables has been considered. These results confirm the differences found between the impacts of the two independent variables.

A Managerial Overview of the Results

The researchers for HATCO performed a series of ANOVAs and MANOVAs in an attempt to understand how the usage (X_9) and satisfaction (X_{10}) levels with HATCO varied across characteristics of the firms involved (i.e., specification buying (X_{11}), type of buying situation (X_{14}), and industry type (X_{13})). For purposes

TABLE 6.16 3×2 Factorial Design MANOVA Summary Table: Main Effect of Buying Situation

MAIN EFFECT: BUYING SITUATION (X_{14})

MULTIVARIATE TESTS OF SIGNIFICANCE

Test Name	Value	Approximate F	Degrees of Freedom		Significance of F Statistic
			Between Group	*Within Group*	
Pillai's criterion	.786	30.411	4	188	.000
Hotelling's trace	2.852	65.601	4	184	.000
Wilks' lambda	.250	45.452	4	186	.000
Roy's *gcr*	.737				

STATISTICAL POWER OF MANOVA TESTS

	Effect Size	Power
Pillai's criterion	.393	1.00
Hotelling's trace	.588	1.00
Wilks' lambda	.500	1.00

UNIVARIATE *F* TESTS

Variable	Between-Groups Sum of Squares	Within-Groups Sum of Squares	Degrees of Freedom	Between-Groups Mean Square	Within-Groups Mean Square	F Statistic	Significance
X_9	5580.664	2361.764	2 and 94	2790.332	25.125	111.057	.000
X_{10}	39.250	32.435	2 and 94	19.625	.345	56.876	.000

ROY-BARGMAN STEPDOWN *F* TESTS

Variable	Between-Groups Mean Square	Within-Groups Mean Square	Stepdown F	Degrees of Freedom		Significance of Stepdown F
				Between Group	*Within Group*	
X_9	2790.332	25.125	111.057	2	94	.000
X_{10}	2.793	.319	8.752	2	93	.000

of this discussion, we focus on the multivariate results as they overlap with the univariate results.

The first MANOVA analysis is very direct: Does the buying method have an effect on satisfaction and usage levels? In this case the researcher tests whether the vectors of mean scores for each group are equivalent. After assuring that all assumptions are met, we find that the results reveal there was a significant difference in that firms using total value analysis had higher usage and satisfaction scores when compared to firms using specification buying. Along with the overall results, management also needed to know if this difference exists not only for the variate but also for the individual variables. Univariate tests and a step-down analysis revealed that there were significant univariate differences for both usage and satisfaction levels. The significant univariate results indicate to management that buying specification uniquely impacts both satisfaction and usage levels. Thus, managers can create unique strategies for each specification buying type to more effectively increase usage and satisfaction.

The next MANOVA follows the same approach, but substitutes a new independent variable, type of buying situation, which has three groups (new buy, modified rebuy, and straight rebuy). Once again, management focuses on usage and satisfaction levels, with significant differences again found. Group profiles and post hoc tests are conducted to determine if this difference exists for usage and satisfaction levels individually. The post hoc tests are significant, indicating that there are unique differences in satisfaction level, depending on the buying situation, when controlling for usage level. From a managerial perspective, the greater the segmentation of customers on variables such as these, the more focused the strategies can become, as the manager has greater knowledge of the groups and if significant differences exist between them. Although this model allows the researcher to consider three groups, it is rather basic compared to the various situations found business settings.

The third example addresses the issue of several independent variables and their potential impact on the dependent variables. This is a more complex design with main effects (the impact on satisfaction and usage levels) and interactions. Using two independent variables also allows the creation of separate categories within a treatment (group). In this example, the independent variables are buying situation (X_{14}) and industry type (X_{13}). The three categories of buying situation are combined with industry type to form six groups. The first step is to review the results for significant interactions; none were found. Given the nonsignificant interactions, one can proceed to interpret the main effects. When doing so, the manager must look for not only statistical but also practical significance. When considered together, both independent variables have an impact on usage and satisfaction levels. But in regards to practical significance, buying situation by far has the greatest impact. Thus, the manager can determine not only whether a variable has an impact but its relative importance as compared to other variables. This identifies for managers the key characteristics that distinguish customers as to their usage and evaluation of HATCO. Strategies that attempt to influence usage and satisfaction should be oriented toward both buying situation and industry type, but the most important factor to consider is buying situation. Thus, if market differences can be incorporated, that will only add to the impact of the strategies that are differentiated by buying situation.

Summary

It may be unrealistic to assume that a difference between experimental treatments will be manifested only in a single measured dependent variable. Unfortunately, many researchers handle multiple-criterion situations by repeated application of individual univariate tests until all the dependent variables have been analyzed. This approach can seriously inflate Type I error rates, and it ignores the possibility that some composite of the dependent variables may provide the strongest evidence of reliable group differences. Appropriate use of MANOVA provides solutions to both these problems.

Questions

1. What are the differences between MANOVA and discriminant analysis? What situations best suit each multivariate technique?
2. Design a two-way factorial MANOVA experiment. What are the different sources of variance in your experiment? What would a significant interaction tell you?

3. Besides the overall, or global, significance, there are at least three approaches to doing follow-up tests: (a) use of Scheffé contrast procedures; (b) stepdown analysis, which is similar to stepwise regression in that each successive F statistic is computed after eliminating the effects of the previous dependent variables; and (c) examination of the discriminant functions. Name the practical advantages and disadvantages of each of these approaches.

4. How is the statistical power affected by statistical and research design decisions? How would you design a study to ensure adequate power?

5. Describe some data analysis situations in which MANOVA and MANCOVA would be appropriate in your areas of interest. What types of uncontrolled variables or covariates might be operating in each of these situations?

References

1. Anderson, T. W. (1958), *Introduction to Multivariate Statistical Analysis.* New York: Wiley.
2. Cattell, R. B., ed. (1966), *Handbook of Multivariate Experimental Psychology.* Chicago: Rand McNally.
3. Cohen, J. (1977), *Statistical Power Analysis for the Behavioral Sciences.* New York: Academic Press.
4. Cole, D. A., S. E. Maxwell, R. Avery, and E. Salas (1994), "How the Power of MANOVA Can Both Increase and Decrease as a Function of the Intercorrelations among Dependent Variables." *Psychological Bulletin* 115: 465–74.
5. Cooley, W. W., and P. R. Lohnes (1971), *Multivariate Data Analysis.* New York: Wiley.
6. Green, P. E. (1978), *Analyzing Multivariate Data.* Hinsdale, Ill.: Holt, Rinehart, & Winston.
7. Green, P. E., and J. Douglas Carroll (1978), *Mathematical Tools for Applied Multivariate Analysis.* New York: Academic Press.
8. Green, P. E., and D. S. Tull (1979), *Research for Marketing Decisions*, 3d ed. Upper Saddle River, N.J.: Prentice Hall.
9. Hand, D. J., and C. C. Taylor (1987), *Multivariate Analysis of Variance and Repeated Measures.* London: Chapman and Hall.
10. Harris, R. J. (1975), *A Primer of Multivariate Statistics.* New York: Academic Press.
11. Hubert, C. J., and J. D. Morris (1989), "Multivariate Analysis versus Multiple Univariate Analyses." *Psychological Bulletin* 105: 302–8.
12. Huitema, B. (1980), *The Analysis of Covariance and Alternatives.* New York: Wiley.
13. Kirk, R. E. (1982), *Experimental Design: Procedures for the Behavioral Sciences*, 2d ed. Belmont, Calif.: Brooks/Cole.
14. Koslowsky, M., and T. Caspy (1991), "Stepdown Analysis of Variance: A Refinement." *Journal of Organizational Behavior* 12: 555–59.
15. Läuter, J. (1978), "Sample Size Requirements for the T^2 Test of MANOVA (Tables for One-Way Classification)." *Biometrical Journal* 20: 389–406.
16. Meyers, J. L. (1975), *Fundamentals of Experimental Design.* Boston: Allyn & Bacon.
17. Morrison, D. F. (1967), *Multivariate Statistical Methods.* New York: McGraw-Hill.
18. Rao, C. R. (1978), *Linear Statistical Inference and Its Application*, 2d ed. New York: Wiley.
19. Stevens, J. P. (1972), "Four Methods of Analyzing between Variations for the k-Group MANOVA Problem." *Multivariate Behavioral Research* 7 (October): 442–54.
20. Stevens, J. P. (1980), "Power of the Multivariate Analysis of Variance Tests." *Psychological Bulletin* 88: 728–37.
21. Tatsuoka, M. M. (1971), *Multivariate Analysis: Techniques for Education and Psychological Research.* New York: Wiley.
22. Wilks, S. S. (1932), "Certain Generalizations in the Analysis of Variance." *Biometrika* 24: 471–94.
23. Winer, B. J. (1962), *Statistical Principles in Experimental Design.* New York: McGraw-Hill.

Annotated Articles

The following annotated articles are provided as illustrations of the application of MANOVA and ANOVA to substantive research questions of both a conceptual and managerial nature. The reader is encouraged to review the complete articles for greater detail on any of the specific issues regarding methodology or findings.

Bello, Daniel C., and Nicholas C. Williamson (1985), "The American Export Trading Company: Designing a New International Marketing Institution." *Journal of Marketing* 49(4): 60–69.

In this article, the authors have used a combination of MANOVA and ANOVA to determine which services should be offered by export management companies. MANOVA allows the researcher to examine the equality of several groups across a number of independent variables in one step. This avoids compounding the Type II error that is associated with obtaining the same information through the use of multiple *t* tests or ANOVAs. MANOVA also allows for the examination of interactions among the variables in order to assess any effect that might be present when multiple service strategies are proposed. Based on importance ratings from a sample of export firm presidents or owners, four export services were identified: promotion, technical export, market contact, and consolidation services. These ratings were examined to determine their effects on product (differentiated or undifferentiated), exporter role (commercial agent or merchant distributor), and supplier sales volume (over or under $1 million). By examining the differences and direction of the mean service importance ratings across the variations in product, exporter role, and supplier sales volume, export management companies should obtain a list of variables for use in developing an appropriate marketing mix. MANOVA, for multivariate comparison, and ANOVA, for univariate comparison, are used to provide information detailing where these differences exist and the probability that these differences could have been obtained by chance. With MANOVA's ability to allow for the examination of interactions, the researchers were provided with information about the possible effects of supplying multiple services.

The dependent variables (four export services scales) are analyzed in a $2 \times 2 \times 2$ (product type $\times$ exporter role $\times$ supplier sales volume) MANOVA design. The authors take into account sampling requirements and ensure that the dependent measures are significantly correlated. The MANOVA results indicate a significant overall main effect for product, exporter role, and supplier sales volume factors with no significant two- or three-way interactions. The authors follow up the MANOVA results with three separate univariate ANOVA tests. The ANOVA results indicate significant individual differences in export services across product, exporter role, and supplier sales volume. Through the use of multivariate analysis of variance (MANOVA), the researchers are able to examine the difference in importance perception of various export services based upon the characteristics of the export intermediary. The authors conclude that the product type, exporter role, and supplier sales volume directly impact importance perception of services provided by the exporter.

Urbany, Joel E., William O. Bearden, and Dan C. Weilbaker (1988), "The Effect of Plausible and Exaggerated Reference Prices on Consumer Perceptions and Price Search." *Journal of Consumer Research* 15(1): 95–110.

This article seeks to empirically test the effect of external reference prices (e.g., advertised prices) on the price perceptions of consumers. Specifically, it addresses whether references prices that are higher than the consumer's initial highest price estimates will affect (1) perceptions of market prices, (2) perceived offer value and search benefit, and (3) purchase behavior. The independent variable, reference price, involves four manipulations (1) no reference price, to serve as a control; (2) average reference price; (3) above-average reference price; and (4) well above market reference price. The authors offer two hypotheses: one examines the effect of a plausible reference price on consumer perceptions and behaviors, and the other examines the effect of an exaggerated reference price on consumer perceptions and behaviors.

To test their hypotheses, the authors perform two experiments. In experiment one, a four-group MANOVA, the subjects are exposed to a sales price and one of four possible reference prices (none, average, above-average, or a well above market reference price). From the results the authors conclude that perceptions of perceived offer value and estimates of typical prices are influenced by both plausible and exaggerated reference prices. Experiment two replicates and extends the earlier findings in a two-group MANOVA. First, the authors vary sales price along two dimensions: a sales price and a lower-than-expected sales price. One manipulation of reference price is dropped in the second experiment, maintaining the control and exaggerated reference prices (above average and well above market). With the addition of the lower-than-expected sales price, the experiment calls for a 2 × 3 (sales price × reference price) factorial design. The results support the hypothesis that exaggerated reference prices can increase consumer estimates of prices and perceived offer value, and reduce search benefits. MANOVA, along with univariate ANOVA, enables the authors to test both hypotheses. In sum, the results demonstrate that higher plausible reference prices give the appearance that the offered price is a better value than if no reference price is indicated.

Brewer, Neil, Lynne Socha, and Rob Potter (1996), "Gender Differences in Supervisors' Use of Performance Feedback." *Journal of Applied Social Psychology* 26(9): 786–803.

This article examines whether supervisor gender affects the nature of performance feedback toward subordinates. Feedback delivery varies based upon frequency, timing, sign (positive/negative/neutral), and specificity. Through the use of MANOVA, the authors are able to measure the impact of variations in supervisor and subordinate gender as well as subordinate performance (above or below average) on the delivery of performance feedback. The authors hypothesize that male supervisors use specific and general negative feedback more frequently and earlier for poor subordinate performance; whereas female supervisors use more specific positive feedback for good performance and more general positive and neutral (encouragement) feedback for both good and poor performance. Testing the hypotheses requires eight treatment conditions in a 2 × 2 × 2 (supervisor gender × subordinate gender × performance) factorial design. Using a sample of 30 male and 30 female undergraduate students acting in supervisory roles, the authors collected data on the delivery of performance feedback in a controlled laboratory setting.

Through multivariate (MANOVA) and univariate (ANOVA) techniques, the authors find that frequencies of general negative, specific positive, general positive, and neutral feedback are unaffected by supervisor gender. To assess the individual differences of the dependent variables and to account for the unique contribution of each dependent variable, the authors also perform a stepdown analysis. Similar to stepwise regression, this approach computes a univariate F statistic for a dependant variable after eliminating the effect of the other dependent variables in the preceding analysis. Although the MANOVA results do not indicate a significant main effect for supervisor gender, the stepdown analysis supports the hypothesis that males use negative feedback in response to poor performance both sooner and more frequently. In sum, the results confirm prior laboratory and field research, which has found males to have a more directive leadership style.

Conjoint Analysis

LEARNING OBJECTIVES

Upon completing this chapter, you should be able to do the following:

- Explain the many managerial uses of conjoint analysis.
- Understand the guidelines for selecting the variables to be examined by conjoint analysis, as well as their values.
- Formulate the experimental plan for simple conjoint analysis, including how to create factorial designs and understanding the impact of choosing from rank choice versus ratings as the measure of preference.
- Assess the relative importance of the predictor variables and each of their levels in affecting consumer judgments.
- Apply a choice stimulator to conjoint results for the prediction of consumer judgments of new attribute combinations.
- Examine the implications for selecting a main effects model versus a model involving interaction terms and demonstrate approaches for establishing the validity of one model versus the other.
- Recognize the limitations of traditional conjoint analysis and select the appropriate alternative methodology (e.g., choice-based or adaptive conjoint) when necessary.

CHAPTER PREVIEW

Since the mid-1970s, conjoint analysis has attracted considerable attention as a method that portrays consumers' decisions realistically as trade-offs among multiattribute products or services [23]. Conjoint analysis gained widespread acceptance and use in many industries, with usage rates increasing up to tenfold in the 1980s [66]. During the 1990s, the application of conjoint analysis increased even further, spreading to many fields of study. Marketing's widespread utilization of conjoint in new product development for consumers led to its adoption in many other areas, such as industrial marketing [41]. This rise in usage in the United States has been paralleled in other parts of the world as well, particularly in Europe [70].

Coincident with this continued growth was the development of alternative methods of constructing the choice tasks for consumers and estimating the conjoint models. Many of the multivariate techniques we discuss in this text have become established in the statistical field; conjoint analysis, on the other hand, has continued and will continue to develop in terms of its design, estimation, and applications within many areas of research [11].

Accelerated use of conjoint analysis has coincided with the widespread introduction of computer programs that integrate the entire process, from generating the combinations of independent variable values to be evaluated to creating choice simulators for predicting consumer choices across a wide number of alternative product and service formulations. Today several widely employed packages can be accessed by any researcher with a personal computer [6, 7, 8, 28, 50, 51, 52, 53, 56, 57]. Moreover, the conversion of research developments into available PC-based programs is continuing [11], and interest in these software programs is increasing [10, 45, 46].

Conjoint analysis is closely related to traditional experimentation. For example, a chemist in a soap manufacturing plant may want to know the effect of the temperature and pressure in the soap-making vats on the density of the resulting bar of soap. The chemist could conduct laboratory experiments to measure these relationships. Once the experiments were conducted, they could be analyzed with ANOVA (analysis of variance) procedures such as those discussed in chapter 6. In situations involving human behavior, we often also need to conduct "experiments" with the factors we control. For example, should the bar of soap be slightly or highly fragranced? Should it be promoted as a cosmetic or a cleaner and/or deodorizer? Which of three prices should be charged? The conjoint analysis technique developed from the need to analyze the effects of the factors we control (independent variables) that are often qualitatively specified or weakly measured [19, 21]. Conjoint analysis is actually a family of techniques and methods, all theoretically based on the models of information integration and functional measurement [38].

In terms of the basic dependence model discussed in chapter 1, conjoint analysis can be expressed as

$$Y_1 = X_1 + X_2 + X_3 + \ldots + X_N$$
(nonmetric or metric)   (nonmetric)

Conjoint analysis is best suited for understanding consumers' reactions to and evaluations of predetermined attribute combinations that represent potential products or services. While maintaining a high degree of realism, it provides the researcher with insight into the composition of consumer preferences. The flexibility and uniqueness of conjoint analysis arise primarily from (1) its ability to accommodate either a metric or a nonmetric dependent variable, (2) the use of categorical predictor variables, and (3) the quite general assumptions about the relationships of independent variables with the dependent variable.

KEY TERMS

Before starting the chapter, review the key terms to develop an understanding of the concepts and terminology to be used. Throughout the chapter the key terms appear in **boldface**. Other points of emphasis in the chapter are *italicized*. Also, cross-references within the Key Terms appear in *italics*.

Adaptive conjoint Methodology for conducting a conjoint analysis that relies on information from the respondents (e.g., importance of attributes) to adapt the conjoint design to make the task even simpler. Examples are the *self-explicated* and *adaptive* or *hybrid models*.

Adaptive model Technique for simplifying conjoint analysis by combining the *self-explicated* and traditional conjoint models. The most common example is *Adaptive Conjoint* from Sawtooth Software.

Additive model Model based on the additive *composition rule*, which assumes that individuals just "add up" the *part-worths* to calculate an overall or "total worth" score indicating *utility* or preference. It is also known as a *main effects model*. It is the simplest conjoint model in terms of the number of evaluations and the estimation procedure required.

Balanced design Stimuli *design* in which each level within a *factor* appears an equal number of times.

Bridging design Stimuli *design* for a large number of *factors* (attributes) in which the attributes are broken into a number of smaller groups. Each attribute group has some attributes contained in other groups, so the results from each group can be combined, or bridged.

Choice-based conjoint approach Alternative form of collecting responses and estimating the conjoint model. The primary difference is that respondents select a single *full profile stimulus* from a set of stimuli (known as a *choice set*) instead of rating or ranking each stimulus separately.

Choice set Set of full profile stimuli constructed through experimental design principles and used in the *choice-based approach*.

Choice simulator Procedure that allows the researcher to assess many "what-if" scenarios, including the preference for possible product or service configurations or the competitive interactions among stimuli assumed to constitute a market. Once the conjoint *part-worths* have been estimated for each respondent, the choice simulator analyzes a set of *full profile stimuli* and predicts both individual and aggregate choices for each stimulus in the set.

Composition rule Rule used in combining attributes to produce a judgment of relative value or *utility* for a product or service. For illustration, let us suppose a person is asked to evaluate four objects. The person is assumed to evaluate the attributes of the four objects and to create some overall relative value for each. The rule may be as simple as creating a mental weight for each perceived attribute and adding the weights for an overall score (*additive model*), or it may be a more complex procedure involving *interaction effects*.

Compositional model Class of multivariate models that base the dependence relationship on observations from the respondent regarding both the dependent and the independent variables. Such models calculate or "compose" the dependent variable from the respondent-supplied values for the independent variables. Principal among such methods are regression analysis and discriminant analysis. These models are in direct contrast to *decompositional models*.

Conjoint variate Combination of variables (known as *factors*) specified by the researcher that constitute the total worth or *utility* of the *stimuli*. The researcher also specifies all the possible values for each factor, with these values known as *levels*.

Decompositional model Class of multivariate models that "decompose" the respondent's preference. This class of models presents the respondent with a predefined set of independent variables, usually in the form of a hypothetical or actual product or service, and then asks for an overall evaluation or preference of the product or service. Once given, the preference is "decomposed" by relating the known attributes of the product (which become the independent variables) to the evaluation (dependent variable). Principal among such models is conjoint analysis and some forms of multidimensional scaling (see chapter 10).

Design Specific set of conjoint stimuli created to exhibit the specific statistical properties of *orthogonality* and *balance*.

Design efficiency Degree to which a *design* matches an *orthogonal* design. This measure is primarily used to evaluate and compare *nearly orthogonal* designs. Design efficiency values range from 0 to 100, which denotes an *optimal design*.

Environmental correlation. See *interattribute correlation*.

Factor Variable the researcher manipulates that represents a specific attribute. In conjoint analysis, the factors (independent variables) are nonmetric. Factors must be represented by two or more values (also known as *levels*), which are also specified by the researcher.

Factorial design Method of designing *stimuli* for evaluation by generating *all* possible combinations of *levels*. For example, a three-factor conjoint analysis with three levels per factor ($3 \times 3 \times 3$) would result in 27 combinations that could act as stimuli.

Fractional factorial design Approach, as an alternative to a *factorial design*, that uses only a subset of the possible stimuli needed to estimate the results based on the assumed composition rule. Its primary task is to reduce the number of evaluations collected while still maintaining *orthogonality* among the *levels* and subsequent *part-worth* estimates. The simplest design is an *additive model*, in which only *main effects* are estimated. If selected *interaction terms* are included, then additional stimuli are created. The design can be created either by referring to published sources or by using computer programs that accompany most conjoint analysis packages.

Full-profile method Method of presenting *stimuli* to the respondent for evaluation that consists of a complete description of the stimuli across all attributes. For example, let us assume that a candy was described by three factors with two levels each: price (15 or 25 cents), flavor (citrus or butterscotch), and color (white or red). A full profile stimulus would be defined by one level of each factor. One such full profile stimulus would be a red butterscotch candy costing 15 cents.

Holdout stimuli See *validation stimuli*.

Hybrid model See *adaptive model*.

Interaction effects Effects of a combination of related features, also known as interaction terms. In assessing value, a person may assign a unique value to specific combinations of features that runs counter to the *additive composition rule*. For example, let us assume a person is evaluating mouthwash products described by the *factors* (attributes) of color and brand. Let us further assume that this person has an average preference for red and brand X. When this specific combination of levels (red and brand X) is evaluated with the same additive composition rule as all other combinations, the red brand X product would have an expected overall preference rating somewhere in the middle of all possible stimuli. If, however, the person actually prefers the red brand X mouthwash more than any other stimuli, even above other combinations of attributes (color and brand) that had higher evaluations of the individual features, then an interaction is found to exist. This unique evaluation of a combination that is greater (or could be less) than expected based on the separate judgments indicates a two-way interaction. Higher-order (three-way or more) interactions can occur among more combinations of levels.

Interattribute correlation   Correlation among attributes, also known as *environmental correlation*, that makes combinations of attributes unbelievable or redundant. A negative correlation depicts the situation in which two attributes are naturally assumed to operate in different directions, such as horsepower and gas mileage. As one increases, the other is naturally assumed to decrease. Thus, because of this correlation, all combinations of these two attributes (e.g., high gas mileage and high horsepower) are not believable. The same effects can be seen for positive correlations, where perhaps price and quality are assumed to be positively correlated. It may not be believable to find a high-price, low-quality product in such a situation. The presence of strong interattribute correlations requires that the researcher closely examine the stimuli presented to respondents and avoid unbelievable combinations that are not useful in estimating the part-worths.

Level   Specific value describing a *factor*. Each factor must be represented by two or more levels, but the number of levels typically never exceeds four or five. If the factor is metric, it must be reduced to a small number of levels. For example, the many possible values of size and price may be represented by a small number of levels: size (10, 12, or 16 ounces); or price ($1.19, $1.39, or $1.99). If the variable is nonmetric, the original values can be used as in these examples: Color (red or blue); brand (X, Y, or Z); or fabric softener additive (present or absent).

Main effects   Direct effect of each factor (independent variable) on the dependent variable. May be complemented by interaction effects in specific situations.

Nearly orthogonal   Characteristic of a stimuli *design* that is not *orthogonal*, but the deviations from orthogonality are slight and carefully controlled in the generation of the stimuli. This type of design can be compared with other stimuli designs with measures of *design efficiency*.

Optimal design   Stimuli *design* that is *orthogonal* and *balanced*.

Orthogonal   Mathematical constraint requiring that the *part-worth* estimates be independent of each other. In conjoint analysis, orthogonality refers to the ability to measure the effect of changing each attribute level and to separate it from the effects of changing other attribute levels and from experimental error.

Pairwise comparison method   Method of presenting a pair of *stimuli* to a respondent for evaluation, with the respondent selecting one stimuli as preferred.

Part-worth   Estimate from conjoint analysis of the overall preference or *utility* associated with each *level* of each *factor* used to define the product or service.

Preference structure   Representation of both the relative importance or worth of each *factor* and the impact of individual *levels* in affecting utility.

Self-explicated model   *Compositional* technique for performing conjoint analysis in which the respondent provides the *part-worth estimates* directly without making choices.

Stimulus   Specific set of *levels* (one per *factor*) evaluated by respondents (also known as a *treatment*). One method of defining stimuli (*factorial design*) is achieved by taking all combinations of all levels. For example, three factors with two levels each would create eight ($2 \times 2 \times 2$) stimuli. However, in many conjoint analyses, the total number of combinations is too large for a respondent to evaluate them all. In these instances, some subsets of stimuli are created according to a systematic plan, most often a *fractional factorial design.*

Trade-off method   Method of presenting stimuli to respondents in which attributes are depicted two at a time and respondents rank all combinations of the levels in terms of preference.

Traditional conjoint analysis   Methodology that employs the "classic" principles of conjoint analysis, using an *additive* model of consumer preference and *pairwise* comparison or *full-profile* methods of presentation.

Treatment   See *stimulus.*

Utility   A subjective preference judgment by an individual representing the holistic value or worth of a specific object. In conjoint analysis, utility is assumed to be formed by the combination of *part-worth estimates* for any specified set of *levels* with the use of an *additive model*, perhaps in conjunction with *interaction effects.*

Validation stimuli   Set of *stimuli* that are not used in the estimation of part-worths. Estimated part-worths are then used to predict preference for the validation stimuli to assess validity and reliability of the original estimates. Similar in concept to the validation sample of respondents in discriminant analysis.

What Is Conjoint Analysis?

Conjoint analysis is a multivariate technique used specifically to understand how respondents develop preferences for products or services. It is based on the simple premise that consumers evaluate the value of a product/service/idea (real or hypothetical) by combining the separate amounts of value provided by each attribute. **Utility,** which is the conceptual basis for measuring value in conjoint analysis, is a subjective judgment of preference unique to each individual. It encompasses all product or service features, both tangible and intangible, and as such is a measure of overall preference. In conjoint analysis, utility is assumed to be based on the value placed on each of the levels of the attributes and expressed in a relationship reflecting the manner in which the utility is formulated for any combination of attributes. For example, we might sum the utility values associated with each feature of a product or service to arrive at an overall utility. Then we would assume that products or services with higher utility values are more preferred and have a better chance of choice.

Conjoint analysis is unique among multivariate methods in that the researcher first constructs a set of real or hypothetical products or services by combining selected levels of each attribute. These combinations are then presented to respondents, who provide only their overall evaluations. Thus, the researcher is asking the respondent to perform a very realistic task—choosing among a set of products. Respondents need not tell the researcher anything else, such as how important an individual attribute is to them or how well the product performs on any specific attribute. Because the researcher constructed the hypothetical products or services in a specific manner, the influence of each attribute and each value of each attribute on the utility judgment of a respondent can be determined from the respondents' overall ratings.

To be successful, the researcher must be able to describe the product or service in terms of both its attributes and all relevant values for each attribute. We use the term **factor** when describing a specific attribute or other characteristic of the product or service. The possible values for each factor are called **levels**. In conjoint terms, we describe a product or service in terms of *its level on the set of factors* characterizing it. For example, brand name and price might be two factors in a conjoint analysis. Brand name might have two levels (brand X and brand Y), whereas price might have four levels (39 cents, 49 cents, 59 cents, and 69 cents). When the researcher selects the factors and the levels to describe a product or service according to a specific plan, the combination is known as a **treatment** or **stimulus.** Therefore, a stimulus for our simple example might be brand X at 49 cents.

A Hypothetical Example of Conjoint Analysis

As an illustration of conjoint analysis, let us assume that HATCO is trying to develop a new industrial cleanser. After discussion with sales representatives and focus groups, management decides that three attributes are important: cleaning ingredients, convenience of use, and brand name. To operationalize these attributes, the researchers create three factors with two levels each:

Factor	*Level*	
Ingredients	Phosphate-free	Phosphate-based
Form	Liquid	Powder
Brand name	HATCO	Generic brand

A hypothetical cleaning product can be constructed by selecting one level of each attribute. For the three attributes (factors) with two values (levels), eight (2 × 2 × 2) combinations can be formed. Three examples of the eight possible combinations (stimuli) are:

- HATCO phosphate-free powder
- Generic phosphate-based liquid
- Generic phosphate-free liquid

HATCO customers are then asked either to rank-order the eight stimuli in terms of preference or to rate each combination on a preference scale (perhaps a 1-to-10 scale). We can see why conjoint analysis is also called "trade-off analysis," because in making a judgment on a hypothetical product, respondents must consider both the "good" and "bad" characteristics of the product in forming a preference. Thus, respondents must weigh all attributes simultaneously in making their judgments.

By constructing specific combinations (stimuli), the researcher is attempting to understand a respondent's **preference structure.** The preference structure "explains" not only how important each factor is in the overall decision, but also how the differing levels within a factor influence the formation of an overall preference (utility). In our example, conjoint analysis would assess the relative impact of each brand name (HATCO versus generic), each form (powder versus liquid), and the different cleaning ingredients (phosphate-free versus phosphate-based) in determining a person's utility. This utility, which represents the total "worth" or overall preference of an object, can be thought of as based on the **part-worths** for each level. The general form of a conjoint model can be shown as

$$\text{(Total worth for product)}_{ij \ldots n} = \text{Part-worth of level } i \text{ for factor 1}$$
$$+ \text{ Part-worth of level } j \text{ for factor 2} + \ldots$$
$$+ \text{ Part-worth of level } n \text{ for factor } m$$

where the product or service has m attributes, each having n levels. The product consists of level i of factor 2, level j of factor 2, and so forth, up to level n for factor m.

In our example, a simple additive model would represent the preference structure for the industrial cleanser as based on the three factors (utility = brand effect + ingredient effect + form effect). The preference for a specific cleanser product can be directly calculated from the part-worth values. For example, the preference for HATCO phosphate-free powder is

Utility = Part-worth of HATCO brand

+ Part-worth of phosphate-free cleaning ingredient

+ Part-worth of powder

With the part-worth estimates, the preference of an individual can be estimated for any combination of factors. Moreover, the preference structure would reveal the factor(s) most important in determining overall utility and product choice. The choices of multiple respondents could also be combined to represent the competitive environment faced in the "real world."

An Empirical Example

To illustrate a simple conjoint analysis, assume that the industrial cleanser experiment was conducted with respondents who purchased industrial supplies. Each respondent was presented with eight descriptions of cleanser products (stimuli) and asked to rank them in order of preference for purchase (1 = most preferred, and 8 = least preferred). The eight stimuli are described in Table 7.1, along with the rank orders given by two respondents.

TABLE 7.1 Stimuli Descriptions and Respondent Rankings for Conjoint Analysis of Industrial Cleanser

	Stimuli Descriptions			*Respondent Rankings*	
	Form	*Ingredients*	*Brand*	*Respondent 1*	*Respondent 2*
1	Liquid	Phosphate-free	HATCO	1	1
2	Liquid	Phosphate-free	Generic	2	2
3	Liquid	Phosphate-based	HATCO	5	3
4	Liquid	Phosphate-based	Generic	6	4
5	Powder	Phosphate-free	HATCO	3	7
6	Powder	Phosphate-free	Generic	4	5
7	Powder	Phosphate-based	HATCO	7	8
8	Powder	Phosphate-based	Generic	8	6

As we examine the responses for respondent 1, we see that the ranks for the stimuli with the phosphate-free ingredients are the highest possible (1, 2, 3, and 4), whereas the phosphate-based ingredient has the four lowest ranks (5, 6, 7, and 8). Thus, the phosphate-free ingredient is much more preferred to the phosphate-based cleanser. This can be contrasted to the ranks for the two brands, which show a mixture of high and low ranks for each brand. Assuming that the basic model (an additive model) applies, we can calculate the impact of each level as differences (deviations) from the overall mean ranking. (Readers may note that this is analogous to multiple regression with dummy variables or ANOVA.) For example, the average ranks for the two cleanser ingredients (phosphate-free versus phosphate-based) for respondent 1 are:

$$\text{Phosphate-free:} \quad (1 + 2 + 3 + 4)/4 = 2.5$$
$$\text{Phosphate-based:} \quad (5 + 6 + 7 + 8)/4 = 6.5$$

With the average rank of the eight stimuli of 4.5 [$(1 + 2 + 3 + 4 + 5 + 6 + 7 + 8)/8 = 36/8 = 4.5$], the phosphate-free level would then have a deviation of -2.0 ($2.5 - 4.5$) from the overall average, whereas the phosphate-based level would have a deviation of $+2.0$ ($6.5 - 4.5$). The average ranks and deviations for each factor from the overall average rank (4.5) for respondents 1 and 2 are given in Table 7.2. In our example, we use smaller numbers to indicate higher ranks and a more preferred stimulus (e.g., 1 = most preferred). When the preference measure is inversely related to preference, such as here, we reverse the signs of the deviations in the part-worth calculations so that positive deviations will be associated with part-worths indicating greater preference.

TABLE 7.2 Average Ranks and Deviations for Respondents 1 and 2

Factor Level	Ranks across Stimuli	Average Rank of Level	Deviation from Overall Average Rank[a]
Respondent 1			
Form			
Liquid	1, 2, 5, 6	3.5	−1.0
Powder	3, 4, 7, 8	5.5	+1.0
Ingredients			
Phosphate-free	1, 2, 3, 4	2.5	−2.0
Phosphate-based	5, 6, 7, 8	6.5	+2.0
Brand			
HATCO	1, 3, 5, 7	4.0	−.5
Generic	2, 4, 6, 8	5.0	+.5
Respondent 2			
Form			
Liquid	1, 2, 3, 4	2.5	−2.0
Powder	5, 6, 7, 8	6.5	+2.0
Ingredients			
Phosphate-free	1, 2, 5, 7	3.75	−.75
Phosphate-based	3, 4, 6, 8	5.25	+.75
Brand			
HATCO	1, 3, 7, 8	4.75	+.25
Generic	2, 4, 5, 6	4.25	−.25

[a]Deviation calculated as: deviation = average Rank of Level − overall Average Rank (4.5). Note that negative deviations imply more preferred rankings.

The part-worths of each level are calculated in four steps:

Step 1: Square the deviations and find their sum across all levels.
Step 2: Calculate a standardizing value that is equal to the total number of levels divided by the sum of squared deviations.
Step 3: Standardize each squared deviation by multiplying it by the standardizing value.
Step 4: Estimate the part-worth by taking the square root of the standardized squared deviation.

Let us examine how we would calculate the part-worth of the first level of ingredients (phosphate-free) for respondent 1. The deviations from 2.5 are squared. The squared deviations are summed (10.5). The number of levels is six (3 factors with 2 levels apiece). Thus, the standardizing value is calculated as .571 (6/10.5 = .571). The squared deviation for phosphate-free (2^2; remember that we reverse signs) is then multiplied by .571 to get 2.284 ($2^2 \times .571 = 2.284$). Finally, to calculate the part-worth for this level, we then take the square root of 2.284, for a value of 1.1511. This process yields part-worths for each level for respondents 1 and 2, as shown in Table 7.3.

Because the part-worth estimates are on a common scale, we can compute the relative importance of each factor directly. The importance of a factor is represented by the range of its levels (i.e., the difference between the highest and lowest values) divided by the sum of the ranges across all factors. For example, for respondent 1, the ranges are 1.512 [.756 − (−.756)], 3.022 [1.511 − (−1.511)], and .756 [.378 − (−.378)]. The sum total of ranges is 5.290. The relative importance for form, ingredients, and brand is calculated as 1.512/5.290, 3.022/5.290, and .756/5.290, or 28.6 percent, 57.1 percent, and 14.3 percent, respectively. We can follow the same procedure for the second respondent and calculate the importance of each factor, with the results of form (66.7 percent), ingredients (25 percent), and brand (8.3 percent). These calculations for respondents 1 and 2 are also shown in Table 7.3.

To examine the ability of this model to predict the actual choices of the respondents, we predict preference order by summing the part-worths for the different combinations of factor levels and then rank-ordering the resulting scores. The calculations for both respondents for all eight stimuli are shown in Table 7.4. Comparing the predicted preference order to the respondent's actual preference order assesses predictive accuracy. Note that the total part-worth values have no real meaning except as a means of developing the preference order and, as such, are not compared across respondents. The predicted and actual preference orders for both respondents are also given in Table 7.4 (p. 398).

The estimated part-worths predict the preference order perfectly for respondent 1. This indicates that the preference structure was successfully represented in the part-worth estimates and that the respondent made choices consistent with the preference structure. The need for consistency is seen when the rankings for respondent 2 are examined. For example, the average rank for the generic brand is lower than the HATCO (refer to Table 7.2), meaning that, all things being equal, the stimuli with the generic brand will be more preferred. Yet, examining the actual rank orders, this is not always seen. Stimuli 1 and 2 are equal except for brand name, yet HATCO is more preferred. This also occurs for stimuli 3 and 4. However, the correct ordering (generic preferred over HATCO) is seen for the stimuli pairs

TABLE 7.3 Estimated Part-Worths and Factor Importance for Respondents 1 and 2

Factor Level	Estimated Part-Worths				Calculating Factor Importance	
	Reversed Deviation[a]	Squared Deviation	Standardized Deviation[b]	Estimated Part-Worth[c]	Range of Part-Worths	Factor Importance[d]
Respondent 1						
Form						
Liquid	+1.0	1.0	+.571	+.756	1.512	28.6%
Powder	−1.0	1.0	−.571	−.756		
Ingredients						
Phosphate-free	+2.0	4.0	+2.284	+1.511	3.022	57.1%
Phosphate-based	−2.0	4.0	−2.284	−1.511		
Brand						
HATCO	+.5	.25	+.143	+.378	.756	14.3%
Generic	−.5	.25	−.143	−.378		
Sum of squared deviations		10.5				
Standardizing value[e]		.571				
Sum of part-worth ranges					5.290	
Respondent 2						
Form						
Liquid	+2.0	4.0	+2.60	+1.612	3.224	66.7%
Powder	−2.0	4.0	−2.60	−1.612		
Ingredients						
Phosphate-free	+.75	.5625	+.365	+.604	1.208	25.0%
Phosphate-based	−.75	.5625	−.365	−.604		
Brand						
HATCO	−.25	.0625	−.04	−.20	.400	8.3%
Generic	+.25	.0625	+.04	+.20		
Sum of squared deviations		9.25				
Standardizing value		.649				
Sum of part-worth ranges					4.832	

[a]Deviations are reversed to indicate higher preference for lower ranks. Sign of deviation used to indicate sign of estimated part-worth.
[b]Standardized deviation equal to the squared deviation times the standardizing value.
[c]Estimated part-worth equal to the square root of the standardized deviation.
[d]Factor importance equal to the range of a factor divided by the sum of the ranges across all factors, multiplied by 100 to yield a percentage.
[e]Standardizing value equal to the number of levels (2 + 2 + 2 = 6) divided by the sum of the squared deviations.

of 5–6 and 7–8. Thus, the preference structure of the part-worths will have a difficult time predicting this choice pattern. When we compare the actual and predicted rank orders (see Table 7.4), we see that respondent 2's choices are many times mispredicted, but most often just miss by one position due to the brand effect. Thus, we would conclude that the preference structure is an adequate representation of the choice process for the more important factors, but that it does not predict choice perfectly for respondent 2, as it does for respondent 1.

TABLE 7.4 Predicted Part-Worth Totals and Comparison of Actual and Estimated Preference Rankings

Stimuli Description			Part-Worth Estimates				Preference Rankings	
Size	Ingredients	Brand	Size	Ingredients	Brand	Total	Estimated	Actual
Respondent 1								
Liquid	Phosphate-free	HATCO	.756	1.511	.378	2.645	1	1
Liquid	Phosphate-free	Generic	.756	1.511	−.378	1.889	2	2
Liquid	Phosphate-based	HATCO	.756	−1.511	.378	−.377	5	5
Liquid	Phosphate-based	Generic	.756	−1.511	−.378	−1.133	6	6
Powder	Phosphate-free	HATCO	−.756	1.511	.378	1.133	3	3
Powder	Phosphate-free	Generic	−.756	1.511	−.378	.377	4	4
Powder	Phosphate-based	HATCO	−.756	−1.511	.378	−1.889	7	7
Powder	Phosphate-based	Generic	−.756	−1.511	−.378	−2.645	8	8
Respondent 2								
Liquid	Phosphate-free	HATCO	1.612	.604	−.20	2.016	2	1
Liquid	Phosphate-free	Generic	1.612	.604	.20	2.416	1	2
Liquid	Phosphate-based	HATCO	1.612	−.604	−.20	.808	4	3
Liquid	Phosphate-based	Generic	1.612	−.604	.20	1.208	3	4
Powder	Phosphate-free	HATCO	−1.612	.604	−.20	−1.208	6	7
Powder	Phosphate-free	Generic	−1.612	.604	.20	−.808	5	5
Powder	Phosphate-based	HATCO	−1.612	−.604	−.20	−2.416	8	8
Powder	Phosphate-based	Generic	−1.612	−.604	.20	−2.016	7	6

The Managerial Uses of Conjoint Analysis

Before discussing the statistical basis of conjoint analysis, we should understand the technique in terms of its role in decision making and strategy development. The simple example we have just discussed presents some of the basic benefits of conjoint analysis. The flexibility of conjoint analysis gives rise to its application in almost any area in which decisions are studied. Conjoint analysis assumes that any set of objects (e.g., brands, companies) or concepts (e.g., positioning, benefits, images) is evaluated as a bundle of attributes. Having determined the contribution of each factor to the consumer's overall evaluation, the marketing researcher could then

1. Define the object or concept with the optimum combination of features.
2. Show the relative contributions of each attribute and each level to the overall evaluation of the object.
3. Use estimates of purchaser or customer judgments to predict preferences among objects with differing sets of features (other things held constant).
4. Isolate groups of potential customers who place differing importance on the features to define high and low potential segments.
5. Identify marketing opportunities by exploring the market potential for feature combinations not currently available.

The knowledge of the preference structure for each individual allows the researcher almost unlimited flexibility in examining both individual and aggregate reactions to a wide range of product- or service-related issues. We examine some of the most popular applications later in this chapter.

Comparing Conjoint Analysis with Other Multivariate Methods

Conjoint analysis differs from other multivariate techniques in three distinct areas: (1) its decompositional nature, (2) the fact that estimates can be made at the individual level, and (3) its flexibility in terms of relationships between dependent and independent variables.

Compositional versus Decompositional Techniques

Conjoint analysis is termed a **decompositional model** because the researcher needs to know only a respondent's overall preference for an object created by the researcher through specifying the values (levels) of each attribute (factor). In this way conjoint analysis can *decompose* the preference to determine the value of each attribute. Conjoint analysis differs from **compositional models** such as discriminant analysis and many regression applications, in which the researcher collects ratings from the respondent on many product characteristics (e.g., favorability toward color, style, specific features) and then relates these ratings to some overall preference rating to develop a predictive model. The researcher does not know beforehand the ratings on the product characteristics, but collects them from the respondent. With regression and discriminant analysis the respondent's ratings and overall preferences are analyzed to "compose" the overall preference from the respondent's evaluations of the product on *each* attribute.

Specifying the Conjoint Variate

Conjoint analysis employs a variate quite similar in form to what we have seen in other multivariate techniques. The **conjoint variate** is a linear combination of effects of the independent variables (factors) on a dependent variable. The important difference is that in the conjoint variate, the researcher specifies both the independent variables (factors) *and* their values (levels). The only information provided by the respondent is the dependent measure. The levels specified by the researcher are then used by conjoint analysis to decompose the respondent's response into effects for each level, much as is done in regression analysis for each independent variable. This feature illustrates the common characteristics shared by conjoint analysis and experimentation, whereby designing the project is a critical step to success. For example, if a variable or effect was not anticipated in the research design, then it will not be available for analysis. For this reason, a researcher may be tempted to include a number of variables that *might* be relevant. However, conjoint analysis is limited in the number of variables it can include, so the researcher cannot just include additional questions to compensate for a lack of clear conceptualization of the problem.

Separate Models for Each Individual

Conjoint analysis differs from almost all other multivariate methods in that it can be carried out at the individual level, meaning that the researcher generates a separate "model" for predicting preference for *each* respondent. Most other multivariate methods take a single measure of preference (observation) from each respondent and then perform the analysis using all respondents simultaneously. In fact, many methods *require* that a respondent provide only a single observation (the assumption of independence) and then develop a common model for all respondents, fitting each respondent with varying degrees of accuracy (represented by the errors of prediction for each observation, such as residuals in regression). In conjoint analysis, however, estimates can be made for the individual (disaggregate) or groups of individuals representing a market segment or the entire market (aggregate). At the disaggregate level, each respondent rates enough stimuli for the analysis to be performed separately for each person. Predictive accuracy is calculated for each person, rather than only for the total sample. The individual results can then be aggregated to portray an overall model as well.

Many times, however, the researcher selects an aggregate analysis method that performs the estimation of part-worths for the group of respondents as a whole. Aggregate analysis can provide (1) a means for reducing the data collection task through more complex designs (discussed in later sections), (2) methods for estimating interactions (e.g., choice-based conjoint), and (3) greater statistical efficiency by using more observations in the estimation. In selecting between aggregate and disaggregate conjoint analyses, the researcher must balance the benefits gained by aggregate methods versus the insights provided by the separate models obtained by disaggregate methods.

Types of Relationships

Conjoint analysis is not limited at all in the types of relationships required between the dependent and independent variables. As discussed in earlier chapters in section 2, most dependence methods assume that a linear relationship exists when the dependent variable increases (or decreases) in equal amounts for each unit change in the independent variable. Conjoint analysis, however, can make separate predictions for the effects of each level of the independent variable and does not assume they are related at all. Conjoint analysis can easily handle nonlinear relationships—even the complex curvilinear relationship, in which one value is positive, the next negative, and the third positive again. As we discuss later, however, the simplicity and flexibility of conjoint analysis compared with the other multivariate methods is based on a number of assumptions made by the researcher.

Designing a Conjoint Analysis Experiment

Although conjoint analysis places the fewest demands on the respondent in terms of both the number and types of responses needed, the researcher must make a number of key decisions in designing the experiment and analyzing the results. Figure 7.1 (stages 1–3) and Figure 7.4 (stages 4–7, p. 419) show the general steps followed in the design and execution of a conjoint analysis experiment. The discussion follows the model-building paradigm introduced in chapter 1. The decision process is initiated with a specification of the objectives of conjoint analysis. Because conjoint analysis is very similar to an experiment, the conceptualization of the research is critical to its success. After the objectives have been defined, the issues related to the actual research design are addressed, and the assumptions are evaluated. The decision process then considers the actual estimation of the conjoint results, the interpretation of the results, and the methods used to validate the results. The discussion ends with an examination of the use of con-

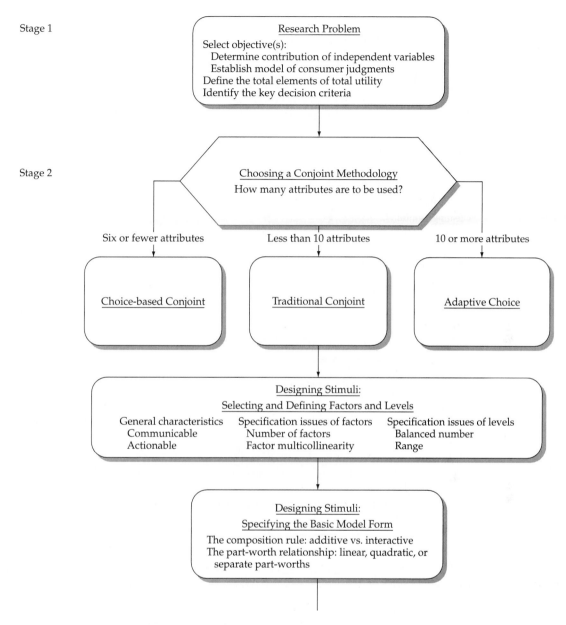

FIGURE 7.1 Stages 1–3 of the Conjoint
Analysis Decision Diagram
(figure continued on next page)

joint analysis results in further analyses such as market segmentation and choice simulators. Each of these decisions stems from the research question and the use of conjoint analysis as a tool in understanding the respondent's preferences and judgment process. We follow this discussion of the model-building approach by examining two alternative conjoint methodologies (choice-based and adaptive conjoint), which are then compared to the issues addressed here for traditional conjoint analysis.

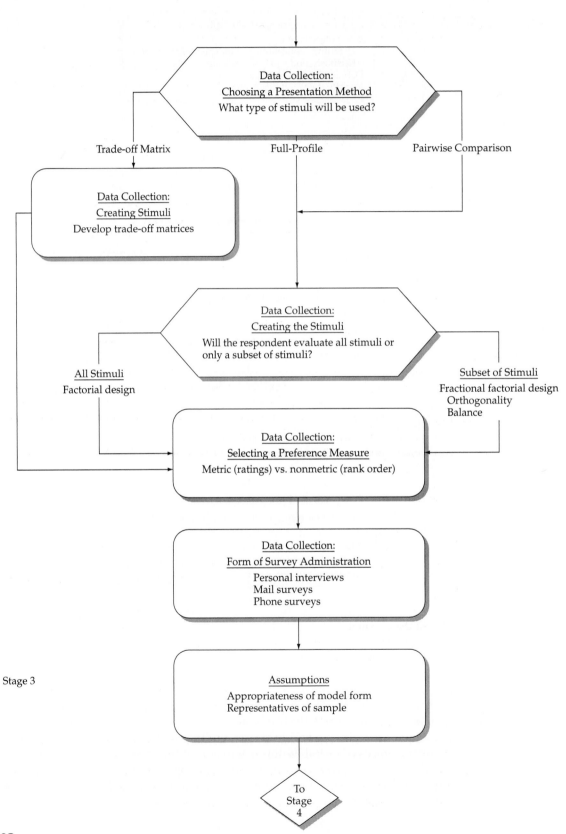

Stage 3

Stage 1: The Objectives of Conjoint Analysis

As with any statistical analysis, the starting point is the research question. In conjoint analysis, experimental design in the analysis of consumer decisions has two objectives:

1. *To determine the contributions of predictor variables and their levels in the determination of consumer preferences.* For example, how much does price contribute to the willingness to buy a product? Which price level is the best? How much change in the willingness to buy soap can be accounted for by differences between the levels of price?

2. *To establish a valid model of consumer judgments.* Valid models allow us to predict the consumer acceptance of any combination of attributes, even those not originally evaluated by consumers. In doing so, the issues addressed include the following: Do the respondent's choices indicate a simple linear relationship between the predictor variables and choices? Is a simple model of "adding up" the value of each attribute sufficient, or do we need to add more complex evaluations of preference to mirror the judgment process adequately?

The respondent reacts only to what the researcher provides in terms of stimuli (attribute combinations). Are these the actual attributes used in making a decision? Are other attributes, particularly attributes of a more qualitative nature such as emotional reactions, important as well? These and other considerations require the research question to be framed around two major issues: (1) Is it possible to describe all the attributes that give utility or value to the product or service being studied? (2) What are the key decision criteria involved in the choice process for this type of product or service? These questions need to be resolved before proceeding into the design phase of a conjoint analysis because they provide critical guidance for key decisions in each stage.

Defining the Total Utility of the Object

The researcher must first be sure to define the total utility of the object. To represent the respondent's judgment process accurately, all attributes that potentially *create* or *detract* from the overall utility of the product or service should be included. It is essential that both positive and negative factors be considered, because (1) focusing on only positive factors will seriously distort the respondents' judgments, and (2) respondents can subconsciously employ the negative factors, even though not provided, and thereby render the experiment invalid. For example, if exploratory focus groups are employed to assess the types of characteristics considered when evaluating the object, the researcher must be sure to address what makes the object unattractive, as well as attractive. Fortunately, the omission of a single factor has only a small impact on the estimates for other factors when using an additive model [49], but may impact the results if important.

Specifying the Determinant Factors

In addition, the researcher must be sure to include all determinant factors (drawn from the concept of determinant attributes [4]). The goal is to include the factors that best *differentiate* between the objects. Many attributes may be considered important but also may not differentiate in making choices because they do not vary substantially between objects. For example, safety in automobiles is a very important attribute, but it would not be determinant in most cases because all cars meet strict government standards and thus are considered safe, at least at an acceptable level. However, other features, such as gas mileage, performance, or price, are both important *and* much more likely to be used to decide among different car models. Thus the researcher should always strive to identify the key determinant variables because they are pivotal in the actual judgment decision.

Stage 2: The Design of a Conjoint Analysis

Having resolved the issues stemming from the research objectives, the researcher shifts attention to the particular issues involved in designing and executing the conjoint analysis experiment. First, which of several alternative conjoint methods should be chosen? With the model type selected, there are specific design issues to resolve. For example, how does one decide which specific combinations of attribute levels to present to the respondent for evaluation? In addition to specifying the combinations (stimuli), the researcher must also decide on issues such as which attributes to include, how many levels of each, how to measure preference and collect data, and what estimation procedure to use. We should note that the design issues are perhaps the most important phase in conjoint analysis, as a poorly designed study cannot be "saved" after administration if design flaws are discovered. Thus, the researcher must pay particular attention to the issues surrounding construction and administration of the conjoint experiment.

Selecting a Conjoint Analysis Methodology

After the researcher has determined the basic attributes that constitute the utility of the product or service (object), a fundamental question must be resolved: Which of the three basic conjoint methodologies (traditional conjoint, adaptive conjoint, or choice-based conjoint) should be used? The choice of conjoint methodologies revolves around three basic characteristics of the proposed research: number of attributes handled, level of analysis, and the permitted model form. Table 7.5 compares the three methodologies on these considerations. **Traditional conjoint analysis,** portrayed in the earlier example, is characterized by a simple additive model containing up to nine factors estimated for each individual. Although this has been the mainstay of conjoint studies for many years, two additional methodologies have been developed in an attempt to deal with certain design issues. The **adaptive conjoint** method was developed to accommodate a large number of factors (many times up to 30) that would not be feasible in traditional conjoint analy-

TABLE 7.5 A Comparison of Alternative Conjoint Methodologies

	Conjoint Methodology		
Characteristic	*Traditional Conjoint*	*Adaptive Conjoint*	*Choice-Based Conjoint*
Maximum number of attributes	9	30	6
Level of analysis	Individual	Individual	Aggregate
Model form	Additive	Additive	Additive + interaction effects

sis. The **choice-based approach** method not only employs a unique form of presenting stimuli in sets rather than one-by-one, but also differs in that it directly includes interactions and must be estimated at the aggregate level. Many times the research objectives create situations not handled well by traditional conjoint analysis, but these alternative methodologies can be employed. The issues of establishing the number of attributes and selecting the model form is discussed in greater detail in the following section, focusing on traditional conjoint analysis. Then, the unique characteristics of the two other methodologies are addressed in subsequent sections. The researcher should note that the basic issues discussed in this section apply to the two other methodologies as well.

Designing Stimuli: Selecting and Defining Factors and Levels

The experimental foundations of conjoint analysis place great importance on the design of the stimuli evaluated by respondents. The design involves specifying the conjoint variate by selecting the factors and levels to be included in constructing the stimuli. In defining the factors and levels, issues must be addressed that relate to the general character of either measure, whereas other considerations are specific to factors and levels. These design issues are important because they affect the effectiveness of the stimuli in the task, the accuracy of the results, and ultimately their managerial relevance.

General Characteristics of Both Factors and Levels

Before discussing the specific issues relating to factors or levels, characteristics applicable to the specification of both factors and levels should be addressed. When operationalizing either factors or levels, the researcher should ensure that the measures are both communicable and actionable.

Communicable Measures   First, the factors and levels must be easily communicated for a realistic evaluation. Traditional methods of administration (pencil and paper or computer) limit the types of factors that can be included. For example, it is difficult to describe the actual fragrance of a perfume or the "feel" of a hand lotion. Written descriptions do not capture sensory effects well unless the respondent sees the product firsthand, smells the fragrance, or uses the lotion. In an attempt to bring a more realistic portrayal of sensory characteristics that may have been excluded in the past, specific forms of conjoint have been developed to employ virtual reality [48] or employ the entire range of sensory and multimedia effects in describing the product or service [37, 54].

Actionable Measures   The factors and levels must also be capable of being put into practice, meaning the attributes must be distinct and represent a concept that can be precisely implemented. They must not be "fuzzy" attributes such as overall quality or convenience. Moreover, levels should not be specified in imprecise terms such as low, moderate, or high. Specifications such as these are imprecise because of the perceptual differences among individuals as to what they actually mean (as compared with actual differences as to how they feel about them). Moreover, concepts are many times not easily implemented; thus the researcher is not sure if the product or service finally developed is actually what was being evaluated by the respondent. If these "fuzzy" factors cannot be defined more precisely, the researcher may use a two-stage process. A preliminary conjoint study defines what determines judgments of the "fuzzy" factors (quality or convenience). Then the factors identified as important in the preliminary study are included in the larger study in more precise terms.

Specification Issues Regarding Factors

Having selected the attributes to be included as factors and ensured that the measures will be communicable and actionable, the researcher still must address three issues specific to defining factors: the number of factors to be included, multicollinearity among the factors, and the unique role of price as a factor.

Number of Factors   The number of factors included in the analysis directly affects the statistical efficiency and reliability of the results. As factors and levels are added, the increased number of parameters to be estimated requires either a larger number of stimuli or a reduction in the reliability of parameters. The minimum number of stimuli that must be evaluated by a respondent if the analysis is performed at the individual level is

Minimum number of stimuli = Total number of levels across all factors
− Number of factors + 1

For example, a conjoint analysis with five factors with three levels each (a total of 15 levels) would need a minimum of eleven ($15 - 5 + 1$) stimuli. This problem is similar to those we encountered in regression when the number of observations was insufficient to estimate valid coefficients. It is especially important in conjoint analysis because each respondent generates the required number of observations, and therefore the problem cannot be "solved" by adding more respondents.

Factor Multicollinearity   Multicollinearity among the factors is a problem that must be remedied. The correlation among factors (known as **interattribute** or **environmental correlation**) denotes a lack of conceptual independence among the factors. In such cases, the parameter estimates are affected just as in regression (chapter 4 contains a discussion of multicollinearity and its impact). Moreover, interattribute collinearity usually results in unbelievable combinations of two or more factors. For example, horsepower and gas mileage are generally thought to be negatively correlated. As a result, how believable is an automobile with the highest levels of both horsepower and gas mileage? The problem lies not in the levels themselves but in the fact that they cannot realistically be paired in all combinations, which is required for parameter estimation.

If multicollinearity creates unrealistic stimuli, the researcher has two options. The most direct is to create "superattributes" that combine the aspects of correlated attributes. In our example of horsepower and gas mileage, perhaps a factor of "performance" could be substituted. As an example of positively correlated attributes, factors of store layout, lighting, and decor may be better addressed by a single concept, such as "store atmosphere." In all cases, when these superattributes are added, they should be made as actionable and specific as possible. If it is not possible to define the broader factors, then the researchers may be forced to eliminate one of them.

The second option involves two possible modifications to the methodology underlying conjoint analysis. The first modification involves refined experimental designs and estimation techniques, which create "near" orthogonal stimuli, which can be used to eliminate any unbelievable stimuli resulting from interattribute correlation [61]. The second modification is to constrain the estimation of part-worths to conform to a prespecified relationship. These constraints can be between factors as well as pertaining to the levels within any single factor [59, 64]. Either of these two modifications to the methodology should be considered only after the more direct remedies have been considered because they add considerable complexity to the design and estimation of the conjoint analysis.

The Unique Role of Price as a Factor Price is a factor that is included in many conjoint studies because it represents a distinct component of value for many products or services being studied. Price, however, is not like other factors in its relationship to other factors [33]. In many, if not most, instances, price has a high degree of interattribute correlation with other factors. For many attributes, an increase in the amount of the attribute is associated with an increase in price, and a decreasing price level may be unrealistic. Second, the price–quality relationship may be operant between certain factors, such that certain combinations may be unrealistic or have the unintended perceptions. Third, many other "positive" factors (e.g., quality, reliability) may be included in defining the utility of the product or service. However, when defining what is "given up" for this utility (i.e., price), only one factor is included. This may inherently decrease the importance of price. Finally, price may interact with other factors, particularly more intangible factors such as brand name. The impact of an interaction in this situation is that a certain price level has different meanings for different brands—one that may be a "premium" brand and another a "discount" brand. We discuss the concept of interactions later in this chapter.

All of these unique features of price as a factor should not cause a researcher to avoid the use of price, but instead to anticipate the impacts and adjust the design and interpretation as required. First, explicit forms of conjoint analysis, such as conjoint value analysis (CVA) have been developed for occasions in which the focus is on price [53]. Moreover, if interactions of price and other factors are considered important, methods such as choice-based conjoint or multistage analyses [47] provide quantitative estimates of these relationships. Even if no specific adjustment is made, the researcher should consider these issues in the definition of the price levels and in the interpretation of the results.

Specification Issues Regarding Levels

The definition of levels is a critical aspect of conjoint analysis because the levels are the actual measures used to form the stimuli. Thus, in addition to being actionable and communicable, research has shown that the number of levels, the balance in levels between factors, and the range of the factors all have distinct effects on the evaluations by respondents.

Balanced Number of Levels Researchers should attempt as best possible to balance or equalize the number of levels across factors. It has been found that the estimated relative importance of a variable increases as the number of levels increases, even if the end points stay the same [68, 69]. It is conjectured that the refined categorization calls attention to the attribute and causes consumers to focus on that factor more than on others. If the relative importance of factors is known a priori, then the researcher may wish to expand the levels of the more important factors to avoid a dilution of importance and to capture additional information on the more important factors [67].

Range of the Factor Levels   The range (low to high) of the levels should be set somewhat outside existing values but not at an unbelievable level. This has a tendency to reduce interattribute correlation, but it also can reduce believability, so the levels should not be too extreme. Completely unacceptable levels can also cause substantial problems and should be eliminated. Before excluding a level, however, the researcher must ensure that it is truly unacceptable, because many times people select products or services that have what they term unacceptable levels. If an unacceptable level is found after the experiment has been administered, the recommended solutions are either to eliminate all stimuli that have unacceptable levels or to reduce part-worth estimates of the offending level to such a low level that any objects containing that level will not be chosen.

The researcher must also apply the criteria of practical relevance and feasibility in defining the levels. Levels that are impractical or would never be used in realistic situations can artificially affect the results. For example, assume that in the normal course of business activity, the range of prices varies about 10 percent around the average market price. If a price level 20 percent lower were included, but it would not realistically be offered, its inclusion would markedly distort the results. Respondents would logically be most favorable to such a price level. When the part-worth estimates are made and the importance of price is calculated, price will artificially appear more important than it would actually be in day-to-day decisions. The researcher must apply the criteria of feasibility and practical relevance to all attribute levels to ensure that stimuli are not created that will be favorably viewed by the respondent but never have a realistic chance of occurring.

Specifying the Basic Model Form

For conjoint analysis to explain a respondent's preference structure only from overall evaluations of a set of stimuli, the researcher must make two key decisions regarding the underlying conjoint model. These decisions affect both the design of the stimuli and the analysis of respondent evaluations.

The Composition Rule: Selecting an Additive versus an Interactive Model

The most wide-ranging decision by the researcher involves the specification of the respondent's **composition rule.** The composition rule describes how the respondent combines the part-worths of the factors to obtain overall worth.

The Additive Model   The most common, basic composition rule is an **additive model,** with which the respondent simply "adds up" the values for each attribute (the part-worths) to get the total value for a combination of attributes (products or services). For example, let us assume that a product has two factors with part-worths 3 and 4. Then the total worth would simply be 7. The additive model accounts for the majority (up to 80 or 90 percent) of the variation in preference in almost all cases, and it suffices for most applications. It is also the basic model underlying both traditional and adaptive conjoint analysis (see Table 7.5).

Adding Interaction Effects   The composition rule using **interaction effects** is similar to the additive form in that it assumes the consumer sums the part-worths to get an overall total across the set of attributes. It differs in that it allows for certain combinations of levels to be more or less than just their sum. By using our previous example, an interactive model would allow for the sum of the two levels to be either more or less than 7, the result of the additive model. In our industrial cleanser example, a respondent may really like a certain brand (generic), but only with a certain type of ingredient (phosphate-based). In this case, the brand has a low part-worth except when combined with another specific level (phosphate-based) of cleaning ingredients. We say that brand and ingredients are interacting as well as using the additive effects for each factor. The interactive form corresponds to the statement, "The whole is greater (or less) than the sum of its parts."

Many times, adding interaction terms to models decreases predictive power because the reduction in statistical efficiency (more part-worth estimates) is not offset by increases in predictive power gained from the interactions. The interactions predict substantially less variance than the additive effects, most often not exceeding a 5 to 10 percent increase in explained variance. Interaction terms are most likely to be substantial in cases for which attributes are less tangible, particularly when aesthetic or emotional reactions play a large role. For example, we discussed earlier the interaction effects that many times occur between price and brand name, which is less tangible but does have specific perceptions. The increased importance of interaction terms comes from the inability to depict the actual differences between certain attributes, with the "unexplained" portions associated with only certain levels of an attribute.

An Example of Interaction Effects on Part-Worth Estimates
Returning to our earlier example of an industrial cleanser, we can posit a situation where the respondent makes choices in which interactions appear to influence the choices. Assume a third respondent made the following preference ordering:

Brand	Ingredients	Form	
		Liquid	Powder
HATCO	Phosphate-free	1	2
	Phosphate-based	3	4
Generic	Phosphate-free	7	8
	Phosphate-based	5	6

These ranks were formed assuming that this respondent prefers HATCO and normally prefers phosphate-free over phosphate-based cleansers. However, a bad experience with a generic cleanser made the respondent select phosphate-based over phosphate-free only if it was a generic brand. This choice is termed an interaction effect between the factors of brand and ingredients. If we consider only an additive model, we obtain the following coefficients:

Form		Ingredients		Brand	
Liquid	Powder	Phosphate-free	Phosphate-based	HATCO	Generic
.42	−.42	0.0	0.0	1.68	1.68

By using these coefficients to calculate preference orders for the combinations, we get the following:

Actual rank	1	2	3	4	5	6	7	8
Predicted	1.5	3.5	1.5	3.5	5.5	7.5	5.5	7.5

The predictions are obviously less accurate, given that we know interactions exist. Also, the coefficients are misleading because the main effects of brand and ingredients are confounded by the interactions. If we were to proceed with only an additive model, we would be violating one of the principal assumptions and making potentially quite inaccurate predictions.

Examining for first-order interactions is a reasonably simple task. For the previous preference data, we can form three two-way matrices of preference order. For example, in the first box, the two preference orders (one for each brand) for each combination of form and ingredients are listed and then summed. To check for interactions, the diagonal values are then added and the difference is calculated. If the total is zero, then no interaction exists. As seen in the example, the only interaction found is between brand and ingredients. It is unusual not to see some slight differences indicating some interactions. As the difference gets greater, the impact of the interaction increases, and it is up to the researcher to decide when the interactions pose enough problems in prediction to warrant the increased complexity of estimating coefficients for the interaction terms.

Selecting the Model Type   The choice of a composition rule determines the types and number of treatments or stimuli that the respondent must evaluate, along with the form of estimation method used. An additive form requires fewer evaluations from the respondent, and it is easier to obtain estimates for the part-worths. However, the interactive form may be a more accurate representation of how respondents actually value a product or service. The researcher does not know with certainty the best model form, but must instead understand the implications of either choice on both the study design and the results obtained. If an additive model form is selected, it is not possible to estimate interactive effects. This does not mean that the researcher should always include interactive effects, as they add substantial complexity to the estimation process and in most cases cause the analysis to be performed at the aggregate rather than individual level. We examine the need for making this choice and the trade-offs associated with choosing either form at various points in our discussion.

Selecting the Part-Worth Relationship: Linear, Quadratic, or Separate Part-Worths

The flexibility of conjoint analysis in handling different types of variables comes from the assumptions the researcher makes regarding the relationships of the part-worths within a factor. In making decisions about the composition rule, the researcher decides how factors relate to one another in the respondent's decision process. In defining the type of part-worth relationship, the researcher is focusing on how the levels of a factor are related.

Types of Part-Worth Relationships Conjoint analysis gives the researcher three alternatives, ranging from the most restrictive (a linear relationship) to the least restrictive (separate part-worths), with the ideal point, or quadratic model, falling in between. Figure 7.2 illustrates the differences among the three types of relationships. The linear model is the simplest yet most restricted form, because we estimate only a single part-worth (similar to a regression coefficient), which is multiplied by the level's value to arrive at separate part-worth values for each level. In the quadratic form, also known as the ideal model, the assumption of strict linearity is relaxed, so that we have a simple curvilinear relationship. The curve can turn either upward or downward. Finally, the separate part-worth form (often referred to simply as the part-worth form) is the most general, allowing for separate estimates for each level. When using separate part-worths, the number of estimated values is the highest and increases quickly as we add factors and levels because each new level has a separate part-worth estimate.

The form of part-worth relationship can be specified for each factor separately, and a mixture of forms is possible if needed. This choice does not affect how the treatments or stimuli are created, and part-worth values are still calculated for each level. It does, however, impact how and what types of part-worths are estimated by conjoint analysis. If the linear or quadratic forms are specified, then the part-worth values for each level are estimated from the relationship, with separate part-worth estimates made directly. If we can reduce the number of part-worths estimated for any given set of stimuli by using a more restricted part-worth relationship (e.g., linear or quadratic form), the calculations will be more efficient and reliable from a statistical estimation perspective. But we must consider the trade-off between these gains and the possibly more accurate representation of how the consumer actually forms overall preference if we employ less restrictive part-worth relationships.

Selecting a Part-Worth Relationship The researcher has several approaches to deciding on the type of relationship for each factor. First, the researcher may rely on prior research or conceptual models to dictate the type of relationship. In an empirical approach, the conjoint model can be estimated first as a part-worth model, and the different part-worth estimates can be examined visually to detect whether a linear or a quadratic form is appropriate. In many instances, the general form is apparent, and the model can be reestimated with relationships

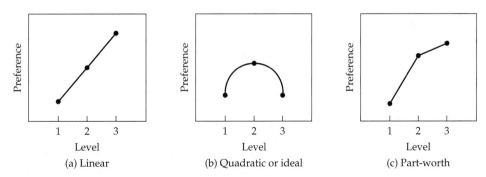

FIGURE 7.2 The Three Basic Types of Relationships between Factor Levels in Conjoint Analysis

specified for each variable as justified. Alternatively, the researcher can assess the changes in predictive ability under different combinations of relationships for one or more variables, but this is not recommended without at least some theoretical or empirical evidence for the possible type of relationship considered (e.g., prior estimates of part-worths). Without such support, the results may have high predictive ability but little use in decision making. In all instances, the researcher must balance predictive ability with the intended use of the study, the conceptual background available and the degree of managerial relevance and interpretation needed.

Data Collection

Having specified the factors and levels, plus the basic model form, the researcher next decides on the type of presentation of the stimuli (trade-off, full-profile, or pairwise comparison), the type of response variable, and the method of data collection. The objective is to convey to the respondent the attribute combinations (stimuli) in the most realistic and efficient manner possible. Most often the stimuli are presented in written descriptions, although physical or pictorial models can be quite useful for aesthetic or sensory attributes.

Choosing a Presentation Method

The **trade-off, full-profile,** and **pairwise comparison methods** are the three methods of stimulus presentation most widely associated with conjoint analysis. Although they differ markedly in the form and amount of information presented to the respondent (see Figure 7.3), they all are acceptable within the traditional conjoint model. The choice between presentation methods focuses on the assumptions as to the extent of consumer processing being performed during the conjoint task and the type of estimation process being employed.

The Trade-Off Presentation Method   The trade-off method compares attributes two at a time by ranking all combinations of levels (see Figure 7.3). It has the advantages of being simple for the respondent and easy to administer, and it avoids information overload by presenting only two attributes at a time. However, usage of this method has decreased dramatically in recent years owing to several limitations: (1) a sacrifice in realism by using only two factors at a time, (2) the large number of judgments necessary for even a small number of levels, (3) a tendency for respondents to get confused or follow a routinized response pattern because of fatigue, (4) the inability to employ pictorial or other nonwritten stimuli, (5) the sole use of nonmetric responses, and (6) its inability to use fractional factorial stimuli designs to reduce the number of comparisons made. Recent studies have shown that the third approach, pairwise comparisons, has now displaced trade-off methods for second place in commercial applications [69].

The Full-Profile Presentation Method   The most popular presentation method is full-profile, principally because of its perceived realism and its ability to reduce the number of comparisons through the use of fractional factorial designs. In this approach, each stimulus is described separately, most often on a profile card (see Figure 7.3 for an example). This approach elicits fewer judgments, but each is more complex and the judgments can be either ranked or rated. Among its advantages are (1) a more realistic description achieved by defining a stimulus in terms of a level for each factor, (2) a more explicit portrayal of the trade-offs among

CONJOINT ANALYSIS

TRADE-OFF APPROACH

Factor 1: Price

	Level 1: $1.19	Level 2: $1.39	Level 3: $1.49	Level 4: $1.69
Level 1: Generic				
Level 2: KX-19				
Level 3: Clean-All				
Level 4: Tidy-Up				

Factor 2: Brand Name

FULL-PROFILE APPROACH

Brand name: KX-19
Price: $1.19
Form: Powder
Color brightener: Yes

PAIRWISE COMPARISON

Brand name: KX-19
Price: $1.19
Form: Powder

VERSUS

Brand name: Generic
Price: $1.49
Form: Liquid

FIGURE 7.3 Examples of the Trade-off and Full-Profile Methods of Presenting Stimuli

all factors and the existing environmental correlations among the attributes, and (3) possible use of more types of preference judgments, such as intentions to buy, likelihood of trial, and chances of switching—all difficult to answer with the trade-off method.

The full-profile method is not flawless and faces two major limitations. First, as the number of factors increases, so does the possibility of information overload. The respondent is tempted to simplify the process by focusing on only a few factors, when in an actual situation all factors would be considered. Second, the order in which factors are listed on the stimulus card may have an impact on the evaluation. Thus the researcher needs to rotate the factors across respondents when possible to minimize order effects. The full-profile method is recommended when the number of factors is six or less. When the number of factors ranges from seven to ten, then the trade-off approach becomes a possible compromise along with the profile method. If the number of factors exceeds 10, then alternative methods (adaptive conjoint) are suggested [20].

The Pairwise Combination Presentation Method   The third presentation method, the pairwise combination, combines the two other methods. The pairwise combination is a comparison of two profiles (see Figure 7.3), with the respondent most often using a rating scale to indicate strength of preference for one profile over the other [31]. The distinguishing characteristic of the pairwise comparison is that the profile typically does not contain all the attributes, as does the full-profile method, but instead only a few attributes at a time are selected in constructing profiles. It is similar to the trade-off method in that pairs are evaluated, but in the case of the trade-off method the pairs being evaluated are attributes, whereas in the pairwise comparison method the pairs are profiles with multiple attributes. The pairwise comparison method is also instrumental in many specialized conjoint designs, such as adaptive conjoint analysis (ACA) [51], which is used in conjunction with a large number of attributes (a more detailed discussion of dealing with a large number of attributes appears later in this chapter).

Creating the Stimuli

Once the factors and levels have been selected and the presentation method chosen, the researcher turns to the task of creating the treatments or stimuli for evaluation by the respondents. For any presentation method, the researcher is always faced with an increasing burden on the respondent as the number of factors and levels increases. The researcher must weigh the benefits of increased task effort versus the additional information gained. The following sections detail the issues involved in creating stimuli for each presentation method.

The Trade-Off Presentation Method   In the case of the trade-off method, all possible combinations of attributes are used. The number of trade-off matrices is strictly based on the number of factors and can be calculated as:

$$\text{Number of trade-off matrices} = \frac{N(N - 1)}{2}$$

where N is the number of factors. For example, five factors would result in 10 trade-off matrices ($5 \times 4/2 = 10$). The researcher should remember, however, that each trade-off matrix involves a number of responses equal to the product of the factors' levels. For example, a trade-off matrix with factors of 3 levels each requires nine (3×3) evaluations in that single matrix. If the five factors in our example each had three levels, then the respondent would evaluate 10 trade-off matrices, each with 9 evaluations, for a total of 90 evaluations overall. As we can see, this presentation method can quickly lead to heavy burdens on the respondent as the number of attributes or levels increases. However, this method does keep the task simple by asking the respondent to evaluate only two factors at a time, whereas the other presentation methods can become quite involved in terms of stimuli complexity.

The Full-Profile or Pairwise Combination Presentation Methods   The two remaining methods—full-profile and pairwise comparison—involve the evaluation of one stimulus at a time (full-profile) or pairs of stimuli (pairwise comparison). In a simple conjoint analysis with a small number of factors and levels (such as those discussed earlier for which three factors with two levels each resulted in eight combinations), the respondent may evaluate all possible stimuli. This is known as a **factorial design** when all combinations are used. But as the number of factors and levels increases, this design becomes impractical in a manner similar to that shown for the trade-off method. If the researcher is interested in assessing the impact of four variables with four levels for each variable, 256 stimuli (4 levels × 4 levels × 4 levels × 4 levels) would be created in a full factorial design for the full-profile method. This is obviously too many for one respondent to evaluate and still give consistent, meaningful answers. An even greater number of pairs of stimuli would be created for the pairwise combinations of profiles with differing numbers of attributes. What is needed is a method for developing a subset of the total stimuli that can be evaluated and still provide the information needed for making accurate and reliable part-worth estimates.

Defining Subsets of Stimuli

A **fractional factorial design** is the most common method for defining a subset of stimuli for evaluation. The fractional factorial design selects a sample of possible stimuli, with the number of stimuli depending on the type of composition rule assumed to be used by respondents. Using the additive model, which assumes only main effects for each factor with no interactions, a study using the full-profile method with four factors at four levels requires only 16 stimuli to estimate the main effects. Table 7.6 shows two possible sets of 16 stimuli. The 16 stimuli must be carefully constructed to ensure the correct estimation of the main effects. The two designs in Table 7.6 are **optimal designs,** as they are **orthogonal** (there is no correlation among levels across attributes) and **balanced** (each level in a factor appears the same number of times).

TABLE 7.6 Alternative Fractional Factorial Designs for an Additive Model (Main Effects Only) with Four Factors at Four Levels Each

	Design 1: Levels for[a]				Design 2: Levels for[a]			
Stimulus	Factor 1	Factor 2	Factor 3	Factor 4	Factor 1	Factor 2	Factor 3	Factor 4
1	3	2	3	1	2	3	1	4
2	3	1	2	4	4	1	2	4
3	2	2	1	2	3	3	2	1
4	4	2	2	3	2	2	4	1
5	1	1	1	1	1	1	1	1
6	4	3	4	1	1	4	4	4
7	1	3	2	2	4	2	1	3
8	2	1	4	3	2	4	2	3
9	2	4	2	1	3	2	3	4
10	3	3	1	3	3	4	1	2
11	1	4	3	3	4	3	4	2
12	3	4	4	2	1	3	3	3
13	1	2	4	4	2	1	3	2
14	2	3	3	4	3	1	4	3
15	4	4	1	4	1	2	2	2
16	4	1	3	2	4	4	3	1

[a]The numbers in the columns under factor 1 through factor 4 are the levels of each factor. For example, the first stimulus in design 1 consists of level 3 for factor 1, level 2 for factor 2, level 3 for factor 3, and level 1 for factor 4.

The creation of an optimal design, with orthogonality and balance, does not mean, however, that all of the stimuli in that design will be acceptable for evaluation. There are several reasons for the occurrence of unacceptable stimuli. The first is the creation of "obvious" stimuli—stimuli whose evaluation is obvious because of their combination of levels. The most common examples are stimuli with all levels at either the highest or lowest values. In these cases, the stimuli really provide little information about choice and can create a perception of unbelievability on the part of the respondent. The second occurrence is the creation of unbelievable stimuli due to interattribute correlation, which can create stimuli with combinations of levels (high gas mileage, high acceleration) that are not realistic. Finally, constraints may be placed on the combinations of attributes. In any of these instances, the unacceptable stimuli present unrealistic choices to the respondent and should be eliminated to ensure a valid estimation process as well as a perception of credibility of the choice task among the respondents.

There are several courses of action for eliminating the unacceptable stimuli. First, the researcher can generate another fractional factorial design and assess the acceptability of its stimuli. If all of the designs contain unacceptable stimuli and a better alternative design cannot be found, then the unacceptable stimulus can be deleted. Although the design will not be totally orthogonal (i.e., it will be somewhat correlated and is termed to be **nearly orthogonal**), it will not violate any assumptions of conjoint analysis. It will create problems similar to multicollinearity in regression (i.e., instability of the estimates when levels are slightly changed and a lessened ability to assess the unique impact of each attribute). All nearly orthogonal designs should be assessed for **design efficiency,** which is a measure of the correspondence of the design in terms of orthogonality and balance to an optimal design [36]. Typically measured on a 100-point scale (optimal designs = 100), alternative nonorthogonal designs can be assessed, and the most efficient design with all acceptable stimuli selected. Most conjoint programs for developing nearly orthogonal designs assess the efficiency of the designs.

Interattribute correlations must be dealt with on conceptual terms. Unacceptable stimuli due to interattribute correlations can occur in optimal or orthogonal designs, and the researcher must accommodate these within the development of designs. In practical terms, interattribute correlations should be minimized but do not need to be zero if small correlations (.20 or less) will add to realism. Most problems are found in the case of negative correlations, as between gas mileage and horsepower. Adding uncorrelated factors can reduce the average interattribute correlation, so that with a realistic number of factors (e.g., 6 factors), the average intercorrelation would be close to .20, which has inconsequential effects. But the researcher should always assess the believability of the stimuli as a measure of practical relevance.

The remaining 240 possible stimuli in our example that are not in the selected fractional factorial design are needed if all 11 interaction terms are to be estimated. As discussed earlier, the researcher may decide that selected interactions are important and should be included in the model estimation. In this case, the fractional factorial design must include additional stimuli to accommodate the interactions. Published guides for fractional factorial designs or conjoint program components will design the subsets of stimuli to maintain orthogonality or maximize efficiency in "near" orthogonal designs, making the generation of full-profile stimuli quite easy [1, 12, 22, 43].

If the number of factors becomes too large and the adaptive conjoint methodology is not acceptable, a **bridging design** can be employed [5]. In this design, the factors are divided in subsets of appropriate size, with some attributes overlapping between the sets so that each set has a factor(s) in common with other sets of factors. The stimuli are then constructed for each subset so that the respondents never see the original number of factors in a single profile. When the part-worths are estimated, the separate sets of profiles are combined, and a single set of estimates is provided. Computer programs handle the division of the attributes, creation of stimuli, and their recombination for estimation [9]. When using pairwise comparisons, the number may be quite large and complex, so that most often interactive computer programs are used that select the optimal sets of pairs as the questioning proceeds.

Selecting a Measure of Consumer Preference

The researcher must also select the measure of preference: rank-ordering versus rating (i.e., a 1-to-10 scale). Although the trade-off method employs only ranking data, the pairwise comparison method can evaluate preference either by obtaining a rating of preference of one stimulus over the other or just a binary measure of which is preferred. The full-profile method also accommodates both ranking and rating methods.

Each preference measure has certain advantages and limitations. Obtaining a rank-order preference measure (i.e., rank-ordering the stimuli from most to least preferred) has two major advantages: (1) it is likely to be more reliable because ranking is easier than rating with a reasonably small number (20 or fewer) of stimuli, and (2) it provides more flexibility in estimating different types of composition rules. It has, however, one major drawback: it is difficult to administer, because the ranking process is most commonly performed by sorting stimulus cards into the preference order, and this sorting can be done only in a personal interview setting.

The alternative is to obtain a rating of preference on a metric scale. Metric measures are easily analyzed and administered, even by mail, and allow conjoint estimation to be performed by multivariate regression. However, respondents can be less discriminating in their judgments than when they are rank-ordering. Also, given the large number of stimuli evaluated, it is useful to expand the number of response categories over that found in most consumer surveys. A rule of thumb is to have 11 categories (i.e., rating from 0 to 10 or 0 to 100 in increments of 10) for 16 or fewer stimuli and expand to 21 categories for more than 16 stimuli [38].

Survey Administration

In the past, the complexity of the conjoint analysis task has led most often to the use of personal interviews to obtain the conjoint responses. Personal interviews enable the interviewer to explain the sometimes more difficult tasks associated with conjoint analysis. Recent developments in interviewing methods, however, have now made conducting conjoint analyses feasible both through the mail (with both pencil-and-paper questionnaires and computer-based surveys) and by telephone. If the survey is designed to ensure that the respondent can assimilate and process the stimuli properly, then all of the interviewing methods produce relatively equal predictive accuracy [2]. The use of disk-by-mail computerized interviewing has greatly simplified the conjoint task demands on the respondent and made such developments as adaptive conjoint analysis [51] widely available.

One concern in any conjoint study is the burden placed on the respondent due to the number of conjoint stimuli evaluated. Obviously, the respondent could not evaluate all 256 stimuli in our earlier factorial design, but what is the appropriate number of tasks in a conjoint analysis? A recent review of commercial conjoint studies found that respondents can easily complete up to 20 conjoint evaluations [34]. After that many evaluations, the responses start to become less reliable and less representative of the underlying reference structure. Although there is no absolute minimum or maximum number of stimuli evaluations, the researcher should always strive to use the fewest possible while maintaining efficiency in the estimation process. Also, there is no substitute for pretesting a conjoint study to assess the respondent burden, the method of administration, and the acceptability of the stimuli.

Stage 3: Assumptions of Conjoint Analysis

Conjoint analysis has the least restrictive set of assumptions involving the estimation of the conjoint model. The structured experimental design and the generalized nature of the model make most of the tests performed in other dependence methods unnecessary. Therefore, the statistical tests for normality, homoscedasticity, and independence that were performed for other dependence techniques are not necessary. The use of statistically based stimuli designs also ensures that the estimation is not confounded and that the results are interpretable under the assumed composition rule.

Yet, even though there are fewer statistical assumptions, the conceptual assumptions are perhaps greater than with any other multivariate technique. As mentioned earlier, the researcher must specify the general form of the model (main effects versus interactive model) before the research is designed. This "builds in" this decision and makes it impossible to test alternative models once the research is designed and the data are collected. Conjoint analysis is not like regression, for example, where additional effects (interaction or nonlinear terms) can be easily evaluated. The researcher must make this decision concerning model form and must design the research accordingly. Thus, conjoint analysis, while having few statistical assumptions, is very theory-driven in its design, estimation, and interpretation.

Stage 4: Estimating the Conjoint Model and Assessing Overall Fit

The options available to the researcher in terms of estimation techniques have increased dramatically in recent years. The development of techniques in conjunction with specialized methods of stimulus presentation (e.g., the adaptive or choice-based conjoint) is just one improvement of this type. The researcher, in obtaining the results of a conjoint analysis study, however, must address the issues of selecting the estimation method and evaluating the results (see Figure 7.4).

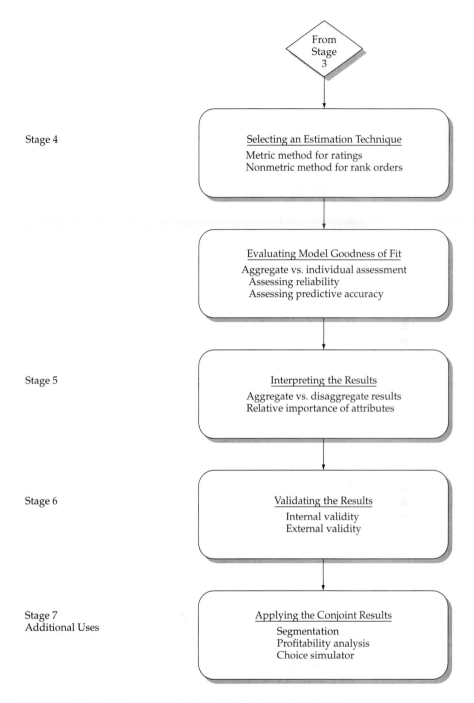

FIGURE 7.4 Stages 4–7 of the Conjoint Analysis Decision Diagram

Selecting an Estimation Technique

Rank-order evaluations require a modified form of analysis of variance specifically designed for ordinal data. Among the most popular and best-known computer programs are MONANOVA (Monotonic Analysis of Variance) [31, 35] and LINMAP [55]. These programs give estimates of attribute part-worths, so that the rank order of their sum (total worth) for each treatment is correlated as closely as possible to the observed rank order. If a metric measure of preference is obtained (e.g., ratings rather than rankings), then many methods, even multiple regression, can estimate the part-worths for each level. Most computer programs available today can accommodate either type of evaluation (ratings or rankings), as well as estimate any of the three types of relationships (linear, ideal point, and part-worth).

Evaluating Model Goodness-of-Fit

Conjoint analysis results are assessed for accuracy at both the individual and aggregate levels. The objective is to ascertain how consistently the model predicts the set of preference evaluations given by each person. This assessment can be for both metric and nonmetric responses. For the rank-order data, correlations based on the actual and the predicted ranks (e.g., Spearman's rho or Kendall's tau) are used. If a metric rating is obtained, then a simple Pearson correlation, just like that used in regression, is appropriate along with a comparison of actual and predicted ranks. In cases of prediction at the individual level, the actual and predicted preferences are correlated for each person and tested for statistical significance.

In most conjoint experiments, however, the number of stimuli does not substantially exceed the number of parameters, and there is always the potential for "overfitting" the data. Researchers are thus strongly encouraged to measure model accuracy not only on the original stimuli but also with a set of **validation** or **holdout stimuli.** In a procedure similar to a holdout sample in discriminant analysis, the researcher prepares more stimulus cards than needed for estimation of the part-worths, and the respondent rates all of them at the same time. Parameters from the estimated conjoint model are then used to predict preference for the new set of stimuli, which is compared with actual responses to assess model reliability. Individuals who have very poor predictive fit for the holdout sample can be deleted from the analysis. The holdout sample also gives the researcher an opportunity for a direct evaluation of stimuli of interest to the research study.

If an aggregate estimation technique is used, then researchers can use a holdout sample of respondents in each group to assess predictive accuracy. This method is not feasible for disaggregate results because there is no "generalized" model to apply to the holdout sample, as each respondent in the estimation sample has individualized part-worth estimates.

Stage 5: Interpreting the Results

Aggregate versus Disaggregate Analysis

The customary approach to interpreting conjoint analysis is disaggregate; that is, each respondent is modeled separately, and the results of the model are examined for each respondent. The most common method of interpretation is an examination of the part-worth estimates for each factor, assessing their magnitude and pattern for both practical relevance as well as correspondence to any theory-based relationships among levels. The higher the part-worth (either positive or negative), the more impact it has on overall utility. Part-worth values can be plotted graphically to identify patterns. Many programs convert the part-worth estimates to some common scale (e.g., maximum of 100 points) to allow for comparison both across factors for an individual and across individuals. This conversion provides a means of using the part-worths in other multivariate techniques such as cluster analysis.

Interpretation can also take place with aggregate results. Whether the model estimation is made at the individual level and then aggregated, or aggregate estimates are made for a set of respondents, the analysis fits one model to the aggregate of the responses. As one might expect, this process generally yields poor results when one is trying to predict what any single respondent would do or when trying to interpret the part-worths for any single respondent. Unless the researcher is dealing with a population definitely exhibiting homogeneous behavior with respect to the factors, aggregate analysis should not be used as the only method of analysis. However, many times aggregate analysis more accurately predicts aggregate behavior, such as market share. Thus the researcher must identify the primary purpose of the study and employ the appropriate level of analysis or a combination of the levels of analysis.

Assessing the Relative Importance of Attributes

In addition to portraying the impact of each level with the part-worth estimates, conjoint analysis can assess the relative importance of each factor. Because part-worth estimates are typically converted to a common scale, the greatest contribution to overall utility—and hence the most important factor—is the factor with the greatest range (low to high) of part-worths. The importance values of each factor can be converted to percentages summing to 100 percent by dividing each factor's range by the sum of all range values. This allows for comparison across respondents in a common scale as well as giving meaning to the magnitude of the importance score. The researcher must always consider the impact on the importance values of an extreme or practically infeasible level. If such a level is found, it should be deleted from the analysis or the importance values should be reduced to reflect only the range of feasible levels.

Stage 6: Validation of the Conjoint Results

Conjoint results can be validated both internally and externally. Internal validation involves confirmation that the selected composition rule (i.e., additive versus interactive) is appropriate. The researcher is typically limited to empirically assessing the validity of only the selected model form in a full study, owing to the high demands of collecting data to test both models. This is most efficiently accomplished by comparing alternative models (additive versus interactive) in a pretest study to confirm which model is appropriate. We have already discussed the use of holdout stimuli to assess the predictive accuracy for each individual or a holdout sample of respondents if the analysis is performed at the aggregate level.

External validation involves in general the ability of conjoint analysis to predict actual choices, and in specific terms the issue of sample representativeness. While conjoint has been employed in numerous studies in the past 20 years, relatively little research has focused on its true external validity. One study did confirm that conjoint analysis closely corresponds to the results from traditional concept testing, an accepted methodology for predicting customer preference [63]. Although there is no evaluation of sampling error in the individual-level models, the researcher must always ensure the sample is representative of the population of study. This becomes especially important when the conjoint results are used for segmentation or choice simulation purposes (see the next section for a more detailed discussion of these uses of conjoint results).

Managerial Applications of Conjoint Analysis

Typically, conjoint models estimated at the individual level (separate model per individual) are used in one or more of the following areas of decision support by representing the decision processes of individuals. With individual-level results, the preferences for each individual can be portrayed empirically. Aggregate conjoint results can represent groups of individuals and also provide a means of predicting their decisions for any number of situations. The most common applications of conjoint analysis in conjunction with its portrayal of the consumer's preference structure include segmentation, profitability analysis, and conjoint simulators.

Segmentation

One of the most common uses of individual-level conjoint analysis results is to group respondents with similar part-worths or importance values to identify segments. The estimated conjoint part-worth utilities can be used solely or in combination with other variables (e.g., demographics) to derive respondent groupings that are most similar in their preferences [17]. In the industrial cleanser example, we might find one group for which brand is the most important feature, whereas another group might value ingredients most highly. For the researcher interested in knowing the presence of such groups and their relative magnitude, a number of different approaches to segmentation are available, all with differing strengths and weaknesses [65].

Profitability Analysis

A complement to the product design decision is a marginal profitability analysis of the proposed product design. If the cost of each feature is known, the cost of each "product" can be combined with the expected market share and sales volume to predict its viability. This process might point to a combination of attributes with a smaller share as the most profitable because of an increased profit margin resulting from the low cost of particular components. An adjunct to profitability analysis is assessing price sensitivity, which can be addressed through either specific research designs [47] or specialized programs [53]. Both individual and aggregate results can be used in this analysis.

Conjoint Simulators

At this point, the researcher still understands only the relative importance of the attributes and the impact of specific levels. But how does conjoint analysis achieve its other primary objective—using "what-if" analysis to predict the share of preferences that a stimulus (real or hypothetical) is likely to capture in various competitive scenarios of interest to management? This is the role played by **choice simulators,** which follow a three-step process:

1. Estimate and validate conjoint models for each respondent (or group).
2. Select the sets of stimuli to test according to possible competitive scenarios.
3. Simulate the choices of all respondents (or groups) for the specified sets of stimuli and predict share of preference for each stimulus by aggregating their choices.

After the conjoint model has been estimated, the researcher can specify any number of sets of stimuli for simulation of consumer choices. Among the possible uses are assessing (1) the impact of adding a product to an existing market; (2) the increased potential from a multiproduct or multibrand strategy, including estimates of cannibalism; or (3) the impact of deleting a product or brand from the market. In each case, the researcher provides the set of stimuli representing the market, and the choices of respondents are then simulated.

Choice simulators typically use two types of rules in predicting choice of a stimulus [16]. The first is the maximum utility model, which assumes that the respondent chooses the stimulus with the highest predicted utility score. This is most appropriate in cases of markets with individuals of widely different preferences and in situations involving sporadic, nonroutine purchases. The alternative choice rule is a purchase probability measure, in which predictions of choice probability sum to 100 percent over the set of stimuli tested. This approach is best suited to repetitive purchasing situations, for which purchases may be more tied to usage situations over time. The two most common methods of making these predictions are the BTL (Bradford-Terry-Luce) and logit models, which make quite similar predictions in almost all situations [24].

The researcher is cautioned, however, in assuming that the share of preference in a conjoint simulation directly translates to market share. The conjoint simulation represents only the product and perhaps price aspects of marketing management, omitting all of the other marketing factors (e.g., advertising and promotion, distribution, competitive responses) that ultimately impact market share. The conjoint simulation does present a view of the product market and the dynamics of preferences that may be seen in the sample under study.

Alternative Conjoint Methodologies

Up to this point we have dealt with conjoint analysis applications involving the traditional conjoint methodology. But real-world applications many times involve 20 to 30 attributes or require a more realistic choice task than used in our earlier discussions. Recent research has been directed toward overcoming these problems encountered in many conjoint studies, with two new conjoint methodologies being developed: (a) an adaptive conjoint for dealing with a large number of attributes, and (b) a choice-based conjoint for providing more realistic choice tasks. These areas represent the primary focus of current research in conjoint analysis [11,20].

Adaptive Conjoint: Conjoint with a Large Number of Factors

The full-profile and trade-off methods start to become unmanageable with between 6 and 9 attributes, yet many conjoint studies need to incorporate up to 20 or 30 attributes. In these cases, some adapted or reduced form of conjoint analysis is used to simplify the data collection effort and still represent a realistic choice decision. The two options are the self-explicated models and adaptive or hybrid models.

Self-Explicated Conjoint Models

In the **self-explicated model,** the respondent provides a rating of the desirability of each level of an attribute and then rates the relative importance of the attribute overall. Part-worths are then calculated by a combination of the two values [58]. This is a compositional approach where ratings are made on the components of utility rather than just overall preference. As a major variant of conjoint analysis and closer to traditional multiattribute models, this model form raises several concerns. First, can respondents assess the relative importance of attributes accurately when research shows that they can be underestimated in multiattribute models because respondents want to give socially desirable answers? Second, interattribute correlations may play a greater role and cause substantial biases in the results due to "double counting" of correlated factors. Finally, respondents never perform a choice task (rating the set of hypothetical combinations of attributes), and this lack of realism is a critical limitation in new-product applications. Recent research, however, has demonstrated that this method may have suitable predictive ability when compared to traditional conjoint methods [18]. Thus, if the number of factors cannot be reduced to a manageable level acceptable for traditional conjoint methods, then a self-explicated model may be a viable alternative method.

Adaptive or Hybrid Conjoint Models

A second approach is the **adaptive,** or **hybrid model,** so termed because it combines the self-explicated and part-worth conjoint models [14, 15]. It utilizes self-explicated values in creating a small (three to nine) subset of stimuli selected from a fractional factorial design. The stimuli are then evaluated in a manner similar to traditional conjoint analysis. The sets of stimuli differ among respondents, and although each respondent evaluates only a small number, collectively all stimuli are evaluated by a portion of the respondents. The approach of integrating information from the respondent to simplify or augment the choice tasks has led to a number of recent research efforts aimed at differing aspects of the research design [3, 30, 60, 64]. One of the most popular variants of this approach is ACA, a computer-administered conjoint program developed by Sawtooth Software [51]. ACA employs self-explicated ratings to reduce the factorial design size and make the process more manageable. Its relative predictive ability has been shown to be comparable to traditional conjoint analysis, and it is a suitable alternative when the number of attributes is large [18, 32, 63, 70].

When faced with a number of factors that cannot be accommodated in the conjoint methods discussed to this point, the self-explicated and adaptive or hybrid models preserve at least a portion of the underlying principles of conjoint analysis. In comparing these two extensions, the self-explicated methods have a slightly lower reliability, although recent developments may provide improvement. When the hybrid models and self-explicated methods are compared with full-profile methods, the results are mixed, with slightly better performance by the hybrid or adaptive method, particularly ACA [25]. Although more research is needed to confirm the comparisons across methods, the empirical studies indicate that the adaptive or hybrid methods and the newer forms of self-explicated models both offer viable alternatives to traditional conjoint when dealing with a large number of factors.

Choice-Based Conjoint: Adding Another Touch of Realism

In recent years, many researchers in the area of conjoint analysis have directed their efforts toward a new conjoint methodology that provides increased realism in the choice task. With the overriding objective of understanding the respondent's decision-making process and predicting behavior in the marketplace, traditional conjoint analysis assumes that the judgment task, based on ranking or rating, captures the choices of the respondent. Yet researchers have argued that this is not the most realistic way of depicting a respondent's actual decision process, and others have pointed to the lack of formal theory linking these measured judgments to choice [39]. What has emerged is an alternative conjoint methodology, known as choice-based conjoint, with the inherent face validity of asking the respondent to choose a full-profile stimulus from a set of alternative stimuli known as a **choice set.** This is much more representative of the actual process of selecting a product from a set of competing products. Moreover, choice-based conjoint provides an option of *not* choosing any of the presented stimuli by including a no-choice option in the choice set. Whereas traditional conjoint assumes respondents' preferences will always be allocated among the set of stimuli, the choice-based approach allows for market contraction if all the alternatives in a choice set are unattractive.

A Simple Illustration of Full-Profile Versus Choice-based Conjoint

Before discussing some of the more technical details of choice-based conjoint and how it differs from the other conjoint methodologies, let us examine a simple example for illustration purposes. A cellular phone company wishes to estimate the market potential for three service options that can be added to the base service fee of $14.95 per month and $0.50 per minute of calling time:

ICA—itemized call accounting with a $2.75-per-month service charge

CW—call waiting with a $3.50-per-month service charge

TWC—three-way calling with a $3.50-per-month service charge

Traditional conjoint analysis is performed with full-profile stimuli representing the various combinations of service, ranging from just the base service to the base service and all three options. The complete set of profiles (factorial design) is shown in Table 7.7 (p. 426). Stimulus 1 represents the base service with no options, stimulus 2 is the base service plus itemized call accounting, and so forth up to stimulus 8 being the base service plus all three options (itemized call accounting, call waiting, and three-way calling). The respondent is asked to rate or rank each of these eight profiles.

In a choice-based approach, the respondent is shown a series of choice sets, each choice set having several full-profile stimuli. A choice-based design is also shown

TABLE 7.7 Comparison of Stimuli Designs Used in Traditional and Choice-Based Conjoint

| Traditional Conjoint Analysis | | | | Choice-Based Conjoint | |
| | Levels of Factors[a] | | | | |
Stimulus	ICA	CW	TWC	Choice Set	Stimuli in Choice Set[b]
1	0	0	0	1	1, 2, 4, 5, 6 and no choice
2	1	0	0	2	2, 3, 5, 6, 7 and no choice
3	0	1	0	3	1, 3, 4, 6, 7, 8 and no choice
4	0	0	1	4	2, 4, 5, 7, 8 and no choice
5	1	1	0	5	3, 5, 6, 8 and no choice
6	1	0	0	6	4, 6, 7 and no choice
7	0	1	1	7	1, 5, 7, 8 and no choice
8	1	1	1	8	1, 2, 6, 8 and no choice
				9	1, 2, 3, 7 and no choice
				10	2, 3, 4, 8 and no choice
				11	1, 3, 4, 5 and no choice

[a]Levels: 1 = service option included; 0 = service option not included.
[b]Stimuli used in choice sets are those defined in the design for the traditional conjoint analysis.

in Table 7.7. The first choice set consists of five of the full-profile stimuli (stimuli 1, 2, 4, 5, and 6) and a "none of these" option. The respondent then chooses only one of the profiles in the choice set ("most preferred" or "most liked") or the "none of these" option. An example choice set task for choice set 6 is shown in Table 7.8. The preparation of stimuli and choice sets is based on experimental design principles [30, 39] and is the subject of considerable research effort to refine and improve on the choice task [3, 11, 27, 30, 47].

The number of profiles varies across choice sets. Also, the number of choices made (one choice for each of 11 choice sets) is actually more in this case than required in the factorial design. But as the number of factors and levels increases, the choice-based design requires considerably fewer evaluations (remember our earlier example of four factors with four levels each generating 256 stimuli). The advantages of the choice-based approach are the additional realism and the ability to estimate interaction terms, which are not possible with traditional conjoint analysis. After each respondent has chosen a stimuli for each choice set, the data are aggregated across respondents (segments or some other homogeneous groupings of respondents) to estimate the conjoint part-worths for each level and the interaction terms. From these results, we can assess the contributions of each factor and factor–level interaction and estimate the likely market shares of competing profiles.

Unique Characteristics of Choice-Based Conjoint

The basic nature of choice-based conjoint and its background in the theoretical field of information integration [38] has led to a somewhat more technical perspective than found in the other conjoint methodologies. While the other methodologies are based on sound experimental and statistical principles, the additional complexity in both stimuli designs and estimation has focused a great deal of developmental efforts in these areas. From these efforts, researchers now have a clearer understanding of the issues involved at each stage. The following sections

TABLE 7.8 Example of a Choice Set in Choice-Based Conjoint

Which Calling System Would You Choose?

1	2	3	4
Base system at $14.95/month and $.50/minute plus: ◆TWC—three-way calling for only $3.50/month	Base system at $14.95/month and $.50/minute plus: ◆ICA—itemized call accounting for only $2.75/month	Base system at $14.95/month and $.50/minute plus: ◆CW—call waiting for only $3.50/month and ◆TWC—three-way calling for only $3.50/month	None of these

detail some of the areas and issues in which choice-based conjoint is unique among the conjoint methodologies.

Type of Decision-Making Process Portrayed   Traditional conjoint has always been associated with an information-intensive approach to decision making, characterized by the scrutiny of the full-profile stimuli composed of levels from each attribute. Each attribute is equally represented and considered in a single profile. But in choice-based conjoint, researchers are coming to the conclusion that the choice task may invoke a different type of decision-making process. In making choices among stimuli, consumers seem to choose among a smaller subset of factors upon which comparisons, and ultimately choice, are made [26]. This parallels the types of decisions associated with time-constrained or simplifying strategies, each characterized by a lower depth of processing. Thus, each conjoint methodology provides different insights into the decision-making process. Because researchers may not be willing at this time to select only one methodology, an emerging strategy is to employ both methodologies and draw unique perspectives from each [26].

Stimuli Design   Perhaps the greatest advantage of choice-based conjoint is the realistic choice process portrayed by the choice set. Two developments have further enhanced the choice task. First, there is a more realistic and informative choice among closely comparable alternatives, rather than the situation in which one or more stimuli are markedly inferior or superior. However, the stimuli design process is focused on achieving orthogonality and balance among the attributes. A recent effort has shown how the choice set can be created to ensure balance not among factor levels, but instead among the utilities of the stimuli [27]. This provides a more realistic task, which can increase consumer involvement and provide better results. Second, in a method involving additional information from the respondents, choice sets are created that fit the unique preferences of each individual and achieve better predictive accuracy in market-based situations [9]. These two developments are characteristic of the efforts aimed at improving upon the choice task to make it an even more realistic and efficient method of assessing consumer preference.

Estimation Technique   The conceptual foundation of choice-based conjoint is psychology [40, 62], but it was the development of the multinomial logit estimation technique [42] that provided an operational method for estimating these types of choice models. Although considerable efforts have refined and made the technique widely available, it still represents a more complex methodology than those associated with the other conjoint methodologies. One particular aspect that is still unresolved is the property of IIA (independence of irrelevant alternatives), an assumption that makes the prediction of very similar alternatives problematic. Although exploring all of the issues underlying IIA is beyond the scope of this discussion, the researcher is cautioned when using choice-based conjoint to understand the ramifications of this assumption.

Some Advantages and Limitations of Choice-Based Conjoint

The growing acceptance of choice-based conjoint analysis among marketing research practitioners is primarily due to the belief that obtaining preferences by having respondents choose a single preferred stimuli from among a set of stimuli is more realistic—and thus a better method—for approximating actual decision processes. Yet the added realism of the choice task is accompanied with a number of "trade-offs" that the researcher must consider before selecting choice-based conjoint.

The Choice Task   Each choice set contains several stimuli and each stimulus contains several factors at different levels, similar to the full-profile stimuli. Therefore, the respondent must process a considerably greater amount of information than the other conjoint methodologies in making a choice in each choice set. Sawtooth Software, developers of the choice-based conjoint (CBC) system, believe those choices involving more than six attributes are likely to confuse and overwhelm the respondent [52]. Although the choice-based method does mimic actual decisions more closely, the inclusion of too many attributes creates a formidable task that ends up with less information than would have been gained through the rating of each stimulus individually.

Predictive Accuracy

In practice, all three conjoint methodologies allow for similar types of analyses, simulations, and reporting, even though the estimation processes are different. Although choice-based models still have to be subjected to more thorough empirical tests, some researchers believe they have an advantage in predicting choice behavior. However, empirical tests indicate little difference between individual-level ratings-based models adjusted to take nonchoice into account and the generalized multinomial logit choice-based models [44]. The conclusion was that both ratings-based and choice-based models predicted equally well. Other research compared the two approaches to conjoint (ratings-based or choice-based) in terms of the ability to predict shares in a holdout sample [13]. Both approaches predict holdout sample choices well, with neither approach dominant and the results mixed in different situations. Ultimately, the decision to use one method over the other is dictated by the objectives and scope of the study, the researcher's familiarity with each method, and the available software to properly analyze the data.

Managerial Applications   Choice-based models estimated at the aggregate level provide the values and statistical significance of all estimates, easily produce realistic market-share predictions for new stimuli [30], and provide the added assurances that "choices" among stimuli were used to calibrate the model. However, aggregate choice models hinder segmentation of the market. Choice-based conjoint is not capable of estimating a separate conjoint model for each respondent, making it impossible to group respondents according to their conjoint model results as discussed earlier. In contrast, the ratings-based models described earlier are well suited to segmentation studies but face the problem of cumbersome tests for the statistical significance of the part-worth estimates. Thus, the results may be difficult to summarize and the simulation of choice shares can be problematic.

Availability of Computer Programs   The good news is that there are several choice-based programs now available for researchers that assist in all phases of research design, model estimation, and interpretation [29, 52]. The bad news is that the recent research by academicians and applied researchers is slowly being integrated into these commercially available programs. Most research advances are still found only in a limited domain and not available for widespread use. These improvements and enhanced capabilities, after rigorous validation by the research community, should become a standard part of all choice-based programs.

Summary

Choice-based conjoint is an emerging conjoint methodology that holds great promise for increasing the interpretative and predictive capabilities of conjoint analysis. The widespread interest and research into improvements in almost all areas of the methodology will provide the necessary foundation for the continued growth, availability, and acceptance of this method. It will add a distinctive component to the researcher's "toolkit" of methods for understanding consumer preferences.

Overview of the Three Conjoint Methodologies

Conjoint analysis has evolved past its origins of what we now know as traditional conjoint to develop two additional methodologies, which address two substantive issues—dealing with large numbers of attributes and making the choice task more realistic. Each methodology has distinctive features that help define those situations in which it is most applicable (see our earlier discussion in stage 2). Yet, in many situations two or more methodologies are feasible and the researcher has the option of selecting one or, more increasingly, combining the methodologies. Only by being knowledgeable about the strengths and weaknesses of each methodology can the researcher make the more appropriate choice. Researchers interested in conjoint analysis are encouraged to continue to monitor the developments of this widely employed multivariate technique.

An Illustration of Conjoint Analysis

In the following sections we examine the steps in an application of conjoint analysis to the industrial cleanser example. The discussion follows the model-building process introduced in chapter 1 and focuses on (1) design of the stimuli, (2) estimation and interpretation of the conjoint part-worths, and (3) application of a conjoint simulator to predict market shares for a new product formulation. The CATEGORIES option of SPSS is used in the design, analysis, and choice simulator phases of this example [57]. Comparable results are obtained with the other conjoint analysis programs available for commercial and academic use.

Stage 1: Objectives of the Conjoint Analysis

In developing a new industrial cleanser, HATCO wanted a more thorough understanding of the needs and preferences of its industrial customers. In an adjunct study to the one described in chapter 1, HATCO commissioned a conjoint analysis experiment among 100 industrial customers. Marketing research and consultation with the product development group identified five factors as the determinant attributes in the targeted segment of the industrial cleaner market. The five attributes are shown in Table 7.9.

Stage 2: Design of the Conjoint Analysis

The decisions at this phase are (1) selecting the conjoint methodology to be used, (2) designing the stimuli to be evaluated, (3) specifying the basic model form, and (4) selecting the method of data collection.

Selecting a Conjoint Methodology

The first issue to be resolved is the selection of the conjoint methodology from among the three options—traditional conjoint, adaptive conjoint, or choice-based conjoint. Given the small number of factors (five), all three methodologies would be appropriate. Because the emphasis was on a thorough understanding of the preference structure, the traditional conjoint methodology was chosen as suitable in terms of response burden on the respondent and depth of information portrayed. Choice-based conjoint was also strongly considered, but the absence of proposed interactions and the desire for reducing the task complexity led to the selection of the traditional conjoint method.

Designing Stimuli

Focus group research established specific levels for each attribute (see Table 7.2) that were deemed actionable and communicable through a small-scale pretest and evaluation study. The price levels were taken from the range found in existing products to ensure they were not unrealistic. The product type did not suggest

TABLE 7.9 Attributes and Levels for the HATCO Conjoint Analysis Experiment

Description	Level
Form of the product	1. Premixed liquid
	2. Concentrated liquid
	3. Powder
Number of applications per container	1. 50
	2. 100
	3. 200
Addition of disinfectant to cleaner	1. Yes
	2. No
Cleaner in biodegradable formulation	1. No
	2. Yes
Price per typical application	1. 35 cents
	2. 49 cents
	3. 79 cents

intangible factors that would contribute to interattribute correlation, and the attributes were specifically defined to minimize interattribute correlation.

Specifying the Basic Model Form

After careful consideration, HATCO researchers felt confident in assuming that an additive composition rule was appropriate. Although research has shown that price often has interactions with other factors, it was assumed that all of the other factors were reasonably orthogonal and that interaction terms were not needed. To allow for the most generalizable results applicable across a broad range of application, all factors were estimated with separate part-worth estimates for each level.

Selecting the Method of Data Collection

To ensure realism and allow for the use of ratings rather than rankings, HATCO decided to use the full-profile method of obtaining respondent evaluations. In choosing the additive rule, they were also able to use a fractional factorial design to avoid the evaluation of all 108 possible combinations ($3 \times 3 \times 2 \times 2 \times 3$). The stimulus design component of the computer program generated a set of 18 full-profile descriptions (see Table 7.10), allowing for the estimation of the orthogonal main effects for each factor. Four additional stimuli were generated to serve as the validation stimuli.

The conjoint analysis experiment was administered during a personal interview. After collecting some preliminary data, the respondents were handed a set of 22

TABLE 7.10 Set of 18 Full-Profile Stimuli Used in the HATCO Conjoint Analysis Experiment

Card Number	Product Form	Number of Applications	Disinfectant Quality	Biodegradability	Price per Application
Stimuli used in estimation of part-worths					
1	Concentrate	200	Yes	No	35 cents
2	Powder	200	Yes	No	35 cents
3	Premixed	100	Yes	Yes	49 cents
4	Powder	200	Yes	Yes	49 cents
5	Powder	50	Yes	No	79 cents
6	Concentrate	200	No	Yes	79 cents
7	Premixed	100	Yes	No	79 cents
8	Premixed	200	Yes	No	49 cents
9	Powder	100	No	No	49 cents
10	Concentrate	50	Yes	No	49 cents
11	Powder	100	No	No	35 cents
12	Concentrate	100	Yes	No	79 cents
13	Premixed	200	No	No	79 cents
14	Premixed	50	Yes	No	35 cents
15	Concentrate	100	Yes	Yes	35 cents
16	Premixed	50	No	Yes	35 cents
17	Concentrate	50	No	No	49 cents
18	Powder	50	Yes	Yes	79 cents
Holdout–validation stimuli					
19	Concentrate	100	Yes	No	49 cents
20	Powder	100	No	Yes	35 cents
21	Powder	200	Yes	Yes	79 cents
22	Concentrate	50	No	Yes	35 cents

cards, each containing one of the full-profile stimulus descriptions. They were also presented with a foldout form that had seven response categories, ranging from "not at all likely to buy" to "certain to buy." Respondents were instructed to place each card in the response category best describing their purchase intentions. After initially placing the cards, they were asked to review their placements and rearrange any cards, if necessary. The validation stimuli were rated at the same time as the other stimuli but withheld from the analysis at the estimation stage. Upon completion, the interviewer recorded the category for each card and proceeded with the interview.

Stage 3: Assumptions in Conjoint Analysis

The relevant assumption in conjoint analysis is the specification of the composition rule and thus the model form used to estimate the conjoint results. In this situation, the nature of the product, the tangibility of the attributes, and the lack of intangible or emotional appeals justifies the use of an additive model. HATCO felt confident in using an additive model for this industrial decision-making situation. Moreover, it simplified the design of the stimuli and facilitated the data collection efforts.

Stage 4: Estimating the Conjoint Model and Assessing Overall Model Fit

The estimation of part-worths and the relative importance of each attribute was first performed for each respondent separately, and the results were then aggregated to obtain an overall result. Separate part-worth estimates were made for all levels initially, with examination of the individual estimates undertaken to examine the possibility of placing constraints on a factor relationship form (i.e., employ a linear or quadratic relationship form). Table 7.11 shows the results for the overall sample, as well as for five respondents. Examination of the overall results suggests that perhaps a linear relationship could be estimated for the price variable. However, review of the individual results shows that only two of the five respondents had part-worth estimates for the price factors that were of a generally linear pattern. Thus application of a linear form for the price factor would severely distort the relationship among levels, and the estimation of separate part-worth values was justified.

The measures of predictive accuracy for the estimation of part-worth were all within the acceptable range for both the aggregate results and the five individuals (see Table 7.11). Although the holdout stimuli generally had a lower level of predictive accuracy, all of these values were acceptable as well. Thus, the researcher would retain all of these respondents as being suitably represented by the conjoint results.

Stage 5: Interpreting the Results

Figure 7.5 (p. 434) shows a diversity of part-worth estimates for every factor. For example, the overall results show little impact for the biodegradable feature. However, when the individual results are shown, we see that respondent 2 values the biodegradable feature, whereas respondent 1 does not. Contrasting results are also seen for the disinfectant feature and the number of applications per container. These differences illustrate the value of conjoint analysis performed for each respondent instead of relying solely on aggregate results.

TABLE 7.11 Conjoint Analysis Experiment Results for the Overall Sample and Selected Respondents

Part-Worth Estimates

	Product Form			Number of Applications			Disinfectant		Biodegradable		Price per Application			Relative Importance of Factors					Predictive Accuracy[a]	
	Premixed	Concentrate	Powder	50	100	200	Yes	No	No	Yes	$.35	$.49	$.79	1	2	3	4	5	Estimation	Holdout
Overall Sample	−.311	.192	.139	−.416	.021	.396	.486	−.486	−.077	.077	.962	.069	−1.031	11.75	18.23	21.82	3.44	44.76	.988	.667
Selected Respondents																				
	−.056	.611	−.556	.444	.611	−1.056	−.208	.208	.542	−.542	1.444	.944	−2.389	14.29	20.41	5.10	13.27	46.94	.929	.707
	−.444	.389	.056	−.944	−.278	1.222	1.083	−1.083	−.417	.417	.389	.889	−1.278	10.20	26.53	26.53	10.20	26.53	.949	.707
	−1.111	1.222	−.111	−.611	.556	.056	.083	−.083	−.417	.417	.556	1.056	−1.611	32.56	16.28	2.33	11.63	37.21	.776	.183
	−1.722	.944	.778	−1.056	−.056	1.111	.792	−.792	−.083	.083	1.944	−.222	−1.722	26.02	21.14	15.45	1.63	35.77	.907	.913
	−.611	.389	.222	−.444	.222	.222	−.417	.417	−.542	.542	2.556	.056	−2.611	11.43	7.62	9.52	12.38	59.05	.851	.707

[a]Spearman's rho is reported for the estimation's predictive accuracy, and Kendall's tau is reported for the predictive accuracy of the holdout sample.

433

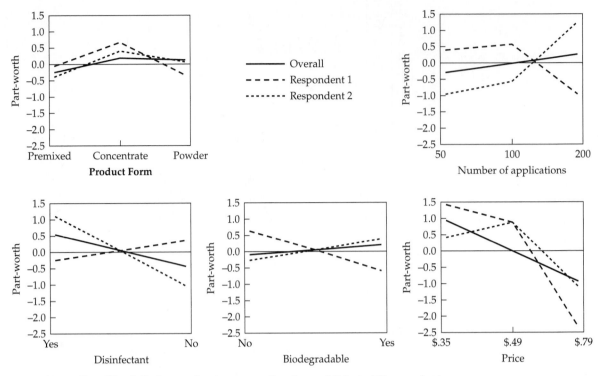

FIGURE 7.5 Part-Worth Estimates for Aggregate Results and Selected Respondents

Figure 7.6 compares the derived importance values of each factor for both the aggregate results and the disaggregate results of two respondents. Although we see a general consistency in the results, each respondent has unique aspects differing from each other and from the aggregate results. The greatest differences are seen for the features of biodegradability and disinfectant quality. Both respondents place more importance on biodegradability than is reflected in the aggregate results. Respondent 1 has little interest in the disinfectant feature, whereas respondent 2 is slightly above average. Finally, respondent 2 is much less price sensitive than either the aggregate results or respondent 1. In summary, the results show two distinct and unique individuals, pointedly illustrating the benefits of conjoint results estimated for each individual instead of for the aggregate. One extension of conjoint analysis is to define groups of respondents with similar part-worth estimates or importance values of the factors using cluster analysis. These segments may then be profiled and assessed for their unique preference structures and market potential.

Stage 6: Validation of the Results

The high levels of predictive accuracy for both the estimation and holdout stimuli across respondents confirm the additive composition rule for this set of respondents. The major validity issue is the representativeness of the sample. In this situation, HATCO would most likely proceed to a larger-scale project with greater coverage of its customer bases to ensure representativeness. Another consideration is the inclusion of noncustomers, especially if the goal is to understand the entire market, not just HATCO customers.

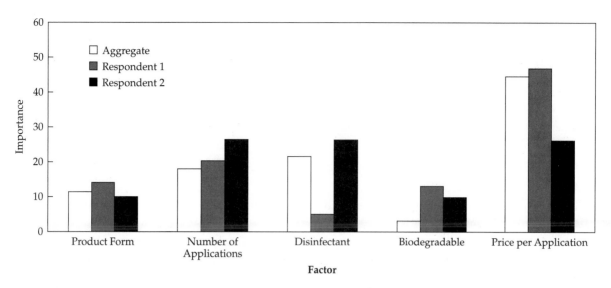

FIGURE 7.6 Factor Importance for Aggregate Results and Selected Respondents

A Managerial Application: Use of a Choice Simulator

In addition to understanding the aggregate and individual preference structures of the respondents, HATCO also used the conjoint results to simulate choices among three possible products. The three products tested were as follows:

Product 1. A premixed cleaner in a handy-to-use size (50 applications per container) that was environmentally safe (biodegradable) and still met all sanitary standards (disinfectant) at only 79 cents per application.

Product 2. An industrial version of product 1 with the environmental and sanitary features, but in a concentrate form in large containers (200 applications) at the low price of 49 cents per application.

Product 3. A real cleanser value in powder form in economical sizes (200 applications per container) for the lowest feasible price of 35 cents per application.

The choice simulator then calculated the preference estimates for the products for each respondent. Predictions of the expected market shares were made with two choice models: the maximum utility model and a probabilistic model. The maximum utility model counts the number of times each of the three products had the highest utility across the set of respondents. As seen in Table 7.12 (p. 436), product 1 was preferred (it had the highest predicted preference value) for only 5 percent of the respondents. Product 2 was next, preferred by 25.5 percent, and the most preferred was product 3, with 69.5 percent. The fractional percentages are due to tied predictions among products 2 and 3.

A second approach to predicting market shares is a probability model, either the BTL or logit model. Both models assess the relative preference of each product and estimate the proportion of times a respondent or the set of respondents will purchase a product. In the HATCO analysis, the aggregated predicted preference values for the products were 2.4, 4.8, and 5.7 for products 1, 2, and 3,

TABLE 7.12 Choice Simulator Results for the Three-Product Formulations

Product Formulation	Aggregate Predicted Preference Scores	Market Share Predictions		
		Maximum Utility Model (%)	Probabilistic Models	
			BTL (%)	Logit (%)
1	2.4	5.0	18.6	9.5
2	4.8	25.5	37.2	32.6
3	5.7	69.5	44.2	57.9

respectively. The predicted market shares using the BTL model are then calculated by

$$\text{Market share}_{\text{product 1}} = 2.4/(2.4 + 4.8 + 5.7) = .186, \text{ or } 18.6\%$$
$$\text{Market share}_{\text{product 2}} = 4.8/(2.4 + 4.8 + 5.7) = .372, \text{ or } 37.2\%$$
$$\text{Market share}_{\text{product 3}} = 5.7/(2.4 + 4.8 + 5.7) = .442, \text{ or } 44.2\%$$

Similar results are obtained using the logit probabilistic model, and are shown in Table 7.12 as well. Using the model recommended in situations involving repetitive choices (probability models), as is the case with an industrial cleaner, HATCO has market share estimates indicating an ordering of product 3, product 2, and finally product 1. It should be remembered that these results are aggregated across the entire sample, and the market shares may differ within specific segments of the respondents.

Summary

Conjoint analysis places more emphasis on the ability of the researcher or manager to theorize about the behavior of choice than it does on analytical technique. The appropriateness of the experimental design and the assumptions concerning the model form and types of relationships among variables are more critical than the choice of estimation technique. As such, it should be viewed primarily as exploratory, because many of its results are directly attributable to basic assumptions made during the course of the design and the execution of the study. Moreover, although conjoint analysis has strong theoretical foundations, its applications have increased dramatically without the corresponding theoretical development. The critical interplay between the assumed conceptual model of decision making and the appropriate elements of the conjoint analysis makes this a unique multivariate method. The researcher must accurately assess many facets of the decision-making process to employ conjoint analysis. Our focus has been on providing a better understanding of the principles of conjoint analysis experiments and how they represent the consumer's choice process. This understanding should enable researchers to avoid misapplication of this relatively new and powerful technique whenever faced with the need to understand choice judgments and preference structures.

Questions

1. Ask three of your classmates to evaluate choice combinations based on these variables and levels relative to their preferred textbook style for a class, and specify the compositional rule you think they will use. Collect information with both the trade-off and full-profile methods.

Factor	Level
Depth	a. Goes into great depth on each subject
	b. Introduces each subject in a general overview
Illustrations	a. Each chapter includes humorous pictures
	b. Illustrative topics are presented
	c. Each chapter includes graphics to illustrate the numeric issues
References	a. General references are included at the end of the textbook
	b. Each chapter includes specific references for the topics covered

   How difficult was it for respondents to handle the wordy and slightly abstract concepts they were asked to evaluate? How would you improve on the descriptions of the factors or levels? Which presentation method was easier for the respondents?

2. Using either the simple numerical procedure discussed earlier or a computer program, analyze the data from the experiment in question 1.

3. Design a conjoint analysis experiment with at least four variables and two levels of each variable that is appropriate to a marketing decision. In doing so, define the compositional rule you will use, the experimental design for creating stimuli, and the analysis method. Use at least five respondents to support your logic.

4. What are the practical limits of conjoint analysis in terms of variables or types of values for each variable? What type of choice problems are best suited to analysis with conjoint analysis? Which are least well served by conjoint analysis?

5. How would you advise a market researcher to choose among the three types of conjoint methodologies? What are the most important issues to consider, along with each methodology's strengths and weaknesses?

References

1. Addelman, S. (1962), "Orthogonal Main-Effects Plans for Asymmetrical Factorial Experiments." *Technometrics* 4: 21–46.

2. Akaah, I. (1991), "Predictive Performance of Self-Explicated, Traditional Conjoint, and Hybrid Conjoint Models under Alternative Data Collection Modes." *Journal of the Academy of Marketing Science* 19: 309–14.

3. Allenby, G. M., N. Arora, and J. L. Ginter (1995), "Incorporating Prior Knowledge into the Analysis of Conjoint Studies." *Journal of Marketing Research* 32 (May): 152–62.

4. Alpert, M. (1971), "Definition of Determinant Attributes: A Comparison of Methods." *Journal of Marketing Research* 8(2): 184–91.

5. Baalbaki, I. B., and N. K. Malhotra (1995), "Standardization versus Customization in International Marketing: An Investigation Using Bridging Conjoint Analysis." *Journal of the Academy of Marketing Science* 23(3): 182–94.

6. Bretton-Clark (1988), *Conjoint Analyzer.* New York: Bretton-Clark.

7. Bretton-Clark (1988), *Conjoint Designer.* New York: Bretton-Clark.

8. Bretton-Clark (1988), *Simgraf.* New York: Bretton-Clark.

9. Bretton-Clark (1988), *Bridger.* New York: Bretton-Clark.

10. Carmone, F. J., Jr., and C. M. Schaffer (1995), "Review of Conjoint Software." *Journal of Marketing Research* 32 (February): 113–20.

11. Carroll, J. D., and P. E. Green (1995), "Psychometric Methods in Marketing Research: Part 1, Conjoint Analysis." *Journal of Marketing Research* 32 (November): 385–91.

12. Conner, W. S., and M. Zelen (1959), *Fractional Factorial Experimental Designs for Factors at Three Levels, Applied Math Series S4.* Washington, D.C.: National Bureau of Standards.

13. Elrod, T., J. J. Louviere, and K. S. Davey (1992), "An Empirical Comparison of Ratings-Based and Choice-Based Conjoint Models." *Journal of Marketing Research* 29: 368–77.

14. Green, P. E. (1984), "Hybrid Models for Conjoint Analysis: An Exploratory Review." *Journal of Marketing Research* 21 (May): 155–69.

15. Green, P. E., S. M. Goldberg, and M. Montemayor (1981), "A Hybrid Utility Estimation Model for Conjoint Analysis." *Journal of Marketing* 45 (Winter): 33–41.

16. Green, P. E., and A. M. Kreiger (1988), "Choice Rules and Sensitivity Analysis in Conjoint Simulators." *Journal of the Academy of Marketing Science* 16 (Spring): 114–27.

17. Green, P. E., and A. M. Kreiger (1991), "Segmenting Markets with Conjoint Analysis." *Journal of Marketing* 55 (October): 20–31.

18. Green, P. E., A. M. Kreiger, and M. K. Agarwal (1991), "Adaptive Conjoint Analysis: Some Caveats and Suggestions." *Journal of Marketing Research* 28 (May): 215–22.

19. Green, P. E., and V. Srinivasan (1978), "Conjoint Analysis in Consumer Research: Issues and Outlook." *Journal of Consumer Research* 5 (September): 103–23.

20. Green, P. E., and V. Srinivasan (1990), "Conjoint Analysis in Marketing: New Developments with Implications for Research and Practice." *Journal of Marketing* 54(4): 3–19.

21. Green, P. E., and Y. Wind (1975), "New Way to Measure Consumers' Judgments." *Harvard Business Review* 53 (July–August): 107–17.

22. Hahn, G. J., and S. S. Shapiro (1966), *A Catalog and Computer Program for the Design and Analysis of Orthogonal Symmetric and Asymmetric Fractional Factorial Experiments*, Report No. 66-C-165. Schenectady, N.Y.: General Electric Research and Development Center.

23. Huber, J. (1987), "Conjoint Analysis: How We Got Here and Where We Are," In *Proceedings of the Sawtooth Conference on Perceptual Mapping, Conjoint Analysis and Computer Interviewing*, M. Metegrano, ed., Ketchum, Idaho: Sawtooth Software, pp. 2–6.

24. Huber, J., and W. Moore (1979), "A Comparison of Alternative Ways to Aggregate Individual Conjoint Analyses," In *Proceedings of the AMA Educator's Conference*, L. Landon, ed., pp. 64–68. Chicago: American Marketing Association.

25. Huber, J., D. R. Wittink, J. A. Fielder, and R. L. Miller (1993), "The Effectiveness of Alternative Preference Elicitation Procedures in Predicting Choice." *Journal of Marketing Research* 30 (February): 105–14.

26. Huber, J., D. R. Wittink, R. M. Johnson, and R. Miller (1992), "Learning Effects in Preference Tasks: Choice-Based versus Standard Conjoint," *Sawtooth Software Conference Proceedings*, M. Metegrano, ed. Ketchum, ID: Sawtooth Software, pp. 275–82.

27. Huber, J., and K. Zwerina (1996), "The Importance of Utility Balance in Efficient Choice Designs." *Journal of Marketing Research* 33 (August): 307–17.

28. Intelligent Marketing Systems, Inc. (1993), *CONSURV—Conjoint Analysis Software, Version 3.0.* Edmonton, Alberta: Intelligent Marketing Systems.

29. Intelligent Marketing Systems, Inc. (1993), *NTELOGIT, Version 3.0.* Edmonton, Alberta: Intelligent Marketing Systems.

30. Jedidi, K., R. Kohli, and W. S. DeSarbo (1996), "Consideration Sets in Conjoint Analysis." *Journal of Marketing Research* 33 (August): 364–72.

31. Johnson, R. M. (1975), "A Simple Method for Pairwise Monotone Regression." *Psychometrika* 40 (June): 163–68.

32. Johnson, R. M. (1991), "Comment on Adaptive Conjoint Analysis: Some Caveats and Suggestions." *Journal of Marketing Research* 28 (May): 223–25.

33. Johnson, R. M., and K. A. Olberts (1991), "Using Conjoint Analysis in Pricing Studies: Is One Price Variable Enough?" In *Advanced Research Techniques Forum Conference Proceedings*, pp. 164–73. Beaver Creek, Colo.: American Marketing Association, pp. 12–18.

34. Johnson, R. M., and B. K. Orme (1996), "How Many Questions Should You Ask in Choice-Based Conjoint Studies?" In *Advanced Research Techniques Forum Conference Proceedings*, Beaver Creek, Colo.: American Marketing Association, pp. 42–49.

35. Kruskal, J. B. (1965), "Analysis of Factorial Experiments by Estimating Monotone Transformations of the Data." *Journal of the Royal Statistical Society* B27: 251–63.

36. Kuhfeld, W. F., R. D. Tobias, and M. Garrath (1994), "Efficient Experimental Designs with Marketing Research Applications." *Journal of Marketing Research* 31 (November): 545–57.

37. Loosschilder, G. H., E. Rosbergen, M. Vriens, and D. R. Wittink (1995), "Pictorial Stimuli in Conjoint Analysis to Support Product Styling Decisions." *Journal of the Marketing Research Society* 37(1): 17–34.

38. Louviere, J. J. (1988), *Analyzing Decision Making: Metric Conjoint Analysis.* Sage University Paper Series on Quantitative Applications in the Social Sciences, vol. 67. Beverly Hills, Calif.: Sage.

39. Louviere, J. J., and G. Woodworth (1983), "Design and Analysis of Simulated Consumer Choice or Alloca-

tion Experiments: An Approach Based on Aggregate Data." *Journal of Marketing Research* 20: 350–67.

40. Luce, R. D. (1959), *Individual Choice Behavior: A Theoretical Analysis.* New York: Wiley.

41. Mahajan, V., and J. Wind (1991), *New Product Models: Practice, Shortcomings and Desired Improvements—Report No. 91-125.* Cambridge, Mass: Marketing Science Institute.

42. McFadden, D. L. (1974), "Conditional Logit Analysis of Qualitative Choice Behavior." In *Frontiers in Econometrics*, P. Zarembka, ed., pp. 105–42. New York: Academic Press.

43. McLean, R., and V. Anderson (1984), *Applied Factorial and Fractional Designs.* New York: Marcel Dekker.

44. Oliphant, K., T. C. Eagle, J. J. Louviere, and D. Anderson (1992), "Cross-Task Comparison of Ratings-Based and Choice-Based Conjoint," In *Sawtooth Software Conference Proceedings*, M. Metegrano, ed., pp. 383–404. Ketchum, Idaho: Sawtooth Software.

45. Oppewal, H. (1995), "A Review of Conjoint Software." *Journal of Retailing and Consumer Services* 2(1): 55–61.

46. Oppewal, H. (1995), "A Review of Choice-Based Conjoint Software: CBC and MINT." *Journal of Retailing and Consumer Services* 2(4): 259–64.

47. Pinnell, J. (1994), "Multistage Conjoint Methods to Measure Price Sensitivity," In *Advanced Research Techniques Forum*, pp. 65–69. Beaver Creek, Colo.: American Marketing Association.

48. Research Triangle Institute (1996), *Trade-Off VR.* Research Triangle Park, N.C.: Research Triangle Institute.

49. Reibstein, D., J. E. G. Bateson, and W. Boulding (1988), "Conjoint Analysis Reliability: Empirical Findings." *Marketing Science* 7 (Summer): 271–86.

50. SAS Institute, Inc. (1992) *SAS Technical Report R-109: Conjoint Analysis Examples.* Cary, N.C.: SAS Institute, Inc.

51. Sawtooth Software (1993), *Adoptive Conjoint Analysis.* Evanston, Ill.: Sawtooth Software.

52. Sawtooth Software (1993), *Choice-Based Conjoint.* Evanston, Ill.: Sawtooth Software.

53. Sawtooth Software (1993), *Conjoint Value Analysis.* Evanston, Ill.: Sawtooth Software.

54. Sawtooth Technologies (1997), *SENSUS, Version 2.0.* Evanston IL.: Sawtooth Technologies.

55. Schocker, A.D., and V. Srinivasan (1977), "LINMAP (Version II): A Fortran IV Computer Program for Analyzing Ordinal Preference (Dominance) Judgments Via Linear Programming Techniques for Conjoint Measurement." *Journal of Marketing Research* 14 (February): 101–103.

56. Smith, Scott M. (1989), *PC-MDS: A Multidimensional Statistics Package.* Provo, Utah: Brigham Young University Press.

57. SPSS, Inc. (1990), *SPSS Categories.* Chicago: SPSS, Inc.

58. Srinivasan, V. (1988), "A Conjunctive-Compensatory Approach to the Self-Explication of Multiattitudinal Preference." *Decision Sciences* 19 (Spring): 295–305.

59. Srinivasan, V., A. K. Jain, and N. Malhotra (1983), "Improving Predictive Power of Conjoint Analysis by Constrained Parameter Estimation." *Journal of Marketing Research* 20 (November): 433–38.

60. Srinivasan, V., and C. S. Park (1997), "Surprising Robustness of the Self-Explicated Approach to Customer Preference Structure Measurement." *Journal of Marketing Research* 34 (May): 286–91.

61. Steckel, J., W. S. DeSarbo, and V. Mahajan (1991), "On the Creation of Acceptable Conjoint Analysis Experimental Design." *Decision Sciences* 22(2): 435–42.

62. Thurstone, L. L. (1927), "A Law of Comparative Judgment." *Psychological Review* 34: 276–86.

63. Tumbush, J. J. (1991), "Validating Adaptive Conjoint Analysis (ACA) Versus Standard Concept Testing," In *Sawtooth Software Conference Proceedings*, M. Metegrano, ed. Ketchum, Idaho: Sawtooth Software, pp. 177–184.

64. van der Lans, I. A., and W. Heiser (1992), "Constrained Part-Worth Estimation in Conjoint Analysis Using the Self-Explicated Utility Model." *International Journal of Research in Marketing* 9: 325–44.

65. Veiens, M., M. Wedel, and T. Wilms (1996), "Metric Conjoint Segmentation Methods: A Monte Carlo Comparison." *Journal of Marketing Research* 33 (February): 73–85.

66. Wittink, D. R., and P. Cattin (1989), "Commercial Use of Conjoint Analysis: An Update." *Journal of Marketing* 53 (July): 91–96.

67. Wittink, D. R., J. Huber, P. Zandan, and R. M. Johnson (1992), "The Number of Levels Effect in Conjoint: Where Does It Come From, and Can It Be Eliminated?" In *Sawtooth Software Conference Proceedings*, M. Metegrano, ed. Ketchum, Idaho: Sawtooth Software, pp. 355–64.

68. Wittink, D. R., L. Krishnamurthi, and J. B. Nutter (1982), "Comparing Derived Importance Weights Across Attributes." *Journal of Consumer Research* 8 (March): 471–74.

69. Wittink, D. R., L. Krishnamurthi, and D. J. Reibstein (1990), "The Effect of Differences in the Number of Attribute Levels on Conjoint Results." *Marketing Letters* 1(2): 113–29.

70. Wittink, D. R., M. Vriens, and W. Burhenne (1994), "Commercial Use of Conjoint Analysis in Europe: Results and Critical Reflections." *International Journal of Research in Marketing* 11: 41–52.

Annotated Articles

The following annotated articles are provided as illustrations of the application of conjoint analysis to substantive research questions of both a conceptual and managerial nature. The reader is encouraged to review the complete articles for greater detail on any of the specific issues regarding methodology or findings.

Rosko, Michael D., Michael DeVita, William F. McKenna, and Lawrence R. Walker (1985), "Strategic Marketing Applications of Conjoint Analysis: An HMO Perspective." *Journal of Health Care Marketing* 5(4): 27–38.

Conjoint analysis is used in the examination of variables that are important in the consumer's decision to participate in a Health Maintenance Organization (HMO). The relevant variables (health system attributes) are examined for their relative importance and perceived utility in making such a decision. A total sample of 97 HMO and non-HMO members are used to determine whether differences exist in HMO attribute importance and utility measures for each of the two groups. This knowledge is then used to examine the impact of changes in the marketing mix on HMO membership, to determine market segments, and to develop a product offering that is more attractive to the chosen segment.

From prior literature and current focus groups, the authors identify 10 actionable factors (five with four levels and five with two levels). An additive conjoint model is used, with each card's rating serving as the dependent variable and the 10 attributes as the independent variables. To collect the data, the authors employ a full-profile method with an orthogonal factorial design to generate 41 profile cards for the respondents to evaluate. This approach reduces the number of comparisons needed (from 4,096) but does not allow for the determination of interaction effects among the variables. Although the validity of their results is verified by testing the model's ability to predict behavior, sampling concerns may inhibit generalizability (i.e., size and representativeness). However, the authors do demonstrate the versatility of the results through segmentation, profitability analysis, and conjoint simulation applications.

Thus, the use of conjoint analysis allows for the examination of all the variables thought to be involved in consumer HMO decision-making. Through this and other multivariate techniques, the authors illustrate how these methods may be employed in the selection of target markets and marketing mix variables (i.e., product design, pricing, placement of facilities, and promotional programs).

Barich, Howard, and V. Srinivasan (1993), "Prioritizing Marketing Image Goals under Resource Constraints." *Sloan Management Review* 34(4): 69–76.

In this article, the authors explain and demonstrate the multiple uses of conjoint analysis as an aid in managerial decision making by exploring marketing image in a high-end department store. Marketing image is how the customer perceives the organization's products and services and has an impact on organizational performance. By understanding how the customer views the organization's marketing image, managers can determine the resources necessary to improve that image. Thus, conjoint provides a means for prioritizing marketing image goals. The authors apply the technique to determine customer importance of marketing image goals and then compare those measures to manager perceptions. They then perform a cost-benefit analysis based on the results in order to prioritize objectives that will bring the greatest return. The authors accomplish this by explaining the technique and then by offering a multistep approach to prioritize marketing image goals.

This application uses a full-profile method with six factors (convenience, customer service, product quality, product variety, reasonable prices, and store attractiveness), each containing three levels (fair, good, and excellent). A fractional factorial design is used to reduce the 729 possible combinations, which results in an additive model of 18 hypothetical combinations. Each of the six part-worth functions is examined for diminishing returns, indicating that the increase in utility from fair to good is larger than the increase from good to excellent. Management perceptions of each attribute's importance and ratings on the degree of difficulty (effort and resources) to make a unit improvement for each of the attributes are also gathered. With this information, the authors are able to compare management perceptions with those of the customer and compute the relative advantage of improving each attribute. Priority is given to improvements with high attribute importance but fewer resource demands. By applying the conjoint results, the authors demonstrate the method's ability to aid in decision making by accounting for customer needs and resource demands.

Money, Arthur, David Tromp, and Trevor Wegner (1988), "The Quantification of Decision Support Benefits within the Context of Value Analysis." *MIS Quarterly* 12(2): 223–36.

The authors use conjoint analysis as a more rigorous approach to evaluate decision support systems (DSS). By identifying and measuring the intangible benefits of these systems, the authors establish a means of evaluating the effectiveness of information technology in adding value to the decision-making process. Using a literature review and the Delphi method, the authors generate a number of benefits. They then use cluster analysis to confirm these as three separate product benefits (operational, managerial–organizational, and personal). The benefit groupings (used as the factors

in the conjoint analysis) contain a total of 12 levels. A sample of 15 DSS workshop attendees is used to obtain the utility scores (raising concerns as to the adequacy of sample size). After determining which benefits are important for end-users, the next step is to apply these results toward judging whether a proposed system adds value.

The authors apply a fractional factorial design, reducing the 48 possible combinations to 12 benefit combinations. The respondents are instructed to rank the combinations of benefits from most important to least important. The results indicate that personally derived benefits, particularly the improvement of decision-making capabilities, are more important than either operational or managerial–organizational benefits. To evaluate the potential of the DSS, the authors perform a t test to compare the differences of the value benefits to a null benefits set (i.e., provides no benefits, which acts as a threshold level). The results, from this example, indicate support for the proposed DSS project. Additionally, the authors demonstrate that self-stated importance ratings offer comparable results. By allowing for the quantification of end-user perceptions of intangible benefits, conjoint analysis has provided a value-based framework for DSS evaluation.

Canonical Correlation Analysis

LEARNING OBJECTIVES

Upon completing this chapter, you should be able to do the following:

- State the similarities and differences between multiple regression, factor analysis, discriminant analysis, and canonical correlation.
- Summarize the conditions that must be met for application of canonical correlation analysis.
- State what the canonical root measures and point out its limitations.
- State how many independent canonical functions can be defined between the two sets of original variables.
- Compare the advantages and disadvantages of the three methods for interpreting the nature of canonical functions.
- Define redundancy and compare it with multiple regression's R^2.

CHAPTER PREVIEW

Until recent years, canonical correlation analysis was a relatively unknown statistical technique. As with almost all of the multivariate techniques, the availability of computer programs has facilitated its increased application to research problems. It is particularly useful in situations in which multiple output measures such as satisfaction, purchase, or sales volume are available. If the independent variables were only categorical, multivariate analysis of variance could be used. But what if the independent variables are metric? Canonical correlation is the answer, allowing for the assessment of the relationship between metric independent variables and multiple dependent measures. As discussed in chapter 1, canonical correlation is considered to be the general model on which many other multivariate techniques are based because it can use both metric and nonmetric data for either the dependent or independent variables. We express the general form of canonical analysis as

$$Y_1 + Y_2 + Y_3 + \ldots + Y_n = X_1 + X_2 + X_3 + \ldots + X_n$$

(metric, nonmetric)   (metric, nonmetric)

This chapter introduces the researcher to the multivariate statistical technique of canonical correlation analysis. Specifically, we (1) describe the nature of canonical correlation analysis, (2) illustrate its application, and (3) discuss its potential advantages and limitations.

KEY TERMS

Before starting the chapter, review the key terms to develop an understanding of the concepts and terminology used. Throughout the chapter the key terms appear in **boldface**. Other points of emphasis in the chapter are *italicized*. Also, cross-references within the Key Terms appear in *italics*.

Canonical correlation Measure of the strength of the overall relationships between the linear composites (*canonical variates*) for the independent and dependent variables. In effect, it represents the bivariate correlation between the two canonical variates.

Canonical cross-loadings Correlation of each observed independent or dependent variable with the opposite *canonical variate*. For example, the independent variables are correlated with the dependent canonical variate. They can be interpreted like *canonical loadings*, but with the opposite canonical variate.

Canonical function Relationship (correlational) between two linear composites (*canonical variates*). Each canonical function has two canonical variates, one for the set of dependent variables and one for the set of independent variables. The strength of the relationship is given by the *canonical correlation*.

Canonical loadings Measure of the simple linear correlation between the independent variables and their respective *canonical variates*. These can be interpreted like factor loadings, and are also known as canonical structure correlations.

Canonical roots Squared *canonical correlations*, which provide an estimate of the amount of shared variance between the respective optimally weighted *canonical variates* of dependent and independent variables. Also known as *eigenvalues*.

Canonical variates Linear combinations that represent the weighted sum of two or more variables and can be defined for either dependent or independent variables. Also referred to as *linear composites*, linear compounds, and linear combinations.

Eigenvalues See *canonical roots*.

Linear composites See *canonical variates*.

Orthogonal Mathematical constraint specifying that the canonical functions are independent of each other. In other words, the canonical functions are derived so that each is at a right angle to all other functions when plotted in multivariate space, thus ensuring statistical independence between the canonical functions.

Redundancy index Amount of variance in a *canonical variate* (dependent or independent) explained by the other canonical variate in the *canonical function*. It can be computed for both the dependent and the independent canonical variates in each canonical function. For example, a redundancy index of the dependent variate represents the amount of variance in the dependent variables explained by the independent canonical variate.

What Is Canonical Correlation?

Chapter 4 discussed multiple regression analysis, which can predict the value of a single (metric) dependent variable from a linear function of a set of independent variables. For some research problems, interest may not center on a single dependent variable; rather, the researcher may be interested in relationships between sets of multiple dependent and multiple independent variables. **Canonical correlation** analysis is a multivariate statistical model that facilitates the study of interrelationships among sets of multiple dependent variables and multiple independent variables [5, 6]. Whereas multiple regression predicts a single dependent variable from a set of multiple independent variables, canonical correlation simultaneously predicts multiple dependent variables from multiple independent variables.

Canonical correlation places the fewest restrictions on the types of data on which it operates. Because the other techniques impose more rigid restrictions, it is generally believed that the information obtained from them is of higher quality and may be presented in a more interpretable manner. For this reason, many researchers view canonical correlation as a last-ditch effort, to be used when all other higher-level techniques have been exhausted. But in situations with multiple dependent and independent variables, canonical correlation is the most appropriate and powerful multivariate technique. It has gained acceptance in many fields and represents a useful tool for multivariate analysis, particularly as interest has spread to considering multiple dependent variables.

Hypothetical Example of Canonical Correlation

To clarify further the nature of canonical correlation, let us consider an extension of the example used in chapter 4. Recall that the HATCO survey results used family size and income as predictors of the number of credit cards a family would hold. That problem involved examining the relationship between two independent variables and a single dependent variable.

Suppose HATCO was interested in the broader concept of credit usage. To measure credit usage, HATCO considered not only the number of credit cards held by the family but also the family's average monthly dollar charges on all credit cards. These two measures were felt to give a much better perspective on a family's credit card usage. Readers interested in the approach of using multiple indicators to represent a concept are referred to chapter 3 (Factor Analysis) and chapter 11 (Structural Equation Modeling). The problem now involves predicting two dependent measures simultaneously (number of credit cards and average dollar charges). Multiple regression is capable of handling only a single dependent variable. Multivariate analysis of variance could be used, but only if all of the independent variables were nonmetric, which is not the case in this problem.

TABLE 8.1 Canonical Correlation of Credit Usage (Number of Credit Cards and Usage Rate) with Customer Characteristics (Family Size and Family Income)

Measures of Credit Usage	*Measures of Customer Characteristics*
• Number of credit cards held • Average monthly dollar expenditures on credit cards	• Family size • Family income

Composite of Dependent Variables	*Canonical Correlation*	*Composite of Independent Variables*
Dependent canonical variate	R_c	Independent canonical variate

Canonical correlation represents the only technique available for examining the relationship with multiple dependent variables.

The problem of predicting credit usage is illustrated in Table 8.1. The two dependent variables used to measure credit usage—number of credit cards held by the family and average monthly dollar expenditures on all credit cards—are listed at the left. The two independent variables selected to predict credit usage—family size and family income—are shown on the right. By using canonical correlation analysis, HATCO creates a composite measure of credit usage that consists of both dependent variables, rather than having to compute a separate regression equation for each of the dependent variables. The result of applying canonical correlation is a measure of the strength of the relationship between two sets of multiple variables (canonical variates). The measure of the strength of the relationship between the two variates is expressed as a canonical correlation coefficient (R_c). The researcher now has two results of interest: the canonical variates representing the optimal linear combinations of dependent and independent variables; and the canonical correlation representing the relationship between them.

Analyzing Relationships with Canonical Correlation

Canonical correlation analysis is the most generalized member of the family of multivariate statistical techniques. It is directly related to several dependence methods. Similar to regression, canonical correlation's goal is to quantify the strength of the relationship, in this case between the two sets of variables (independent and dependent). It corresponds to factor analysis in the creation of composites of variables. It also resembles discriminant analysis in its ability to determine independent dimensions (similar to discriminant functions) for *each* variable set, in this situation with the objective of producing the maximum correlation between the dimensions. Thus, canonical correlation identifies the optimum structure or dimensionality of each variable set that maximizes the relationship between independent and dependent variable sets.

Canonical correlation analysis deals with the association between composites of sets of multiple dependent and independent variables. In doing so, it develops a number of independent **canonical functions** that maximize the correlation between the **linear composites,** also known as **canonical variates,** which are sets of dependent and independent variables. Each canonical function is actually based on the correlation between two canonical variates, one variate for the dependent variables and one for the independent variables. Another unique feature of canonical correlation is that the variates are derived to maximize their correlation. Moreover, canonical correlation does not stop with the derivation of a single relationship between the sets of variables. Instead, a number of canonical functions (pairs of canonical variates) may be derived.

Our discussion of canonical correlation analysis is organized around the model-building process described in chapter 1. Figure 8.1 (stages 1–3) and Figure 8.2 (stages 4–6; p. 449) depict the stages for canonical correlation analysis, which include (1) specifying the objectives of canonical correlation, (2) developing the analysis plan, (3) assessing the assumptions underlying canonical correlation, (4) estimating the canonical model and assessing overall model fit, (5) interpreting the canonical variates, and (6) validating the model.

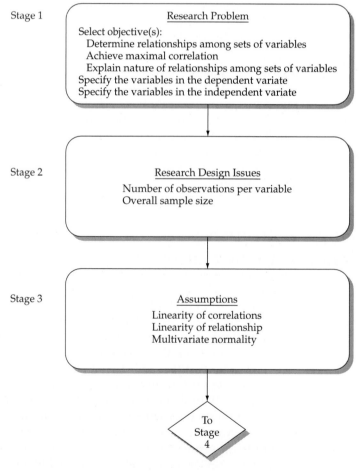

FIGURE 8.1 Stages 1–3 in the Canonical Correlation Decision Diagram

Stage 1: Objectives of Canonical Correlation Analysis

The appropriate data for canonical correlation analysis are two sets of variables. We assume that each set can be given some theoretical meaning, at least to the extent that one set could be defined as the independent variables and the other as the dependent variables. Once this distinction has been made, canonical correlation can address a wide range of objectives. These objects may be any or all of the following:

1. Determining whether two sets of variables (measurements made on the same objects) are independent of one another or, conversely, determining the magnitude of the relationships that may exist between the two sets.
2. Deriving a set of weights for each set of dependent and independent variables so that the linear combinations of each set are maximally correlated. Additional linear functions that maximize the remaining correlation are independent of the preceding set(s) of linear combinations.
3. Explaining the nature of whatever relationships exist between the sets of dependent and independent variables, generally by measuring the relative contribution of each variable to the canonical functions (relationships) that are extracted.

The inherent flexibility of canonical correlation in terms of the number and types of variables handled, both dependent and independent, makes it a logical candidate for many of the more complex problems addressed with multivariate techniques.

Stage 2: Designing a Canonical Correlation Analysis

As the most general form of multivariate analysis, canonical correlation analysis shares basic implementation issues common to all multivariate techniques. Discussions in other chapters (particularly multiple regression, discriminant analysis, and factor analysis) on the impact of measurement error, the types of variables, and their transformations that can be included are relevant to canonical correlation analysis as well.

The issues of the impact of sample size (both small and large) and the necessity for a sufficient number of observations per variable are frequently encountered with canonical correlation. Researchers are tempted to include many variables in both the independent and dependent variable set, not realizing the implications for sample size. Sample sizes that are very small will not represent the correlations well, thus obscuring any meaningful relationships. Very large samples will have a tendency to indicate statistical significance in all instances, even where practical significance is not indicated. The researcher is also encouraged to maintain at least 10 observations per variable to avoid "overfitting" the data.

The classification of variables as dependent or independent is of little importance for the statistical estimation of the canonical functions, because canonical correlation analysis weights both variates to maximize the correlation and places no particular emphasis on either variate. Yet because the technique produces variates to maximize the correlation between them, a variable in either set relates to all other variables in both sets. This allows the addition or deletion of a single variable to affect the entire solution, particularly the other variate. The composition of each variate, either independent or dependent, becomes critical. A researcher must have conceptually linked sets of the variables before applying canonical correlation analysis. This makes the specification of dependent versus independent variates essential to establishing a strong conceptual foundation for the variables.

Stage 3: Assumptions in Canonical Correlation

The generality of canonical correlation analysis also extends to its underlying statistical assumptions. The assumption of linearity affects two aspects of canonical correlation results. First, the correlation coefficient between any two variables is based on a linear relationship. If the relationship is nonlinear, then one or both variables should be transformed, if possible. Second, the canonical correlation is the linear relationship between the variates. If the variates relate in a nonlinear manner, the relationship will not be captured by canonical correlation. Thus, while canonical correlation analysis is the most generalized multivariate method, it is still constrained to identifying linear relationships.

Canonical correlation analysis can accommodate any metric variable without the strict assumption of normality. Normality is desirable because it standardizes a distribution to allow for a higher correlation among the variables. But in the strictest sense, canonical correlation analysis can accommodate even nonnormal variables if the distributional form (e.g., highly skewed) does not decrease the correlation with other variables. This allows for transformed nonmetric data (in the form of dummy variables) to be used as well. However, multivariate normality is required for the statistical inference test of the significance of each canonical function. Because tests for multivariate normality are not readily available, the prevailing guideline is to ensure that each variable has univariate normality. Thus, although normality is not strictly required, it is highly recommended that all variables be evaluated for normality and transformed if necessary.

Homoscedasticity, to the extent that it decreases the correlation between variables, should also be remedied. Finally, multicollinearity among either variable set will confound the ability of the technique to isolate the impact of any single variable, making interpretation less reliable. Readers unfamiliar with these statistical assumptions, the tests for their diagnosis, or the alternative remedies when the assumptions are not met should refer to chapter 2.

Stage 4: Deriving the Canonical Functions and Assessing Overall Fit

The first step of canonical correlation analysis is to derive one or more canonical functions (see Figure 8.2). Each function consists of a pair of variates, one representing the independent variables and the other representing the dependent vari-

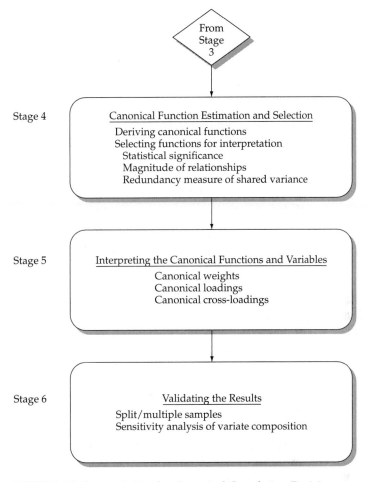

FIGURE 8.2 Stages 4–6 in the Canonical Correlation Decision Diagram

ables. The maximum number of canonical variates (functions) that can be extracted from the sets of variables equals the number of variables in the smallest data set, independent or dependent. For example, when the research problem involves five independent variables and three dependent variables, the maximum number of canonical functions that can be extracted is three.

Deriving Canonical Functions

The derivation of successive canonical variates is similar to the procedure used with unrotated factor analysis (see chapter 3). The first factor extracted accounts for the maximum amount of variance in the set of variables, then the second factor is computed so that it accounts for as much as possible of the variance not accounted for by the first factor, and so forth, until all factors have been extracted. Therefore, successive factors are derived from residual or leftover variance from earlier factors. Canonical correlation analysis follows a similar procedure but focuses on accounting for the maximum amount of the relationship between the two sets of variables, rather than within a single set. The result is that the first pair of canonical variates is derived so as to have the highest intercorrelation possible between the two sets of variables. The second pair of canonical variates is then derived so that it exhibits the maximum relationship between the two sets of variables (variates) not accounted for by the first pair of variates. In short, successive pairs of canonical variates are based on residual variance, and their respective canonical correlations (which reflect the interrelationships between the variates) become smaller as each additional function is extracted. That is, the first pair of canonical variates exhibits the highest intercorrelation, the next pair the second-highest correlation, and so forth.

One additional point about the derivation of canonical variates: as noted, successive pairs of canonical variates are based on residual variance. Therefore, each of the pairs of variates is **orthogonal** and independent of all other variates derived from the same set of data.

The strength of the relationship between the pairs of variates is reflected by the canonical correlation. When squared, the canonical correlation represents the amount of variance in one canonical variate accounted for by the other canonical variate. This also may be called the amount of shared variance between the two canonical variates. Squared canonical correlations are called **canonical roots** or **eigenvalues.**

Which Canonical Functions Should Be Interpreted?

As with research using other statistical techniques, the most common practice is to analyze functions whose canonical correlation coefficients are statistically significant beyond some level, typically .05 or above. If other independent functions are deemed insignificant, these relationships among the variables are not interpreted. Interpretation of the canonical variates in a significant function is based on the premise that variables in each set that contribute heavily to shared variances for these functions are considered to be related to each other.

The authors believe that the use of a single criterion such as the level of significance is too superficial. Instead, they recommend that three criteria be used in conjunction with one another to decide which canonical functions should be interpreted. The three criteria are (1) level of statistical significance of the function, (2) magnitude of the canonical correlation, and (3) redundancy measure for the percentage of variance accounted for from the two data sets.

Level of Significance

The level of significance of a canonical correlation generally considered to be the minimum acceptable for interpretation is the .05 level, which (along with the .01 level) has become the generally accepted level for considering a correlation coefficient statistically significant. This consensus has developed largely because of the availability of tables for these levels. These levels are not necessarily required in all situations, however, and researchers from various disciplines frequently must rely on results based on lower levels of significance. The most widely used test, and the one normally provided by computer packages, is the F statistic, based on Rao's approximation [3].

In addition to separate tests of each canonical function, a multivariate test of all canonical roots can also be used for evaluating the significance of canonical roots. Many of the measures for assessing the significance of discriminant functions, including Wilks' lambda, Hotelling's trace, Pillai's trace, and Roy's gcr, are also provided. See chapter 5 for a discussion of these measures.

Magnitude of the Canonical Relationships

The practical significance of the canonical functions, represented by the size of the canonical correlations, also should be considered when deciding which functions to interpret. No generally accepted guidelines have been established regarding suitable sizes for canonical correlations. Rather, the decision is usually based on the contribution of the findings to better understanding of the research problem being studied. It seems logical that the guidelines suggested for significant factor loadings (see chapter 3) might be useful with canonical correlations, particularly when one considers that canonical correlations refer to the variance explained in the canonical variates (linear composites), not the original variables.

Redundancy Measure of Shared Variance

Recall that squared canonical correlations (roots) provide an estimate of the shared variance between the canonical variates. Although this is a simple and appealing measure of the shared variance, it may lead to some misinterpretation because the squared canonical correlations represent the variance shared by the linear composites of the sets of dependent and independent variables, and not the variance extracted from the sets of variables [1]. Thus, a relatively strong canonical correlation may be obtained between two linear composites (canonical variates), even though these linear composites may not extract significant portions of variance from their respective sets of variables.

Because canonical correlations may be obtained that are considerably larger than previously reported bivariate and multiple correlation coefficients, there may be a temptation to assume that canonical analysis has uncovered substantial relationships of conceptual and practical significance. Before such conclusions are warranted, however, further analysis involving measures other than canonical correlations must be undertaken to determine the amount of the dependent variable variance accounted for or shared with the independent variables [7].

To overcome the inherent bias and uncertainty in using canonical roots (squared canonical correlations) as a measure of shared variance, a **redundancy index** has been proposed [8]. It is the equivalent of computing the squared multiple correlation coefficient between the total independent variable set and each variable in the dependent variable set, and then averaging these squared coefficients to arrive at an average R^2. This index provides a summary measure of the ability of a set of independent variables (taken as a set) to explain variation in the dependent variables (taken one at a time). As such, the redundancy measure is perfectly analogous to multiple regression's R^2 statistic, and its value as an index is similar.

The Stewart-Love index of redundancy calculates the amount of variance in one set of variables that can be explained by the variance in the other set. This index serves as a measure of accounted-for variance, similar to the R^2 calculation used in multiple regression. The R^2 represents the amount of variance in the dependent variable explained by the regression function of the independent variables. In regression, the total variance in the dependent variable is equal to 1, or 100 percent. Remember that canonical correlation is different from multiple regression in that it does not deal with a single dependent variable but has a composite of the dependent variables, and this composite has only a portion of each dependent variable's total variance. For this reason, we cannot assume that 100 percent of the variance in the dependent variable set is available to be explained by the independent variable set. The set of independent variables can be expected to account only for the shared variance in the dependent canonical variate. For this reason, the calculation of the redundancy index is a three-step process. The first step involves calculating the amount of shared variance from the set of dependent variables included in the dependent canonical variate. The second step involves calculating the amount of variance in the dependent canonical variate that can be explained by the independent canonical variate. The final step is to calculate the redundancy index, found by multiplying these two components.

Step 1: The Amount of Shared Variance.   To calculate the amount of shared variance in the dependent variable set included in the dependent canonical variate, let us first consider how the regression R^2 statistic is calculated. R^2 is simply the square of the correlation coefficient R, which represents the correlation between the actual dependent variable and the predicted value. In the canonical case, we are concerned with correlation between the dependent canonical variate and each of the dependent variables. Such information can be obtained from the canonical loadings (L_1), which represent the correlation between each input variable and its own canonical variate (discussed in more detail in the following section). By squaring each of the dependent variable loadings (L_i^2), one may obtain a measure of the amount of variation in each of the dependent variables explained by the dependent canonical variate. To calculate the amount of shared variance explained by the canonical variate, a simple average of the squared loadings is used.

Step 2: The Amount of Explained Variance.   The second step of the redundancy process involves the percentage of variance in the dependent canonical variate that can be explained by the independent canonical variate. This is simply the squared correlation between the independent canonical variate and the dependent canonical variate, which is otherwise known as the canonical correlation. The squared canonical correlation is commonly called the canonical R^2.

Step 3: The Redundancy Index.   The redundancy index of a variate is then derived by multiplying the two components (shared variance of the variate multiplied by the squared canonical correlation) to find the amount of shared variance that can be explained by each canonical function. To have a high redundancy index, one must have a high canonical correlation and a high degree of shared variance explained by the dependent variate. A high canonical correlation alone does not ensure a valuable canonical function. Redundancy indices are calculated for both the dependent and the independent variates, although in most instances the researcher is concerned only with the variance extracted from the dependent variable set, which provides a much more realistic measure of the predictive ability of canonical relationships. The researcher should note that while the canonical correlation is the same for both variates in the canonical function, the redundancy index will most likely vary between the two variates, as each will have a differing amount of shared variance.

What is the minimum acceptable redundancy index needed to justify the interpretation of canonical functions? Just as with canonical correlations, no generally accepted guidelines have been established. The researcher must judge each canonical function in light of its theoretical and practical significance to the research problem being investigated to determine whether the redundancy index is sufficient to justify interpretation. A test for the significance of the redundancy index has been developed [2], although it has not been widely utilized.

Stage 5: Interpreting the Canonical Variate

If the canonical relationship is statistically significant and the magnitudes of the canonical root and the redundancy index are acceptable, the researcher still needs to make substantive interpretations of the results. Making these interpretations involves examining the canonical functions to determine the relative importance of each of the original variables in the canonical relationships. Three methods have been proposed: (1) canonical weights (standardized coefficients), (2) canonical loadings (structure correlations), and (3) canonical cross-loadings.

Canonical Weights

The traditional approach to interpreting canonical functions involves examining the sign and the magnitude of the canonical weight assigned to each variable in its canonical variate. Variables with relatively larger weights contribute more to the variates, and vice versa. Similarly, variables whose weights have opposite signs exhibit an inverse relationship with each other, and variables with weights of the same sign exhibit a direct relationship. However, interpreting the relative importance or contribution of a variable by its canonical weight is subject to the same criticisms associated with the interpretation of beta weights in regression techniques. For example, a small weight may mean either that its corresponding variable is irrelevant in determining a relationship or that it has been partialed out of the relationship because of a high degree of multicollinearity. Another problem with the use of canonical weights is that these weights are subject to considerable instability (variability) from one sample to another. This instability occurs because the computational procedure for canonical analysis yields weights that maximize the canonical correlations for a particular sample of observed dependent and independent variable sets [7]. These problems suggest considerable caution in using canonical weights to interpret the results of a canonical analysis.

Canonical Loadings

Canonical loadings have been increasingly used as a basis for interpretation because of the deficiencies inherent in canonical weights. **Canonical loadings,** also called canonical structure correlations, measure the simple linear correlation between an original observed variable in the dependent or independent set and the set's canonical variate. The canonical loading reflects the variance that the observed variable shares with the canonical variate and can be interpreted like a factor loading in assessing the relative contribution of each variable to each canonical function. The methodology considers each independent canonical function separately and computes the within-set variable-to-variate correlation. The larger the coefficient, the more important it is in deriving the canonical variate. Also, the criteria for determining the significance of canonical structure correlations are the same as with factor loadings (see chapter 3).

Canonical loadings, like weights, may be subject to considerable variability from one sample to another. This variability suggests that loadings, and hence the relationships ascribed to them, may be sample-specific, resulting from chance or extraneous factors [7]. Although canonical loadings are considered relatively more valid than weights as a means of interpreting the nature of canonical relationships, the researcher still must be cautious when using loadings for interpreting canonical relationships, particularly with regard to the external validity of the findings.

Canonical Cross-Loadings

The computation of **canonical cross-loadings** has been suggested as an alternative to canonical loadings [4]. This procedure involves correlating each of the original observed dependent variables directly with the independent canonical variate, and vice versa. Recall that conventional loadings correlate the original observed variables with their respective variates after the two canonical variates (dependent and independent) are maximally correlated with each other. This may also seem similar to multiple regression, but it differs in that each independent variable, for example, is correlated with the dependent variate instead of a single dependent variable. Thus cross-loadings provide a more direct measure of the dependent–independent variable relationships by eliminating an intermediate step involved in conventional loadings. Some canonical analyses do not compute correlations between the variables and the variates. In such cases the canonical weights are considered comparable but not equivalent for purposes of our discussion.

Which Interpretation Approach to Use

Several different methods for interpreting the nature of canonical relationships have been discussed. The question remains, however: Which method should the researcher use? Because most canonical problems require a computer, the researcher frequently must use whichever method is available in the standard statistical packages. The cross-loadings approach is preferred, and it is provided by many computer programs, but if the cross-loadings are not available, the researcher is forced either to compute the cross-loadings by hand or to rely on the other methods of interpretation. The canonical loadings approach is somewhat more representative than the use of weights, just as was seen with factor analysis and discriminant analysis. Therefore, whenever possible the loadings approach is recommended as the best alternative to the canonical cross-loadings method.

Stage 6: Validation and Diagnosis

As with any other multivariate technique, canonical correlation analysis should be subjected to validation methods to ensure that the results are not specific only to the sample data and can be generalized to the population. The most direct procedure is to create two subsamples of the data (if sample size allows) and perform the analysis on each subsample separately. Then the results can be compared for similarity of canonical functions, variate loadings, and the like. If marked differences are found, the researcher should consider additional investigation to ensure that the final results are representative of the population values, not solely those of a single sample.

Another approach is to assess the sensitivity of the results to the removal of a dependent and/or independent variable. Because the canonical correlation procedure maximizes the correlation and does not optimize the interpretability, the canonical weights and loadings may vary substantially if one variable is removed from either variate. To ensure the stability of the canonical weights and loading, the researcher should estimate multiple canonical correlations, each time removing a different independent or dependent variable.

Although there are few diagnostic procedures developed specifically for canonical correlation analysis, the researcher should view the results within the limitations of the technique. Among the limitations that can have the greatest impact on the results and their interpretation are the following:

1. The canonical correlation reflects the variance shared by the linear composites of the sets of variables, not the variance extracted from the variables.
2. Canonical weights derived in computing canonical functions are subject to a great deal of instability.
3. Canonical weights are derived to maximize the correlation between linear composites, not the variance extracted.
4. The interpretation of the canonical variates may be difficult because they are calculated to maximize the relationship, and there are no aids for interpretation, such as rotation of variates, as seen in factor analysis.
5. It is difficult to identify meaningful relationships between the subsets of independent and dependent variables because precise statistics have not yet been developed to interpret canonical analysis, and we must rely on inadequate measures such as loadings or cross-loadings [7].

These limitations are not meant to discourage the use of canonical correlation. Rather, they are pointed out to enhance the effectiveness of canonical correlation as a research tool.

An Illustrative Example

To illustrate the application of canonical correlation, we use variables drawn from the database introduced in chapter 1. Recall that the data consisted of a series of measures obtained on a sample of 100 HATCO customers. The variables included ratings of HATCO on seven attributes (X_1 to X_7) and two measures reflecting the effects of HATCO's efforts (X_9, usage of HATCO products, and X_{10}, customer satisfaction with HATCO).

As with previous chapters, the discussion of this application of canonical correlation analysis follows the six-stage process discussed earlier in the chapter. At each stage the results illustrating the decisions in that stage are examined.

Stage 1: Objectives of Canonical Correlation Analysis

In demonstrating the application of canonical correlation, we use all nine variables as input data. The HATCO ratings (X_1 through X_7) are designated as the set of independent variables. The measures of usage level and satisfaction level (variables X_9 and X_{10}) are specified as the set of dependent variables. The statistical problem involves identifying any latent relationships (relationships between composites of variables rather the individual variables themselves) between a customer's perceptions about HATCO and the customer's level of usage and satisfaction.

Stages 2 and 3: Designing a Canonical Correlation Analysis and Testing the Assumptions

The designation of the variables includes two metric-dependent and seven metric-independent variables. The conceptual basis of both sets is well established, so there is no need for alternative model formulations testing different sets of variables. The seven variables resulted in a 13-to-1 ratio of observations to variables, exceeding the guideline of 10 observations per variable. The sample size of 100 is not felt to affect the estimates of sampling error markedly and thus should have no impact on the statistical significance of the results. Finally, both dependent and independent variables were assessed in chapter 2 for meeting the basic distributional assumptions underlying multivariate analyses and passed all statistical tests.

Stage 4: Deriving the Canonical Functions and Assessing Overall Fit

The canonical correlation analysis was restricted to deriving two canonical functions because the dependent variable set contained only two variables. To determine the number of canonical functions to include in the interpretation stage, the analysis focused on the level of statistical significance, the practical significance of the canonical correlation, and the redundancy indices for each variate.

Statistical and Practical Significance

The first statistical significance test is for the canonical correlations of each of the two canonical functions. In this example, both canonical correlations are statistically significant (see Table 8.2). In addition to tests of each canonical function separately, multivariate tests of both functions simultaneously are also performed. The test statistics employed are Wilks' lambda, Pillai's criterion, Hotelling's trace, and Roy's gcr. Table 8.2 also details the multivariate test statistics, which all indicate that the canonical functions, taken collectively, are statistically significant at the .01 level.

In addition to statistical significance, the canonical correlations were both of sufficient size to be deemed practically significant. The final step was to perform redundancy analyses on both canonical functions.

Redundancy Analysis

A redundancy index is calculated for the independent and dependent variates of the first function in Table 8.3. As can be seen, the redundancy index for the dependent variate is substantial (.751). The independent variate, however, has a

TABLE 8.2 Canonical Correlation Analysis Relating Levels of Usage and Satisfaction with HATCO to Perceptions of HATCO

	Measures of Overall Model Fit for Canonical Correlation Analysis			
Canonical Function	Canonical Correlation	Canonical R^2	F Statistic	Probability
1	.937	.878	30.235	.000
2	.510	.260	5.391	.000

	Multivariate Tests of Significance		
Statistic	Value	Approximate F Statistic	Probability
Wilks' lambda	.090	30.235	.000
Pillai's trace	1.138	17.348	.000
Hotelling's trace	7.535	48.441	.000
Roy's ger	.878		

TABLE 8.3 Calculation of the Redundancy Indices for the First Canonical Function

Variate/Variables	Canonical Loading	Canonical Loading Squared	Average Loading Squared	Canonical R^2	Reundancy Index[a]
Dependent variables					
X_9 Usage level	.913	.834			
X_{10} Satisfaction level	.936	.876			
Dependent variate		1.710	.855	.878	.751
Independent variables					
X_1 Delivery speed	.764	.584			
X_2 Price level	.061	.004			
X_3 Price flexibility	.624	.389			
X_4 Manufacturer image	.414	.171			
X_5 Overall service	.765	.585			
X_6 Salesforce image	.348	.121			
X_7 Product quality	−.278	.077			
Independent variate		1.931	.276	.878	.242

[a]The redundancy index is calculated as the average loading squared times the canonical R^2.

markedly lower redundancy index (.242), although in this case, because there is a clear delineation between dependent and independent variables, this lower value is not unexpected or problematic. The low redundancy of the independent variate results from the relatively low shared variance in the independent variate (.276), not the canonical R^2. From the redundancy analysis and the statistical significance tests, the first function should be accepted.

The redundancy analysis for the second function produces quite different results (see Table 8.4, p. 458). First, the canonical R^2 is substantially lower (.260). Moreover, both variable sets have low shared variance in the second function (.145 for the dependent variate and .082 for the independent variate). Their combination with the canonical root in the redundancy index produces values of .038 for the dependent variate and .021 for the independent variate. Thus, although the second function is statistically significant, it has little practical significance. With such a small percentage, one must question the value of the function. This is an excellent example of a statistically significant canonical function that does not have practical significance because it does not explain a large proportion of the dependent variables' variance.

The interested researcher should review chapter 3 with attention to the discussion of scale development. Canonical correlation is in some ways a form of scale development, as the dependent and independent variates represent dimensions of the variable sets similar to the scales developed with factor analysis. The primary difference is that these dimensions are developed to maximize the relationship between them, whereas factor analysis maximizes the explanation (shared variance) of the variable set.

Stage 5: Interpreting the Canonical Variates

With the canonical relationship deemed statistically significant and the magnitude of the canonical root and the redundancy index acceptable, the researcher proceeds to making substantive interpretations of the results. Although the second function could be considered practically nonsignificant, owing to the low redundancy value, it is included in the interpretation phase for illustrative reasons.

TABLE 8.4 Redundancy Analysis of Dependent and Independent Variates for Both Canonical Functions

Standardized Variance of the Dependent Variables Explained by

	Their Own Canonical Variate (Shared Variance)			The Opposite Canonical Variate (Redundancy)	
Canonical Function	Percentage	Cumulative Percentage	Canonical R^2	Percentage	Cumulative Percentage
1	.855	.855	.878	.751	.751
2	.145	1.000	.260	.038	.789

Standardized Variance of the Independent Variables Explained by

	Their Own Canonical Value (Shared Variance)			The Opposite Canonical Variate (Redundancy)	
Canonical Function	Percentage	Cumulative Percentage	Canonical R^2	Percentage	Cumulative Percentage
1	.276	.276	.878	.242	.242
2	.082	.358	.260	.021	.263

These interpretations involve examining the canonical functions to determine the relative importance of each of the original variables in deriving the canonical relationships. The three methods for interpretation are (1) canonical weights (standardized coefficients), (2) canonical loadings (structure correlations), and (3) canonical cross-loadings.

Canonical Weights

Table 8.5 contains the standardized canonical weights for each canonical variate for both dependent and independent variables. As discussed earlier, the magnitude of the weights represents their relative contribution to the variate. Based on the size of the weights, the order of contribution of independent variables to the first variate is X_3, X_5, X_4, X_1, X_2, X_6, and X_7, and the dependent variable order on the first variate is X_{10}, then X_9. Similar rankings can be found for the variates of the second canonical function. Because canonical weights are typically unstable, particularly in instances of multicollinearity, owing to their calculation solely to optimize the canonical correlation, the canonical loading and cross-loadings are considered more appropriate.

Canonical Loadings

Table 8.6 (p. 460) contains the canonical loadings for the dependent and independent variates for both canonical functions. The objective of maximizing the variates for the correlation between them results in variates "optimized" not for interpretation, but instead for prediction. This makes identification of relationships more difficult. In the first dependent variate, both variables have loadings exceeding .90, resulting in the high shared variance (.855). This indicates a high

TABLE 8.5 Canonical Weights for the Two Canonical Functions

	Canonical Weights	
	Function 1	*Function 2*
Standardized canonical coefficients for the independent variables		
X_1 Delivery speed	.225	−.965
X_2 Price level	.103	−.868
X_3 Price flexibility	.569	.160
X_4 Manufacturer image	.348	−1.456
X_5 Overall service	.445	1.530
X_6 Salesforce image	−.051	.736
X_7 Product quality	.001	.478
Standardized canonical coefficients for the dependent variables		
X_9 Usage level	.501	1.330
X_{10} Satisfaction level	.580	−1.298

degree of intercorrelation among the two variables and suggests that both, or either, measures are representative of the effects of HATCO's efforts.

The first independent variate has a quite different pattern, with loadings ranging from .061 to .765, with one independent variable (X_7) even having a negative loading, although it is rather small and not of substantive interest. The three variables with the highest loadings on the independent variate are X_5 (overall service), X_1 (delivery speed), and X_3 (price flexibility). This variate does not correspond to the dimensions extracted in factor analysis (see chapter 3), but it would not be expected to because the variates in canonical correlation are extracted only to maximize predictive objectives. As such, it should correspond more to the results from other dependence techniques. There is a close correspondence to multiple regression (see chapter 4). Two of these variables (X_3 and X_5) were included in the stepwise regression analysis in which X_9 (one of the two variables in the dependent variate) was the dependent variable. Thus, the first canonical function closely corresponds to the multiple regression results, with the independent variate representing the set of variables best predicting the two dependent measures. The researcher should also perform a sensitivity analysis of the independent variate in this case to see whether the loadings change when an independent variable is deleted (see stage 6).

The second variate's poor redundancy values are exhibited in the substantially lower loadings for both variates on the second function. Thus, the poorer interpretability as reflected in the lower loadings, coupled with the low redundancy values, reinforce the low practical significance of the second function.

Canonical Cross-Loadings

Table 8.6 also includes the cross-loadings for the two canonical functions. In studying the first canonical function, we see that both independent variables (X_9 and X_{10}) exhibit high correlations with the independent canonical variate (function 1): .855 and .877, respectively. This reflects the high shared variance between these two variables. By squaring these terms, we find the percentage of the variance for each of the variables explained by function 1. The results show that 73 percent of the variance in X_9 and 77 percent of the variance in X_{10} is explained by function 1. Looking at the independent variables' cross-loadings, we see that variables X_1

TABLE 8.6 Canonical Structure for the Two Canonical Functions

	Canonical Loadings	
	Function 1	*Function 2*
Correlations between the independent variables and their canonical variates		
X_1 Delivery speed	.764	.109
X_2 Price level	.061	.141
X_3 Price flexibility	.624	.123
X_4 Manufacturer image	.414	−.626
X_5 Overall service	.765	.222
X_6 Salesforce image	.348	−.199
X_7 Product quality	−.278	.219
Correlations between the dependent variables and their canonical variates		
X_9 Usage level	.913	.408
X_{10} Satisfaction level	.936	−.352

	Canonical Cross-Loadings[a]	
	Function 1	*Function 2*
Correlations between the independent variables and dependent canonical variates		
X_1 Delivery speed	.716	.056
X_2 Price level	.058	.072
X_3 Price flexibility	.584	.063
X_4 Manufacturer image	.388	−.319
X_5 Overall service	.717	.113
X_6 Salesforce image	.326	−.102
X_7 Product quality	−.261	.112
Correlations between the dependent variables and independent canonical variates		
X_9 Usage level	.855	.208
X_{10} Satisfaction level	.877	−.180

[a]The canonical cross-loadings are provided by SAS because SPSS does not report the cross-loadings.

and X_5 both have high correlations of roughly .72 with the dependent canonical variate. From this information, approximately 52 percent of the variance in each of these two variables is explained by the dependent variate (the 52 percent is obtained by squaring the correlation coefficient, .72). The correlation of X_3 (.584) may appear high, but after squaring this correlation, only 34 percent of the variation is included in the canonical variate.

The final issue of interpretation is examining the signs of the cross-loadings. All independent variables except X_7 (product quality) have a positive, direct relationship. For the second function, two independent variables (X_4 and X_6), plus a dependent variable (X_{10}), are negative. The three highest cross-loadings of the first independent variate correspond to the variables with the highest canonical loadings as well. Thus all the relationships are direct except for one inverse relationship in the first function.

Stage 6: Validation and Diagnosis

The last stage should involve a validation of the canonical correlation analyses through one of several procedures. Among the available approaches would be (1) splitting the sample into estimation and validation samples, or (2) sensitivity analysis of the independent variable set. Table 8.7 contains the result of such a

TABLE 8.7 Sensitivity Analysis of the Canonical Correlation Results to Removal of an Independent Variable

	Complete Variate	Results after Deletion of		
		X_1	X_2	X_7
Canonical correlation (R)	.937	.936	.937	.937
Canonical root (R^2)	.878	.876	.878	.878
INDEPENDENT VARIATE				
Canonical loadings				
X_1 Delivery speed	.764	omitted	.765	.764
X_2 Price level	.061	.062	omitted	.061
X_3 Price flexibility	.624	.624	.624	.624
X_4 Manufacturer image	.414	.413	.414	.415
X_5 Overall service	.765	.766	.766	.765
X_6 Salesforce image	.348	.348	.348	.348
X_7 Product quality	−.278	−.278	−.278	omitted
Shared variance	.276	.225	.322	.309
Redundancy	.242	.197	.282	.271
DEPENDENT VARIATE				
Canonical loadings				
X_9 Usage level	.913	.915	.914	.913
X_{10} Satisfaction level	.936	.934	.935	.936
Shared variance	.855	.855	.855	.855
Redundancy	.750	.749	.750	.750

sensitivity analysis in which the canonical loadings are examined for stability when individual independent variables are deleted from the analysis. As seen, the canonical loadings in our example are remarkably stable and consistent in each of the three cases where an independent variable (X_1, X_2, or X_7) is deleted. The overall canonical correlations also remain stable. But the researcher examining the canonical weights (not presented in the table) would find widely varying results, depending on which variable was deleted. This reinforces the procedure of using the canonical loading and cross-loading for interpretation purposes.

A Managerial Overview

The canonical correlation analysis addresses two primary objectives: (1) the identification of dimensions among the dependent and independent variables that (2) maximize the relationship between the dimensions. From a managerial perspective, this provides the researcher with some insight into the structure of the different variable sets as they relate to a dependence relationship. First, the results indicate only a single relationship exists, supported by the low practical significance of the second canonical function. In examining this relationship, we first see that the two dependent variables are quite closely related and create a well-defined dimension for representing the outcomes of HATCO's efforts. Second, this outcome dimension is fairly well predicted by the set of independent variables when acting as a set. The redundancy value of .750 would be a quite acceptable R^2 for a comparable multiple regression. When interpreting the independent variate, we see that three variables, X_5 (overall service), X_1 (delivery speed), and X_3 (price flexibility) provide the substantive contributions and thus are the key predictors of the outcome dimension. These should be the focal points in the development of any strategy directed toward impacting the outcomes of HATCO.

Summary

Canonical correlation analysis is a useful and powerful technique for exploring the relationships among multiple dependent and independent variables. The technique is primarily descriptive, although it may be used for predictive purposes. Results obtained from a canonical analysis should suggest answers to questions concerning the number of ways in which the two sets of multiple variables are related, the strengths of the relationships, and the nature of the relationships defined.

Canonical analysis enables the researcher to combine into a composite measure what otherwise might be an unmanageably large number of bivariate correlations between sets of variables. It is useful for identifying overall relationships between multiple independent and dependent variables, particularly when the data researcher has little a priori knowledge about relationships among the sets of variables. Essentially, the researcher can apply canonical correlation analysis to a set of variables, select those variables (both independent and dependent) that appear to be significantly related, and run subsequent canonical correlations with the more significant variables remaining, or perform individual regressions with these variables.

Questions

1. Under what circumstances would you select canonical correlation analysis instead of multiple regression as the appropriate statistical technique?
2. What three criteria should you use in deciding which canonical functions should be interpreted? Explain the role of each.
3. How would you interpret a canonical correlation analysis?
4. What is the relationship among the canonical root, the redundancy index, and multiple regression's R^2?
5. What are the limitations associated with canonical correlation analysis?
6. Why has canonical correlation analysis been used much less frequently than the other multivariate techniques?

References

1. Alpert, Mark I., and Robert A. Peterson (1972), "On the Interpretation of Canonical Analysis." *Journal of Marketing Research* 9 (May): 187.
2. Alpert, Mark I., Robert A. Peterson, and Warren S. Marti (1975), "Testing the Significance of Canonical Correlations." *Proceedings, American Marketing Association* 37: 117–19.
3. Bartlett M. S. (1941), "The Statistical Significance of Canonical Correlations." *Biometrika* 32: 29.
4. Dillon, W. R., and M. Goldstein (1984), *Multivariate Analysis: Methods and Applications.* New York: Wiley.
5. Green, P. E. (1978), *Analyzing Multivariate Data.* Hinsdale, Ill.: Holt, Rinehart, & Winston.

6. Green, P. E., and J. Douglas Carroll (1978), *Mathematical Tools for Applied Multivariate Analysis*. New York: Academic Press.

7. Lambert, Z., and R. Durand (1975), "Some Precautions in Using Canonical Analysis." *Journal of Marketing Research* 12 (November): 468–75.

8. Stewart, Douglas, and William Love (1968), "A General Canonical Correlation Index." *Psychological Bulletin* 70: 160–63.

Annotated Articles

The following annotated articles are provided as illustrations of the application of canonical correlation analysis to substantive research questions of both a conceptual and managerial nature. The reader is encouraged to review the complete articles for greater detail on any of the specific issues regarding methodology or findings.

Schul, Patrick L., William M. Pride, and Taylor L. Little (1983), "The Impact of Channel Leadership Behavior on Intrachannel Conflict." *Journal of Marketing* 47(3): 21–34.

This article uses canonical correlation analysis, a technique that allows for the investigation of multiple independent variable effects upon multiple dependent variables. In this article, the multivariate technique is applied to determine the effects of leadership style upon perceptions of intrachannel conflict. By examining the strength and direction of the relationship between leadership style and overall intrachannel conflict, canonical correlation analysis provides the researcher with information to improve distribution channel transactions. Findings of this nature also provide managers of channel member organizations with a means of structuring their conduct with other channel members. Three leadership styles—participative, directive, and supportive—which have been theorized to exhibit an inverse relationship with two forms of intrachannel conflict—administrative and product-service—are defined and measured. It is through this technique that researchers are able to explore the relationship between such multifaceted constructs as leadership and conflict.

Data from a sample of 349 franchised real estate brokers were analyzed to test for an association between channel leadership (as the predictor, or independent variables) and intrachannel conflict (as the criterion, or dependent variables). The authors are able to confirm that there exists a strong relationship between the type of channel leadership and intrachannel conflict by examining the redundancy index (indicates the amount of variance explained in a canonical variate by the other canonical variate) and the correlation between the two variates. Through individual analysis, all three leadership styles are shown to reduce conflict. Participative and supportive techniques facilitate understanding and acceptance of policies and procedures, whereas a directive style reduces role ambiguity. From the large cross-loading values, the results indicate that supportive leadership has a stronger relationship with reducing channel conflict than participative or directive leadership. For validation purposes, the authors run a canonical correlation analysis on an analysis and hold-out sample. Canonical weights are compared across the two samples, thereby providing an indication of the stability of this measure for the combined sample. These results indicate that a channel leader should implement a style that best fits the needs of the franchisee in order to minimize channel conflict.

Luthans, Fred, Dianne H. B. Welsh, and Lewis A. Taylor III (1988), "A Descriptive Model of Managerial Effectiveness." *Group and Organization Studies* 13(2): 148–62.

By conducting a canonical correlation analysis, this study seeks to determine which specific managerial activities relate to organizational effectiveness. The authors identify nine managerial activities and eight items on organizational subunit effectiveness. The activities of 78 managers are observed and recorded in order to measure engagement in the identified behaviors. To eliminate same-source bias that may be introduced if the managers rate subunit effectiveness, their subordinates (278 in all) rate subunit effectiveness. Whereas other studies examine the activities of successful managers (i.e., those on a fast promotion track), this study seeks to identify those behaviors that contribute to organizational effectiveness. The canonical correlation analysis between the frequency of the managerial activities and subordinate-reported subunit effectiveness is used in order to reveal the presence and strength of the relationship between the two sets of variables.

Results indicate a significant canonical variate (canonical correlation = .44); however, the strength of the relationship is not assessed (i.e., the redundancy index, which is a better measure of the ability of the predictor variables to explain variation in the criterion variables, is not reported). Interpretation of the results suggests a continuum of management orientation from quantity-oriented human resources to quality-oriented traditional. Quantity-oriented human resource managers are understood to focus on staffing and motivating or reinforcing activities and are perceived as having quantity performance in their units. These managers, however, have limited outside interaction, engagement in controlling and planning activities, or the perception of quality performance in their units. On the other hand, quality-oriented traditional managers are understood as having quality performance in their units, interacting with outsiders, and engaging in controlling and planning activities. Although the findings are not validated, the authors maintain that the results should assist organizational planners in identifying the necessary managerial skills for the desired organizational outcome (i.e., human resource activities may aid in the attainment of more output whereas traditional management activities may improve quality).

Van Auken, Howard E., B. Michael Doran, and Kil-Jhin Yoon (1993), "A Financial Comparison between Korean and U.S. Firms: A Cross–Balance Sheet Canonical Correlation Analysis." *Journal of Small Business Management* 31(3): 73–83.

In this article, the authors seek to examine cross–balance sheet relationships and general financing strategies of small- to medium-sized Korean firms through the use of canonical correlation analysis. From a random sample of 45 Korean firms, financial position statements from 1988 were obtained for various asset, liability, and equity accounts. The relationships between the assets (cash, accounts receivable, inventories, and long-term assets) and liabilities and equities (accounts payable, other current liabilities, long-term debt, and equity) were explored using canonical correlation analysis. The study's design allowed the authors to compare the results of previously published works on Mexican and U.S. small- and medium-sized firms. This and previous studies have demonstrated that the financial strategies of small firms are influenced by economic conditions and cultural elements. The results of the study should aid in the decision making of small business owners who are developing financing strategies in similar economies.

The analysis resulted in all four canonical functions being significant. As a further assessment of the canonical functions, the authors calculate a redundancy index. The proportion of the asset variance accounted for by the liability variance is .60. The liability variance shared with the asset variance is .24. Because the canonical relationships were acceptable, the authors proceed to make the following interpretations about small- to medium-sized Korean firms: (1) they experience hedging, (2) they use collateral for loans, (3) their inventories are associated with accounts payable, and (4) they manage risk with the simultaneous use of lower leverage and greater liquidity balances. Compared to U.S. firms, Korean firms rely heavily on the use of current debt. The findings extend earlier studies, which suggest that the financial strategies of small firms in other countries are dependent on the marketing constraints of the country in which the firm operates.

Mahmood, Mo Adam, and Gary J. Mann (1993), "Measuring the Organizational Impact of Information Technology Investment: An Exploratory Study." *Journal of Management Information Systems* 10(1): 97–122.

This article seeks to determine whether a relationship exists between information technology (IT) investment and the strategic and economic performance of the firm. From past research measuring the impact of IT on the organization, the authors determine which measures to include in the study. The predictor variables consist of five IT investment measures, and the criterion variables include six organizational strategic and economic performance measures. Canonical analysis is used for exploratory purposes with no specific hypotheses offered. The technique enables the researchers to determine the presence and magnitude of the association between multiple IT and organizational performance measures. From the results, the authors offer hypotheses and a model depicting the interrelationships between IT investment and performance.

The canonical correlation analysis is performed with a sample of 100 firms. The results indicate a significant relationship with 10.4 percent of the variation in the organizational performance measures explained by IT investment. Although only one of the five canonical functions is significant, the authors interpret the two functions that account for nearly 86 percent of the total explained variation (i.e., of the 10.4 percent). By examining the canonical loadings of the two functions, the authors are able to determine the relative importance of each variable. Altogether, the findings are interpreted as indicating that IT investment contributes to organizational performance when the firm invests in both equipment and employee IT training. These conclusions lead the authors to call for further research to test the interdependencies between IT investment and firm performance using different methods and samples.

INTERDEPENDENCE TECHNIQUES

Overview

The dependence methods described in section 2 provide the researcher with several methods for assessing relationships between one or more dependent variables and a set of independent variables. Many methods were discussed that accommodated all types (metric and nonmetric) and potentially large numbers of both dependent and independent variables that could be applied to sets of observations. But what if the variables or observations are related in ways not captured by the dependence relationships? What if the assessment of interdependence (i.e., structure) is missing? One of the most basic abilities of human beings is to classify and categorize objects and information into simpler schema, such that we can characterize the objects within the groups in total rather than having to deal with each individual object. That is the objective of the methods in this section: to identify the structure among a defined set of variables, observations, or objects. The identification of structure offers not only simplicity, but also a means of description and even discovery.

Interdependence techniques, however, are focused solely on the definition of structure, assessing interdependence without any associated dependence relationships. None of the interdependence techniques will define structure to optimize or maximize a dependence relationship. It is the researcher's task to first utilize these methods in identifying structure and then to employ it where appropriate. The objectives of dependence relationships are not "mixed" in these interdependence methods—they assess structure for its own sake and no other.

Chapters in Section 3

Section 3 comprises only two chapters, but there are really three interdependence techniques. The first interdependence technique, factor analysis (chapter 3), was discussed in section 2 on preparing for a multivariate analysis because it provides us with a tool for understanding the relationships among variables, a knowledge fundamental to all of our multivariate analyses. The issues of multicollinearity and model parsimony are reflective of the underlying structure of the variables, and factor analysis provides an objective means of assessing the groupings of variables and the ability to incorporate composite variables reflecting these variable groupings into other multivariate techniques.

But it is not only variables that have structure. Although we assume independence among the observations and variables in our estimation of relationships, we also know that most populations have subgroups sharing general characteristics. Marketers look for target markets of differentiated groups of homogeneous consumers, strategy researchers look for groups of similar firms to identify common strategic elements, and financial modelers look for stocks with similar fundamentals to create stock portfolios. These and many other situations require techniques that find these groups of similar objects based on a set of characteristics.

This is the goal of chapter 9, Cluster Analysis. Cluster analysis is ideally suited for defining groups of objects with maximal homogeneity within the groups while also having maximum heterogeneity between the groups—determining the most similar groups that are also most different from each other. As we show, cluster analysis has a rich tradition of application in almost every area of inquiry. But its ability to define groups of similar objects is countered by its rather subjective nature and the instrumental role played by the researcher's judgment in several key decisions. This does not reduce the usefulness of the technique, but it does place a greater burden on the researcher to fully understand the technique and the impact of certain decisions on the ultimate cluster solution.

But what if we know only how similar objects are and don't know the source of that similarity or how to best group the objects? This situation is addressed in chapter 10, Multidimensional Scaling. Multidimensional scaling is a technique that starts out as a univariate analysis—a single measure of similarity among objects—and infers the dimensionality of the similarities among the objects. It attempts to answer this basic question: Can the objects be grouped in one, two, three, or *n*-dimensional space in such a way as to adequately represent the similarities among the objects by their proximity? As such, multidimensional scaling is a form of decompositional analysis, somewhat like conjoint analysis (see chapter 7), but in this case there are not any known characteristics of the objects, only their similarities. A special form of multidimensional scaling is correspondence analysis, which analyzes a distinct form of data— cross-tabulated categorical variables. From these data, correspondence analysis is able to portray the relationships between rows and columns (e.g., products and attributes) in a dimensional perspective in which proximity represents similarity.

Cluster analysis, factor analysis, and multidimensional scaling provide the researcher with methods that bring "order" to the data in the form of structure among the observations or variables. In this way, the researcher can better understand the basic structures of the data, not only facilitating the description of the data, but also providing a foundation for a more refined analysis of the dependence relationships.

CHAPTER **9**

Cluster Analysis

LEARNING OBJECTIVES

Upon completing this chapter, you should be able to do the following:

- Define the appropriate research questions addressed by cluster analysis.
- Understand how interobject similarity is measured.
- Distinguish between the various distance measures.
- Differentiate between the clustering algorithms and their appropriate applications.
- Understand the differences between hierarchical and nonhierarchical clustering techniques.
- Understand how to select the number of clusters to be formed.
- Follow the guidelines for cluster validation.
- Construct profiles for the derived clusters and assess managerial significance.
- State the limitations of cluster analysis.

CHAPTER PREVIEW

Academicians and market researchers often encounter situations best resolved by defining groups of homogeneous objects, whether they are individuals, firms, products, or even behaviors. Strategy options based on identifying groups within the population, such as segmentation and target marketing would not be possible without an objective methodology. This same need is encountered in other areas, ranging from the physical sciences (e.g., creating a biological taxonomy for the classification of various animal groups—insects versus mammals versus reptiles) to the social sciences (e.g., analyzing various psychiatric profiles). In all instances, the researcher is searching for a "natural" structure among the observations based on a multivariate profile.

The most commonly used technique for this purpose is cluster analysis. Cluster analysis groups individuals or objects into clusters so that objects in the same cluster are more similar to one another than they are to objects in other clusters. The attempt is to maximize the homogeneity of objects within the clusters while also maximizing the heterogeneity between the clusters. This chapter explains the nature and purpose of cluster analysis and guides the researcher in the selection and use of various cluster analysis approaches.

KEY TERMS

Before starting the chapter, review the key terms to develop an understanding of the concepts and terminology used. Throughout the chapter the key terms appear in **boldface**. Other points of emphasis in the chapter are *italicized*. Also, cross-references within the Key Terms appear in *italics*.

Agglomerative methods   *Hierarchical procedure* that begins with each *object* or observation in a separate cluster. In each subsequent step, the two object clusters that are most similar are combined to build a new aggregate cluster. The process is repeated until all objects are finally combined into a single cluster.

Algorithm   Set of rules or procedures; similar to an equation.

Average linkage   *Algorithm* used in *agglomerative methods* that represents *similarity* as the average distance from all objects in one cluster to all objects in another. This approach tends to combine clusters with small variances.

Centroid   Average or mean value of the objects contained in the cluster on each variable, whether used in the *cluster variate* or in the validation process.

Centroid method   *Agglomerative algorithm* in which *similarity* between clusters is measured as the distance between *cluster centroids*. When two clusters are combined, a new centroid is computed. Thus cluster centroids migrate, or move, as the clusters are combined.

City-block approach   Method of calculating distances based on the sum of the absolute differences of the coordinates for the objects. This method assumes the variables are uncorrelated and unit scales are compatible.

Cluster centroid   Average value of the objects contained in the cluster on all the variables in the *cluster variate.*

Cluster seeds   Initial *centroids* or starting points for clusters. These values are selected to initiate *nonhierarchical clustering* procedures, in which clusters are built around these prespecified points.

Cluster variate   Set of variables or characteristics representing the *objects* to be clustered and used to calculate the *similarity* between objects.

Complete linkage   *Agglomerative algorithm* in which *interobject similarity* is based on the maximum distance between *objects* in two clusters (the distance between the most dissimilar members of each cluster). At each stage of the agglomeration, the two clusters with the smallest maximum distance (most similar) are combined.

Criterion validity   Ability of clusters to show the expected differences on a variable not used to form the clusters. For example, if clusters were formed on performance ratings, marketing thought would suggest that clusters with higher performance ratings should also have higher satisfaction scores. If this was found to be so in empirical testing, then criterion validity is supported.

Dendrogram   Graphical representation (tree graph) of the results of a *hierarchical procedure* in which each *object* is arrayed on one axis, and the other axis portrays the steps in the hierarchical procedure. Starting with each object represented as a separate cluster, the dendrogram shows graphically how the clusters are combined at each step of the procedure until all are contained in a single cluster.

Divisive method   Clustering procedure that begins with all *objects* in a single cluster, which is then divided at each step into two clusters that contain the most dissimilar objects. This method is the opposite of the *agglomerative method.*

Entropy group   Group of *objects* independent of any cluster (i.e., they do not fit into any cluster) that may be considered outliers and possibly eliminated from the cluster analysis.

Euclidean distance   Most commonly used measure of the *similarity* between two *objects.* Essentially, it is a measure of the length of a straight line drawn between two objects.

Hierarchical procedures   Stepwise clustering procedures involving a combination (or division) of the *objects* into clusters. The two alternative procedures are the *agglomerative* and *divisive methods.* The result is the construction of a hierarchy, or treelike structure (*dendrogram*), depicting the formation of the clusters. Such a procedure produces $N - 1$ cluster solutions, where N is the number of objects. For example, if the agglomerative procedure starts with five objects in separate clusters, it will show how four clusters, then three, then two, and finally one cluster are formed.

Interobject similarity   The correspondence or association of two *objects* based on the variables of the *cluster variate.* Similarity can be measured in two ways. First is a measure of association, with higher positive correlation coefficients representing greater similarity. Second, "proximity," or "closeness," between each pair of objects can assess similarity, where measures of distance or difference are used, with smaller distances or differences representing greater similarity.

Mahalanobis distance (D^2)   Standardized form of *Euclidean distance.* Scaling responses in terms of standard deviations that standardizes the data, with adjustments made for intercorrelations between the variables.

Multicollinearity   Extent to which a variable can be explained by the other variables in the analysis. As multicollinearity increases, it complicates the interpretation of the *variate* because it is more difficult to ascertain the effect of any single variable, owing to the variables' interrelationships.

Nonhierarchical procedures   Procedures that produce only a single cluster solution for a set of cluster seeds. Instead of using the treelike construction process found in the *hierarchical procedures, cluster seeds* are used to group *objects* within a prespecified distance of the seeds. For example, if four cluster seeds are specified, only four clusters are formed. Nonhierarchical procedures do not produce results for all possible numbers of clusters as is done with a *hierarchical procedure.*

Normalized distance function   Process that converts each raw data score to a standardized variate with a mean of 0 and a standard deviation of 1, to remove the bias introduced by differences in scales of several variables.

Object   Person, product or service, firm, or any other entity that can be evaluated on a number of attributes.

Optimizing procedure   *Nonhierarchical clustering* procedure that allows for the reassignment of *objects* from the originally assigned cluster to another cluster on the basis of an overall optimizing criterion.

Parallel threshold method   *Nonhierarchical clustering* procedure that selects the cluster seeds simultaneously in the beginning. *Objects* within the threshold distances are assigned to the nearest seed. Threshold distances can be adjusted to include fewer or more objects in the clusters. This method is the opposite of the *sequential threshold method.*

Predictive validity   See *criterion validity.*

Profile diagram   Graphical representation of data that aids in screening for outliers or the interpretation of the final cluster solution. Typically, the variables of the *cluster variate* or those used for validation are listed along the horizontal axis, and the value scale is used for the vertical axis. Separate lines depict the scores (original or standardized) for individual *objects* or cluster *centroids* in a graphic plane.

Response-style effect   Series of systematic responses by a respondent that reflect a "bias" or consistent pattern. Examples include responding that an object always performs excellently or poorly across all attributes with little or no variation.

Row-centering standardization   See *within-case standardization.*

Sequential threshold method   *Nonhierarchical clustering* procedure that begins by selecting one *cluster seed.* All *objects* within a prespecified distance are then included in that cluster. Subsequent cluster seeds are selected until all objects are grouped in a cluster.

Similarity   See *interobject similarity.*

Single-linkage   *Hierarchical clustering* procedure in which *similarity* is defined as the minimum distance between any *object* in one cluster and any object in another. This simply means the distance between the closest objects in two clusters. This procedure has the potential for creating less compact, or even chainlike, clusters. This differs from the *complete linkage* method, which uses the maximum distance between objects in the cluster.

Stopping rule   *Algorithm* for determining the final number of clusters to be formed. With no stopping rule inherent in cluster analysis, researchers have developed several criteria and guidelines for this determination. Two classes of rules exist that are applied post hoc and calculated by the researcher are (1) measures of *similarity* and (2) adapted statistical measures.

Taxonomy   Empirically derived classification of actual *objects* based on one or more characteristics. Typified by the application of cluster analysis or other grouping procedures. This classification can be contrasted to a *typology.*

Typology   Conceptually based classification of *objects* based on one or more characteristics. A typology does not usually attempt to group actual observations, but instead provides the theoretical foundation for the creation of a *taxonomy,* which groups actual observations.

Vertical icicle diagram   Graphical representation of clusters. The separate *objects* are shown horizontally across the top of the diagram, and the hierarchical clustering process is depicted in combinations of clusters vertically. This diagram is similar to an inverted *dendrogram* and aids in determining the appropriate number of clusters in the solution.

Ward's method *Hierarchical clustering* procedure in which the *similarity* used to join clusters is calculated as the sum of squares between the two clusters summed over all variables. This method has the tendency to result in clusters of approximately equal size due to its minimization of within-group variation.

Within-case standardization Method of standardization in which a respondent's responses are not compared to the overall sample but instead to their own responses. Also known as ipsitizing, the respondents' average responses are used to standardize their own responses.

What Is Cluster Analysis?

Cluster analysis is the name for a group of multivariate techniques whose primary purpose is to group objects based on the characteristics they possess. Cluster analysis classifies **objects** (e.g., respondents, products, or other entities) so that each object is very similar to others in the cluster with respect to some predetermined selection criterion. The resulting clusters of objects should then exhibit high internal (within-cluster) homogeneity and high external (between-cluster) heterogeneity. Thus, if the classification is successful, the objects within clusters will be close together when plotted geometrically, and different clusters will be far apart.

In cluster analysis, the concept of the variate is again a central issue, but in a quite different way from other multivariate techniques. The **cluster variate** is the set of variables representing the characteristics used to compare objects in the cluster analysis. Because the cluster variate includes only the variables used to compare objects, it determines the "character" of the objects. Cluster analysis is the only multivariate technique that does not estimate the variate empirically but instead uses the variate as specified by the researcher. The focus of cluster analysis is on the comparison of objects based on the variate, not on the estimation of the variate itself. This makes the researcher's definition of the variate a critical step in cluster analysis.

Cluster analysis has been referred to as Q analysis, typology construction, classification analysis, and numerical taxonomy. This variety of names is due in part to the usage of clustering methods in such diverse disciplines as psychology, biology, sociology, economics, engineering, and business. Although the names differ across disciplines, the methods all have a common dimension: classification according to natural relationships [1, 2, 3, 6, 12, 16]. This common dimension represents the essence of all clustering approaches. As such, the primary value of cluster analysis lies in the classification of data, as suggested by "natural" groupings of the data themselves. Cluster analysis is comparable to factor analysis (see chapter 3) in its objective of assessing structure. But cluster analysis differs from factor analysis in that cluster analysis groups objects, whereas factor analysis is primarily concerned with grouping variables.

Cluster analysis is a useful data analysis tool in many different situations. For example, a researcher who has collected data by means of a questionnaire may be faced with a large number of observations that are meaningless unless classified into manageable groups. Cluster analysis can perform this data reduction procedure objectively by reducing the information from an entire population or sample to information about specific, smaller subgroups. For example, if we can understand the attitudes of a population by identifying the major groups within the population, then we have reduced the data for the entire population into profiles of a number of groups. In this fashion, the researcher has a more concise, understandable description of the observations, with minimal loss of information.

Cluster analysis is also useful when a researcher wishes to develop hypotheses concerning the nature of the data or to examine previously stated hypotheses. For example, a researcher may believe that attitudes toward the consumption of diet versus regular soft drinks could be used to separate soft-drink consumers into logical segments or groups. Cluster analysis can classify soft-drink consumers by their attitudes about diet versus regular soft drinks, and the resulting clusters, if any, can be profiled for demographic similarities and differences.

These examples are just a small fraction of the types of applications of cluster analysis. Ranging from the derivation of taxonomies in biology for grouping all living organisms, to psychological classifications based on personality and other personal traits, to segmentation analyses of marketers, cluster analysis has always had a strong tradition of grouping individuals. This tradition has been extended to classifying objects, including the market structure, analyses of the similarities and differences among new products, and performance evaluations of firms to identify groupings based on the firms' strategies or strategic orientations. The result has been an explosion of applications in almost every area of inquiry, creating not only a wealth of knowledge on the use of cluster analysis but also the need for a better understanding of the technique to minimize its misuse.

Yet, along with the benefits of cluster analysis come some caveats. Cluster analysis can be characterized as descriptive, atheoretical, and noninferential. Cluster analysis has no statistical basis upon which to draw statistical inferences from a sample to a population, and it is used primarily as an exploratory technique. The solutions are not unique, as the cluster membership for any number of solutions is dependent upon many elements of the procedure, and many different solutions can be obtained by varying one or more elements. Moreover, cluster analysis will always create clusters, regardless of the "true" existence of any structure in the data. Finally, the cluster solution is totally dependent upon the variables used as the basis for the similarity measure. The addition or deletion of relevant variables can have a substantial impact on the resulting solution. Thus, the researcher must take particular care in assessing the impact of each decision involved in performing a cluster analysis.

How Does Cluster Analysis Work?

The nature of cluster analysis can be illustrated by a simple bivariate example. Suppose a marketing researcher wishes to determine market segments in a small community based on their patterns of loyalty to brands and stores. A small sample of seven respondents is selected as a pilot test of how cluster analysis is applied. Two measures of loyalty—V_1 (store loyalty) and V_2 (brand loyalty)—were measured for each respondent on a 0-to-10 scale. The values for each of the seven respondents are shown in Figure 9.1, along with a scatter diagram depicting each observation on the two variables.

Data Values

Clustering Variable	Respondents						
	A	B	C	D	E	F	G
V_1	3	4	4	2	6	7	6
V_2	2	5	7	7	6	7	4

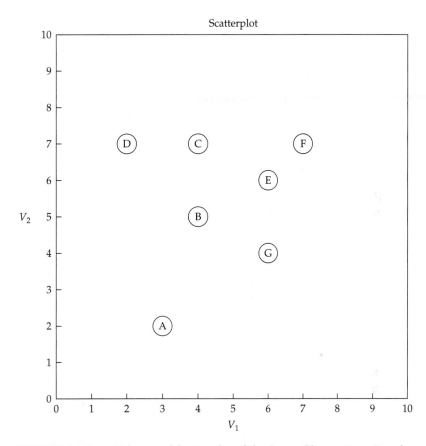

FIGURE 9.1 Data Values and Scatterplot of the Seven Observations Based on the Two Clustering Variables (V_1 and V_2)

The primary objective of cluster analysis is to define the structure of the data by placing the most similar observations into groups. But to accomplish this task, we must address three basic questions. First, how do we measure similarity? We require a method of simultaneously comparing observations on the two clustering variables (V_1 and V_2). Several methods are possible, including the correlation between objects, a measure of association used in other multivariate techniques, or perhaps measuring their proximity in two-dimensional space such that the distance between observations indicates similarity. Second, how do we form clusters? No matter how similarity is measured, the procedure must group those observations that are most similar into a cluster. This procedure must determine the group membership of each observation. Third, how many groups do we form?

Any number of "rules" might be used, but the fundamental task is to assess the "average" similarity across clusters such that as the average increases, the clusters become less similar. The researcher then faces a trade-off: fewer clusters versus less homogeneity. Simple structure, in striving toward parsimony, is reflected in as few clusters as possible. Yet as the number of clusters decreases, the homogeneity within the clusters necessarily decreases. Thus, a balance must be made between defining the most basic structure (fewer clusters) that still achieves the necessary level of similarity within the clusters. Once we have procedures for addressing each issue, we can perform a cluster analysis.

Measuring Similarity

We will illustrate a cluster analysis for the seven observations (respondents A–G) using simple procedures for each of the issues. First, similarity will be measured according to the **Euclidean** (straight-line) **distance** between each pair of observations. Table 9.1 contains measures of proximity between each of the seven respondents. In using distance as the measure of proximity, we must remember that smaller distances indicate greater similarity, such that observations E and F are the most similar (1.414), and A and F are the most dissimilar (6.403).

Forming Clusters

Once we have the similarity measure, we must next develop a procedure for forming clusters. As shown later in this chapter, many methods have been proposed. But for our purposes here, we use this simple rule: Identify the two most similar (closest) observations not already in the same cluster and combine their clusters. We apply this rule repeatedly, starting with each observation in its own "cluster" and combining two clusters at a time until all observations are in a single cluster. This is termed a **hierarchical procedure** because it moves in a stepwise fashion to form an entire range of cluster solutions. It is also an **agglomerative method** because clusters are formed by the combination of existing clusters.

Table 9.2 details the steps of the hierarchical process, first depicting the initial state with all seven observations in single member clusters. Then clusters are joined in the agglomerative process until only one cluster remains. Step 1 identifies the two closest observations (E and F) and combines them into a cluster, moving from seven to six clusters. Next, step 2 finds the next closest pairs of observations. In this case, three pairs have the same distance of 2.000 (E-G, C-D,

TABLE 9.1 Proximity Matrix of Euclidean Distances between Observations

| | *Observation* | | | | | | |
Observation	A	B	C	D	E	F	G
A	—						
B	3.162	—					
C	5.099	2.000	—				
D	5.099	2.828	2.000	—			
E	5.000	2.236	2.236	4.123	—		
F	6.403	3.606	3.000	5.000	1.414	—	
G	3.606	2.236	3.606	5.000	2.000	3.162	—

TABLE 9.2 Agglomerative Hierarchical Clustering Process

| | *Agglomeration Process* | | *Cluster Solution* | | |
Step	*Minimum Distance between Unclustered Observations[a]*	*Observation Pair*	*Cluster Membership*	*Number of Clusters*	*Overall Similarity Measure (Average Within-Cluster Distance)*
	Initial Solution		(A) (B) (C) (D) (E) (F) (G)	7	0
1	1.414	E-F	(A) (B) (C) (D) (E-F) (G)	6	1.414
2	2.000	E-G	(A) (B) (C) (D) (E-F–G)	5	2.192
3	2.000	C-D	(A) (B) (C–D) (E-F–G)	4	2.144
4	2.000	B-C	(A) (B–C-D) (E-F-G)	3	2.234
5	2.236	B-E	(A) (B-C-D–E-F-G)	2	2.896
6	3.162	A-B	(A–B-C-D-E-F-G)	1	3.420

[a]Euclidean distance between observations

and B-C). Let us start with E-G. G is a single member cluster, but E was combined in the prior step with F. So, the cluster formed at this stage now has three members: G, E, and F. Step 3 combines the single member clusters of C and D, and step 4 combines B with the two-member cluster C-D that was formed in step 3. At this point, we now have three clusters: cluster 1 (A), cluster 2 (B, C, and D), and cluster 3 (E, F, and G).

The next smallest distance is 2.236 for three pairs of observations (E-B, B-G, and C-E). We use only one of these distances, however, as each observation pair contains a member from each of the two existing clusters (B, C, and D versus E, F, and G). Thus, step 5 combines the two three-member clusters into a single six-member cluster. The final step (6) is to combine observation A with the remaining cluster (six observations) into a single cluster at a distance of 3.162. You will note that there are distances smaller or equal to 3.162, but they are not used because they are between members of the same cluster.

The hierarchical clustering process can be portrayed graphically in several ways. Figure 9.2 (p. 478) illustrates two such methods. First, because the process is hierarchical, the clustering process can be shown as a series of nested groupings (see Figure 9.2a). This process, however, can represent the proximity of the observations for only two or three clustering variables in the scatterplot or three-dimensional graph. A more common approach is the dendrogram, which represents the clustering process in a tree-like graph. The horizontal axis represents the agglomeration coefficient, in this instance the distance used in joining clusters. This approach is particularly useful in identifying outliers, such as observation A. It also depicts the relative size of varying clusters, although it becomes unwieldy when the number of observations increases.

Determining the Number of Clusters in the Final Solution

A hierarchical method results in a number of cluster solutions—in this case they range from a one-cluster solution to a six-cluster solution. But which one should we choose? We know that as we move from single member clusters, homogeneity

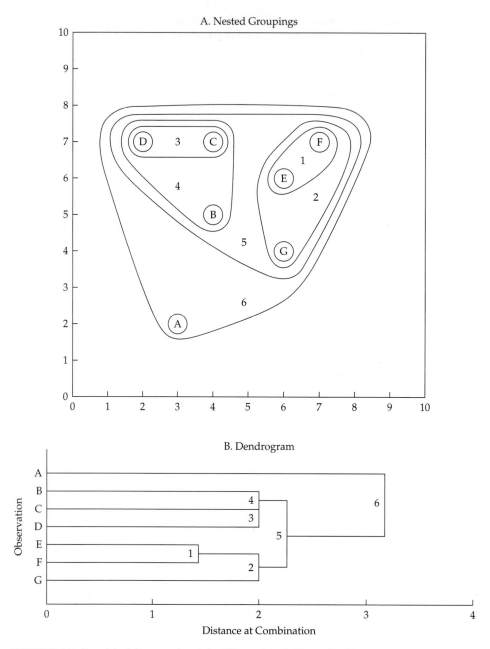

FIGURE 9.2 Graphical Portrayals of the Hierarchical Clustering Process

decreases. So why not stay at seven clusters, the most homogeneous possible? The problem is that we have not defined any structure with seven clusters. So the researcher must view each cluster solution for its description of structure balanced against the homogeneity of the clusters. In this example, we use a very simple measure of homogeneity: the average distances of all observations within clusters. In the initial solution with seven clusters, our overall similarity measure is 0—no observation is paired with another. For the six-cluster solution, the overall similarity is the distance between the two observations (1.414) joined at step 1. Step 2 forms a three-member cluster (E, F, and G), so that the overall similarity measure is the mean of the distances between E and F (1.414), E and G (2.000), and F and G (3.162), for an average of 2.192. In step 3, a new two-member cluster is formed with a distance of 2.000, which causes the overall average to fall slightly to 2.144. We can proceed to form new clusters in this manner until a single-cluster solution is formed (step 6), in which the average of all distances in the distance matrix is 3.420.

Now, how do we use this overall measure of similarity to select a cluster solution? Remember that we are trying to get the simplest structure possible that still represents homogeneous groupings. If we monitor the overall similarity measure as the number of clusters decreases, large increases in the overall measure indicate that two clusters were not that similar. In our example, the overall measure increases when we first join two observations (step 1) and then again when we make our first three-member cluster (step 2). But in the next two steps (3 and 4), the overall measure does not change substantially. This indicates that we are forming other clusters with essentially the same homogeneity of the existing clusters. But when we get to step 5, which combines the two three-member clusters, we see a large increase. This indicates that joining these two clusters resulted in a single cluster that was markedly less homogeneous. We would consider the cluster solution of step 4 much better found in step 5. We can also see that in step 6 the overall measure again increased slightly, indicating that even though the last observation remained separate until the last step, when it was joined it changed the cluster homogeneity. However, given the rather unique profile of observation A compared to the others, it might best be designated as a member of the **entropy group,** those observations that are outliers and independent of the existing clusters. Thus, when reviewing the range of cluster solutions, the three-cluster solution of step 4 seems the most appropriate for a final cluster solution, with two equally sized clusters and the single outlying observation.

As is probably clear by now, the selection of the final cluster solution requires substantial researcher judgment and is considered by many as too subjective. Even though more sophisticated methods have been developed to assist in evaluating the cluster solutions, it still falls to the researcher to make the final decision as to the number of clusters to accept as the final solution. Cluster analysis is rather simple in this bivariate case because the data are two-dimensional. In most marketing research studies, however, more than two variables are measured on each object, and the situation is much more complex with many more observations. In the remainder of this chapter, we discuss how the researcher can employ more sophisticated procedures to deal with the increased complexity of "real world" applications.

Cluster Analysis Decision Process

Cluster analysis, like the other multivariate techniques discussed earlier, can be viewed from the six-stage model-building approach introduced in chapter 1 (see Figure 9.3 for stages 1–3, p. 480, and Figure 9.6 on p. 492 for stages 4–6). Starting with research objectives that can be either exploratory or confirmatory, the design of a cluster analysis deals with partitioning the data set to form clusters,

Stage 1

Stage 2

Stage 3

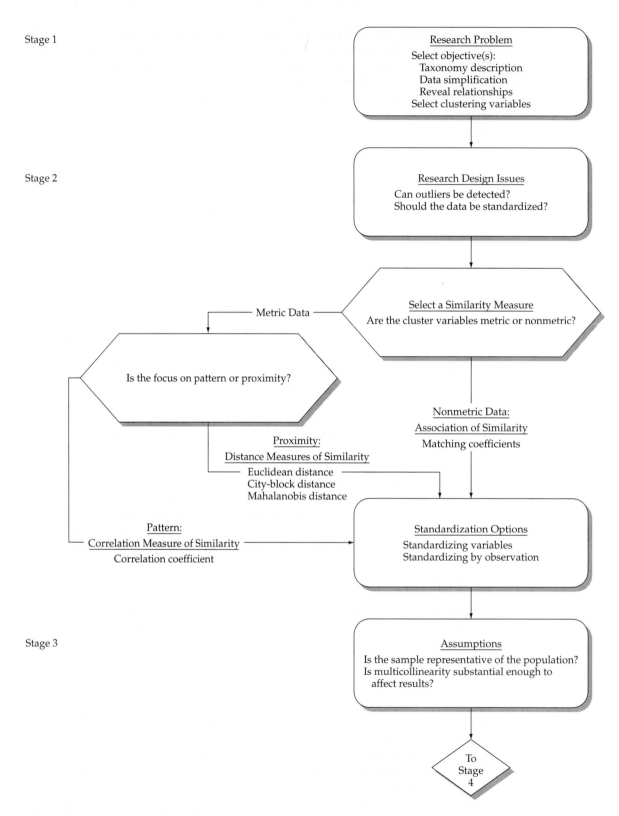

FIGURE 9.3 Stages 1–3 of the Cluster Analysis Decision Diagram

interpreting the clusters, and validating the results. The partitioning process determines how clusters may be developed. The interpretation process involves understanding the characteristics of each cluster and developing a name or label that appropriately defines its nature. The final process involves assessing the validity of the cluster solution (i.e., determining its stability and generalizability), along with describing the characteristics of each cluster to explain how they may differ on relevant dimensions such as demographics. The following sections detail all these issues through the six stages of the model-building process.

Stage 1: Objectives of Cluster Analysis

The primary goal of cluster analysis is to partition a set of objects into two or more groups based on the similarity of the objects for a set of specified characteristics (cluster variate). In forming homogeneous groups, the researcher can achieve any of three objectives:

1. *Taxonomy description.* The most traditional use of cluster analysis has been for exploratory purposes and the formation of a **taxonomy**—an empirically based classification of objects. As described earlier, cluster analysis has been used in a wide range of applications for its partitioning ability. But cluster analysis can also generate hypotheses related to the structure of the objects. Yet, although viewed principally as an exploratory technique, cluster analysis can be used for confirmatory purposes. If a proposed structure can be defined for a set of objects, cluster analysis can be applied, and a proposed **typology** (theoretically based classification) can be compared to that derived from the cluster analysis.
2. *Data simplification.* In the course of deriving a taxonomy, cluster analysis also achieves a simplified perspective on the observations. With a defined structure, the observations can be grouped for further analysis. Whereas factor analysis attempts to provide "dimensions" or structure to variables (see chapter 3), cluster analysis performs the same task for observations. Thus, instead of viewing all of the observations as unique, they can be viewed as members of a cluster and profiled by its general characteristics.
3. *Relationship identification.* With the clusters defined and the underlying structure of the data represented in the clusters, the researcher has a means of revealing relationships among the observations that was perhaps not possible with the individual observations. Whether analyses such as discriminant analysis are used to empirically identify relationships, or the groups are subjected to more qualitative methods, the simplified structure from cluster analysis many times portrays relationships or similarities and differences not previously revealed.

Selection of Clustering Variables

In any application, the objectives of cluster analysis cannot be separated from the selection of variables used to characterize the objects to be clustered. Whether the objective is exploratory or confirmatory, the researcher has effectively constrained the possible results by the variables selected for use. The derived clusters reflect the inherent structure of the data only as defined by the variables.

Selecting the variables to be included in the cluster variate must be done with regard to theoretical and conceptual as well as practical considerations. Any application of cluster analysis must have some rationale upon which variables are selected. Whether the rationale is based on an explicit theory, past research, or supposition, the researcher must realize the importance of including only those variables that (1) characterize the objects being clustered, and (2) relate specifically to the objectives of the cluster analysis. The cluster analysis technique has no means of differentiating relevant from irrelevant variables. It only derives the most consistent, yet distinct, groups of objects across *all* variables. The inclusion of an irrelevant variable increases the chance that outliers will be created on these variables, which can have a substantive effect on the results. Thus, one should never include variables indiscriminately but instead carefully choose the variables with the research objective as the criterion for selection.

In a practical vein, cluster analysis can be dramatically affected by the inclusion of only one or two inappropriate or undifferentiated variables [8]. The researcher is always encouraged to examine the results and to eliminate the variables that are not distinctive (i.e., that do not differ significantly) across the derived clusters. This procedure allows the cluster techniques to maximally define clusters based only on those variables exhibiting differences across the objects.

Stage 2: Research Design in Cluster Analysis

With the objectives defined and variables selected, the researcher must address three questions before starting the partitioning process: (1) Can outliers be detected and, if so, should they be deleted? (2) How should object similarity be measured? (3) Should the data be standardized? Many different approaches can be used to answer these questions. However, none of them has been evaluated sufficiently to provide a definitive answer to any of these questions, and, unfortunately, many of the approaches provide different results for the same data set. Thus cluster analysis, along with factor analysis, is much more an art than a science. For this reason, our discussion reviews these issues in a very general way by providing examples of the most commonly used approaches and an assessment of the practical limitations where possible.

The importance of these issues and the decisions made in later stages becomes apparent when we realize that although cluster analysis is seeking structure in the data, it must actually impose a structure through a selected methodology. Cluster analysis cannot evaluate all the possible partitions because, even for the relatively small problem of partitioning 25 objects into 5 nonoverlapping clusters, there are 2.4×10^{15} possible partitions [2]. Instead, based on the decisions of the researcher, the technique identifies one of the possible solutions as "correct." From this viewpoint, the research design issues and the choice of methodologies made by the researcher have greater impact than perhaps with any other multivariate technique.

Detecting Outliers

In its search for structure, cluster analysis is very sensitive to the inclusion of irrelevant variables. But cluster analysis is also sensitive to outliers (objects that are very different from all others). Outliers can represent either (1) truly "aberrant" observations that are not representative of the general population, or (2) an undersampling of actual group(s) in the population that causes an underrepresentation of the group(s) in the sample. In both cases, the outliers distort the true structure and make the derived clusters unrepresentative of the true population structure. For this reason, a preliminary screening for outliers is always necessary. Probably the easiest way to conduct this screening is to prepare a graphic profile diagram, such as that shown in Figure 9.4. The profile diagram lists the variables along the horizontal axis and the variable values along the vertical axis. Each point on the graph represents the value of the corresponding variable, and the points are connected to facilitate visual interpretation. Profiles for all objects are then plotted on the graph, a line for each object. Outliers are those objects with very different profiles, most often characterized by extreme values on one or more variables.

Obviously, such a procedure becomes cumbersome with a large number of objects (observations) or variables. For the observations shown in Figure 9.4, there is no obvious outlier that has all extremely high or low values. But just as in detecting multivariate outliers in other multivariate techniques, outliers may also be defined as having unique profiles that distinguish them from all of the other observations. For these instances, the procedures for identifying outliers discussed in chapter 2 can be applied. Also, they may emerge in the calculation of similarity. By whatever means used, observations identified as outliers must be assessed for their representativeness of the population and deleted from the analysis if deemed unrepresentative. But, as in other instances of outlier detection, the researcher should exhibit caution in deleting observations from the sample because such deletion may distort the actual structure of the data.

Similarity Measures

The concept of similarity is fundamental to cluster analysis. **Interobject similarity** is a measure of correspondence, or resemblance, between objects to be clustered. In our discussion of factor analysis, we created a correlation matrix between

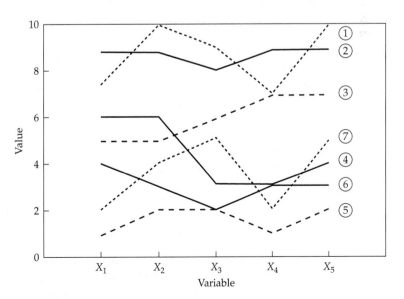

FIGURE 9.4 Profile Diagram

variables that was then used to group variables into factors. A comparable process occurs in cluster analysis. Here, the characteristics defining similarity are first specified. Then, the characteristics are combined into a similarity measure calculated for all pairs of objects, just as we used correlations between variables in factor analysis. In this way, any object can be compared to any other object through the similarity measure. The cluster analysis procedure then proceeds to group similar objects together into clusters.

Interobject similarity can be measured in a variety of ways, but three methods dominate the applications of cluster analysis: correlational measures, distance measures, and association measures. Each of the methods represents a particular perspective on similarity, dependent on both its objectives and type of data. Both the correlational and distance measures require metric data, whereas the association measures are for nonmetric data.

Correlational Measures

The interobject measure of similarity that probably comes to mind first is the correlation coefficient between a pair of objects measured on several variables. In effect, instead of correlating two sets of variables, we invert the objects' X variables matrix so that the columns represent the objects and the rows represent the variables. Thus, the correlation coefficient between the two columns of numbers is the correlation (or similarity) between the profiles of the two objects. High correlations indicate similarity and low correlations denote a lack of it. This procedure is followed in the application of Q-type factor analysis (see chapter 3).

Correlational measures represent similarity by the correspondence of patterns across the characteristics (X variables). This is illustrated by the example of seven observations shown in Figure 9.4. A correlational measure of similarity does not look at the magnitude but instead at the patterns of the values. In Table 9.3, which contains the correlations among these seven observations, we can see two distinct groups. First, cases 1, 5, and 7 all have similar patterns and corresponding high positive intercorrelations. Likewise, cases 2, 4, and 6 also have high positive correlations among themselves but low or negative correlations with the other observations. Case 3 has low or negative correlations with all other cases, thereby perhaps forming a group by itself. Thus, correlations represent patterns across the variables much more than the magnitudes. Correlational measures, however, are rarely used because emphasis in most applications of cluster analysis is on the magnitudes of the objects, not the patterns of values.

Distance Measures

Even though correlational measures have an intuitive appeal and are used in many other multivariate techniques, they are not the most commonly used measure of similarity in cluster analysis. Distance measures of similarity, which represent similarity as the proximity of observations to one another across the variables in the cluster variate, are the similarity measure most often used. Distance measures are actually a measure of dissimilarity, with larger values denoting lesser similarity. Distance is converted into a similarity measure by using an inverse relationship. A simple illustration of this was shown in our hypothetical example, in which clusters of observations were defined based on the proximity of observations to one another when each observation's scores on two variables were plotted graphically (see Figure 9.2).

TABLE 9.3 Calculating Correlational and Distance Measures of Similarity

ORIGINAL DATA

			Variables		
Case	X_1	X_2	X_3	X_4	X_5
1	7	10	9	7	10
2	9	9	8	9	9
3	5	5	6	7	7
4	6	6	3	3	4
5	1	2	2	1	2
6	4	3	2	3	3
7	2	4	5	2	5

SIMILARITY MEASURE: CORRELATION

				Case			
Case	1	2	3	4	5	6	7
1	1.00						
2	−.147	1.00					
3	.000	.000	1.00				
4	.087	.516	−.824	1.00			
5	.963	−.408	.000	−.060	1.00		
6	−.466	.791	−.354	.699	−.645	1.00	
7	.891	−.516	.165	−.239	.963	−.699	1.00

SIMILARITY MEASURE: EUCLIDEAN DISTANCE

				Case			
Case	1	2	3	4	5	6	7
1	nc						
2	3.32	nc					
3	6.86	6.63	nc				
4	10.24	10.20	6.00	nc			
5	15.78	16.19	10.10	7.07	nc		
6	13.11	13.00	7.28	3.87	3.87	nc	
7	11.27	12.16	6.32	5.10	4.90	4.36	nc

nc = distances not calculated.

Comparison to Correlational Measures The difference between correlational and distance measures can be seen by referring again to Figure 9.4. Distance measures focus on the magnitude of the values and portray as similar cases that are close together, but may have very different patterns across the variables. Table 9.3 also contains distance measures of similarity for the seven cases, and we see a very different clustering of cases emerging than that found when using the correlational measures. With smaller distances representing greater similarity, we see that cases 1 and 2 form one group, and cases 4, 5, 6, and 7 make up another group. These groups represent those with higher versus lower values. A third group, consisting of only case 3, differs from the other two groups because it has values that are both low and high. Although the two clusters using distance measures have different members than those using correlations, case 3 is unique in either measure of similarity. The choice of a correlational measure rather than the more traditional distance measure requires a quite different interpretation of the results by the researcher. Clusters based on correlational measures may not have similar values but instead have similar patterns. Distance-based clusters have more similar values across the set of variables, but the patterns can be quite different.

Types of Distance Measures   Several distance measures are available. The most commonly used is **Euclidean distance.** An example of how Euclidean distance is obtained is shown geometrically in Figure 9.5. Suppose that two points in two dimensions have coordinates (X_1, Y_1) and (X_2, Y_2), respectively. The Euclidean distance between the points is the length of the hypotenuse of a right triangle, as calculated by the formula under the figure. This concept is easily generalized to more than two variables. The Euclidean distance is used to calculate specific measures such as the simple Euclidean distance (calculated as described above) and the squared, or absolute, Euclidean distance, which is the sum of the squared differences without taking the square root. The squared Euclidean distance has the advantage of not having to take the square root which speeds computations markedly, and it is the recommended distance measure for the centroid and Ward's methods of clustering.

Several options not based on the Euclidean distance are also available. One of the most widely used alternative measures involves replacing the squared differences by the sum of the absolute differences of the variables. This procedure is called the absolute, or city-block, distance function. The **city-block approach** to calculating distances may be appropriate under certain circumstances, but it causes several problems. One is the assumption that the variables are not correlated with one another; if they are correlated, the clusters are not valid [15]. Other measures that employ variations of the absolute differences or the powers applied to the differences (other than just squaring the differences) are also available in most cluster programs.

Impact of Unstandardized Data Values   A problem faced by all the distance measures that use unstandardized data involves the inconsistencies between cluster solutions when the scale of the variables is changed. For example, suppose three objects, A, B, and C, are measured on two variables, probability of purchasing

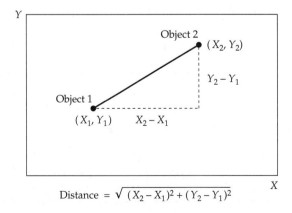

$$\text{Distance} = \sqrt{(X_2 - X_1)^2 + (Y_2 - Y_1)^2}$$

FIGURE 9.5 An Example of Euclidean Distance between Two Objects Measured on Two Variables, X and Y

brand X (in percentages) and amount of time spent viewing commercials for brand X (in minutes or seconds). The values for each observation are shown in Table 9.4.

From this information, distance measures can be calculated. In our example, we calculate three distance measures for each object pair: simple Euclidean distance, the absolute or squared Euclidean distance, and the city-block distance. First, we calculate the distance values based on purchase probability and viewing time in minutes. These distances, with smaller values indicating greater proximity and similarity, and their rank order are shown in Table 9.4. As we can see, the most

TABLE 9.4 Variations in Distance Measures Based on Alternative Data Scales

ORIGINAL DATA

Object	Purchase Probability	Commercial Viewing Time	
		Minutes	Seconds
A	60	3.0	180
B	65	3.5	210
C	63	4.0	240

DISTANCE MEASURES BASED ON PURCHASE PROBABILITY AND MINUTES OF COMMERCIAL VIEWING TIME

Object Pair	Simple Euclidean Distance		Squared or Absolute Euclidean Distance		City-block Distance	
	Value	Rank	Value	Rank	Value	Rank
A-B	5.025	3	25.25	3	5.5	3
A-C	3.162	2	10.00	2	4.0	2
B-C	2.062	1	4.25	1	2.5	1

DISTANCE MEASURES BASED ON PURCHASE PROBABILITY AND SECONDS OF COMMERCIAL VIEWING TIME

Object Pair	Simple Euclidean Distance		Squared or Absolute Euclidean Distance		City-block Distance	
	Value	Rank	Value	Rank	Value	Rank
A-B	30.41	2	925	2	35	3
A-C	60.07	3	3,609	3	63	2
B-C	30.06	1	904	1	32	1

DISTANCE MEASURES BASED ON STANDARDIZED VALUES OF PURCHASE PROBABILITY AND MINUTES OR SECONDS OF COMMERCIAL VIEWING TIME

Object Pair	Standardized Values		Simple Euclidean Distance		Squared or Absolute Euclidean Distance		City-block Distance	
	Purchase Probability	Minutes/Seconds of Viewing Time	Value	Rank	Value	Rank	Value	Rank
A-B	−1.06	−1.0	2.22	2	4.95	2	2.99	2
A-C	.93	0.0	2.33	3	5.42	3	3.19	3
B-C	.13	1.0	1.28	1	1.63	1	1.79	1

similar objects (with the smallest distance) are B and C, followed by A and C, with A and B the least similar (or least proximal). This ordering holds for all three distance measures, but the relative similarity or dispersion between objects is the most pronounced in the squared Euclidean distance measure.

The ordering of similarities can change markedly with only a change in the scaling of one of the variables. If we measure the viewing time in seconds instead of minutes, then the rank orders change (see Table 9.4). Objects B and C are still the most similar, but now pair A-B is next most similar and is almost identical to the similarity of B-C. Yet when we use minutes of viewing time, pair A-B is the least similar by a substantial margin. What has occurred is that the scale of the viewing time variable has dominated the calculations, making purchase probability less significant in the calculations. The reverse is true, however, when we measure viewing time in minutes, as then purchase probability is dominant in the calculations. The researcher should thus note the tremendous impact that variable scaling can have on the final solution. Standardization of the clustering variables, whenever possible conceptually, should be employed to avoid such instances as found in our example.

A commonly used measure of Euclidean distance that directly incorporates a standardization procedure is the **Mahalanobis distance (D^2).** The Mahalanobis approach not only performs a standardization process on the data by scaling in terms of the standard deviations but also sums the pooled within-group variance–covariance, which adjusts for intercorrelations among the variables. Highly intercorrelated sets of variables in cluster analysis can implicitly overweight one set of variables in the clustering procedures. In short, the Mahalanobis generalized distance procedure computes a distance measure between objects comparable to R^2 in regression analysis. Although many situations are appropriate for use of the Mahalanobis distance, not all programs include it as a measure of similarity. In such cases, the researcher usually selects the squared Euclidean distance.

In attempting to select a particular distance measure, the researcher should remember the following caveats. Different distance measures or a change in the scales of the variables may lead to different cluster solutions. Thus, it is advisable to use several measures and compare the results with theoretical or known patterns. Also, when the variables are intercorrelated (either positively or negatively), the Mahalanobis distance measure is likely to be the most appropriate because it adjusts for intercorrelations and weights all variables equally. Of course, if the researcher wishes to weight the variables unequally, other procedures are available [10, 11].

Association Measures

Association measures of similarity are used to compare objects whose characteristics are measured only in nonmetric terms (nominal or ordinal measurement). As an example, respondents could answer yes or no on a number of statements. An association measure could assess the degree of agreement or matching between each pair of respondents. The simplest form of association measure would be the percentage of times there was agreement (both respondents said yes or both said no to a question) across the set of questions. Extensions of this simple matching coefficient have been developed to accommodate multicategory nominal variables and even ordinal measures. Many computer programs, however, have limited support for association measures, and the researcher is many times forced to first calculate the similarity measures and then input the similarity matrix into the cluster program. Reviews of the various types of association measures can be found in several sources [4, 16].

Standardizing the Data

With the similarity measure selected, the researcher must address only one more question: Should the data be standardized before similarities are calculated? In answering this question, the researcher must consider several issues. First, most distance measures are quite sensitive to differing scales or magnitude among the variables. We saw this impact earlier when we changed from minutes to seconds on one of our variables. In general, variables with larger dispersion (i.e., larger standard deviations) have more impact on the final similarity value. Let us consider another example to illustrate this point. Assume that we want to cluster individuals on three variables—attitudes toward a product, age, and income. Now assume that we measured attitude on a seven-point scale of liking–disliking, with age measured in years and income in dollars. If we plotted this on a three-dimensional graph, the distance between points (and their similarity) would be almost totally based on the income differences. The possible differences on attitude range from 1 to 7, whereas income may have a range perhaps a thousand times greater. Thus, graphically we would not be able to see any difference on the dimension associated with attitude. For this reason the researcher must be aware of the implicit weighting of variables based on their relative dispersion, which occurs with distance measures.

Standardizing by Variables

The most common form of standardization is the conversion of each variable to standard scores (also known as Z scores) by subtracting the mean and dividing by the standard deviation for each variable. This is an option in all computer programs and many times is even directly included in the cluster analysis procedure. This is the general form of a **normalized distance function,** which utilizes a Euclidean distance measure amenable to a normalizing transformation of the raw data. This process converts each raw data score into a standardized value with a mean of 0 and a standard deviation of 1. This transformation, in turn, eliminates the bias introduced by the differences in the scales of the several attributes or variables used in the analysis.

The benefits of standardization can be seen in the last section of Table 9.4, in which two variables (purchase probability and viewing time) have been standardized before computing the three distance measures. First, it is much easier to compare between variables as they are on the same scale (a mean of 0 and standard deviation of 1). Positive values are above the mean, and negative values are below; the magnitude represents the number of standard deviations the original value is from the mean. Second, there is no difference in the standardized values when only the scale changes. For example, when viewing time in minutes and then seconds is standardized, the values are the same. Thus, using standardized variables truly eliminates the effects due to scale differences not only across variables, but for the same variable as well. However, the researcher should not always apply standardization without consideration for its consequences. There is no reason to absolutely accept the cluster solution using standardized variables versus unstandardized variables. If there is some "natural" relationship reflected in the scaling of the variables, then standardization may not be appropriate. The decision to standardize has both empirical and conceptual impacts and should always be made with careful consideration.

Standardizing by Observation

Up to now we have discussed standardizing only variables. What about "standardizing" respondents or cases? Why would we ever do this? Let us take a simple example. Suppose we had collected a number of ratings on a 10-point scale from respondents on the importance of several attributes in their purchase decisions for a product. We could apply cluster analysis and obtain clusters, but one very distinct possibility is that what we would get are clusters of people who said everything was important, some who said everything had little importance, and perhaps some clusters in between. What we are seeing are **response-style effects** in the clusters. Response-style effects are the systematic patterns of responding to a set of questions, such as yea-sayers (answer very favorably to all questions) or nay-sayers (answer unfavorably to all questions).

If we want to identify groups according to their response style, then standardization is not appropriate. But in most instances what is desired is the *relative* importance of one variable to another. In other words, is attribute 1 more important than the other attributes, and can clusters of respondents be found with similar patterns of importance? In this instance, standardizing by respondent would standardize each question not to the sample's average but instead to that respondent's average score. This **within-case** or **row-centering standardization** can be quite effective in removing response effects and is especially suited to many forms of attitudinal data [14]. We should note that this is similar to a correlational measure in highlighting the pattern across variables, but the proximity of cases still determines the similarity value.

Stage 3: Assumptions in Cluster Analysis

Cluster analysis, like multidimensional scaling (see chapter 10), is not a statistical inference technique in which parameters from a sample are assessed as possibly being representative of a population. Instead, cluster analysis is an objective methodology for quantifying the structural characteristics of a set of observations. As such, it has strong mathematical properties but not statistical foundations. The requirements of normality, linearity, and homoscedasticity that were so important in other techniques really have little bearing on cluster analysis. The researcher must focus, however, on two other critical issues: representativeness of the sample and multicollinearity.

Representativeness of the Sample

Rarely does the researcher have a census of the population to use in the cluster analysis. Usually, a sample of cases is obtained and the clusters derived in the hope that they represent the structure of the population. The researcher must therefore be confident that the obtained sample is truly representative of the population. As mentioned earlier, outliers may really be only an undersampling of divergent groups that, when discarded, introduce bias in the estimation of structure. The researcher must realize that cluster analysis is only as good as the representativeness of the sample. Therefore, all efforts should be taken to ensure that the sample is representative and the results are generalizable to the population of interest.

Impact of Multicollinearity

Multicollinearity was an issue in other multivariate techniques because of the difficulty in discerning the "true" impact of multicollinear variables. But in cluster analysis the effect is different because those variables that are multicollinear are implicitly weighted more heavily. Let us start with an example that illustrates the effect of multicollinearity. Suppose that respondents are being clustered on 10 variables, all attitudinal statements concerning a service. When multicollinearity is examined, we see that there are really two sets of variables, the first made up of eight statements and the second consisting of the remaining two statements. If our intent is to really cluster the respondents on the dimensions of the product (in this case represented by the two groups of variables), then use of the original 10 variables will be quite misleading. Because each variable is weighted equally in cluster analysis, the first dimension will have four times as many chances (eight items compared to two items) to affect the similarity measure, and so will the second dimension.

Multicollinearity acts as a weighting process not apparent to the observer but affecting the analysis nonetheless. For this reason, the researcher is encouraged to examine the variables used in cluster analysis for substantial multicollinearity and, if found, either reduce the variables to equal numbers in each set or use one of the distance measures, such as the Mahalanobis distance, that compensates for this correlation. There is debate over the use of factor scores in cluster analysis, as some research has shown that the variables that truly discriminate among the underlying groups are not well represented in most factor solutions. Thus, when factor scores are used, it is quite possible that a poor representation of the true structure of the data will be obtained [13]. The researcher must deal with both multicollinearity and discriminability of the variables to arrive at the best representation of structure.

Stage 4: Deriving Clusters and Assessing Overall Fit

With the variables selected and the similarity matrix calculated, the partitioning process begins (see Figure 9.6, p. 492). The researcher must first select the clustering algorithm used for forming clusters and then make the decision on the number of clusters to be formed. Both decisions have substantial implications not only on the results that will be obtained but also on the interpretation that can be derived from the results.

Clustering Algorithms

The first major question to answer in the partitioning phase is, What procedure should be used to place similar objects into groups or clusters? That is, What clustering algorithm or set of rules is the most appropriate? This is not a simple question because hundreds of computer programs using different algorithms are available, and more are always being developed. The essential criterion of all the

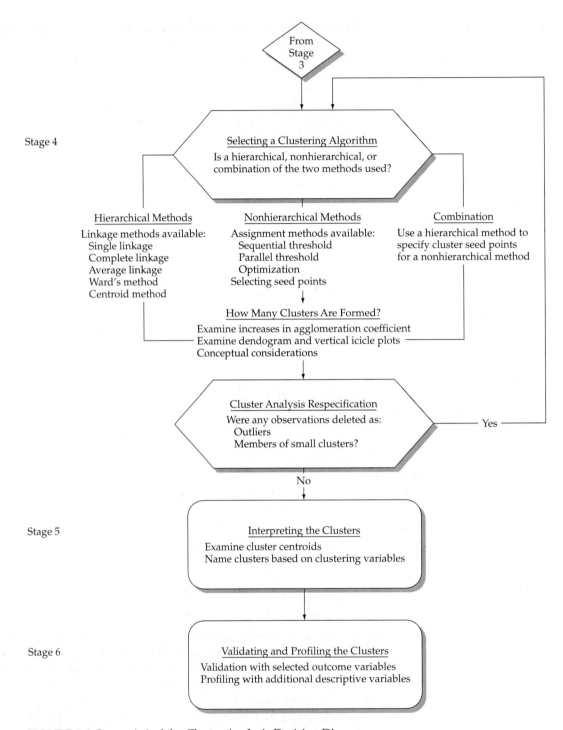

FIGURE 9.6 Stages 4–6 of the Cluster Analysis Decision Diagram

algorithms, however, is that they attempt to maximize the differences between clusters relative to the variation within the clusters, as shown in Figure 9.7. The ratio of the between-cluster variation to the average within-cluster variation is then comparable to (but not identical to) the *F* ratio in analysis of variance.

Most commonly used clustering algorithms can be classified into two general categories: (1) hierarchical and (2) nonhierarchical. We discuss the hierarchical techniques first.

Hierarchical Cluster Procedures

Hierarchical procedures involve the construction of a hierarchy of a treelike structure. There are basically two types of hierarchical clustering procedures—agglomerative and divisive. In the **agglomerative methods,** each object or observation starts out as its own cluster. In subsequent steps, the two closest clusters (or individuals) are combined into a new aggregate cluster, thus reducing the number of clusters by one in each step. In some cases, a third individual joins the first two in a cluster. In others, two groups of individuals formed at an earlier stage may join together in a new cluster. Eventually, all individuals are grouped into one large cluster; for this reason, agglomerative procedures are sometimes referred to as buildup methods.

An important characteristic of hierarchical procedures is that the results at an earlier stage are always nested within the results at a later stage, creating a similarity to a tree. For example, a six-cluster solution is obtained by joining two of the clusters found at the seven-cluster stage. Because clusters are formed only by joining existing clusters, any member of a cluster can trace its membership in an unbroken path to its beginning as a single observation. This process is shown in Figure 9.8; the representation is referred to as a **dendrogram** or tree graph. Another popular graphical method is the **vertical icicle diagram.**

When the clustering process proceeds in the direction opposite to agglomerative methods, it is referred to as a **divisive method.** In divisive methods, we begin

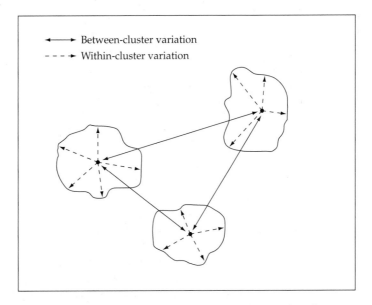

FIGURE 9.7 Cluster Diagram Showing Between- and Within-Cluster Variation

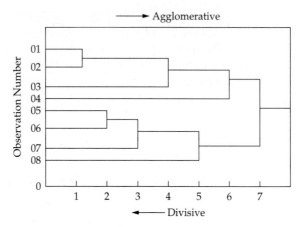

FIGURE 9.8 Dendrogram Illustrating Hierarchical Clustering

with one large cluster containing all the observations (objects). In succeeding steps, the observations that are most dissimilar are split off and made into smaller clusters. This process continues until each observation is a cluster in itself. In Figure 9.8, agglomerative methods move from left to right, and divisive methods move from right to left. Because most commonly used computer packages use agglomerative methods, and divisive methods act almost as agglomerative methods in reverse, we focus here on the agglomerative methods.

Five popular agglomerative algorithms used to develop clusters are (1) single linkage, (2) complete linkage, (3) average linkage, (4) Ward's method, and (5) centroid method. These algorithms differ in how the distance between clusters is computed.

Single Linkage   The **single-linkage** procedure is based on minimum distance. It finds the two objects separated by the shortest distance and places them in the first cluster. Then the next-shortest distance is found, and either a third object joins the first two to form a cluster, or a new two-member cluster is formed. The process continues until all objects are in one cluster. This procedure has also been called the nearest-neighbor approach.

The distance between any two clusters is the shortest distance from any point in one cluster to any point in the other. Two clusters are merged at any stage by the single shortest or strongest link between them. This was the rule applied in the example at the beginning of this chapter. Problems occur, however, when clusters are poorly delineated. In such cases, single linkage procedures can form long, snakelike chains, and eventually all individuals are placed in one chain. Individuals at opposite ends of a chain may be very dissimilar.

An example of this arrangement is shown in Figure 9.9. Three clusters (A, B, and C) are to be joined. The single linkage algorithm, focusing on only the closest points in each cluster, would link clusters A and B because of their short distance at the extreme ends of the clusters. Joining clusters A and B creates a cluster that encircles cluster C. Yet in striving for within-cluster homogeneity, it would be much better to join cluster C with either A or B. This is the principal disadvantage of the single linkage algorithm.

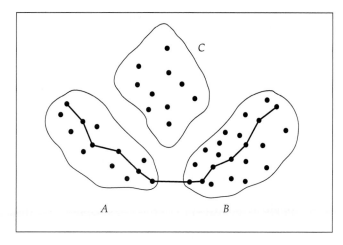

FIGURE 9.9 Example of Single Linkage Joining Dissimilar
Clusters A and B

Complete Linkage   The **complete linkage** procedure is similar to single linkage
except that the cluster criterion is based on maximum distance. For this reason,
it is sometimes referred to as the farthest-neighbor approach or as a diameter
method. The maximum distance between individuals in each cluster represents
the smallest (minimum-diameter) sphere that can enclose all objects in both clusters. This method is called complete linkage because all objects in a cluster are
linked to each other at some maximum distance or by minimum similarity. We
can say that within-group similarity equals group diameter. This technique eliminates the snaking problem identified with single linkage.
    Figure 9.10 shows how the shortest (single linkage) and longest (complete link-
age) distances represent similarity between groups. Both measures reflect only

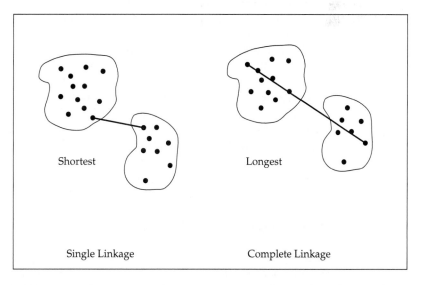

FIGURE 9.10 Comparison of Distance Measures for Single Linkage and
Complete Linkage

one aspect of the data. The use of the shortest distance reflects only a single pair of objects (the closest), and the complete linkage also reflects a single pair, this time the two most extreme. It is thus useful to visualize the measures as reflecting the similarity of most similar pair or least similar pair of objects.

Average Linkage   The **average linkage** method starts out the same as that of single linkage or complete linkage, but the cluster criterion is the average distance from all individuals in one cluster to all individuals in another. Such techniques do not depend on extreme values, as do single linkage or complete linkage, and partitioning is based on all members of the clusters rather than on a single pair of extreme members. Average linkage approaches tend to combine clusters with small within-cluster variation. They also tend to be biased toward the production of clusters with approximately the same variance.

Ward's Method   In **Ward's method,** the distance between two clusters is the sum of squares between the two clusters summed over all variables. At each stage in the clustering procedure, the within-cluster sum of squares is minimized over all partitions (the complete set of disjoint or separate clusters) obtainable by combining two clusters from the previous stage. This procedure tends to combine clusters with a small number of observations. It is also biased toward the production of clusters with approximately the same number of observations.

Centroid Method   In the **centroid method** the distance between two clusters is the distance (typically squared Euclidean or simple Euclidean) between their centroids. **Cluster centroids** are the mean values of the observations on the variables in the cluster variate. In this method, every time individuals are grouped, a new centroid is computed. Cluster centroids migrate as cluster mergers take place. In other words, there is a change in a cluster centroid every time a new individual or group of individuals is added to an existing cluster. These methods are the most popular with biologists but may produce messy and often confusing results. The confusion occurs because of reversals, that is, instances when the distance between the centroids of one pair may be less than the distance between the centroids of another pair merged at an earlier combination. The advantage of this method is that it is less affected by outliers than are other hierarchical methods.

Nonhierarchical Clustering Procedures

In contrast to hierarchical methods, **nonhierarchical procedures** do not involve the treelike construction process. Instead, they assign objects into clusters once the number of clusters to be formed is specified. Thus, the six-cluster solution is not just a combination of two clusters from the seven-cluster solution, but is based only on finding the best six-cluster solution. In a simple example, the process works this way. The first step is to select a **cluster seed** as the initial cluster center, and all objects (individuals) within a prespecified threshold distance are included in the resulting cluster. Then another cluster seed is chosen, and the assignment continues until all objects are assigned. Objects then may be reassigned if they are closer to another cluster than the one originally assigned. There are several different approaches for selecting cluster seeds and assigning objects that we discuss in the next section. Nonhierarchical clustering procedures are frequently referred to as K-means clustering, and they typically use one of the following three approaches for assigning individual observations to one of the clusters [5, p. 428].

Sequential Threshold The **sequential threshold method** starts by selecting one cluster seed and includes all objects within a prespecified distance. When all objects within the distance are included, a second cluster seed is selected, and all objects within the prespecified distance are included. Then a third seed is selected, and the process continues as before. When an object is clustered with a seed, it is no longer considered for subsequent seeds.

Parallel Threshold In contrast, the **parallel threshold method** selects several cluster seeds simultaneously in the beginning and assigns objects within the threshold distance to the nearest seed. As the process evolves, threshold distances can be adjusted to include fewer or more objects in the clusters. Also, in some variants of this method, objects remain unclustered if they are outside the prespecified threshold distance from any cluster seed.

Optimization The third method, referred to as the **optimizing procedure,** is similar to the other two nonhierarchical procedures except that it allows for reassignment of objects. If, in the course of assigning objects, an object becomes closer to another cluster that is not the cluster to which it is currently assigned, then an optimizing procedure switches the object to the more similar (closer) cluster.

Selecting Seed Points Nonhierarchical procedures are available in a number of computer programs, including all major statistical packages. The sequential threshold procedure (e.g., FASTCLUS program in SAS) is an example of a nonhierarchical clustering program designed for large data sets. After the researcher specifies the maximum number of clusters allowed, the procedure begins by selecting cluster seeds, which are used as initial guesses of the means of the clusters. The first seed is the first observation in the data set with no missing values. The second seed is the next complete observation (no missing data) that is separated from the first seed by a specified minimum distance. The default option is a zero minimum distance. After all seeds have been selected, the program assigns each observation to the cluster with the nearest seed. The researcher can specify that the cluster seeds be revised (updated) by calculating seed cluster means each time an observation is assigned. In contrast, the parallel threshold methods (e.g., QUICK CLUSTER in SPSS) establishes the seed points as user-supplied points or selects them randomly from all observations.

The major problem faced by all nonhierarchical clustering procedures is how to select the cluster seeds. For example, with a sequential threshold option, the initial and probably the final cluster results depend on the order of the observations in the data set, and shuffling the order of the data is likely to affect the results. Specifying the initial cluster seeds, as is done in the sequential threshold procedure, can reduce this problem. But even selecting the cluster seeds randomly will produce different results for each set of random seed points. Thus, the researcher must be aware of the impact of the cluster seed selection process on the final results.

Should Hierarchical or Nonhierarchical Methods Be Used?

A definitive answer to this question cannot be given for two reasons. First, the research problem at hand may suggest one method or the other. Second, what we learn with continued application to a particular context may suggest one method over the other as more suitable for that context.

Pros and Cons of Hierarchical Methods   In the past, hierarchical clustering techniques were more popular, with Ward's method and average linkage probably being the best available [8]. Hierarchical procedures do have the advantage of being fast and therefore taking less computer time. But with the computing power of today, even personal computers can handle large data sets quite easily. Hierarchical methods can be misleading, however, because undesirable early combinations may persist throughout the analysis and lead to artificial results. Of specific concern is the substantial impact of outliers on hierarchical methods, particularly with the complete linkage method. To reduce this possibility, the researcher may wish to cluster analyze the data several times, each time deleting problem observations or outliers. The deletion of cases, however, even those not found to be outliers, can many times distort the solution. Thus, the researcher must employ extreme care in the deletion of observations for any reason.

Also, although computations of the clustering process are relatively fast, hierarchical methods are not amenable to analyzing very large samples. As sample size increases, the data storage requirements increase dramatically. For example, a sample of 400 cases requires storage of approximately 80,000 similarities, and this increases to almost 125,000 for a sample of 500. Even with today's technological advances, problems of this size exceed the capacity of most personal computers, thus limiting the application in many instances. The researcher may take a random sample of the original observations to reduce sample size but must now question the representativeness of the sample taken from the original sample.

Emergence of Nonhierarchical Methods   Nonhierarchical methods have gained increased acceptability and are applied increasingly. Their use, however, depends on the ability of the researcher to select the seed points according to some practical, objective, or theoretical basis. In these instances, nonhierarchical methods have several advantages over hierarchical techniques. The results are less susceptible to the outliers in the data, the distance measure used, and the inclusion of irrelevant or inappropriate variables. These benefits are realized, however, only with the use of nonrandom (i.e., specified) seed points; thus, the use of nonhierarchical techniques with random seed points is markedly inferior to the hierarchical techniques. Even a nonrandom starting solution does not guarantee an optimal clustering of observations. In fact, in many instances the researcher will get a different final solution for each set of specified seed points. How is the researcher to select the "correct" answer? Only by analysis and validation can the researcher select what is considered the "best" representation of structure, realizing there are many alternatives that may be as acceptable.

A Combination of Both Methods   Another approach is to use *both* methods (hierarchical and nonhierarchical) to gain the benefits of each [8]. First, a hierarchical technique can establish the number of clusters, profile the cluster centers, and identify any obvious outliers. After outliers are eliminated, the remaining observations can then be clustered by a nonhierarchical method with the cluster centers from the hierarchical results as the initial seed points. In this way, the advantages of the hierarchical methods are complemented by the ability of the nonhierarchical methods to "fine-tune" the results by allowing the switching of cluster membership.

How Many Clusters Should Be Formed?

Perhaps the most perplexing issue for the researcher using cluster analysis is determining the final number of clusters to be formed (also known as the **stopping rule**). Unfortunately, no standard, objective selection procedure exists. Because there is no internal statistical criterion used for inference, such as the statistical significance tests of other multivariate methods, researchers have developed many criteria and guidelines for approaching the problem. The principal drawback is that these are ad hoc procedures and must be computed by the researcher, and many times this involves fairly complex procedures [1, 9]. One class of stopping rules that is relatively simple examines some measure of similarity or distance between clusters at each successive step, with the cluster solution defined when the similarity measure exceeds a specified value or when the successive values between steps makes a sudden jump. A simple example of this was used in the example at the beginning of the chapter, which looked for large increases in the average within-cluster distance. When a large increase occurs, the researcher selects the prior cluster solution on the logic that its combination caused a substantial decrease in similarity. This stopping rule has been shown to provide fairly accurate decisions in empirical studies [9]. A second general class of stopping rules attempt to apply some form of statistical rule or adapt a statistical test, such as the point-biserial/tau correlations or the likelihood ratio. Although some of these—such as the cubic clustering criterion (CCC) [9] contained in SAS—have been shown to have notable success, many seem overly complex for the improvement they provide over simpler measures. There are a number of other specific procedures that have been proposed, but none have been found to be substantially better in all situations.

Also, the researcher should complement the strictly empirical judgment with any conceptualization of theoretical relationships that may suggest a natural number of clusters. One might start this process by specifying some criteria based on practical considerations, such as saying, "My findings will be more manageable and easier to communicate if I have three to six clusters," and then solving for this number of clusters and selecting the best alternative after evaluating all of them. In the final analysis, however, it is probably best to compute a number of different cluster solutions (e.g., two, three, four) and then decide among the alternative solutions by using a priori criteria, practical judgment, common sense, or theoretical foundations. The cluster solutions will be improved by restricting the solution according to conceptual aspects of the problem.

Should the Cluster Analysis Be Respecified?

When an acceptable cluster analysis solution is identified, the researcher should examine the fundamental structure represented in the defined clusters. Of particular note are widely disparate cluster sizes or clusters of only one or two observations. Researchers must examine widely varying cluster sizes from a conceptual perspective, comparing the actual results with the expectations formed in the research objectives. More troublesome are single-member clusters, which may be outliers not detected in earlier analyses. If a single-member cluster (or one of very small size compared with other clusters) appears, the researcher must decide if it represents a valid structural component in the sample or if it should be deleted as unrepresentative. If any observations are deleted, especially when hierarchical solutions are employed, the researcher should rerun the cluster analysis and start the process of defining clusters anew.

Stage 5: Interpretation of the Clusters

The interpretation stage involves examining each cluster in terms of the cluster variate to name or assign a label accurately describing the nature of the clusters. To clarify this process, let us refer to the example of diet versus regular soft drinks. Let us assume that an attitude scale was developed that consisted of statements regarding consumption of soft drinks, such as "diet soft drinks taste harsher," "regular soft drinks have a fuller taste," "diet drinks are healthier," and so forth. Further, let us assume that demographic and soft drink consumption data were also collected.

When starting the interpretation process, one measure frequently used is the cluster's centroid. If the clustering procedure was performed on the raw data, this would be a logical description. If the data were standardized or if the cluster analysis was performed using factor analysis (component factors), the researcher would have to go back to the raw scores for the original variables and compute average profiles using these data.

Continuing with our soft drink example, in this stage we examine the average score profiles on the attitude statements for each group and assign a descriptive label to each cluster. Many times discriminant analysis is applied to generate score profiles, but we must remember that statistically significant differences would not indicate an "optimal" solution because statistical differences are expected, given the objective of cluster analysis. Examination of the profiles allows for a rich description of each cluster. For example, two of the clusters may have favorable attitudes about diet soft drinks and the third cluster negative attitudes. Moreover, of the two favorable clusters, one may exhibit favorable attitudes toward only diet soft drinks, whereas the other may display favorable attitudes toward both diet and regular soft drinks. From this analytical procedure, one would evaluate each cluster's attitudes and develop substantive interpretations to facilitate labeling each. For example, one cluster might be labeled "health- and calorie-conscious," whereas another might be labeled "get a sugar rush."

The profiling and interpretation of the clusters, however, achieve more than just description. First, they provide a means for assessing the correspondence of the derived clusters to those proposed by prior theory or practical experience. If used in a confirmatory mode, the cluster analysis profiles provide a direct means of assessing the correspondence. Second, the cluster profiles provide a route for making assessments of practical significance. The researcher may require that substantial differences exist on a set of clustering variables and the cluster solution be expanded until such differences arise. In assessing either correspondence or practical significance, the researcher compares the derived clusters to a preconceived typology.

Stage 6: Validation and Profiling of the Clusters

Given the somewhat subjective nature of cluster analysis about selecting an "optimal" cluster solution, the researcher should take great care in validating and ensuring practical significance of the final cluster solution. Although no single method exists to ensure validity and practical significance, several approaches have been proposed to provide some basis for the researcher's assessment.

Validating the Cluster Solution

Validation includes attempts by the researcher to assure that the cluster solution is representative of the general population, and thus is generalizable to other objects and is stable over time. The most direct approach in this regard is to cluster analyze separate samples, comparing the cluster solutions and assessing the correspondence of the results. This approach, however, is often impractical because of time or cost constraints or the unavailability of objects (particularly consumers) for multiple cluster analyses. In these instances, a common approach is to split the sample into two groups. Each is cluster analyzed separately, and the results are then compared. Other approaches include (1) a modified form of split sampling whereby cluster centers obtained from one cluster solution are employed to define clusters from the other observations and the results are compared [7], and (2) a direct form of cross-validation [12].

The researcher may also attempt to establish some form of **criterion** or **predictive validity.** To do so, the researcher selects variable(s) *not used to form the clusters* but known to vary across the clusters. In our example, we may know from past research that attitudes toward diet soft drinks vary by age. Thus, we can statistically test for the differences in age between those clusters that are favorable to diet soft drinks and those that are not. The variable(s) used to assess predictive validity should have strong theoretical or practical support as they become the benchmark for selecting among the cluster solutions.

Profiling the Cluster Solution

The profiling stage involves describing the characteristics of each cluster to explain how they may differ on relevant dimensions. This typically involves the use of discriminant analysis (see chapter 5). The procedure begins after the clusters are identified. The researcher utilizes *data not previously included* in the cluster procedure to profile the characteristics of each cluster. These data typically are demographic characteristics, psychographic profiles, consumption patterns, and so forth. Although there may not be a theoretical rationale for their difference across the clusters, such as required for predictive validity assessment, they should at least have practical importance. Using discriminant analysis, the researcher compares average score profiles for the clusters. The categorical dependent variable is the previously identified clusters, and the independent variables are the demographics, psychographics, and so on. From this analysis, assuming statistical significance, the researcher could conclude, for example, that the "health- and calorie-conscious" cluster from our previous example consists of better-educated, higher-income professionals who are moderate consumers of soft drinks. In short, the profile analysis focuses on describing not what directly determines the clusters but the characteristics of the clusters after they have been identified. Moreover, the emphasis is on the characteristics that differ significantly across the clusters and those that could predict membership in a particular cluster.

Summary of the Decision Process

Cluster analysis provides researchers with an empirical and objective method for performing one of the most inherent tasks for humans—classification. Whether for purposes of simplification, exploration, or confirmation, cluster analysis is a potent analytical tool that has a wide range of applications. But with this technique comes a responsibility on the part of the researcher to apply the underlying principles appropriately. As mentioned in the introduction to this chapter, cluster analysis has many caveats, which cause even the experienced researcher to apply it with caution. Yet when used appropriately, it has the potential to reveal structures within the data that could not be discovered by any other means. Thus, this powerful technique addresses a fundamental need of researchers in all fields, yet it can be applied with the knowledge that it can be as easily abused as used wisely.

An Illustrative Example

To illustrate the application of cluster analysis techniques, let us turn to the HATCO database. The seven perceptions of HATCO provide a basis for illustrating one of the most common uses of cluster analysis—the formation of customer segments. In our example, we follow the stages of the model-building process, starting with setting objectives, then addressing research design issues, and finally actually partitioning respondents into clusters and the interpretation and validation of the results. The following sections detail these procedures through each of the stages.

Stage 1: Objectives of the Cluster Analysis

We begin by cluster analyzing the ratings by HATCO customers as to the performance of HATCO on the seven attributes (X_1 through X_7). Our objective is to segment objects (customers) into groups with similar perceptions of HATCO. Once identified, HATCO can then formulate strategies with different appeals for the separate groups—the requisite basis for market segmentation. A primary concern is that the seven attributes used to form the clusters be adequate in scope and detail. From the examples in other chapters with the various multivariate techniques, we have found that these variables have sufficient predictive power to justify their use as the basis for segmentation.

Stage 2: Research Design of the Cluster Analysis

The first step is to identify any outliers in the sample before partitioning begins. The sample of 100 observations has been examined for outliers and found to have no strong candidates for deletion regarding the issues raised in chapter 2 and the discussion on outlier detection presented earlier in this chapter. But the researcher should also examine the cluster solutions in later stages to assess if outliers have emerged during the clustering process. The next issue involves the choice of a similarity measure. Given that the set of seven variables is metric, squared Euclidean distance is chosen as the similarity measure. If multicollinearity is deemed substantial or to have the effect of uneven weighting of the variables, then the Mahalanobis distance (D^2) would be appropriate. Also, correlational measures are not employed because the derivation of segments should consider the magnitude of the perceptions (favorable versus unfavorable) as well as the pattern. This is best accomplished with a distance measure of similarity. Finally, no form of standardization is used. The standardization of variables is not undertaken because all variables are on the same scale, and within-case standardization is not appropriate because the magnitude of the perceptions is an important element of the segmentation objectives.

Stage 3: Assumptions in Cluster Analysis

For purposes of illustration, the sample is considered a representative sample of HATCO customers. Still left to be resolved is the impact of multicollinearity on the implicit weighting of the results. The analysis of multicollinearity detailed in chapters 3 and 4 identifies only minimal levels that should not impact the cluster analysis in any substantial manner.

Stage 4: Deriving Clusters and Assessing Overall Fit

In our example, we follow the approach of employing hierarchical and nonhierarchical methods in combination. The first step in the partitioning stage uses the hierarchical procedure to identify the appropriate number of clusters. Then, in step 2 we use nonhierarchical procedures to "fine-tune" the results even further by utilizing the hierarchical results as a basis for generating the seed points. The hierarchical and nonhierarchical procedures from SPSS are used in this analysis, although comparable results would be obtained with any of the clustering programs.

Step 1: Hierarchical Cluster Analysis

The first question to ask is, Which clustering algorithm should we use? Ward's method is chosen to minimize the within-cluster differences and to avoid problems with "chaining" of the observations found in the single linkage method. Table 9.5 (pp. 504–505) contains the results of the cluster analysis, including the cases being combined at each stage of the process and the agglomeration coefficient. The agglomeration coefficient (fourth column) is the within-cluster sum of squares. For the other linkage methods, the agglomeration coefficient is the squared Euclidean distance between the two cases of clusters being combined.

Selecting a Cluster Solution   How many clusters should we have? Because the data involve profiles of HATCO customers and our interest is in identifying types or profiles of these customers that may form the bases for differing strategies, a manageable number of clusters would be in the range of two to five. The researcher must now select the final cluster solution from these possible cluster solutions.

This agglomeration coefficient is particularly amenable for use in a stopping rule that evaluates the changes in the coefficient at each stage of the hierarchical process. Small coefficients indicate that fairly homogeneous clusters are being merged. Joining two very different clusters results in a large coefficient or a large percentage change in the coefficient. The researcher looks for large increases in the value, similar to the scree test in factor analysis. This test has been shown to be a fairly accurate algorithm, although it has the tendency to indicate too few clusters [9]. Another popular stopping rule is the cubic clustering criterion (CCC) popularized by SAS, which is not available in this output. Although CCC has been

TABLE 9.5 Agglomeration Schedule of Hierarchical Cluster Analysis Using
Ward Method

Stage	Cluster Combined		Agglomeration Coefficient	Stage Cluster First Appears		Next Stage
	Cluster 1	Cluster 2		Cluster 1	Cluster 2	
1	15	20	0.000	0	0	60
2	5	42	0.005	0	0	94
3	24	27	0.010	0	0	74
4	47	61	0.020	0	0	78
5	19	28	0.040	0	0	60
6	67	90	0.070	0	0	39
7	18	92	0.105	0	0	65
8	51	77	0.140	0	0	72
9	33	62	0.175	0	0	63
10	36	41	0.210	0	0	45
11	85	87	0.260	0	0	69
12	65	79	0.310	0	0	68
13	43	46	0.360	0	0	76
14	25	44	0.410	0	0	63
15	38	63	0.475	0	0	54
16	69	81	0.555	0	0	52
17	94	98	0.650	0	0	73
18	56	91	0.745	0	0	66
19	50	72	0.840	0	0	52
20	75	99	0.950	0	0	62
21	1	95	1.060	0	0	72
22	16	73	1.170	0	0	61
23	37	48	1.280	0	0	58
24	11	100	1.405	0	0	69
25	4	89	1.545	0	0	62
26	84	88	1.685	0	0	45
27	2	83	1.825	0	0	82
28	29	78	1.965	0	0	61
29	3	71	2.105	0	0	75
30	23	32	2.245	0	0	66
31	17	64	2.435	0	0	83
32	12	76	2.650	0	0	67
33	8	68	2.865	0	0	70
34	9	74	3.130	0	0	55
35	52	60	3.420	0	0	57
36	10	34	3.755	0	0	43
37	26	59	4.105	0	0	64
38	49	97	4.525	0	0	81
39	7	67	4.995	0	6	77
40	13	21	5.515	0	0	51
41	82	93	6.040	0	0	91
42	40	54	6.565	0	0	53
43	10	30	7.097	36	0	50
44	66	80	7.632	0	0	59
45	36	84	8.189	10	26	70
46	22	55	8.749	0	0	71
47	6	70	9.409	0	0	57
48	45	86	10.239	0	0	53
49	39	96	11.079	0	0	68

TABLE 9.5 (Continued)

Stage	Cluster Combined		Agglomeration Coefficient	Stage Cluster First Appears		Next Stage
	Cluster 1	Cluster 2		Cluster 1	Cluster 2	
50	10	53	11.965	43	0	56
51	13	35	13.025	40	0	71
52	50	69	14.468	19	16	65
53	40	45	15.970	42	48	73
54	14	38	17.558	0	15	59
55	9	58	19.213	34	0	67
56	10	31	21.261	50	0	58
57	6	52	23.516	47	35	88
58	10	37	25.869	56	23	75
59	14	66	28.244	54	44	80
60	15	19	30.704	1	5	77
61	16	29	33.179	22	28	78
62	4	75	35.714	25	20	74
63	25	33	38.537	14	9	64
64	25	26	41.568	63	37	84
65	18	50	44.879	7	52	76
66	23	56	48.546	30	18	87
67	9	12	52.279	55	32	80
68	39	65	56.214	49	12	89
69	11	85	60.252	24	11	87
70	8	36	64.364	33	45	83
71	13	22	68.580	51	46	90
72	1	51	73.083	21	8	84
73	40	94	77.887	53	17	85
74	4	24	82.785	62	3	82
75	3	10	88.133	29	58	79
76	18	43	93.522	65	13	92
77	7	15	98.977	39	60	86
78	16	47	104.835	61	4	90
79	3	57	111.625	75	0	91
80	9	14	118.530	67	59	81
81	9	49	126.007	80	38	86
82	2	4	134.773	27	74	85
83	8	17	143.875	70	31	88
84	1	25	156.719	72	64	92
85	2	40	170.259	82	73	89
86	7	9	185.590	77	81	94
87	11	23	201.110	69	66	93
88	6	8	218.441	57	83	93
89	2	39	236.111	85	68	96
90	13	16	258.731	71	78	95
91	3	82	281.428	79	41	97
92	1	18	305.027	84	76	95
93	6	11	333.081	88	87	96
94	5	7	364.898	2	86	98
95	1	13	398.082	92	90	98
96	2	6	446.283	89	93	97
97	2	3	522.981	96	91	99
98	1	5	614.954	95	94	99
99	1	2	994.752	98	97	0

shown to be quite successful in identifying the correct number of clusters, it has the tendency to select cluster solutions that have too many clusters [9]. The researcher should consider a range of cluster solutions, with emphasis on examining the tendencies of the particular stopping rule used in the analysis.

The clustering (agglomeration) coefficient shows rather large increases in going from four to three clusters ($523.0 - 446.3 = 76.6$), three to two clusters ($615.0 - 523.0 = 92.0$), and two to one cluster ($994.8 - 615.0 = 379.8$). To help identify large relative increases in the cluster homogeneity, we calculate the percentage change in the clustering coefficient for 10 to 2 clusters (see Table 9.6). The largest percentage increase by far occurs in going from two to one cluster, and the next noticeable change in the percentage increase occurs in combining four into three clusters. Thus, for illustrative purposes as well as the assessment of the practical significance of a four-cluster solution, both the two- and four-cluster solutions are examined.

Identifying Outliers in the Cluster Solutions   The dendrogram (Figure 9.11) and agglomeration schedule (Table 9.5) both provide a means of identifying outliers in the sample. The dendrogram permits a visual inspection for outliers, where an outlier would be a "branch" that did not join until very late. The researcher can also readily identify small clusters, as they exhibit a long "branch" for only a small number of observations. In the agglomeration schedule, the researcher can ascertain the presence of single-member clusters quite easily with many of the computer programs. The example in Table 9.5 from SPSS shows on the left side of the agglomeration coefficient the clusters being combined. In the right-hand columns, the steps at which each cluster was formed are noted. An observation that has never been joined into a cluster has a stage of 0. In the first 38 stages, single observations are being joined together. Only at stage 39 does the cluster analysis first join a cluster formed at another stage. We can use this information also to identify single observations that are joined very late in the clustering process—potential outliers. Looking backward from stage 99 in Table 9.5, we see that at stage 94 (six clusters) a cluster formed at stage 2 was joined. This means that if we selected a seven-cluster solution, one of the clusters would have only two observations. We can also see that the last single-member cluster to be joined occurs at stage 79. Thus, if analysis is confined to a smaller number of clusters (say, 10 or fewer), then the researcher has only one potential problem (the two-member clus-

TABLE 9.6 Analysis of Agglomeration Coefficient for Hierarchical Cluster Analysis

Number of Clusters	Agglomeration Coefficient	Percentage Change in Coefficient to Next Level
10	258.7	8.8
9	281.4	8.4
8	305.0	9.2
7	333.1	9.6
6	364.9	9.1
5	398.1	12.1
4	446.3	17.2
3	523.0	17.6
2	615.0	61.8
1	994.8	—

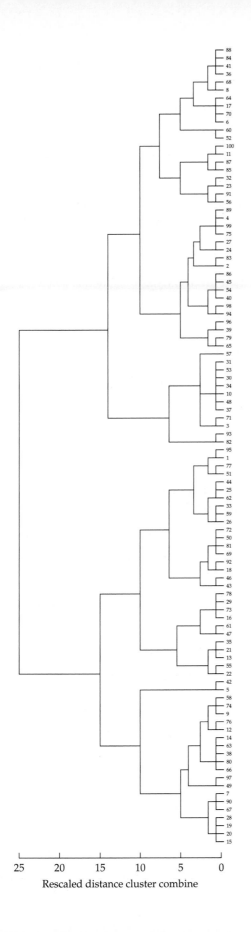

FIGURE 9.11 Dendrogram for Hierarchical Cluster Analysis Using Ward's Method

ter) to deal with. In this case, the selection of less than seven clusters eliminates the need for any further respecification of the cluster analysis.

Profiling the Two-Cluster and Four-Cluster Solutions With no need for respecification, the next step is to profile the clusters in both solutions to assist in the selection of the final cluster solution. Table 9.7 contains the clustering variable profiles for both the two- and four-cluster solutions. The focus at this stage is not to provide an interpretation of the cluster, but to ensure that they are truly distinctive. Examination of the two-cluster profiles reveals two clusters that are almost mirror images of each other. Cluster 1 has higher values on X_1 and X_3, whereas cluster 2 has higher values for X_2, X_4, X_6, and X_7. Only X_5 showed no significant differences between the clusters. If the results were restricted to only this solution, the results would suggest two quite different segments. The clusters cannot be characterized as all low or all high values, but instead must be evaluated from both a conceptual and practical standpoint.

TABLE 9.7 Clustering Variable Profiles for the Two-Cluster and Four-Cluster Solutions from the Hierarchical Cluster Analysis

CLUSTERING VARIABLE PROFILES

| | | | | Clustering Variable Mean Values | | | | |
Cluster	X_1 Delivery Speed	X_2 Price Level	X_3 Price Flexibility	X_4 Manufacturer Image	X_5 Overall Service	X_6 Salesforce Image	X_7 Product Quality	Cluster Size
Two-cluster solution								
1	4.460	1.576	8.900	4.926	2.992	2.510	5.904	50
2	2.570	3.152	6.888	5.570	2.840	2.820	8.038	50
Four-cluster solution								
1	4.207	1.624	8.597	4.372	2.879	2.014	5.124	29
2	2.213	2.834	7.166	5.358	2.505	2.689	7.968	38
3	3.700	4.158	6.008	6.242	3.900	3.233	8.258	12
4	4.810	1.510	9.319	5.690	3.148	3.195	6.981	21

SIGNIFICANCE TESTING OF DIFFERENCES BETWEEN CLUSTER CENTERS

Variable	Cluster Mean Square	Degrees of Freedom	Error Mean Square	Degrees of Freedom	F value	Significance
Two-cluster solution						
X_1 Delivery speed	89.302	1	.851	98	104.95	.000
X_2 Price level	62.094	1	.811	98	76.61	.000
X_3 Price flexibility	101.204	1	.909	98	111.30	.000
X_4 Manufacturer image	10.368	1	1.187	98	8.73	.004
X_5 Overall service	.578	1	.564	98	1.02	.314
X_6 Salesforce image	2.402	1	.576	98	4.17	.044
X_7 Product quality	113.849	1	1.377	98	82.68	.000
Four-cluster solution						
X_1 Delivery speed	37.962	3	.613	96	61.98	.000
X_2 Price level	26.082	3	.659	96	39.56	.000
X_3 Price flexibility	39.927	3	.735	96	54.34	.000
X_4 Manufacturer image	12.884	3	.917	96	14.04	.000
X_5 Overall service	6.398	3	.382	96	16.75	.000
X_6 Salesforce image	7.367	3	.383	96	19.26	.000
X_7 Product quality	52.203	3	.960	96	54.37	.000

Also shown in Table 9.7 are the profiles for the four-cluster solution. The increased number of clusters provides for a more well-defined structure and more variation in terms of the clustering variables. Examining the four-cluster profiles reveals a number of patterns of high versus low values. Another aspect that varies from the two-cluster solution is that all of the clustering variables differ in a statistically significant manner across the four groups. This compares to the five variables in the two-cluster solution that were significantly different at the .01 level. The increased number of clusters does exhibit an improvement in representing distinct groups that may reflect an underlying structure.

Assessing the Nonhierarchical Results All of the indicators at this stage (stopping rule, absence of outliers, and distinctive profiles) support either the two- or four-cluster solutions. Thus, both solutions will be carried forward into the nonhierarchical analysis (step 2) to obtain final cluster solutions. The selection between the two-cluster and four-cluster solutions will then be based on evaluations made in stages five and six.

Step 2: Nonhierarchical Cluster Analysis

The second step uses nonhierarchical techniques to adjust or "fine-tune" the results from the hierarchical procedures. In performing the cluster analysis, the researcher should take the initial seed points from the results in step 1; in this case, these are the cluster centroids on the seven clustering variables (X_1 to X_7) for both the two- and four-cluster solutions. The results (centroid values and cluster size) are shown in Table 9.8 (p. 510) for both cluster solutions. Just as seen in the hierarchical methods, the two-cluster solution resulted in groups of almost equal size (52 versus 48 observations), and the profiles correspond quite well with the cluster profiles from the hierarchical procedure. Again, only X_5 showed no significant differences between the clusters. For the four-group solution, the cluster sizes were quite similar to those from the hierarchical procedure, varying in size at the most by only four observations. Also, the cluster profiles matched quite well. The correspondence and stability of the two cluster solutions between the nonhierarchical and hierarchical methods confirms the results subject to theoretical and practical acceptance.

Stage 5: Interpretation of the Clusters

Information essential to the interpretation and profiling stages is provided in Table 9.8. For each cluster, the centroid on each of the seven clustering variables is provided along with the univariate F ratios and levels of significance comparing the differences between the cluster means. Whereas in stage four we examined the clusters for distinctiveness, we now consider the practical significance of the clusters in meeting the objectives of market segmentation. Which is a better segmentation of the market—two segments or four?

In evaluating the profiles on the clustering variables, we should remember some of the results from the factor analysis performed on these seven variables (see chapter 3). In defining the factors that are "dimensions" of variables that highly intercorrelate, two factors emerge. The first factor contains X_1, X_2, X_3, and X_7, and the second factor contains the two image items X_4 and X_6. In the first factor, X_2 and X_7 are inversely related to X_1 and X_3, meaning as one increases the other should decrease. For our purposes, this suggests that high scores on X_1 and X_3 should be associated with low scores on X_2 and X_7. Thus, to expect a cluster to

TABLE 9.8 Two-Cluster and Four-Cluster Solutions of the Nonhierarchical Cluster Analysis with Initial Seed Points from the Hierarchical Results

TWO-CLUSTER SOLUTION

	Mean Values							
Cluster	X_1 Delivery Speed	X_2 Price Level	X_3 Price Flexibility	X_4 Manufacturer Image	X_5 Overall Service	X_6 Salesforce Image	X_7 Product Quality	Cluster Size
Final cluster centers								
1	4.383	1.581	8.900	4.925	2.958	2.525	5.904	52
2	2.575	3.212	6.804	5.598	2.871	2.817	8.127	48
Statistical significance of cluster differences								
F value	87.72	86.75	133.18	9.60	.33	3.67	96.40	
Significance	.000	.000	.000	.003	.566	.058	.000	

	Profiling the Clusters			
	Cluster			
Predictive Validity	1	2	F Value	Significance
X_9 Usage level	49.212	42.729	14.789	.000
X_{10} Satisfaction	5.133	4.379	23.826	.000

FOUR-CLUSTER SOLUTION

	Mean Values							
Cluster	X_1 Delivery Speed	X_2 Price Level	X_3 Price Flexibility	X_4 Manufacturer Image	X_5 Overall Service	X_6 Salesforce Image	X_7 Product Quality	Cluster Size
Final cluster centers								
1	4.094	1.621	8.630	4.415	2.830	2.079	5.273	33
2	2.171	2.846	7.123	5.403	2.489	2.683	8.194	35
3	3.662	4.200	5.946	6.123	3.900	3.177	7.946	13
4	4.884	1.511	9.368	5.811	3.179	3.300	7.000	19
Statistical significance of cluster differences								
F Value	56.35	46.71	67.86	14.60	18.60	19.71	57.60	
Significance	.000	.000	.000	.000	.000	.000	.000	

	Profiling the Clusters					
	Cluster					
Predictive Validity	1	2	3	4	F Value	Significance
X_9 Usage level	46.333	41.229	46.769	54.211	11.304	.000
X_{10} Satisfaction	4.839	4.134	5.038	5.642	22.212	.000

be all high or all low is inappropriate, so that when we speak of a cluster being "high" on this set of variables, it means high scores for X_1 and X_3 and low scores for X_2 and X_7. Another possibility is to calculate composite scores for each dimension (see chapter 3 for more details), but in our example we profile only the individual variables.

Two-Cluster Solution

We start by examining the levels of significance for the differences across clusters and note that six of the seven are statistically significant (see Table 9.8). Only X_5, overall service, is not significantly different between the two clusters. In forming a profile of the clusters (see Figure 9.12), we see that cluster 1 is significantly higher on that first "dimension" of variables, whereas cluster 2 has higher perceptions of HATCO on the two image variables (X_4 and X_6). One should note, however, that the differences are much more distinctive on the first set of variables and that the image variables have substantially less delineation between the clusters, even though they are statistically significant. This should focus managerial attention on the four variables in the first set and relegate the image variables to a secondary role, although they are the key descriptors for the second cluster.

Four-Cluster Solution

The four-cluster solution actually represents the splitting of each cluster in the two-cluster solution. Cluster 1 is split into clusters 1 and 4, whereas cluster 2 splits into clusters 2 and 3. In formulating an interpretation of the profiles, several

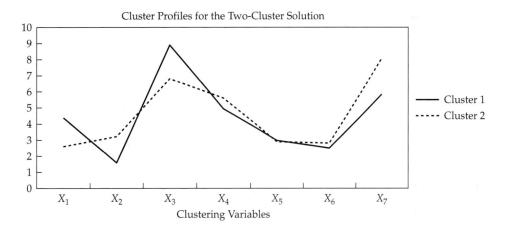

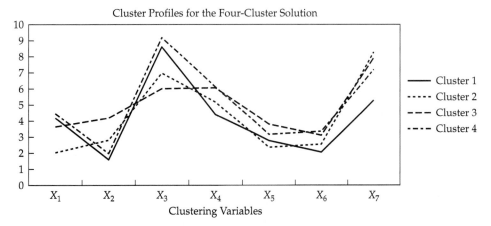

FIGURE 9.12 Graphical Profiles of the Two-Cluster and Four-Cluster Solution of the Nonhierarchical Cluster Analysis

patterns emerge. Because clusters 1 and 4 emerged from cluster 1 in the two-cluster solution, they should share at least some similar patterns. They do so on the first variable set (X_1, X_2, X_3, and X_7), maintaining their distinctiveness from clusters 2 and 3. Their profiles do vary slightly, with cluster 1 higher on X_2 and cluster 4 higher on X_1, X_3, and X_7. The marked difference comes on the image variables, where cluster 1 follows the pattern seen in the two-cluster solution with lower scores whereas cluster 4 counters this trend by having high positive image scores. This represents the most marked difference between clusters 1 and 4. Cluster 4 has relatively high perceptions of HATCO on all variables, whereas cluster 1 is high only on the first set of variables.

The remaining clusters, 2 and 3, follow the pattern of having relatively high scores on the image variables and low scores elsewhere. But they also differentiate themselves, as there is separation between them on the first set of variables, particularly X_1, X_2, and X_3. Identifying these patterns among the clustering variables allows the researcher to develop more in-depth profiles and develop tactics directed toward each group's specific perceptions.

Stage 6: *Validation and Profiling of the Clusters*

In this final stage, the processes of validation and profiling are critical due to the exploratory and often atheoretical basis for the cluster analysis. It is essential that the researcher perform all possible tests to confirm the validity of the cluster solution while also ensuring the cluster solution has practical significance. Researchers who minimize or skip this step are at risk for accepting a cluster solution that is specific only to the sample and has limited generalizability or even little use beyond its mere description of the data on the clustering variables.

Validation of the Cluster Solutions

The process of validation is accomplished in two steps. First, validity is assessed by applying alternative cluster methods and comparing the solutions. Then the clusters are assessed for predictive validity on two additional measures (X_9, usage level, and X_{10}, satisfaction level) that are indicative of the potential for differentiated strategies between the clusters (see Table 9.8).

Application of a Second Hierarchical Analysis As a first validity check for stability of the cluster solution, a second nonhierarchical analysis is performed, this time allowing the procedure to randomly select the initial seed points for both cluster solutions. The results in Table 9.9 confirm a consistency in the results for both cluster solutions. The cluster sizes are comparable for each solution, and the cluster profiles are very similar. Another approach, although not undertaken here, is to employ different similarity measures and linkage methods in the hierarchical analysis to assess the consistency of the results. But given the stability of the results between the specified seed points and random selection, management should feel confident that "true" differences do exist among customers in terms of their needs in a supplier and that the structure depicted in the cluster analysis is supported empirically.

Assessing Predictive Validity To assess predictive validity, we focus on variables that have a theoretically based relationship to the seven clustering variables, but were not included in the cluster solution. In this example, we consider variables X_9 (usage level) and X_{10} (satisfaction level) as shown in Table 9.8. These two variables have been shown to have a definite relationship with the clustering variables in prior mul-

TABLE 9.9 Clustering Variable Profiles for the Two-Cluster and Four-Cluster Solutions from the Nonhierarchical Cluster Analysis with Random Seed Points

CLUSTERING VARIABLE PROFILES
Clustering Variable Mean Values

Cluster	X_1 Delivery Speed	X_2 Price Level	X_3 Price Flexibility	X_4 Manufacturer Image	X_5 Overall Service	X_6 Salesforce Image	X_7 Product Quality	Cluster Size
Two-Cluster Solution								
Initial seed points								
1	4.100	.600	6.900	4.700	2.400	2.300	5.200	
2	1.800	3.000	6.300	6.600	2.500	4.000	8.400	
Final cluster solution								
1	4.383	1.581	8.900	4.925	2.958	2.525	5.904	52
2	2.575	3.212	6.804	5.598	2.871	2.817	8.127	48
Four-Cluster Solution								
Initial seed points								
1	4.100	.600	6.900	4.700	2.400	2.300	5.200	
2	1.800	3.000	6.300	6.600	2.500	4.000	8.400	
3	3.400	5.200	5.700	6.000	4.300	2.700	8.200	
4	2.700	1.000	7.100	5.900	1.800	2.300	7.800	
Final cluster solution								
1	4.144	1.579	8.635	4.418	2.835	2.088	5.315	34
2	2.024	2.766	6.941	5.162	2.366	2.555	8.269	29
3	3.614	4.129	5.950	6.064	3.843	3.164	7.950	14
4	4.404	1.943	9.183	6.087	3.165	3.352	7.187	23

tivariate techniques, notably canonical correlation and multiple regression. Given this relationship, we should see significant differences in these variables across the clusters. If significant differences do exist on these two variables, we can draw the conclusion that the clusters do depict groups that have predictive validity.

For the two-cluster solution, the univariate F ratios show that the cluster means for both variables are significantly different. The profiling process here shows that cluster 1 customers, who rated HATCO higher on the first set of variables, had higher levels of usage and satisfaction with HATCO, as would be expected. Likewise, cluster 2 customers had lower ratings on the first set of variables and lower ratings on the two additional variables. Although it is not the purpose of cluster analysis to identify the relationship between the clustering variables and the additional variables, the results here do support the relationships found with other multivariate techniques.

The four-cluster solution exhibits a similar pattern, with the clusters showing statistically significant differences on these two additional variables. Although it is beyond the scope of this analysis to relate the clustering variables to these usage and satisfaction levels, the patterns of differences among the clusters lend sufficient evidence to draw the conclusion that the four-cluster solution also has an adequate level of predictive validity.

Profiling the Cluster Solutions on Additional Variables

The final task is to profile the clusters on a set of additional variables not included in the clustering procedure or used to assess predictive validity. In this example, a number of characteristics of the HATCO customers are available. These include X_8 (firm size), X_{11} (specification buying), X_{12} (structure of procurement), X_{13} (industry type), and X_{14} (buying situation). Table 9.10 (p. 514) provides a descriptive

TABLE 9.10 Profile of the Two- and Four-Cluster Solutions on Additional Characteristics (X_8, X_{11}, X_{12}, X_{13}, and X_{14})

	Profiling the Clusters			
	Cluster		Pearson	
X_8 Firm Size	1	2	Chi-Square	Significance
Small	50	10		
Large	2	38	59.001	.000
X_{11} Specification Buying				
Specification Buying	2	38		
Total Value Analysis	50	10	59.001	.000
X_{12} Structure of Procurement				
Decentralized	50	0		
Centralized	2	48	92.308	.000
X_{13} Type of Industry				
SIC Category One	28	22		
SIC Category Two	24	26	.641	.423
X_{14} Type of Buying Situation				
New Task	10	24		
Modified Rebuy	10	22		
Straight Rebuy	32	2	36.634	.000

	Cluster				Pearson	
X_8 Firm Size	1	2	3	4	Chi-Square	Significance
Small	31	6	4	19		
Large	2	29	9	0	59.919	.000
X_{11} Specification Buying						
Specification Buying	2	29	9	0		
Total Value Analysis	31	6	4	19	59.919	.000
X_{12} Structure of Procurement						
Dencentralized	31	0	0	19		
Centralized	2	35	13	0	92.485	.000
X_{13} Type of Industry						
SIC Category One	23	17	5	5		
SIC Category Two	10	18	8	14	10.105	.018
X_{14} Type of Buying Situation						
New Task	8	23	1	2		
Modified Rebuy	8	10	12	2		
Straight Rebuy	17	2	0	15	62.189	.000

profile of both cluster solutions on these characteristics. As we can see, both the two-cluster and four-cluster solutions have distinctive profiles on this set of additional variables. Only X_{13} shows no difference across the two-cluster solution, with all other variables having significant differences for the two-cluster and four-cluster solutions. For example, cluster 1 in the two-cluster solution can be characterized as small firms engaged in straight rebuys with HATCO that primarily use total value analysis in a decentralized procurement system. Similar profiles can be developed for each cluster.

The importance of identifying unique profiles on these sets of additional variables is in assessing both their practical significance and theoretical basis. In assessing practical significance, the researcher may require that the clusters exhibit differences of a set of additional variables. In our example, a successful segmentation analysis not only requires the identification of the homogeneous groups (clusters), but that they also be identifiable (uniquely described on other variables). When cluster analysis is used to verify a typology or other proposed grouping of objects, associated variables, either antecedents or outcomes, may be profiled to ensure correspondence of the groupings within a larger theoretical model.

A Managerial Overview

The cluster analysis of the 100 HATCO customers was successful in performing a market segmentation of HATCO customers. It not only created homogeneous groupings of customers based on their perceptions of HATCO, but also found that these clusters met the tests of predictive validity and distinctiveness on additional sets of variables, all necessary for achieving practical significance. The segments represent quite different customer perspectives of HATCO, varying in both the types of variables that are viewed most positively as well as the magnitude of the perceptions.

One issue unresolved to this point is the selection of the final cluster solution between the two- and four-clusters. Both cluster solutions provide viable market segments, as they are of substantial size and do not have any small cluster sizes caused by outliers. Moreover, they meet all of the criteria for a successful market segmentation. In either instance, the clusters (market segments) represent sets of consumers with homogeneous perceptions that can be uniquely identified, thus being prime candidates for differentiated marketing programs. The researcher can utilize the two-cluster solution to provide a basic delineation of customers that vary in perceptions and buying behavior, or employ the four-cluster solution for a more complex segmentation strategy that provides a highly differentiated mix of customer perceptions as well as targeting options.

Summary

Cluster analysis can be a very useful data-reduction technique. But because its application is more an art than a science, it can easily be abused or misapplied by researchers. Different interobject measures and different algorithms can and do affect the results. The selection of the final cluster solution in most cases is based on both objective and subjective considerations. The prudent researcher, therefore, considers these issues and always assesses the impact of all decisions.

Cluster analysis, along with multidimensional scaling, due to their lack of a statistical basis for inference to the population, are most in need of replication under varying conditions. If the researcher proceeds cautiously, however, cluster analysis can be an invaluable tool in identifying latent patterns by suggesting useful groupings (clusters) of objects that are not discernible through other multivariate techniques.

Questions

1. What are the basic stages in the application of cluster analysis?
2. What is the purpose of cluster analysis, and when should it be used instead of factor analysis?
3. What should the researcher consider when selecting a similarity measure to use in cluster analysis?
4. How does the researcher know whether to use hierarchical or nonhierarchical cluster techniques? Under which conditions would each approach be used?
5. How does a researcher decide the number of clusters to have in your solution?
6. What is the difference between the interpretation stage and the profiling stage?
7. How do researchers use the graphical portrayals of the cluster procedure?

References

1. Aldenderfer, Mark S., and Roger K. Blashfield (1984), *Cluster Analysis.* Thousand Oaks, Calif.: Sage Publications.
2. Anderberg, M. (1973), *Cluster Analysis for Applications.* New York: Academic Press.
3. Bailey, Kenneth D. (1994), *Typologies and Taxonomies: An Introduction to Classification Techniques.* Thousand Oaks, Calif.: Sage Publications.
4. Everitt, B. (1980), *Cluster Analysis,* 2d ed. New York: Halsted Press.
5. Green, P. E. (1978), *Analyzing Multivariate Data.* Hinsdale, Ill.: Holt, Rinehart & Winston.
6. Green, P. E., and J. Douglas Carroll (1978), *Mathematical Tools for Applied Multivariate Analysis.* New York: Academic Press.
7. McIntyre, R. M., and R. K. Blashfield (1980), "A Nearest-Centroid Technique for Evaluating the Minimum-Variance Clustering Procedure." *Multivariate Behavioral Research* 15: 225–38.
8. Milligan, G. (1980), "An Examination of the Effect of Six Types of Error Perturbation on Fifteen Clustering Algorithms." *Psychometrika* 45 (September): 325–42.
9. Milligan, Glenn W., and Martha C. Cooper (1985), "An Examination of Procedures for Determining the Number of Clusters in a Data Set." *Psychometrika* 50(2): 159–79.
10. Morrison, D. (1967), "Measurement Problems in Cluster Analysis." *Management Science* 13 (August): 775–80.
11. Overall, J. (1964), "Note on Multivariate Methods for Profile Analysis." *Psychological Bulletin* 61(3): 195–98.
12. Punj, G., and D. Stewart (1983), "Cluster Analysis in Marketing Research: Review and Suggestions for Application." *Journal of Marketing Research* 20 (May): 134–48.
13. Rohlf, F. J. (1970), "Adaptive Hierarchical Clustering Schemes." *Systematic Zoology* 19: 58.
14. Schaninger, C. M., and W. C. Bass (1986), "Removing Response-Style Effects in Attribute-Determinance Ratings to Identify Market Segments." *Journal of Business Research* 14: 237–52.
15. Shephard, R. (1966), "Metric Structures in Ordinal Data." *Journal of Mathematical Psychology* 3: 287–315.
16. Sneath, P. H. A., and R. R. Sokal (1973), *Numerical Taxonomy.* San Francisco: Freeman Press.

Annotated Articles

The following annotated articles are provided as illustrations of the application of cluster analysis to substantive research questions of both a conceptual and managerial nature. The reader is encouraged to review the complete articles for greater detail on any of the specific issues regarding methodology or findings.

Singh, Jagdip (1990), "A Typology of Consumer Dissatisfaction Response Styles," *Journal of Retailing* 66(1): 57–99.

The author utilized cluster analysis, as well as factor analysis and discriminant analysis, to develop a categorization system of consumer complaint behavior (CCB) styles. The research design closely follows the recommended six-step process in the text. The author identifies three dimensions of complaint intentions/behaviors that differ based on the type of response (actions directed at the seller, negative word-of-mouth, or complaints to third parties). Based on these three complaint-intentioned behaviors, the author uses cluster analysis to develop groups of similar individuals. Testing whether the response styles would reproduce differences in actual behavior is offered as support for the result's validity. Finally, a number of demographic, personality/attitudinal, and situational variables—identified in prior research as important to understanding consumer complaints—are used to profile the CCB styles.

To form the clusters, the author splits the sample in two using the analysis sample to derive the number of clusters and their centroids and the holdout sample to confirm the stability of the clusters. Using the initial centroids obtained from Ward's clustering method and the Euclidean measurement of distance, a K-means nonhierarchical approach is used. This results in four unique styles of consumer groups: (1) no-action, (2) voice actions only, (3) voice and private actions, and (4) voice, private, and third-party actions. This study uses a coefficient of agreement between the two subsamples to determine the optimal number of clusters. The validity of these results is supported by actual behaviors correctly identified by response styles. The final step involves interpretation and profiling of the cluster membership.

Additionally, the author uses multiple discriminant analysis to determine the relative importance of each of the factor and demographic variables for each of the clusters. This study extends previous research and demonstrates the multifaceted nature of complaint styles. Such findings should be of interest to retail managers by increasing knowledge about customers and improving the handling of customer complaints.

Larwood, Laurie, Cecilia M. Falbe, Mark P. Kriger, and Paul Miesing (1995), "Structure and Meaning of Organizational Vision," *Academy of Management Journal* 38(3): 740–769.

In this application, cluster analysis is used for confirmatory purposes. The authors offer a theoretically derived organization value structure that is then compared to the cluster analysis results. The proposed theoretical four-value structure is based on the nature of organization's value systems, specifically, the four-value structures—elite, leadership, meritocratic, or collegial—were created based on how the organization addresses the problem of distributive justice (the balance of equality and equity concerns). To measure the presence of a prespecified set of nine individual values, the authors performed a content analysis on the organizational documents of 88 organizations over a period of five years. The organizations are clustered based on the nine espoused values (affiliation, authority, commitment, leadership, normative, participation, performance, reward, and teamwork) to form value profiles that are both distinct and interpretable. To ensure the result's validity, the authors compared perceptions of organizational change across the four types of value structures to assess whether differences were consistent with the proposed theoretical interpretation.

A content analysis of organizational documents was done to identify the frequency with which each of the nine values occurred to create a database. For the cluster analysis, the authors specified the initial seed points for each of the nine clustering variables for each of the four-value structures based on their theoretical profiles. Using a squared Euclidean distance measure, organizations were assigned to the nearest cluster that caused the centroids to change based on the average of the value profiles selected. The stability of the cluster solution was examined by repeating the analysis in several different ways and comparing the cluster membership from each cluster procedure. Using MANOVA, the authors established that significant differences exist between the clusters. They also achieve good predictive results by correctly classifying group membership using discriminant analysis. As a final validity test, the authors established that different organizational change themes were fairly consistent with the theoretical value profiles. From the cluster analysis results, the authors are able to confirm the existence of organizational value structures and their consistency with value-related processes, like organizational change.

Kabanoff, Boris, Robert Waldersee, and Marcus Cohen (1995), "Espoused Values and Organizational Change Themes," *Academy of Management Journal* 38(4): 1075–1104.

Through clustering techniques, this study explores the content and structure of organizational visions of top business executives. Executives were asked to articulate their visions and to evaluate them using a 26-item self-evaluation list. Factor analysis of the self-evaluation items reveals seven factors: vision formulation, implementation, innovative realism, general, detailed, risk taking, and profit-oriented. The authors use the 26 items that created the seven factors as clustering variables. The resultant three-cluster solution allows the researchers to examine whether relationships exist between patterns of vision and individual and organizational characteristics. From the findings, the study enhances understanding of the structure of organizational vision, and whether these patterns of vision are related to organizational and individual differences.

In summary, a three-cluster solution was formed using all 26 items on a sample of 331 respondents. A fourth cluster was excluded because it contained only one organization. Each cluster is then profiled based on mean differences in the self-evaluation items. From the results, the authors determined whether relationships exist between cluster membership and individual and organizational characteristics. Comparing the clusters to organizational characteristics (industry type and size) revealed no significant difference; however, membership did differ based on individual characteristics (length of vision horizon, perception of firm and industry change, and perceptions of and need for control). As an extension to their findings, the authors contrasted the current results with those obtained from a previous study of business school deans. Although organizational differences were not apparent, significant differences between executive and business school dean self-evaluations of their respective organizational visions warrants caution as to the universality of the concept of vision.

Furse, David H., Girish N. Punj, and David W. Stewart (1984), "A Typology of Individual Search Strategies among Purchasers of New Automobiles," *Journal of Consumer Research* 10(4): 417–431.

Using cluster analysis, this study seeks to identify a typology of new car buyers based on aspects of the purchaser's search activities. The authors identify 24 search activities that, through factor analysis, are reduced to five dimensions. The dimensions are dealership visits, level of personal participation, participation and involvement of others, interpersonal search factors, and the amount of search activity. Using these five factors, the cluster technique reveals six groups. The results are then validated using data from sales personnel and a proposed theoretical framework.

Data are collected from a sample of 1,031 recent car purchasers, which is divided into two groups. From the first sample, the authors use Ward's hierarchical clustering method with Euclidean distance to obtain the initial seed points and number of clusters. The holdout sample, using K-means nonhierarchical clustering procedure, is then used to confirm the stability of the hierarchical cluster solution. Although other methods are available to determine the optimal numbers of cluster, this study used a coefficient of agreement (Kappa) between the two subsamples to determine the appropriate number. Based on a variety of descriptive variables, the six-cluster solution is interpreted and comparisons made among the clusters. To validate the findings, the authors collect similar information from 48 salespersons. Using a similar procedure, the cluster analysis reveals six similar groups based on seller perceptions of buyer search strategies. These results are then related to current theory in consumer research.

CHAPTER **10**

Multidimensional Scaling

LEARNING OBJECTIVES

Upon completing this chapter, you should be able to do the following:

- Appreciate how spatial representation of data can clarify underlying relationships.
- Determine the number of dimensions represented in the data.
- Interpret the spatial maps so the dimensions can be understood.
- Understand the differences between similarity data and preference data as used in multidimensional scaling.
- Distinguish between the various approaches to applying multidimensional scaling.

CHAPTER PREVIEW

Multidimensional scaling (MDS) refers to a series of techniques that help the researcher to identify key dimensions underlying respondents' evaluations of objects. For example, multidimensional scaling is often used in marketing to identify key dimensions underlying customer evaluations of products, services, or companies. Other common applications include the comparison of physical qualities (e.g., food tastes or various smells), perceptions of political candidates or issues, and even the assessment of cultural differences between distinct groups. Multidimensional scaling techniques can infer the underlying dimensions from a series of similarity or preference judgments provided by respondents about objects. Once the data are in hand, multidimensional scaling can help determine (1) what dimensions respondents use when evaluating objects, (2) how many dimensions they may use in a particular situation, (3) the relative importance of each dimension, and (4) how the objects are related perceptually.

KEY TERMS

Before starting the chapter, review the key terms to develop an understanding of the concepts and terminology used. Throughout the chapter the key terms appear in **boldface**. Other points of emphasis in the chapter are *italicized*. Also, cross-references within the Key Terms appear in *italics.*

Aggregate analysis   Approach to MDS in which a *perceptual map* is generated for a group of respondents' evaluations of *objects.* This composite perceptual map may be created by a computer program or by the researcher to find a few "average" or representative subjects.

Chi-square   Method of standardizing data in a *contingency table* by comparing the actual cell frequencies to an expected cell frequency. The expected cell frequency is based on the marginal probabilities of its row and column (probability of a row and column among all rows and columns).

Compositional method   Alternative approach to the more traditional *decompositional* methods of perceptual mapping that derive overall *similarity* or *preference* evaluations from evaluations of separate attributes by each respondent. These separate attribute evaluations are combined (composed) for an overall evaluation. The most common examples of compositional methods are the techniques of factor analysis and discriminant analysis.

Confusion data   Procedure to obtain respondents' perceptions of *similarities data.* Respondents indicate the similarities between pairs of stimuli. The pairing (or "confusing") of one stimulus with another is taken to indicate similarity. Also known as *subjective clustering.*

Contingency table   Cross-tabulation of two nonmetric or categorical variables in which the entries are the frequencies of responses that fall into each "cell" of the matrix. For example, if three brands were rated on four attributes, the brand-by-attribute contingency table would be a three-row by four-column table. The entries would be the number of times a brand was rated as having an attribute.

Correspondence analysis   *Compositional approach* to perceptual mapping that relates categories of a *contingency table.* Most applications involve a set of *objects* and attributes, with the results portraying both objects and attributes in a common *perceptual map.*

Cross-tabulation table   See *contingency table.*

Decompositional method   Perceptual mapping method associated with MDS techniques in which the respondent provides only an overall evaluation of *similarity* or *preference* between *objects.* This set of overall evaluations is then "decomposed" into a set of "dimensions" that best represent the objects' differences.

Degenerate solution   MDS solution that is invalid because of (1) inconsistencies in the data or (2) too few objects compared with the dimensionality of the solution. Even though the computer program may indicate a valid solution, the researcher should disregard the degenerate solution and examine the data for the cause. This type of solution is typically portrayed as a circular pattern with illogical results.

Derived measures   Procedure to obtain respondents' perceptions of *similarities data.* Derived similarities are typically based on a series of "scores" given to stimuli by respondents, which are then combined in some manner. The semantic differential scale is frequently used to elicit such "scores."

Dimensions   Features of an *object*. A particular object can be thought of as possessing both *perceived*, or *subjective*, dimensions (e.g., expensive, fragile) and *objective* dimensions (e.g., color, price, features).

Disaggregate analysis   Approach to MDS in which the researcher generates *perceptual maps* on a respondent-by-respondent basis. The results may be difficult to generalize across respondents. Therefore, the researcher may attempt to create fewer maps by some process of *aggregate analysis*, in which the results of respondents are combined.

Disparities   Differences in the computer-generated distances representing *similarity* and the distances provided by the respondent.

Ideal point   Point on a perceptual map that represents the most preferred combination of perceived attributes (according to the respondents). A major assumption is that the position of the ideal point (relative to the other objects on the perceptual map) would define relative *preference* such that objects farther from the ideal point should be preferred less.

Importance–performance grid   Two-dimensional approach for assisting the researcher in labeling dimensions. The first axis is the respondents' perceptions of the importance (e.g., as measured on a scale of "extremely important" to "not at all important"). The opposing axis is performance (e.g., as measured on a scale of "highly likely to perform" to "highly unlikely to perform") for each brand, or product or service, on various attributes. Each object is represented by its values on importance and performance.

Index of fit   Squared correlation index (R^2) that may be interpreted as indicating the proportion of variance of the *disparities* (optimally scaled data) that can be accounted for by the MDS procedure. It measures how well the raw data fit the MDS model. This index is an alternative to the *stress measure* for determining the number of dimensions. Similar to measures of covariance in other multivariate techniques, measures of .60 or greater are considered acceptable.

Initial dimensionality   A starting point in selecting the best spatial configuration for data. Before beginning an MDS procedure, the researcher must specify how many *dimensions* or features are represented in the data.

Multiple correspondence analysis   Form of *correspondence analysis* that involves three or more categorical variables related in a common perceptual space.

Object   Any stimulus, including tangible entities (product or physical object), actions (service), sensory perceptions (smell, taste, sights), or even thoughts (ideas, slogans), that can be compared and evaluated by the respondent.

Objective dimension   Physical or tangible characteristics of an *object* that have an objective basis of comparison. For example, a product has size, shape, color, weight, and so on.

Perceived dimension   A respondent's subjective attachment of features to an *object* that represents its intangible characteristics. Examples include "quality," "expensive," and "good-looking." These perceived dimensions are unique to the individual respondent and may bear little correspondence to actual *objective dimensions.*

Perceptual map   Visual representation of a respondent's perceptions of *objects* on two or more dimensions. Usually this map has opposite levels of *dimensions* on the ends of the X- and Y-axes, such as "sweet" to "sour" on the ends of the X-axis and "high-priced" to "low-priced" on the ends of the Y-axis. Each *object* then has a spatial position in the perceptual map that reflects the relative *similarity* or *preference* to other objects with regard to the dimensions of the perceptual map.

Preference   Implies that *objects* are judged by the respondent in terms of dominance relationships; that is, the stimuli are ordered in preference with respect to some property. Direct ranking, paired comparisons, and preference scales are frequently used to determine respondent preferences.

Projections   Points defined by perpendicular lines from an object to a vector. Projections are used in determining the *preference* order with *vector* representations.

Similarities data   Data used to determine which *objects* are the most similar to each other and which are the most dissimilar. Implicit in similarities measurement is the ability to compare all pairs of objects. Three procedures to obtain similarities data are paired comparison of objects, *confusion data*, and *derived measures*.

Similarity   See *similarities data*.

Similarity scale   Arbitrary scale, for example, from −5 to +5 that allows the representation of an ordered relationship between objects from the most similar (closest) to the least similar (farthest apart). This type of scale is appropriate only for representing a single dimension.

Spatial map   See *perceptual map*.

Stress measure   Proportion of the variance of the *disparities* (optimally scaled data) not accounted for by the MDS model. This type of measurement varies according to the type of program and the data being analyzed. The stress measure helps to determine the appropriate number of *dimensions* to include in the model.

Subjective clustering   See *confusion data*.

Subjective dimension   See *perceived dimension*.

Subjective evaluation   Method of determining how many *dimensions* are represented in the MDS model. The researcher makes a "subjective inspection" of the spatial maps and asks whether the configuration looks reasonable. The objective is to obtain the best fit with the least number of dimensions.

Unfolding   Representation of an individual respondent's *preferences* within a common (aggregate) stimulus space derived for all respondents as a whole. The individual's preferences are "unfolded" and portrayed as the best possible representation within the aggregate analysis.

Vector   Method of portraying an ideal point or attribute in a perceptual map. Involves the use of *projections* to determine an *object's* order on the vector.

What Is Multidimensional Scaling?

Multidimensional scaling (MDS), also known as perceptual mapping, is a procedure that allows a researcher to determine the perceived relative image of a set of objects (firms, products, ideas, or other items associated with commonly held perceptions). The purpose of MDS is to transform consumer judgments of similarity or preference (e.g., preference for stores or brands) into distances represented in multidimensional space. Assume that objects A and B are judged by respondents to be the most similar compared with all other possible pairs of objects. MDS techniques will position objects A and B so that the distance between them in multidimensional space is smaller than the distance between any other two pairs of objects. The resulting **perceptual map,** also known as a **spatial map,** shows the relative positioning of all objects, as shown in Figure 10.1.

Multidimensional scaling is based on the comparison of **objects.** Any object (e.g., product, service, image, aroma) can be thought of as having both perceived and objective **dimensions.** For example, HATCO's management may see their product (a lawn mower) as having two color options (red and green), a two-horse-power motor, and a 24-inch blade. These are the **objective dimensions.** On the other hand, customers may (or may not) see these attributes. Customers may also perceive the HATCO mower as expensive-looking or fragile. These are **perceived dimensions,** also known as **subjective dimensions.** Two products may have the same physical characteristics (objective dimensions) but be viewed differently because the different brands are perceived to differ in quality (a perceived dimension) by many customers. Thus, the following two differences between objective and perceptual dimensions are very important:

1. *Individual Differences:* The dimensions perceived by customers may not coincide with (or may not even include) the objective dimensions assumed by the researcher. We expect that each individual may have different perceived dimensions, but the researcher must also accept that the objective dimensions may also vary substantially. Individuals may consider different sets of objective characteristics as well as vary the importance they attach to each dimension.
2. *Interdependence:* The evaluations of the dimensions (even if the perceived dimensions are the same as the objective dimensions) may not be independent and may not agree. Both perceived and objective dimensions may interact with one another to create unexpected evaluations. For example, one soft drink may

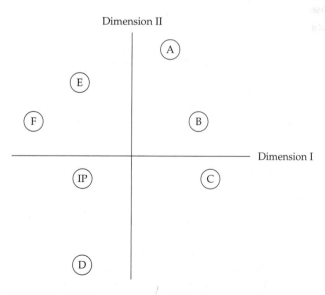

FIGURE 10.1 Illustration of a Multidimensional "Map" of Perceptions of Six Industrial Suppliers (A to F) and the Ideal Point (IP)

be judged sweeter than another because the first has a fruitier aroma, although both contain the same amount of sugar.

The challenge to the researcher is first to understand the perceived dimensions and then to relate them to objective dimensions, if possible. Additional analysis is needed to assess which attributes predict the position of each object in both perceptual and objective space.

A note of caution must be raised, however, concerning the interpretation of dimensions. Because this process is more an art than a science, the researcher must resist the temptation to allow personal perception to affect the qualitative dimensionality of the perceived dimensions. Given the level of researcher input, caution must be taken to be as objective as possible in this critical, yet still rudimentary, area.

A Simplified Look at How MDS Works

To facilitate a better understanding of the basic procedures in multidimensional scaling, we first present a simple example to illustrate the basic concepts underlying MDS and the procedure by which it transforms similarity judgments into the corresponding spatial positions.

Market researchers are interested in understanding consumers' perceptions of six candy bars that are currently on the market. Instead of trying to gather information about consumer's evaluations of the candy bars on a number of attributes, the researchers will instead gather only perceptions of overall similarities or dissimilarities. The data are typically gathered by having respondents give simple global responses to statements such as these:

- Rate the similarity of products A and B on a 10-point scale.
- Product A is more similar to B than to C.
- I like product A better than product B.

From these simple responses, a perceptual map can be drawn that best portrays the overall pattern of similarities among the six candy bars. We will illustrate the process of creating a perceptual map with the data from a single respondent, although this process could also be applied to multiple respondents or to the aggregate responses of a group of consumers.

The data are gathered by first creating a set of 15 unique pairs of the six candy bars ($6 \times 5/2 = 15$ pairs). Respondents are then asked to rank the following 15 candy bar pairs, where a rank of 1 is assigned to the pair of candy bars that is most similar and a rank of 15 indicates the pair that is least alike. The results (rank orders) for all pairs of candy bars for one respondent are shown in Table 10.1.

The respondent whose ranks are shown in Table 10.1 thought that candy bars D and E were the most similar, candy bars A and B were the next most similar, and so forth until candy bars E and F were the least similar. If we want to illustrate the similarity among candy bars graphically, a first attempt would be to draw a single **similarity scale** and fit all the candy bars to it. This represents a one-dimensional portrayal of similarity, with distance representing similarity. Thus, objects closer together on the scale are more similar and those farther away less similar. The objective is to position the candy bars on the scale so that the rank

TABLE 10.1 Similarity Data (Rank Orders) for Candy Bar Pairs

Candy Bar	A	B	C	D	E	F
A	—	2	13	4	3	8
B		—	12	6	5	7
C			—	9	10	11
D				—	1	14
E					—	15
F						—

Note: Lower values indicate greater similarity; 1 indicates the most similar pair and 15 the least similar pair.

orders are best represented (rank order of 1 is closest, rank order of 2 is next closest, and so forth).

Let us try to see how we would place some of the objects. Positioning two or three candy bars is fairly simple; the first real test comes with four objects. We choose candy bars A, B, C, and D. Table 10.1 shows that the rank order of the pairs is as follows: $\overline{AB} < \overline{AD} < \overline{BD} < \overline{CD} < \overline{BC} < \overline{AC}$ (the line over each pair of letters indicates that the expression refers to the distance [similarity] between the pair). From these values, we must place the four candy bars on a single scale so that the most similar ($\overline{AB}$) are the closest and the least similar ($\overline{AC}$) are the farthest apart. Figure 10.2 (p. 526, part A) contains a one-dimensional perceptual map that matches the orders of pairs described above. If the person judging the similarity between the candy bars had been thinking of a simple rule of similarity that involved only one attribute (dimension), such as amount of chocolate, then all the pairs could be placed on a single scale that reproduces the similarity values.

Although a one-dimensional map can be accomplished with four objects, the task becomes increasingly difficult as the number of objects increases. The interested reader is encouraged to attempt this task with six objects. When a single dimension is employed with the six objects, the actual ordering varies substantially from the respondent's original rank orders. Because one-dimensional scaling does not fit the data well, a two-dimensional solution should be attempted. This would allow for another scale (dimension) to be used in configuring the six candy bars. The procedure is quite tedious to attempt by hand; the two-dimensional solution produced by an MDS program is shown in Figure 10.2 (part B). This configuration matches the rank orders of Table 10.1 exactly, supporting the notion that the respondent most probably used two dimensions in evaluating the candy bars. The conjecture that at least two attributes (dimensions) were considered is based on the inability to represent the respondent's perceptions in one dimension. However, we are still not aware of what attributes the respondent used in this evaluation.

Although we have no information as to what these dimensions are, we may be able to look at the relative positions of the candy bars and infer what attribute(s) the dimensions represent. For example, suppose that candy bars A, B, and F were a form of combination bar (e.g., chocolate and peanuts, chocolate and peanut butter), and C, D, and E were strictly chocolate bars. We could then infer that dimension I represents the type of candy bar (chocolate versus combination). When we look at the position of the candy bars on the vertical dimension, other attributes may emerge as the descriptors of that dimension as well.

A. One-Dimensional Perceptual Map of Four Observations

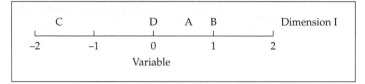

B. Two-Dimensional Perceptual Map of Six Observations

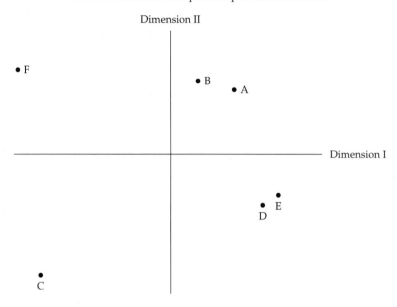

FIGURE 10.2 One- and Two-Dimensional Perceptual Maps

MDS allows researchers to understand the similarity between the six candy bars by asking only for overall similarity perceptions. The procedure may also assist in determining which attributes actually enter into the perceptions of similarity. Although we do not directly incorporate the attribute evaluations into the MDS procedure, we can use them in subsequent analyses to assist in interpreting the dimensions and the impacts each attribute has on the relative positions of the candy bars.

Comparing MDS to Other Interdependence Techniques

Multidimensional scaling can be compared to the other interdependence techniques such as factor and cluster analysis based on its approach to defining structure. Factor analysis groups variables into variates that define underlying dimensions in the original set of variables. Variables that highly correlate are grouped together. Cluster analysis groups observations according to their profile on a set of variables (the cluster variate) in which observations in close proximity to each other are grouped together. MDS differs from cluster analysis in two key aspects: (1) a solution can be obtained for each individual, and (2) it does not use a variate.

Individual as the Unit of Analysis

In MDS, each respondent provides evaluations of all objects being considered, so that a solution can be obtained for each individual that is not possible in cluster analysis or factor analysis. As such, the focus is not on the objects themselves but instead on how the individual perceives the objects. The structure being defined is the perceptual dimensions of comparison for the individual(s). Once the perceptual dimensions are defined, the relative comparisons among objects can also be made.

Lack of a Variate

Multidimensional scaling, unlike the other multivariate techniques, does not use a variate. Instead, the "variables" that would make up the variate (i.e., the perceptual dimensions of comparison) are inferred from global measures of similarity among the objects. In a simple analogy, this is like providing the dependent variable (similarity among objects) and figuring out what the independent variables (perceptual dimensions) must be. MDS has the advantage of reducing the influence of the researcher by not requiring the specification of the variables to be used in comparing objects, as was required in cluster analysis. But it also has the disadvantage that the researcher is not really sure what variables the respondent is using to make the comparisons.

A Decision Framework for Perceptual Mapping

Perceptual mapping encompasses a wide range of possible methods, including MDS, and all these techniques can be viewed through the model-building process introduced in chapter 1. These steps represent a decision framework, depicted in Figure 10.3 (stages 1–3, p. 528) and 10.4 (stages 4–6, p. 538) within which all perceptual mapping techniques can be applied and the results evaluated.

Stage 1: Objectives of MDS

Perceptual mapping, and MDS in particular, is most appropriate for achieving two objectives:

1. as an exploratory technique to identify unrecognized dimensions affecting behavior
2. as a means of obtaining comparative evaluations of objects when the specific bases of comparison are unknown or undefined

In MDS, it is not necessary to specify the attributes of comparison for the respondent. All that is required is to specify the objects and make sure that the objects share a common basis of comparison. This flexibility makes MDS particularly

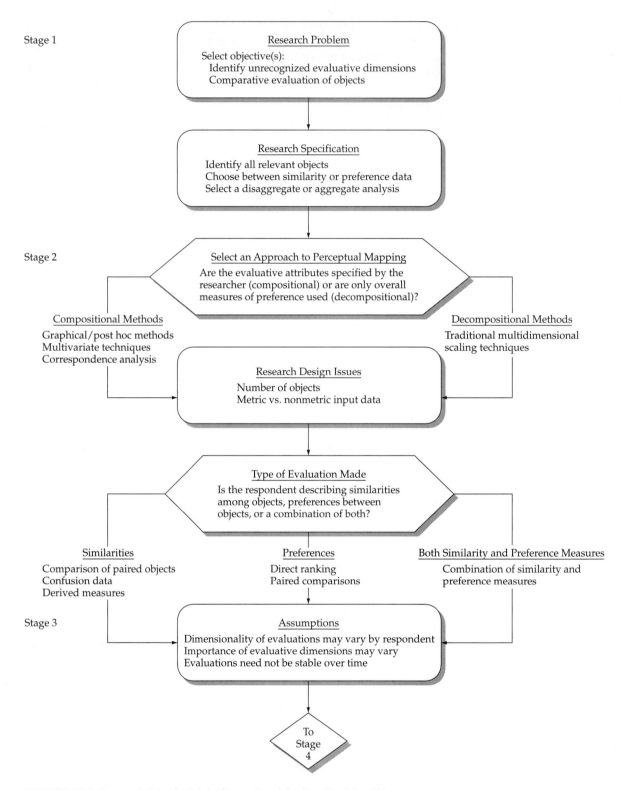

FIGURE 10.3 Stages 1–3 in the Multidimensional Scaling Decision Diagram

suited to image and positioning studies in which the dimensions of evaluation may be too global or too emotional and affective to be measured by conventional scales. Some MDS methods combine the positioning of objects and subjects in a single overall map. In these techniques, the relative positions of objects and consumers make segmentation analysis much more direct.

Key Decisions in Setting Objectives

A common characteristic of each objective is the lack of specificity in defining the standards of evaluation of the objects. The strength of perceptual mapping is its ability to "infer" dimensions without the need for defined attributes. The flexibility and inferential nature of MDS places a greater responsibility on the researcher to "correctly" define the analysis. Conceptual as well as practical considerations are essential for MDS to achieve its best results. To ensure this success, the researcher must define an MDS analysis through three key decisions: selecting the objects that will be evaluated, deciding whether similarities or preferences are to be analyzed, and choosing whether the analysis will be performed at the group or individual level.

Identification of All Relevant Objects to Be Evaluated

The most basic, but important, issue in perceptual mapping is the defining of the objects to be evaluated. The researcher must ensure that all "relevant" firms, products, services, or other objects are included, because perceptual mapping is a technique of relative positioning. Relevancy is determined by the research questions to be addressed. The perceptual maps resulting from any of the methods can be greatly influenced by either the omission of objects or the inclusion of inappropriate ones [7, 20]. If "irrelevant" or noncomparable objects are included, the researcher is forcing the technique not only to infer the perceptual dimensions that distinguish among comparable objects but also to infer those dimensions that distinguish among noncomparable objects as well. This task is beyond the scope of MDS and results in a solution that addresses neither question well.

Similarities versus Preference Data

Having selected the objects for study, the researcher must next select the basis of evaluation: similarity versus preference. To this point, we have discussed perceptual mapping and MDS mainly in terms of similarity judgments. In providing **similarities data,** the respondent does not apply any "good–bad" aspects of evaluation in the comparison. The assessment "good–bad" is done, however, within **preference** data, which assumes that differing combinations of perceived attributes are valued more highly than other combinations. Both bases of comparison can be used to develop perceptual maps, but with differing interpretations. Similarity-based perceptual maps represent attribute similarities and perceptual dimensions of comparison but do not reflect any direct insight into the determinants of choice. Preference-based perceptual maps do reflect preferred choices but may not correspond in any way to the similarity-based positions, because respondents may base their choices on entirely different dimensions or criteria from those on which they base comparisons. There is no optimal base for evaluation, but the decision between similarities and preference data must be made with the ultimate research question in mind, because they are fundamentally different in what they represent.

Aggregate versus Disaggregate Analysis

In considering similarities or preference data, we are taking respondent's perceptions of stimuli and creating outputs of representations of stimulus proximity in t-dimensional space (where the number of dimensions t is less than the number of stimuli). The researcher can generate this output on a subject-by-subject basis (producing as many maps as subjects), a method known as a **disaggregate analysis.** One of the distinctive characteristics of MDS techniques is their ability to estimate solutions for each respondent, which can then be represented separately. The advantage is the representation of the unique elements of each respondent's perceptions. The disadvantage is that the researcher must identify the common elements across respondents.

MDS techniques can also combine respondents and create fewer perceptual maps by some process of **aggregate analysis.** The aggregation may take place either before or after scaling the subjects' data. Before scaling, the simplest approach is for the researcher to find the "average" evaluations for all respondents and obtain a single solution for the group of respondents as a whole. To identify groups of similar respondents, the researcher may cluster analyze the subjects' responses to find a few average or representative subjects and then develop maps for the cluster's "average respondent." Alternatively, the researcher may develop maps for each individual and cluster the maps according to the coordinates of the stimuli on the maps. It is recommended that "average" evaluations be used rather than clustering the individual perceptual maps because minor rotations of essentially the same map can cause problems in creating reasonable clusters by the second approach.

A specialized form of disaggregate analysis is available with INDSCAL (individual differences scaling) [4] and its variants, which have characteristics of both disaggregate and aggregate analyses. INDSCAL assumes that all individuals share a common or group space (an aggregate solution) but that the respondents individually weight the dimensions, including zero weights when totally ignoring a dimension. As a first step, INDSCAL derives the perceptual space shared by all individuals, just as do other aggregate solutions. However, individuals are also portrayed in a special group space map. Here the respondents' position is determined by their weights for each dimension. Respondents positioned closely together employ similar combinations of the dimensions from the common group space. Moreover, the distance of the individual from the origin is an approximate measure of the proportion of variance for that subject accounted for by the solution. Thus a position farther from the origin indicates better fit. Being at the origin means "no fit" because all weights are zeros. If two or more subjects or groups of subjects are at the origin, separate group spaces need to be configured for each of them. In this analysis, the researcher is presented with not only an overall representation of the perceptual map but also the degree to which each respondent is represented by the overall perceptual map. These results for each respondent can then be used to group respondents and even identify different perceptual maps in subsequent analyses.

The choice of aggregate or disaggregate analysis is based on the study objectives. If the focus is on an understanding of the overall evaluations of objects and the dimensions employed in those evaluations, an aggregate analysis is the most suitable. But if the objective is to understand variation among individuals, then a disaggregate approach is the most helpful.

Stage 2: Research Design of MDS

Although MDS looks quite simple computationally, the results, as with other multivariate techniques, are heavily influenced by a number of key issues that must be resolved before the research can proceed. We cover four of the major issues, ranging from discussions of research design (selecting the approach and objects or stimuli for study) to specific methodological concerns (metric versus nonmetric methods) and data collection methods.

Selection of Either a Decompositional (Attribute-Free) or Compositional (Attribute-Based) Approach

Perceptual mapping techniques can be classified by the nature of the responses obtained from the individual concerning the object. One approach, the **decompositional method,** measures only the overall impression or evaluation of an object and then attempts to derive spatial positions in multidimensional space that reflect these perceptions. This technique is typically associated with MDS. The **compositional method** is an alternative approach, which employs several of the multivariate techniques already discussed that are used in forming an impression or evaluation based on a combination of specific attributes. Each approach has advantages and disadvantages. Our discussion here centers on the distinctions between the two approaches and then we focus primarily on the decompositional methods.

Decompositional or Attribute-Free Approach

Commonly associated with the techniques of MDS, decompositional methods rely on global or overall measures of similarity, from which the perceptual maps and relative positioning of objects are formed. They have two distinct advantages. First, they require only that respondents give their overall perceptions of objects; respondents do not detail the attributes used in this evaluation. Second, because each respondent gives a full assessment of similarities among all objects, perceptual maps can be developed for individual respondents or aggregated to form a composite map.

Decompositional methods have disadvantages as well. First, the researcher has no objective basis provided by the respondent on which to identify the basic "dimensions" of evaluation of the objects (i.e., the correspondence of perceptual and objective dimensions). In many instances, the usefulness to managers of attribute-free studies is restricted because the studies provide little guidance for specific action. For example, the inability to develop a direct link between actions by the firm (the objective dimension) and market positions of their products (the perceptual dimension) many times diminishes the value of perceptual mapping. Moreover, the researcher has little guidance, other than generalized guidelines or a priori beliefs, in determining both the dimensionality of the perceptual map and the representativeness of the solution. Although some overall measures of fit are available, they are nonstatistical, and thus decisions about the final solution involve substantial researcher judgment.

Characterized by the generalized category of MDS techniques, a wide range of possible decompositional techniques is available. Selection of a specific method requires decisions regarding the nature of the respondent's input (rating versus ranking), whether similarities or preferences are obtained, and whether individual or composite perceptual maps are derived. Among the most common multidimensional scaling programs are KYST, MDSCAL, PREFMAP, MDPREF, INDSCAL, ALSCAL, MINISSA, POLYCON, and MULTISCALE. Detailed descriptions of the programs and sources for obtaining them are available in Schiffman et al. [23, 24].

Compositional or Attribute-Based Approach

Compositional methods include some of the more traditional multivariate techniques (e.g., discriminant analysis or factor analysis), as well as methods specifically designed for perceptual mapping, such as correspondence analysis. A principle common to all of these methods, however, is the assessment of similarity in which a defined set of attributes is considered in developing the similarity between objects.

One advantage of this approach is the explicit description of the dimensions of perceptual space. Because the respondent provides detailed evaluations across numerous attributes for each object, the evaluative criteria represented by the dimensions of the solution are much easier to ascertain. Second, these methods provide a direct method of representing both attributes and objects on a single map, with several methods providing the additional positioning of respondent groups. This information provides unique managerial insight into the competitive marketplace.

There are four primary disadvantages to compositional techniques. First, the similarity between objects is limited to only the attributes rated by the respondents. If salient attributes are omitted, there is no opportunity for the respondent to incorporate them, as there would be if a single overall measure were provided. Second, the researcher must *assume* some method of combining these attributes to represent overall similarity, and the chosen method may not represent the respondents' thinking. Third, the data collection effort is substantial, especially as the number of choice objects increases. Finally, results are not typically available for the individual respondent.

Techniques of compositional methods can be grouped into one of three basic groups:

1. *Graphical or post hoc approaches*—included in this set are analyses such as semantic differential plots or **importance–performance grids,** which rely on researcher judgment and univariate or bivariate representations of the objects.
2. *Conventional multivariate statistical techniques*—these techniques, especially *factor analysis* and *discriminant analysis*, are particularly useful in developing a dimensional structure among numerous attributes and then representing objects on these dimensions.
3. *Specialized perceptual mapping methods*—notable in this class is *correspondence analysis*, developed specifically to provide perceptual mapping with only qualitative or nominally scaled data as input.

Selecting between Compositional and Decompositional Techniques

Perceptual mapping can be performed with both compositional and decompositional techniques, but each technique has specific advantages and disadvantages that must be considered in view of the research objectives. If perceptual mapping is undertaken in the "spirit" of either of the two basic objectives discussed earlier, the decompositional or attribute-free approaches are the most appropriate. If, however, the research objectives shift to the portrayal among objects on a defined set of attributes, then the compositional techniques become the preferred alternative. Our discussion of the compositional methods in past chapters has illustrated their uses and application along with their strengths and weaknesses. The researcher must always remember the alternatives that are available in the event that the objectives of the research change. Thus, we focus here on the decompositional approaches, followed by a discussion of correspondence analysis, a widely used compositional technique particularly suited to perceptual mapping. As such, we also consider as synonymous the terms *perceptual mapping* and *multidimensional scaling* unless necessary distinctions are made.

Objects: Their Number and Selection

Before beginning any perceptual mapping study, the researcher must address several questions about the objects being evaluated. First and foremost, are the objects really comparable? An implicit assumption in perceptual mapping is that there are common characteristics, either objective or perceived, that the respondent could use for evaluations. It is not possible for the researcher to "force" the respondent to make comparisons by creating pairs of noncomparable objects. Even if responses are given in such a forced situation, their usefulness is questionable.

A second question deals with the number of objects to be evaluated. In deciding how many objects to include, the researcher must balance two desires: a smaller number of objects to ease the effort on the part of the respondent versus the required number of objects to obtain a stable multidimensional solution. A suggested guideline for stable solutions is to have more than four times as many objects as dimensions desired [9]. Thus at least five objects are required for a one-dimensional perceptual map, nine objects for a two-dimensional solution, and so on. When using the method of evaluating pairs of objects for similarity, the respondent must make 36 comparisons of the nine objects—a substantial task. And a three-dimensional solution suggests at least 13 objects be evaluated, necessitating the evaluation of 78 pairs of objects. Therefore, a trade-off must be made between the dimensionality accommodated by the objects (and the implied number of underlying dimensions that can be identified) and the effort required on the part of the respondent.

The number of objects also affects the determination of an acceptable level of fit. Many times, having less than the suggested number of objects for a given dimensionality causes an inflated estimate of fit. Similar to the overfitting problem we found in regression, falling below the recommended guidelines of at least four objects per dimension greatly increases the chances of a misleading solution. For example, an empirical study demonstrated that when seven objects are fit to three dimensions with *random similarity* values, acceptable stress levels and apparently valid perceptual maps are generated more than 50 percent of the time. If the seven objects with random similarities were fit to four dimensions, the stress values decreased to zero, indicating perfect fit, in half the cases [18]. Yet in both instances, there was no real pattern of similarity among the objects. Thus we must be aware of violating the guidelines for the number of objects per dimension and the impact this has on both the measures of fit and the validity of the resulting perceptual maps.

Nonmetric versus Metric Methods

The original MDS programs were truly nonmetric, meaning that they required only nonmetric input but they also provided only nonmetric (rank-order) output. The nonmetric output, however, limited the interpretability of the perceptual map. Therefore, all MDS programs used today produce metric output. The metric multidimensional positions can be rotated about the origin, the origin can be changed by adding a constant, the axes can be flipped (reflection), or the entire solution can be uniformly stretched or compressed, all without changing the relative positions of the objects.

Because all programs today produce metric output, the distinction is based on the *input measures of similarity.* Nonmetric methods, distinguished by the non-metric input typically generated by rank-ordering pairs of objects, are more flexible in that they do not assume any specific type of relationship between the calculated distance and the similarity measure. However, because nonmetric methods contain less information for creating the perceptual map, they are more likely to result in degenerate or suboptimal solutions. This is a particular problem when there are wide variations in the perceptual maps between respondents or the perceptions between objects are not distinct or well defined. Metric methods assume that input as well as output is metric. This assumption allows us to strengthen the relationship between the final output dimensionality and the input data. Rather than assuming that only the ordered relationships are preserved in the input data, we can assume that the output preserves the interval and ratio qualities of the input data. Even though the assumptions underlying metric programs are more difficult to support conceptually in many cases, the results of non-metric and metric procedures applied to the same data are often very similar. Thus, selection of the input data type must consider both the research situation (variations of perceptions among respondents and distinctiveness of objects) and the preferred mode of data collection.

Collection of Similarity or Preference Data

As already noted, the primary distinction among MDS programs is the type of data (metric versus nonmetric) used to represent similarity and preferences. Here we address issues associated with making similarity-based and preference judgments. For many of the data collection methods, either metric (ratings) or non-metric (rankings) data may be collected. In some instances however, the responses are limited to only one type of data.

Similarities Data

When collecting **similarities data,** the researcher is trying to determine which items are the most similar to each other and which are the most dissimilar. The terms of dissimilarities and similarities often are used interchangeably to represent measurement of the differences between objects. Implicit in similarity measurement is the ability to compare all pairs of objects. If, for example, all pairs of objects of the set A, B, C (i.e., $\overline{AB}$, $\overline{AC}$, $\overline{BC}$) are rank-ordered, then all pairs of objects can also be compared. Assume that the pairs were ranked $\overline{AB} = 1$, $\overline{AC} = 2$, and $\overline{BC} = 3$ (where 1 denotes most similar). Clearly, pair $\overline{AB}$ is more similar than pair $\overline{AC}$, pair $\overline{AB}$ is more similar than pair $\overline{BC}$, and pair $\overline{AC}$ is more similar than pair $\overline{BC}$.

Several procedures are commonly used to obtain respondents' perceptions of the similarities among stimuli. Each procedure is based on the notion that the relative differences between any pair of stimuli must be measured so that the researcher can determine whether the pair is more or less similar to any other pair. We discuss three procedures commonly used to obtain respondents' perceptions of similarities: comparison of paired objects, confusion data, and derived measures.

Comparison of Paired Objects   By far the most widely used method of obtaining *similarity judgments* is that of paired objects, in which the respondent is asked simply to rank or rate the similarity of all pairs of objects. If we have stimuli A, B, C, D, and E, we could rank pairs AB, AC, AD, AE, BC, BD, BE, CD, CE, and DE from most similar to least similar. If, for example, pair AB is given the rank of 1, we would assume that the respondent sees that pair as containing the two stimuli that are the most similar, in contrast to all other pairs. This procedure would provide a nonmetric measure of similarity. Metric measures of similarity would involve a rating of similarity (e.g., from 1 "very similar" to 10 "not at all similar"). Either form (metric or nonmetric) can be used in most MDS programs.

Confusion Data   The pairing (or "confusing") of stimulus I with stimulus J is taken to indicate similarity. Also known as **subjective clustering,** the typical procedure for gathering these data is to place the objects whose similarity is to be measured (e.g., 10 candy bars) on small cards, either descriptively or with pictures. The respondent is asked to sort the cards into stacks so that all the cards in a stack represent similar candy bars. Some researchers tell the respondents to sort into a fixed number of stacks; others say to sort into as many stacks as the respondent wants. In either situation, the data result in an aggregate similarities matrix similar to a cross-tabulation table. These data then indicate which products appeared together most often and are therefore considered the most similar. Collecting data in this manner allows only for the calculation of aggregate similarity, because the responses from all individuals are combined to obtain the similarities matrix.

Derived Measures   Derived measures of similarity are typically based on scores given to stimuli by respondents. For example, subjects are asked to evaluate three stimuli (cherry, strawberry, and lemon-lime soda) on a number of attributes (diet versus nondiet, sweet versus tart, and light tasting versus heavy tasting) using semantic differential scales. The responses would be evaluated for each respondent (e.g., correlation, index of agreement) to create similarity measures between the objects. There are three important assumptions here:

1. The researcher has selected the appropriate dimensions to measure.
2. The scales can be weighted (either equally or unequally) to achieve the similarities data for a subject or group of subjects.
3. Even if the weighting of scales can be determined, all individuals have the same weights.

Of the three procedures we have discussed, the derived measure is the least desirable in meeting the "spirit" of MDS—that the evaluation of objects be made with minimal influence by the researcher.

Collecting Preference Data
Preference implies that stimuli should be judged in terms of dominance relationships; that is, the stimuli are ordered in terms of the preference for some property. For example, brand A is preferred over brand C. The two most common

procedures for obtaining preference data are direct ranking and paired comparisons.

Direct Ranking   Each respondent ranks the objects from most preferred to least preferred. This is a very popular method of gathering nonmetric similarity data because it is easy to administer for a small to moderate number of objects. It is quite similar in concept to the subjective clustering procedure discussed earlier, only in this case each object must be given a unique rank (no ties).

Paired Comparisons   A respondent is presented with all possible pairs and asked to indicate which member of each pair is preferred. In this way, the researcher gathers explicit data for each comparison, which is much more detailed than just the direct rankings. The principal drawback to this method is the large number of tasks involved with even a relatively small number of objects. For example, 10 objects result in 90 paired comparisons, which are too many tasks for most research situations. Note that paired comparisons are also used in collecting similarity data, as noted in the example at the beginning of the chapter.

Preference Data versus Similarity Data

Preference data allow the researcher to view the location of objects in a perceptual map for which distance implies differences in preference. This procedure is useful because an individual's perception of objects in a preference context may be different from that in a similarity context; that is, a particular dimension may be very useful in describing the differences between two objects but may not be consequential in determining preference. Therefore, two objects could be perceived as different in a similarity-based map but be similar in a preference-based spatial map. This would result in two quite different maps, such as two different brands of candy bars being far apart in a similarity-based map but, with equivalent preference, being positioned close to each other on a preference map.

In summary, the collection procedures for both similarity and preference data have the common purpose of obtaining a series of unidimensional responses that represent the respondents' judgments. These judgments then serve as inputs to the many MDS procedures that define the underlying multidimensional pattern leading to these judgments.

Stage 3: Assumptions of MDS Analysis

Multidimensional scaling, while having no restraining assumptions on the methodology, type of data, or form of the relationships among the variables, does require that the researcher accept several tenets about perception, including the following:

1. *Variation in dimensionality*—each respondent will not perceive a stimulus to have the same dimensionality (although it is thought that most people judge in terms of a limited number of characteristics or dimensions). For example, some might evaluate a car in terms of its horsepower and appearance, whereas others do not consider these factors at all but instead assess it in terms of cost and interior comfort.

2. *Variation in importance*—respondents need not attach the same level of importance to a dimension, even if all respondents perceive this dimension. For example, two respondents perceive a cola drink in terms of its level of carbonation, but one may consider this dimension unimportant whereas the other may consider it very important.

3. *Variation over time*—judgments of a stimulus in terms of either dimensions or levels of importance need not remain stable over time. In other words, one may not expect respondents to maintain the same perceptions for long periods of time.

In spite of these assumptions and the differences we can expect between individuals, MDS attempts to represent perceptions spatially so that any common underlying relationships can be examined. The purpose of employing an MDS technique lies not only in understanding each separate individual but also in identifying the shared perceptions and evaluative dimensions within the sample of respondents.

Stage 4: Deriving the MDS Solution and Assessing Overall Fit

The variety of computer programs for MDS is rapidly expanding, particularly for use on the personal computer. Today the basic MDS programs are available in all of the major statistical programs. We employ a series of MDS applications to illustrate the types of input data, the differing types of spatial representations, and the interpretational alternatives. Our objective here is to provide an overview of MDS to allow a ready understanding of the differences among these programs. However, as with other multivariate techniques, there is continual development in both application and knowledge. Thus, we refer the user interested in specific program applications to other texts devoted solely to multidimensional scaling [9, 10, 16, 18, 23].

Determining an Object's Position in the Perceptual Map

The first task of stage 4 involves the positioning of objects to best reflect the similarity evaluations provided by the respondents (see Figure 10.4, p. 538). MDS programs follow a common process for determining the optimal positions. This process can be described in four steps:

Step 1: Select an initial configuration of stimuli (S_k) at a desired **initial dimensionality** (*t*). A number of options for obtaining the initial configuration are available. The two most widely used are configurations either applied by the researcher based on previous data or generated by selecting pseudorandom points from an approximately normal multivariate distribution.

Step 2: Compute the distances between the stimuli points and compare the relationships (observed versus derived) with a measure of fit. Once a configuration is found, the interpoint distances between stimuli (d_{ij}) in the starting configurations are compared with distance measures ($\hat{d}_{ij}$) derived from the similarity judgments (s_{ij}). The two distance measures are then

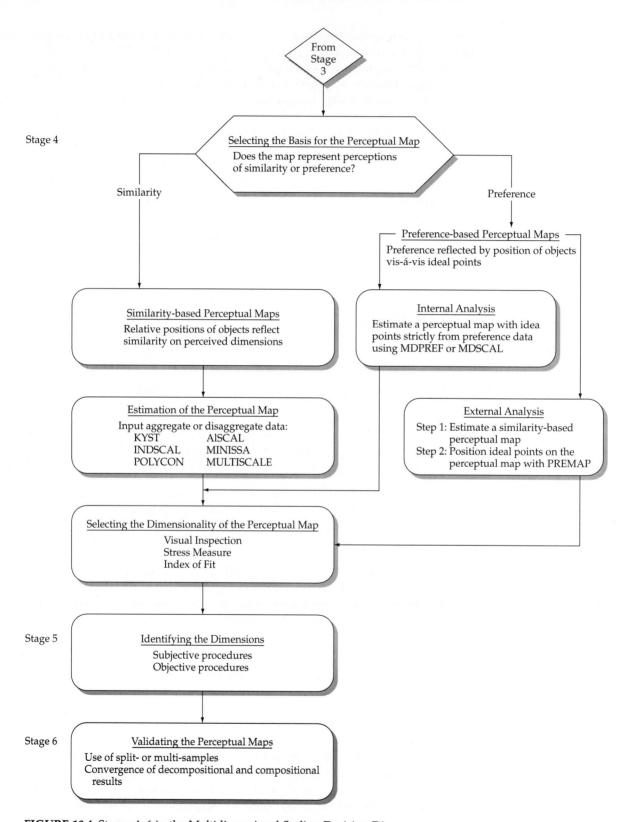

FIGURE 10.4 Stages 4–6 in the Multidimensional Scaling Decision Diagram

compared by a measure of fit, typically a measure of stress. (Fit measures are discussed in a later section.)

Step 3: If the measure of fit does not meet a selected predefined stopping value, find a new configuration for which the measure of fit is further minimized. The program determines the directions in which the best improvement in fit can be obtained and then moves the points in the configuration in those directions in small increments.

Step 4: Once satisfactory stress has been achieved, the dimensionality is reduced by one, and the process is repeated until the lowest dimensionality with an acceptable measure of fit has been reached.

The need for a computer program versus hand calculations becomes apparent as the number of objects and dimensions increases. For example, if we have 10 products to be evaluated, each respondent must now rank all 45 pairs of products from most similar (1) to least similar (45). We place the 10 points (representing the 10 products) randomly on a sheet of graph paper and then measure the distances between every pair of points (45 distances). We then must calculate the stress of the solution, a measure that shows the rank-order agreement between the Euclidean (straight-line) distances of the plotted objects and the original 45 ranks. If the straight-line distances do not agree with the original ranks, we must move the 10 points and try again. The process becomes intractable as the number of objects increases and the differences in perception and dimensions used in the evaluation increase. The computer is used only to replace the manual calculations and allow for a more accurate and detailed solution.

The primary criterion in all instances for finding the best representation of the data is preservation of the ordered relationship between the original rank data and the derived distances between points. The stress measure is simply a measure of how well (or poorly) the ranked distances on the map agree with the ranks given by the respondents.

In evaluating a perceptual map, the researcher should always be aware of **degenerate solutions,** which are derived perceptual maps that are not accurate representations of the similarity responses. Most often they are caused by inconsistencies in the data or an inability of the MDS program to reach a stable solution. They are characterized most often by either a circular pattern, in which all objects are shown to be equally similar, or a clustered solution, in which the objects are grouped at two ends of a single dimension. In both cases, the computer program is unable to differentiate among the objects for some reason. The researcher should then reexamine the research design to see where the inconsistencies occur.

Selecting the Dimensionality of the Perceptual Map

The objective of this step is the selection of a spatial configuration in a specified number of dimensions. The determination of how many dimensions are actually represented in the data is generally reached through one of three approaches: subjective evaluation, scree plots of the stress measures, or an overall index of fit.

The spatial map is a good starting point for the evaluation. The number of maps necessary for interpretation depends on the number of dimensions. A map is produced for each combination of dimensions. One objective of the researcher should be to obtain the best fit with the smallest possible number of dimensions. Interpretation of solutions derived in more than three dimensions is extremely difficult and usually is not worth the improvement in fit. The researcher typically makes a **subjective evaluation** of the perceptual maps and determines whether the configuration looks reasonable. This evaluation is important because at a later stage the dimensions will need to be interpreted and explained.

A second approach is to use a **stress measure,** which indicates the proportion of the variance of the **disparities** not accounted for by the MDS model. This measurement varies according to the type of program and the data being analyzed. Kruskal's [17] stress is the most commonly used measure for determining a model's goodness of fit. It is defined by:

$$\text{Stress} = \sqrt{\frac{(d_{ij} - \hat{d}_{ij})^2}{(d_{ij} - \bar{d})^2}}$$

where

$\bar{d}$ = the average distance ($\Sigma d_{ij}/n$) on the map
$\hat{d}_{ij}$ = derived distance from similarity data
d_{ij} = original distances provided by respondents

The stress value becomes smaller as the estimated $\hat{d}_{ij}$ approaches the original d_{ij}. Stress is minimized when the objects are placed in a configuration so that the distances between the objects best match the original distances.

A problem found in using stress, however, is analogous to that of R^2 in multiple regression in that stress always improves with increased dimensions. (Remember that R^2 always increases with additional variables.) A trade-off must then be made between the fit of the solution and the number of dimensions. As was done for the extraction of factors in factor analysis, we can plot the stress value against the number of dimensions to determine the best number of dimensions to utilize in the analysis [18]. For example, in the scree plot in Figure 10.5, the elbow indicates that there is substantial improvement in the goodness of fit when the number of dimensions is increased from 1 to 2. Therefore, the best fit is obtained with a relatively low number (2) of dimensions.

A squared correlation index is sometimes used as an **index of fit.** It can be interpreted as indicating the proportion of variance of the disparities (optimally scaled data) accounted for by the MDS procedure. In other words, it is a measure

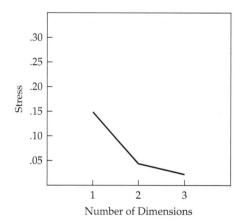

FIGURE 10.5 Use of a Scree Plot to Determine the Appropriate Dimensionality

of how well the raw data fit the MDS model. The R^2 measure on multidimensional scaling represents essentially the same measure of variance as it does in other multivariate techniques. Therefore, it is possible to use similar measurement criteria; that is, measures of .60 or better are considered acceptable. Of course, the higher the R^2, the better the fit.

Incorporating Preferences into MDS

Up to this point, we have concentrated on developing perceptual maps based on similarity judgments. However, perceptual maps can also be derived from preferences. The objective is to determine the preferred mix of characteristics for a set of stimuli that predicts preference, given a set configuration of objects [8, 9]. In doing so, a joint space is developed portraying both the objects (stimuli) and the subjects (ideal points). A critical assumption is the homogeneity of perception across individuals for the set of objects. This allows all differences to be attributed to preferences, not perceptual differences.

Ideal Points

The term **ideal point** has been misunderstood or misleading at times. We can assume that if we locate (on the derived perceptual map) the point that represents the most preferred combination of perceived attributes, we have identified the position of an ideal object. Equally, we can assume that the position of this ideal point (relative to the other products on the derived perceptual map) defines relative preferences so that products farther from the ideal should be less preferred. An ideal point is positioned so that the distance from the ideal conveys changes in preference. Consider, for example, Figure 10.6. When preference data on the six candy bars were obtained from the person indicated by the dot (·), the point (·) was positioned so that increasing the distance from it indicated declining preference. One may assume that this person's preference order is C, F, D, E, A, B. To imply that the ideal candy bar is exactly at the point (·) or even beyond (in the

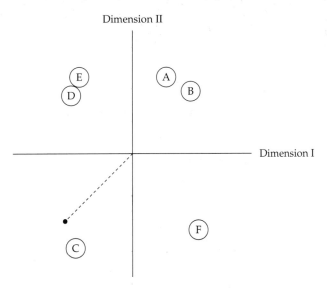

• Indicates respondent's ideal point

FIGURE 10.6 A Respondent's Ideal Point within the
Perceptual Map

direction shown by the dashed line from the origin) can be misleading. The ideal point simply defines the ordered preference relationship among the set of six candy bars for that respondent. Although ideal points individually may not offer much insight, clusters of them can be very useful in defining segments. Many respondents with ideal points in the same general area represent potential market segments of persons with similar preferences, as indicated in Figure 10.7.

Two approaches have generally been used to determine ideal points: explicit and implicit estimation. Explicit estimation proceeds from the direct responses of subjects. This procedure may involve asking the subject to rate a hypothetical ideal on the same attributes on which the other stimuli are rated. Alternatively, the respondent is asked to include among the stimuli used to gather similarities data a hypothetical ideal stimulus (e.g., brand, image).

When asking respondents to conceptualize an ideal of anything, we typically run into problems. Often the respondent conceptualizes the ideal at the extremes of the explicit ratings used or as being similar to the most preferred product from among those with which the respondent has had experience. Also, the respondent must think in terms not of similarities but of preferences, which is often difficult with relatively unknown objects. Often these perceptual problems lead the researcher to use implicit ideal point estimation.

There are several procedures for implicitly positioning ideal points (see the following section for a more detailed description). The basic assumption underlying most procedures is that derived measures of ideal points' spatial positions are maximally consistent with individual respondents' preferences. Srinivasan and Schocker [25] assume that the ideal point for all pairs of stimuli is determined so that it violates with least harm the constraint that it be closer to the most preferred in each pair than it is to the least preferred.

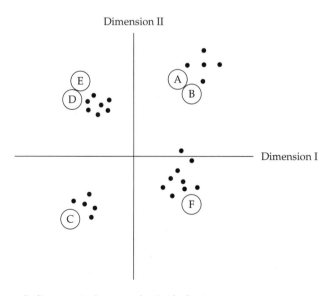

• Indicates a single respondent's ideal point

FIGURE 10.7 Incorporating Multiple Ideal Points in the
          Perceptual Map

In summary, there are many ways to approach ideal point estimation, and no one best method has been demonstrated. The choice depends on the researcher's skills and the MDS procedure selected.

Positioning the Ideal Point

Implicit positioning of the ideal point from preference data can be accomplished through either an internal or an external analysis. Internal analysis of preference data refers to the development of a spatial map shared by both stimuli and subject points (or vectors) solely from the preference data. External analysis of preference uses a prespecified configuration of objects and then attempts to place the ideal points within this perceptual map. Each approach has advantages and disadvantages, which are discussed in the following sections.

Internal Analysis   Internal analyses must make certain assumptions in deriving both the perceptual map of stimuli and ideal points. The objects' positions are calculated based on **unfolding** preference data for each individual. The results reflect perceptual dimensions that are "stretched" and weighted to predict preference. One characteristic of internal estimation methods is that they typically employ a vector representation of the ideal point (see the following section for a discussion of vector versus point representations), whereas external models can estimate either vector or point representations.

As an example of this approach, MDPREF [5] or MDSCAL [17], two of the more widely used programs of this type, allow the user to find configurations of stimuli and ideal points. In doing so, the researcher must assume (1) no difference between subjects exists, (2) separate configurations for each subject, or (3) a single configuration with individual ideal points. By gathering preference data, the researcher can represent both stimuli and respondents on a single perceptual map.

External Analysis   External analysis of preference data refers to fitting ideal points (based on preference data) to stimulus space developed from similarities data obtained from the same subjects. For example, we might scale similarities data individually, examine the individual maps for commonality of perception, and then scale the preference data for any group identified in this fashion. If this procedure is followed, the researcher has to gather both preference and similarities data to achieve external analysis.

PREFMAP [6] was developed solely to perform external analysis of preference data. Because the similarity matrix defines the objects in the perceptual map, the researcher can now define both attribute descriptors (assuming that the perceptual space is the same as the evaluative dimensions) and ideal points for individuals. PREFMAP provides estimates for a number of different types of ideal points, each based on different assumptions as to the nature of preferences (e.g., vector versus point representations, or equal versus differential dimension weights).

Choosing between Internal and External Analysis   It is generally accepted [9, 10, 23] that external analysis is clearly preferable to internal analysis. This conclusion is based on computational difficulties with internal analysis procedures and on the confounding of differences in preference with differences in perception. In addition, the saliences of perceived dimensions may change as one moves from perceptual space (are the stimuli similar or dissimilar?) to evaluative space (which stimulus is preferred?).

We illustrate the procedure of external estimation in our example of perceptual mapping with MDS at the end of this chapter.

Vector versus Point Representations

The discussion of perceptual mapping of preference data has emphasized an ideal point that portrays the relationship of an individual's preference ordering for a set of stimuli. The most easily understood method of portraying the ideal point is to use the straight-line (Euclidean) distance measure of preference ordering from the ideal point to all the points representing the objects. We are assuming that the direction of distance from the ideal point is not critical, only the relative distance. An example is shown in Figure 10.8. Here, the ideal point as positioned indicates that the most preferred object is E, followed by C, then B, D, and finally A.

The ideal point can also be shown as a **vector.** To calculate the preferences in this approach, perpendicular lines (also known as **projections**) are drawn from the objects to the vector. Preference increases in the direction the vector is pointing. The preferences can be read directly from the order of the projections. Figure 10.9 illustrates the vector approach for two subjects with the same set of stimuli positions. For subject 1, the vector has the direction of lower preference in the bottom left-hand corner to higher preference in the upper right-hand corner. When the projection for each object is made, the preference order (highest to lowest) is A, B, C, E, and D. However, the same objects have a quite different preference order for subject 2. For subject 2 the preference order ranges from the most preferred, E, to the least preferred, C. In this manner, a separate vector can represent each subject. In the vector approach, there is no single ideal point, but it is assumed that the ideal point is at an infinite distance outward on the vector.

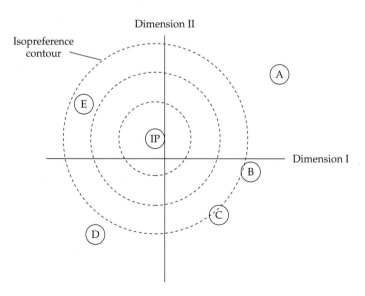

Preference order (highest to lowest): E > C > B > D > A

Ⓐ Object   ⒤Ⓟ Ideal Point

FIGURE 10.8 Point Representation of an Ideal Point

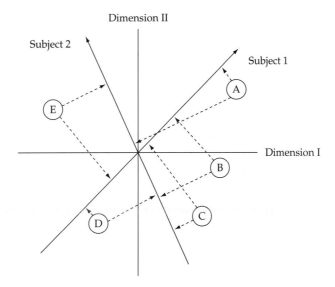

Preference order (highest to lowest): Subject 1: A > B > C > E > D
Subject 2: E > A > D > B > C

FIGURE 10.9 Vector Representations of Two Ideal Points:
Subjects 1 and 2

Although either the point or vector representations can indicate what combinations of attributes are more preferred, these observations are often not borne out by further experimentation. For example, Raymond [22] cites an example in which the conclusion was drawn that people would prefer brownies on the basis of degree of moistness and chocolate content. When the food technicians applied this result in the laboratory, they found that their brownies made to the experimental specification became chocolate milk. One cannot always assume that the relationships found are independent or linear, or that they hold over time, as noted previously. However, MDS is a beginning in understanding perceptions and choice that will expand considerably as applications extend our knowledge of both the methodology and human perception.

Summary
Preference data are best examined using external analysis as a means to better understand both the perceptual differences between objects based on similarity judgments and the preference choices made within this perceptual map of objects. In this manner, the researcher can distinguish between both types of perceptual evaluations and more accurately understand the perceptions of individuals in the "true spirit" of multidimensional scaling.

Stage 5: Interpreting the MDS Results

Once the perceptual map is obtained, the two approaches—compositional and decompositional—again diverge in their interpretation of the results. For compositional methods, the perceptual map must be validated against other measures of perception, because the positions are totally defined by the attributes specified by the researcher. For example, discriminant analysis results might be applied to a new set of objects or respondents, assessing the ability to differentiate with these new observations.

For decompositional methods, the most important issue is the description of the perceptual dimensions and their correspondence to attributes. A number of descriptive techniques to "label" the dimensions, as well as to integrate preferences (for objects and attributes) with the similarity judgments, are discussed later. Again, in line with their objectives, the decompositional methods provide an initial look into perceptions from which more formalized perspectives may emerge.

Because other chapters in this text have dealt with many of the compositional techniques, the remainder of this chapter focuses on decompositional methods, primarily the various techniques used in multidimensional scaling. A notable exception is the discussion of a compositional approach—correspondence analysis—that, to a degree, bridges the gap between the two approaches in its flexibility and methods of interpretation.

Identifying the Dimensions

As discussed in chapter 3, identifying underlying dimensions is often a difficult task. Multidimensional scaling techniques have no built-in procedure for labeling the dimensions. The researcher, having developed the maps with a selected dimensionality, can adopt several procedures, either subjective or objective.

Subjective Procedures

Interpretation must always include some element of researcher or respondent judgment, and in many cases this proves adequate for the questions at hand. A quite simple, yet effective, method is labeling (by visual inspection) the dimensions of the perceptual map by the respondent. Respondents may be asked to interpret the dimensionality subjectively by inspecting the maps, or a set of "experts" may evaluate and identify the dimensions. Although there is no attempt to quantitatively link the dimensions to attributes, this approach may be the best available if the dimensions are believed to be highly intangible, or affective or emotional, in content, so that adequate descriptors cannot be devised.

In a similar manner, the researcher may describe the dimensions in terms of known (objective) characteristics. In this way, the correspondence is made between objective and perceptual dimensions directly, although these relationships are not a result of respondent feedback but of the researcher's judgment.

Objective Procedures

As a complement to the subjective procedures, a number of more formalized methods have been developed. The most widely used method, PROFIT (PROperty FITting) [3], collects attribute ratings for each object and then finds the best correspondence of each attribute to the derived perceptual space. The attempt is to identify the determinant attributes in the similarity judgments made by individuals. Measures of fit are given for each attribute, as well as their correspondence with the dimensions. The researcher can then determine which attributes best describe the perceptual positions and are illustrative of the dimensions. The need for correspondence between the attributes and the defined dimensions diminishes with the use of metric output, as the dimensions can be rotated freely without any changes in interpretation.

Selecting between Subjective and Objective Procedures

For either subjective or objective procedures, the researcher must remember that although a dimension can represent a single attribute, it usually does not. A more common procedure is to collect data on several attributes, associate them either subjectively or empirically with the dimensions where applicable, and determine labels for each dimension using multiple attributes, similar to factor analysis. Many researchers suggest that using attribute data to help label the dimensions is the best alternative. The problem, however, is that the researcher may not include all the important attributes in the study. Thus, the researcher can never be totally assured that the labels represent all relevant attributes.

Both subjective and objective procedures illustrate the difficulty of labeling the axes. This task cannot be left until completion, as the dimensional labels are essential for further interpretation and use of the results. The researcher must select the type of procedure that best suits both the objectives of the research and the available information. Thus, the researcher must plan for the derivation of the dimensional labels as well as the estimation of the perceptual map.

Stage 6: Validating the MDS Results

Validation in MDS is as important as in any other multivariate technique. Owing to the highly inferential nature of MDS, this effort should be directed toward ensuring the generalizability of the results both across objects and to the population. But validation efforts are problematic. The only output of MDS that can be used for comparative purposes involves the relative positions of the objects. Thus, although the positions can be compared, the underlying dimensions have no basis for comparison. If the positions vary, the researcher cannot determine whether the objects are viewed differently, the perceptual dimensions vary, or both. Moreover, systematic methods of comparison have not been developed and integrated into the statistical programs. The researcher is left to improvise with procedures that may address general concerns but are not specific to MDS results.

What options are available? The most direct approach is a split- or multisample comparison, in which either the original sample is divided or a new sample is collected. In either instance, the researcher must then find a means of comparing the results. Most often the comparison between results is done visually or with a simple correlation of coordinates. Some matching programs are available, such as FMATCH [24], but the researcher must still determine how many of the disparities are due to differences in object perceptions, differing dimensions, or both.

Another method is to obtain a convergence of MDS results by applying both decompositional and compositional methods to the same sample. The decompositional method(s) could be applied first, along with interpretation of the dimensions to identify key attributes. Then one or more compositional methods, particularly correspondence analysis, could be applied to confirm the results. The researcher must realize that this is not true validation of the results as being generalizable but does confirm the interpretation of the dimension. From this point, validation efforts with other samples and other objects could be undertaken to demonstrate generalizability to other samples.

Correspondence Analysis

Up to this point we have discussed the traditional decompositional approaches to MDS, but what about compositional techniques? In the past, compositional approaches have relied on traditional multivariate techniques such as discriminant and factor analysis. But recent developments have combined aspects of both methods and MDS to form potent new tools for perceptual mapping.

Correspondence analysis (CA) is an interdependence technique that has become increasingly popular for dimensional reduction and perceptual mapping [1, 2, 11, 13, 19]. It is a compositional technique because the perceptual map is based on the association between objects and a set of descriptive characteristics or attributes specified by the researcher. Among the compositional techniques, factor analysis is the most similar, but correspondence analysis extends beyond factor analysis. Its most direct application is portraying the "correspondence" of categories of variables, particularly those measured in nominal measurement scales. This correspondence is then the basis for developing perceptual maps. The benefits of CA lie in its unique abilities for representing rows and columns, for example, brands *and* attributes, in joint space.

We first examine a simple example of CA to gain some perspective on its basic principles. Then we discuss each of the six stages of the decision-making process introduced in chapter 1. The emphasis is on those unique elements of CA as compared to the decompositional methods of MDS discussed earlier.

A Simple Example of CA

Let us examine a simple situation as an introduction to CA. In its most basic form, CA examines the relationships between categories of nominal data in a **contingency table,** the cross-tabulation of two categorical variables. For example, assume that sales figures for products A, B, and C are broken down by three age categories (young adults, who are 18 to 35 years old; middle age, who are 36 to 55 years old; and senior citizens, who are 56 or older). The cross-tabulated data are shown in Table 10.2.

The data show that unit sales vary substantially across products (product C has the highest total sales, product B the lowest) and age groups (middle age buys the most units, young adults the least). But we want to identify any pattern to the

TABLE 10.2 Cross-Tabulated Data Detailing Product
Sales by Age Category

	Product Sales			
Age Category	A	B	C	Total
Young adults (18–35 years old)	20	20	20	60
Middle age (36–55 years old)	40	10	40	90
Senior citizens (56+ years old)	20	10	40	70
Total	80	40	100	220

sales, such that we can state that young adults buy more of product X or senior citizens buy more of product Z. To do this, we need a standardized measure of unit sales that simultaneously considers the differences in sales for a specific product–age category combination. Then, if we still see that a certain age group buys more units of a product than expected, we can associate that age group with that product. In a graphical portrayal, age groups would be located closer to products with which they are highly associated and farther away from products with lower associations. Likewise, we want to be able to view any product and see its associations with various age groups.

Calculating a Measure of Association

Correspondence analysis uses one of the most basic statistical concepts, chi-square, to standardize the sales (frequency values) and form the basis for associations. **Chi-square** is a standardized measure of actual cell frequencies compared to expected cell frequencies. In our cross-tabulated data, each cell contains the sales for a product–age group combination. The chi-square procedure then proceeds in three steps to calculate a chi-square value for each cell:

1. *Calculate expected sales.* The first step is to calculate the expected sales for a cell as if no association existed. The expected sales are defined as the joint probability of the column (product) and row (age group) combination. This is calculated as the marginal probability for the product (sales of that product across all age groups ÷ total sales across all age groups and products) times the marginal probability for the age group (sales of that age group across all products ÷ total sales across all age groups and products). This value is then multiplied by the total sales across all age groups and products. It can be simplified by canceling terms so that the equation is:

$$\text{Expected sales} = \frac{\text{Total sales of age category} \times \text{Total sales of product type}}{\text{Overall total sales}}$$

   In our simple example, the expected sales for the young adults buying product A is 21.82 units, as shown in the following calculation:

$$\text{Expected sales}_{\text{young adults, product A}} = \frac{60 \times 80}{220} = 21.82$$

   This calculation is performed for each cell, with the results shown in Table 10.3 (p. 550).

2. *Difference in expected and actual sales.* The next step is to calculate the difference between the expected and actual sales as follows: Difference = Expected sales − Actual sales. Again, in our example of the cell for the young adults purchasing product A, the difference is 1.82 (21.82 − 20.00). Larger positive differences would mean that that age group–product combination had fewer sales than would be expected (a negative association) and large negative differences would indicate positive associations (the cell actually bought more than expected). The differences for each cell are also shown in Table 10.3.

3. *Calculate the chi-square value.* The final step is to standardize the differences across cells so that comparisons can be easily made. Standardization is required because it would be much easier for differences to occur if the cell frequency (sales) was very high compared to a cell with only a few sales. So, we standardize the differences to form a chi-square value by dividing each squared difference by the expected sales value. Thus, the chi-square value for a cell is calculated as:

$$\text{Chi-square value } (\chi^2) \text{ of a cell} = \frac{(\text{Difference})^2}{\text{Expected sales}}$$

For our example cell, the chi-square value would be:

$$\text{Chi-Square } (\chi^2)_{\text{young adults, product A}} = \frac{(1.82)^2}{21.82} = .15$$

The chi-square values can then be converted to similarity measure by applying the opposite sign of their difference. Thus, for our aforementioned example cell, the chi-square value of .15 would be stated as a similarity value of $-.15$ because the difference was positive. This is necessary because the chi-square calculation

TABLE 10.3 Calculating Chi-Square Similarity Values for Cross-Tabulated Data

Age Category	A	B	C	Total
		Product Sales		
Young adults				
Sales	20	20	20	60
Column percentage	25%	50%	20%	27%
Row percentage	33%	33%	33%	100%
Expected sales[a]	21.82	10.91	27.27	60
Difference[b]	1.82	−9.09	7.27	—
Chi-square value[c]	.15	7.58	1.94	9.67
Similarity	−.15	7.58	−1.94	
Middle age				
Sales	40	10	40	90
Column percentage	50%	25%	40%	41%
Row percentage	44%	11%	44%	100%
Expected sales	32.73	16.36	40.91	90
Difference	−7.27	6.36	.91	—
Chi-square value	1.62	2.47	.02	4.11
Similarity	1.62	−2.47	−.02	
Senior citizens				
Sales	20	10	40	70
Column percentage	25%	25%	40%	32%
Row percentage	29%	14%	57%	100%
Expected sales	25.45	12.73	31.82	70
Difference	5.45	2.73	−8.18	—
Chi-square value	1.17	.58	2.10	3.85
Similarity	−1.17	−.58	2.10	
Total				
Sales	80	40	100	220
Column percentage	100%	100%	100%	100%
Row percentage	36%	18%	46%	100%
Expected sales	80	40	100	220
Difference	—	—	—	—
Chi-square value	2.94	10.63	4.06	17.63

[a]Expected sales = (Row total × Column total) ÷ Overall total
  Example: Cell$_{\text{young adults, product A}}$ = (60 × 80) ÷ 220 = 21.82
[b]Difference = Expected Sales − Actual Sales
  Example: Cell$_{\text{young adults, product A}}$ = 21.82 − 20.00 = 1.82
[c]Chi-square value = $\dfrac{(\text{Difference})^2}{\text{Expected sales}}$
  Example: Cell$_{\text{young adults, product A}}$ = $(1.82)^2$ ÷ 21.82 = .15

squares the difference and negative signs are eliminated. The result is a measure that acts just like the similarity measures used in earlier examples. Negative values indicate less association (similarity) and positive values indicate greater association. The chi-square values for each cell are also shown in Table 10.3.

The cells with large positive similarity values (indicating a positive association) are young adults–product B (+7.58), middle age–product A (+1.62), and senior citizens–product C (+2.10). Each of these pairs of categories should be close together on a perceptual map. Cells with large negative similarity values (meaning that expected sales exceeded actual sales, or a negative association) were young adults–product C (−1.94), middle age–product B (−2.47), and senior citizens–product A (−1.17). Where possible, these categories should be far apart on the map.

Creating the Perceptual Map

The similarity (signed chi-square) values provide a standardized measure of association, much like the similarity judgments in the earlier candy bar example. With these association measures, CA creates a metric distance measure and creates orthogonal dimensions upon which the categories can be placed to best account for the strength of association represented by the chi-square distances. As we did in the MDS example, we can consider first a one-dimensional solution, then expand to two dimensions and continue until we reach the maximum number of dimensions. The maximum is one less than the smaller of the number of rows or columns. In this example, we can only have two dimensions (number of rows or columns minus one = 3 − 1 = 2). The two-dimensional perceptual map is shown in Figure 10.10.

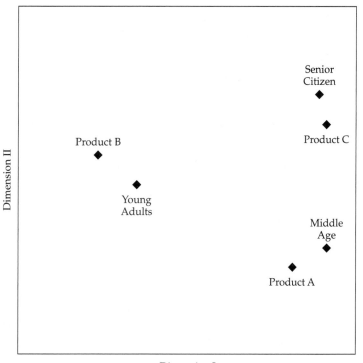

FIGURE 10.10 Perceptual Map from Correspondence Analysis

Corresponding to our examination of the chi-square distances, the age group of young adults is closest to product B, middle age is closest to product A, and senior citizens is closest to product C. Likewise, the negative associations are also represented in the positions of products and age groups. The researcher can examine the perceptual map to understand the product preferences among age groups based on their sales patterns. But, as with MDS, we do not know why the sales patterns existed, but only how to identify these patterns.

Stage 1: Objectives of CA

Researchers are constantly faced with the need to "quantify the qualitative data" found in nominal variables. CA differs from other interdependence techniques in its ability to accommodate both nonmetric data and nonlinear relationships. It performs dimensional reduction similar to multidimensional scaling or factor analysis and a type of perceptual mapping, in which categories are represented in the multidimensional space. Proximity indicates the level of association among row or column categories. CA can address either of two basic objectives:

1. *Association among row or column categories.* CA can be used to examine the association among the categories of just a row or column. A typical use is the examination of the categories of a scale, such as the Likert scale (five categories from "strongly agree" to "strongly disagree") or other qualitative scales (e.g., excellent, good, poor, bad). The categories can be compared to see if two can be combined (i.e., they are in close proximity on the map) or if they do provide discrimination (i.e., they are located separately in the perceptual space).

2. *Association between row <u>and</u> column categories.* In this application, interest lies is portraying the association between categories of the rows and columns, such as our example of product by age group. This use is most similar to the previous example of MDS and has propelled CA into more widespread use across many research areas.

The researcher must determine the specific objectives of the analysis because certain decisions are based on which type of objective is chosen. CA provides a multivariate representation of interdependence for nonmetric data not possible with other methods. But as a compositional method, the researcher must ensure that all of the relevant variables appropriate for the research question have been included. This is in contrast to the decompositional MDS procedures described earlier, which require only the overall measure of similarity.

Stage 2: Research Design of CA

Correspondence analysis requires only a rectangular data matrix (cross-tabulation) of nonnegative entries. The rows and columns do not have predefined meanings (i.e., attributes do not always have to be rows, and so on) but instead represent the responses to one or more categorical variables. The categories for a row or column need not be a single variable but can represent any set of relationships. A prime example is the "pick any" method [14, 15], in which respondents are given a set of objects and characteristics. The respondents then indicate which objects, if any, are described by the characteristics. Note that the respondent may choose any number of objects for each characteristic, rather than a prespecified number (i.e., choose only the object best described or the top two objects). In this situation, the cross-tabulation table would be the total number of times each object was described by each characteristic.

The cross-tabulation of more than two variables in a multiway matrix form is known as **multiple correspondence analysis.** In a procedure quite similar to two-way analysis, the additional variables are "fitted" so that all the categories are placed in the same multidimensional space.

Stage 3: Assumptions in CA

Correspondence analysis shares with the more traditional MDS techniques a relative freedom from assumptions. The use of strictly nonmetric data in its simplest form (cross-tabulated data) represents linear and nonlinear relationships equally well. The lack of assumptions, however, must not cause the researcher to neglect the efforts to ensure the comparability of objects and, because this is a compositional technique, the completeness of the attributes used.

Stage 4: Deriving CA Results and Assessing Overall Fit

With a cross-tabulation table, the frequencies for any row–column combination of categories are related to other combinations based on the marginal frequencies. This procedure yields a conditional expectation (a chi-square value). Once obtained, these chi-square values are standardized and converted to a distance metric, and then a process much like multidimensional scaling defines lower-dimensional solutions. These "factors" simultaneously relate the rows and columns in a single joint plot. The result is a representation of categories of rows and/or columns (e.g., brands *and* attributes) in the same plot. A number of computer programs are available to perform correspondence analysis. Among the more popular programs are ANACOR and HOMALS, available with SPSS; CA from BMDP; CORRAN and CORRESP from PC-MDS [24]; and MAPWISE [21].

To assess overall fit, the researcher must first identify the appropriate number of dimensions and their importance. The maximum number of dimensions that can be estimated is one less than the smaller of the number of rows or columns. For example, with six columns and eight rows, the maximum number of dimensions would be five, which is six (the number of columns) minus one. Eigenvalues, also known as singular values, are derived for each dimension and indicate the relative contribution of each dimension in explaining the variance in the categories. Some programs, such as those of SPSS, introduce a measure termed "inertia," which also measures explained variation and is directly related to the eigenvalue. The researcher selects the number of dimensions based on the overall level of explained variance desired and the incremental explanation gained by adding another dimension. A rule of thumb is that dimensions with inertia (eigenvalues) greater than .2 should be included in the analysis. As discussed with regard to perceptual mapping, using a three-dimensional or lower representation facilitates interpretation.

Stage 5: Interpretation of the Results

Once the dimensionality has been established, the researcher can identify a category's association with other categories by proximity after the appropriate normalization. The researcher must select the type of normalization, and determine whether comparisons are to be made between row categories, column categories, or row and column categories. In most instances, the researcher wishes to compare between row and column categories. There may be instances, however, in which the focus is on only rows or columns, such as when examining the categories of a scale to see if they can be combined. At this time, there is debate on the appropriateness of comparing between row and column categories. Some computer programs provide for a normalization procedure to allow for this direct comparison. If only a row or column normalization is available, alternative procedures are proposed to make all categories comparable [2, 21], but there is still disagreement as to their success [12]. In the cases for which direct comparisons are not possible, the general correspondence still holds and specific patterns can be distinguished.

If the researcher is interested in defining the character of one or more dimensions in terms of the row or column categories, there are descriptive measures that indicate the association of each category with a specific dimension. Similar in character to factor loadings, these measures detail the extent of association individually for each dimension as well as collectively. From the collective measures, an assessment of the fit for each category can also be made.

Stage 6: Validation of the Results

The compositional nature of correspondence analysis provides more specificity for the researcher to validate the results. As with all MDS techniques, an emphasis must be made to ensure generalizability through split- or multisample analyses. However, as with other perceptual mapping techniques, the generalizability of the objects (individually and as a set) must also be established. The sensitivity of the results to the addition or deletion of an object can be evaluated, as well as the addition or deletion of an attribute. The goal is to assess whether the analysis is dependent on only a few objects and/or attributes. In either instance, the researcher must understand the "true" meaning of the results in terms of the objects and attributes.

Overview of Correspondence Analysis

Correspondence analysis presents the researcher with a number of advantages. First, the simple cross-tabulation of multiple categorical variables, such as product attributes versus brands, can be represented in a perceptual space. This approach allows the researcher either to analyze existing responses or to gather responses at the least restrictive measurement type, the nominal or categorical level. For example, the respondent need rate only yes or no for a set of objects on a number of attributes. These responses can then be aggregated in a cross-tabulation table and analyzed. Other techniques, such as factor analysis, require interval ratings of each attribute for each object.

Second, CA portrays not only the relationships between the rows and columns, but also the relationships between the categories of either the rows or the columns. For example, if the columns were attributes, multiple attributes in close proximity would all have similar profiles across products. This forms a group of attributes quite similar to a factor from principal components analysis.

Finally, and most important, CA can provide a joint display of row and column categories in the same dimensionality. Certain program modifications allow for interpoint comparisons in which relative proximity is directly related to higher association among separate points [1, 21]. When these comparisons are possible, they allow row and column categories to be examined simultaneously. An analysis of this type would enable the researcher to identify groups of products characterized by attributes in close proximity.

With the advantages of CA, however, come a number of disadvantages or limitations. First, the technique is descriptive and not at all appropriate for hypothesis testing. If the quantitative relationship of categories is desired, methods such as log-linear models are suggested. CA is best suited for exploratory data analysis. Second, CA, as is the case with many dimensionality-reducing methods, has no method for conclusively determining the appropriate number of dimensions. As with similar methods, the researcher must balance interpretability versus parsimony of the data representation. Finally, the technique is quite sensitive to outliers, in terms of either rows or columns (e.g., attributes or brands). Also, for purposes of generalizability, the problem of omitted objects or attributes is critical.

Illustration of MDS and CA

To demonstrate the use of MDS techniques, we examine data gathered in a series of interviews with company representatives from a cross section of potential customers. In the course of the perceptual mapping analysis, we apply both decompositional and compositional methods. The discussion first examines the initial three stages of the model-building process that are common to both methods. The discussion then focuses on the next two stages for decompositional methods. This is followed by a discussion of a compositional method, CA, applied to the same sample of respondents. The sixth stage addresses the validation of the analysis through comparison of the results from both types of methods. Finally, an overview of MDS results is presented.

Stage 1: Objectives of Perceptual Mapping

The purpose of the research is to explore HATCO's image and competitiveness. This exploration includes addressing the perceptions in the market of HATCO and nine major competitors, as well as investigating preferences, among potential customers. The data are analyzed in a two-phase plan: (1) identification of the position of HATCO in a perceptual map of major competitors in the market with an understanding of the dimensions comparison used by potential customers, and (2) assessment of the preferences toward HATCO relative to major competitors. Before proceeding with a discussion of the results, we briefly describe the data collection process.

Stage 2: Research Design of the Perceptual Mapping Study

The HATCO image study is comprised of in-depth interviews with 18 midlevel management personnel from different firms selected as representative of the potential customer base existing in the market. The nine competitors, plus HATCO, represent all the major firms in this industry and collectively have more than 85 percent of total sales. In the course of the interviews, three types of data were collected: similarity judgments, attribute ratings of firms, and preferences for each firm in different buying situations.

Similarity Data

The starting point for data collection was obtaining the perceptions of the respondents concerning the similarity or dissimilarity of HATCO and nine competing firms in the market. Similarity judgments were made with the comparison-of-paired-objects approach. The 45 pairs of firms [(10 × 9)/2] were presented to the respondents, who indicated how similar each was on a nine-point scale, with 1 being "not at all similar" and 9 being "very similar." Note that the values have to be transformed because increasing values for the similarity ratings indicate greater similarity, the opposite of a distance measure of similarity.

Attribute Ratings

In addition to the similarity judgments, ratings of each firm for eight attributes were obtained by two methods. The attributes included product quality, management orientation, service quality, delivery speed, price level, salesforce image, price flexibility, and manufacturing image. In the first method, each firm was rated on a six-point scale for each attribute. In the second method, each respondent was asked to pick the firms best characterized by each attribute. As with the "pick any" method [14, 15], the respondent could pick any number of firms for each attribute.

Preference Evaluations

The final data assessed the preferences of each respondent for the 10 firms in three different buying situations: a straight-rebuy, a modified-rebuy, and a new-buy situation. In each situation, the respondents ranked the firms in order of preference for that particular type of purchase. For example, in the straight-rebuy situation, the respondent indicated the most preferred firm for the simple reordering of products (rank order = 1), the next most preferred (rank order = 2), and so on. Similar preferences were gathered for the remaining two buying situations.

Stage 3: Assumptions in Perceptual Mapping

The assumptions of MDS and CA deal primarily with the comparability and representativeness of the objects being evaluated and the respondents. With regard to the sample, the sampling plan emphasized obtaining a representative sample of HATCO customers. Moreover, care was taken to obtain respondents of comparable position and market knowledge. Because HATCO and the other firms serve a fairly distinct market, all the firms evaluated in the perceptual mapping should be known, ensuring that positioning discrepancies can be attributed to perceptual differences among respondents.

Multidimensional Scaling: Stages 4 and 5

Having specified the 10 firms to be included in the image study, HATCO's management specified that both decompositional (MDS) and compositional (CA) approaches were to be employed in constructing the perceptual maps. We first discuss a series of decompositional techniques, then examine a compositional approach to perceptual mapping.

Stage 4: Deriving MDS Results and Assessing Overall Fit

INDSCAL [4] was used to develop both a composite, or aggregate, perceptual map and the measures of the differences between respondents in their perception. The 45 similarity judgments from the 18 respondents were input as separate matrices, but a matrix of mean scores was calculated to illustrate the general pattern of similarities (see Table 10.4). The table also details the high similarities (greater than 6.0) as well as the lowest similarity for each firm. With these relationships, the basic patterns can be identified and available for comparison to the resulting map.

TABLE 10.4 Mean Similarity Ratings for HATCO and Nine Competing Firms

					Firm					
Firm	HATCO	A	B	C	D	E	F	G	H	I
HATCO	0.00									
A	6.61	0.00								
B	5.94	5.39	0.00							
C	2.33	2.61	3.44	0.00						
D	2.56	2.56	4.11	6.94	0.00					
E	4.06	2.39	2.17	4.06	2.39	0.00				
F	2.50	3.50	4.00	2.22	2.17	4.06	0.00			
G	2.33	2.39	3.72	2.67	2.61	3.67	2.28	0.00		
H	2.44	4.94	6.61	2.50	7.06	5.61	2.83	2.56	0.00	
I	6.17	6.94	2.83	2.50	2.50	3.50	6.94	2.44	2.39	0.00

Maximum and Minimum Similarity Ratings for Each Firm

Similarities > 6.0	A, I	HATCO I	H	D	C, H	None	I	None	B, D	HATCO A, F
Lowest similarities	C, G	E, G	E	F	F	B	C	F	I	H

Note: Similarity ratings are on a nine-point scale (1 = not at all similar, 9 = very similar)

The first analysis of the MDS results is to determine the appropriate dimensionality and portray the results in a perceptual map. To do so, the researcher should consider both the indices of fit at each dimensionality and the researcher's ability to interpret the solution. Table 10.5 shows the indices of fit for solutions of two to five dimensions (a one-dimensional solution was not considered a viable alternative for 10 firms). As the table shows, there is substantial improvement in moving from two to three dimensions, after which the improvement diminishes somewhat and remains consistent as we increase in the number of dimensions. Balancing this improvement in fit against the increasing difficulty of interpretation, the two- or three-dimensional solutions seem the most appropriate. For purposes of illustration, the two-dimensional solution is selected for further analyses,

TABLE 10.5 Assessing Overall Model Fit and Determining the Appropriate Dimensionality

Dimensionality of the Solution	Average Measures of Fit[a]			
	Stress[b]	Percentage Change	R Squared[c]	Percentage Change
5	.20068	—	.6303	—
4	.21363	6.4	.5557	11.8
3	.23655	10.7	.5007	9.9
2	.30043	27.0	.3932	21.5

[a]Average across 18 individual solutions
[b]Kruskal's stress formula
[c]Proportion of original similarity ratings accounted for by scaled data (distances) from the perceptual map

but the methods we discuss here could just as easily be applied to the three-dimensional solution. The researcher is encouraged to explore other solutions to assess whether any substantively different conclusions would be reached based on the dimensionality selected.

The two-dimensional aggregate perceptual map is shown in Figure 10.11. HATCO is most closely associated with firm A, with respondents considering them almost identical. Other pairs of firms considered highly similar based on their proximity are E and G, D and H, and F and I. Comparisons can also be made between these firms and HATCO. HATCO differs from C, E, and G primarily on dimension II, whereas dimension I differentiates HATCO most clearly from firms B, C, D, and H in one direction and firms F and I in another direction. All of these differences are reflected in their relative positions in the perceptual map. Similar comparisons can be made among all sets of firms. To understand the sources of these differences, however, the researcher must interpret the dimensions.

The researcher can also look at the fit of the solution in a scatterplot of actual distances (scaled similarity values) versus fitted distances from the perceptual map (see Figure 10.12). This plot can identify true outliers that are not well represented by the current solution. If a consistent set of objects or individuals is identified as outliers, they can be considered for deletion. In this instance, no firm exhibits a large number of outlying points that would make it a candidate for elimination from the analysis.

In addition to developing the composite perceptual map, INDSCAL also provides the means for assessing one of the assumptions of MDS, the homogeneity of respondents' perceptions. Weights are calculated for each respondent indicating the

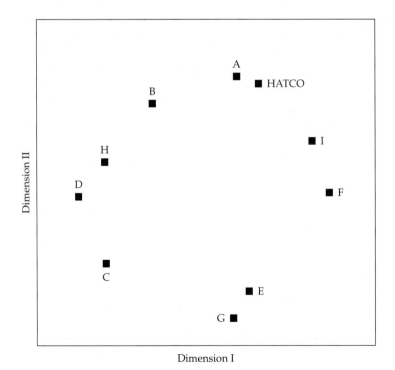

FIGURE 10.11 Perceptual Map of HATCO and Major Competitors

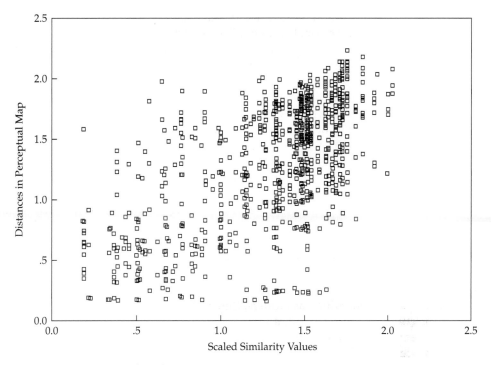

FIGURE 10.12 Scatterplot of Linear Fit

correspondence of their own perceptual space and the aggregate perceptual map. These weights provide a measure of comparison among respondents because respondents with similar weights have similar individual perceptual maps. INDSCAL also provides a measure of fit for each subject by correlating the computed scores and the respondent's original similarity ratings. Table 10.6 (p. 560) contains the weights and measures of fit for each respondent, and Figure 10.13 (p. 560) is a graphical portrayal of the individual respondents based on their weights. Examination of the weights (Table 10.6) and Figure 10.12 reveals that the respondents are quite homogeneous in their perceptions, because the weights show few substantive differences on either dimension, and no distinctive "clusters" of individuals emerge. This is shown in Figure 10.12 by all of the individual weights falling roughly on a straight line, indicating a consistent weight between dimensions I and II. The distance of each individual weight from the origin indicates its level of fit with the solution. Better fits are shown by farther distances from the origin. Thus, respondents 4, 7, and 10 have the highest fit, and respondents 1 and 9 have the lowest fit. The fit values show relative consistency in both the stress and R^2 measures, with mean values of .300 (stress) and .393 (R^2). Moreover, all respondents are well represented by the composite perceptual map, with the lowest measure of fit being .27. Thus, no individual should be eliminated due to a poor fit in the two-dimensional solution.

Incorporating Preferences in the Perceptual Map   Up to this point we have dealt only with judgments of firms based on similarity, but many times we may wish to extend the analysis to the decision-making process and to understand the respondent's preferences for the objects (in this case, firms). To do so, we can employ

TABLE 10.6 Measures of Individual Differences in Perceptual Mapping: Respondent-Specific Measures of Fit and Dimension Weights

Subject	Measures of Fit		Dimension Weights	
	Stress[a]	R Squared[b]	Dimension I	Dimension II
1	.358	.274	.386	.353
2	.297	.353	.432	.408
3	.302	.378	.395	.472
4	.237	.588	.572	.510
5	.308	.308	.409	.375
6	.282	.450	.488	.461
7	.247	.547	.546	.499
8	.302	.332	.444	.367
9	.320	.271	.354	.382
10	.280	.535	.523	.511
11	.299	.341	.397	.429
12	.301	.343	.448	.378
13	.292	.455	.497	.456
14	.302	.328	.427	.381
15	.290	.371	.435	.426
16	.311	.327	.418	.390
17	.281	.433	.472	.458
18	.370	.443	.525	.409
Average[c]	.300	.393		

[a]Kruskal's stress formula
[b]Proportion of original similarity ratings accounted for by scaled data (distances) from the perceptual map
[c]Average across 18 individual solutions

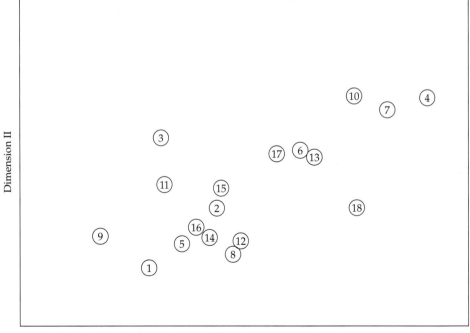

FIGURE 10.13 Individual Subject Weights

several MDS techniques that allow for the estimation of ideal points, from which preferences for the objects can be determined. In this example, we employ an external method of preference formation (PREFMAP [6]) that utilizes the aggregate perceptual maps derived in the prior section.

Preferences were measured by asking respondents to detail their preferences for firms in purchasing situations. Here we examine the preferences for firms in the new-buy situation. The program inputs include the coordinates of firms in the aggregate perceptual map and the preferences of five respondents. The preference rankings for these five respondents are given in Table 10.7.

The program can estimate both point- and vector-based ideal points. In this situation, HATCO management decided on the point representations, which resulted in the derivation of ideal points for the five respondents, plus an ideal point for the "average" subject. The results are shown in Figure 10.14 (p. 562). The distances of each firm to the ideal points are provided in Table 10.7. Lower values indicate a closer proximity to the ideal point.

All of the respondents form a general group somewhat clustered around the average. However, we can still detect differences in proximity for the group as a whole as well as for individual firms. First, the group as a whole is closer to firms C, D, F, and H, whereas HATCO, A, B, E, and G are somewhat farther away. Note that in this case both proximity and dimensionality are important. The assumption of an external analysis is that as you change your perceptual map position on the dimensions, you can change your proximity to the ideal points and your preference ordering. In terms of the individual respondents, there are some close associations. Respondent 1 has a relatively close association with firm F and respondents 3 and 5 have close associations with firms C, D, and H. Although this group of respondents is relatively homogeneous in its preferences, as indicated by its clustering together, Figure 10.13 still portrays each firm's relative position not only in perception, but now also in preference.

TABLE 10.7 Preference Data of the New-Buy Purchasing Situation for Selected Respondents

					Firm						
Subject	*HATCO*	*A*	*B*	*C*	*D*	*E*	*F*	*G*	*H*	*I*	*Fit[a]*
1	2	3	5	6	7	4	10	8	1	9	
	−.867	−.972	−.920	−1.096	−1.095	−.636	−.264	−1.054	−.854	−.371	.787
2	5	2	7	6	9	3	4	1	10	8	
	−1.049	−1.056	−.622	−.906	−.642	−1.111	−.879	−1.596	−.413	−.825	.961
3	4	1	8	7	6	9	3	5	10	2	
	−.894	−.868	−.448	−.133	−.106	−.449	−.726	−.576	−.132	−.779	.855
4	4	3	10	2	7	8	6	1	9	5	
	−1.098	−1.128	−.736	−1.060	−.813	−1.136	−.822	−1.672	−.544	−.790	.884
5	4	1	8	7	9	3	5	2	10	6	
	−.905	−.868	−.401	−.362	−.188	−.769	−.870	−1.019	−.126	−.838	.977
Average	NA	NA	NA	NA	NA	NA	NA	NA	NA	NA	
	−.916	−.931	−.580	−.668	−.525	−.776	−.666	−1.140	−.370	−.674	.990

Note: Top values in each cell are original preference rankings, and bottom entries are signed squared distances of firms to the ideal point. NA indicates average rankings are not available.
[a]Fit is the squared correlation between preferences and signed distance values.

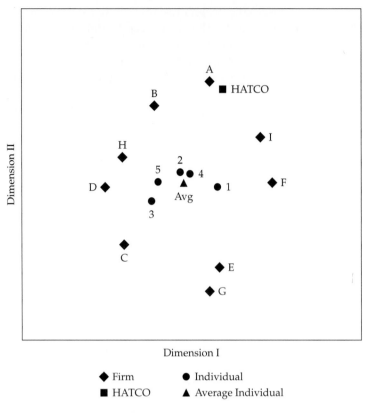

FIGURE 10.14 Map of Ideal Points for Selected and Average
Respondents: New-Buy Purchasing Situation

Stage 5: Interpretation of the Results

Once the perceptual map has been established, we can begin the process of interpretation. Because the INDSCAL procedure uses only the overall similarity
judgments, HATCO also gathered ratings of the firms on eight attributes—the
seven evaluations used before and a new variable, X_{15}, representing strategic orientation—descriptive of typical strategies followed in this industry. The ratings
for each firm were then averaged across the respondents for a single overall rating. To provide an objective means of interpretation, PROFIT [3], a vector model,
was used to match the ratings to the firm positions in the perceptual map. The
results of applying the ratings data to the composite perceptual map are shown
in Figure 10.15. This shows that there are three distinct "groups" or dimensions
of attributes. The first involves X_1 (delivery speed), X_2 (price level), and X_3 (price
flexibility), which are all pointed in the same direction, and X_5 (overall service),
which is in the direction opposite to that of the three price-oriented variables. This
directional difference indicates a negative correspondence of service versus the
three other variables. The second set of variables reflects more global evaluations,
consisting of the two image variables, X_6 (salesforce image) and X_4 (manufacturer
image), along with a new variable, X_{15} (strategic orientation). Finally, X_7 (product quality) runs almost perpendicular to the price and service dimension, indicating a separate and distinct evaluative dimension.

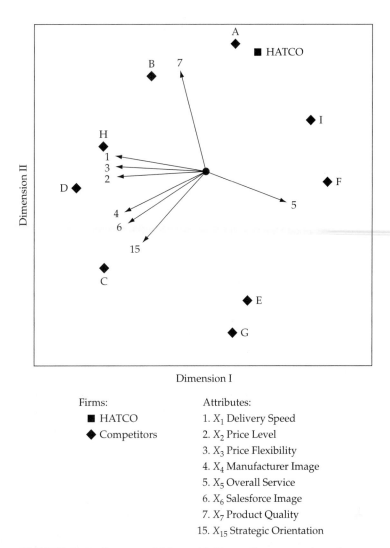

FIGURE 10.15 Perceptual Map with Vector Representation of Attributes

Firms:
- ■ HATCO
- ◆ Competitors

Attributes:
1. X_1 Delivery Speed
2. X_2 Price Level
3. X_3 Price Flexibility
4. X_4 Manufacturer Image
5. X_5 Overall Service
6. X_6 Salesforce Image
7. X_7 Product Quality
15. X_{15} Strategic Orientation

To interpret the dimensions, the researcher looks for attributes closely aligned with the axis. In this instance, the groups of attributes are slightly angled from the original axes. However, because the perceptual map is a point representation, the axes can be rotated without any impact on the relative positions. If we rotate the axes slightly (much as is done in factor analysis in chapter 3), we would now have a dimension of price and service versus a second dimension of product quality. Although it is not necessary to actually perform the rotation because firms can be compared directly on the attribute vectors, many times rotation can aid in gaining a more fundamental understanding of the perceived dimension. Rotation is especially helpful in solutions involving more than two dimensions.

To determine the values for any firm on an attribute vector, we need to calculate the projections from the firm to the vector. To assist in interpretation, the PROFIT program provides the projection values for each attribute, which are listed as the second row of values for each variable in Table 10.8. Also included are the original ratings (values in the first row) to see how well the vector represents the respondent's actual perceptions. For example, the order of the firm's ratings on product quality (from highest to lowest) were H, B, A, HATCO, F, E, D, C, I, and G. Using the vector projections, we see that the order of firms is A, B, HATCO, H, D, I, F, C, E, and G. There is a fairly close correspondence between the original and calculated values, particularly among the top four firms. A statistical measure of fit for each attribute is the correlation between the original ratings and the vector projections. In the case of product quality, the correlation is .710. The researcher should not expect a perfect fit for several reasons. First, the perceptual map is based on the overall evaluation, which may not be directly comparable to the ratings. Second, the ratings are averaged across respondents, so their values are determined by differences between individuals as well as differences between firms. Given these factors, the level of fit for the attributes individually and collectively is acceptable.

Overview of the Decompositional Results

The decompositional methods employed in this image study illustrate both the advantages and disadvantages of this approach. The use of overall similarity judgments provides a perceptual map based on only the relevant criteria chosen by each respondent. However, the attribute-free techniques also demonstrate the notable difficulty of interpreting the perceptual map in terms of specific attributes. The researcher is required to infer the bases for comparison among objects without direct confirmation from the respondent.

The results do provide insight into the relative perceptions of HATCO and the other nine firms. In terms of perceptions, HATCO is most associated with firm A

TABLE 10.8 Interpreting the Perceptual Map with PROFIT

Original Attribute Ratings and Projections on Fitted Vectors

					Firm						
Variables	HATCO	A	B	C	D	E	F	G	H	I	Fit[a]
X_7 Product quality	6.94	7.17	7.67	3.22	4.78	5.11	6.56	1.61	8.78	3.17	.710
	1.0038	1.1529	1.1057	−.6284	.2474	−1.4128	−.5099	−1.6815	.5717	.1511	
X_{15} Strategic orientation	4.0	1.83	6.33	7.67	6.00	5.78	5.50	6.11	7.50	4.17	.785
	−1.2912	−1.1645	−.2293	1.5350	1.1914	.5743	−.9644	.9370	.6718	−1.2600	
X_5 Overall service	6.94	5.67	3.39	3.67	3.67	6.94	6.44	7.22	4.94	6.11	.842
	.1994	−.0704	−.9018	−.8079	−1.3700	.8879	1.4219	.8185	−1.2092	1.0317	
X_1 Delivery speed	4.00	3.39	7.33	6.11	7.50	4.22	7.17	4.33	8.22	5.56	.510
	−.4202	−.1535	.7696	1.0137	1.4560	−.6928	−1.4604	−.5627	1.2144	−1.1641	
X_2 Delivery level	5.16	3.47	6.41	5.88	6.06	4.94	5.29	4.82	8.35	4.65	.653
	−.5571	−.2965	.6725	1.1322	1.4903	−.5561	−1.4639	−.3885	1.1998	−1.2328	
X_6 Salesforce image	5.11	1.22	5.78	7.89	6.56	3.83	4.28	6.94	8.67	4.72	.720
	−1.1133	−.9244	.1007	1.5053	1.4050	.1842	−1.2369	.5012	.9313	−1.3530	
X_3 Price flexibility	5.33	3.72	6.33	5.56	6.39	4.72	5.28	5.22	7.33	5.11	.651
	−.5138	−.2509	.7045	1.0956	1.4812	−.6007	−1.4666	−.4450	1.2060	−1.2122	
X_4 Manufacturer image	4.17	1.56	6.06	8.22	7.72	4.28	3.89	6.33	7.72	5.06	.829
	−1.0905	−.8962	.1333	1.4956	1.4201	.1443	−1.2587	.4553	.9531	−1.3563	

Note: Top values in the cells are the original attributes ratings; bottom numbers are projections for the fitted vectors.
[a]Fit is measured as the correlation between the original attribute ratings and the vector projections.

and somewhat with firms B and I. There do exist some competitive groupings (e.g., F and I, E and G), which must also be considered. None of the firms are markedly distinct so as to be considered an outlier. HATCO can be considered "average" on several attributes (X_1, X_2, and X_3), but has lower scores on several attributes (X_4, X_6, and X_{15}) countered by a high score on attribute X_7. Finally, HATCO has no real advantage in terms of proximity to respondent ideal points, with other firms, such as D, H, and F, being located in much closer proximity to the ideal points for several respondents. These results provide HATCO insight into not only its perceptions, but also the perceptions of the other major competitors in the market.

Correspondence Analysis: Stages 4 and 5

An alternative to attribute-free perceptual mapping is CA, a compositional method based on eight binary firm ratings provided for each firm (i.e., the yes–no ratings of each firm on each attribute). In this attribute-based method, the perceptual map is a joint space, showing both attributes and firms in a single representation. Moreover, the positions of firms are relative not only to the other firms included in the analysis but also to the attributes selected as well.

Stage 4: Deriving the CA

Preparing the data for analysis involves creating a cross-tabulation matrix relating the attributes (represented as rows) to the ratings of firms (the columns). The individual entries in the matrix are the number of times a firm is rated as possessing a specific attribute. Thus, simple frequencies are provided for each firm across the entire set of attributes (see Table 10.9).

Correspondence analysis is based on a transformation of the chi-square value into a metric measure of distance. The chi-square value is calculated as the actual frequency of occurrence minus the expected frequency. Thus a negative value indicates, in this case, that a firm was rated less often than would be expected. The expected value for a cell (any firm–attribute combination in the cross-tabulation table) is based on how often the firm was rated on other attributes and how often other firms were rated on that attribute. (In statistical terms, the expected value is based on the row [attribute] and column [firm] marginal probabilities.) Table 10.10 (p. 566) contains the transformed (metric) chi-square distances for each cell of cross-tabulation from Table 10.9. High positive values indicate a strong degree of "correspondence" between the attribute and firm, and negative values have an

TABLE 10.9 Cross-Tabulated Frequency Data of Attribute Descriptors for HATCO and Nine Competing Firms

					Firms					
Variables	*HATCO*	*A*	*B*	*C*	*D*	*E*	*F*	*G*	*H*	*I*
X_7 Product quality	4	3	1	13	9	6	3	18	2	10
X_{15} Strategic orientation	15	16	15	11	11	14	16	12	14	14
X_5 Overall service	15	14	6	4	4	15	14	13	7	13
X_1 Delivery speed	16	13	8	13	9	17	15	16	6	12
X_2 Price level	14	14	10	11	11	14	12	13	10	14
X_6 Salesforce image	7	18	13	4	9	16	14	5	4	16
X_3 Price flexibility	6	6	14	10	11	8	7	4	14	4
X_4 Manufacturer image	15	18	9	2	3	15	16	7	8	8

TABLE 10.10 Measures of Similarity in CA: Chi-Square Distances

					Firms					
Variables	HATCO	A	B	C	D	E	F	G	H	I
X_7 Product quality	−1.27	−1.83	−2.08	3.19	1.53	−.86	−1.73	4.07	−1.42	.97
X_{15} Strategic orientation	.02	−.13	.76	−.01	.04	−.73	.07	−.60	1.07	−.20
X_5 Overall service	1.08	.40	−1.10	−1.52	−1.48	.57	.59	.65	−.36	.53
X_1 Delivery speed	.68	−.51	−.95	.95	−.27	.40	.20	.86	−1.15	−.37
X_2 Price level	.19	−.19	−.30	.37	.42	−.30	−.54	.08	.20	.23
X_6 Salesforce image	−1.32	−1.49	1.15	−1.54	.23	.81	.55	−1.80	−1.44	1.39
X_3 Price flexibility	−1.02	−1.28	2.37	1.27	1.71	−.73	−.83	−1.59	2.99	−1.66
X_4 Manufacturer image	1.24	1.69	−.01	−2.14	−1.76	.72	1.32	−1.07	.10	−.85

opposite interpretation. For example, the high values for HATCO and firms A and F with the manufacturer image attribute (X_4) indicate that they should be located close together on the perceptual map if at all possible. Likewise, the high negative values for firms C and D on the same variable would indicate that their position should be far from the attribute's location.

CA tries to satisfy all of these relationships simultaneously by producing dimensions representing the chi-square distances. To determine the dimensionality of the solution, the researcher examines the cumulative percent of variation explained, much as in factor analysis, and determines the appropriate dimensionality. Table 10.11 contains the eigenvalues and cumulative and explained percentages of variation for each dimension up to the maximum of seven. Again, the researcher balances the desire for increased explanation versus interpretability. A two-dimensional solution in this situation explains 86 percent of the variation, whereas increasing to a three-dimensional solution adds only an additional 10 percent. The researcher must compare and contrast the additional variance explained in relation to the increased complexity in interpreting the results. Thus a two-dimensional solution is deemed adequate for further analysis.

Stage 5: Interpreting CA Results

The attribute-based perceptual map shows the relative proximities of both firms and attributes (see Figure 10.16). If we focus on the firms first, we see that the pattern of firm groups is similar to that found in the MDS results. Firms A, E, F, and I, plus HATCO form one group; firms C and D and firms H and B form two

TABLE 10.11 Determining the Appropriate Dimensionality in CA

Dimension	Eigenvalue (Singular Value)	Inertia (Normalized Chi-Square)	Percentage Explained	Cumulative Percentage
1	.27666	.07654	53.1	53.1
2	.21866	.04781	33.2	86.3
3	.12366	.01529	10.6	96.9
4	.05155	.00266	1.8	98.8
5	.02838	.00081	.6	99.3
6	.02400	.00058	.4	99.7
7	.01951	.00038	.3	100.0

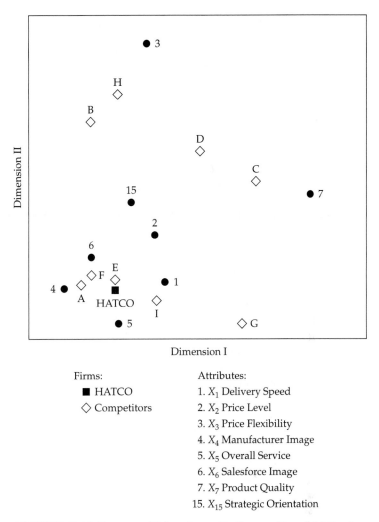

FIGURE 10.16 Perceptual Mapping with Compositional Methods: Correspondence Analysis

other similar groups. However, the relative proximities of the members in each group differ somewhat from the MDS solution. Also, firm G is more isolated and distinct, and firms F and E are now seen as more similar to HATCO.

It may be helpful to interpret the dimensions if row or column normalizations are used. For these purposes, the inertia (explained variation) of each dimension can be attributed among rows and columns. Table 10.12 (p. 568) provides the contributions of both sets of categories to each dimension. For the attributes, we can see that X_7 (product quality) is the primary contributor to dimension I, and X_4 (manufacturer image) is a secondary contributor. Note that these are the two extreme attributes in terms of their location on dimension I (i.e., highest or lowest values on dimension I). Between these two attributes, 86 percent of dimension I is accounted for. A similar pattern follows for dimension II, for which X_3 (price flexibility) is the primary contributor followed by X_5 (overall service). If we shift our focus to the 10 firms, we see a somewhat more balanced situation, with three

TABLE 10.12 Interpreting the Dimensions and Their Correspondence to Firms and Attributes

Object	Coordinates		Contribution to Inertia[a]		Explanation by Dimension[b]		
	I	II	I	II	I	II	Total
Attributes							
X_7 Product quality	.044	1.235	.001	.689	.002	.989	.991
X_{15} Strategic orientation	−.676	−.285	.196	.044	.789	.111	.901
X_5 Overall service	.115	.046	.007	.001	.469	.058	.527
X_1 Delivery speed	1.506	0.298	.665	.033	.961	.030	.991
X_2 Price level	−.081	.245	.004	.045	.093	.678	.772
X_6 Salesforce image	−.440	−.099	.087	.006	.358	.014	.372
X_3 Price flexibility	−.202	−.502	.018	.142	.138	.677	.816
X_4 Manufacturer image	.204	−.245	.022	.040	.289	.330	.619
Firms							
HATCO	−.247	−.293	.024	.042	.206	.228	.433
A	−.537	−.271	.125	.040	.772	.156	.928
B	−.444	.740	.063	.224	.294	.648	.942
C	1.017	.371	.299	.050	.882	.093	.975
D	.510	.556	.074	.111	.445	.418	.863
E	−.237	−.235	.025	.031	.456	.356	.812
F	−.441	−.209	.080	.023	.810	.144	.954
G	.884	−.511	.292	.123	.762	.201	.963
H	−.206	.909	.012	.289	.049	.748	.797
I	.123	−.367	.006	.066	.055	.390	.446

[a]Proportion of dimension's inertia attributable to each category
[b]Proportion of category variation accounted for by dimension

firms (A, C, and G) contributing above the average of 10 percent. For the second dimension, four firms (B, D, G, and H) have contributions above average. Although the comparisons in this example are between both sets of categories and not restricted to a single set of categories (either row or column), these measures of contribution demonstrate the ability to interpret the dimension when desired.

One final measure provides an assessment of fit for each category. Comparable to squared factor loadings in factor analysis (see chapter 3 for a more detailed discussion), these values represent the amount of variation in the category accounted for by the dimension. A total value represents the total amount of variation across all dimensions, with the maximum possible being 100 percent. Table 10.12 contains fit values for each category on each dimension. As we can see the fit values range from a high of 99.1 for X_1 (delivery speed) and X_7 (product quality) to a low of .372 for X_6 (salesforce image). Among the attributes, only X_6 (salesforce image) has a value below 50 percent and only two firms (HATCO and firm I) fall below this value. Even though these are somewhat low, they still represent a substantial enough explanation to retain them in the analysis and deem the analysis of sufficient practical significance.

Overview of CA

These and other comparisons highlight the differences between MDS and CA methods and their results. CA results provide a means for directly comparing the similarity or dissimilarity of firms and the associated attributes, whereas MDS allows only for the comparison of firms. But the CA solution is conditioned on the set of attributes included. It assumes that all attributes are appropriate for all firms and that the same dimensionality applies to each firm. Thus the resulting perceptual map should always be viewed only in the context of both the firms and attributes included in the analysis.

Correspondence analysis is a quite flexible technique applicable to a wide range of issues and situations. The advantages of the joint plot of attributes and objects must always be weighed against the inherent interdependencies that exist and the potentially biasing effects of a single inappropriate attribute or firm, or perhaps more important, the omitted attribute of a firm. Yet CA still provides a powerful tool for gaining managerial insight into the relative position of firms and the attributes associated with those positions.

Stage 6: Validation of the Results

Perhaps the strongest internal validation of this analysis is to assess the convergence between the results from the separate decompositional and compositional techniques. Each technique employs different types of consumer responses, but the resulting perceptual maps are representations of the same perceptual space and should correspond. If the correspondence is high, the researcher can be assured that results reflect the problem as depicted. The researcher should note that this type of convergence does not address the generalizability of the results to other objects or samples of the population.

Comparison of the decompositional and compositional methods, shown in Figures 10.11 and 10.16, can take two approaches: examining the relative positioning of objects and interpreting the axes. Let us start by examining the positioning of firms. When Figures 10.11 and 10.16 are rotated to obtain the same perspective, they show quite similar patterns of firms reflecting two groups: firms B, H, D, and C versus firms E, F, G, and I. While the relative distances among firms do vary between the two perceptual maps, we still see HATCO associated strongly with firms A and I in each perceptual map. CA produces more distinction between the firms, but its objective is to define firm positions as a result of differences; thus it will generate more distinctiveness in its perceptual maps.

The interpretation of axes and distinguishing characteristics also shows similar patterns in the two perceptual maps. For the decompositional method shown in Figure 10.15, we noted in the earlier discussion that by rotating the axes we would obtain a clearer interpretation. If we rotate the axes, then dimension I becomes associated with delivery speed, price level, and price flexibility (X_1, X_2, and X_3), whereas dimension II reflects product quality (X_7). The remaining attributes are not associated strongly with either axis. In the correspondence analysis (Figure 10.15), we see that there are really three attribute groups representing evaluative dimensions: product quality (X_7), price flexibility (X_3), and all of the remaining attributes in a third group. This compares quite favorably with the decompositional results except in the case of price flexibility, which was not distinguished as a separate dimension.

Overall, although some differences do exist, owing to the characteristics of each approach, the convergence of the two results does provide some internal validity to the perceptual maps. Perceptual differences may exist for a few attributes, but the overall patterns of firm positions and evaluative dimensions are supported by both approaches. The disparity of the price flexibility attribute illustrates the differences in the two approaches. The decompositional method determines position based on overall judgments, with attributes applied only as an attempt to explain the positions. The compositional method positions firms according to the selected set of attributes, thus creating positions based on the attributes. Moreover, each attribute is weighted equally, thus potentially distorting the map with irrelevant attributes. These differences do not make either approach better or optimal but instead must be understood by the researcher to ensure selection of the method most suited to the research objectives.

A Managerial Overview of MDS Results

Perceptual mapping is a unique technique providing overall comparisons not readily possible with any other multivariate method. As such, its results present a wide range of perspectives for managerial use. The most common application of the perceptual maps is for the assessment of image for any firm or group of firms. As a strategic variable, image can be quite important as an overall indicator of market presence or position. In this study, we found that HATCO is most closely associated with firms A and I, and most dissimilar from firms C, E, and G. Thus, when serving the same product markets, HATCO can identify those firms considered similar to or different from its image. With the results based not on any set of specific attributes, but instead on respondents' overall judgments, they present the benefit of not being subject to a researcher's subjective judgments as to attributes to include or how to weight the individual attributes. This is in the "true spirit" of assessing image. However, MDS technologies are less useful in guiding strategy because they are less helpful in prescribing how to change image. The global responses that are advantageous for comparison now work against us in explanation. Although MDS techniques can augment the explanation of the perceptual maps, they must be viewed as supplemental and thus expect greater inconsistencies than if integral to the process. Thus, additional research may assist in helping to "explain" the relative positions.

To this end, CA results are a compromise approach in trying to portray perceptual maps from a compositional perspective. The comparison of CA results to those from the classical MDS solution reveals a number of consistencies, but some discrepancies as well. These should be expected, as the techniques are based on two different approaches. What we can see in these two solutions is some general pattern of associations between firms (such as HATCO and firms A and I) and between attributes, particularly the distinctiveness of X_3 (price flexibility) and X_7 (product quality). HATCO management can use these results not only as a guide to overall policy, but also as a framework for further investigation with other multivariate techniques into more specific research questions.

Summary

Multidimensional scaling is a set of procedures that may be used to display the relationships tapped by data representing similarity or preference. It has been used successfully (1) to illustrate market segments based on preference judgments, (2) to determine which products are more competitive with each other (i.e., are more similar), and (3) to deduce what criteria people use when judging objects (e.g., products, companies, advertisements). MDS can reveal relationships that appear to be obscured when one examines only the numbers resulting from a study. A visual perceptual map does much to emphasize the relationships between the stimuli under study. However, great care must be taken when attempting to use this technique. Misuse is common. The researcher should become very familiar with the technique before using it and should view the output as only the first step in the determination of perceptual information.

Questions

1. How does MDS differ from other interdependence techniques (cluster analysis and factor analysis)?
2. What is the difference between preference data and similarities data, and what impact does it have on the results of MDS procedures?
3. How are ideal points used in MDS procedures?
4. How do metric and nonmetric MDS procedures differ?
5. How can the researcher determine when the "best" MDS solution has been obtained?
6. How does the researcher identify the dimensions in MDS? Compare this procedure with the procedure for factor analysis.
7. Compare and contrast CA and MDS techniques.
8. Describe how "correspondence" or association is derived from a contingency table.

References

1. Carroll, J. Douglas, Paul E. Green, and Catherine M. Schaffer (1986), "Interpoint Distance Comparisons in Correspondence Analysis." *Journal of Marketing Research* 23 (August): 271–80.
2. Carroll, J. Douglas, Paul E. Green, and Catherine M. Schaffer (1987), "Comparing Interpoint Distances in Correspondence Analysis: A Clarification." *Journal of Marketing Research* 24 (November): 445–50.
3. Chang, J. J., and J. Douglas Carroll (1968), "How to Use PROFIT, a Computer Program for Property Fitting by Optimizing Nonlinear and Linear Correlation." Unpublished paper, Bell Laboratories, Murray Hill, N.J.
4. Chang, J. J., and J. Douglas Carroll (1969), "How to use INDSCAL, a Computer Program for Canonical Decomposition of *n*-way Tables and Individual Differences in Multidimensional Scaling." Unpublished paper, Bell Laboratories, Murray Hill, N.J.
5. Chang, J. J., and J. Douglas Carroll (1969), "How to Use MDPREF, a Computer Program for Multidimensional Analysis of Preference Data." Unpublished paper, Bell Laboratories, Murray Hill, N.J.
6. Chang, J. J., and J. Douglas Carroll (1972), "How to Use PREFMAP and PREFMAP2—Programs Which Relate Preference Data to Multidimensional Scaling Solution." Unpublished paper, Bell Laboratories, Murray Hill, N.J.

7. Green, P. E. (1975), "On the Robustness of Multidimensional Scaling Techniques." *Journal of Marketing Research* 12 (February): 73–81.
8. Green, P. E., and F. Carmone (1969), "Multidimensional Scaling: An Introduction and Comparison of Nonmetric Unfolding Techniques." *Journal of Marketing Research* 7 (August): 33–41.
9. Green, P. E., F. Carmone, and Scott M. Smith (1989), *Multidimensional Scaling: Concept and Applications.* Boston: Allyn & Bacon.
10. Green, P. E., and Vithala Rao (1972), *Applied Multidimensional Scaling.* New York: Holt, Rinehart and Winston.
11. Greenacre, Michael J. (1984), *Theory and Applications of Correspondence Analyses.* London: Academic Press.
12. Greenacre, Michael J. (1989), "The Carroll-Grenn-Schaffer Scaling in Correspondence Analysis: A Theoretical and Empirical Appraisal." *Journal of Marketing Research* 26 (August): 358–65.
13. Hoffman, Donna L., and George R. Franke (1986), "Correspondence Analysis: Graphical Representation of Categorical Data in Marketing Research." *Journal of Marketing Research* 23 (August): 213–27.
14. Holbrook, Morris B., William L. Moore, and Russell S. Winer (1982), "Constructing Joint Spaces from Pick-Any Data: A New Tool for Consumer Analysis." *Journal of Consumer Research* 9 (June): 99–105.

15. Levine, Joel H. (1979), "Joint-Space Analysis of 'Pick-Any' Data: Analysis of Choices from an Unconstrained Set of Alternatives." *Psychometrika* 44 (March): 85–92.

16. Lingoes, James C. (1972), *Geometric Representations of Relational Data*. Ann Arbor, Mich.: Mathesis Press.

17. Kruskal, Joseph B., and Frank J. Carmone (1967), "How to Use MDSCAL. Version 5-M, and Other Useful Information." Unpublished paper, Bell Laboratories, Murray Hill, N.J.

18. Kruskal, Joseph B., and Myron Wish (1978), *Multidimensional Scaling*. Sage University Paper Series on Quantitative Applications in the Social Sciences, 07–011, Beverly Hills, Calif.: Sage.

19. Lebart, Ludovic, Alain Morineau, and Kenneth M. Warwick (1984), *Multivariate Descriptive Statistical Analysis: Correspondence Analysis and Related Techniques for Large Matrices*. New York: Wiley.

20. Maholtra, Naresh (1987), "Validity and Structural Reliability of Multidimensional Scaling." *Journal of Marketing Research* 24 (May): 164–73.

21. Market ACTION Research Software, Inc. (1989), MAPWISE: *Perceptual Mapping Software*. Peoria, Ill.: Business Technology Center, Bradley University.

22. Raymond, Charles (1974), *The Art of Using Science in Marketing*. New York: Harper & Row.

23. Schiffman, Susan S., M. Lance Reynolds, and Forrest W. Young (1981), *Introduction to Multidimensional Scaling*. New York: Academic Press.

24. Smith, Scott M. (1989), PC–MDS: *A Multidimensional Statistics Package*. Provo, Utah: Brigham Young University.

25. Srinivasan, V., and A. D. Schocker (1973), "Linear Programming Techniques for Multidimensional Analysis of Preferences." *Psychometrica* 38 (September): 337–69.

Annotated Articles

The following annotated articles are provided as illustrations of the application of multidimensional scaling to substantive research questions of both a conceptual and managerial nature. The reader is encouraged to review the complete articles for greater detail on any of the specific issues regarding methodology or findings.

Wind, Yoram, and Patrick J. Robinson (1972), "Product Positioning: An Application of Multidimensional Scaling." In *Attitude Research in Transition*, ed. R. I. Haley (Chicago: American Marketing Association). 155–75.

To illustrate the applicability of multidimensional scaling in measuring product positioning, the authors present five different usages of the technique. The article maintains that in order to effectively plan a product or organizational strategy the manager needs to know consumers' subjective perceptions and preferences for the product or firm as well as relevant objective information about these items. Use of multiple variables in the development of preference and similarity mapping provided by MDS techniques allows researchers or managers to better interpret perceptions of their product, service, or firm relative to others. From the results, the authors are able to apply the findings toward identifying discrepancies between subjective and objective product evaluation. Additionally, they measure how the perceptions of organizations and products change over time. The results indicate that a firm may improve product positioning, segment markets, compare objective attributes and subjective consumer evaluations, and monitor changing perceptions by employing the findings of MDS.

The objective of each of the five applications of MDS is to provide the researcher with some form of diagram or map to facilitate the examination of the product (e.g., service, firm) in question relative to other products. The authors specify the objective of each particular study, stating its sample size, the number of dimensions, the procedure used, and the amount of variance accounted for by the model. Additionally, the authors identify the dimensions and describe the implications of the results. In all, each application demonstrates the measurement of product positioning by accounting for the perception and preferences of buyers for a given product in relation to its competitors. By demonstrating the ability of the technique to measure product positioning for differing products (calculators, new diet products, medical journals, financial services, and retail stores) and to make multiple comparisons (objective and subjective evaluations, products and market segments, and perceptions over time), the article explores the range of applications for which MDS is suited.

Robinson, Sandra L., and Rebecca J. Bennett (1995), "A Typology of Deviant Workplace

Behaviors: A Multidimensional Scaling Study." *Academy of Management Journal* 38(2):555–72.

This article seeks to develop a typology of deviant workplace behaviors using a multidimensional scaling technique. The authors define employee deviance as voluntary behavior that violates organizational norms and is destructive to the organization and/or its employees. To capture the full dimensionality of the construct, the authors derive a list of 45 deviant workplace behaviors from prior research. They then conduct a pretest with 70 respondents. Based on this pool of items, the MDS technique is empirically applied to establish a typology of deviant workplace behaviors. This classification schema identifies the underlying dimensions of deviant workplace behaviors and provides a basis of comparison among these behaviors. The results should aid in further theory development and provide managers with a means for fairly allocating punishments for deviant behaviors.

Similarities data are gathered from a sample of 180 respondents who evaluated the 45 deviant workplace behaviors as to their similarity to or difference from a target behavior. From these ratings, the spatial configurations of objects are derived based on the perceived difference between the objects. The authors determine that two dimensions best represent the data based on an evaluation of stress measures of a scree plot. With the aid of four independent judges, the researchers identify the dimensions. The first dimension is based on the severity of the act and the second is based on whether the behaviors are directed at individuals or the organization. The authors validate their interpretation through further independent judging. In all, the results classify deviant workplace behaviors into four types, which contain implications for both theory development and managerial action.

Green, Paul E., Catherine M. Schaffer, and Karen M. Patterson (1988), "A Reduced-Spaced Approach to the Clustering of Categorical Data in Market Segmentation." *Journal of the Market Research Society* 30(3): 267–88.

This article provides three different market segmentation applications to illustrate how correspondence analysis can be used to facilitate the matching of a market to product attributes. Used as a data reduction tool, CA reduces the space representation of respondents and categories and provides proximity measures that are suitable for use in such grouping procedures as cluster analysis. The objective of the procedures outlined in the article is to provide the researcher with a diagram or map that indicates how items such as products or services, or multiple variations of these items, may be matched with a particular market. From this information, a profile of the type of consumer served is derived. While the authors do not elaborate on many of the technical details, they do indicate the purpose, design, results, and implication of each of the three examples.

In the first example, the authors use CA to provide a two-dimensional map that indicates how modes of advertising can be more matched to a specific target type in joint space. The second and third examples use CA as a data-reduction technique by representing individual responses as proximity measures that are suitable for use in cluster analysis. These results demonstrate how CA provides a multidimensional representation based upon respondent classification of multiple variables. These results may then be applied for segmentation purposes by matching items such as product attributes and consumer type.

Javalgi, Rajshekhar, Thomas Whipple, Mary McManamon, and Vicki Edick (1992), "Hospital Image: A Correspondence Analysis Approach." *Journal of Health Care Marketing* 12(4): 34–41.

Using correspondence analysis, this study seeks to uncover the unobservable dimensions of hospital image. CA allows the researcher to determine the presence of relations between categorical variables, which may then be represented in perceptual space. The authors identify 13 hospital features (rows) and 16 hospitals (columns) for the study. CA portrays not only how the features are related to other features and hospitals to other hospitals but also how specific features are related to certain hospitals. By exposing these associations, the technique promotes an understanding of the complexity of the hospital image. This facilitates image management that aids in the realization of strategic objectives.

The results, derived from a sample of 503 health care consumers, indicate that two dimensions adequately account for the variation in the original variables. This is ascertained through an examination of eigenvalues much as in factor analysis.

The two dimensions in this study account for 82 percent of the variance, whereas a three-dimensional solution adds only another 6.6 percent. The authors then examine the similarity and dissimilarity of features and the relative contribution of each to the overall solution. They also construct a joint map that illustrates interdependencies and competitive positions. As an aid to strategic planning, these results may then be applied toward assessing the hospital's strengths and weaknesses.

ADVANCED AND EMERGING TECHNIQUES

Overview

This section provides a simple and concise introduction to some of the "cutting edge" techniques that are now emerging in multivariate analysis. Too often the adoption of a new technique is slowed by a "mystique" that is fostered by "experts" who don't or can't pass on their knowledge to others. Yet today's researchers are being offered a range of new and exciting techniques that is wider than ever before and can extend their capability to deal with entirely new sets of problems. Thus, the real need today is an introduction to these techniques that provides not only a general understanding of the procedures, but also the knowledge of when they can be applied and to which problems they are uniquely suited. By no means have we been able to cover all that is new in multivariate analysis. Instead, we have selected one major advance in dealing with multiple dependence relationships (structural equation modeling) and several techniques that have emerged from the applied domain. We hope that this balance will provide adequate coverage of the developments occurring in both the academic and applied communities.

Chapters in Section 4

Section 4 contains two chapters. Chapter 11 discusses structural equation modeling (SEM), a procedure for accommodating measurement error directly into the estimation of a series of dependence relationships. SEM provides two distinct advantages not found in the multivariate techniques discussed thus far: (1) the ability to directly incorporate measurement error in the estimation process, and (2) the simultaneous estimation of several interrelated dependence relationships. Structural equation modeling has been widely accepted in the academic community, but has gained little use elsewhere due in part to the tremendous "learning curve" perceived to be necessary. We do not wish to underestimate the effort involved, but no researcher should avoid SEM only for this reason because the principles of factor analysis and multiple regression form the underlying methods used in SEM. Chapter 12 introduces a range of emerging applications that have primarily been developed in the applied domain. Issues such as data warehousing and data mining have been spurred by the tremendous amounts of information being collected in organizations today and the need for timely and concise analyses. Today's researchers face enormous problems in dealing with large databases that may contain hundreds of thousands or millions of cases while trying to address research questions that require an exploratory component. How do we identify potential solutions from among the possibly thousands of alternatives? Moreover, how do we leverage these large databases and "mine them" for the "nuggets" of information they might provide? These are issues addressed in chapter 12.

CHAPTER **11**

Structural Equation Modeling

LEARNING OBJECTIVES

Upon completing this chapter, you should be able to do the following:

- Understand the role of causal relationships in statistical analysis.
- Represent a series of causal relationships in a path diagram.
- Translate a path diagram into a set of equations for estimation.
- Appreciate the role and influence of a variable's measurement properties on the results of a statistical analysis.
- Differentiate between observable and unobservable variables and their roles in structural equation modeling.
- Evaluate the results of a structural equation modeling analysis for its support of the proposed relationships and the possible areas of improving the results.
- Apply structural equation modeling techniques to such problems as confirmatory factor analysis, path analysis, and simultaneous equation estimation.

CHAPTER PREVIEW

One of the primary objectives of multivariate techniques is to expand the researcher's explanatory ability and statistical efficiency. Multiple regression, factor analysis, multivariate analysis of variance, discriminant analysis, and the other techniques discussed in previous chapters all provide the researcher with powerful tools for addressing a wide range of managerial and theoretical questions. But they all share one common limitation: each technique can examine only a single relationship at a time. Even the techniques allowing for multiple dependent variables, such as multivariate analysis of variance and canonical analysis, still represent only a single relationship between the dependent and independent variables.

All too often, however, the researcher is faced with a set of interrelated questions. For example, what variables determine a store's image? How does that image combine with other variables to affect purchase decisions and satisfaction at the store? And finally, how does satisfaction with the store result in long-term loyalty to it? This series of issues has both managerial and theoretical importance. Yet none of the multivariate techniques we have examined allow us to address all these questions with a single comprehensive technique. For this reason, we now examine the technique of **structural equation modeling (SEM),** an extension of several multivariate techniques we have already studied, most notably multiple regression and factor analysis.

As briefly described in chapter 1, structural equation modeling examines a series of dependence relationships simultaneously. It is particularly useful when one dependent variable becomes an independent variable in subsequent dependence relationships. This set of relationships, each with dependent and independent variables, is the basis of SEM. The basic formulation of SEM in equation form is

$$Y_1 = X_{11} + X_{12} + X_{13} + \ldots + X_{1n}$$
$$Y_2 = X_{21} + X_{22} + X_{23} + \ldots + X_{2n}$$
$$Y_m = X_{m1} + X_{m2} + X_{m3} + \ldots + X_{mn}$$

(metric)   (metric, nonmetric)

Structural equation modeling has been used in almost every conceivable field of study, including education, marketing, psychology, sociology, management, testing and measurement, health, demography, organizational behavior, biology, and even genetics. The reasons for its attractiveness to such diverse areas is twofold: (1) it provides a straightforward method of dealing with multiple relationships simultaneously while providing statistical efficiency, and (2) its ability to assess the relationships comprehensively and provide a transition from **exploratory** to **confirmatory analysis.** This transition corresponds to greater efforts in all fields of study toward developing a more systematic and holistic view of problems. Such efforts require the ability to test a series of relationships constituting a large-scale model, a set of fundamental principles, or an entire theory. These are tasks for which structural equation modeling is well suited.

KEY TERMS

Before starting the chapter, review the key terms to develop an understanding of the concepts and terminology used. Throughout the chapter the key terms appear in **boldface.** Other points of emphasis in the chapter are *italicized.* Also, cross-references within the Key Terms appear in *italics.*

Absolute fit measure   Measure of overall *goodness-of-fit* for both the *structural* and *measurement models* collectively. This type of measure does not make any comparison to a specified *null model* (*incremental fit measure*) or adjust for the number of parameters in the estimated model (*parsimonious fit measure*).

Biserial correlation   Correlation measure used to replace the product-moment correlation when a metrically measured variable is associated with a nonmetric binary (0, 1) measure. Also see *polyserial correlation.*

Bootstrapping Form of resampling in which the original data are repeatedly sampled with replacement for model estimation. Parameter estimates and standard errors are no longer calculated with statistical assumptions, but instead are based on empirical observations.

Causal relationship Dependence relationship of two or more variables in which the researcher clearly specifies that one or more variables "cause" or create an outcome represented by at least one other variable. Must meet the requirements for *causation.*

Causation Principle by which cause and effect are established between two variables. It requires that there be a sufficient degree of association (correlation) between the two variables, that one variable occurs before the other, (that one variable is clearly the outcome of the other), and that there be no other reasonable causes for the outcome. Although in its strictest terms causation is rarely found, in practice strong theoretical support can make empirical estimation of causation possible.

Competing models strategy Strategy that compares the proposed model with a number of alternative models in an attempt to demonstrate that no better-fitting model exists. This is particularly relevant in structural equation modeling because a model can be shown only to have acceptable fit, but acceptable fit alone does not guarantee that another model will not fit better or equally well.

Confirmatory analysis Use of a multivariate technique to test (confirm) a prespecified relationship. For example, suppose we hypothesize that only two variables should be predictors of a dependent variable. If we empirically test for the significance of these two predictors and the nonsignificance of all others, this test is a confirmatory analysis. It is the opposite of *exploratory analysis.*

Confirmatory modeling strategy Strategy that statistically assesses a single model for its fit to the observed data. This approach is actually less rigorous than the *competing models strategy* because it does not consider alternative models that might fit better or equally well than the proposed model.

Construct Concept that the researcher can define in conceptual terms but cannot be directly measured (e.g., the respondent cannot articulate a single response that will totally and perfectly provide a measure of the concept) or measured without error (see *measurement error*). Constructs are the basis for forming *causal relationships*, as they are the "purest" possible representation of a concept. A construct can be defined in varying degrees of specificity, ranging from quite narrow concepts, such as total household income, to more complex or abstract concepts, such as intelligence or emotions. No matter what its level of specificity, however, a construct cannot be measured directly and perfectly but must be approximately measured by *indicators.*

Cronbach's alpha Commonly used measure of *reliability* for a set of two or more *construct* indicators. Values range between 0 and 1.0, with higher values indicating higher reliability among the *indicators.*

Degrees of freedom The number of bits of information available to estimate the sampling distribution of the data after all model parameters have been estimated. In practical terms, degrees of freedom are the number of nonredundant correlations or covariances in the input matrix minus the number of estimated coefficients. The researcher attempts to maximize the degrees of freedom available while still obtaining the best-fitting model. Each estimated coefficient "uses up" a degree of freedom. A model can never estimate more coefficients than the number of nonredundant correlations or covariances, meaning that zero is the lower bound for the degrees of freedom for any model.

Direct estimation   Process whereby a model is estimated directly with a selected estimation procedure and the confidence interval (and standard error) of each parameter estimate is based on sampling error.

Endogenous construct   *Construct* or variable that is the dependent or outcome variable in at least one *causal relationship*. In terms of a path diagram, there are one or more arrows leading *into* the endogenous construct or variable.

Equivalent models   Comparable models that have the same number of degrees of freedom (*nested models*) but differ in one or more paths. The number of equivalent models expands very quickly as model complexity increases.

Exogenous construct   *Construct* or variable that acts only as a predictor or "cause" for other constructs or variables in the model. In path diagrams, the exogenous *constructs* have only causal arrows leading out of them and are not predicted by any other *constructs* in the model.

Exploratory analysis   Analysis that defines possible relationships in only the most general form and then allows the multivariate technique to estimate relationship(s). The opposite of *confirmatory analysis*, the researcher is not looking to "confirm" any relationships specified prior to the analysis, but instead lets the method and the data define the nature of the relationships. An example is stepwise multiple regression, in which the method adds predictor variables until some criterion is met.

Goodness-of-fit   Degree to which the actual or observed input matrix (covariances or correlations) is predicted by the estimated model. *Goodness-of-fit* measures are computed only for the total input matrix, making no distinction between exogenous and endogenous constructs or indicators.

Heywood cases   A common type of *offending estimate*, which occurs when the estimated error term for an indicator becomes negative, which is a nonsensical value. The problem is remedied either by deleting the indicator or by constraining the measurement error value to be a small positive value.

Identification   Degree to which there is a sufficient number of equations to "solve for" each of the coefficients (unknowns) to be estimated. Models can be *underidentified* (cannot be solved), *just-identified* (number of equations equals number of estimated coefficients with no degrees of freedom), or *overidentified* (more equations than estimated coefficients and degrees of freedom greater than zero). The researcher desires to have an overidentified model for the most rigorous test of the proposed model. Also see *degrees of freedom*.

Incremental fit measure   Measure of *goodness-of-fit* that compares the current model to a specified *null model* to determine the degree of improvement over the null model. Complements the other two types of goodness-of-fit measures, the *absolute fit* and *parsimonious fit measures*.

Indicator   Observed value (*manifest variable*) used as a measure of a concept or *latent construct* that cannot be measured directly. The researcher must specify which indicators are associated with each latent construct.

Influential observation   Any observation that has a disproportionate influence on the estimated parameters.

Jackknife   Procedure for drawing repeated samples based on omitting one observation when creating each sample. Allows for the empirical estimation of parameter confidence levels rather than parametric estimation.

Just-identified model   Structural model with zero *degrees of freedom* and exactly meeting the *order condition*. This corresponds to a perfectly fitting model, but has no generalizability.

Latent construct or **variable**   Operationalization of a *construct* in structural equation modeling. A latent variable cannot be measured directly but can be represented or measured by one or more variables (*indicators*). For example, a person's attitude toward a product can never be measured so precisely that there is no uncertainty, but by asking various questions we can assess the many aspects of the person's attitude. In combination, the answers to these questions give a reasonably accurate measure of the latent construct (attitude) for an individual.

Manifest variable   Observed value for a specific item or question, obtained either from respondents in response to questions (as in a questionnaire) or from observations by the researcher. Manifest variables are used as the *indicators* of *latent constructs* or *variables.*

Maximum likelihood estimation (MLE)   Estimation method commonly employed in structural equation models, including LISREL and EQS. An alternative to ordinary least squares used in multiple regression, MLE is a procedure which iteratively improves parameter estimates to minimize a specified fit function.

Measurement error   Degree to which the variables we can measure (the *manifest variables*) do not perfectly describe the *latent construct(s)* of interest. Sources of measurement error can range from simple data entry errors to definition of constructs (e.g., abstract concepts such as patriotism or loyalty that mean many things to different people) that are not perfectly defined by any set of manifest variables. For all practical purposes, all constructs have some measurement error, even with the best *indicator variables.* However, the researcher's objective is to minimize the amount of measurement error. SEM can take measurement error into account in order to provide more accurate estimates of the *causal relationships.*

Measurement model   Submodel in SEM that (1) specifies the *indicators* for each *construct*, and (2) assesses the *reliability* of each construct for estimating the *causal relationships.* The measurement model is similar in form to factor analysis; the major difference lies in the degree of control provided the researcher. In factor analysis, the researcher can specify only the number of factors, but all variables have loadings (i.e., they act as indicators) for each factor. In the measurement model, the researcher specifies which variables are indicators of each construct, with variables having no loadings other than those on its specified construct.

Model   Specified set of dependence relationships that can be tested empirically—an operationalization of a *theory.* The purpose of a model is to concisely provide a comprehensive representation of the relationships to be examined. The model can be formalized in a path diagram or in a set of structural equations.

Model development strategy   Structural modeling strategy incorporating *model respecification* as a theoretically driven method of improving a tentatively specified model. This allows exploration of alternative model formulations that may be supported by theory. It does not correspond to an *exploratory approach* in which model respecifications are made atheoretically.

Model respecification   Modification of an existing model with estimated parameters to correct for inappropriate parameters encountered in the estimation process or to create a *competing model* for comparison.

Modification indices   Values calculated for each unestimated relationship possible in a specified model. The modification index value for a specific unestimated relationship indicates the improvement in overall model fit (the reduction in the chi-square statistic) that is possible if a coefficient is calculated for that untested relationship. The researcher should use modification indices only as a guideline for model improvements of those relationships that can theoretically be justified as possible modifications.

Nested models   Models that have the same constructs but differ in terms of the number or types of *causal relationships* represented. The most common form of nested model occurs when a single relationship is added to or deleted from another model. Thus the model with fewer estimated relationships is "nested" within the more general model.

Nonrecursive   Relationship in a path diagram indicating a mutual or reciprocal relationship between two constructs. Each construct in the pair has a causal relationship with the other construct. Depicted by a two-headed straight arrow between both constructs.

Null model   Baseline or comparison standard used in *incremental fit indices*. The null model is hypothesized to be the simplest model that can be theoretically justified. The most common example is a single *construct* model related to all *indicators* with no *measurement error.*

Offending estimates   Any value that exceeds its theoretical limits. The most common occurrences are *Heywood cases* with negative error variances (the minimum value should be zero, indicating no measurement error) or very large standard errors. The researcher must correct the offending estimate with one of a number of remedies before the results can be interpreted for overall model fit and the individual coefficients can be examined for statistical significance.

Order condition   Requirement for *identification* that the model's *degrees of freedom* must be greater than or equal to zero.

Overidentified model   Structural model with a positive number of *degrees of freedom*, indicating that some level of generalizability may be possible. The objective is to achieve the maximum model fit with the largest number of degrees of freedom.

Parsimonious fit measure   Measure of overall *goodness-of-fit* representing the degree of model fit per estimated coefficient. This measure attempts to correct for any "overfitting" of the model and evaluates the *parsimony* of the model compared to the goodness-of-fit.

Parsimony   Degree to which a model achieves *goodness-of-fit* for each estimated coefficient. The objective is not to minimize the number of coefficients or to maximize the fit, but to maximize the amount of fit per estimated coefficient and avoid "overfitting" the model with additional coefficients that achieve only small gains in model fit.

Path analysis   Method that employs simple bivariate correlations to estimate the relationships in a system of structural equations. The method is based on specifying the relationships in a series of regression-like equations (portrayed graphically in a *path diagram*) that can then be estimated by determining the amount of correlation attributable to each effect in each equation simultaneously. When employed with multiple relationships among *latent constructs* and a *measurement model,* it is then termed *structural equation modeling.*

Path diagram   Graphical portrayal of the complete set of relationships among the model's constructs. *Causal relationships* are depicted by straight arrows, with the arrow emanating from the predictor variable and the arrowhead "pointing" to the dependent *construct* or variable. Curved arrows represent correlations between constructs or indicators, but no causation is implied.

Polychoric correlation   Measure of association employed as a replacement for the product–moment correlation when both variables are ordinal measures with three or more categories.

Polyserial correlation   Measure of association used as a substitute for the product–moment correlation when one variable is measured on an ordinal scale and the other variable is metrically measured.

Rank condition   Requirement for *identification* that each estimated parameter be defined algebraically. Much more laborious than the *order condition* or empirical estimation of identification.

Reciprocal   See *nonrecursive.*

Reliability   Degree to which a set of *latent construct indicators* are consistent in their measurements. In more formal terms, reliability is the extent to which a set of two or more indicators "share" in their measurement of a construct. The indicators of highly reliable constructs are highly intercorrelated, indicating that they all are measuring the same latent construct. As reliability decreases, the indicators become less consistent and thus are poorer indicators of the latent construct. Reliability can be computed as 1.0 minus the *measurement error.* Also see *validity.*

Simulation   Creation of multiple data input matrices based on specified parameters that reflect variation in the distribution of the input data.

Specification error   Lack of model goodness-of-fit resulting from the omission of a relevant variable from the proposed model. Tests for specification error are quite complicated and involve numerous trials among alternative models. The researcher can avoid specification error to a high degree by using only theoretical bases for constructing the proposed model. In this manner, the researcher is less likely to "overlook" a relevant construct for the model.

Starting value   Initial parameter estimate used for incremental or iterative estimation processes, such as LISREL.

Structural equation modeling (SEM)   Multivariate technique combining aspects of multiple regression (examining dependence relationships) and factor analysis (representing unmeasured concepts—factors—with multiple variables) to estimate a series of interrelated dependence relationships simultaneously.

Structural model   Set of one or more dependence relationships linking the hypothesized model's *constructs.* The structural model is most useful in representing the interrelationships of variables between dependence relationships.

Tetrachoric correlation   Measure of association used for relating two binary measures. Also see *polychoric correlation.*

Theory   A systematic set of *causal relationships* providing a consistent and comprehensive explanation of a phenomenon. In practice, a theory is a researcher's attempt to specify the entire set of dependence relationships explaining a particular set of outcomes. A theory may be based on ideas generated from one or more of three principal sources: (1) prior empirical research; (2) past experiences and observations of actual behavior, attitudes, or other phenomena; and (3) other theories that provide a perspective for analysis. Thus, theory building is not the exclusive domain of academic researchers; it has an explicit role for practitioners as well. For any researcher, theory provides a means to address the "big picture" and assess the relative importance of various concepts in a series of relationships.

Underidentified model   *Structural model* with a negative number of *degrees of freedom*. This indicates an attempt to estimate more parameters than possible with the input matrix.

Unidimensionality   Characteristic of a set of *indicators* that has only one underlying trait or concept in common. From the match between the chosen indicators and the theoretical definition of the unidimensional *construct*, the researcher must establish both conceptually and empirically that the indicators are reliable and valid measures of only the specified construct before establishing unidimensionality. Similar to the concept of *reliability*.

Unobserved concept or **variable**   See *latent construct* or *variable.*

Validity   Ability of a *construct's indicators* to measure accurately the concept under study, such as household income or intelligence. Validity is determined to a great extent by the researcher, because the original definition of the construct or concept is proposed by the researcher and must be matched to the selected indicators or measures. Validity does not guarantee reliability, and vice versa. A measure may be accurate (valid) but not consistent (reliable). Also, it may be quite consistent but not accurate. Thus validity and reliability are two separate but interrelated conditions.

Variance extracted measure   Amount of "shared" or common variance among the *indicators* or *manifest variables* for a *construct*. Higher values represent a greater degree of shared representation of the indicators with the construct.

What Is Structural Equation Modeling?

Structural equation modeling (SEM) encompasses an entire family of models known by many names, among them covariance structure analysis, latent variable analysis, confirmatory factor analysis, and often simply LISREL analysis (the name of one of the more popular software packages). Resulting from an evolution of multiequation modeling developed principally in econometrics and merged with the principles of measurement from psychology and sociology, SEM has emerged as an integral tool in both managerial and academic research [7, 11, 12, 20, 25, 39, 41, 51, 61, 63, 71, 80, 96]. SEM can also be used as a means of estimating other multivariate models, including regression, principal components [37], canonical correlation [40], and even MANOVA [9].

As might be expected for a technique with such widespread use and so many variations in applications, many researchers are uncertain about what constitutes structural equation modeling. Yet all SEM techniques are distinguished by two characteristics: (1) estimation of multiple and interrelated dependence relationships, and (2) the ability to represent **unobserved concepts** in these relationships and account for measurement error in the estimation process.

Accommodating Multiple Interrelated Dependence Relationships

The most obvious difference between SEM and other multivariate techniques is the use of separate relationships for each of a set of dependent variables. In simple terms, SEM estimates a series of separate, but interdependent, multiple regression equations simultaneously by specifying the **structural model** used by the statistical program. First, the researcher draws upon theory, prior experience, and the research objectives to distinguish which independent variables predict each dependent variable. In our example from earlier chapters, we first wanted to predict store image. We then wanted to use store image to predict satisfaction, both of which in turn were used to predict store loyalty. Thus, some dependent variables become independent variables in subsequent relationships, giving rise to the interdependent nature of the structural model. Moreover, many of the same variables affect each of the dependent variables, but with differing effects. The structural model expresses these relationships among independent and dependent variables, even when a dependent variable becomes an independent variable in other relationships.

The proposed relationships are then translated into a series of structural equations (similar to regression equations) for each dependent variable. This feature sets SEM apart from techniques discussed previously that accommodate multiple dependent variables—multivariate analysis of variance and canonical correlation—in that they allow only a *single* relationship between dependent and independent variables.

Incorporating Variables that We Do Not Measure Directly

The estimation of multiple interrelated dependence relationships is not the only unique element of structural equation modeling. SEM also has the ability to incorporate **latent variables** into the analysis. A latent variable is a hypothesized and unobserved concept that can only be approximated by observable or measurable variables. The observed variables, which we gather from respondents through various data collection methods (e.g., surveys, tests, observations), are known as **manifest variables.** Yet why would we want to use a latent variable that we did not measure instead of the exact data (manifest variables) the respondents provided? Although this may sound like a nonsensical or "black box" approach, it has both practical and theoretical justification by improving statistical estimation, better representing theoretical concepts, and accounting for measurement error.

Improving Statistical Estimation

Statistical theory tells us that a regression coefficient is actually composed of two elements: the "true" or structural coefficient between the dependent and independent variable and the **reliability** of the predictor variable. Reliability is the degree to which the independent variable is "error-free" [19]. In all the multivariate techniques to this point, we have assumed we had no error in our variables. But we know from both practical and theoretical perspectives that we cannot perfectly measure a concept and that there is always some degree of **measurement error.** For example, when asking about something as straightforward as household income, we know some people will answer incorrectly, either overstating or understating the amount or not knowing it precisely. The answers provided have some measurement error and thus affect the estimation of the "true" structural coefficient [87].

The impact of measurement error (and the corresponding lowered reliability) can be shown from an expression of the regression coefficient as

$$\beta_{y \cdot x} = \beta_s \times \rho_x$$

where $\beta_{y \cdot x}$ is the observed regression coefficient, β_s is the "true" structural coefficient, and ρ_x is the reliability of the predictor variable. Unless the reliability is 100%, the observed correlation will always understate the "true" relationship. Because all dependence relationships are based on the observed correlation (and resulting regression coefficient) between variables, we would hope to "strengthen" the correlations used in the dependence models and make them more accurate estimates of the structural coefficients by first accounting for the correlation attributable to any number of measurement problems.

Representing Theoretical Concepts

Measurement error is not just caused by inaccurate responses but occurs when we use more abstract or theoretical concepts, such as attitude toward a product or motivations for behavior. With concepts such as these, the researcher tries to design the best questions to measure the concept [82]. The respondents also may be somewhat unsure about how to respond or may interpret the questions in a way that is different from what the researcher intended. Both situations can give rise to measurement error. But if we know the magnitude of the problem, we can incorporate the reliability into the statistical estimation and improve our dependence model.

Specifying Measurement Error

How do we account for measurement error? SEM provides the **measurement model,** which specifies the rules of correspondence between manifest and latent variables. The measurement model allows the researcher to use one or more variables for a single independent or dependent concept and then estimate (or specify) the reliability. For example, the dependent variable might be a concept represented by a set of questions (like the summated scale introduced in chapter 3). In the measurement model the researcher can assess the contribution of each scale item as well as incorporate how well the scale measures the concept (its reliability) into the estimation of the relationships between dependent and independent variables. This procedure is similar to performing a factor analysis of the scale items and using the factor scores in the regression. These similarities and specific details are discussed in a later section of this chapter.

A Simple Example of SEM

Structural equation modeling provides the researcher with the ability to accommodate multiple interrelated dependence relationships in a single model. Its closest analogy is multiple regression, which can estimate a single relationship (equation). But SEM can estimate many equations at once, and they can be interrelated, meaning that the dependent variable in one equation can be an independent variable in other equation(s). This allows the researcher to model complex relationships that are not possible with any of the other multivariate techniques we have discussed in this text.

The following example illustrates how SEM works with multiple relationships. We should note, however, that it will not depict one of SEM's other strengths, the ability to employ multiple measures of a concept in a manner similar to factor analysis. Chapter 3 introduced the concept of forming scales to represent concepts or using factor scores as replacements for sets of variables. SEM can perform a similar procedure when the researcher specifies the set of variables to represent each concept. For simplicity, we assume that each concept in the following example is measured by a single variable. We address the benefit of multiple indicators later in this chapter, but now focus only on the basic principles of estimating multiple relationships.

The Research Question

HATCO managers have long hoped to increase employee retention and productivity through a better understanding of their motivations and attitudes toward HATCO. To this end, the personnel department identified three employee attitudes they felt most important: job satisfaction, organizational commitment, and probability of employee turnover. They then developed relationships linking each attitude with its antecedents in a dependence relationship:

Dependent Variable	Independent Variables
Job satisfaction	= Coworker attitude + Work environment
Organizational commitment	= Job satisfaction + Pay level
Probability of employee turnover	= Job satisfaction + Organizational commitment

Setting Up the Structural Equation Model for Path Analysis

When the personnel department brought their work to the research department, the researchers realized that multiple regression would not estimate the relationships, and that instead they must use SEM. To better portray the interrelated relationships easily, they created a pictorial portrayal of the relationships (see Figure 11.1) which they termed **path diagrams.** Straight arrows depict the impact of independent variables on the dependent variables and curved arrows depict the correlation among variables, just like multicollinearity in multiple regression.

What was the purpose of developing the path diagram? Path diagrams are the basis for **path analysis,** the procedure for empirical estimation of the strength of each relationship (paths) depicted in the path diagram. Path analysis calculates the strength of the relationships using only a correlation or covariance matrix as input. The simple (bivariate) correlation between any two variables can be

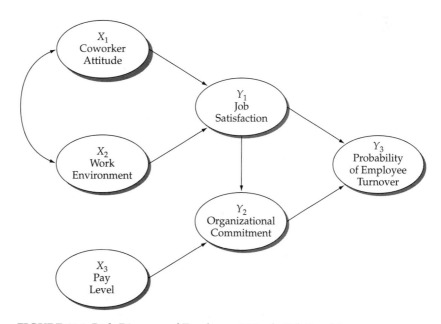

FIGURE 11.1 Path Diagram of Employee Attitude Relationships

represented as the sum of the compound paths connecting these points. A compound path is a path along the arrows that follows three rules:

1. After going forward on an arrow, the path cannot go backward again; but the path can go backward as many times as necessary before going forward.
2. The path cannot go through the same variable more than once.
3. The path can include only one curved arrow (correlated variable pair).

An Application of Path Analysis

These may seem to be quite complicated rules, but they are really quite simple. We can illustrate how they work by applying them to a small part of our path diagram. In Figure 11.2 we have chosen a simple dependence relationship for job satisfaction that is part of the larger model. The two independent variables of coworker attitude (X_1) and work environment (X_2) are correlated and predict the dependent variable of job satisfaction (Y_1). This can be stated simply as:

$$Y_1 = b_1X_1 + b_2X_2$$

The path analysis rules allow us to use the simple correlations between constructs to estimate the causal relationships represented by the coefficients b_1 and b_2. For ease in referring to the paths, the causal paths are labeled A, B, and C. Causal path A is X_1 correlated with X_2, path B is the effect of X_1 predicting Y_1, and path C shows the effect of X_2 predicting Y_1.

Path analysis uses the simple correlations as shown in Figure 11.2 to estimate the causal paths using the three rules given earlier. For example, the correlation of X_1

Path Diagram

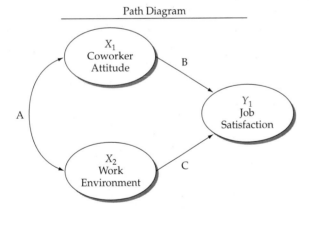

Bivariate Correlations

	X_1	X_2	Y_1
X_1	1.0		
X_2	.50	1.0	
Y_1	.60	.70	1.0

Correlations as Compound Paths

$$\text{Corr}_{X_1X_2} = A$$
$$\text{Corr}_{X_1Y_1} = B + AC$$
$$\text{Corr}_{X_2Y_1} = C + AB$$

Solving for the Structural Coefficients

$$.50 = A$$
$$.60 = B + AC$$
$$.70 = C + AB$$

Substituting A = .50

$$.60 = B + .50C$$
$$.70 = C + .50B$$

Solving for B and C

$$B = .33$$
$$C = .53$$

FIGURE 11.2 Calculating Structural Coefficients with Path Analysis

and Y_1 ($r_{x_1 y_1} = .60$) can be represented as two causal paths: B and AC. The symbol B represents the direct path from X_1 to Y_1, and the other path (a compound path) follows the curved arrow from X_1 to X_2 and then to Y_1. Likewise, the correlation of X_2 and Y_1 can be shown to be composed of two causal paths: C and AB. Finally, the correlation of X_1 and X_2 is equal to A. This relationship forms three equations:

$$r_{x_1 x_2} = A$$
$$r_{x_1 y_1} = B + AC$$
$$r_{x_2 y_1} = C + AB$$

We know that A equals .50, so we can substitute this value into the other equations. By solving these two equations, we get values of B (b_1) = .33 and C (b_2) = .53. The actual calculations are shown in Figure 11.2. This approach allows path analysis to solve for any causal relationship based only on the correlations among the constructs and the specified causal model. As you can see from this simple example, if we change the path model in some way, the causal relationships will change as well. Such a change provides the basis for modifying the model to achieve better fit, if theoretically justified.

With these simple rules, the larger model can now be modeled simultaneously, using correlations or covariances as the input data. We should note that when used in a larger model, we can solve for the interrelated equations. For example, solving the relationship

Probability of employee turnover = Job satisfaction + Organizational commitment

is done with the same rules as our simple example earlier. Thus, dependent variables in one relationship can easily be independent variables in another relationship. No matter how large the path diagram gets or how many relationships are included, path analysis provides a way to analyze the set of relationships.

Summary

Path analysis can be extended to any system of relationships. With these procedures, all of the relationships in any path diagram can be estimated to quantify the effects between the dependent and independent variables, even if interrelated. The researcher now has a technique for analyzing a set of relationships in a simultaneous manner. We have presented only a simple illustration of the process; interested readers are encouraged to examine other treatments of this topic [33, 38, 68, 99].

The Role of Theory in Structural Equation Modeling

Throughout the discussion of SEM, we refer to the need for theoretical justification for specification of the dependence relationships, modifications to the proposed relationships, and many other aspects of estimating a **model.** "Theory" provides the rationale for almost all aspects of SEM. For purposes of this chapter, **theory** can be defined as a systematic set of relationships providing a consistent and comprehensive explanation of a phenomenon. From this definition, we see that theory is not the exclusive domain of academia but can be rooted in experience and practice obtained by observation of real-world behavior. Theory is often a primary objective of academic research, but practitioners may develop or propose a set of relationships that are as complex and interrelated as any academically based theory. Thus researchers from both academia and industry can benefit from the unique analytical tools provided by SEM.

From a practical perspective, a theory-based approach to SEM is a necessity because the technique must be almost completely specified by the researcher. Whereas with other multivariate techniques the researcher may have been able to specify a basic model and allow default values in the statistical programs to "fill in" the remaining estimation issues, SEM has none of these features. Although the seven stage process we discuss makes these decisions straightforward, each component of the structural and measurement models must be explicitly defined. Moreover, any model modifications must be made through specific actions by the researcher. The need for a "theoretical" model to guide the estimation process becomes especially critical when making model modifications. Because of the flexibility of SEM, the chances for "overfitting" the model or developing a model with little generalizability are quite high. Thus, when we stress the need for theoretical justification, our intent is to gain recognition by the researcher that SEM is a confirmatory method, guided more by theory than by empirical results.

Developing a Modeling Strategy

One of the most important concepts a researcher must learn regarding multivariate techniques is that there is no single "correct" way to apply them. Instead, the researcher must formulate the objectives of the research and apply the appropriate technique in the most suitable manner to achieve the desired objectives. In some instances, the relationships are strictly specified and the objective is a confirmation of the relationship. At other times, the relationships are loosely recognized and the objective is the discovery of relationships. In each extreme instance and points in between, the researcher must formulate the use of the technique in accordance with the research objectives.

The application of SEM follows this same tenet. Its flexibility provides the researcher with a powerful analytical tool appropriate for many research objectives. But the researcher must define these objectives as guidelines in a modeling strategy. The use of the term *strategy* is designed to denote a plan of action toward a specific outcome. In the case of SEM, the ultimate outcome is always the assessment of a series of relationships. However, this can be achieved through many avenues. For our purposes, we define three distinct strategies in the application of SEM: confirmatory modeling strategy, competing models strategy, and model development strategy.

Confirmatory Modeling Strategy

The most direct application of structural equation modeling is a **confirmatory modeling strategy,** wherein the researcher specifies a single model, and SEM is used to assess its statistical significance. Here the researcher is saying, "It either works or it doesn't." Although this may seem to be the most rigorous application, it actually is not the most stringent test of a proposed model. Research has even shown that the techniques developed for assessing structural equation models have a *confirmation bias*, which tends to confirm that the model fits the data [88]. Thus, if the proposed model has acceptable fit by whatever criteria are applied, the researcher has not "proved" the proposed model but only confirmed that it is one of several possible acceptable models. Several different models might have equally acceptable model fits. Thus, the more rigorous test is achieved by comparing alternative models.

Competing Models Strategy

Obtaining an acceptable level of fit for both the overall model and the measurement and structural models does not assure the researcher that the "best" model has been found. Numerous alternative models may provide equal or even better fits. As a means of evaluating the estimated model with alternative models, overall model comparisons can be performed in a **competing models strategy.** The strongest test of a proposed model is to identify and test competing models that represent truly different hypothetical structural relationships. When comparing these models, the researcher comes much closer to a test of competing "theories," which is a much stronger test than just a slight modification of a single "theory."

How does the researcher generate this set of competing models? One possible source of competing models is alternative formulations of the underlying theory. For example, in one formulation, trust may precede commitment, yet in another commitment may precede trust. This could be the basis for two competing models. **Equivalent models** provide a second perspective on developing a set of competing models. It has been shown that for any structural equation model, there is at least one other model with the same number of parameters and the same level of model fit that varies in the relationships portrayed. This implies that no model is unique in the level of fit achieved, and for any model with acceptable fit there are any number of alternative models with the same level of model fit. A series of generalized rules have been defined to identify the equivalent models for any structural model [69]. As a general rule of thumb, the more complex a model, the more equivalent models exist. A third approach, the TETRAD program, is an emerging empirical method that systematically examines a structural model and identifies additional relationships that are supported by the data [50]. With the input data matrix and initial model specification supplied by the researcher, the program examines the patterns of relationships (tetrads) and isolates those relationships that could be empirically supported. The TETRAD program does not actually estimate model parameters, but instead identifies relationships to be included in the original model to form competing models. The program works best by starting with a simple model and adding relationships. Many researchers object to this method as being atheoretical and too mechanistic or "black box," but it may provide some researchers insights into their models that could not be obtained in any other fashion.

One common example of a competing models strategy is the process of assessing factorial invariance, the equality of factor models across groups. There is an established procedure to assess the degree of invariance, starting with the most loosely constrained models and slowly adding additional constraints until the most restrictive model is tested [75]. Restrictions are added to represent invariance across groups, loadings, and even factor intercorrelations. This is also an example of a **nested model** approach, in which the number of constructs and indicators remains constant, but the number of estimated relationships changes. Although competing models are typically nested models, they also can be nonnested (differ in number of constructs or indicators), but this requires specialized measures of model fit for comparing between models.

Model Development Strategy

The **model development strategy** differs from the prior two strategies in that although a model is proposed, the purpose of the modeling effort is to improve the model through modifications of the structural and/or measurement models. In many applications, theory can provide only a starting point for development of a theoretically justified model that can be empirically supported. Thus, the researcher must employ SEM not just to test the model empirically but also to provide insights into its respecification. One note of caution must be made. The researcher must be careful not to employ this strategy to the extent that the final model has acceptable fit but cannot be generalized to other samples or populations. Moreover, the respecification of a model must always be made with theoretical support rather than just empirical justification.

Stages in Structural Equation Modeling

The true value of SEM comes from the benefits of using the structural and measurement models simultaneously, each playing distinct roles in the overall analysis. To ensure that both models are correctly specified and the results are valid, we now discuss a seven-stage process (see Figure 11.3, which shows stages 1–3, and Figure 11.6 on p. 602, which shows stages 4–7). This process differs from the six-stage model-building approach introduced in chapter 1 and used in our discussions of the other multivariate methods. The introduction of this separate process for SEM does not invalidate the model-building approach for other multivariate techniques, but just accentuates the uniqueness of SEM.

The seven stages in structural equation modeling are (1) developing a theoretically based model, (2) constructing a path diagram of causal relationships, (3) converting the path diagram into a set of structural and measurement models, (4) choosing the input matrix type and estimating the proposed model, (5) assessing the identification of the structural model, (6) evaluating goodness-of-fit criteria, and (7) interpreting and modifying the model, if theoretically justified.

Stage 1: Developing a Theoretically Based Model

Structural equation modeling is based on **causal relationships,** in which the change in one variable is assumed to result in a change in another variable [57]. We encountered this type of statement when we defined a dependence relationship, such as is found in regression analysis. Causal relationships can take many forms and meanings, from the strict causation found in physical processes, such as a chemical reaction, to the less well-defined relationships encountered in behavioral research, such as the "causes" of educational achievement or the "reasons" why we purchase one product rather than another. The strength and conviction with which the researcher can assume **causation** between two variables lies not in the analytical methods chosen but in the theoretical justification provided to support the analyses. The "requirements" for asserting causation have deep roots in various views of philosophy of science [8, 27, 60]. There is general agreement with at least four established criteria for making causal assertions: (1) sufficient association between the two variables, (2) temporal antecedence of the cause versus the effect, (3) lack of alternative causal variables, and (4) a theoretical basis for the relationship. Although in many instances all of the established

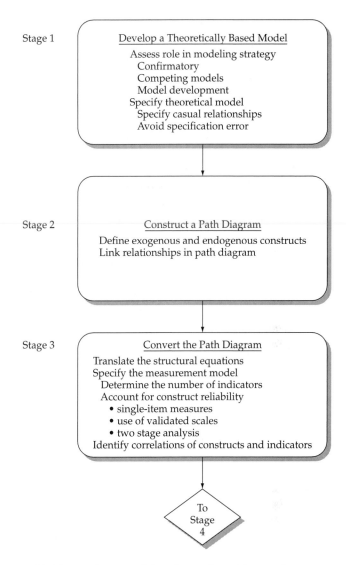

FIGURE 11.3 Stages 1–3 in a Seven-Stage Process for Structural Equation Modeling

criteria for making causal assertions are not strictly met, causal assertions can possibly be made if the relationships are based on a theoretical rationale. But we caution any researcher against assuming that the techniques discussed in this chapter will by themselves provide a means of "proving" causation without having some guiding theoretical perspective. Using these techniques in an "exploratory" manner is invalid and misleads the researcher more often than it provides appropriate results.

The most critical error in developing theoretically based models is the omission of one or more key predictive variables, a problem known as **specification error.** The implication of omitting a significant variable is to bias the assessment of the importance of other variables [72]. For example, assume that two variables (*a* and *b*) were predictors of *c*. If we included both *a* and *b* in our analysis, we would make the correct assessment of their relative importance as shown by their estimated coefficients. But if we left variable b out of the analysis, the coefficient for variable a would be different. This difference, or bias, is the result of the coefficient for variable a reflecting not only its effect on c but the effect it shared with b as well. This shared effect, however, is controlled for when both variables are included in the analysis. We encourage readers who still do not fully understand the impact of omitted variables also to review the material in chapter 4 concerning specification error and its effects.

The desire to include all variables, however, must be balanced against the practical limitations of SEM. Although no theoretical limit on the number of variables in the models exists, practical concerns occur even before the limits of most computer programs are met. Most often, interpretation of the results, particularly statistical significance, becomes quite difficult as the number of concepts becomes large (exceeding 20 concepts). The researcher should never omit a concept solely because the number of variables is becoming large, but should recognize the benefits of parsimonious and concise theoretical models.

Stage 2: Constructing a Path Diagram of Causal Relationships

Up to this point, we have expressed causal relationships only in terms of equations. But there is another method of portraying these relationships called **path diagrams,** which are especially helpful in depicting a series of causal relationships. A path diagram is more than just a visual portrayal of the relationships because it allows the researcher to present not only the predictive relationships among constructs (i.e., the dependent–independent variable relationships), but also associative relationships (correlations) among constructs and even indicators. We discuss the implications of each type of relationship in later sections of this chapter. Here we discuss the path diagram, which presents a concise method for expressing each of these types of relationships.

Elements of a Path Diagram

Before examining path diagrams, we must define two basic elements used in their construction. The first is the concept of a **construct,** which is a theoretically based concept that acts as a "building block" used to define relationships. A construct can represent a concept as simple as age, income, or gender, or as complex as socioeconomic status, knowledge, preference, or attitude. A researcher defines path diagrams in terms of constructs and then finds variables to measure each construct. For example, we may ask someone's age and use this as a measure of the construct age. Likewise, we may ask a series of questions about a person's opinions and use this as a measure of attitude. Both sets of questions provide numerical values for the constructs. We can then assess the questions for the amount of measurement error they possess and include this in the estimation process. From now on, we use the term "construct" to represent a particular concept, no matter how it is measured. A construct is typically represented in a path diagram by an oval.

The second basic element is the arrow, used to represent specific relationships between constructs. A straight arrow indicates a direct causal relationship from one construct to another. A curved arrow (or a line without arrowheads) between constructs indicates simply a correlation between constructs. Finally, a straight

Then the predictor variables are all constructs at the ends, or "tails," of the straight arrows leading into the endogenous variable. It is that simple.

Figure 11.5 illustrates this translation process for each of the path diagrams in Figure 11.4. As we see, each endogenous variable (Y_j) can be predicted either by exogenous variable(s) (X_j) or by other endogenous variable(s). For each hypothesized effect, we estimate a structural coefficient (b_{jm}). Also, because we know that we will have prediction errors, just as in multiple regression, we include an error term (ϵ_i) for each equation as well. The error term represents the sum of the effects due to specification error and random measurement error. It is not possible to separate these two sources of error except in special situations.

Measurement Model

We have referred to the measurement model in general terms up to this point, but we now define it in specific terms. We discuss not only the basic procedures of specifying the measurement model, but also issues regarding the number of indicators per construct and the process of specifying the reliability of the construct rather than by its estimation. But before doing so, let us review the foundations of factor analysis (chapter 3), which are quite analogous to the measurement model.

Correspondence to Factor Analysis   In factor analysis, each individual variable was "explained" by its loading on each factor. The objective was to best represent all the variables in a small number of factors. The factors related to "underlying dimensions" in the data, which we then had to interpret and label. Factor analysis as discussed in chapter 3 is often termed an exploratory technique because there are no constraints on the variable loadings. Each variable has a loading on each factor. We represent these relationships mathematically as shown in Table 11.1 (p. 598).

The value for each factor (factor score) is calculated by the loadings on each variable (e.g., for factor $1 = L_{11}V_1 + L_{21}V_2 + L_{31}V_3 + L_{41}V_4 + L_{51}V_5$, where V_1 to V_5 are the actual data values for each variable). Also, the predicted value for each variable is calculated by the loadings of the variable on each factor. However, every variable has a factor loading on each factor; thus, each factor is always a

	ENDOGENOUS VARIABLE	=	EXOGENOUS VARIABLES			+	ENDOGENOUS VARIABLES			+	ERROR
	Y_1		X_1	X_2	X_3		Y_1	Y_2	Y_3		ε_i
Path Diagram											
Figure 11.4(a)	Y_1	$=$	$b_1X_1 + b_2X_2$							+	ε_1
Figure 11.4(b)	Y_1	$=$	$b_1X_1 + b_2X_2$							+	ε_1
	Y_2	$=$	b_3X_2			+	b_4Y_1			+	ε_2
Figure 11.4(c)	Y_1	$=$	$b_1X_1 + b_2X_2$							+	ε_1
	Y_2	$=$	$b_3X_2 + b_4X_3$			+	b_5Y_1		$+ b_6Y_3$	+	ε_2
	Y_3	$=$				+	$b_7Y_1 + b_8Y_2$			+	ε_3

FIGURE 11.5 Translation of Path Diagrams into Structural Equations

TABLE 11.1 Comparison of Factor Analysis Loadings and Indicator Loadings of the Measurement Model

	Factor Analysis: Factor Loadings on Factors			Measurement Model: Indicator Loadings on Constructs		
Variable	Factor 1	Factor 2	Factor 3	Construct A	Construct B	Construct C
V_1	L_{11}	L_{12}	L_{13}	L_1		
V_2	L_{21}	L_{22}	L_{23}	L_2		
V_3	L_{31}	L_{32}	L_{33}		L_3	
V_4	L_{41}	L_{42}	L_{43}		L_4	
V_5	L_{51}	L_{52}	L_{53}			L_5

composite of all variables, although their loadings vary in magnitude. Therefore, a factor is actually a latent construct, defined by the loadings of all the variables.

Specifying the Measurement Model   To specify the measurement model, we make the transition from factor analysis, in which the researcher had no control over which variables describe each factor, to a **confirmatory** mode, in which the researcher specifies which variables define each construct (factor). The manifest variables we collect from the respondents are termed **indicators** in the measurement model, because we use them to measure, or "indicate," the latent constructs (factors). Assume in our example that V_1 and V_2 are now hypothesized to be indicators of construct A, V_3, and V_4 are indicators of construct B, and V_5 is a single indicator of construct C. The measurement model would be expressed as shown in Table 11.1.

How and why does this configuration differ from the factor analysis loadings we discussed before? The most obvious difference is the much smaller number of loadings. In the exploratory mode of factor analysis, the researcher cannot control the loadings. In the measurement model, however, the researcher has complete control over which variables describe each construct. In the example, each variable was an indicator of only one construct; thus, there was a lower number of loadings. Although a variable may be an indicator for more than one construct, this method is not recommended except in specific situations with strong theoretical rationale. The researcher specifies a measurement model for both the exogenous constructs and the endogenous constructs in exactly this manner.

Determining the Number of Indicators   We have already discussed the rationale and justification for employing multiple indicators (variables) to represent a construct, but a fundamental question remains: How many indicators should be used per construct? The minimum number of indicators for a construct is one, but the use of only a single indicator requires the researcher to provide estimates of reliability. A construct can be represented with two indicators, but three is the preferred minimum number of indicators, because using only two indicators increases the chances of reaching an infeasible solution [36]. Apart from the theoretical basis that should be used to select variables as indicators of a construct, there is no upper limit in terms of the number of indicators. As a practical matter, however, five to seven indicators should represent most constructs. The notable exception is the use of preexisting scales, which may contain many items, each acting as an indicator of the construct. In these instances, the researcher should assess the **unidimensionality** of the construct and the possibility of multiple subdimensions that can be represented in a second-order factor model (see further discussion in stage 6).

Accounting for Construct Reliability   Once the measurement model has been specified, the researcher must then provide for the reliability of the indicators. We provide a much more detailed discussion of reliability in stage 6, but at this stage the researcher must determine the basic method in which reliability will be established for each construct. There are two principal methods to establish reliability: (1) empirical estimation or (2) specification by the researcher.

Empirically Estimating Reliabilities
Empirical estimation of reliability is possible only if the construct has two or more indicators. For a construct with only one indicator, the researcher must specify the reliability. For empirical estimation, the researcher specifies the loading matrix as described, along with an error term for each indicator variable (because we do not expect to predict each indicator perfectly). When the structural and measurement models are estimated, the loading coefficients provide estimates of the reliabilities of the indicators and the overall construct. In this approach, the researcher has no impact on the reliability value used in the estimation of the model except by the sets of indicators included. We illustrate the exact steps required for this approach in stage 6.

Specifying the Reliabilities
In some instances it is appropriate for the researcher to specify, or "fix," the reliabilities. The specification of reliabilities for the indicator(s) of any latent construct may seem to be counter to the objectives of structural equation modeling; however, in at least three situations it is justified and strongly recommended. In one instance, empirical estimation of the reliability is not possible, yet the researcher may know that measurement error still exists. In another, the indicators may have been used extensively; therefore, the reliabilities are known before use. And finally, a two-step approach in which the reliabilities are first assessed and then specified in the estimation process. This two-step approach explicitly separates the two empirical processes and provides insight into each separately.

Single-Item Measures   With single-item measures, it is not possible to empirically estimate reliability; thus the researcher is faced with two possibilities. First, set ("fix") the reliability at 1.0, indicating that there is no measurement error in the indicator. Yet, as discussed before, we know this is erroneous in almost all instances, if for no other reason than reliability is affected by the quality of data collection. For example, gender may be "perfect" or very close (99 percent), with error due only to coding errors. However, income may have a higher level of error (e.g., 10 percent) owing to reporting bias and the level of measurement. Most often, therefore, the researcher should make some estimate of the reliability and specify the value for single-item indicators. A number of recommended approaches are provided by Hayduk [54].

Use of Validated Scales or Measures   Many times a researcher employs a scale or measure that has been extensively tested in previous research. If the objective in its use is replication of the effects found in prior studies, then the reliability of the scale or measure should be fixed at previously established levels. This is an example of the researcher specifying reliabilities to maintain "control" over the meaning of the constructs. By fixing the reliability, the researcher "forces" an indicator to have the amount of variance appropriate for the construct and maintains a specific meaning for the construct.

Two-Step Analysis Many researchers propose a two-step process of structural equation modeling in which the measurement model is first estimated, much like factor analysis, and then the measurement model is "fixed" in the second stage when the structural model is estimated [4, 61, 68, 77, 102]. The rationale of this approach is that accurate representation of the reliability of the indicators is best accomplished in two steps to avoid the interaction of measurement and structural models. Although we cannot truly evaluate the measurement and structural models in isolation, we must consider the potential for within-construct versus between-construct effects in estimation, which can be substantial and result in what Burt [28] terms "interpretational confounding."

A single-step analysis with the simultaneous estimation of both structural and measurement models is the best approach when the model possesses both strong theoretical rationale and highly reliable measures, resulting in more accurate relationships and decreasing the possibility for the structure or measurement interaction. However, when faced with measures that are less reliable, or theory that is only tentative, the researcher should consider a staged approach to maximize the interpretability of both measurement and structural models. Considerable debate has emerged on the appropriateness of this approach and about those instances in which it is justified, on both conceptual and empirical grounds [5, 44, 45, 55].

Methods of Specifying the Reliability To "fix" the reliability of an indicator in a correlation matrix, the researcher specifies the loading value as the square root of the desired or estimated reliability, or specifies the error term of that variable as 1.0 minus the desired reliability value. If a covariance matrix is used, then the error term or loading value is multiplied by the variable's variance value. In specifying the reliabilities, the researcher may specify the loading value, the error term, or both. Because specifying either the loading or the error terms automatically determines the other value, we recommend that both be set for the greatest model parsimony and that a coefficient not be used for estimating a value that could be specified. This procedure can be simply done in statistical programs through a single statement for each variable.

A variation from setting the reliability of the entire scale is an approach advocated by Hayduk [54, 55] that involves a staged process. First, the researcher selects the single indicator judged to be the best representative of the construct. For this indicator, the reliability (loading and error term) is fixed. Then, additional indicators can be added and their loadings and error terms empirically estimated, allowing for calculating the reliability of the entire construct. The rationale behind this approach is for the researcher to specify the desired meaning of the construct through this indicator and then allow other indicators to add meaning to the concept as already defined. The contention is that when all loadings and error terms are empirically estimated simultaneously, the researcher is not sure exactly what the construct represents except some underlying concept common to all of the indicators.

Correlations among Constructs and Indicators

In addition to the structural and measurement models, the researcher also specifies any correlations between the exogenous constructs or between the endogenous constructs. Many times exogenous constructs are correlated, representing a "shared" influence on the endogenous variables. Correlations among the endogenous constructs, however, have fewer appropriate applications and are not recommended for typical use because they represent correlations among the structural equations that confound their interpretation. Finally, the indicators in the measurement model can also be correlated separately from the construct correlations. This method is to be avoided except in specific situations, such as a study in which there are known effects from the measurement or data collection process on two or more indicators, or a longitudinal study in which the same indicator is collected in two periods of time [8, 89]. For a more complete discussion of the distorting effects of correlated indicators, see Gerbing and Anderson [47].

Stage 4: Choosing the Input Matrix Type and Estimating the Proposed Model

As shown with the prior stages, much more is required of the researcher using SEM in terms of specifying the model to be estimated than with any other multivariate technique, with the possible exception of conjoint analysis. Now the researcher must address the actual process of estimating the specified model, including the issues of inputting the data in the appropriate form and selecting the estimation procedure (see Figure 11.6, p. 602). The decisions made in these areas have profound impacts on the results achieved.

Inputting Data

SEM differs from other multivariate techniques in that it uses only the variance–covariance or correlation matrix as its input data. Individual observations can be input into the programs, but they are converted into one of these two types of matrices before estimation. The focus of SEM is not on individual observations but on the pattern of relationships across respondents. Input for the program is a correlation or variance–covariance matrix of all indicators used in the model. The measurement model then specifies which indicators correspond to each construct, and the latent construct scores are then employed in the structural model.

Assumptions SEM shares three assumptions with the other multivariate methods we have studied: independent observations, random sampling of respondents, and the linearity of all relationships. In addition, SEM is more sensitive to the distributional characteristics of the data, particularly the departure from multivariate normality (critical in the use of LISREL) or a strong kurtosis (skewness) in the data. Some software, such as EQS, are less sensitive to nonnormal data, but the data should be evaluated no matter which program is used. Generalized least squares, an alternative estimation method, can adjust for these violations, but this method quickly becomes impractical as the model size and complexity increase; thus its use is limited [98]. A lack of multivariate normality is particularly troublesome because it substantially inflates the chi-square statistic and creates upward bias in critical values for determining coefficient significance [79, 98].

Because the programs accept only the correlation or variance–covariance matrices, the researcher must perform all of the diagnostic tests on the data before

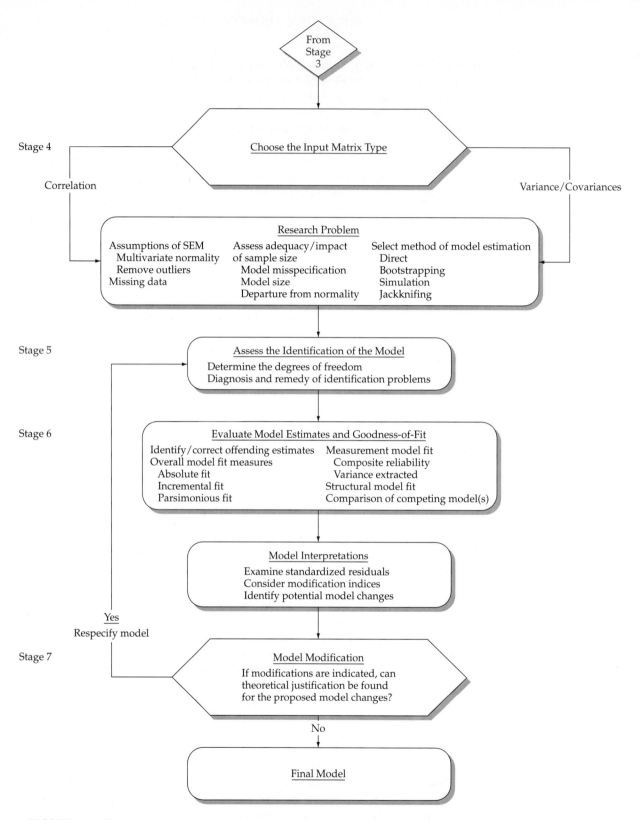

FIGURE 11.6 Stages 4–7 in a Seven-Stage Process for Structural Equation Modeling

they are used in the estimation procedure. Although the structural equation programs do not have built-in diagnostic procedures for testing these assumptions, they can be tested with conventional methods (see a more detailed discussion in chapter 2) or through programs such as PRELIS [66]. The researcher should also identify any outliers in the data before they are converted to matrix form.

Missing Data Missing data can have a profound effect on calculating the input data matrix and its ability to be used in the estimation process. There are two ways in which missing data can be incorporated into SEM. The first is the direct method, in which model parameters are estimated with both the complete and incomplete data [2, 67]. This approach is rarely used, however, given the complexity of the resulting model. More common is an indirect method, whereby an input data matrix is estimated using some or all of the available information. As discussed in chapter 2, there are many methods available for "solving" the missing data problem, ranging from listwise deletion to imputation methods. Recent research has shown that the EM method introduces the least bias into the estimated models, but that the pairwise and listwise options perform adequately if the proportion of missing data is not too great [26]. One drawback of the listwise method is that it may seriously reduce sample size, a key concern as we will see in a later discussion. The pairwise approach can introduce irregularities into the data matrix that later cause serious problems in the estimation process. Thus, there is no single approach that always produces the best results, and if possible the researcher should employ several approaches to assess the stability of the results.

Covariances versus Correlations An important issue in interpreting the results is the use of the variance–covariance matrix versus the correlation matrix. SEM was initially formulated for use with the variance–covariance matrix (hence its common designation as covariance structure analysis). The covariance matrix has the advantage of providing valid comparisons between different populations or samples, a feature not possible when models are estimated with a correlation matrix. Interpretation of the results, however, is somewhat more difficult when using covariances because the coefficients must be interpreted in terms of the units of measure for the constructs.

The correlation matrix has gained widespread use in many applications. Correlation matrices have a common range that makes possible direct comparisons of the coefficients within a model, because it is simply a "standardized" variance–covariance matrix in which the scale of measurement of each variable is removed by dividing the variances or covariances by the product of the standard deviations. Use of correlations is appropriate when the objective of the research is only to understand the pattern of relationships between constructs, but not to explain the total variance of a construct. Another appropriate use is to make comparisons across different variables, because the scale of measurement affects the covariances. Coefficients obtained from the correlation matrix are always in standardized units, similar to beta weights in regression, and range between -1.0 and $+1.0$. Moreover, research has shown that the correlation matrix provides more conservative estimates of the significance of coefficients and is not upwardly biased, as previously thought [35].

In summary, the researcher should employ the variance–covariance matrix any time a true "test of theory" is being performed, as the variances and covariances satisfy the assumptions of the methodology and are the appropriate form of the data for validating causal relationships. However, often the researcher is concerned only with patterns of relationships, not with total explanation as needed in theory testing, and the correlation matrix is acceptable. Any time the correlation matrix is used, the researcher should cautiously interpret the results and generalizability to different situations.

Types of Correlations or Covariances Used   The most widely used means of computing the correlations or covariances between manifest variables is the Pearson product–moment correlation. This is also the most common form of correlation used in multivariate analysis, making it quite easy for the researcher to compute the correlation or covariance matrices. An assumption of the product–moment correlation, however, is that both variables are metrically measured. This makes the product–moment correlation inappropriate for use with nonmetric (ordinal or binary) measures. To allow for incorporation of nonmetric measures into structural equation models, the researcher must employ different types of correlation measures. If both variables are ordinal with three or more categories (polychotomous), then **polychoric correlation** is appropriate. If both nonmetric measures are binary, then **tetrachoric correlation** is used. For instances in which a metric measure is related to a polychotomous ordinal measure, **polyserial correlation** represents the relationship. If a binary measure is related to a metric measure, **biserial correlation** is used.

Sample Size   Even though individual observations are not needed, as with all other multivariate methods, the sample size plays an important role in the estimation and interpretation of SEM results. Sample size, as in any other statistical method, provides a basis for the estimation of sampling error. The critical question in SEM involves how large a sample is needed. Although there is no single criterion that dictates the necessary sample size, there are at least four factors that impact the sample size requirements [83]: (1) model misspecification, (2) model size, (3) departures from normality, and (4) estimation procedure.

Model Misspecification
Model misspecification refers to the extent that the model suffers from specification error. As discussed earlier, specification error is the omission of relevant variables from the specified model. All structural equation models suffer from specification error to some extent in that every potential construct and indicator cannot be included. The impact of the omitted constructs and indicators, however, should be negligible if the researcher has included all those relevant to the theory. Sample size impacts the ability of the model to be correctly estimated and identify specification error if desired. Thus, if the researcher has concerns about the impact of specification error, sample size requirements should be increased over what would otherwise be required.

Model Size
The absolute minimum sample size must be at least greater than the number of covariances or correlations in the input data matrix. However, more typical is a minimum ratio of at least five respondents for each estimated parameter, with a ratio of 10 respondents per parameter considered most appropriate. Thus, as model complexity increases, so do the sample size requirements. Note that this requirement differs from the concept of degrees of freedom (discussed later) and

concerns the number of original respondents used to calculate the covariance or correlation matrix.

Departures from Normality

As the data violate the assumptions of multivariate normality, the ratio of respondents to parameters needs to increase with a generally accepted ratio of 15 respondents for each parameter. Although some estimation procedures are specifically designed to deal with nonnormal data, the researcher is always encouraged to provide sufficient sample size to allow for the sampling error's impact to be minimized, especially for nonnormal data [98].

Estimation Procedure

Maximum likelihood estimation (MLE), the most common estimation procedure, has been found to provide valid results with sample sizes as small as 50, but a sample this small is *not recommended*. It is generally accepted that the minimum sample size to ensure appropriate use of MLE is 100 to 150 [36]. As we increase the sample size above this value, the MLE method increases in its sensitivity to detect differences among the data. As the sample size becomes large (exceeding 400 to 500), the method becomes "too sensitive" and almost any difference is detected, making all goodness-of-fit measures indicate poor fit [30, 73, 94]. Although there is no correct sample size, recommendations are for a size ranging between 100 to 200. One approach is always to test a model with a sample size of 200, no matter what the original sample size was, because 200 is proposed as being the "critical sample size" [58]. If an asymptotically distribution-free estimation procedure (see next section for more detail) is selected, the sample size requirements are markedly increased, as it requires large samples to offset the reliance on the distributional assumptions of other methods.

Summary

There are many factors impacting the required sample size. As a matter of course, we recommend a sample size of 200, with increases occurring if misspecification is suspected, the model is overly large or complex, the data exhibit nonnormal characteristics, or an alternative estimation procedure is used. One separate assessment is the critical N diagnostic, which is the sample size that would make the obtained model fit (measured by χ^2) significant at the stated level of significance [58]. The chi-square measure is provided by most programs and is a basis of comparison for sample size. The researcher is encouraged, if in doubt, to determine the model's sensitivity to variations in sample size by estimating the model under varying sample size specifications.

Model Estimation

Once the structural and measurement models are specified and the input data type is selected, the researcher must choose how the model will be estimated. In the case of SEM, the researcher has several options for both the estimation procedure and the computer program to be used.

Estimation Technique Early attempts at structural equation model estimation were performed with ordinary least squares (OLS) regression. But these efforts were quickly supplanted by maximum likelihood estimation, which is efficient and unbiased when the assumption of multivariate normality is met. As such, it was used in the early versions of LISREL and became a widely employed technique in most computer programs. The sensitivity of MLE to nonnormality, however, created a need for alternative estimation techniques, and such methods as weighted least squares (WLS), generalized least squares (GLS), and asymptotically distribution free (ADF) became available [55]. The ADF technique has received particular attention recently due to its insensitivity to nonnormality of the data. Its primary drawback is the increased sample size required. All of the alternative estimation techniques have become more widely available as the computing power of the personal computer has increased making them feasible for typical problems.

Estimation Processes   In addition to the estimation technique employed, the researcher can also choose among several estimation processes. These processes range from the direct estimation of the model, which is similar to what we have seen in past multivariate techniques, to methods that generate thousands of model estimations from which final model results are obtained. We discuss four basic processes—direct estimation, bootstrapping, simulation, and jackknifing—in this chapter. The reader should also refer to chapter 12 for a more detailed discussion of several of these processes.

Direct Estimation
The most common estimation process is that of **direct estimation,** in which a model is estimated directly with a selected estimation procedure. In this process, like the multivariate models in past chapters, we estimate the parameter, and then the confidence interval (and standard error) of each parameter estimate is based on sampling error. Both the parameter estimate and its confidence interval come from the model estimated on a single sample.

Bootstrapping
Several alternatives exist, however, that do not rely on a single model estimation, but instead calculate parameter estimates and their confidence intervals based on multiple estimations [24]. The first option is **bootstrapping,** which is accomplished in four basic steps. First, the original sample is designated to act as the population for sampling purposes. In the second step, the original sample is resampled a specified number of times (perhaps up to several thousand) to generate a large number of new samples, each a random subset of the original sample. In the third step, the model is estimated for each new sample and the estimated parameters are saved. In the last step, the final parameter estimates are calculated as the average of the parameter estimates across all of the samples. The confidence interval is not estimated by sampling error, but instead is directly observed by examining the actual distribution of the parameter estimates around the mean. In this manner, the final parameter estimates and their confidence estimates are derived directly from multiple model estimations across separate samples and do not rely on assumptions as to the statistical distribution of the parameters.

Simulation
The researcher may also employ **simulation** techniques, which also rely on multiple samples and estimated models. Simulation processes differ from bootstrapping in that during the process of generating the new samples, the simulation program may change certain characteristics of the sample to meet the researcher's

objectives. For example, the degree of correlation between variables may be varied across the samples in some systematic manner. In this way, the researcher has not only random sampling variation between the samples, but also a systematic pattern specified in the simulation procedure. With these samples, models are again estimated for each sample and the results compiled, as in the bootstrapping process.

Jackknifing

The final estimation process is the **jackknife** procedure, where again repeated samples are created from the original sample. The jackknife differs from the simulation and bootstrapping procedures, however, in the method of creating the new samples. Instead of creating a large number of new random samples, the jackknife process creates N new samples, where N is the original sample size. Each time a new sample is created, one different observation is omitted. Thus, each new sample has a sample size of $N - 1$ with a different observation omitted in each sample. The advantage of this process is the ease of identifying **influential observations** by examining the changes in parameter estimates. If desired, the final parameter estimate can be calculated as the average parameter, but in instances of small sample sizes there are not enough new samples to adequately calculate the confidence interval.

Computer Programs Once the estimation procedure is selected, the next step is to select the computer program used for actually estimating the model. The most widely used program is LISREL (LInear Structural RELations) [64, 65, 67], a truly flexible model for a number of research situations (cross-sectional, experimental, quasi-experimental, and longitudinal studies). LISREL has found applications across all fields of study [8] and has become almost synonymous with structural equation modeling. However, a number of alternative programs exist, among them EQS [13, 14, 18], AMOS [6], PROC CALIS of SAS [53], COSAN [46], or LVPLS [104]. EQS places less stringent assumptions on the multivariate normality of the data, and LVPLS is better suited to prediction yet limited for interpretation purposes. AMOS has gained increased popularity in recent years due to its simple interface for the user, and it has been compared recently to LISREL and EQS [59]. All of these programs are available in versions that can be run on personal computers.

Solving the "Not Positive Definite" Problem

A problem familiar to all users of SEM is the computer error message, "The _____ matrix is not positive definite." What has happened is that either the input data matrix or the estimated data matrix is singular, meaning there is a linear dependency or inconsistency among some set of variables. There are many sources of this problem, but some general findings have emerged [105]. If the error occurs in the input data matrix, the most likely sources are (1) the missing data approach used, especially pairwise deletion [6]; or (2) a linear dependency among the variables, such as including all of the scale items and the scale total in the input matrix. In these two instances, the researcher should generate a new data matrix, which employs an alternative missing data process or eliminates the offending variables. If the problem occurs in the estimated data matrix, then the researcher should correct any negative error variances (known as Heywood cases and described in stage 6) or try different starting values.

Stage 5: Assessing the Identification of the Structural Model

During the estimation process, the most likely cause of the computer program "blowing up" or producing meaningless or illogical results is a problem in the identification of the structural model. An **identification** problem, in simple terms, is the inability of the proposed model to generate unique estimates. It is based on the principle that we must have a separate and unique equation to estimate each coefficient, reflected in the dictum "You must have more equations than unknowns" that we learned in algebra when defining a series of equations. However, as structural models become more complex, there is no guaranteed approach for ensuring that the model is identified [23].

Degrees of Freedom

For purposes of identification, the researcher is concerned with the size of the covariance or correlation matrix relative to the number of estimated coefficients. This difference between the number of correlations or covariances and the actual number of coefficients in the proposed model is termed **degrees of freedom.** Similar to the degrees of freedom we encountered in multiple regression or MANOVA, a degree of freedom is an unconstrained element of the data matrix. The number of degrees of freedom for a proposed model is calculated as

$$df = \frac{1}{2}[(p + q)(p + q + 1)] - t$$

where:

    p = the number of endogenous indicators,

    q = the number of exogenous indicators,

    t = the number of estimated coefficients in the proposed model.

The first portion of the equation calculates the nonredundant size of the correlation or covariance matrix (i.e., the lower or upper half of the matrix plus diagonal). Then each estimated coefficient "uses up" a degree of freedom. The primary difference in the degrees of freedom used in SEM compared to other multivariate techniques is that the number of estimated parameters is compared to the number of elements in the data matrix, not the sample size. In SEM the sample size is used to estimate sampling error, but *does not affect the degrees of freedom.*

Rules for Identification

Although there is no single rule that will establish the identification of a model, the researcher does have a number of "rules" or heuristics available [10, 34, 84, 85]. The two most basic rules are the rank and order conditions. The **order condition** states that the model's degrees of freedom must be greater than or equal to zero. This corresponds to what are termed just-identified or overidentified models. A **just-identified** model has exactly zero degrees of freedom. Although this will provide a perfect fit of the model, the solution is uninteresting because it has no generalizability. An **overidentified model** is the goal for all structural equation models. It has more information in the data matrix than the number of parameters to be estimated, meaning that there is a positive number of degrees of freedom. Just as in other multivariate techniques, the researcher is striving to achieve acceptable fit with the largest number of degrees of freedom possible. This ensures that the model is as generalizable as possible. A model failing to meet the order condition is known as an **underidentified model.** This model has negative degrees of freedom, meaning that it tries to estimate more parameters than there is information available. The model cannot be estimated until some parameters are fixed or constrained.

The order condition is a necessary, but not sufficient, condition for identification. The model must also meet the **rank condition,** which requires that the researcher algebraically determine if each parameter is uniquely identified (estimated). For all but the simplest models, this is too complex an exercise to be undertaken directly by the researcher. Instead, several heuristics are available. First is the three-measure rule, which asserts that any construct with three or more indicators will always be identified. There is also the recursive model rule, which states that recursive models with identified constructs (three-measure rule) will always be identified [85]. A recursive model has no nonrecursive or reciprocal relationships in the structural model.

Diagnosing Identification Problems

The structural equation programs will also perform empirical tests to diagnose identification problems. LISREL performs a simple test for identification during the estimation process by examining the information matrix, whereas EQS performs the Wald rank test. Although these tests will identify most identification problems, they may not assess the uniqueness of each estimated parameter, as required for the rank test.

The researcher can also perform tests when the equation is identified to see whether the results are unstable because of the level of identification. First, the model can be reestimated several times, each time with a different **starting value.** The researcher can specify an initial value for any estimated parameter, a "starting point" for the estimation process. If the starting value is not provided, the programs automatically compute them by one of several methods. If the results do not converge at the same point for different starting values, the identification should be examined more thoroughly. The second test to assess the identification's effect on a single coefficient is first to estimate the model and obtain the coefficient estimate. Then "fix" the coefficient to its estimated value and reestimate the equation (fixing coefficients is discussed in a later section). If the overall fit of the model varies markedly, identification problems are indicated.

Another approach is to look for possible symptoms of an identification problem. These include (1) very large standard errors for one or more coefficients, (2) the inability of the program to invert the information matrix, (3) wildly unreasonable estimates or impossible estimates such as negative error variances, or (4) high correlations ($\pm.90$ or greater) among the estimated coefficients. Stage 6 addresses interpreting these results in more detail.

Sources and Remedies of Identification Problems

If an identification problem is indicated, the researcher should first look to three common sources: (1) a large number of estimated coefficients relative to the number of covariances or correlations, indicated by a small number of degrees of freedom—similar to the problem of overfitting the data found in many other multivariate techniques; (2) the use of reciprocal effects (two-way causal arrows between two constructs); or (3) failure to fix the scale of a construct. We discuss this procedure later in our analysis of the example data.

The only solution for an identification problem is to define more constraints on the model—that is, to eliminate some of the estimated coefficients. The researcher should follow a structured process, gradually adding more constraints (deleting paths from the path diagram) until the problem is remedied. In doing so, the researcher is attempting to achieve an overidentified model that has degrees of freedom available with which to assess, if possible, the amount of sampling and measurement error and thus provide better estimates of the "true" causal relationships. To achieve this end, the following process is recommended by Hayduk [54]: (1) Build a theoretical model with the minimum number of coefficients (unknowns) that can be justified. If identification problems are encountered, proceed with remedies in this order: (2) fix the measurement error variances of constructs if possible, (3) fix any structural coefficients that are reliably known, and (4) eliminate troublesome variables. If identification problems still exist, the researcher must reformulate the theoretical model to provide more constructs relative to the number of causal relationships examined.

Stage 6: *Evaluating Goodness-of-Fit Criteria*

The first step in evaluating the results is an initial inspection for "offending estimates." Once the model is established as providing acceptable estimates, the goodness-of-fit must then be assessed at several levels: first for the overall model and then for the measurement and structural models separately.

Offending Estimates

The results are first examined for **offending estimates.** These are estimated coefficients in either the structural or measurement models that exceed acceptable limits. The most common examples of offending estimates are (1) negative error variances or nonsignificant error variances for any construct, (2) standardized coefficients exceeding or very close to 1.0, or (3) very large standard errors associated with any estimated coefficient. If offending estimates are encountered, the researcher must first resolve each occurrence before evaluating any specific results of the model, as changes in one portion of the model can have significant effects on other results.

Several approaches for resolution have already been examined in the discussion of identification issues. If identification problems are corrected and problems still exist, several other remedies are available. In the case of negative error variances (also known as **Heywood cases**), one possibility is to fix the offending error variances to a very small positive value (.005) [17, 35]. Although this remedy meets the practical requirements of the estimation process, it only masks the underlying problem and must be considered when interpreting the results. If correlations in the standardized solution exceed 1.0, or two estimates are correlated highly, then the researcher should consider elimination of one of the constructs or should ensure that true discriminant validity has been established among the constructs. In many instances, such situations are the result of atheoretical models, established without sufficient theoretical justification or modified solely on the basis of empirical considerations.

Overall Model Fit

Once the researcher has established that there are no offending estimates, the next step is to assess the overall model fit with one or more goodness-of-fit measures. **Goodness-of-fit** measures the correspondence of the actual or observed input (covariance or correlation) matrix with that predicted from the proposed model.

In developing any statistical model, the researcher must guard against "over-fitting" the model to the data. In discussing regression, we showed that a certain ratio (perhaps five to one) should be maintained between the number of estimated coefficients and the number of respondents. This ratio should be maintained in SEM as well. The researcher should strive for a larger number of degrees of freedom, all other things being equal. In doing so, the model achieves **parsimony**—the achievement of better or greater model fit for each estimated coefficient. The better fit we can achieve with fewer coefficients, the better the test of the model and the more confidence we can have that the results are not a result of overfitting the data.

Goodness-of-fit measures are of three types: (1) **absolute fit measures,** (2) incremental fit measures, or (3) **parsimonious fit measures. Absolute fit measures** assess only the overall model fit (both structural and measurement models collectively), with no adjustment for the degree of "overfitting" that might occur. **Incremental fit measures** compare the proposed model to another model specified by the researcher. Finally, **parsimonious fit measures** "adjust" the measures of fit to provide a comparison between models with differing numbers of estimated coefficients, the purpose being to determine the amount of fit achieved by each estimated coefficient.

The researcher is faced with the question of which measures to choose. No single measure or set of measures emerges as the only measures needed. As SEM has evolved in recent years, goodness-of-fit measures have been continually developed, and additional measures (not discussed here) have been proposed [78]. The researcher is encouraged to employ one or more measures from each type. Assessing the goodness-of-fit of a model is more a relative process than one with absolute criteria. The application of multiple fit measures will enable the researcher to gain a consensus across types of measures as to the acceptability of the proposed model. Appendix 11B describes several measures within each class of goodness-of-fit measures in more detail, including how to calculate measures not provided by the computer program. An excellent review of the various goodness-of-fit measures and their application in a number of situations is contained in various sources [22, 67]. An acceptable level of overall goodness-of-fit does not guarantee that all constructs will meet the requirements for measurement model fit, nor is the structural model certain to be fully supported. The researcher must assess each of these areas separately to confirm their meeting the requirements or as a means of identifying potential problems that affected overall goodness-of-fit.

Measurement Model Fit

Once the overall model fit has been evaluated, the measurement of each construct can then be assessed for unidimensionality and reliability. **Unidimensionality** is an assumption underlying the calculation of reliability and is demonstrated when the indicators of a construct have acceptable fit on a single-factor (one-dimensional) model. This topic is beyond the scope of this discussion, but the interested reader can consult several sources [3, 4, 5, 44, 45]. The use of reliability measures, such as **Cronbach's alpha** [32], does not ensure unidimensionality but instead assumes it exists. The researcher is encouraged to perform unidimensionality tests on all multiple-indicator constructs before assessing their reliability. The next step is to examine the estimated loadings and to assess the statistical significance of each one. If statistical significance is not achieved, the researcher may wish to eliminate the indicator or attempt to transform it for better fit with the construct.

Composite Reliability   Beyond examination of the loadings for each indicator, a principal measure used in assessing the measurement model is the composite reliability of each construct. Reliability is a measure of the internal consistency of the construct indicators, depicting the degree to which they "indicate" the common latent (unobserved) construct. More reliable measures provide the researcher with greater confidence that the individual indicators are all consistent in their measurements. A commonly used threshold value for acceptable reliability is .70, although this is not an absolute standard, and values below .70 have been deemed acceptable if the research is exploratory in nature.

We should note, however, that reliability does not ensure **validity.** Validity is the extent to which the indicators "accurately" measure what they are supposed to measure. For example, several measures of how and why consumers purchase products may be quite reliable, but the researcher may mistakenly assume they measure brand loyalty when in fact they are indicators of purchase intentions. In this instance, the indicators are a reliable set of measures but an invalid measure of brand loyalty. The issue of validity rests on the researcher's specification of indicators for a latent construct. The means for assessing validity in its many forms are reviewed in Bollen [22].

The reliability and variance extracted (see next section) for a latent construct must be computed separately for each multiple indicator construct in the model. Although LISREL and other programs do not compute them directly, all the necessary information is readily provided. The composite reliability of a construct is calculated as

$$\text{Construct reliability} = \frac{(\Sigma \text{ standardized loading})^2}{(\Sigma \text{ standardized loading})^2 + \Sigma \varepsilon_j}$$

where the standardized loadings are obtained directly from the program output, and ε_j is the measurement error for each indicator [43]. The measurement error is 1.0 minus the reliability of the indicator, which is the square of the indicator's standardized loading. The indicator reliabilities should exceed .50, which roughly corresponds to a standardized loading of .7.

Variance Extracted   Another measure of reliability is the **variance extracted measure.** This measure reflects the overall amount of variance in the indicators accounted for by the latent construct. Higher variance extracted values occur when the indicators are truly representative of the latent construct. The variance extracted measure is a complementary measure to the construct reliability value. The variance extracted measure is calculated as

$$\text{Variance extracted} = \frac{\Sigma (\text{standardized loading}^2)}{\Sigma (\text{standardized loading})^2 + \Sigma \varepsilon_j}$$

This measure is quite similar to the reliability measure but differs in that the standardized loadings are squared before summing them [43]. Guidelines suggest that the variance extracted value should exceed .50 for a construct. Actual examples of the calculations for both the reliability and variance extracted measures are provided in the confirmatory factor analysis example later in this chapter.

Structural Model Fit

The most obvious examination of the structural model involves the significance of estimated coefficients. Structural equation modeling methods provide not only estimated coefficients but also standard errors and calculated t values for each coefficient. If we can specify the significance level we deem appropriate (e.g., .05), then each estimated coefficient can be tested for statistical significance (i.e., that it is different from zero) for the hypothesized causal relationship. However, given the statistical properties of MLE and its characteristics at smaller sample sizes, the researcher is encouraged to be conservative in specifying a significance level, choosing smaller levels (.025 or .01) instead of the traditional .05 level.

The selection of a critical value also depends on the theoretical justification for the proposed relationships. If a positive or negative relationship is hypothesized, then a one-tailed test of significance can be employed. However, if the researcher cannot prespecify the direction of the relationship, then a two-tailed significance test must be used. The difference is in the critical t values used to assess significance. For example, for the .05 significance level, the critical value for a one-tailed test is 1.645, but it increases to 1.96 for a two-tailed test. Thus the researcher can more accurately detect differences if stronger theory can be utilized in model specification.

As a measure of the entire structural equation, an overall coefficient of determination (R^2) is calculated, similar to that found in multiple regression. Although no test of statistical significance can be performed, it provides a relative measure of fit for each structural equation.

The results of SEM can be affected by multicollinearity, just as was found in regression. Here the researcher must be aware of the correlations among construct estimates in the SEM results. Computer programs provide a correlation matrix of the estimated values for the latent constructs. If large values appear, then corrective action should be taken. This action may include the deletion of one construct or the reformulation of causal relationships. Although no limit has been set that defines what are considered as high correlations, values exceeding .90 should always be examined, and many times correlations exceeding .80 can be indicative of problems.

Comparison of Competing or Nested Models

The more common modeling strategies—a competing models or model development strategy—involve the comparison of model results to determine the best-fitting model from a set of models. In a competing models strategy, the researcher postulates a number of alternative models. The objective is to fit the "best" from among the set of models. In a model development strategy, the researcher starts with an initial model and engages in a series of model respecifications, each time hoping to improve the model fit while maintaining accordance with the underlying theory.

To assist in comparing models, a large number of measures have been developed to assess model fit. One class of measures assesses the overall model fit in absolute terms, providing specific measures of the fit. One drawback to these measures is that they do not account for the number of relationships used in obtaining the model fit. To measure model parsimony, a series of parsimonious fit measures have been proposed. Their objective is to determine the "fit per coefficient," because the absolute fit will always improve as estimated coefficients are added.

A comprehensive procedure for this purpose was proposed by Anderson and Gerbing [4], in which a series of competing models are specified. Differences between models can be shown to be simply the difference in the chi-square values for the different models. This chi-square difference can then be tested for statistical significance with the appropriate degrees of freedom being the difference in the number of estimated coefficients for the two models. The only requirement is that the number of constructs and indicators remains the same, so that the null model is the same for both models (i.e., they are nested models). The effect of adding or deleting one or more causal relationships can also be tested in this manner by making comparisons between models with and without the relationships. If the models become nonnested (have a different number of indicators or constructs), the researcher must rely on the parsimonious fit measures described earlier, as the chi-square difference test is not appropriate in this instance.

Stage 7: Interpreting and Modifying the Model

Once the model is deemed acceptable, the researcher should first examine the results for their correspondence to the proposed theory. Are the principal relationships in the theory supported and found to be statistically significant? Do the competing models add insight in alternative formulations of the theory that can be supported? Are all of the relationships in the hypothesized direction (positive or negative)? All of these and many more questions can be addressed from the empirical results. In the course of addressing these questions, the researcher may find the need to consider two issues of interpretation: the use of the standardized versus unstandardized solutions and model respecification.

Standardized versus Unstandardized Solutions

One aspect of evaluating an estimated relationship is the assessment of the actual size of the parameter. But just as for other multivariate techniques, such as multiple regression, there is a marked difference in the standardized and unstandardized solutions in terms of their interpretation and use. In structural equation models, the standardized coefficients all have equal variances and a maximum value of 1.0, thus closely approximating effect sizes, as was shown by beta weights in regression. Coefficients near zero have little, if any, substantive effect, whereas an increase in value corresponds to increased importance in the causal relationships. The standardized coefficients are useful for determining relative importance, but are sample specific and not comparable across samples.

The unstandardized coefficients correspond to the regression weights in multiple regression in that they are expressed in terms of the construct's scale, in this case its variance. This makes these coefficients comparable across samples and retains their scale effect. Because the scale varies for each construct, however, comparison between coefficients is more difficult than with the standardized coefficients.

Model Respecification

Once model interpretation is complete, the researcher most likely is looking for methods to improve model fit and/or its correspondence to the underlying theory. In such cases, the researcher may engage in **model respecification,** the process of adding or deleting estimated parameters from the original model. Before addressing some approaches for identifying model modification, we caution the researcher to make such modifications with care and only after obtaining theoretical justification for what empirically is deemed significant. Modifications to the original model should be made only after deliberate consideration. If modifications are made, the model should be cross-validated (i.e., estimated on a separate set of data) before the modified model can be accepted.

A Process of Model Respecification Before identifying any possible model respecification, the researcher should classify all relationships (estimated or not) into one of two categories: theoretical or empirical [88]. The theoretical relationships are essential to the underlying theory and cannot be modified. They are "off-limits" to respecification. The empirical category contains relationships that are added to provide fit to the model. These can be respecified. The objective is to define a nesting of theoretical models, where the set of models is viewed as a set of different levels of parsimony for the same underlying theory. In this way, the models become a series of competing models lending various levels of support to the theory.

Empirical Indicators of Possible Respecification Where can the researcher look for model improvements? The first indication comes from examination of the residuals of the predicted covariance or correlation matrix. The standardized residuals (also called normalized residuals) represent the differences between the observed correlation or covariance and the estimated correlation or covariance matrix. With the latest versions of LISREL (version 7 and higher), the calculation of residuals was improved and the standard for assessing "significant" residuals has changed (the prior threshold was ± 2.0). Residual values greater than ± 2.58 are now to be considered statistically significant at the .05 level. Significant residuals indicate a substantial prediction error for a pair of indicators (i.e., one of the correlations or covariances in the original input data). A standardized residual indicates only that a difference exists but lends no insight into how it may be reduced. The researcher must identify a remedy by the addition or modification of the causal relationships [31, 49].

Another aid in assessing the fit of a specified model involves **modification indices,** which are calculated for each nonestimated relationship. The modification index value corresponds approximately to the reduction in chi-square that would occur if the coefficient were estimated. A value of 3.84 or greater suggests that a statistically significant reduction in the chi-square is obtained when the coefficient is estimated. Although modification indices can be useful in assessing the impact of theoretically based model modifications, the researcher should never make model changes based solely on the modification indices. This atheoretical approach is totally contrary to the "spirit" of the technique and should be avoided in all instances. Model modification must have a theoretical justification before being considered, and even then the researcher should be quite skeptical about the changes [72].

In addition to the modification index, LISREL also provides an expected change parameter, which denotes the magnitude and direction of each fixed (not estimated) parameter. This parameter differs from the modification index in that it does not indicate the change in overall model fit (χ^2), instead it depicts the change in the actual parameter value. EQS provides the Lagrange multiplier and Wald statistic, which assess the effect of freeing a set of parameters simultaneously.

As model modifications are made, the researcher must return to stage 4 of the seven-stage process and reevaluate the modified models. If extensive model modifications are anticipated, the data should be divided into two samples, one providing the basis for model estimation and modification, and the other providing for validation of the final model.

A Recap of the Seven-Stage Process

Structural equation modeling gives the researcher more flexibility than any of the other multivariate methods we have discussed. But along with this flexibility comes the potential for inappropriate use of the method. One overriding concern in any application of SEM should have a steadfast reliance on a theoretically based foundation for the proposed model and any modifications. SEM, when applied correctly, provides a strong confirmatory test to a series of causal relationships. However, when the method is applied in an "exploratory" manner, the researcher is faced with the rather high probability of falling prey to "data snooping" or "fishing" and identifying relationships that have little generalizability by simply capitalizing on the relationships specific to the sample data being studied.

Two Illustrations of Structural Equation Modeling

Structural equation modeling can address a wide variety of causal relationships. Among the two most common types of analyses performed are confirmatory factor analysis and the estimation of a series of structural equations. We now illustrate both of these analyses by following the seven-stage process for SEM. First, a confirmatory factor analysis examines a two-factor solution developed from the factor analysis performed in chapter 3. Then a series of structural relationships are proposed from a new database dealing with an expanded study of supplier characteristics that customers deem important.

Confirmatory Factor Analysis

Factor analysis, as discussed in chapter 3, is concerned with exploring the patterns of relationships among a number of variables. These patterns are represented by what are termed principal components or, more commonly, factors. As variables load highly on a factor, they become descriptors of the underlying dimension. Only on examination of the loadings of the variables on the factors does the researcher identify the character of the underlying dimension.

At this point, the reader may now see the similarity between the objectives of factor analysis and the measurement model in SEM. The factors are, in measurement model terms, the latent variables. Each variable acts as an indicator of each factor (because every variable has a loading for each factor). Used in this manner, factor analysis is primarily an exploratory technique because of the researcher's limited control over which variables are indicators of which latent construct (i.e., which variables "load" on each factor). SEM, however, can play a confirmatory role because the researcher has complete control over the specification of indicators for each construct. Moreover, SEM allows for a statistical test of the goodness-of-fit for the proposed confirmatory factor solution, which is not possible with principal components or factor analysis. Confirmatory factor analysis (CFA) is particularly useful in the validation of scales for the measurement of specific constructs [92].

Stage 1: Developing a Theoretically Based Model

The confirmatory use of SEM can be illustrated by a synthesis of the principal components and common factor analysis examples in chapter 3. In review, six variables measured the respondents' perceptions of HATCO on the supplier characteristics of X_1, delivery speed; X_2, price level; X_3, price flexibility; X_4, manufacturer image; X_6, salesforce image; and X_7, product quality. Variable X_5, overall service, was omitted from the factor analyses and will be deleted in this example as well to maintain comparability.

The factor analyses in chapter 3 indicated the existence of two dimensions (factors). First, the four objective measures (delivery speed, price level, price flexibility, and product quality) were the principal descriptors of one dimension. The second dimension was consistently characterized by the two image variables (salesforce and manufacturer image). These two dimensions of supplier perceptions are characterized as specific strategy actions versus more global or affective evaluations. Thus the hypothesized model posits two factors (*strategy* and *image*), with each set of variables now acting as indicators of the separate constructs. There is no reason to expect uncorrelated perceptions; thus the factors will be allowed to correlate as well.

Recent research has demonstrated the benefits of using factor analysis as a complement to theory in specifying the appropriate factor loadings in the measurement model [48]. A more detailed discussion of the underlying estimation procedure of using LISREL for estimating principal components is also available [37].

Stage 2: Constructing a Path Diagram of Causal Relationships

The next stage is to portray the relationships in a path diagram. In this case, the two hypothesized factors are considered exogenous constructs. The path diagram, including the variables measuring each construct, is shown in Figure 11.7. The correlation between perceptions is represented by the curved line connecting the two constructs.

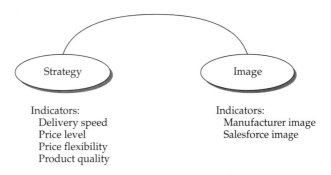

Indicators:
  Delivery speed
  Price level
  Price flexibility
  Product quality

Indicators:
  Manufacturer image
  Salesforce image

FIGURE 11.7 Path Diagram for CFA

TABLE 11.2 Two-Construct Measurement Model

Variables	Indicator Loadings on Constructs	
	Strategy	Image
X_1 Delivery speed	L_1	
X_2 Price level	L_2	
X_3 Price flexibility	L_3	
X_4 Manufacturer image		L_4
X_6 Salesforce image		L_5
X_7 Product quality	L_6	

Stage 3: Converting the Path Diagram into a Set of Structural and Measurement Models

Because all the constructs in the path diagram are exogenous, we need only consider the measurement model and the associated correlation matrices for exogenous constructs and indicators. With no structural model, the measurement model constitutes the entire structural equation modeling effort (hence we refer to *confirmatory* factor analysis).

The measurement model can be represented simply by a two-construct (strategy and image) model, as shown in Table 11.2. In addition, the two constructs are hypothesized to be correlated, and no within-construct correlated measurements are proposed.

For the interested reader, the appropriate LISREL notation (as discussed in appendix 11A) is as shown in Table 11.3.

TABLE 11.3 LISREL Notation for the Measurement Model

Exogenous Indicator		Exogenous Constructs		Error
X_1	=	$\lambda_{11}^x \xi_1$	+	δ_1
X_2	=	$\lambda_{21}^x \xi_1$	+	δ_2
X_3	=	$\lambda_{31}^x \xi_1$	+	δ_3
X_4	=	$\lambda_{42}^x \xi_2$	+	δ_4
X_5	=	$\lambda_{52}^x \xi_2$	+	δ_5
X_6	=	$\lambda_{61}^x \xi_1$	+	δ_6

Correlation among Exogenous Constructs (ϕ)

	ξ_1	ξ_2
ξ_1	—	
ξ_2	ϕ_{21}	—

Stage 4: Choosing Input Matrix Type and Estimating the Proposed Model

Inputting Data

Structural equation modeling will accommodate either a covariance or a correlation matrix. For purposes of CFA, either type of input matrix can be employed. However, because the objective is an exploration of the pattern of interrelationships, correlations are the preferred input data type. The correlation matrix of the six variables is shown in Table 11.4.

The basic assumptions of structural modeling, similar to other multivariate methods, were examined in chapter 2. For purposes of illustration, the variables are deemed as meeting the assumptions, and we move now to examining the results for offending estimates.

Model Estimation

LISREL [64, 67] is used for estimation of the measurement model and the construct correlations. Regarding estimation of the measurement model for constructs with more than one variable: Because of the estimation procedure, the construct must be made "scale invariant," meaning that the indicators of a construct must be "standardized" in a way to make constructs comparable [65, 71, 81]. There are two common approaches to this procedure. First, one of the loadings in each construct can be set to the fixed value of 1.0. The second approach is to estimate the construct variance directly. Either approach results in the same estimates, but for purposes of theory testing, the second approach (estimating construct variance) is recommended [4]. In this example, the second approach is employed. The complete set of commands used for the LISREL analysis is contained in appendix A.

Stage 5: Assessing the Identification of the Structural Model

Identification is a relatively simple matter in confirmatory factor analysis, and the diagnostic procedures of the program are sufficient to detect identification problems. The most common problem may occur if multiple variables are hypothesized to be indicators for two or more constructs. It is possible for identification problems to arise in such instances, but they are not encountered in most confirmatory factor analyses, and their chances of occurrence are minimized by the use of strong theoretical foundations for specification of the measurement model.

TABLE 11.4 Correlation Matrix for CFA

Variables	X_1	X_2	X_3	X_4	X_6	X_7
X_1 Delivery speed	1.000					
X_2 Price level	−.349	1.000				
X_3 Price flexibility	.509	−.487	1.000			
X_4 Manufacturer image	.050	.272	−.116	1.000		
X_6 Salesforce image	.077	.186	−.034	.788	1.000	
X_7 Product quality	−.483	.470	−.448	.200	.177	1.000

Stage 6: Evaluating Goodness-of-Fit Criteria

Offending Estimates

Table 11.5 contains the LISREL estimates for the measurement model and the construct correlations. The occurrence of a loading for manufacturer image (X_4) is greater than 1.0 (it is therefore an offending estimate). The reader should also note a corresponding negative error measurement value (known as a Heywood case) for the same variable. Such estimates are theoretically inappropriate and must be corrected before the model can be interpreted and the goodness-of-fit assessed.

Model Respecification

Several remedies are possible for correcting the offending estimate, including dropping the offending variable. In this situation, the variable will be retained and the corresponding error variance ($-.325$ on the diagonal of the measurement error matrix) will be set to a small value (.005) to ensure that the loading will now be less than 1.0. The model is then reestimated.

The results of the respecified model are shown in Table 11.6. In examining the results, no offending estimates are present; thus we can proceed to assessing the goodness-of-fit of the confirmatory factor analysis.

TABLE 11.5 CFA Results: Initial Model

CONSTRUCT LOADINGS

| | Exogenous Construct | |
Variables	Strategy	Image
X_1 Delivery speed	.644	.000
X_2 Price level	−.654	.000
X_3 Price flexibility	.722	.000
X_4 Manufacturer image	.000	1.151
X_6 Salesforce image	.000	.685
X_7 Product quality	−.689	.000

CORRELATIONS AMONG LATENT
CONSTRUCTS (FACTORS)

	Strategy	Image
Strategy	1.000	
Image	−.175	1.000

MEASUREMENT ERROR FOR INDICATORS

Variables	X_1	X_2	X_3	X_4	X_6	X_7
X_1 Delivery speed	.585					
X_2 Price level	.000	.573				
X_3 Price flexibility	.000	.000	.479			
X_4 Manufacturer image	.000	.000	.000	−.325		
X_6 Salesforce image	.000	.000	.000	.000	.531	
X_7 Product quality	.000	.000	.000	.000	.000	.526

TABLE 11.6 CFA Results: Revised Model

CONSTRUCT LOADINGS (*t* VALUES IN PARENTHESES)

	Strategy	*Image*
Variables	*Exogenous Construct*	
X_1 Delivery speed	.643	.000
	(6.263)	
X_2 Price level	−.654	.000
	(−6.381)	
X_3 Price flexibility	.718	.000
	(7.118)	
X_4 Manufacturer image	.000	.997
		(14.001)
X_6 Salesforce image	.000	.790
		(9.467)
X_7 Product quality	−.692	.000
	(−6.821)	

CORRELATIONS AMONG LATENT
CONSTRUCTS (*t* VALUE IN PARENTHESES)

	Strategy	*Image*
Strategy	1.000	
Image	−.202	1.000
	(−1.826)	

MEASUREMENT ERROR FOR INDICATORS

Variables	X_1	X_2	X_3	X_4	X_6	X_7
X_1 Delivery speed	.586					
X_2 Price level	.000	.573				
X_3 Price flexibility	.000	.000	.484			
X_4 Manufacturer image	.000	.000	.000	.005		
X_6 Salesforce image	.000	.000	.000	.000	.376	
X_7 Product quality	.000	.000	.000	.000	.000	.521

Overall Model Fit: Revised Model

The first assessment of model fit must be done for the overall model. In confirmatory factor analysis, overall model fit portrays the degree to which the specified indicators represent the hypothesized constructs. The latest version of LISREL (version 8) now provides a full range of goodness-of-fit measures, as described in appendix 11C. For purposes of the confirmatory factor analysis, we focus on only a limited number of measures from each type. A more detailed discussion of all the available measures occurs in the second example of estimating a full structural model and evaluating competing models. The three types of overall model fit measures useful in CFA can be represented by absolute, incremental, and parsimonious fit measures.

Absolute Fit Measures   LISREL provides absolute goodness-of-fit measures, as shown in Table 11.7 (p. 622). The first measure is the likelihood ratio chi-square statistic. The value ($\chi^2 = 15.745$, nine degrees of freedom) has a statistical significance level of .072, above the minimum level of .05, but not above the more conservative

TABLE 11.7 LISREL Goodness-of-Fit Measures for
CFA: Revised and Null Models

Revised Model

Chi-square (χ^2)	15.745
Degrees of freedom[a]	9
Significance level	.072
Goodness-of-fit index (GFI)	.949
Adjusted Goodness-of-fit index (AGFI)	.881
Root mean square residual (RMSR)	.075

Null Model

Chi-square (χ^2)	212.033
Degrees of freedom	15
Significance level	.000

[a]The expected number of degrees of freedom for the revised
model is 8 [(.5)(6)(7) − 13]. However, because the measurement error of X_4 was set at .005 and not estimated, an additional degree of freedom is available.

levels of .10 or .20. This statistic shows some support for believing that the differences of the predicted and actual matrices are nonsignificant, indicative of acceptable fit. Moreover, the sample size of 100 is within the acceptable range for application of this measure. It also meets the criteria of five observations per estimated parameter. Note, however, that the potential exists for the χ^2 with sample sizes of 100 or less to denote no differences even when the model has no significant relationships, and it would always be advisable in situations such as this to increase the sample size. Thus additional measures of fit must be employed. The goodness-of-fit index (GFI) has a value of .949, which is quite high, but it is not adjusted for model parsimony. Finally, the root mean square residual (RMSR) indicates that the average residual correlation is .075, deemed acceptable given the rather strong correlations in the original correlation matrix. Even though all the measures fall within acceptable levels, the incremental fit and parsimonious fit indices are needed to ensure acceptability of the model from other perspectives.

Incremental Fit Measures   The next type of goodness-of-fit measure assesses the incremental fit of the model compared to a **null model.** In this case, the null model is hypothesized as a single-factor model with no measurement error. The estimation results for the null model are also shown in Table 11.7. (The control cards for null model estimation are shown in appendix A.)

The null model has a χ^2 value of 212.033 with 15 degrees of freedom. With this information, we can now calculate the two incremental fit measures, the Tucker–Lewis Index and the normed fit index.

Tucker–Lewis Index (TLI)

$$\text{TLI} = \frac{(\chi^2_{\text{null}}/df_{\text{null}}) - (\chi^2_{\text{proposed}}/df_{\text{proposed}})}{(\chi^2_{\text{null}}/df_{\text{null}}) - 1}$$

$$= \frac{(212.033/15) - (15.75/9)}{(212.033/15) - 1} = .943$$

Normed Fit Index (NFI)

$$\text{NFI} = \frac{\chi^2_{\text{null}} - \chi^2_{\text{proposed}}}{\chi^2_{\text{null}}}$$

$$= \frac{(212.033 - 15.75)}{212.033} = .926$$

Both incremental fit measures exceed the recommended level of .90, further supporting acceptance of the proposed model.

Parsimonious Fit Measures The final measures of the overall model assess the parsimony of the proposed model by evaluating the fit of the model versus the number of estimated coefficients (or conversely, the degrees of freedom) needed to achieve that level of fit. Two measures appropriate for direct model evaluations are the adjusted goodness-of-fit index (AGFI) and the normed chi-square. The AGFI is provided by the LISREL program. The AGFI value of .881 is close to the recommended level of .90; thus marginal acceptance can be given on this measure. The normed chi-square (χ^2/df) has a value of 1.75 (15.75/9). This falls well within the recommended levels of 1.0 to 2.0. Combined with the AGFI, this result allows conditional support to be given for model parsimony.

In summary, the various measures of overall model goodness-of-fit lend sufficient support to deeming the results an acceptable representation of the hypothesized constructs.

Measurement Model Fit

Now that the overall model has been accepted, each of the constructs can be evaluated separately by (1) examining the indicator loadings for statistical significance and (2) assessing the construct's reliability and variance extracted. First, for each variable the t values associated with each of the loadings exceed the critical values for the .05 significance level (critical value = 1.96) and the .01 significance level as well (critical value = 2.576). Thus all variables are significantly related to their specified constructs, verifying the posited relationships among indicators and constructs.

We now need estimates of the reliability and variance-extracted measures for each construct to assess whether the specified indicators are sufficient in their representation of the constructs. Computations for each measurement are shown in Table 11.8 (p. 624). Both constructs (.772 and .893) exceed the recommended level of .70, although we must remember that the reliability of construct 2 (image) is inflated as a result of respecification of the measurement error for variable X_4 to almost zero to eliminate the negative error variance value.

For the variance-extracted measures, construct 1 (strategy) has a value of .459, falling somewhat short of the recommended 50 percent. Construct 2 (image), with a value of .809, again exceeds the recommended level substantially. The lower level of variance extracted for construct 1 indicates that more than half of the variance for the specified indicators is not accounted for by the construct. This finding may lead the researcher to explore additional loadings for these indicators on the other construct if theoretically justified.

Another coefficient estimated in the measurement model is the correlation between the two constructs. The t value for the obtained correlation of $-.202$ is 1.826 (see Table 11.6) which falls below the critical value for the .05 significance

TABLE 11.8 Reliability and Variance-Extracted Estimates for Constructs in CFA

Reliability

$$\text{Construct reliability} = \frac{(\text{Sum of standardized loadings})^2}{(\text{Sum of standardized loadings})^2 + \text{Sum of indicator measurement error}}$$

Sum of Standardized Loadings[a]
Strategy = .643 + .654 + .718 + .692 = 2.707
Image = .997 + .790 = 1.787

Sum of Measurement Error[b]
Strategy = .586 + .573 + .484 + .521 = 2.164
Image = .005 + .376 = .381

Reliability Computation
$$\text{Strategy} = \frac{(2.707)^2}{(2.707)^2 + 2.164} = .772$$
$$\text{Image} = \frac{(1.787)^2}{(1.787)^2 + .381} = .893$$

Variance Extracted

$$\text{Variance extracted} = \frac{\text{Sum of squared standardized loadings}}{\text{Sum of squared standardized loadings} + \text{Sum of indicator measurement error}}$$

Sum of Squared Standardized Loadings
Strategy = $.643^2 + .654^2 + .718^2 + .692^2 = 1.836$
Image = $.997^2 + .790^2 = 1.618$

Variance Extracted Computation
$$\text{Strategy} = \frac{1.836}{1.836 + 2.164} = .459$$
$$\text{Image} = \frac{1.618}{1.618 + .381} = .809$$

---

[a]For purposes of computing the reliability of a construct, the signs of a loading can be ignored. However, if the researcher wishes to compute a summed measure (such as a scale total) for the indicators, the variables with negative loadings must be reversed in magnitude. This ensures that the negative indicators do not offset the positive indicators.

[b]Indicator measurement error can be calculated as $1 - (\text{standardized loading})^2$ or the diagonal of the measurement error correlation matrix (theta-delta matrix) in the LISREL output.

level, although only slightly. This indicates that there may be weak support at best (only at the .10 significance level) for believing that the constructs are correlated.

Summary

The overall model goodness-of-fit results and the measurement model assessments lend substantial support for confirmation of the proposed two-factor model. Although acceptable results are achieved with this model, the next stage explores possible modifications that may improve on the model results if theoretically justified.

Stage 7: Interpreting and Modifying the Model

The objectives of CFA are (1) to verify the proposed factor structure and (2) to explore if any significant modifications are needed. In doing so, we first examine the factor loadings and factor correlations and then look for possible model respecifications.

Interpretation

Interpretation of CFA is quite similar to that undertaken for factor analysis in chapter 3. For our example, all four variables have significant loadings on the first construct and both loadings are also significant on the second construct. In comparing these results to the factor analysis of the same variables in chapter 3, we find that the conclusions are similar, but with a couple of differences. First, the signs of the loadings are reversed from that found in chapter 3. This does not impact the interpretation of the construct, but does impact the relationship to the second construct. This is shown in the correlation between constructs, which is now negative (versus positive in chapter 3) because the two variables (X_1 and X_3) with negative loadings in the chapter 3 results now have positive loadings in CFA. Second, the correlation between the constructs is increased. This is due to the estimation process accounting for the measurement error of the two factors, which will increase the estimated parameters. Remember, this occurs because if the observed correlation is calculated with measurement error (which should reduce it somewhat), then the "true" relationship will be greater.

Model Respecification

Possible modifications to the proposed model may be indicated through examination of the normalized residuals and the modification indices. However, in any instance the proposed modifications must first have theoretical justification before a respecified model can be tested. In this example, the only possible modifications would be the possibility of a variable acting as an indicator for both constructs.

The normalized residuals are shown in Table 11.9 (p. 626), and examination reveals only one value exceeding 2.58. Thus only one correlation from the original input matrix has a statistically significant residual. This falls within the acceptable range of one in 20 residuals exceeding 2.58 strictly by chance. Table 11.9 also contains the modification indices for the measurement model. Two of the indicators (delivery speed and price level) have modification indicators above the suggested level (3.84) for possible model respecification. In this instance, the modification indices indicate that these variables might be indicators on the second construct as well as the first (i.e., multiple loadings).

Theoretical support for such a structure cannot be found; thus the model is not respecified. If model respecification is based only on the values of the modification indices, the researcher is capitalizing on the uniqueness of these particular data, and the result will most probably be an atheoretical, but statistically significant, model that has little generalizability and limited use in testing causal relationships.

Higher-Order Factor Analysis Models

An additional perspective on the factor analytic structure can be gained with the introduction of higher-order (also known as second-order) factor models [62, 74]. The factor analysis model we have just examined is known as a *first-order factor model.* In this type of CFA model, the researcher specifies just one level of factors (the first order) that are correlated. But this assumes that the factors, although correlated, are separate constructs. What if the researcher had a construct with several facets or dimensions, such as store image or job satisfaction? It would be necessary to see if they are correlated, but what is actually needed is a means of

TABLE 11.9 Normalized Residuals and Modification Indices for CFA

NORMALIZED RESIDUALS

Variables	X_1	X_2	X_3	X_4	X_6	X_7
X_1 Delivery speed	.000					
X_2 Price level	1.826	.000				
X_3 Price flexibility	1.535	−.595	.000			
X_4 Manufacturer image	2.707	2.157	.508	.000		
X_6 Salesforce image	2.234	1.028	1.067	.000	.000	
X_7 Product quality	−1.091	.523	1.959	1.011	.868	.000

MODIFICATION INDICES

	Construct Loadings	
	Strategy	Image
X_1 Delivery speed	.000	7.359
X_2 Price level	.000	4.634
X_3 Price flexibility	.000	.269
X_4 Manufacturer image	.321	.000
X_6 Salesforce image	.321	.000
X_7 Product quality	.000	1.029

	Indicator Measurement Error					
	X_1	X_2	X_3	X_4	X_6	X_7
X_1 Delivery speed	.000					
X_2 Price level	3.333	.000				
X_3 Price flexibility	2.357	.355	.000			
X_4 Manufacturer image	1.712	2.182	.216	.321		
X_6 Salesforce image	.182	.046	.927	.321	.000	
X_7 Product quality	1.191	.274	3.837	.038	1.009	.000

demonstrating the structural relationships between the facets and dimensions. This would allow for a stronger statement by the researcher as to the dimensionality of the store image and job satisfaction constructs.

This is accomplished through the specification of a *second-order factor model*, which posits that the first-order factors estimated are actually subdimensions of a broader and more encompassing construct, in this example, job satisfaction or store image. For illustration purposes, assume that store image is hypothesized to contain five dimensions (atmosphere, price competitiveness, product assortment, product quality, and clientele). By following the procedures described earlier, a five-factor model could be tested. But what would a second-order model represent? Figure 11.8 portrays a second-order factor model in which each of these five constructs are now related (or arise from) the second-order factor.

There are two unique characteristics of the second-order model. First, the second-order factor becomes the exogenous construct, whereas the first-order factors are endogenous. By this, we mean that the second-order factor "causes" the first-order factors. That is why the arrows lead from the second-order factor to the first-order factors. Second, there are no indicators of the second-order factor. Although we can have a number of statements measuring each of the five di-

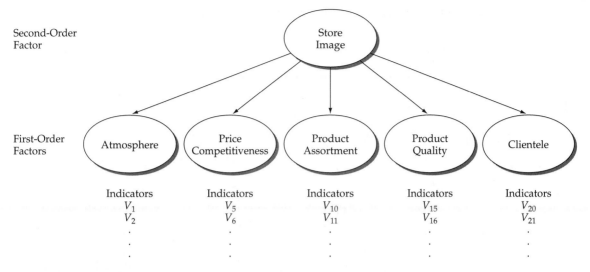

FIGURE 11.8 Path Diagram of Second-Order Factor Analysis of Store Image

mensions, the second-order factor is completely latent, unobservable, and not measurable. To many researchers this seems too abstract or like a "black box" approach. But we must remember that it actually represents an abstract concept that gives rise to latent constructs. It is solely a function of the relationships between the first-order factors. The basic steps and procedures for estimating a higher-order factor model are readily available [48, 75]. The procedures for testing higher-order factor models are also well documented [10].

Summary

Confirmatory factor analysis provides adequate support for the proposed model based on the factor analyses performed in chapter 3. However, because the model has a data inconsistency (the Heywood case for X_4, manufacturer image), the researcher should gather additional data and test the model on a new correlation matrix to ensure generalizability across multiple samples.

Estimating a Path Model with SEM

The next example of SEM examines a set of two causal relationships, relating customers' perceptions of HATCO to their usage level of HATCO products and their satisfaction with HATCO. The structural model allows for an understanding not only of the relative importance of the various perceptions of HATCO in determining purchases but also of possible sources of customers' satisfaction with HATCO.

To avoid duplication of CFA, just examined, an additional survey was conducted among a new set of HATCO customers. A total of 136 customers rated HATCO on a number of possible purchase determinants. This survey is similar to the original database introduced in chapter 1 in that respondents were asked to rate HATCO on a 0 to 10 scale (0 = very poor, and 10 = excellent), indicating how well HATCO performs on that attribute. The causal relationships thus relate to determining what elements of HATCO's performance determine the level of customers' purchases and their satisfaction. It differs in the attributes and the types of relationships examined.

As in the prior example, the discussion follows the seven-stage process, illustrating the issues and interpretation at each stage. The reader is encouraged to refer to the earlier discussions for specific details on issues raised in the example.

Stage 1: Developing a Theoretically Based Model

The purpose of this analysis is to better understand how customers' perceptions of HATCO affect their behavior and attitudes. As already described, 136 customers provided evaluations on 13 attributes across the entire spectrum of ways in which they might interact with and evaluate HATCO. The attributes were derived by a two-stage process. First, all regional managers were participants in a series of focus groups in which these issues, among others, were discussed. From these discussions, 27 attributes were identified as possible evaluative items. This set of 27 items was then presented to the district managers, who rated each item in terms of their perceptions of its importance in customer purchase decisions and satisfaction. Concurrently, a sample of customers of other firms in the industry was also asked for their perceptions as to the importance of each attribute. Analysis of the set of 27 items for both samples resulted in the final set of 13 items composing three general areas of evaluation. The items and evaluation dimensions are shown in Table 11.10.

When the dimensions of evaluation that will act as the independent variables in this analysis have been established, the final step is to determine their relationships to purchase decisions and satisfaction. Because satisfaction is based on a customer's evaluation of past experiences with HATCO, the appropriate causal relationship is for current satisfaction levels to be predicted by level of purchases from HATCO. Although a causal relationship can be posed depicting purchases arising *from* satisfaction, this method would require a greater degree of control over longitudinal aspects of data collection necessary to establish the temporal order. However, we know that purchase level reflects past purchase behavior, so

TABLE 11.10 Evaluative Dimensions and Firm Attributes for the New HATCO Survey

Evaluative Dimension	Firm Attribute
Firm and product factors	X_1 Product quality
	X_2 Invoice accuracy
	X_3 Technical support
	X_4 Introduction of new products
	X_5 Reliable delivery
	X_6 Customer service
Price-based factors	X_7 Product value
	X_8 Low-price supplier
	X_9 Negotiation position (policies)
Buying relationship factors	X_{10} Mutuality of interests
	X_{11} Integrity and honesty
	X_{12} Flexibility
	X_{13} Problem resolution

that it can be a valid predictor of current satisfaction. The researcher must always examine each proposed relationship from a theoretical perspective to ensure that the results are conceptually valid. In this case, the purchase-to-satisfaction link could be easily reversed and estimates of the relationship made, because the correlation works equally well for both relationships. But the results could be potentially misleading because there is a conceptual flaw in the reversed relationship if estimated with these data.

Stage 2: Constructing a Path Diagram of Causal Relationships

Having developed a set of causal relationships, we next portray the relationships in a path diagram (see Figure 11.9). As we see, the three evaluative dimensions act as the exogenous variables, each related to the product usage level, which is one of the two endogenous variables. Product usage is then posited to be the sole predictor of satisfaction. The evaluative dimensions could also be sources of satisfaction; however, these causal relationships are not proposed initially, but instead are explored through the testing of alternative model specifications (i.e., competing models). The path diagram also indicates that the three exogenous dimensions are all proposed to be intercorrelated. Although the evaluative dimensions are proposed to be distinct, it is recognized that some perceptions are shared, and thus there are correlations among the constructs.

Stage 3: Converting the Path Diagram into a Set of Structural and Measurement Models

The path diagram provides the basis for specification of the structural equations and the proposed correlation (1) between exogenous constructs and (2) between structural equations. From the path model, the researcher can construct a series

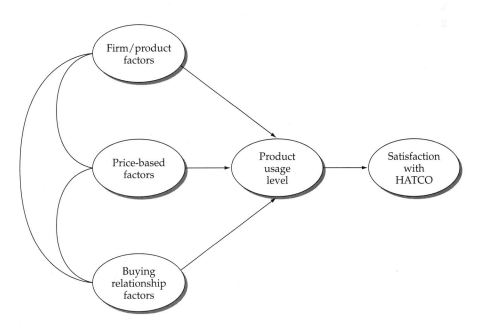

FIGURE 11.9 Path Diagram for SEM

of structural equations (one for each endogenous construct) to constitute the structural model. Then, the measurement model is specified wherein indicators are assigned to each construct (exogenous and endogenous).

Specifying Structural Equations

The specification of a structural equation for each endogenous construct (usage level and satisfaction) must specify the relationships to both the exogenous constructs and other endogenous constructs as well. The four coefficients to be estimated in the structural equations can be expressed as follows:

Endogenous Variable	*Exogenous*			*Endogenous*	
	Firm and Product Factors	Price-Based Factors	Buying Relationship Factors	Usage Level	Satisfaction Level
Usage level	b_1	b_2	b_3		
Satisfaction				b_4	

Note that the error terms for each equation were omitted, but they must be included when the model is estimated.

Specifying the Measurement Model

The measurement model specifies the correspondence of indicators to constructs. For the two endogenous variables, only single indicators are available. Therefore, the researcher has to specify their reliability. Although estimates of the level of measurement error can be made, for illustration purposes the measurement error will be set at zero (indicating perfect measurement of both constructs). The relaxation of this assumption then provides the basis for an alternative model. (This alternative model is not tested in our discussion, but the reader is encouraged to undertake the analysis and examine the effects of varying levels of measurement error.)

The exogenous constructs are measured by 13 indicators reflecting the dimensions described previously. We do not here make a formal specification of the measurement model for the exogenous constructs, and the reader can refer to the prior example for more specific detail on this process if needed. Also, there are no instances for which indicators should be correlated; thus no measurement error correlations are allowed, either in the initial model or as a basis for model modifications.

Correlations among Constructs and Indicators

In addition to the structural equations, the correlations must be specified between the set of exogenous constructs as well as between any of the structural equations. In this example, each exogenous construct (firm and product factors, price-based factors, and buying relationship factors) is allowed to be correlated with the other two exogenous constructs. This represents the realization of shared influences on these external constructs that are not specified in the model, thus causing them to be correlated. There are no correlations, however, between the structural equations.

Stage 4: Choosing Input Matrix Type and Estimating the Proposed Model

With the model completely specified, the next steps are to test the data for meeting the assumptions underlying structural equation modeling, to select the type of input matrix (covariances or correlations) to be used for model estimation, and to estimate the structural and measurement models.

Inputting the Data

Although the design of the study assures independent and random responses, the 15 variables (13 attributes and 2 performance measures) still must be assessed for their distributional characteristics, particularly normality and kurtosis. The procedures described in chapter 2 for assessing these assumptions are employed, along with the procedures for multivariate kurtosis and skewness. No variable is found to have significant departure from normality or pronounced kurtosis, and thus all 15 variables are deemed suitable for use.

When testing a series of causal relationships, covariances are the preferred input matrix type. In this example, correlations are used for both practical and theoretical reasons. From a practical perspective, correlations are more easily interpreted, and diagnosis of the results is more direct. From a theoretical perspective, the primary purpose of the analysis is to examine the *pattern* of relationships among the exogenous and endogenous constructs. For this purpose the correlation matrix is an acceptable input matrix. Moreover, the model most likely suffers from specification error, which is the omission of relevant variables from the model. Both theory and experience would suggest additional predictors of a customer's satisfaction with HATCO. Thus, given this limitation, the researcher should draw conclusions only about the patterns of relationships rather than about the predictive ability of the constructs [22]. For the purposes stated, the example employs the correlation matrix (see Table 11.11, p. 632). The sample size of 136 customers falls within the acceptable limits for use of SEM.

Model Estimation

As in the prior example, the structural equation model was estimated with the LISREL program [67]. As noted in our discussion of CFA, each construct must be made scale invariant through one of two methods. In this analysis we employ the method not used earlier (i.e., we set the loading of one variable per construct to a value of 1.0) to illustrate the procedure. The structural and measurement model coefficients are unaffected by the selected approach. Thus, the variables with fixed loadings of 1.0 are X_1, X_7, and X_{10}.

Stage 5: Assessing the Identification of the Structural Model

Before examining the results, the researcher must be assured that the model is identified. The LISREL program assesses the identification of the model and highlights almost all problems. For this model, no identification problems were indicated. As an example of what might cause an identification problem, the specification of a reciprocal (two-way) relationship between usage level and satisfaction would in most instances result in an identification problem with at least one of the structural equations. Although reciprocal relationships can be

TABLE 11.11 Correlation Matrix for the SEM

Variables	Y_1	Y_2	X_1	X_2	X_3	X_4	X_5	X_6	X_7	X_8	X_9	X_{10}	X_{11}	X_{12}	X_{13}
Y_1 Usage level	1.000														
Y_2 Satisfaction	0.411	1.000													
X_1 Product quality	0.288	0.168	1.000												
X_2 Invoice accuracy	0.359	0.159	0.785	1.000											
X_3 Technical support	0.268	0.141	0.676	0.637	1.000										
X_4 Introduction of new products	0.212	0.081	0.581	0.622	0.627	1.000									
X_5 Reliable delivery	0.250	0.060	0.632	0.644	0.538	0.699	1.000								
X_6 Customer service	0.305	0.127	0.690	0.667	0.551	0.625	0.692	1.000							
X_7 Product value	0.328	0.133	0.293	0.263	0.336	0.290	0.207	0.174	1.000						
X_8 Low-price supplier	0.268	0.046	0.184	0.124	0.230	0.260	0.110	0.254	0.301	1.000					
X_9 Negotiation position	0.142	0.104	0.289	0.249	0.299	0.219	0.155	0.299	0.307	0.676	1.000				
X_{10} Mutuality of interests	0.329	0.063	0.327	0.292	0.112	0.292	0.326	0.253	0.179	0.114	0.173	1.000			
X_{11} Integrity and honesty	0.519	0.077	0.413	0.385	0.265	0.346	0.259	0.262	0.347	0.225	0.205	0.411	1.000		
X_{12} Flexibility	0.510	0.090	0.330	0.272	0.192	0.151	0.210	0.225	0.348	0.259	0.162	0.555	0.532	1.000	
X_{13} Problem resolution	0.341	0.173	0.154	0.173	0.182	0.153	0.107	0.102	0.272	0.052	0.093	0.202	0.411	0.425	1.000

estimated, they require model constraints in other aspects of the structural equations. Interested readers may refer to Bollen [22] or Hayduk [54].

Stage 6: Evaluating Goodness-of-Fit Criteria

Being assured that the model is correctly specified and the estimation process is not constrained by identification problems, the researcher can proceed to evaluate the specific results for the proposed model. If the assumptions underlying SEM are met, the estimated coefficients are evaluated along with the overall model fit.

Offending Estimates

Having met the assumptions of SEM, the results are first examined for nonsensical or theoretically inconsistent estimates. The three most common offending estimates are negative error variances, standardized coefficients exceeding or very close to 1.0, or very large standard errors. Examination of the standardized results in Tables 11.12 and 11.13 (p. 634) reveals no instances of any of these problems.

Overall Model Fit

Before evaluating the structural or measurement models, the researcher must assess the overall fit of the model to ensure that it is an adequate representation of the entire set of causal relationships. Each of the three types of goodness-of-fit measures are used.

Absolute Fit Measures   Three measures of the most basic measures of absolute fit are the likelihood-ratio chi-square (χ^2), the goodness-of-fit index, and the root mean square residual (see Table 11.14, p. 635). The chi-square value of 178.714 with 85 degrees of freedom is statistically significant at the .000 significance level.

TABLE 11.12 SEM Results: Standardized Parameter Estimates for the Structural Model

STRUCTURAL EQUATION COEFFICIENTS (t VALUES IN PARENTHESES)

Endogenous Constructs		Endogenous Constructs		Exogenous Constructs			
		Usage Level	Satisfaction	Firm and Product Factors	Price-Based Factors	Buying Relationship Factors	Structural Equation Fit (R^2)
Usage level	=	.000	.000	.056 (.621)	.038 (.428)	.615 (4.987)	.433
Satisfaction	=	.411 (5.241)	.000	.000	.000	.000	.169

CORRELATIONS AMONG THE EXOGENOUS CONSTRUCTS (t VALUES IN PARENTHESES)

Exogenous Constructs	Firm and Product Factors	Price-Based Factors	Buying Relationship Factors
Firm and product factors	1.000		
Price-based factors	.355 (2.724)	1.000	
Buying relationship factors	.465 (3.746)	.353 (2.489)	1.000

TABLE 11.13 SEM Results: Standardized Parameter Estimates for the Measurement Model

CONSTRUCT LOADINGS (t VALUES IN PARENTHESES)

Indicators	Exogenous Constructs		
	Firm and Product Factors	Price-Based Factors	Buying Relationship Factors
X_1 Product quality	.863 (.000)[a]	.000	.000
X_2 Invoice accuracy	.857 (12.855)	.000	.000
X_3 Technical support	.747 (10.297)	.000	.000
X_4 Introduction of new products	.759 (10.550)	.000	.000
X_5 Reliable delivery	.781 (11.020)	.000	.000
X_6 Customer service	.803 (11.524)	.000	.000
X_7 Product value	.000	.404 (.000)[a]	.000
X_8 Low-price supplier	.000	.811 (4.181)	.000
X_9 Negotiation position	.000	.821 (4.169)	.000
X_{10} Mutuality of interests	.000	.000	.603 (.000)[a]
X_{11} Integrity and honesty	.000	.000	.729 (6.216)
X_{12} Flexibility	.000	.000	.786 (6.447)
X_{13} Problem resolution	.000	.000	.506 (4.786)

[a]Values were not calculated because loading was set to 1.0 to fix construct variance.

Because the sensitivity of this measure is not overly affected by the sample size of 136, the researcher may conclude that significant differences exist. However, we must also note that the chi-square test becomes more sensitive as the number of indicators rises. With this in mind, we examine a number of other measures. The GFI value of .865 is at a marginal acceptance level, as is the RMSR value of .076. The RMSR must be evaluated in light of the input correlation matrix, and in this context the residual value is relatively high.

As a complement to these basic measures, the researcher can also examine other absolute fit measures. The root mean square error of approximation has a value of .090, which falls just outside the acceptable range of .08 or less, but less than the upper threshold of .10. Three other measures—the noncentrality index, the scaled noncentrality index, and the expected cross-validation index—are all used in comparisons among alternative models. There is no established range of acceptable values for these measures. They are examined later, when competing models are analyzed.

All of the absolute fit measures indicate that the model is marginally acceptable at best. This should not halt further examination of the results, however, un-

TABLE 11.14 Goodness-of-Fit Measures for SEM

LISREL-Provided Measures

Absolute Fit Measures

Chi-square (χ^2) of estimated model	178.714[a]
Degrees of freedom	85
Significance level	.000[a]
Noncentrality parameter (NCP)	93.714
Goodness-of-fit index (GFI)	.865[a]
Root mean square residual (RMSR)	.076[a]
Root mean square error of approximation (RMSEA)	.090
P-Value of close fit (RMSEA < .05)	.000
Expected cross-validation index (ECVI)	
ECVI for estimated model	1.842
ECVI for saturated model	1.778
ECVI for independence model	7.927

Incremental Fit Measures

Chi-square (χ^2) of null or independence model	1040.194
Degrees of freedom	105
Adjusted goodness-of-fit index (AGFI)	.810[a]
Tucker–Lewis index (TLI) or Non-normed fit index (NNFI)	.876
Normed fit index (NFI)	.828

Parsimonious Fit Measures

Parsimonious normed fit index (PNFI)	.670
Parsimonious goodness-of-fit index	.613
Akaike information criterion (AIC)	
Estimated model	248.714
Saturated model	240.000
Independence model	1070.194
Comparative fit index (CFI)	.900
Incremental fit index (IFI)	.902
Relative fit index (RFI)	.788
Critical *N* (CN)	90.318
Sample size: 136 respondents	

Calculated Measures of Overall Model Goodness-of-Fit

Normed chi-square	$\dfrac{178.714}{85} = 2.103$
Scaled noncentrality parameter (SNCP)	$\dfrac{93.714}{136} = .689$

[a]Measures provided directly by all versions of LISREL. Other measures provided only by version 8 of LISREL and must be calculated separately for earlier versions.

less the overall model fit was deemed so poor that both the structural and measurement models would be invalid. Moreover, the other types of fit measures will provide different perspectives on the acceptability of the model fit.

Incremental Fit Measures   In addition to the overall measures of fit, a model can be evaluated in comparison to a baseline or null model. In this instance, the null model is a single-factor model with no measurement error. The null model has a chi-square value of 1,040.194 with 105 degrees of freedom. Although we gain a substantial reduction in the chi-square value owing to the estimated coefficients in the model, the incremental fit measures also provide only marginal support. The AGFI, Tucker–Lewis index, and NFI all fall slightly below the desired threshold of .90.

Although the .90 threshold has no statistical basis, practical experience and research have demonstrated its usefulness in distinguishing between acceptable and unacceptable models. However, all the incremental fit measures exceed .80, and the true test comes with the comparison of the proposed model against alternative or competing models.

Parsimonious Fit Measures   This final type of measure provides a basis for comparison between models of differing complexity and objectives. One applicable measure for evaluating a single model is the normed chi-square measure. With a computed value of 2.103, it falls within some threshold limits for this measure but exceeds other limits. Thus, again, only marginal support is provided. Three other parsimonious fit measures are available (the parsimonious normed fit index, the parsimonious goodness-of-fit index, and the Akaike information criterion), but they are suited only for intermodel comparisons.

A review of the three types of overall measures of fit reveals a consistent pattern of marginal support for the overall model as proposed. As noted earlier, the truer test for the overall model is a comparison to a series of alternative models, which is carried out in stage 7.

Measurement Model Fit

Even though only marginal support was found for the overall model, it was deemed sufficient to proceed and assess the measurement model fit as well. The first step is an examination of the loadings, particularly focusing on any non-significant loading. Referring to Table 11.13, we see that all the indicators are statistically significant for the proposed constructs. Because no indicators have loadings so low that they should be deleted and the model reestimated, the reliability and variance extracted measures need to be computed.

Table 11.15 contains the computations for both the reliability and the variance-extracted measures. In terms of reliability, all three exogenous constructs exceed the suggested level of .70. In terms of variance extracted, the firm/product construct exceeds the threshold value of .50. The price-based construct only slightly misses the .50 guideline, and the buying-relationship construct has a somewhat lower value (.442). Thus, for all three constructs, the indicators are sufficient in terms of how the measurement model is now specified.

Structural Model Fit

Having assessed the overall model and aspects of the measurement model, the researcher is now prepared to examine the estimated coefficients themselves for both practical and theoretical implications. Review of Table 11.12 reveals that both structural equations contain statistically significant coefficients. For the causal relationship linking the three evaluative dimensions with usage level, only one dimension, the buying relationship, is statistically significant. Therefore, the more personal aspects of the transaction, such as mutuality, integrity, flexibility in dealing, and problem resolution, have the only distinct and substantive impact on increasing product usage levels. Thus, although HATCO must not ignore the other aspects of the business, emphasis should be placed on the maintenance of existing buying relationships and development of new ones focusing on these features. Moreover, the combined effect of these three factors achieves an R^2 value of .433, or 43.3 percent of the variance in usage levels. For the structural equation pre-

TABLE 11.15 Reliability and Variance-Extracted Estimates for Exogenous Constructs in SEM

Reliability

$$\text{Construct reliability} = \frac{(\text{Sum of standardized loadings})^2}{(\text{Sum of standardized loadings})^2 + \text{Sum of indicator measurement error}}$$

Sum of Standardized Loadings
　Firm and product factors = .863 + .857 + .747 + .759 + .781 + .803 = 4.81
　Price-based factors = .404 + .811 + .821 = 2.036
　Buying relationship = .603 + .729 + .786 + .506 = 2.624

Sum of Measurement Error[a]
　Firm and product factors = .256 + .265 + .442 + .424 + .390 + .355 = 2.132
　Price-based factors = .837 + .342 + .327 = 1.506
　Buying relationship = .636 + .469 + .382 + .744 = 2.231

Reliability Computation

$$\text{Firm and product factors} = \frac{(4.81)^2}{(4.81)^2 + 2.132} = .916$$

$$\text{Price-based factors} = \frac{(2.036)^2}{(2.036)^2 + 1.506} = .734$$

$$\text{Buying relationship} = \frac{(2.624)^2}{(2.624)^2 + 2.231} = .755$$

Variance Extracted

$$\text{Variance Extracted} = \frac{\text{Sum of squared standardized loadings}}{\text{Sum of squared standardized loadings} + \text{Sum of indicator measurement error}}$$

Sum of Squared Standardized Loadings
　Firm and product factors = $.863^2 + .857^2 + .747^2 + .759^2 + .781^2 + .803^2 = 3.868$
　Price-based factors = $.404^2 + .811^2 + .821^2 = 1.495$
　Buying relationship = $.603^2 + .729^2 + .786^2 + .502^2 = 1.769$

Variance Extracted Computation

$$\text{Firm and product factors} = \frac{3.868}{3.868 + 2.132} = .645$$

$$\text{Price-based factors} = \frac{1.495}{1.495 + 1.506} = .498$$

$$\text{Buying relationship} = \frac{1.769}{1.769 + 2.231} = .442$$

[a]Indicator measurement error can be calculated as $1 - (\text{standardized loading})^2$ or the diagonal of the measurement error correlation matrix (theta-delta matrix) in the LISREL output.

dicting satisfaction, the other endogenous construct (usage level) was statistically significant. Thus a significant causal relationship has been identified and may now be used as a basis for the formulation of alternative or competing models. The low R^2 value (.169) is to be expected because it is based only on the correlation between usage level and satisfaction.

Also of interest are the correlations between the evaluative dimensions (exogenous constructs). In this example, each construct is significantly correlated with the other exogenous constructs. This correlation reveals that, whereas the buying relationship is fundamental to increases in usage levels, all three constructs are interwoven, and the firm must not focus exclusively on any single dimension.

As a final means of examining the results, the correlations between estimated constructs (endogenous variables) should be reviewed and high values should be noted as an indication of an unacceptable level of intercorrelated constructs. For the endogenous constructs of this study, the correlation of their estimated values is only .411 (see Table 11.12), too low to indicate troublesome intercorrelation.

Competing Models

The final approach to model assessment is to compare the proposed model with a series of competing models, which act as alternative explanations to the proposed model. In this way, the researcher can determine whether the proposed model, regardless of overall fit (within reasonable limits), is acceptable, because no other similarly formulated model can achieve a higher level of fit. This step is particularly important when the chi-square statistic indicates no significant differences in overall model fit, because there may always be a better-fitting model, even in the case of nonsignificant differences. For a systematic approach to specification of competing models, see Anderson and Gerbing [4].

For the purpose of this example, two alternative models are proposed (see Figure 11.10 for these models expressed as path diagrams). The first model (COMPMOD1) adds the three exogenous constructs as predictors of satisfaction. The second model (COMPMOD2) adds only buying relationship as a predictor of satisfaction because it was already a significant predictor of usage levels. As a means of comparison, a set of goodness-of-fit measures are calculated for each model and then compared to determine which of the three is the most parsimonious.

Table 11.16 compares the three models on all three types of fit measures. For the absolute fit measures, COMPMOD1 has the lowest chi-square value, but this model also has the largest number of estimated parameters (and thus the lowest degrees of freedom). COMPMOD1 also is lowest on the RMSR measure. COMPMOD2 excels on the measures that attempt to minimize sample size and sample-specific estimates to model fit. Table 11.16 shows that NCP, SNCP, and ECVI all have their lowest values with COMPMOD2. The estimated model, although not achieving the best fit on any of these measures, is still very close on all measures, making it a viable alternative for acceptance along with the other two competing models.

Two of the incremental fit measures favor COMPMOD2 (AGFI and NNFI), whereas COMPMOD1 has the best performance on the NFI measure. Again, the estimated model is quite close with no substantive differences. The parsimonious fit measures are the final set to be considered. Here, the estimated model has the best fit as measured by the PNFI and PGFI, and COMPMOD2 excels on the normed chi-square and Akaike information criterion.

The results across all three types of measures show mixed results, sometimes favoring the estimated model or one of the competing models. If the focus is limited to the parsimonious fit indices, which account for the model parsimony, the results are still split between the original estimated model and the second competing model (COMPMOD2). Thus, although the fit of the proposed model did not exceed the recommended guidelines in many instances, we accept the proposed model with reservations until additional constructs can be added, measures refined, or causal relationships respecified. The selection of one of the competing models must be made on both theoretical and empirical bases, just as was done in formulating the original theoretical model that was the basis for the estimated

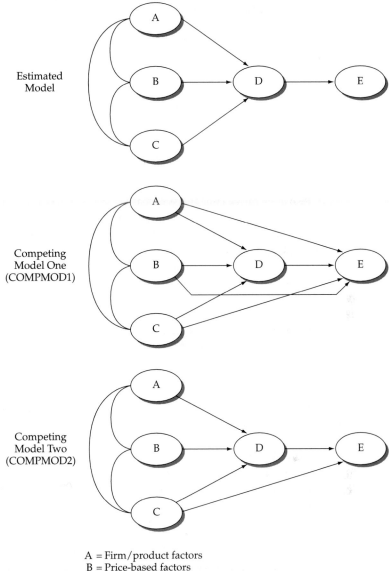

A = Firm/product factors
B = Price-based factors
C = Buying relationship factors
D = Product usage level
E = Satisfaction with HATCO

FIGURE 11.10 Path Diagrams of Estimated Model and Competing
Models (COMPMOD1 and COMPMOD2)

model. This example illustrates the value of testing competing models for all structural equation models to assure the researcher that the proposed or revised model
is truly the "best" model available.

Stage 7: Interpreting and Modifying the Model

The final stage involves interpreting the results in both empirical and practical
terms, as well as examining the results for any potential model modifications.
Although interpretation is the basis for support of the theoretical model, model

TABLE 11.16 Comparison of Goodness-of-Fit Measures for the Estimated and
Competing Models

Goodness-of-Fit Measure	Estimated Model	Competing Models	
		COMPMOD1	COMPMOD2
Absolute Fit Measures			
Likelihood-ratio chi-square (χ^2)	178.714	174.450	175.397
Degrees of freedom	85	82	84
Noncentrality parameter (NCP)	93.714	92.450	91.397
Scaled noncentrality parameter (SNCP)	.689	.680	.672
Goodness-of-fit index (GFI)	.865	.867	.866
Root mean square residual (RMSR)	.076	.074	.075
Root mean square error of approximation (RMSEA)	.090	.091	.090
Expected cross-validation index (ECVI)	1.842	1.855	1.833
Incremental Fit Measures			
Adjusted goodness-of-fit index (AGFI)	.810	.805	.809
Tucker–Lewis index (TLI) or (NNFI)	.876	.873	.878
Normed fit index (NFI)	.828	.832	.831
Parsimonious Fit Measures			
Parsimonious normed fit index (PNFI)	.670	.650	.665
Parsimonious goodness-of-fit index (PGFI)	.613	.592	.606
Normed chi-square	2.103	2.127	2.088
Akaike information criterion (AIC)	248.714	250.450	247.397

respecification procedures are necessary to ensure that the estimated model has
the more appropriate empirical constraints in order to maximize model goodness-
of-fit.

Interpretation

The measurement model provides the researcher with a means of providing meaning to the constructs, in this example the three exogenous constructs, because the
two endogenous constructs were single indicator constructs. Each of the set of the
indicators for the three constructs had fairly comparable indicators, the only exception being the loading for X_7 on the price-based construct (see Table 11.13).
The loadings are all statistically significant, thus supporting the theoretical basis
for assignment of indicators to each construct.

The significance tests for the structural model parameters is the basis for accepting or rejecting the proposed relationships between exogenous and endogenous constructs. In the first basic relationship, the three exogenous constructs
(firm/product factors, price-based factors, and buying relationship factors) were
proposed to be the antecedents of usage. The estimated model, however, found
support only for the role of buying relationship in affecting usage level. Both other
constructs had very low parameter estimates and were deemed not statistically
significant. The correlation of the three exogenous constructs were also significant
and all positive. The second relationship concerned the prediction of satisfaction,

which was deemed to be solely a function of usage level. This relationship was supported by a parameter of .411 (see Table 11.12), which was statistically significant and of the appropriate direction (positive). The earlier discussion of competing models identified some additional effects that could impact satisfaction, but these must be tested in further research.

Model Respecification

Finally, we examine some diagnostic elements (residuals and modification indices) that may indicate potentially significant model modifications. As noted earlier, we are assuming that all model modifications have theoretical support before being implemented.

Normalized Residual Analysis   Table 11.17 (p. 642) contains the normalized residuals from the proposed model when estimated by LISREL. There are 11 potentially significant residuals—residuals exceeding the threshold value of ±2.58. This number exceeds the guideline of having only 5 percent of the normalized residuals exceed the threshold value, which in this case would be six residuals (5 percent of 120). We can note only one consistent pattern among the residuals. The variable X_7 is connected with a majority of the residuals exceeding 2.58. This observation indicates that perhaps the elimination of only this single indicator will create a substantial improvement in fit. If this modification is made, the chi-square value decreases to 146.94 with 71 degrees of freedom. Using the parsimonious fit index, we note that the computed value for this reduced indicator model is .557, which is not greater than the original proposed model. Thus, although model fit will improve, model parsimony will not. If we exclude X_7, the normalized residuals marginally meet the guideline. The researcher's decision regarding the exclusion of X_7 should be based on theoretical grounds.

Modification Indices   Another indication of possible model respecification is the modification index. Table 11.18 (p. 643) contains the modification indices obtained during the estimation of the proposed model. Looking for values exceeding 3.84 reveals that modifications are suggested only in the measurement model. We should note that the two highest modification indices are for X_7, the variable that the residual analysis suggested dropping. Besides X_7, loadings on more than a single construct were suggested for X_8 and X_9, although the anticipated reduction in chi-square is minimal and not appropriate for model respecification.

Overview of the Seven-Stage Process

The seven-stage process has empirically investigated a series of causal relationships with interrelated dependent (endogenous) constructs. The estimated model, even though not achieving the recommended levels of fit, may represent the best available model until further research identifies improvements in theoretical relationships or measurement of the constructs. Although several model modifications improve model fit, they do not increase model parsimony, and little theoretical support can be found for these respecifications of theory. Thus, in situations such as these, the researcher should be very cautious about the type and the extent of model respecifications undertaken.

TABLE 11.17 Normalized Residuals for the SEM

	Y_1	Y_2	X_1	X_2	X_3	X_4	X_5	X_6	X_7	X_8	X_9	X_{10}	X_{11}	X_{12}	X_{13}
Y_1 Usage level	.000														
Y_2 Satisfaction	.000	.000													
X_1 Product quality	-.542	.546	.000												
X_2 Invoice accuracy	1.526	.436	3.138	.000											
X_3 Technical support	.052	.409	1.430	-.154	.000										
X_4 Introduction of new products	-1.196	-.384	-3.423	-1.292	1.839	.000									
X_5 Reliable delivery	-.600	-.693	-2.039	-1.233	-1.495	3.569	.000								
X_6 Customer service	.437	.124	-.120	-1.117	-1.724	.559	2.499	.000							
X_7 Product value	2.972	1.040	2.293	1.887	2.985	2.363	1.243	.780	.000						
X_8 Low-price supplier	1.402	-.593	-1.454	-2.736	.255	.757	-2.176	.438	-1.103	.000					
X_9 Negotiation position	-2.788	.147	.870	-.013	1.469	-.034	-1.383	1.287	-1.058	3.810	.000				
X_{10} Mutuality of interests	-1.582	-1.347	1.396	.851	-1.468	1.199	1.641	.427	1.174	-.941	-.035	.000			
X_{11} Integrity and honesty	1.385	-1.704	2.357	1.830	.188	1.492	-.096	-.181	3.186	.308	-.128	-.847	.000		
X_{12} Flexibility	-.189	-1.794	.310	-.900	-1.450	-2.287	-1.146	-1.323	3.150	.735	-1.478	2.956	-2.310	.000	
X_{13} Problem resolution	.214	.485	-.742	-.436	.085	-.370	-1.105	-1.271	2.467	-1.343	-.775	-1.977	1.063	.836	.000

Note: Underlined values are residuals exceeding the suggested guideline of ± 2.58.

TABLE 11.18 Modification Indices for the SEM

MEASUREMENT MODEL

Exogenous Construct Indicators

	Exogenous Constructs		
Variables	Firm and Product Factors	Price-Based Factors	Buying Relationship Factors
X_1 Product quality	.000	.040	2.224
X_2 Invoice accuracy	.000	1.574	.690
X_3 Technical support	.000	2.009	.872
X_4 Introduction of new products	.000	.543	.276
X_5 Reliable delivery	.000	3.565	.570
X_6 Customer service	.000	1.294	.589
X_7 Product value	6.242	.000	15.289
X_8 Low-price supplier	3.907	.000	.000
X_9 Negotiation position	.489	.000	3.916
X_{10} Mutuality or interests	1.094	.107	.000
X_{11} Integrity and honesty	2.959	.574	.000
X_{12} Flexibility	3.484	.003	.000
X_{13} Problem resolution	.778	.766	.000

STRUCTURAL MODEL

Structural Equations

	Endogenous Constructs		Exogenous Constructs		
Endogenous Construct	Usage Level	Satisfaction	Firm and Product Factors	Price-Based Factors	Buying Relationship Factors
Usage level	.000	2.729	.000	.000	.000
Satisfaction	.000	.000	.016	.026	3.249

Correlations between Exogenous Constructs

	Exogenous Constructs		
	Firm and Product Factors	Price-Based Factors	Buying Relationship Factors
Firm and product factors	.000		
Price-based factors	.000	.000	
Buying relationship factors	.000	.000	.000

Correlations between Endogenous Constructs (Structural Equations)

	Usage Level	Satisfaction
Usage level	.000	
Satisfaction	2.729	.000

Summary

Structural equation modeling is a technique combining elements of both multiple regression and factor analysis that enables the researcher not only to assess quite complex interrelated dependence relationships but also to incorporate the effects of measurement error on the structural coefficients at the same time. Upon completion of this chapter, both the academic researcher and the industry researcher should be able to recognize the benefits afforded by SEM. Although SEM is useful in many instances, it should be used only in a confirmatory mode, leaving exploratory analyses to other multivariate techniques.

Questions

1. What are the similarities between structural equation modeling and the multivariate techniques discussed in earlier chapters?
2. Why should a researcher assess measurement error and incorporate it into the analysis?
3. Using the definition of "theory" presented in this chapter, describe theories that might be of interest to academic and industry researchers.
4. Specify a set of causal relationships and then represent them in a path diagram.
5. What are the criteria by which the researcher should decide whether a model has achieved an acceptable level of fit?
6. Using the path diagram for question 4 (or one prepared for this question), specify at least two alternative models with the supporting theoretical rationale.
7. What is meant by higher-order confirmatory factor analysis? When should it be used and what can it tell you that other factor analyses will not?

A Mathematical Representation in LISREL Notation

The discussion of structural equation modeling to this point has relied on terminology similar to that used in our discussion of multiple regression and other dependence relationships. However, a researcher attempting either (1) to read program manuals, articles, and texts on this topic, or (2) to publish in academic journals using this technique will need a more formal introduction to the mathematical notation used in SEM. Because of its widespread application, LISREL [64, 65, 67] has become the standard for notation. This appendix first presents the complete formulation of the set of causal relationships and measurement relationships in LISREL notation. Then, an example will progress from a path diagram through the structural equations into the appropriate LISREL notation for model specification.

LISREL Notation

The entire LISREL model can be expressed in terms of eight matrices, two defining the structural equations, two defining the correspondence of indicators and constructs, one for the correlation of exogenous constructs, one for the correlation of endogenous constructs, and finally two detailing the correlated errors for the measurement of exogenous and endogenous variables. Table 11A.1 (p. 646) lists each of the matrices as well as constructs and indicators, and provides a brief description and notation for the matrix and its elements.

TABLE 11A.1 Matrices, Constructs and Indicators, and Model Equation Notation of the LISREL Model

LISREL Model Element	Description	Notation Matrix	Notation Element
Matrix			
Structural Model			
Beta	Relationships of endogenous to endogenous constructs	B	β_{nn}
Gamma	Relationships of exogenous to endogenous constructs	Γ	γ_{nm}
Phi	Correlation among exogenous constructs	Φ	ϕ_{mm}
Psi	Correlation of structural equations or endogenous constructs	Ψ	ψ_{nn}
Measurement Model			
Lambda-X	Correspondence (loadings) of exogenous indicators	Λ_x	λ^x_{pm}
Lambda-Y	Correspondence (loadings) of endogenous indicators	Λ_y	λ^y_{qn}
Theta-delta	Correlational matrix of prediction error for exogenous construct indicators	Θ_δ	θ^δ_{pp}
Theta-epsilon	Correlational matrix of prediction error for endogenous construct indicators	Θ_ε	θ^ε_{qq}
Constructs and Indicators			
Construct			
Exogenous	Exogenous construct		ξ
Endogenous	Endogenous construct		η
Indicator			
Exogenous	Exogenous indicator		X
Endogenous	Endogenous indicator		Y
Structural and Measurement Model Equations			
Structural Model	Relationships between exogenous and endogenous constructs		$\eta = \Gamma\xi + \beta\eta + \zeta$
Measurement Model			
Exogenous	Specification of indicators for exogenous constructs		$X = \Lambda_x\xi + \delta$
Endogenous	Specification of indicators for endogenous constructs		$Y = \Lambda_y\eta + \varepsilon$
Size Subscripts for Matrices			
	Number of exogenous constructs		m
	Number of endogenous constructs		n
	Number of exogenous construct indicators		p
	Number of endogenous construct indicators		q

The matrices are used to form the basic equations for both structural and measurement models. The equation for the structural model is

$$\eta = \Gamma\xi + B\eta + \zeta$$

And the equations for the measurement model are as follows:

Exogenous Constructs	Endogenous Constructs
$X = \Lambda_x\xi + \delta$	$Y = \Lambda_y\eta + \varepsilon$

A more straightforward presentation, however, is through the actual equations (structural and measurement). For purposes of illustration, there are two endogenous constructs ($n = 2$) and three exogenous constructs ($m = 3$), with four indicators of both endogenous and exogenous constructs ($p = 4$, $q = 4$). No causal relationships are expressed; rather, the complete equations are shown (see Table 11A.2). An example of specifying a proposed path model in LISREL terms is addressed in the example that follows.

When the structural and measurement equations are actually estimated, the error terms become part of the between-construct and between-indicator matrices described. For example, the δ_p and ε_p values are actually estimated as the diagonal values of the theta-delta and theta-epsilon matrices, respectively. Likewise, the structural equation errors (ζ_n) are the diagonals of the psi matrix.

TABLE 11A.2 Complete Structural and Measurement Model Equations

STRUCTURAL MODEL EQUATIONS

Endogenous Construct		Exogenous Construct		Endogenous Construct		Error
η_1	=	$\gamma_{11}\xi_1 + \gamma_{12}\xi_2 + \gamma_{13}\xi_3$	+	$\beta_{11}\eta_1 + \beta_{12}\eta_3$	+	ζ_1
η_2	=	$\gamma_{21}\xi_1 + \gamma_{22}\xi_2 + \gamma_{23}\xi_3$	+	$\beta_{21}\eta_2 + \beta_{22}\eta_2$	+	ζ_2

MEASUREMENT MODEL EQUATIONS

Exogenous Indicator		Exogenous Construct		Error
X_1	=	$\lambda_{11}^x\xi_1 + \lambda_{12}^x\xi_2 + \lambda_{13}^x\xi_3$	+	δ_1
X_2	=	$\lambda_{21}^x\xi_1 + \lambda_{22}^x\xi_2 + \lambda_{23}^x\xi_3$	+	δ_2
X_3	=	$\lambda_{31}^x\xi_1 + \lambda_{32}^x\xi_2 + \lambda_{33}^x\xi_3$	+	δ_3
X_4	=	$\lambda_{41}^x\xi_1 + \lambda_{42}^x\xi_2 + \lambda_{43}^x\xi_3$	+	δ_4

Endogenous Indicator		Endogenous Constructs		Error
Y_1	=	$\lambda_{11}^y\eta_1 + \lambda_{12}^y\eta_2$	+	ε_1
Y_2	=	$\lambda_{21}^y\eta_1 + \lambda_{22}^y\eta_2$	+	ε_2
Y_3	=	$\lambda_{31}^y\eta_1 + \lambda_{32}^y\eta_2$	+	ε_3
Y_4	=	$\lambda_{41}^y\eta_1 + \lambda_{42}^y\eta_2$	+	ε_4

STRUCTURAL EQUATION CORRELATIONS AMONG CONSTRUCTS

	Among Exogenous Constructs (Phi ϕ)				Among Endogenous Constructs (Psi ψ)	
	ξ_1	ξ_2	ξ_3		η_1	η_2
ξ_1	—			η_1	—	
ξ_2	ϕ_{21}	—		η_2	ψ_{21}	—
ξ_3	ϕ_{31}	ϕ_{32}	—			

MEASUREMENT MODEL (INDICATOR) CORRELATIONS

	Among Exogenous Indicators (Theta-delta θ_δ)					Among Endogenous Indicators (Theta-epsilon ϕ_ϵ)			
	X_1	X_2	X_3	X_4		Y_1	Y_2	Y_3	Y_4
X_1	—				Y_1	—			
X_2	$\phi_{\delta 21}$	—			Y_2	$\phi_{\varepsilon 21}$	—		
X_3	$\phi_{\delta 31}$	$\phi_{\delta 32}$	—		Y_3	$\phi_{\varepsilon 31}$	$\phi_{\varepsilon 32}$	—	
X_4	$\phi_{\delta 41}$	$\phi_{\delta 42}$	$\phi_{\delta 43}$	—	Y_4	$\phi_{\varepsilon 41}$	$\phi_{\varepsilon 42}$	$\phi_{\varepsilon 43}$	—

From a Path Diagram to LISREL Notation

We now illustrate the complete process of moving from the path diagram to the complete LISREL notation, ready for input into any of the SEM programs. The path diagram in Figure 11A.1 describes the set of causal relationships to be examined in this example. There are three endogenous (Y, or dependent) variables related to four exogenous (X, or independent) variables. Each of the seven constructs are measured by two variables (V_1 to V_{14}). Also, the measures of X_1 are correlated, as are the measures of Y_2. Moreover, correlations exist between two independent variable pairs (X_1-X_2 and X_3-X_4) and two dependent variables (Y_1-Y_3). Do not confuse the use of X and Y variables in the path diagram with the designation of exogenous and endogenous indicators as X and Y, respectively, in LISREL notation. Indicators in the path diagram, which are not shown in Figure 11A.1, are the variables V_1 to V_{14}.

Constructing Structural Equations from the Path Diagram

The first step is to translate the path diagram into a series of structural equations for each endogenous variable. The equations are given in Table 11A.3.

Denoting the Correspondence of Indicators and Constructs

When the structural equations have been specified, the measurement of each construct must be defined. In this example, each construct has two indicators. The correspondence of indicators and constructs is shown in Table 11A.4.

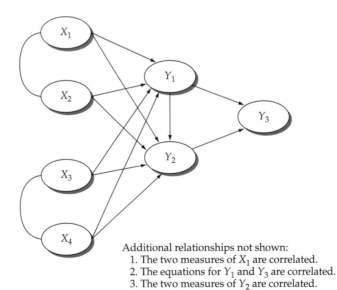

Additional relationships not shown:
1. The two measures of X_1 are correlated.
2. The equations for Y_1 and Y_3 are correlated.
3. The two measures of Y_2 are correlated.
4. There are two measures per construct.

FIGURE 11.A1 Path Diagram of Causal Relationships

TABLE 11A.3 Structural Model Equations for the Path Diagram

Endogenous Variable		Exogenous Variables					Endogenous Variables				Error
		X_1	X_2	X_3	X_4		Y_1	Y_2	Y_3		
Y_1	$=$	$b_1X_1 + b_2X_2 + b_3X_3 + b_4X_4$				$+$				$+$	ε_1
Y_2	$=$	$b_5X_1 + b_6X_2 + b_7X_3 + b_8X_4$				$+$	b_9Y_1			$+$	ε_2
Y_3	$=$						$b_{10}Y_1 + b_{11}Y_2$			$+$	ε_3

Specifying the LISREL Structural and Measurement Model Equations

We now can specify the set of equations for both the structural and measurement models. Only those coefficients to be estimated are included. First, we define the structural and measurement equations (see Tables 11A.5 and 11A.6, p. 650).

The reader should note that the original variables (V_1 to V_{14}) must be identified as either exogenous or endogenous indicators. Variables V_1 to V_8 correspond to X_1 to X_8, respectively (the exogenous construct indicators), and V_9 to V_{14} correspond to Y_1 to Y_6 (the endogenous construct indicators). Using this notation, we present the measurement model equations in Table 11A.6.

Specifying the Structural Equation Correlations

Next, there are two correlation matrices pertaining to the structural equations. The first is the phi matrix, which denotes the correlations among the exogenous constructs. From the path diagram, we note that there are two correlations of this type between constructs 1-2 (ϕ_{21}) and 3-4 (ϕ_{43}). These are noted in the matrix in Table 11A.7 (p. 650). There is also a relationship among the structural equations for two endogenous constructs (η_1 and η_3), represented in the psi matrix (ψ_{31}) in Table 11A.7.

Measurement Model (Indicator) Correlations

The final two matrices portray any within-construct measurement errors among the indicators of the exogenous and endogenous constructs. Note that although the correlations can theoretically be between any exogenous or endogenous indicators, they cannot occur between indicators of different types.

For purposes of illustration, two measurement error correlations are hypothesized. The first is between two exogenous indicators (X_1 and X_2), as shown in the theta-delta matrix (Table 11A.8, p. 651). The second correlation is between two

TABLE 11A.4 Constructs and Indicators for the Path Diagram in LISREL Notation

Exogenous		Endogenous	
Construct	Indicators	Construct	Indicators
X_1	V_1, V_2	Y_1	V_9, V_{10}
X_2	V_3, V_4	Y_2	V_{11}, V_{12}
X_3	V_5, V_6	Y_3	V_{13}, V_{14}
X_4	V_7, V_8		

TABLE 11A.5 Measurement Model Equations for the Path Diagram

Exogenous Indicator		Exogenous Constructs					Error
		ξ_1	ξ_2	ξ_3	ξ_4		
X_1	=	$\lambda_{11}^x \xi_1$				+	δ_1
X_2	=	$\lambda_{21}^x \xi_1$				+	δ_2
X_3	=		$\lambda_{32}^x \xi_2$			+	δ_3
X_4	=		$\lambda_{42}^x \xi_2$			+	δ_4
X_5	=			$\lambda_{53}^x \xi_3$		+	δ_5
X_6	=			$\lambda_{63}^x \xi_3$		+	δ_6
X_7	=				$\lambda_{74}^x \xi_4$	+	δ_7
X_8	=				$\lambda_{84}^x \xi_4$	+	δ_8

Endogenous Indicator		Endogenous Constructs				Error
		η_1	η_2	η_3		
Y_1	=	$\lambda_{11}^y \eta_1$			+	ε_1
Y_2	=	$\lambda_{21}^y \eta_1$			+	ε_2
Y_3	=		$\lambda_{32}^y \eta_2$		+	ε_3
Y_4	=		$\lambda_{42}^y \eta_2$		+	ε_4
Y_5	=			$\lambda_{53}^y \eta_3$	+	ε_5
Y_6	=			$\lambda_{63}^y \eta_3$	+	ε_6

TABLE 11A.6 Structural Model Equations for the Path Diagram

Endogenous Construct		Exogenous Constructs					Endogenous Constructs				Error
		ξ_1	ξ_2	ξ_3	ξ_4		η_1	η_2	η_3		
η_1	=	$\gamma_{11}\xi_1 + \gamma_{12}\xi_2 + \gamma_{13}\xi_3 + \gamma_{14}\xi_4$				+				+	ζ_1
η_2	=	$\gamma_{21}\xi_1 + \gamma_{22}\xi_2 + \gamma_{23}\xi_3 + \gamma_{24}\xi_4$				+	$\beta_{21}\eta_1$			+	ζ_2
η_3	=						$\beta_{31}\eta_1 + \beta_{32}\eta_2$			+	ζ_3

TABLE 11A.7 Correlations among Constructs for the Path Diagram

	Exogenous					Endogenous			
	ξ_1	ξ_2	ξ_3	ξ_4		η_1	η_1	η_2	η_3
ξ_1	—								
ξ_2	ϕ_{21}	—							
ξ_3			—			η_2		—		
ξ_4			ϕ_{43}	—		η_3	ψ_{31}		—	

endogenous indicators (Y_3 and Y_4), which is represented in the theta-epsilon matrix (Table 11A.8).

We have now completed the specification of the LISREL matrices needed for estimation of the proposed model. Figure 11A.2 (p. 651) illustrates the complete model, with all relationships portrayed in a causal model format. In the graphical

TABLE 11A.8 Measurement Error Correlation for Exogenous and Endogenous Indicators for the Path Diagram

Exogenous Indicators

	X_1	X_2	X_3	X_4	X_5	X_6	X_7	X_8
X_1	—							
X_2	$\phi_{\delta_{21}}$	—						
X_3			—					
X_4				—				
X_5					—			
X_6						—		
X_7							—	
X_8								—

Endogenous Indicators

	Y_1	Y_2	Y_3	Y_4	Y_5	Y_6
Y_1	—					
Y_2		—				
Y_3			—			
Y_4			$\phi_{\varepsilon_{43}}$	—		
Y_5					—	
Y_6						—

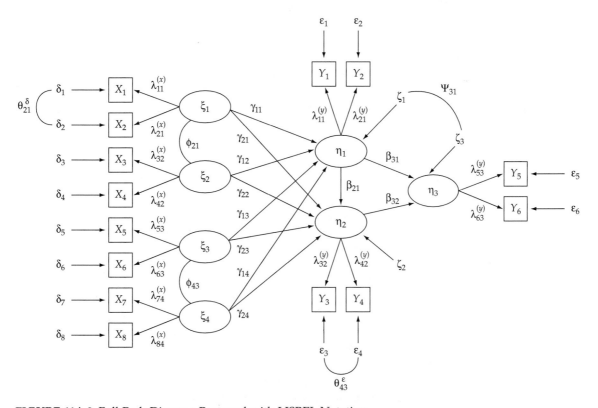

FIGURE 11A.2 Full Path Diagram Portrayal with LISREL Notation

portrayal of the structural and measurement models, some specific conventions are followed. First, as with path diagrams, straight arrows indicate causal relationships, and curved arrows denote correlations. Second, indicators are represented by rectangles; constructs are represented by ellipses.

Summary

Specification of any causal relationship can be incorporated directly into one of the eight matrices in the LISREL notation. Although it is not necessary to specify the models in this notation initially, after this review of the notational form used in SEM, researchers should feel equipped to portray their relationships in either format. Moreover, as the researcher becomes more acquainted with the notation, more advanced relationships can easily be incorporated as well.

Overall Goodness-of-Fit Measures for Structural Equation Modeling

Assessing the overall goodness-of-fit for structural equation models is not as straightforward as with other multivariate dependence techniques such as multiple regression, discriminant analysis, multivariate analysis of variance, or even conjoint analysis. SEM has no single statistical test that best describes the "strength" of the model's predictions. Instead, researchers have developed a number of goodness-of-fit measures that, when used in combination, assess the results from three perspectives: overall fit, comparative fit to a base model, and model parsimony. The discussions that follow present alternative measures for each of these perspectives, along with the methods of calculation for those measures that are not contained in the results and that must be computed separately.

One common question arises in the discussion of each measure: What is an acceptable level of fit? None of the measures (except the chi-square statistic) has an associated statistical test. Although in many instances guidelines have been suggested, no absolute test is available, and the researcher must ultimately decide whether the fit is acceptable. Bollen [22, p. 275] addresses this issue directly: "Overall, selecting a rigid cutoff for the incremental fit indices is like selecting a minimum R^2 for a regression equation. Any value will be controversial. Awareness of the factors affecting the values and good judgment are the best guides to evaluating their size." This advice applies equally well to the other goodness-of-fit measures.

Before examining the various goodness-of-fit measures, it may be useful to review the derivation of degrees of freedom in structural models. The number of unique data values in the input matrix is s (where $s = \frac{1}{2}(k)(k - 1)$ and k is the total number of indicators for both endogenous and exogenous constructs). The degrees of freedom (df) for any estimated model are then calculated as $df = s - t$, where t is the number of estimated coefficients. If the researcher knows the df for an estimated model and the total number of indicators, then t can be calculated directly as $t = s - df$.

The examination and derivation of goodness-of-fit measures for SEM has gained widespread interest among academic researchers in recent years, resulting in the continual development of new goodness-of-fit measures [36, 86, 90, 95, 100]. This is reflected in the statistical programs as they are continually modified to provide the most relevant information regarding the estimated model. In our discussions we have focused our attention on the LISREL program because of its widespread application in the social sciences. It has undergone these changes as well. The newest version of LISREL substantially expands the number and type of fit indices available directly in the output. For this reason, the following discussion and example data detail the calculations of those measures not provided in earlier versions of the program.

Measures of Absolute Fit

Absolute fit measures determine the degree to which the overall model (structural and measurement models) predicts the observed covariance or correlation matrix. No distinction is made as to whether the model fit is better or worse in the structural or measurement models. Among the absolute fit measures commonly used to evaluate SEM are the chi-square statistic, the noncentrality parameter, the goodness-of-fit statistic, the root mean square error, the root mean square error of approximation, and the expected cross-validation index.

Likelihood-Ratio Chi-Square Statistic

The most fundamental measure of overall fit is the likelihood-ratio chi-square (χ^2) statistic, the only statistically based measure of goodness-of-fit available in SEM [67]. A large value of chi-square relative to the degrees of freedom signifies that the observed and estimated matrices differ considerably. Statistical significance levels indicate the probability that these differences are due solely to sampling variations. Thus, low chi-square values, which result in significance levels greater than .05 or .01, indicate that the actual and predicted input matrices are not statistically different. In this instance, the researcher is looking for *nonsignificant differences* because the test is between actual and predicted matrices. The researcher must remember that this method differs from the customary desire to find statistical significance. However, even statistical nonsignificance does not guarantee that the "correct" model has been identified, but only that this proposed model fits the observed covariances and correlations well. It does not assure the researcher that another model would not fit as well or better. The .05 significance level is recommended as the minimum accepted, and levels of .1 or .2 should be exceeded before nonsignificance is confirmed [42].

An important criticism of the chi-square measure is that it is too sensitive to sample size differences, especially for cases in which the sample size exceeds 200 respondents. As sample size increases, this measure has a greater tendency to indicate significant differences for equivalent models. If the sample size becomes large enough, significant differences will be found for any specified model. Moreover, as the sample size nears 100 or goes even lower, the chi-square test will show acceptable fit (nonsignificant differences in the predicted and observed input matrices), even when none of the model relationships are shown to be statistically significant. Thus, the chi-square statistic is quite sensitive in different ways to both small and large sample sizes, and the researcher is encouraged to complement this measure with other measures of fit in all instances. The use of chi-square is appropriate for sample sizes between 100 and 200, with the significance test becoming less reliable with sample sizes outside this range.

The sensitivity of the chi-square measure extends past sample size considerations. For example, it has been shown that this measure varies based on the number of categories in the response variable [52]. Given its sensitivity to many factors, the researcher is encouraged to complement the chi-square measure with other goodness-of-fit measures.

Noncentrality and Scaled Noncentrality Parameters

The noncentrality parameter (NCP) is the result of the statisticians' search for an alternative measure to the likelihood-ratio chi-square statistic that is less affected by or independent of the sample size. Statistical theory suggests that a noncentrality chi-square measure will be less affected by sample size in its representation of the differences between the actual and estimated data matrices [76]. In a LISREL problem, the noncentrality parameter can be calculated as:

$$NCP = \chi^2 - \text{Degrees of freedom}$$

Although this measure adjusts the χ^2 by the degrees of freedom of the estimated model, it is still in terms of the original sample size. To "standardize" the NCP, divide it by the sample size to obtain the scaled noncentrality parameter (SNCP) [76]. This can be calculated as

$$SNCP = \frac{\chi^2 - \text{Degrees of freedom}}{\text{Sample size}}$$

This scaled measure is analogous to the average squared Euclidean distance measure between the estimated model and the unrestricted model [76]. For both the unscaled and the scaled parameters, the objective is to minimize the parameter value. Because there is no statistical test for this measure, it is best used in making comparisons between alternative models.

Goodness-of-Fit Index

The goodness-of-fit index [64, 67] is another measure provided by LISREL. It is a nonstatistical measure ranging in value from 0 (poor fit) to 1.0 (perfect fit). It represents the overall degree of fit (the squared residuals from prediction compared with the actual data), but is not adjusted for the degrees of freedom. Higher values indicate better fit, but no absolute threshold levels for acceptability have been established.

Root Mean Square Residual

The root mean square residual is the square root of the mean of the squared residuals—an average of the residuals between observed and estimated input matrices. If covariances are used, the RMSR is the average residual covariance. If a correlation matrix is used, then the RMSR is in terms of an average residual correlation. The RMSR is more useful for correlations, which are all on the same scale, than for covariances, which may differ from variable to variable, depending on unit of measure. Again, no threshold level can be established, but the researcher can assess the practical significance of the magnitude of the RMSR in light of the research objectives and the observed or actual covariances or correlations [9].

Root Mean Square Error of Approximation

Another measure that attempts to correct for the tendency of the chi-square statistic to reject any specified model with a sufficiently large sample is the root mean square error of approximation (RMSEA). Similar to the RMSR, the RMSEA is the discrepancy per degree of freedom. It differs from the RMSR, however, in that the discrepancy is measured in terms of the population, not just the sample used for estimation [93]. The value is representative of the goodness-of-fit that could be expected if the model were estimated in the population, not just the sample drawn for the estimation. Values ranging from .05 to .08 are deemed acceptable. An empirical examination of several measures found that the RMSEA was best suited to use in a confirmatory or competing models strategy with larger samples [86].

Expected Cross-Validation Index

The expected cross-validation index (ECVI) is an approximation of the goodness-of-fit the estimated model would achieve in another sample of the same size. Based on the sample covariance matrix, it takes into account the actual sample size and the difference that could be expected in another sample. The ECVI also takes into account the number of estimated parameters for both the structural and measurement models. The ECVI is calculated as

$$\text{ECVI} = \frac{\chi^2}{\text{Sample size} - 1} + \frac{2 \times \text{number of estimated parameters}}{\text{Sample size} - 1}$$

The EVCI has no specified range of acceptable values, but it is used in comparing between alternative models.

Cross-Validation Index

The cross-validation index (CVI) assesses goodness-of-fit when an actual cross-validation has been performed. Cross-validation is performed in two steps. First, the overall sample is split into two samples—an estimation sample and a validation sample. The estimation sample is used to estimate a model and create the estimated correlation of covariance matrix. This matrix is then compared to the sample from the validation sample. A double cross-validation process can be performed by comparing the estimated correlation or covariance matrix from each sample to a data matrix from the other sample.

Incremental Fit Measures

The second class of measures compares the proposed model to some baseline model, most often referred to as the null model. The null model should be some realistic model that all other models should be expected to exceed. In most cases, the null model is a single-construct model with all indicators perfectly measuring the construct (i.e., this represents the chi-square value associated with the total variance in the set of correlations or covariances). There is, however, some disagreement over exactly how to specify the null model in many situations [91].

Adjusted Goodness-of-Fit Index

The adjusted goodness-of-fit is an extension of the GFI, adjusted by the ratio of degrees of freedom for the proposed model to the degrees of freedom for the null model. It is quite similar to the parsimonious normed fit index (discussed later), and a recommended acceptance level is a value greater than or equal to .90.

Tucker–Lewis Index

The next incremental fit measure is the Tucker–Lewis index [97], also known as the nonnormed fit index (NNFI). First proposed as a means of evaluating factor analysis, the TLI has been extended to SEM. It combines a measure of parsimony into a comparative index between the proposed and null models, resulting in values ranging from 0 to 1.0. It is expressed as:

$$\text{TLI} = \frac{(\chi^2_{\text{null}}/df_{\text{null}}) - (\chi^2_{\text{proposed}}/df_{\text{proposed}})}{(\chi^2_{\text{null}}/df_{\text{null}}) - 1}$$

A recommended value of TLI is .90 or greater. This measure can also be used for comparing between alternative models by substituting the alternative model for the null model.

Normed Fit Index

One of the more popular measures is the normed fit index [16], which is a measure ranging from 0 (no fit at all) to 1.0 (perfect fit). Again, the NFI is a relative comparison of the proposed model to the null model. The NFI is calculated as:

$$\text{NFI} = \frac{(\chi^2_{\text{null}} - \chi^2_{\text{proposed}})}{\chi^2_{\text{null}}}$$

As with the Tucker–Lewis index, there is no absolute value indicating an acceptable level of fit, but a commonly recommended value is .90 or greater.

Other Incremental Fit Measures

A number of other incremental fit measures have been proposed, and the newer version of LISREL includes three in its output. The relative fit index (RFI), the incremental fit index (IFI), and the comparative fit index (CFI) all represent comparisons between the estimated model and a null or independence model. The values lie between 0 and 1.0, and larger values indicate higher levels of goodness-of-fit. The CFI has been found to be more appropriate in a model development strategy or when a smaller sample is available [86]. The interested reader can find the specific details of each measure in selected readings [15, 21, 22].

Parsimonious Fit Measures

Parsimonious fit measures relate the goodness-of-fit of the model to the number of estimated coefficients required to achieve this level of fit. Their basic objective is to diagnose whether model fit has been achieved by "overfitting" the data with too many coefficients. This procedure is similar to the "adjustment" of the R^2 in multiple regression. However, because no statistical test is available for these measures, their use in an absolute sense is limited in most instances to comparisons between models.

Parsimonious Normed Fit Index

The first measure in this case is the parsimonious normed fit index (PNFI) [61], a modification of the NFI. The PNFI takes into account the number of degrees of freedom used to achieve a level of fit. Parsimony is defined as achieving higher degrees of fit per degree of freedom used (one degree of freedom per estimated coefficient). Thus more parsimony is desirable. The PNFI is defined as:

$$\text{PNFI} = \frac{df_{\text{proposed}}}{df_{\text{null}}} \times \text{NFI}$$

Higher values of PNFI are better, and its principal use is for the comparison of models with differing degrees of freedom. It is used to compare alternative models, and there are no recommended levels of acceptable fit. However, when comparing between models, differences of .06 to .09 are proposed to be indicative of substantial model differences [103].

Parsimonious Goodness-of-Fit Index

The parsimonious goodness-of-fit index (PGFI) modifies the GFI differently from the AGFI. Where the AGFI's adjustment of the GFI was based on the degrees of freedom in the estimated and null models, the PGFI is based on the parsimony of the estimated model. It adjusts the GFI in the following manner:

$$\text{PGFI} = \frac{df_{\text{proposed}}}{\frac{1}{2}(\text{No. of manifest variables})(\text{No. of manifest variables} + 1)} \times \text{GFI}$$

The value varies between zero and 1.0, with higher values indicating greater model parsimony.

Normed Chi-Square

Jöreskog [63] proposed that the chi-square be "adjusted" by the degrees of freedom to assess model fit for various models. This measure can be termed the normed chi-square, and is the ratio of the chi-square divided by the degrees of freedom. This measure provides two ways to assess inappropriate models: (1) a model that may be "overfitted," thereby capitalizing on chance, typified by values less than 1.0; and (2) models that are not yet truly representative of the observed data and thus need improvement, having values greater than an upper threshold, either 2.0 or 3.0 [30] or the more liberal limit of 5.0 [63]. However, because the chi-square value is the major component of this measure, it is subject to the sample size effects discussed earlier with regard to the chi-square statistic.

The normed chi-square, however, has been shown to be somewhat unreliable [56, 100], so researchers should always combine it with other goodness-of-fit measures.

Akaike Information Criterion

Another measure based on statistical information theory is the Akaike information criterion [1]. Similar to the PNFI, the AIC is a comparative measure between models with differing numbers of constructs. The AIC is calculated as:

$$\text{AIC} = \chi^2 + 2 \times \text{Number of estimated parameters}$$

AIC values closer to zero indicate better fit and greater parsimony. A small AIC generally occurs when small chi-square values are achieved with fewer estimated coefficients. This shows not only a good fit of observed versus predicted covariances or correlations, but also a model not prone to "overfitting."

A Review of the Structural Model Goodness-of-Fit Measures

Table 11B.1 (pp. 660–661) presents a summary of the goodness-of-fit measures for the structural model estimated in chapter 11. The table demonstrates the derivation and interpretation of each measure as well. As this example illustrates, the researcher looks to the various measures to evaluate differing aspects of the model, and one hopes that all measures would indicate agreement on the level of model acceptability.

In evaluating the set of measures, some general criteria are applicable and indicate models with acceptable fit:

- Nonsignificant χ^2 (at least $p > .05$, perhaps .10 or .20)
- Incremental fit indices (NFI, TLI) greater than .90
- Low RMSR and RMSEA values based on the use of correlations or covariances
- Parsimony indices that portray the proposed model as more parsimonious than alternative models

As indicated earlier, the researcher should evaluate the proposed models on a series of measures from each type. A consensus should be reached on the acceptability of the model only after examination of the results from the entire set of goodness-of-fit measures.

Summary

The types and number of goodness-of-fit measures are increasing as researchers continually explore the possibilities of structural equation modeling. The user, however, is faced with the task of not only selecting the appropriate measures but also assessing by admittedly subjective standards whether the model is acceptable. The final effect is an uncertainty as to what is acceptable versus unacceptable, leaving the burden of proof on the researcher rather than a statistically based test.

Prevailing thought holds that the strongest test of any proposed model is through the comparison of the model to any number of proposed models. The

TABLE 11B.1 Comparison of Goodness-of-Fit Measures for the Structural Model

Structural Model Data:
   Fifteen indicators for five constructs (three exogenous, two endogenous)
   Total degrees of freedom: $\frac{1}{2}(15 \times 16) = 120$
   Number of estimated parameters (structural and measurement models): 35
   Sample Size: 136
   Proposed Model: $\chi^2 = 178.714$ $df = 85$ $p = .000$
   Null or Independence Model: $\chi^2 = 1040.194$ $df = 105$ $p = .000$

Evaluation of Structural Model with Goodness-of-Fit Measures

Goodness-of-Fit Measure	*Levels of Acceptable Fit*	*Calculation of Measure*	*Acceptability[a]*
Absolute Fit Measures			
Likelihood ratio chi-square statistic (χ^2)	Statistical test of significance provided	$\chi^2 = 178.714$[b] significance level: 000	Marginal
Noncentrality parameter (NCP)	Stated in terms of respecified χ^2, judged in comparison to alternative models	NCP = 93.714	Not applicable
Scaled noncentrality parameter (SNCP)	NCP stated in terms of average difference per observation for comparison between models	$SNCP = \dfrac{93.714}{136} = .689$	Not applicable
Goodness-of-fit index (GFI)	Higher values indicate better fit, no established thresholds	GFI = .865[b]	Marginal
Root mean square residual (RMSR)	Stated in terms of input matrix (covariance or correlation), with acceptable levels set by analyst	RMSR = .076[b]	Marginal
Root mean square error of approximation (RMSEA)	Average difference per degree of freedom expected to occur in the population, not the sample. Acceptable values under .08	RMSEA = .090	Marginal
Expected cross-validation index (ECVI)	The goodness-of-fit expected in another sample of the same size. No established range of acceptable values. Used in comparing between models	$ECVI = \dfrac{178.714}{136 - 1} + \dfrac{2 \times 35}{136 - 1}$ $= 1.842$	Not applicable
Incremental Fit Measures			
Tucker–Lewis index (TLI) or NNFI	Recommended level: 90	$TLI = \dfrac{(1040.194/105) - (178.714/85)}{(1040.194/105) - 1}$ $= .876$	Marginal
Normed fit index (NFI)	Recommended level: 90	$NFI = \dfrac{1040.194 - 178.714}{1040.194}$ $= .828$	Marginal
Adjusted goodness-of-fit index (AGFI)	Recommended level: 90	AGFI = .810[b]	Marginal

TABLE 11B.1 (*Continued*)

Parsimonious Fit Measures

Parsimonious goodness-of-fit index (PGFI)	A respecification of the GFI with higher values reflecting greater model parsimony. Used in comparing between models.	$PGFI = \dfrac{85}{\frac{1}{2}(15 \times 16)} \times .865$ $= .613$	Not applicable
Normed chi-square	Recommended level: Lower limit: 1.0 Upper limit: 2.0/3.0 or 5.0	$\text{Normed } \chi^2 = \dfrac{178.714}{85} = 2.103$	Marginal
Parsimonious normed fit index (PNFI)	Higher values indicate better fit, used only in comparing between alternative models	$PNFI = \dfrac{85}{105} \times .828 = .670$	Not applicable
Akaike information criterion (AIC)	Smaller positive values indicate parsimony, used in comparing alternative models	$AIC = 178.714 + 2(35)$ $= 248.714$	Not applicable

[a]An acceptability level of "not applicable" is applied to measures used only in comparison with alternative models.
[b]Measures provided in all versions of LISREL.

incremental fit measures were an attempt to provide a "standard" alternative model, termed the null or independence model. But the inability of LISREL or any other SEM program to ensure that no other model will have a better fit to the data than the proposed model makes a formalized process of comparison between alternative or competing models the strictest test of theory. Researchers are strongly encouraged to examine alternative models not in a model development approach but in a test of competing models to find the "best" representation of the proposed theoretical model.

References

1. Akaike, H. (1987), "Factor Analysis and AIC." *Psychometrika* 52: 317–32.
2. Allison, Paul D. (1987), "Estimation of Linear Models with Incomplete Data." In C. Clogg (ed.) *Sociological Methodology*, San Francisco: Jossey-Bass, pp. 71–103.
3. Anderson J. C., and D. W. Gerbing (1982), "Some Methods for Respecifying Measurement Models to Obtain Unidimensional Construct Measures." *Journal of Marketing Research* 19 (November): 453–60.
4. Anderson, J. C., and D. W. Gerbing (1988), "Structural Equation Modeling in Practice: A Review and Recommended Two-Step Approach." *Psychological Bulletin* 103: 411–23.
5. Anderson, J. C., and D. W. Gerbing (1990), "Assumptions and Comparative Strengths of the Two-Step Approach: Comment on Fornell and Yi." *Sociological Methods and Research* 20: 321–33.
6. Arbuckle, J. L. (1994), "AMOS—Analysis of Moment Structures," *Psychometrika* 59(1): 135–37.
7. Austin, J. T., and R. F. Calderon (1996), "Theoretical and Technical Contributions to Structural Equation Modeling: An Updated Annotated Bibliography," *Structural Equation Modeling* 3(2): 105–25.
8. Bagozzi, R. P. (1980), *Causal Models in Marketing*. New York: Wiley.
9. Bagozzi, R. P., and Y. Yi (1988), "On the Use of Structural Equation Models in Experimental Designs." *Journal of Marketing Research* 26 (August): 271–84.
10. Becker, Paul A., Arjen Merckens, and Toms J. Wansbrek (1994), *Identification, Equivalent*

Models and Computer Algebra. New York: Academic Press.

11. Bentler, P. M. (1980), "Multivariate Analysis with Latent Variables: Causal Modeling." *Annual Review of Psychology* 31: 419–56.

12. Bentler, P. M. (1986), "Structural Modeling and Psychometrika: A Historical Perspective on Growth and Achievements." *Psychometrika* 51: 35–51.

13. Bentler, P. M. (1988), *Theory and Implementation of EQS, a Structural Equations Program*, Los Angeles: BMDP Statistical Software.

14. Bentler, P. M. (1992), *EQS: Structural Equations Program Manual.* Los Angeles: BMDP Statistical Software.

15. Bentler, P. M. (1990), "Comparative Fit Indexes in Structural Models." *Psychological Bulletin* 107: 238–46.

16. Bentler, P. M., and D. G. Bonnett. (1980), "Significance Tests and Goodness of Fit in the Analysis of Covariance Structures." *Psychological Bulletin* 88: 588–606.

17. Bentler, P. M., and C. Chou (1987), "Practical Issues in Structural Modeling." *Sociological Methods and Research* 16 (August): 78–117.

18. Bentler, P. M., and E. J. C. Wu (1993), *EQS/Windows User's Guide.* Los Angeles: BMDP Statistical Software.

19. Blalock, H. M. (1982), *Conceptualization and Measurement in the Social Sciences.* Beverly Hills, Calif.: Sage.

20. Blalock, H. M. (1985), *Causal Modeling in the Social Sciences.* New York: Academic Press.

21. Bollen, K. A. (1986), "Sample Size and Bentler and Bonnett's Nonnormed Fit Index." *Psychometrika* 51: 375–77.

22. Bollen, K. A. (1989), *Structural Equations with Latent Variables.* New York: Wiley.

23. Bollen, K. A., and K. G. Joreskog (1985), "Uniqueness Does Not Imply Identification." *Sociological Methods and Research* 14: 155–63.

24. Bollen, K. A., and R. A. Stine (1993), "Bootstrapping Goodness-of-Fit Measures in Structural Equation Models." In K. A. Bollen and J. S. Long (eds.), *Testing Structural Equation Models*, Newbury Park, Calif.: Sage.

25. Breckler, S. J. (1990), "Applications of Covariance Structure Modeling in Psychology: Cause for Concern?" *Psychological Bulletin* 107(2): 260–73.

26. Brown, R. L. (1994), "Efficacy of the Indirect Approach for Estimating Structural Equation Models with Missing Data: A Comparison of Five Methods," *Structural Equation Modeling* 1(4): 375–80.

27. Bullock, H. E., L. L. Harlow, and S. Mulaik (1994), "Causation Issues in Structural Equation Modeling." *Structural Equation Modeling* 1(3): 253–67.

28. Burt, R. S. (1976), "Interpretational Confounding of Unidimensional Variables in Structural Equation Modeling." *Sociological Methods and Research* 5: 3–51.

29. Byrne, B. M. (1995), "Strategies in Testing for an Invariant Second-Order Factor Structure: A Comparison of EQS and LISREL," *Structural Equation Modeling* 2(1): 53–72.

30. Carmines, E., and J. McIver (1981), "Analyzing Models with Unobserved Variables: Analysis of Covariance Structures." In G. Bohrnstedt and E. Borgatta (eds.), *Social Measurement: Current Issues*, Beverly Hills, Calif.: Sage.

31. Costner, H. L., and R. Schoenberg (1979), "Diagnosing Indicator Ills in Multiple Indicator Models." In A. Goldberger and O. Duncan (eds.), *Structural Equation Models in the Social Sciences*, New York: Seminar Press.

32. Cronbach, L. J. (1951), "Coefficient Alpha and the Internal Structure of Tests." *Psychometrica* 16: 297–334.

33. Darden, W. R. (1981), "Review of Behavioral Modeling in Marketing." In Ben M. Enis and Kenneth J. Roering (eds.), *Review of Marketing.* Chicago: American Marketing Association.

34. Davis, W. R. (1993), "The FCI Rule of Identification for Confirmatory Factor Analysis: A General Sufficient Condition." *Sociological Methods and Research* 21: 403–37.

35. Dillon, W., A. Kumar, and N. Mulani (1987), "Offending Estimates in Covariance Structure Analysis—Comments on the Causes and Solutions to Heywood Cases." *Psychological Bulletin* 101: 126–35.

36. Ding, L., W. F. Velicer, and L. L. Harlow (1995), "Effects of Estimation Methods, Number of Indicators per Factor and Improper Solutions on Structural Equation Modeling Fit Indices." *Structural Equation Modeling*, 2: 119–43.

37. Dolan, C. (1996), "Principal Component Analysis Using LISREL 8," *Structural Equation Modeling* 3(4): 307–22.

38. Duncan, O. D. (1966), "Path Analysis: Sociological Examples." *American Journal of Sociology* 72: 1–16.

39. Duncan, O. D. (1975), *Introduction to Structural Equation Models.* New York: Academic Press.

40. Fan, Xitao (1997), "Canonical Correlation Analysis and Structural Equation Modeling: What Do They Have in Common?" *Structural Equation Modeling* 4(1): 65–79.

41. Fassinger, R. E. (1987), "Use of Structural Equation Modeling in Counseling Psychology Research." *Journal of Counseling Psychology* 34: 425–36.

42. Fornell, C. (1983), "Issues in the Application of Covariance Structure Analysis: A Comment." *Journal of Consumer Research* 9: 443–48.

43. Fornell, C., and D. F. Larker (1981), "Evaluating Structural Equation Models with Unobservable Variables and Measurement Error." *Journal of Marketing Research* 18 (February): 39–50.

44. Fornell, C., and Y. Yi (1992), "Assumptions of the Two-Step Approach to Latent Variable Modeling." *Sociological Methods and Research* 20: 291–320.

45. Fornell, C., and Y. Yi (1992), "Assumptions of the Two-Step Approach: Reply to Anderson and Gerbing." *Sociological Methods and Research* 20: 334–39.

46. Fraser, C. (1980), *COSAN User's Guide.* Toronto: Ontario Institute for Studies in Education.

47. Gerbing, D. W., and J. C. Anderson (1984), "On the Meaning of Within-Factor Correlated Measurement Errors." *Journal of Consumer Research* 11: 572–80.

48. Gerbing, D. W., and J. G. Hamilton (1996), "Viability of Exploratory Factor Analysis as a Precursor to Confirmatory Factor Analysis," *Structural Equation Modeling* 3(1): 62–72.

49. Glymour, C. (1988), *Discovering Causal Structure.* Orlando, Fla.: Academic Press.

50. Glymour, C., R. Scheines, P. Spirtes, and K. Kelly (1987), "Discovering Casual Structure: Artificial Intelligence." *Philosophy of Science and Statistical Models,* New York: Academic Press.

51. Goldberger, A. S., and O. D. Duncan (1973), *Structural Equation Models in the Social Sciences.* New York: Seminar Press.

52. Green, S. B., T. M. Akey, K. K. Fleming, S. C. Hershberger, and J. G. Marquis (1997), "Effect of the Number of Scale Points on Chi-Square Fit Indices in Confirmatory Factor Analysis." *Structural Equation Modeling* 4(2): 108–20.

53. Hatcher, L. (1996), "Using SAS® PROC CALIS for Path Analysis: An Introduction." *Structural Equation Modeling* 3(2): 176–92.

54. Hayduk, L. A. (1987), *Structural Equation Modeling with LISREL: Essentials and Advances.* Baltimore: Johns Hopkins University Press.

55. Hayduk, L. A. (1996), *LISREL Issues, Debates and Strategies.* Baltimore: Johns Hopkins University Press.

56. Hayduk, L. A. (1987), *Structural Equation Modeling with LISREL.* Baltimore: Johns Hopkins University Press.

57. Heise, D. R. (1975), *Causal Analysis.* New York: Wiley.

58. Hoelter, J. W. (1983), "The Analysis of Covariance Structures: Goodness-of-Fit Indices." *Sociological Methods and Research* 11: 325–44.

59. Hox, J. J. (1995), "AMOS, EQS and LISREL for Windows: A Comparative Review." *Structural Equation Modeling* 2(1): 79–91.

60. Hunt, S. D. (1990), *Marketing Theory: The Philosophy of Marketing Science.* Homewood, Ill.: Irwin.

61. James, L. R., S. A. Muliak, and J. M. Brett (1982), *Causal Analysis: Assumptions, Models and Data.* Beverly Hills, Calif.: Sage.

62. Jöreskog, K. G. (1969), "A General Approach to Confirmatory Maximum Likelihood Factor Analysis." *Psychometrika* 34: 183–202.

63. Jöreskog, K. G. (1970), "A General Method for Analysis of Covariance Structures." *Biometrika* 57: 239–51.

64. Jöreskog, K. G., and D. Sorbom (1988), *LISREL VII: Analysis of Linear Structure Relationships by the Method of Maximum Likelihood.* Mooresville, Ill.: Scientific Software.

65. Jöreskog, K. G., and D. Sorbom (1988), *LISREL 7.* Chicago: SPSS, Inc.

66. Jöreskog, K. G., and D. Sorbom (1993), *PRELIS2: A Program for Multivariate Data Screening and Data Summarization.* Mooresville, Ill.: Scientific Software.

67. Jöreskog, K. G., and D. Sorbom (1993), *LISREL 8: Structural Equation Modeling with the SIMPLIS Command Language.* Mooresville, Ill.: Scientific Software.

68. Kenny, D. A. (1979), *Correlation and Causation.* New York: Wiley.

69. Lee, S., and S. Hershberger (1990), "A Simple Rule for Generating Equivalent Models in Structural Equation Modeling." *Multivariate Behavioral Research* 25: 313–34.

70. Loehlin, J. C. (1987), *Latent Variable Models: An Introduction to Factor, Path and Structural Analysis.* Hillsdale, N.J.: Lawrence Erlbaum.

71. Long, J. S. (1983), *Covariance Structure Models: An Introduction to LISREL.* Beverly Hills, Calif.: Sage.

72. MacCullum, R. (1986), "Specification Searches in Covariance Structure Modeling." *Psychological Bulletin* 100: 107–20.

73. Marsh, H. W., J. R. Balla, and R. P. McDonald (1988), "Goodness-of-Fit Indices in Confirmatory Factor Analysis: The Effect of Sample Size." *Psychological Bulletin* 103: 391–410.

74. Marsh, H. W., and D. Hoceuar (1985), "Application of Confirmatory Factor Analysis to the Study of Self-Concept: First- and Higher-Order Factor Models and Their Invariance Across Groups." *Psychological Bulletin* 97(1): 562–82.

75. Marsh, H. W., and D. Hoceuar (1994), "Confirmatory Factor Analysis Models of Factorial Invariance: A Multifaceted Approach." *Structural Equation Modeling* 1(10): 5–34.

76. McDonald, R. P., and H. W. Marsh (1990), "Choosing a Multivariate Model: Noncentrality and Goodness of Fit." *Psychological Bulletin* 107: 247–55.

77. Mulaik, S. A., L. R. James, J. Van Alstine, N. Bennett, S. Lind, and D. C. Stillwell (1989), "An Evaluation of Goodness of Fit Indices for Structural Equation Models." *Psychological Bulletin* 103: 430–55.

78. Mulaik, S. A., Larry R. James, Judith Van Alstine, Nathan Bennett, Sherri Lind, and C. Dean Stilwell (1989), "Evaluation of Goodness-of-Fit Indices for Structural Equation Models." *Psychological Bulletin* 105: 430–45.

79. Muthen, B., and D. Kaplan (1985), "A Comparison

of Methodologies for the Factor Analysis of Nonnormal Likert Variables." *British Journal of Mathematical and Statistical Psychology* 38: 171–89.

80. Neale, M. C., A. C. Heath, J. K. Kewitt, L. J. Eaves, and D. W. Walker (1989), "Fitting Genetic Models with LISREL: Hypothesis Testing." *Behavior Genetics* 19: 37–49.

81. O'Brien, R. M., and T. Reilly (1995), "Equality in Constraints and Metric-Setting Measurement Models." *Structural Equation Modeling* 2(1): 53–72.

82. Predhazur, E. J., and L. P. Schmelkin (1992), *Measurement Design and Analysis: An Integrated Approach.* Hillsdale, N.J.: Lawrence Erlbaum and Associates.

83. Raykou, T., and K. F. Widaman (1995), "Issues in Applied Structural Equation Modeling Research." *Structural Equation Modeling* 2(4): 289–318.

84. Reilly, T. (1995), "A Necessary and Sufficient Condition for Identification of Confirmatory Factor Analysis Models of Complexity One." *Sociological Methods and Research* 23(4): 421–41.

85. Rigdon, E. E. (1995), "A Necessary and Sufficient Identification Rule for Structural Models Estimated in Practice." *Multivariate Behavior Research* 30(3): 359–83.

86. Rigdon, E. E. (1996), "CFI versus RMSEA: A Comparison of Two Fit Indices for Structural Equation Modeling." *Structural Equation Modeling* 3(4): 369–79.

87. Rigdon, E. E. (1994), "Demonstrating the Effects of Unmodeled Random Measurement Error." *Structural Equation Modeling* 1(4): 375–80.

88. Robles, J. (1996), "Confirmation Bias in Structural Equation Modeling." *Structural Equation Modeling* 3(1): 307–22.

89. Rubio, D. M., and D. E. Gillespie (1995), "Problems with Error in Structural Equation Models," *Structural Equation Modeling* 2(4): 367–78.

90. Satorra, A., and P. Bentler (1994), "Correction to Test Statistics and Standard Errors in Covariance Structure Analysis. In A. Von Eye and C. Clogg (eds.), Latent Variable Analysis: Applications for Developmental Research, Newbury Park, Calif.: Sage, pp. 399–419.

91. Sobel, M. E., and G. W. Bohrnstedt (1985), "The Use of Null Models in Evaluating the Fit of Covariance Structure Models." In Nancy Brandon Tuma (ed.), *Sociological Methodology*, San Francisco: Jossey-Bass.

92. Steenkamp, J. E. M., and H. C. M. van Trijp (1991), "The Use of LISREL in Validating Marketing Constructs." *International Journal of Research in Marketing* 8(4): 283–99.

93. Steiger, J. H. (1990), "Structural Model Evaluation and Modification: An Interval Estimation Approach." *Multivariate Behavioral Research* 25: 173–80.

94. Tanaka, J. (1987), "How Big Is Enough? Sample Size and Goodness-of-Fit in Structural Equation Models with Latent Variables." *Child Development* 58: 134–46.

95. Tanaka, J. (1993), "Multifaceted Conceptions of Fit in Structural Equation Models." In K. A. Bollen and J. S. Long (eds.), *Testing Structural Equation Models*, Newbury Park, Calif.: Sage.

96. Tremblay, P. F., and R. G. Gardner (1996), "On the Growth of Structural Equation Modeling in Psychological Journals." *Structural Equation Modeling* 3(2): 93–104.

97. Tucker, L. R., and C. Lewis (1973), "The Reliability Coefficient for Maximum Likelihood Factor Analysis." *Psychometrika* 38: 1–10.

98. Wang, L. L., X. Fan, and V. L. Wilson (1996), "Effects of Nonnormal Data on Parameter Estimates for a Model with Latent and Manifest Variables: An Empirical Study." *Structural Equation Modeling* 3(3): 228–47.

99. Werts, C. E., and R. L. Linn (1970), "Path Analysis: Psychological Examples." *Psychological Bulletin* 74: 193–212.

100. Wheaton, B. (1987), "Assessment of Fit in Overidentified Models with Latent Variables." *Sociological Methods and Research* 16: 118–54.

101. Wheaton, B., D. Muthen, D. Alwin, and G. Summers (1977), "Assessing Reliability and Stability in Panel Models." In D. Heise (ed.), *Sociological Methodology*, San Francisco: Jossey-Bass.

102. Williams, L. J., and J. T. Hazer (1986), "Antecedents and Consequences of Organizational Turnover: A Reanalysis Using a Structural Equations Model." *Journal of Applied Psychology* 71 (May): 219–31.

103. Williams, L. J., and Patricia J. Holahan (1994), "Parsimony-Based Fit Indices for Multiple-Indicator Models." *Structural Equation Modeling* 1(2): 161–89.

104. Wold, Herman, ed. (1981), *The Fixed Point Approach to Interdependent Systems.* Amsterdam: North Holland.

105. Wothke, W. (1993), "Nonpositive Definite Matrices in Structural Modeling." In K. A. Bolden and J. S. Long (eds.), *Testing Structural Equation Models*, Newbury Park, Calif.: Sage.

Annotated Articles

The following readings have been selected for their illustration of the application of structural equation models to specific research problems.

The annotations of each article are provided to give the reader a sense of the issues involved in each instance and the types of results achieved with this multivariate technique. Interested readers are encouraged to review the original articles in their complete form to gain a deeper appreciation for "real world" applications of structural equation modeling.

Netemeyer, Richard G., James S. Boles, Daryl O. McKee, and Robert McMurrian (1997), "An Investigation into the Antecedents of Organizational Citizenship Behaviors in a Personal Selling Context." *Journal of Marketing* 61(3): 85–98.

Based on prior research and two separate studies, the authors explore predictors of organizational citizenship behaviors (OCBs) of sales personnel. OCBs are discretionary behaviors that directly benefit the organization without necessarily influencing salesperson productivity. Given their ability to enhance organizational performance and increase effectiveness, OCBs are of strategic importance to organizations. In this study, OCB is treated as a higher-order factor containing four first-order factors: sportsmanship, civic virtue, conscientiousness, and altruism. The authors follow a two-step approach in which a measurement model is first estimated before assessing the structural model relationships. The proposed model specifies direct relationships between three exogenous variables (person–organization fit, leadership support, fairness in reward allocation) and job satisfaction. Only job satisfaction is hypothesized as influencing OCBs directly.

The measurement and structural model results were fairly consistent across both studies, with differences attributed to sample-specific characteristics (e.g., maturity level of the two samples). Results for the two studies are based on sample sizes of 91 and 182 respondents. From the goodness-of-fit measures (chi-square GFI, AGFI, TLI, and CFI), support is offered for the measurement model in that the indicators are found to adequately represent the hypothesized constructs. The model's ability to adequately represent the specified causal relationships (structural model) was also supported. General agreement was found for the hypothesized relationships. Additionally, the authors compare alternative models, noting little improvement in model fit. In all, the article's empirical findings have potential implications for both academicians and practitioners. The results facilitate theory development and are of strategic importance for organizations.

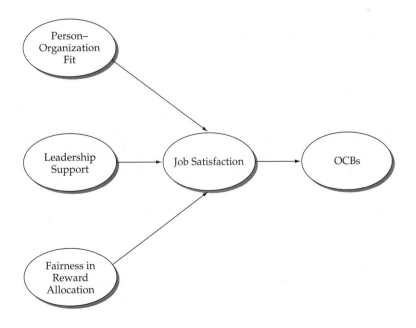

Hughes, Marie Adele, R. Leon Price, and Daniel W. Marrs (1986), "Linking Theory Construction and Theory Testing: Models with Multiple Indicators of Latent Variables." *Academy of Management Review* 11(1): 128–44.

The authors present a description and illustration of the use of structural equation modeling as a tool for integrating theory construction and testing. Data from a previous study is used to confirm a hypothesized measurement model comprised of relationships among latent construct variables and their indicators. The measurement model is tested using confirmatory factor analysis. The authors then specify and estimate a structural model consisting of the relationships among three latent constructs. From this study, the authors demonstrate a technique that allows for empirical tests of theoretical relationships.

To illustrate the applicability of structural equation modeling for the testing of theory, the authors use a sample of 114 respondents from a previous study. In the first phase of the analysis, the authors use confirmatory factor analysis to assess the internal consistency and discriminant validity of the hypothesized model structure containing 14 indicators. In this example, findings from the confirmatory factor analysis are also used to refine the measurement model. The structural model consists of two exogenous variables (knowledge of job enrichment and attitude toward management) predicting a single endogenous variable (attitude toward job enrichment).

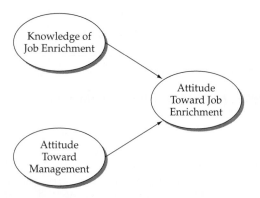

The final model provides an adequate fit for the model, as was indicated by the goodness-of-fit statistics. In all, the technique allows the authors

to confirm the hypothesized measurement model and theoretical relationships. Through the use of multiple indicators of multiple, operationally defined latent variables, the measurement and structural model provides a direct means for testing scientific theory.

Doll, William J., Weidong, Xia, and Gholamreza Torkzadeh (1994), "A Confirmatory Factor Analysis of the End-User Computing Satisfaction Instrument." *MIS Quarterly* 18(4): 453–61.

The authors employ a rigorous cross-validation examination of the multifaceted construct, end-user computing satisfaction (EUCS), using confirmatory factor analysis. Prior research has proposed that user satisfaction is a single construct. From a sample of 409 computer end users, the authors specify and test four hypothesized measurement models. Based on comparisons of model-data fit, the authors chose one model as best representing the dimensionality of the construct in a parsimonious manner. Additionally, the authors assess the reliability of the factors and indicators. From their results, the authors provide researchers and practitioners with a standardized instrument for measuring user satisfaction that is both empirically and conceptually reliable and valid.

Each model is derived from prior research and is comprised of 12 items. The models are assessed using several goodness-of-fit indexes (chi-square, NFI, GFI, AGFI, and RMSR). Model 1, a hypothesized one first-order factor model, consists of all 12 items summed into one construct. Models 2 and 3 each contain five first-order factors to represent end-user satisfaction—content, accuracy, format, ease of use, and timeliness. In Model 2 the factors are uncorrelated, whereas in Model 3 they are correlated. Model 4 consists of five first-order factors, which form a single second-order factor. Results, based on the fit indices indicate that Models 3 and 4 provide satisfactory representations of the underlying data; however, by reducing the number of indicators from 12 to 5, Model 4 is a more parsimonious representation. The authors recommend that future researchers improve upon the construct (i.e., additional tests of reliability and validity) and examine its antecedents and consequences.

CHAPTER 12

Emerging Techniques in Multivariate Analysis

LEARNING OBJECTIVES

Upon completing this chapter, you should be able to do the following:

- Understand the principles and reasons underlying data warehousing
- Identify and describe the techniques associated with data mining
- Describe the unique principles upon which neural networks are based
- Explain the basic operations of neural networks, especially the learning process
- Distinguish between resampling and parametric methods of parameter estimation
- Enumerate the advantages of the bootstrap and jackknife methods

CHAPTER PREVIEW

The transition of our economy to one based on information has had a profound effect not only on the operations of organizations, but also on the environment for business analysis. Researchers face new challenges that require using traditional multivariate techniques in different ways and combining these traditional techniques with new analytical techniques. The tremendous amount of information generated by today's organizations will only increase. As electronic commerce and electronic communication have become more common on the Internet, firms are developing a direct link with customers, another source of information.

Today's researcher must confront the tasks of both data management and data analysis. Data management is benefiting from the concept of data warehousing, which integrates data into information suitable for analytical applications. Data analysis has seen two new procedures complement the traditional multivariate techniques. The first is learning models, typified by neural networks, which are particularly suited for analyzing complex patterns in large databases without prior model specification. The second is resampling, replacing the statistical basis for inference with strict empirical estimation based on repeated sampling of the original sample. The computational power available today allows the researcher to empirically estimate the confidence interval of a parameter instead of relying on statistical assumptions. This chapter contains a brief introduction to each of these topics as well as empirical examples where appropriate.

KEY TERMS

Before starting the chapter, review the key terms to develop an understanding of the concepts and terminology used. Throughout the chapter the key terms appear in **boldface**. Other points of emphasis in the chapter are *italicized*. Also, cross-references within the Key Terms appear in *italics*.

Activation function Mathematical function within the *node* that translates the summated score of the weighted input values into a single output value. Although the function can be any type, the most common form is a *sigmoid function*.

Aggregated data Data created through a process of combination or summarization. It can then be stored in this summarized form in a *data warehouse*.

Analytical data Operational data that has been integrated and combined with existing analytical data through a *data warehousing* operation to form a historical record of a business event or object (e.g., customer, firm, or product).

Artificial intelligence (AI) Generalized area of computer science dealing with the creation of computer programs that attempt to emulate the learning properties of the human brain. AI has evolved into a number of more specialized areas, including *neural networks* and *genetic algorithms*.

Association rules Rules based on the correlation between attributes or *dimensions* of *data elements*.

Backpropagation Most common *learning* process in *neural networks*, in which errors in estimating the output *nodes* are "fed back" through the system and used as indicators as to how to recalibrate the weights for each node.

Bias node Additional *node* to a *hidden layer*, which has a constant value. This node functions in a manner similar to the constant term in multiple regression.

Confidence The proportion of times two or more events occur jointly. For example, if product A is purchased 60 percent of the time a customer also purchases product B, we have a confidence of 60 percent in saying products A and B are purchased jointly.

Data cleaning Removal of random and systematic errors from *data elements* through filtering, merging, and translation.

Data element Level of data stored in a database. Separate data elements may represent different levels of aggregation of summarization, ranging from *primitive data* to *aggregate data*.

Data mining   Extracting valid and previously unknown actionable information from large databases and applying it to business models.

Data scrubbing   See *data cleaning*.

Data visualization   Techniques for portraying data in graphical format, typically used in an exploratory manner to identify basic relationships.

Data warehousing   Assimilation and integration of data from internal and external sources in a format suitable for *on-line analytical processing* and *decision support system* applications.

Decision support systems (DSS)   Interactive systems developed to provide users access to data for (1) ad hoc or ill-defined queries, (2) prespecified reports (e.g., exception reporting), and (3) structured analyses (rules-based procedures, multivariate statistics, neural networks, or other models).

Decision tree   Rule-based model consisting of *nodes* (decision points) and branches (connections between nodes) that reaches multiple outcomes based on passing through two or more nodes.

Dimension   Attribute of a *data element*, such as age or gender of a customer or date of a historical record.

Drill down   Accessing more detailed data used to create *aggregate data*. For example, after reviewing weekly sales figures for a set of stores, the researcher would drill down to view daily sales reports for sales or reports for separate stores. One can drill down to the lowest level of detail (*primitive data*) for available data elements.

Encoding   Conversion of the categories of a nonmetric variable to a series of binary variables, one for each category. Similar to the dummy variable coding discussed in chapter 2 without the deletion of the reference category.

Genetic algorithm   Learning-based models developed on the principles of evolution. Partial solutions to a problem "compete" with one another, and then the best solutions are selected and combined to provide the basis for further problem-solving.

Hidden layer   Series (layer) of *nodes* in a *multilayer perceptron neural network* that are between the input and output nodes. The researcher may or may not control the number of nodes per hidden layer or the number of hidden layers. The hidden layers provide the capability to represent nonlinear functions in the neural network system.

Jitter   Small amounts of random noise added to the values used in *training* a neural network that smooth the error function and assist in obtaining a global optimum solution.

Knowledge discovery in databases (KDD)   Extraction of new information from databases through a variety of knowledge discovery processes.

Kohonen model   *Neural network* model that is designed for clustering problems and operates in an unsupervised *learning* mode.

Learning   Sequential processing of large samples of observations (known as the *training* sample) in which the prediction or classification errors are used to recalibrate the weights to improve estimation.

Metadata   Complete description of a data element, not only defining its storage location and data type, but also any attributes, dimensions, or associations with any other data elements. Simply put, "data about data," which are stored in a centralized *repository* for common access to all database users.

Multidimensional data   *Data element* with multiple *dimensions* (attributes).

Multilayer perceptron   Most popular form of *neural network*, contains at least one *hidden layer* of *nodes* between input and output nodes.

Neural network   Nonlinear predictive model that learns through *training*. It resembles the structure of biological neural systems.

Node   Basic "building block" of *neural networks* that can act as an input, output, or processing/analysis function. Analogous in operation to a neuron in the brain.

On-line analytical processing (OLAP)   Database application that allows users to view, manipulate, and analyze *multidimensional data.*

Operational data   Data used by operational systems (e.g., accounting, inventory management) that reflect the current status of the organization that is necessary for day-to-day business functions.

Operational systems   Information systems that service the basic functions of the organization (accounting, inventory management, order processing) for their day-to-day operations.

Primitive data   Data elements maintained at their lowest level of detail, such as individual transactions.

Query   Directed access to a database requesting specific information. Can be used to *drill down* from *aggregated data.*

Radial basis function   Alternative form of *neural network* that functions much like the *multilayer perceptron* in being appropriate for prediction and classification problems.

Repository   Centralized collection of *metadata* available for access by all functions of the *data warehouse.*

Resampling   Nonstatistical estimation of a parameter's confidence interval through the use of an empirically derived sampling distribution. The empirical distribution is calculated from multiple samples drawn from the original sample.

Search space   Range of potential solutions defined for various learning models, such as *genetic algorithms* or *neural networks.* The solutions can be specified in terms of *dimensions* considered, data elements included, or criteria for feasible or infeasible solutions.

Sigmoid function   Nonlinear function with a general S-shaped distribution. One common example is the logistic function.

Structured query language (SQL)   Method for extracting information from a database according to specified criteria. It differs from data mining in that the only results provided are those meeting the specific request conditions. SQL cannot perform any analytical processing, such as developing classification models or determining levels of association. For example, SQL can extract information for all customers between ages 35 and 55 who bought a product within the last month by credit card, but it cannot group customers into segments without specified criteria.

Supervised   *Learning* process that employs a *training sample* and provides feedback to the *neural network* concerning errors at the output *nodes.*

Support   Percentage of the total sample for which an *association rule* is valid. Support indicates the substantiality of the group.

Training   See *learning.*

Training sample   Observations used in the calibration of a *neural network.* It must contain actual values for the output *node* so that errors in output value prediction can be determined and used in the *learning* process.

Transaction   Action upon an *operational data element* (creation, modification, or deletion) representing a single business event. Examples include the sale of a product to a customer, receipt of a payment, transfer of funds between accounts, or transfer of completed or purchased products to inventory.

Introduction

Until recently, the fundamental character of data analysis had remained basically static since the last major shift in the mid-1970s, when the introduction of the main-frame computer and its evolution to the personal computer caused a revolutionary change in how data analysis was practiced. Researchers then had enough computational power to essentially eliminate all restrictions on the types of statistical techniques available, and the multivariate statistical methods we have discussed in earlier chapters all gained widespread acceptance and use. The educational focus for the researcher was statistical methods, with an emphasis on the theoretical background and applications of multivariate techniques. Little consideration was given to the nature or character of the data, other than assessing its statistical qualities in terms of meeting the assumptions of the statistical techniques.

But the "information age" has now brought a second revolution, particularly outside the academic community. Researchers are again being challenged, this time to reorient the application of their techniques to a new research environment. Large-scale databases with potentially hundreds of thousands or even millions of observations provide very detailed attribute profiles ranging over time spans of years. This data explosion is in stark contrast to the past, when the researcher often was analyzing data collected specifically for a research question. Now data abound, and organizations are looking for new ways in which to extract information. This second revolution in data analysis is primarily based on two trends: an information avalanche and a questioning of statistical inference.

The Information Avalanche

For both academic and commercial communities, the days of limited samples and/or small sample sizes are past. Academicians have access to published databases with topics ranging from scanner data of consumer purchases to historical files of complete financial information for all publicly traded firms. Coupled with the emerging types of information available from governmental sources, the academician in many cases has more than enough data. But in the commercial sector, the scenario is much more troublesome. Organizations of all types have become increasing computerized in all functional areas, facilitating not only their operations, but their ability to collect data. Technological advances provide the capacity to effectively store all information in a common format, and data access is available to all members of the organization, not just a few select decision makers or planners. These trends have combined to create something akin to an "information avalanche," wherein the researcher can easily feel overwhelmed by the continual flow of massive amounts of data. One hundred gigabyte databases are common, and the terabyte (1,000 gigabyte) database is now possible. How can this new information be processed and then analyzed on a timely basis? What techniques are best suited for this new challenge? These are some of the issues faced by all researchers in this new research environment.

Analysis without Statistical Inference

The information avalanche has also drawn attention to alternatives to statistical inference for assessing the "significance" of parameter estimates. But why would we want to abandon this principle—the basis for most of the multivariate techniques discussed in this text? First, there is a move to "return to the data," in which the researcher applies as few assumptions as possible to the analysis and "lets the data talk." The development of artificial intelligence programs and their offspring (e.g., neural networks, genetic algorithms) has led researchers to explore a vast range of analytical models not grounded in statistical inference. These techniques were based on the application of simple learning rules to a data set in as unconstrained a setting as possible. Many researchers initially saw these methods as too much like a "black box," where some mysterious process occurred and then produced results. This is far from the truth because these methods are very structured in their procedures—it is just that they react to the data rather than the data being shaped to fit the statistical assumptions.

Second, the academic community has made a commitment to teach students the underlying principles of statistics without the rote memorization of formulas and statistical theory. Students can readily grasp the concepts of statistical inference when they can actually visualize the process, such as proving the concept of a probability distribution by flipping coins or combinations of coins. So why discard this simple perspective when we move to more sophisticated techniques? Combined with the computational power available today, these methods allow for an empirical estimation of the distribution rather than a statistical assumption. This is akin to flipping the coin many times, but now with decidedly more complex questions and analytical techniques. This has led to the addition of these alternatives, such as neural networks and resampling techniques, to the researcher's toolbox.

Topics Covered in this Chapter

The changes and forces in our research environment have been evidenced in many ways in recent years, and three specific concepts emerge as representative of marked departures from our past ways of thinking about data analysis: data warehousing and data mining, neural networks, and resampling. The purpose of this chapter is not to provide an in-depth discussion of each topic, but instead to discuss the basic concepts, terminology, and issues, using as our reference point the reader's exposure in this text to multivariate statistical techniques. The following section contains brief descriptions of each topic, which will then be followed by more detailed coverage with empirical examples where applicable. The reader is encouraged to first review these descriptions to gain a basic understanding of the concepts and then to judge their benefit in specific situations.

Data Warehousing and Data Mining

No concept has garnered more attention in the information technology (IT) community than the emergence of data warehousing, and now data mining [12, 14]. **Data warehousing** is the attempt to combine all sources of data and information relevant to an organization into a single, unified database with a structure amenable to the support of analytical decision making by all levels of the organization. There are many issues of a quite technical nature that we will not discuss here, but it is essential that the researcher using a data warehouse understand its basic structure and operation as it impacts its usefulness for various types of analysis. **Data mining** is a somewhat new perspective of data analysis with more of an exploratory orientation rather than a confirmatory mode. Given the vast amounts of available data, the research perspective becomes more directed toward delving deeply into the characteristics of the data and not focusing on generalization to other situations. Thus, techniques other than the traditional multivariate methods are employed, many times of a quite simple and qualitative nature. The researcher, given this depth of information, is following a strategy of discovery through examination of the data for all types of relationships.

Neural Networks

Once viewed strictly as a method used by computer scientists in their attempts at **artificial intelligence,** neural networks have become an accepted tool for "everyday" researchers, particularly those outside the academic community. Following the impetus of data warehousing and data mining, neural networks provide a relatively straightforward approach to exploration and discovery [4, 22, 33]. Also, given the more pragmatic nature of this exploration, neural networks provide a powerful prediction tool based on quantification and replication of complex patterns in the data. Less emphasis is placed on what it is doing than on how well it is doing it. Although neural networks can replicate and many times outperform many statistical techniques, such as multiple regression, discriminant analysis, logistic regression, and cluster analysis, it is not easily interpreted due to the complex relationships that are handled "invisibly" by the methodology. Thus, the researcher is encouraged to employ neural networks when exploration and prediction, but not explanation, are the focal points of the research.

Resampling

The final topic—resampling—is one that might surprise many by its inclusion in a multivariate text. But resampling, the analysis approach that substitutes empirical estimation of sampling distributions for the assumed properties of theoretical distributions such as the normal or t-distributions, differs only in its underlying assumptions about the data and its characteristics. All of the multivariate techniques based on classical statistical inference, such as multiple regression, discriminant analysis, or MANOVA, can be performed just as easily with resampling. The computational power available today enables the user to avoid the assumptions of normality by repeatedly drawing samples and empirically defining the actual sampling distribution. In this manner, the researcher is again "returning to the data" and representing the actual characteristics of the data rather than assuming certain characteristics for estimation purposes. As with the topics already discussed, most of the principles and techniques we have learned are still operable; we are just using a somewhat different process of model estimation, inference, and interpretation.

Data Warehousing and Data Mining

The emergence of data warehousing and data mining may be best described by the saying, "Necessity is the mother of invention." Organizations have enthusiastically supported increases in automation and computerization over the years and valued the strategic role of information. But three forces have combined to force a realization that new approaches were necessary for the effective assimilation and use of the information available in today's information-intensive economy [2, 28].

The first factor is the expanding pool of information that almost all organizations have at their disposal. Computerized accounting systems make current information readily available for all levels of the organization. Automation of the manufacturing process, particularly procurement and the rise of inventory control systems, also provides a continual source of information. The participation in a global economy and the need for international operations requires even more cooperation and control, again a reason for increased computerization. The recent emergence of the Internet only promises to increase the information demand and its supply, particularly as electronic commerce provides a direct link to consumers and information that until now was available only to retailers. The automation of the salesforce, with emphasis on computerization and a "virtual office," provides one more source of information that was until just recently in a format that was of no benefit to anyone except the salesman. Organizations also have evolved from relying strictly on operational data to the integration of external information directed toward a specific function, whether it be customer satisfaction feedback for the marketing department or financial reports to assess market performance and return.

A second related force is that many organizations, because of the processes described above, are producing information as fast as products or services. Organizations, in capturing data from all of their processes, are creating a parallel role: constant and immediate evaluation of their operations. This force was countered by the third—techniques that can match the production of information with the production of knowledge. What was needed were analyses that modify themselves without human intervention, freeing the researcher to focus on the results and the model's performance. As these techniques became available and accepted, the three forces were aligned to promote action. What we have seen recently has been an explosion in the use and application of data warehousing and data mining. The following sections detail the principal characteristics of both data warehousing and data mining, ending with a discussion of the implications they have for multivariate techniques.

What Are Data Warehousing and Data Mining?

Data mining and data warehousing are complementary elements in the improvement of data access for decision making. Data warehousing is the facilitating mechanism for **decision support systems (DSS),** storing an organization's data in a single, integrated database and providing a historical perspective. Two key concepts underlying data warehousing are integration and time-invariance [18, 20]. Integration refers to the unified database design that combines all data sources within the organization into a single access point. Time invariant means that it preserves a historical perspective, such that "slices of reality" are available in any retrospective analysis. A data warehouse, however, is not an application, but instead is a facilitator of applications by providing timely data in the required format. Its role as a strategic investment is becoming more widely accepted [6, 28].

Data mining, also known as **knowledge discovery in databases (KDD),** is the search for relationships and data patterns in large databases [3, 12, 13, 26]. As the term suggests, data mining has an exploratory orientation of searching for knowledge obscured by the complex patterns of association and large amounts of data. A few "nuggets" of knowledge may be found only after processing vast amounts of information. An implicit assumption is that by revealing these relationships in the database, benefits will accrue to the extent that the database truly reflects the organization's decision environment. A number of varying types of analytical techniques can be employed in data mining, ranging from the most basic descriptive and graphical approaches to more sophisticated multivariate techniques (e.g., cluster analysis, or multiple or logistic regression) and newer learning-type models (e.g., neural networks and genetic algorithms) [3, 33].

Fundamental Concepts in Data Warehousing

The primary goal of a data warehouse is an enterprise-wide integration of data in a format amenable to analytical scrutiny. Therefore, data warehousing achieves at least three objectives: (1) it provides support for DSS in terms of data access and data organization; (2) it segregates data access to the data warehouse, thus reducing the performance degradation of operational systems due to repeated queries; and (3) it forces a recognition of the differing data structure needs for operational versus analytical purposes.

Operating a Data Warehouse

The operation of a data warehouse has many sophisticated technical aspects involving data organization, data structures, and database design that are beyond the scope of this text. Interested readers may wish to review an introductory text in data warehouse design and architecture [8, 17, 18, 21, 31, 34]. From an operational perspective, the data warehouse encompasses seven basic phases:

1. *Data acquisition.* The first task is to acquire data from all relevant sources, internal as well as external to the organization [2]. The internal data sources are primarily legacy (existing) systems dealing with operational processes, such as transaction processing and inventory management. The external information sources may be customer information (e.g., demographics), research studies (e.g., satisfaction surveys or other marketing research results), or even commercial sources, such as census information or commercial databases (e.g., Dunn and Bradstreet profiles). Note that these data are at differing levels of analysis or aggregation and must be integrated with other information at the same level of aggregation.

2. *Data integration.* This phase integrates all of the data sources, maintaining consistency by matching characteristics, attributes, and level of aggregation. As we discuss in a later section, the data take on their multidimensional qualities during this phase.

3. **Data cleaning.** The integration of many data sources necessitates scrutiny of the data to eliminate errors and poor quality data while also performing consistency checks [27]. The quality of all analyses depends on the underlying quality of the data and this becomes one of the greatest hurdles in the successful implementation of a data warehouse.

4. *Metadata creation.* Metadata comprise a complete description of a **data element,** including not only its attributes, but also its original source and any transformations or summarizations. The metadata description provides a profile of the data element that can be used in any further analysis, transformation, or summarization.

5. *Data import.* At periodic intervals, data are imported into the data warehouse. Because the data warehouse provides a historical archive, these data are merged to the existing database and provide a temporal perspective.
6. *Data warehousing.* This phase includes the organization and processing of the database, including the summarization of data at varying levels in anticipation of user queries.
7. *Decision support.* Decision support refers to user-directed applications from **OLAP (on-line analytical processing)** or data mining procedures that access the database with a specified query. OLAP comprises those "real-time" queries of the database that access stored prespecified multidimensional data summaries, when possible, for immediate response [31]. Data mining techniques are specialized analytical procedures, which will be discussed in more detail in the following section.

In describing the operation of a data warehouse, it is important to distinguish between two systems: **operational systems** and decision support systems (DSS). An operational system controls the basic functions **(transactions)** of the business, such as accounting, inventory, and order processing. Without these systems, the organization could not function. Operational systems require data that reflect the current status of the organization; historical data is of little use. In contrast, decision support systems involve applications dealing with the planning and strategy of the organization [30, 33]. In the short run, an organization could exist without DSS. But given the critical nature of information in the organization's ability to compete and react in today's business environment, an organization without DSS is short-lived. Operational systems function with their own databases based on the efficient processing of multiple single requests. A decision support system, however, requires the existence of a data warehouse to collect and format the data in a format suitable for DSS applications. Without a data warehouse, extensive time and effort are spent in data preparation for each DSS information request.

Data Definitions

The benefits of a data warehouse come not only from its day-to-day operations and storage of data, but also from data definition. Whether defining its basic character or level of aggregation, or forming a complete profile, these data definitions are instrumental in the efficiency and effectiveness of a data warehouse. The three most important definitions are: operational versus analytical data, primitive versus aggregated data, and metadata.

Operational versus Analytical Data   **Operational data** are the foundation of maintaining automated day-to-day operations of the organization, but they are not well suited to further analysis. A primary focus of data warehousing is the creation of **analytical data** by processing and integrating operational data into the data warehouse database in a format suitable for data mining applications [18, 19]. Analytical data reflect an internal structure that combines separate events around a common object, whether it be a customer's history of purchases, the sales of a product item, or some other unit of analysis. This transformation process highlights that data warehousing is not just about storing data, but also about how it is stored.

Primitive versus Aggregated Data   A second characteristic of a data warehouse is its ability to summarize **primitive data** (data at their original level) and store it as **aggregated data** at a higher level of analysis. Operational systems rely on primitive data because they are concerned only with single events, not some composite of events. DSS applications, however, can benefit from the increased speed of access provided by aggregated data. This becomes a critical issue as the database size grows, making queries requiring data summarization much less efficient. Moreover, OLAP's quick response time is predicated on accessing aggregated data and avoiding the computational delays associated with the summarization process. The key is to aggregate only the information that is needed, since aggregated data can greatly add to the storage requirements. As the database size and complexity grows, the number of possible aggregations quickly exceeds any feasible limit, so the data warehousing staff must select the aggregations they feel will best support their users.

This does not mean, however, that primitive data are useful only in aggregated form. The retention of primitive data allows for the **drill down** process, whereby the researcher examines the data underlying selected aggregated data. Assume that after reviewing monthly sales totals, the researcher wants to see total sales by outlet by week and then focus on the sales of particular store by product line. The retention of primitive data allows aggregation at any level and on any dimension. This facilitates the discovery process by allowing exploration in as flexible a manner as possible.

Metadata The final data definition that distinguishes data warehousing is **metadata.** Literally "data about data," metadata provide a complete profile of the data element, including its source, transformations, any summarization, a complete list of dimensions, time frame, and any other pertinent information [2, 18, 19, 20]. Metadata also allow for a standard classification system among data elements. For example, the term *sales* has many meanings within an organization. To marketing, it is transactions with customers. To the inventory managers, it is product to be replaced. For the accounting department, it is revenues minus returns and allowances. Each of these are sales within their functional area, but differ slightly as to their time frame, when they occur, and the actions they require. Metadata allow for a uniform classification that specifies distinctly the character of each data element in the data warehouse.

Summary
The data warehouse, in concept, has been around for as long as researchers have managed data within organizations. As anyone with experience in organizational database systems will tell you, rarely are data structured in a format that provides the type of information needed for decision support systems. Data warehousing emerged in response to the vast amounts of data accumulating in organizations and the need for a systematic process of translation from operational data to analytical data. Its continued development will foster not only the use of DSS applications, but also the recognition of the strategic asset embodied in the organization's database information.

Fundamental Issues in Data Mining
With the data warehouse in place, the researcher is now prepared for the next step: data mining. As described earlier, data mining is the process of extracting information from large databases. Given its analytical nature, we first discuss some general similarities and differences from the multivariate techniques discussed in

earlier chapters. A second discussion focuses on the predominant orientation of data mining—data exploration—and how it changes the researcher's approach to data analysis. Then the discussion centers on the basic types of analytical techniques used in data mining, ranging from the simplest forms (queries) to the more complex techniques (neural networks and genetic algorithms). We conclude by examining some specific recommendations for the researcher in making the transition from the statistically based orientation of this text to this different orientation toward data analysis.

What's Different versus What's the Same in Data Mining?

The reader will note that data mining is not an analytical technique, but instead an approach to data analysis. As such, much of what is discussed in the previous chapters still applies. But data mining also has some unique characteristics, especially in its approach to the analytical process in general.

Some Differences   Up to this point we have dealt with well-defined research problems for which a detailed research plan is developed, including the selection of *the* appropriate multivariate technique. Most often a confirmatory mode was used, in which proposed models were assessed for their explanation of the situation. Data mining, however, presents a somewhat different situation for several reasons.

First, many times the researcher is operating in an exploratory mode structured around a very general research question (e.g., what affects customer retention?). Also, although data mining encompasses a wide range of analytical tools, we have far less guidance in selecting a technique than we did in our earlier discussions, in which the measurement scale and number of variables for the dependent and/or independent variates determined the method to be employed. Because exploration is a major focal point of data mining, many of the techniques, by necessity, are quite generalized and can be employed for any number of different functions and for all types of variables. Thus, the researcher selects the analytical method for its approach to the problem as much as for its analytical sophistication or explicit requirements for data input.

Some Similarities   Many of the issues faced in previous chapters emerge when discussing data mining. Even while differing in form many times, data mining is based on multivariate analyses, with even the simplest methods designed to utilize a fairly large number of variables. The very nature of the data warehouse concept implies a truly multivariate perspective. Perhaps more so than ever, theory can play a large role in a successful data mining effort. The large databases used for data mining not only contain many observations, but often also include a tremendous amount of information about each respondent. Without guidance, however, the exploratory process can quickly become overwhelming, and confirmation is not even possible. Thus, researchers must never "let the machine do the thinking" in terms of formulating the research design. Some methods, such as neural networks or other learning-based methods, may require little specification of the actual relationships, but they will suffer as much as any other analysis when the variables are inappropriately selected or key variables are omitted.

Validation is a key component of any data mining analysis. Given the exploratory nature of data mining and the capability of its methods to represent complex relationships, the researcher should always guard against becoming too "sample-specific," thus losing generalizability. In terms of internal validation, most data mining methods will use a sample of the database (even though a sample can be thousands of records). It is essential to ensure that these results can be extended to the remainder of the database. Although it might be argued that external validation is not needed because often all of a firm's customers are represented in the database, this argument ultimately fails in light of customer turnover and the continual quest to acquire new customers.

Finally, assessing practical significance of the results is critical for several reasons. Many times in multivariate research the researcher can become too enamored with the techniques and forget to scrutinize the results for their applicability and true insight. We raise this issue here because the newness of the field and the use of more "exotic" techniques presents a prime opportunity for this problem. In any research, one must always critically examine the results and determine the true extent to which they address the original research question.

Exploration versus Confirmation

Although earlier discussions of multivariate techniques generally focused on confirmatory applications, data mining generally has an exploratory nature. This exploratory focus does not exclude a confirmatory or verification orientation in data mining as well, but its smaller role can be attributed to several factors. First, by its very nature, data mining involves searching for hidden information. The amount of information available and the uncertain nature of many typical research questions both lead to an exploratory mode with a wide range of possible relationships to be explored. Also, many of the techniques are designed to explore a cross section of options and then pursue the most promising ones. As such, the researcher must manage the exploratory process not so much in methodological terms, but to ensure that the research goals are kept in focus.

But confirmation or verification is not "dead." It still plays an essential role in providing separate and objective support of the discovered relationships. Confirmation can come from the use of validation procedures or from the use of more structured multivariate techniques. As discussed earlier, validation is an essential step because many data mining techniques can easily be "overfitted" or "overtrained" and lose generality. Validation is typically performed by using separate and independent samples, of either an internal or external nature. In this sense, validation is a simple form of confirmation. The more rigorous approach is the application of a specified model—many times one of the multivariate techniques discussed in earlier chapters—to replicate and confirm the results. For example, neural networks can perform tasks quite similar to multiple regression, discriminant analysis, and cluster analysis. Many times these results from a neural network procedure can be reestimated using the corresponding multivariate technique. Although direct correspondence may be difficult to obtain, this does provide the researcher a more structured and controllable means of quantifying the discovered relationships.

Data Mining Techniques

The field of data mining encompasses a broad field of analytical techniques, ranging from the simple to the sophisticated. One characteristic common to all, however, is that most of the techniques are based on very simple principles that are understandable, even to those individuals without mathematical or statistical training.

Although our previous discussions of multivariate techniques never focused exclusively on the statistical underpinnings, any user should appreciate the complex statistical theory that allows for the estimation of these models. As such, texts that focus more on application are a useful complement to the discussions based in mathematical and statistical terms. Yet, grasping the basic principles of even correlation and regression requires some background. In data mining, many of the techniques take basic principles such as matching (association rules) or biological analogies (neural networks and genetic algorithms) as the bases for the procedure. Most people can relate more readily to these principles than to those of a statistical nature. As such, researchers and users of the results may easily feel comfortable with the procedures and the results. We must caution, however, against any notion that their simplicity makes them less "quantitative" or rigorous, as each approach can perform quite specific and detailed analyses.

Query   The first technique used in many data mining analyses is some form of **query.** Queries are considered a separate technique because their use can be quite specific, and many times they constitute the first stage of exploration. One form of query in the data warehouse is a structured query, many times known just as SQL (pronounced see-kwull). This form of search process allows the researcher to formulate very specific requests, such as, "How many purchasers of product A are in the middle-income brackets with young children at home and have made purchases of product B?" By making repeated queries, the researcher can begin to see or identify patterns and relationships. Also, queries allow the researcher to drill down, whereby after identifying an interesting fact in aggregated data, the underlying data and identifying patterns are explored in more detail.

Another more structured form of query is on-line analytical processing (OLAP). This technique specializes in the examining of **multidimensional data** (data with multiple **dimensions** or attributes) that is arranged in a unique manner in the database. This data structure facilitates the examination of the intersections (combinations) of attributes and identification of patterns and relationships. The interactive nature also allows for the exploration of a wide range of possible relationships and then "drilling down" when interesting findings occur.

Visualization   **Data visualization** techniques play an important role in data mining by allowing the researcher to employ one of our most valuable assets—the mind's ability to process and recognize patterns. Whether it be with simple scatterplots or more sophisticated approaches that create multidimensional displays, the researcher can gain great insight into basic relationships through these procedures. Programs such as DIAMOND, developed originally by IBM, specifically deal with visualization techniques. For example, DIAMOND has developed procedures for presenting multidimensional portrayals of association, much as seen in a scatterplot, but instead of representing only two dimensions, it can portray up to nine dimensions. The objective is to portray higher-order relationships in a manner most likely to identify the hidden information that would otherwise be obscured by the more quantitative results.

Multivariate Statistical Tools   Many of the multivariate techniques already discussed are employed in data mining, as evidenced by the heavy involvement of firms such as SPSS and SAS in the data mining field. For example, all of the techniques in the following discussion are extensively employed in data mining.

Multiple regression provides a direct means of confirmation and some types of exploration of dependence relationships. Factor analysis is employed to assess patterns of variables and cluster analysis objectively assesses the association of objects across many variables. Discriminant analysis and logistic regression are frequently employed for classification purposes. Even MDS is used to present overall associations and generate leads for inquiry in some instances. Consequently, the multivariate researcher will find extensive usage of familiar techniques in data mining, the primary difference being the research orientation, as discussed earlier.

Association Rules **Association rules** are much like the binary matching procedure in cluster analysis in that they quantify the joint occurrence of two events. For example, consider these questions concerning two events: (1) How often do you buy toothpaste when you buy shampoo? and (2) How often do you buy a toothbrush when you buy toothpaste? Each of these are the basis for an association rule, calculated as the joint occurrence of two events (the percentage of times they occur). An association rule is not restricted to just two events, as it could as easily calculate how often you buy all three products (toothpaste, toothbrush, and shampoo). Obviously, it would be simple to calculate thousands of rules if the number of events (product purchases in this example) becomes large.

But how can we evaluate these associations? Two measures are normally used. The first is **confidence,** measured as the likelihood of event A happening when event B occurs. For example, what is the association between purchases of brand X's toothpaste and toothbrushes. We find that a customer buying a brand X toothbrush has an 80 percent chance of also buying brand X toothpaste. Thus, we have a confidence of 80 percent that a customer purchasing brand X toothpaste will also purchase brand X toothbrushes. In addition, we must also examine how often this occurs. Termed **support,** this is the percentage of time that the joint event occurs out of the total population. If we find that only 5% of the customers buy brand X toothpaste as well as toothbrushes, then we probably will not consider this association very important.

Association rules are a valuable tool for profiling objects, particularly in a marketing context, for which the objects are consumers. From our earlier example, we could state that 5 percent of the sample bought brand X toothpaste and toothbrushes together 80 percent of the time. The researcher can then set acceptable levels of support and confidence to identify groups for further consideration.

Decision Trees The next method, decision trees, has a familiar look (like the dendrogram in cluster analysis), but is constructed and interpreted in a totally different manner. **Decision trees** are a sequential partitioning of the dataset in order to maximize differences on a dependent variable. The two most widely used programs are CHAID (Chi Square Interaction Detector) and CART (Classification and Regression Trees). Let us look at the simple example of partitioning buyer versus nonbuyer by three categorical independent variables: age, gender, and income (shown in Figure 12.1, p. 682). Because buyer–nonbuyer is dichotomous, the objective is to identify which of the three variables gives the best split of buyers versus nonbuyers (because the dependent variable is dichotomous, we will consider just the percentage of buyers). Overall, 40 percent of the sample buys the product. The best split of the entire group is with gender, with 30 percent of males buying the product and 60 percent of the females buying the product. The procedure then takes each of these two groups (males and females) and finds the variable that best splits each group. Note that the same variable does not have to

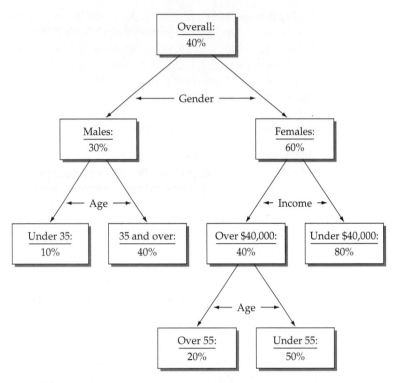

Legend: The top value in the box is the group label, and the bottom value is the
percentage of buyers. The variable below a group is the variable used to
split that group. For example, gender was used to split the overall group.

FIGURE 12.1 A Decision-Tree Model

split each group, as one variable could be used for males and another for females.
For example, males are split on age, whereas females are split on income. The
procedure continues until there are no independent variables left or there are no
significant splits left to make. The final result is a set of mutually exclusive groups
of customers that all vary in their percentage of buyers.

The results in our example show that the highest percentage of buyers is found
in females with an income of under $40,000. The group with the lowest percent-
age of buyers is males under the age of 35. We should note that the second lowest and second highest groups both come from the female split, where further
splits identified a group of females actually lower than one of the male groups.
Each group can be profiled by reading the tree diagram "back up." Decision trees
provide a rather concise way to develop groups that are consistent in their attributes but vary in terms of the dependent variable.

Neural Networks   Neural networks are one of the tools most likely to be associated with data mining. Patterned after the workings of the neural system of the
brain, the **neural network** tries to "learn" by repeated trials how to best organize
itself to achieve maximal prediction. We discuss this method in more detail later
in this chapter, but let us now just examine its basic operation. The model is composed of **nodes,** which act as inputs, outputs, or intermediate processors. Each
node connects to the next set of modes by a series of weighted paths (similar to

the weights in a regression model). Based on a learning paradigm, the model takes the first case, inputs its data, and makes an initial decision based on the weights. The prediction error is assessed, and then the model makes its best effort to modify the weights to improve prediction and then moves onto the next case. This cycle repeats itself for each case in what is termed the **training** phase, when the model is being calibrated. After calibration, the model can be used on a separate sample to assess its external validity.

Many researchers view neural networks as too much of a "black box" in that the researcher does not control the structure of the model, that is, which nodes are connected or the weights of the paths between nodes. Instead, this is all accomplished through the learning process. But this structure provides great flexibility. The neural network system can represent very complex, even nonlinear, relationships—something that is very difficult to do with most multivariate models. Moreover, in most cases it can achieve greater predictive accuracy than the comparable statistical method. However, the researcher must guard against becoming too sample-specific and losing generalizability. Neural networks have gained more widespread use in applied areas than in the academic area because, although they achieve very good predictive results, which is needed in applied areas, they fall short of the academic area's need for explanation. Continued research should encourage a greater appreciation of neural networks in both communities.

Genetic Algorithms   The final type of data mining technique is also a learning-based model, but based on a different biological analogy. **Genetic algorithms** mimic the evolutionary process by using natural selection. We start with a number of possible solutions to a problem. The "survivors" of this first "generation" form a new generation. Some will be more successful than the last generation, and some less successful. The survivors in each successive generation move on and compete for survival in the next generation. Slowly, over time, natural selection "weeds out" the inferior solutions and results in overall improvement. This process continues until acceptable rates of prediction are achieved. In operationalizing this process, two issues must be resolved. First, there must be a way for the "survivors" to be determined. There are many ways to measure success, but they are all based on some form of a dependent variable. Second, there must be a way for the survivors to be formed for the next generation. This can be achieved by some combination of the survivors (using a cross-over function) or mutation (a random variation of each survivor).

One advantage of the genetic algorithms is that they generally converge on the optimal solution. The drawbacks are that this may take many generations and a very large number of individuals are required. Consequently, this method is not particularly efficient, but advances in computational power have made it feasible for a wider range of applications.

A Multivariate Researcher's Perspective

In our discussion so far, the multivariate researcher might feel left out. The emphasis has not been on statistical inference, but on a host of alternative methods of varying approaches. And even when there is a statistical basis (e.g., association rules or decision trees), the statistical significance we seek in our other multivariate methods seems somewhat secondary. Does this mean that statistics has little to offer those in data mining? Is a new "field" of data analysis being formed with totally unique characteristics?

The answer is a resounding "No!" to both questions. Multivariate researchers will still play a key role, but they must understand what they can uniquely add to the development of data mining. Their contributions will not be through more elaborate or sophisticated models, but instead in fostering at least five basic statistical principles [23]. First, retain a clarity of goals for the research design. Each data mining method should be evaluated as critically as any multivariate technique for its appropriate application. This is particularly relevant when choosing between exploration, explanation, and prediction. Second, ensure that the selected technique is reliable in the research context. Will the method act as expected given the sample size, number of variables, and research objectives? Third, remember that uncertainties still exist. Even though the statistical inference of parameter estimates may not be emphasized, what about the probability associated with model search? If the data mining method has tested literally hundreds or thousands of models, what reliance should we place on the final "optimal" model? Could it occur by chance? What about the second-best model? Does it provide the same results and interpretations? Fourth, be wary of nonrandom data. Even though the data warehouse or database may hold data sets too large to even analyze, there may still be certain nonrandom biases present. For example, is the model valid for noncustomers if the results come from a sample of customers only? Even if the results are to be used internally, generalizability is essential. Finally, because the data are retrospective (historical) and nonexperimental, causal statements are quite difficult. Avoid the temptation to assign causality based solely on the technique used or the assertion of causality. The requirements for causal statements are addressed in chapter 11.

Summary

The fields of data warehousing and data mining are a research context quite different from that normally encountered by a multivariate researcher. But these differences should not preclude contributions from both areas, and perhaps a "symbiosis of statistics" can occur where each field contributes equally [23]. The collaboration would aid each party in developing the field overall as well as their own particular interests. In a broader perspective, this interaction could provide a prototype for such cooperation in other areas with distinct conceptual and applied constituencies.

Neural Networks

Neural networks are a totally different approach to data analysis from any of the multivariate techniques. Instead of conceptualizing the problem as a mathematical one, neural networks use the human brain and its structure to develop a processing strategy. Although we will never be able to construct neural networks as complex as a human brain, we can use its basic principles of multiple parallel processing units engaged in pattern recognition. Neural networks differ not only in their structure, but also in process. A key element in a neural network is learning (another analogy to the human brain), by which output errors (prediction or classification) are fed back into the system and it is adjusted accordingly. It then proceeds again, learning from each set of output errors. This is a sequential process, one case at a time, as compared to our multivariate technique, which considers an entire set of cases simultaneously.

Neural networks were first operationalized in the late 1950s and appeared to offer great promise. However, in the late 1960s, research demonstrated that the neural networks of that time were really quite limited in capability, and the field in general suffered a setback. Interest resumed in the 1980s as theoretical developments were coupled with increased computational power. A major improvement was the addition of hidden layers (a later section has more details), which allowed neural networks to portray much more complex systems. Today neural networks are used in almost every discipline or area of analysis. The flexible nature of system specification makes them adaptable to a wide range of problems, ranging from prediction to classification and even to time series analysis.

First, let us clarify the relationship between neural networks and multivariate statistical techniques. Neural networks can address many of the same problems as the multivariate techniques of multiple regression, discriminant analysis, and cluster analysis. In most instances the neural networks produce comparable results [1, 24], and thus it is up to the researcher to select between the methods based on research objectives. Also, neural networks do have an underlying statistical basis, for example, the impact of input distributions (nonnormal) on the estimation of weights. The primary difference from the multivariate techniques is the absence of any statistical inference tests for model weights of overall model fit. But researchers should not look on neural networks as any less rigorous, just a variation in approach.

The discussion in this chapter provides only a brief introduction to the field of neural networks. The interested reader is referred to any of a number of introductions to the use of neural networks in a wide range of applications [5, 7, 11, 15, 16, 29, 35]. Our discussion focuses first on some basic concepts of the neural network and then explores the steps needed to estimate a neural network. Finally, a brief example is presented comparing a neural network model to the discriminant analysis results from chapter 5.

Basic Concepts of Neural Networks

Neural networks have a simple structure and operation that can be described by four concepts: (1) the type of neural network model; (2) the individual processing units (nodes) that collect information, process that information, and create an output value; (3) the system of nodes (network) arranged to transfer signals from the input nodes to the output nodes, with some intermediate nodes in between; and (4) the learning function by which the system "feeds back" errors in prediction to recalibrate the model. We discuss each of these concepts and how they interrelate in more detail in the following section.

Types of Neural Network Models

There are three basic types of neural networks: the multilayer perceptron, the radial basis function, and the Kohonen networks. The **multilayer perceptron** model is the most commonly used and is the type used in the following example. The **radial basis function** is a more recently developed method that can be used for the same tasks as the multilayer model, but works in a somewhat different manner. The **Kohonen model** is appropriate only for clustering problems. The interested reader can find more detailed discussions of the strengths and weaknesses of each network in a number of resources [5, 11, 15, 29, 35]. All of our discussions focus on the multilayer perceptron model.

Nodes

The most basic element in a neural network is a **node,** a self-contained processing unit that acts in parallel with other nodes in the neural network. The node is analogous to the neuron of the human brain, which accepts inputs and then creates an output. A simple representation of the node and its operation is shown in Figure 12.2, which shows that nodes accept a number of inputs from other sources (nodes). Each connection from another node has an assigned weight. The first task of the node is to process the incoming data by creating a summated value in which each input value is multiplied by its respective weight. (Note that this is the exact operation performed in our multivariate methods when calculating a variate value.) This summed value is then processed by an **activation function** to generate an output value, which is sent to the next node in the system. Activation functions are generally a nonlinear function, such as the **sigmoid function,** which is a general class of S-shaped curves that includes the logistic function (see chapter 5 for a more detailed discussion of logistic functions).

Neural Network

The neural network is a sequential arrangement of three basic types of nodes or layers: input nodes, output nodes, and intermediate (hidden) nodes (see Figure 12.2). The input nodes receive initial data values from each case and transmit them to the neural network. An input node represents a single variable or pattern. Metric variables require only one node for each variable. Nonmetric variables must be **encoded,** meaning that each category is represented by a binary variable (just like the creation of binary dummy variables in chapter 2, only none of them are deleted as a reference category). Therefore, a three-category nonmetric variable would be represented by three binary input nodes. The first category would have the values of 1,0,0 across the three variables, the second category would be 0,1,0, and the third would be 0,0,1. An output node receives input and calculates an output value, but instead of going to another node, this is the final value. If this is a predictive model, then this is the predicted value. If the model is used for classification, then this is the value used in the classification process (you may wish to review chapter 5 regarding classification processes in discriminant analysis).

In almost every neural network there is a third type of node contained in the **hidden layer.** This is a set of nodes used by the neural network to represent more complex relationships than just a one-to-one relationship from input to output. It is the hidden layer(s) and the activation function that allow neural networks to easily represent nonlinear relationships, which are very problematic for multivariate techniques. This design of the neural network allows each node to act independently, but in parallel, with all of the other nodes. This provides the neural network with great flexibility in the types of input–output relationships that it can handle. We discuss issues regarding the number of nodes and hidden layers in a later section.

Learning

The feature of a neural network that truly sets it apart from the other multivariate techniques is its ability to "learn," or to "correct itself" based on its errors. Earlier, we discussed the concepts of weights on the connections between nodes. In using a biological analogy, the weights represent a state of memory, the model's "best guess" as to how to make the prediction of the output nodes.

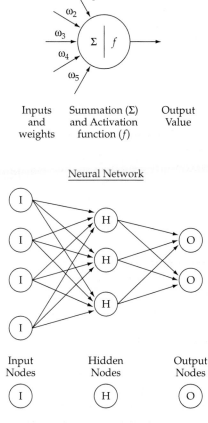

FIGURE 12.2 Nodes and the Neural
Network

Once the input for a case is processed through the system, it can be compared to the actual output value. This is termed training the system in a **supervised** learning mode. The actual and output values are compared. If there is any difference between the two values (similar to a residual value), then we would like to adjust the model in the hope of improving it. The most common form of training is **backpropagation.** In this approach, the error in the output value is calculated and then distributed backward through the system. As it works its way through the system of nodes, the weights are changed proportionally, increasing or decreasing depending on the direction of the error. Once all of the weights have been recalibrated, the input for another case is entered into the network and the process starts all over. The objective is to process a large number of cases through the neural network in the training phase so that it can make the best predictions across all of the input data patterns. There is also an unsupervised mode in which no feedback is given as to what is the correct output value. This approach is used only in clustering problems because there is no way to know actual cluster solutions.

Estimating a Neural Network Model

Now that we have described the basic operations and components of the neural network, we look at five fundamental issues involved in actually estimating a neural network model: data preparation, defining the model structure, estimating the model, evaluating the model results, and model validation.

Data Preparation

Neural networks are like all other statistical methods in one regard: garbage in will get you garbage out. Neural networks do not have the ability to transform poor quality or ill-conditioned data into a successful model. The researcher, therefore, must examine the data just as thoroughly as with any other statistical method. In preparing data for use in a neural network model, the researcher must consider (1) sample size, and (2) distributions of the data, transformations, and encoding.

Sample Size   The first task is to create the datasets for estimating your neural network. You need a calibration sample, known as the **training sample,** used to estimate the weights, and a separate validation sample to independently assess the predictive ability of the model. The split of the original sample is based primarily on the required sample size for the calibration and validation samples.

The sample size used to calibrate the neural network model can have as much impact on the results as with any of the other multivariate techniques. First, a model can be perfectly fitted if the sample size is less than the number of estimated weights. But as the number of cases approaches the number of weights, overfitting occurs and the model becomes too sample-specific, losing generalizability [30]. The number of weights in a simple neural net model can add up quickly. The neural net model in Figure 12.2 has 12 weights between the inputs and the hidden layer and another 6 weights between the hidden layer and the output nodes for a total of 18 weights. Thus, the number of weights is related to both the number of layers and the number of nodes at each layer. Adding one more input node would add three weights, whereas adding a second hidden layer of three nodes would add nine weights.

The model requires more than just one more case than the number of weights. The rule of thumb is to have 10 to 30 cases in the calibration dataset for each estimated weight. Although this is just a recommendation, it should be strongly considered in designing the network. We should note that you can never really have too many cases (the opposite of statistically based tests, in which the power becomes too great and the statistical inference test always shows significance). Cases may be randomly sampled if the original sample size is judged too large to process or unwieldy by the researcher.

Examining the Data   Even though we are not using a statistical inference test, the researcher still should examine the data with the techniques presented in chapter 2 to assess skewness, nonnormality, and outliers. Skewness can have an impact on both metric and nonmetric variables. For metric variables, techniques are available (see chapter 2) to assess these characteristics quite easily and to perform the transformations if necessary. Nonmetric variables can be just as problematic, particularly if the output node is a skewed nonmetric variable. If there are two categories, it does little good to have 98 percent of the cases in one category and

2 percent in the other. The neural network model will concentrate on the large (98 percent) value because the error in missing the other category is small. The researcher should look for a somewhat equal size for each category. If this does not occur naturally, the researcher is encouraged to equalize the categories through sampling. The problem is that the overall sample size is limited by the smallest output group.

Besides skewness and normality, the data should also be examined for missing data points, following the recommended solutions in chapter 2. Outliers should be carefully considered for deletion because they can seriously affect the overall results and training procedures. The final issue deals with standardization of the variables. The scale used to measure the variables can cause the estimation of larger weights for higher scaled values. Although there are many suggestions as to when to standardize, standardization carries no real risk, and we suggest that all variables be standardized. One last note, it is possible to add some **jitter** (random noise) to the input data to ease the estimation process and smooth out the fit function so as to have a better chance at a global optimum solution (see later section for a more detailed discussion).

Defining the Model Structure

After the data have been examined and any transformations have been made, the researcher must specify the model structure. Because the inputs and outputs are already selected, the decisions at this stage are the number of hidden layers and the number of nodes in each layer. Although there may be multiple hidden layers, the consensus is to use only one hidden layer. An additional hidden layer may improve estimation slightly, but at the greater risk of overfitting and finding only a suboptimal solution, while also greatly increasing the time needed for estimation. If a second layer is added, rigorous validation procedures should be used. As for the number of nodes in a hidden layer, this is as much a trial-and-error decision as any. The researcher should just vary the number of nodes in the hidden layer to find the best model fit. If two solutions have equal fit, take the simpler (smaller number of nodes) structure. One final element is a **bias node,** which is like a hidden node but has a constant value of 1.0. This may be added, to act like a constant term in a regression equation.

Model Estimation

The primary goal in the estimation process is to achieve the best possible model fit by finding the global optimum solution and not overtraining the model. The global optimum solution is the best possible solution across all possible solutions, termed the **search space.** Because the estimation process is an iterative procedure and it searches for ways to decrease the fit incrementally, it is possible to not always find the global optimum. For example, Figure 12.3 (p. 690) portrays a possible suboptimal solution and the global optimum solution. If the starting point were to the far right, the solution may have improved as it moved to the left, but when it reached the bottom of the "valley," it could not make any more improvement. The procedure does not have any way to "jump out" of the valley so that it can proceed to the global optimum. However, if the starting point were on the left, the global optimum would be reached. This illustrates the necessity for using multiple starting points to ensure that you have reached the global optimum.

We know that all of our data has some random error or noise in it, so we should not expect perfect fit (just as we wouldn't expect perfect fit in our multivariate

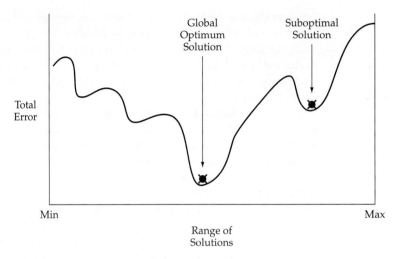

FIGURE 12.3 Global Optimum and Suboptimal Solutions

models). But if a model is trained too long with too many cases, it is possible for it to be overtrained, meaning that it has explained all of the basic relationships among the nodes and begins to represent the random error in the model. The researcher can avoid this in one of two ways. The first is to set a lower limit of error and stop training when this is reached. Although arbitrary, this directly avoids the problem. A second approach is to monitor the error rate for both the calibration and validation samples. When the neural network is being initially trained, the fit will improve for both samples as the weights are calibrated. But at some point, the validation sample error will level off and even start to increase, diverging from the calibration sample. This is the point at which the calibration sample is becoming overtrained, as it is becoming too sample specific (not generalizable). This also causes a poorer fit to the validation sample. The researcher should stop the training process when the best possible solution has been found.

Evaluating Model Results

Evaluating a neural network model consists primarily of assessing the level of prediction or classification of the output variables. For example, in a classification problem, the classification matrix cross-tabulates the actual and predicted values with correct predictions on the diagonal. From this number, the percent correctly classified can be calculated and compared to differing criteria (see the discussion in chapter 5 on predictive accuracy in discriminant analysis for more detail). If the output is a metric variable, then the common measure of fit is mean squared error.

Even though the neural network model can perform comparable tasks to other multivariate techniques, such as multiple regression and discriminant analysis, it does not provide information on the relative importance of the input variables. Although there are weights for each input variable to the hidden layer, they are not directly interpretable because they must also be combined with the weights from the hidden layer to the output variable. At this time there is no simple method of interpreting the solution with regard to the importance or impact of a single input variable.

Model Validation

The final step is to validate the solution to ensure that it is the global optimum and that it is as generalizable as possible. As mentioned earlier, it is essential that a validation sample be created to provide an independent assessment of model fit other than the calibration sample. Whenever possible, the researcher should employ a new sample of cases for an additional assessment of fit. The solution's stability is assessed by providing different starting points for the weights and re-arranging the order of the cases in the calibration sample. Finally, the researcher should also vary the number of nodes to ensure that a better solution is not possible.

Summary

The process of estimating a neural network model is not accomplished in a single step, but requires that the researcher constantly assess the model's solutions for predictive accuracy, overtraining or overfitting, and generalizability. The researcher should "experiment" with the solution as much as possible because the procedure does not always guarantee a global optimum solution or generalizability. Any researcher who wishes to use this procedure should be aware of the considerations at each stage, because this is not a method in which you can just use the "default" values and be assured of an acceptable model.

Using a Neural Network for Classification

One of the most common applications of neural networks is with classification problems: deciding to which group an observation belongs. This corresponds to a discriminant analysis or logistic regression problem. To provide a comparison between the abilities of discriminant analysis and neural networks, the two-group discriminant analysis problem from chapter 6 is revisited here. In addition to the discriminant results, we also estimate four neural network models. The first three have one hidden layer with four, six, and eight nodes. The fourth neural network model has two hidden layers with four nodes each. The alternative neural network models are estimated to ascertain the predictive accuracy gained through increasing model complexity.

Table 12.1 (p. 692) contains the neural network model descriptions and the predictive results. Classification accuracy is measured by percent correctly classified (see chapter 6 for a more detailed description, if needed). First, the discriminant analysis model predicts very well at a rate of 97.5 percent accuracy. But the neural network models matched the discriminant analysis for higher degrees of model complexity. The levels of classification accuracy for the neural network models ranged from 80 percent for the simplest model to 100 percent for the two most complex models. Note that although it was not possible to gain in predictive accuracy for the two highest models, the addition of the second hidden layer actually increased the training error rate. This supports the contention that a single hidden layer will suffice in almost all instances.

Researchers should consider the application of neural networks to prediction and classification problems, especially when the emphasis is on classification accuracy and not interpretation of the variate. As we have seen, neural networks can match and even surpass the predictive ability of the appropriate multivariate techniques. The growing availability of user-friendly software, such as NEURAL CONNECTION from SPSS, greatly facilitates the adoption of this technique.

TABLE 12.1 Comparison of Neural Network and Discriminant Analysis Models: Two Group Classification

Neural Network Model

Model	Structure[a]	Number of Weights	Training Error[b]	Classification Accuracy[c]
NN1	7-4-1	32	.196	80.0%
NN2	7-6-1	48	.142	90.0%
NN3	7-8-1	64	.026	100.0%
NN4	7-4-4-1	56	.034	100.0%

Discriminant Analysis

Two Group Model	97.5%

[a]First value is number of input nodes. Last value is number of output nodes. Middle values represent nodes in hidden layers.
[b]Root mean square (RMS) error.
[c]Percent correctly classified in testing sample (neural network) or validation sample (discriminant analysis).

Summary

Neural networks present an approach to data analysis that is markedly different from that covered elsewhere in the text. Its ability to handle complex relationships, particularly those of a nonlinear nature, provides an analytical tool with great flexibility in the types of problems that can be addressed. This flexibility provides the basis for superior estimation results in many predictive and classification problems. It does not provide, however, interpretation as to the relative importance of the input variables, nor their intercorrelation. Thus, the researcher should employ neural networks in those situations in which (1) multivariate techniques can not address the complex relationships, and (2) prediction and classification are the primary objectives.

Resampling

The kind of nonparametric statistics that has been taught in the past played an important role in analyzing data that are not continuous and thus cannot employ the normal probability distribution in making parameter and confidence interval estimates. But there is a new perspective on nonparametric estimation that also relates to parameter and confidence interval estimation for metric variables. With it, we do not have to assume that the confidence interval for a parameter follows a normal distribution. We can even generate confidence intervals for parameters such as the median, which currently is difficult to assess with the traditional parametric inference techniques. This nonparametric approach, known generally as resampling, has been gaining support as an alternative to our classical parametric inference methods. **Resampling** discards the assumed sampling distribution of a parameter and calculates an empirical distribution—the actual distribution of the parameter—across hundreds or thousands of samples. With resampling, we don't have to rely on the assumed distribution or be cautious about violating one of the underlying assumptions. We can calculate an actual distribution of sample parameters and can now see where the 95th or 99th percentiles actually are.

But where do the multiple samples come from? Is it necessary to gather separate samples, markedly raising the expense of data collection? Over the years statisticians have developed several procedures for creating the multiple samples needed for resampling *from the original sample.* Now one sample can generate a large number of other samples that can be used to generate the empirical sampling distribution. In this section we briefly review parametric estimation and then discuss some of these methods, focusing mainly on the jackknife and bootstrapping methods. After examining some of the issues involved in their use, we present a simple example of resampling applied to a multiple regression model.

A Brief Review of Parametric Inference

First, let us review exactly what is meant by parametric inference. First, we know that when we estimate a parameter (e.g., the mean) from a sample that the calculated value is only one estimate and that we will have other values for other samples. As the samples get larger the variation of the sample means gets smaller, but still there is some variation. So no matter what sample size we have, we need to know not only what the estimated parameter value is, but how much we can expect it to vary across all possible samples of this size.

Classical statistical theory provides a simple way of making these estimates. If we assume that the means will follow a normal distribution, also meaning that the underlying values of the sample are normally distributed, then we can directly calculate the standard error of the mean (where the standard error is the expected standard deviation of estimated means across samples). Thus, ± 1.96 standard errors should give us the 95 percent confidence interval for our calculated mean. We now have the calculated mean and a measure of how much it would vary.

Statistical theory provides some assurance in that the central limit theorem shows that the means will be normally distributed for large samples even if the underlying values are not. But we cannot be assured that it will hold in smaller samples, and the confidence intervals of many parameters are difficult to calculate.

Basic Concepts in Resampling

Resampling, however, does not use the assumed probability distribution, but instead calculates an empirical distribution of estimated parameters. By creating multiple samples from the original sample, resampling now only needs computational power to estimate a parameter value from each sample. Once they are all calculated, we can examine a histogram of the values and even calculate confidence intervals from the actual distribution of estimated parameters. The underlying theory for resampling is not covered in our discussion here, but is available in other sources [9, 10, 25]. Resampling actually encompasses several well-known methods. We now discuss two of the most widely used: the jackknife and the bootstrap.

Resampling Methods

A key difference between the various resampling methods is whether the samples are drawn with or without replacement. Sampling with replacement draws an observation from the sample and then places it back in the sample to possibly be sampled again. Sampling without replacement draws observations from the sample, but once drawn they are not available to be sampled again. The true power of resampling comes from sampling with replacement. Research has shown that this method provides direct estimates of the confidence intervals, although there are several modifications to the simple methods for deriving the confidence intervals. Let us review how the jackknife and bootstrap differ.

Jackknife versus Bootstrap   The jackknife and bootstrap methods differ in how they derive the sample. The jackknife method [32] computes n subsets (n = sample size) by sequentially eliminating one case from each sample. Thus each sample has a sample size of $n - 1$ and differs only by the single omitted case in each sample. Although the jackknife method has been surpassed by the bootstrap as an efficient estimator of confidence intervals, it still remains a viable measure of influential observations (see chapter 4 for a more detailed discussion of influential observations) and an option for many statistical packages.

The bootstrap method derives its sample by sampling with replacement from the original sample. The key is the replacement of the observations after sampling, which allows the researcher to create as many samples as needed and never worry about duplicating samples except by chance. Each sample can be analyzed independently and the results compiled across samples. For example, the best estimate of the mean is just the average of all the estimated means across the samples. The confidence interval can also be directly calculated. The two simplest approaches either (1) calculate the standard error as simply the standard deviation of the estimated means, or (2) literally rank order the estimated means and define the values that contain the outer 5 percent (or 1 percent) of the estimated mean values. Other approaches add some correction factors to this process, and are markedly more complex in their calculations [25].

Limitations

While resampling procedures are not bound by any parametric assumptions, they still have certain limitations [25]. First, and perhaps most important, the sample must be large enough and drawn (supposedly randomly) in a manner to be representative of the complete population. Resampling techniques cannot overcome any biases due to a nonrepresentative sample, and in many cases will exacerbate the problem. Second, parametric methods are better in many cases for making point estimates, such as the overall mean. Thus, the resampling procedures can complement the point estimates of parametric methods by providing the confidence interval estimates. Finally, resampling techniques are not suitable for identifying parameters that have a very narrow sampling range, such as the minimum or maximum values. Resampling works best when the entire distribution is to be considered.

An Example of Resampling and Multiple Regression

As an illustration of the insights provided by resampling, multiple regression is performed using both the parameter estimates and also the bootstrap estimates. The problem is similar to the example used in chapter 4, except that a simultaneous solution (all variables retained in the solution) is performed. Seven independent variables (X_1 to X_7) are used to predict a metric dependent variable (X_9). A more detailed explanation of the variables is available in chapter 1.

Table 12.2 contains the regression estimates (coefficients and standard errors) for each variable. Only two of the variables (X_3 and X_5) are statistically signifi-

TABLE 12.2 Comparison of Regression Coefficients and Standard Errors Obtained by Parametric Estimation and Bootstrapping

Variable	Regression Estimate		Bootstrap Estimates[a] for Sample Size				
	Coefficient[a]	Significance	50	100	500	1000	2500
X_1	−.058	.977	−.050	−.720	−.323	−.512	−.527
	(2.013)		(1.820)	(2.799)	(2.351)	(2.458)	(2.409)
X_2	−.697	.740	−.751	−1.491	−.971	−1.148	−1.196
	(2.090)		(1.831)	(2.743)	(2.431)	(2.492)	(2.460)
X_3	3.368	.000	3.425	3.287	3.374	3.376	3.352
	(.411)		(.406)	(.384)	(.376)	(.381)	(.384)
X_4	−.042	.950	−.105	.078	.033	.039	.043
	(.667)		(.533)	(.624)	(.636)	(.591)	(.598)
X_5	8.369	.035	8.535	9.916	8.996	9.301	9.372
	(3.918)		(3.255)	(5.348)	(4.567)	(4.743)	(4.677)
X_6	1.281	.180	1.418	1.099	1.183	1.215	1.169
	(.947)		(.923)	(1.135)	(1.116)	(1.044)	(1.089)
X_7	.567	.114	.664	.599	.558	.581	.587
	(.355)		(.363)	(.301)	(.322)	(.365)	(.347)

[a]Standard errors are shown in parentheses.

cant. The bootstrap estimates are also shown in Table 12.2 for sample sizes of 50, 100, 500, 1,000, and 2,500. The sample sizes refer to the number of separate estimates made for each coefficient. Each estimate requires a separate sample, so to achieve the sample size of 2,500 required the generation of 2,500 bootstrap samples (each with a size of 100) and then the estimation of the regression equation for each sample. Standard errors for the bootstrap samples were calculated with the normal approximation method.

Comparison of the coefficients and the standard errors between the regression and bootstrap samples indicates that assessments of statistical significance would be identical for both methods. Although the nonsignificant variables show a fair amount of variability across sample sizes and between the regression and bootstrap estimates, this would be expected because they have less correlation with the dependent variables and/or multicollinearity with the other independent variables. The two statistically significant variables (X_3 and X_5) are quite similar, although the coefficients for X_5 are typically greater in the bootstrap estimates.

Figure 12.4 (p. 696) portrays the bootstrap sampling distribution for each of the seven coefficients for a sample size of 1,000. Four of the coefficients (X_3, X_4, X_6, and X_7) approximate quite closely the normal distribution assumed in parametric estimation. The other three estimates, however, differ markedly. Both X_1 and X_2 have a left-skewness, whereas X_5 has the opposite orientation. Moreover, all three of these skewed distributions are much more peaked than the normal distribution. This skewness generates substantially larger standard errors as the sample sizes increase for all three variables. This would indicate that the parametric estimates are smaller than actual and thus the statistical tests are liberal.

Figure 12.5 (p. 697) demonstrates the changes in the form of the empirical distribution as the sample size increases. At small sample sizes the distribution is less well formed, but as the sample sizes increase to 1,000 and above, the distribution becomes very well formed and in this instance approximates the normal distribution. This suggests that bootstrap samples should be based on at least

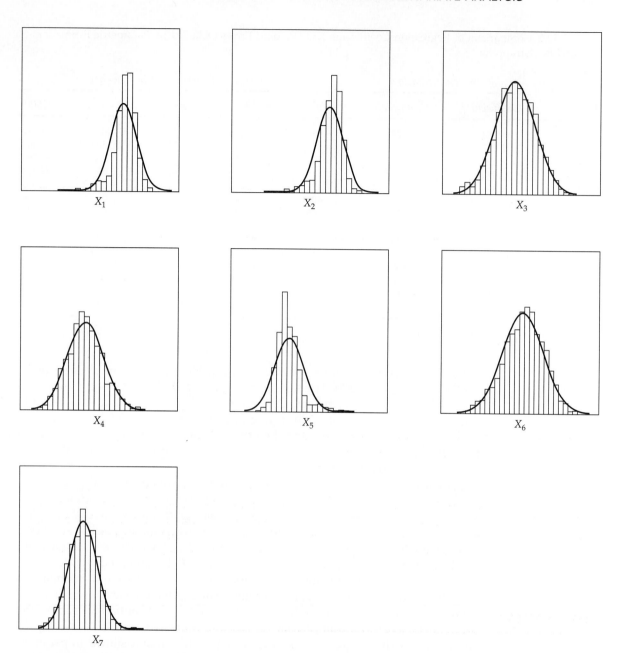

FIGURE 12.4 Bootstrap Sampling Distributions for the Seven Regression Coefficients (X_1 to X_7).

1,000 samples. Although this may at first seem to be problematic, personal computers can easily handle problems of this size.

Summary

The resampling procedures we have discussed provide an alternative perspective on one of the key assessments made in data analysis, the variability of the estimated parameter. This is the basis of hypothesis testing and assessing statistical

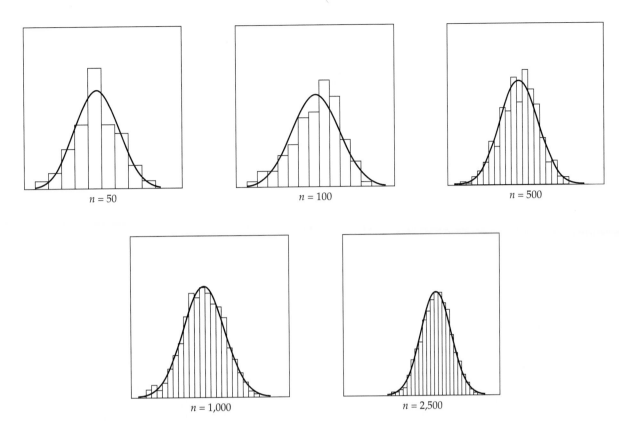

FIGURE 12.5 Changes in the Bootstrap Sampling Distribution of the Regression Coefficient (X_3) across Varying Sample Sizes

significance. Resampling techniques enhance the ability of the researcher to examine the actual distribution of the estimated parameters instead of relying on an assumed distribution. Techniques such as these provide a direct means to "know your data" and avoid the all-too-common pitfall of becoming too reliant on statistical techniques rather than reasoned judgment based on knowledge of the data.

Summary

This final chapter has attempted to provide a broader perspective on data analysis than that achieved just by study of the multivariate techniques. We are in a period of immense change. Information abounds and organizations are recognizing its increasing value in making strategic and tactical decisions. But as the pace of decision making increases, researchers must also become more aware of all of the available techniques and be well schooled in their application. An interesting observation is that many of the techniques and issues described in this chapter evolved from needs identified primarily in the applied fields of multivariate research. We hope that this will signal the beginning of an era of closer cooperation between academic and applied users to create a true partnership to the benefit of both parties.

Questions

1. Define the roles data mining and data warehousing play in today's research environment.
2. Explain the logic underlying decision trees and how they might be used in addressing a research problem.
3. Explain how neural networks learn and train themselves. Why can't multivariate techniques do the same?
4. What are the major differences between the resampling techniques of bootstrapping and the jackknife?
5. Describe how genetic algorithms work. What are their advantages and disadvantages?
6. Describe the differences in the procedures by which a parametric method and a nonparametric (resampling) method estimate confidence intervals.

References

1. Balakrishanan, P. V., M. C. Cooper, V. S. Jacob, and P. A. Lewis (1994), "A Study of the Classification Capabilities of Neural Networks Using Unsupervised Learning: A Comparison with K-means Clustering." *Psychometrika* 59: 509–25.
2. Banquin, R., and H. Edelstein, eds. (1996), *Planning and Designing the Data Warehouse*. Upper Saddle River, N.J.: Prentice Hall.
3. Berry, M., and G. Linoff (1997), *Data Mining Techniques for Marketing, Sales and Customer Support*. New York: Wiley.
4. Bigus, J. (1996), *Data Mining with Neural Networks*. New York: McGraw-Hill.
5. Bigus, J. (1996), *Data Mining with Neural Networks: Solving Business Problems—from Application Development to Decision Support*. New York: McGraw-Hill.
6. Boar, Bernard (1996), *Understanding Data Warehousing Strategically*. Lincroft, N.J.: NCR Corporation.
7. Chester, M. (1993), *Neural Networks: A Tutorial*. Upper Saddle River, N.J.: Prentice Hall.
8. Devlin, B. (1977), *Data Warehouse: From Architecture to Implementation*. New York: Addison-Wesley.
9. Efron, B. (1982), *The Jackknife, the Bootstrap and Other Resampling Plans*, vol. 38 of the CBSM-NSF Regional Conference Series in Applied Mathematics. Philadelphia: Society for Industrial and Applied Mathematics.
10. Efron, B., and R. J. Tibshiran (1993), *An Introduction to the Bootstrap*. New York: Chapman and Hall.
11. Fausett, L. (1994), *Fundamental of Neural Networks: Architectures, Algorithms and Applications*. Upper Saddle River, N.J.: Prentice Hall.
12. Fayyad, U. M., G. Piatetsky-Shapiro, P. Smyth, and R. Uthurusamy (1996), *Advances in Knowledge Discovery and Data Mining*. Cambridge, Mass.: AAAI Press/MIT Press.
13. Groth, R. (1997), *Data Mining: A Hands-On Approach for Business Professionals*. Upper Saddle River, N.J.: Prentice Hall.
14. Hackathorn, D. (1995), "Reinventing Enterprise Systems via Data Warehousing." The Data Warehousing Institute Annual Conference, Washington, D.C.
15. Hertz, J. (1991), *Introduction to the Theory of Neural Computing*. Reading, Mass.: Addison-Wesley.
16. Hinton, G. E. (1992), "How Neural Networks Learn from Experience." *Scientific American* 267 (September): 144–51.
17. Holsheimer, M., and M. Kersten (1994), *Architectural Support for Data Mining*. Technical Report CS–R9429. Amsterdam: CWI.
18. Inmon, W. H. (1996), *Building the Data Warehouse*. London: QED Publishing Group.
19. Inmon, W. H., and R. D. Hackathorn (1994), *Using the Data Warehouse*. New York: Wiley.
20. Inmon, W. H., and J. D. Welch, and K. Glassey (1987), *Managing the Data Warehouse*. New York: Wiley.
21. Kimball, R. (1996), *The Data Warehouse Toolkit*. New York: Wiley.
22. Medsker, L., and J. Liebowitz (1994), *Design and Development of Expert Systems and Neural Networks*. New York: Macmillan.
23. Michalewicz, Z. (1994), *Genetic Algorithms + Data Structures = Evolution Programs*. New York: Springer-Verlag.
24. Michie, D., D. J. Spiegelhalter, and C. C. Taylor (1994), *Machine Learning, Neural and*

Statistical Classification. New York: Ellis Horwood.

25. Mooney, C. Z., and R. D. Duval (1993), *Bootstrapping: A Nonparametric Approach to Statistical Inference.* Newbury Park, Calif.: Sage.

26. Piatetsky-Shapiro, G., and W. Frawley (1991), *Knowledge Discovery in Databases.* Cambridge, Mass.: AAAI Press/MIT Press.

27. Redman, T. C. (1996), *Data Quality for the Information Age.* New York: Artech House.

28. Simon, A. R. (1995), *Strategic Database Technology: Management for the Year 2000.* New York: Morgan Kaufman.

29. Smith, M. (1993), *Neural Networks for Statistical Modeling.* New York: Van Nostrand Reinhold.

30. Sprauge, R. H., and H. Watson (1994), *Decision Support for Management.* Upper Saddle River, N.J.: Prentice Hall.

31. Thomsen, Erik (1997), *OLAP Solutions: Building Multidimensional Information Systems.* New York: Wiley.

32. Tukey, J. W. (1958), "Bias and Confidence in Not Quite Large Samples." *Annals of Mathematical Statistics* 29: 614.

33. Turban, E. (1995), *Decision Support Systems and Expert Systems.* Upper Saddle River, N.J.: Prentice Hall.

34. Ulman, J. D. (1989), *Principles of Database Systems.* Santa Clara, Calif.: Computer Science Press.

35. Weiss, K. M., and C. A. Kulikuwski (1991), *Computer Systems that Learn.* New York: Morgan Kaufman.

Applications of Multivariate Data Analysis

APPENDIX PREVIEW

As we have seen for each multivariate technique, statistical programs are needed to handle the computational demands for all but the most trivial problems. And as noted many times in the text, the acceptance and widespread use of multivariate data analysis was in large part due to the availability of computer programs and access to computers for both academic researchers and business analysts. Over the years, researchers have continually sought new and more sophisticated programs to improve and extend their multivariate "toolkit." Luckily, today even the most sophisticated techniques are available through the major statistical packages (e.g., SPSS, SAS, SYSTAT, or BMDP). These programs, which used to be the exclusive domain of mainframe computers, all have PC-based versions that have the same capabilities as their mainframe predecessors. With the continual increase in computational power, the analyses that required a mainframe computer only a few years ago is now available to any user with even minimal access to a relatively new computer.

In our attempt to foster a working knowledge of multivariate data techniques, this appendix is devoted to providing the control commands necessary to execute the techniques illustrated in the chapters. Whenever possible, the commands for the various analyses are provided for the two most popular statistical packages: SPSS and SAS. These statistical packages have gained widespread usage in all types of computers, ranging from mainframes to personal computers. Fortunately, there are few, if any, differences between the control commands necessary for each type of computer. Our examples will be oriented toward personal computer applications, but only slight modifications are necessary for usage on other systems.

Two exceptions to the use of SPSS or SAS for our analyses should be noted. First, several of the multidimensional scaling analyses are performed with a popular MDS package available for the personal computer. These control commands are documented in a separate section. The other exception is in the area of SEM. We employ the popular LISREL program, and the necessary control commands for its execution will be provided as well in a separate section. In both situations, the necessary software is widely available at quite affordable prices.

We assume that the reader has some working knowledge of the programs being used. It is beyond the scope of this text to provide a tutorial on the general features of the statistical packages or the commands necessary for general operation of the packages. Although we are more familiar with SPSS, we have made every effort to provide the comparable control commands where possible.

We should note that the usage of command syntax is lessened dramatically with the advent of PC-based programs, especially those of a Windows-like nature. The ability of the user to specify the analysis through a series of menus greatly facilitates the use of the programs across a wide range of analysts. The analyst is no longer required to learn the syntax before a single analysis can be performed. Instead, it may be that only in the most advanced instances will the analyst have to deal with the command syntax. We provide the command syntax for these programs to illustrate the actual commands for experienced users and to provide guidance to new users if the need arises to utilize command syntax. For each analysis, we present an annotated set of control commands, in which specific commands are explained and optional commands detailed.

The continued advances in these statistical programs will require some updating of the control commands as the years pass. We encourage any interested researcher to visit our Web site at http://www.prenhall.com. We will maintain current control commands in addition to suggestions from other researchers involved with these or other statistical programs.

Software Used and Presented

Joreskog, K. G., and D. Sorbom (1993), *LISREL VIII*. Mooresville, Ind.: Scientific Software, Inc.

SAS Institute, Inc. (1990), *SAS User's Guide: Basics, Version 6*. Cary, N.C.: SAS Institute, Inc.

SAS Institute, Inc. (1990), *SAS User's Guide: Statistics, Version 6*. Cary, N.C.: SAS Institute, Inc.

Smith, Scott M. (1989), *PC-MDS: A Multidimensional Statistics Package*. Provo, Utah: Brigham Young University.

SPSS, Inc. (1997), *SPSS for Windows, Version 7.5*. Chicago: SPSS, Inc.

SPSS, Inc. (1997), *SPSS for Windows, Missing Data Module*. Chicago: SPSS, Inc.

SPSS, Inc. (1993), *SPSS Categories Module*. Chicago: SPSS, Inc.

ANNOTATED SPSS for Windows CONTROL COMMANDS

CHAPTER 2: EXAMINING YOUR DATA

Creating the SPSS System File

```
DATA LIST /ID 4-6 X1 10-12 X2 16-18
   X3 21-24 X4 28-30 X5 34-36 X6 40-42
   X7 45-48 X8 51 X9 54-57 X10 61-63          Identifies variables and column location.
   X11 66 X12 69 X13 72 X14 75.

FORMATS ID X1 X2 X4 X5 X6 X10 (F3.1).
FORMATS X3 X7 X9 (F4.1).                       Specifies the format of each variable.
FORMATS X8 X11 X12 X13 X14 (F1.0).

VARIABLE LABELS ID 'ID'
 /X1 'Delivery Speed'
 /X2 'Price Level'
 /X3 'Price Flexibility'
 /X4 'Manufacturer Image'
 /X5 'Overall Service'
 /X6 'Salesforce Image'
 /X7 'Product Quality'                         Specifies a label for each variable.
 /X8 'Firm Size'
 /X9 'Usage Level'
 /X10 'Satisfaction Level'
 /X11 'Specification Buying'
 /X12 'Structure of Procurement'
 /X13 'Industry Type'
 /X14 'Buying Situation'

VALUE LABELS X8 0 'SMALL' 1 'LARGE'
 /X11 0 'SPECIFICATION BUYING'
     1 'TOTAL VALUE ANALYSIS'                  Specifies a label for variable values.
 /X12 0 'DECENTRALIZED' 1 'CENTRALIZED'
 /X13 0 'FIRM TYPE ONE' 1 'FIRM TYPE TWO'
 /X14 1 'NEW TASK' 2 'MODIFIED REBUY'
     3 'STRAIGHT REBUY'

BEGIN DATA
     1   4.1    .6   6.9   4.7   2.4   2.3   5.2   0   32.0   4.2   1   0   1   1
     2   1.8   3.0   6.3   6.6   2.5   4.0   8.4   1   43.0   4.3   0   1   0   1
     3   3.4   5.2   5.7   6.0   4.3   2.7   8.2   1   48.0   5.2   0   1   1   2
     .    .     .     .     .     .     .     .     .    .     .    .   .   .   .
     .    .     .     .     .     .     .     .     .    .     .    .   .   .   .
```

A complete listing of the dataset is provided at the end of this appendix.

```
     .    .     .     .     .     .     .     .     .    .     .    .   .   .   .
     .    .     .     .     .     .     .     .     .    .     .    .   .   .   .
    98   2.0   2.8   5.2   5.0   2.4   2.7   8.4   1   38.0   3.7   0   1   0   1
    99   3.1   2.2   6.7   6.8   2.6   2.9   8.4   1   42.0   4.3   0   1   0   1
   100   2.5   1.8   9.0   5.0   2.2   3.0   6.0   0   33.0   4.4   1   0   0   1
END DATA.

SAVE OUTFILE='HATCO.SAV'.                       Saves data as a system file.
```

Missing Data Analysis

```
MVA
  X1 X2 X3 X4 X5 X6 X7 X9 X10
  /ID = ID
```
Specifies the metric variables for inclusion in the missing data analysis.

```
/TTEST PROB PERCENT=5
```
Performs *t*-test of mean differences between respondents with valid and missing data.

```
/CROSSTAB PERCENT=5
```
Cross-tabulation of any categorical variables between groups formed by respondents with missing and valid data.

```
/MISMATCH PERCENT=5
```
Matrix of percentages of cases that are mismatched on variable pairs (i.e., one variable has valid/missing and other variable has opposite).

```
/DPATTERN DESCRIBE=X1 X2 X3 X4 X5 X6 X7 X9
  X10
/MPATTERN DESCRIBE=X1 X2 X3 X4 X5 X6 X7 X9
  X10
/TPATTERN PERCENT=1 DESCRIBE=X1 X2 X3 X4
  X5 X6 X7 X9 X10
```
Three different types of portrayals of missing data patterns.

```
/LISTWISE
/PAIRWISE
```
Computation of correlation matrix among matrix variables using pairwise and listwise deletion practices.

```
/EM (TOLERANCE=0.001 CONVERGENCE=0.0001
  ITERATIONS=200)
```
Employs an EM (expectation-maximization) procedure to impute values and then estimate means and correlations.

```
/REGRESSION (TOLERANCE=0.001 FLIMIT=4.0
  ADDTYPE=RESIDUAL).
```
Employs a regression-based procedure to impute values and then estimate means and correlations.

Descriptive Statistics

```
EXAMINE
  VARIABLES=X1 X2 X3 X4 X5 X6 X7 X9 X10
```
Selects variables for calculation of descriptive statistics.

```
/PLOT NONE
```
Cancels all plots.

```
/STATISTICS DESCRIPTIVES
/CINTERVAL 95
/MISSING LISTWISE
  /NOTOTAL.
```
Selects descriptive statistics and use of pairwise deletion for missing data handling.

Testing for Homoscedasticity

EXAMINE
  VARIABLES=X1 X2 X3 X4 X5 X6 X7 X9 X10 BY
  X8 X11 X12 X13 X14 /ID=ID
	Selects metric variables (X_1, X_2, X_3, X_4, X_5, X_6, X_7, X_9, X_{10}) for comparison across categorical variables (X_8, X_{11}, X_{12}, X_{13}, X_{14}).

/PLOT SPREADLEVEL(1)	Performs spread-level analysis and Levene test for homoscedasticity.

/STATISTICS DESCRIPTIVES
  /CINTERVAL 95
  /MISSING LISTWISE
  /NOTOTAL.
	Selects descriptive statistics and use of listwise deletion for missing data handling.

CHAPTER 3: FACTOR ANALYSIS

Principal Components Analysis with VARIMAX Rotation (without X_5)

FACTOR /VARIABLES X1 X2 X3 X4 X6 X7	Specifies the factor analysis procedure of X_1 to X_4, X_6, and X_7. X_5 is omitted after initial analysis.
/CRITERIA ITERATE(50)	Specifies the maximum number of iterations for the factor solution; default is 25.
/FORMAT BLANK(0)	Controls the displayed format of the factor matrices. BLANK(.30) would not show any variable loading below .30. In this example, all values are shown.
/PRINT ALL	Prints all available statistics.
/PLOT EIGEN ROTAT(1,2)	Plots eigenvalues in descending order in scree plot. Generates a factor loading plot with factors 1 and 2 as the axes.
/EXTRACTION PC	Specifies the method of extraction to be principal components.
/ROTATION VARIMAX.	Requests the rotation method available. The default is VARIMAX. Other rotation methods (EQUAMAX, QUARTIMAX, OBLIMIN) available with additional /ROTATION commands.

Principal Components Analysis with Oblique Rotation (without X_5)

	The syntax for the principal components analysis with oblique rotation remains the same except for the rotation that is specified as oblique (OBLIMIN).

```
FACTOR
  /VARIABLES X1 X2 X3 X4 X6 X7
  /CRITERIA ITERATE(50)
  /FORMAT BLANK(0)
  /PRINT ALL
  /PLOT EIGEN ROTAT(1,2)
  /EXTRACTION PC
  /ROTATION OBLIMIN.
```

Validation of Components Factor Analysis by Split-Sample Estimation with VARIMAX Rotation

	The sample is split into two equal samples of 50 respondents. The factor model is then reestimated.
`SET SEED=34567.`	Specifies the random number seed.
`COMPUTE SPLIT=UNIFORM(1)>.52.`	Computes the variable SPLIT with a uniform distribution between 0 and 1.
`EXECUTE.` `SORT CASES BY SPLIT.`	Reorders the sequence of cases based on the values of the SPLIT variable.
`SPLIT FILE`	Splits the file into two subgroups which are analyzed separately.
`LAYERED BY SPLIT.` `FACTOR`	Same syntax as principal components analysis with VARIMAX rotation above.

`/VARIABLES X1 X2 X3 X4 X6 X7`
`  /CRITERIA ITERATE(50)`
`  /FORMAT BLANK(0)`
`/PRINT ALL`
`  /PLOT EIGEN ROTAT(1,2)`
`  /EXTRACTION PC`
`  /ROTATION VARIMAX.`	

Common Factor Analysis (without X_5)

	The primary difference between principal components factor analysis and common factor analysis is the specification of the extraction method, which is now PAF. For this dataset, the variable X_5 was dropped in the common factor analysis.
`FACTOR` `  /VARIABLES X1 X2 X3 X4 X6 X7`	Variables to be analyzed without X_5.
`  /PRINT INITIAL EXTRACTION ROTATION FSCORE` `  /CRITERIA MINEIGEN(1)  ITERATE(150)`	
`  /EXTRACTION PAF`	Specifies the common factor extraction method (PAF).
`  /CRITERIA ITERATE(150)`	Must increase the number of possible iterations in common factor analysis to ensure convergence.
`  /ROTATION VARIMAX.`	Requests the rotation method available.

CHAPTER 4: MULTIPLE REGRESSION ANALYSIS

Multiple Regression

`REGRESSION`	Initiates the regression procedure.
`/DESCRIPTIVES ALL`	Requests all descriptive statistics.
`/STATISTICS ALL`	Prints all summary statistics.

`/CRITERIA=PIN(.05) POUT(.10)`	Specifies the statistical criteria used in building the regression equation: (1) PIN = probability of F-to-enter. (2) POUT = probability of F-to-remove.
`/DEPENDENT X9`	Specifies the dependent variable as X_9.
`/METHOD = STEPWISE X1 X2 X3 X4 X5 X6 X7`	Specifies the variable selection method as stepwise and the variables for analysis as X_1 to X_7.
`/PARTIALPLOT ALL`	Specifies partial regression plots using all independent variables.
`/SCATTERPLOT=(*ZRESID,X9)`	Specifies variables for scatterplots.
`/RESIDUALS HIST(ZRESID) NORM(ZREZID) ID(ID)`	Specifies output of information on outliers, statistics, histograms, and normal probability plots. ID(ID) specifies id number as the label for casewise or outlier plots.
`/CASEWISE ALL SRE MAH SDR COOK LEVER`	Specifies the inclusion of all cases in the casewise plot and names the diagnostic variables to be used. Other diagnostic variables are available.

`/SAVE PRED ZPRED MAHAL COOK LEVER RESID`
`  ZRESID SRESID DRESID SDRESID DFBETA`
`  SDBETA DFFIT SDFIT COVRATIO.`
	Generates new variables, which serve as diagnostic measures for identifying influential observations.

Split-Sample Validation of the Stepwise Estimation

`SET SPEED=34567.`	Specifies the random number seed.
`COMPUTE SPLIT=UNIFORM(1)>.52.`	Computes the variable SPLIT with a uniform distribution between 0 and 1.
`EXECUTE.` `SORT CASES BY SPLIT.`	Reorders the sequence of cases based on the values of the SPLIT variable.
`SPLIT FILE`	Splits the file into two subgroups, which are analyzed separately.
`LAYERED BY SPLIT.` `REGRESSION`	Initiates the regression procedure.
`  /DESCRIPTIVES ALL`	Requests all descriptive statistics.
`  /STATISTICS ALL`	Prints all summary statistics.
`  /CRITERIA=PIN(.05) POUT(.10)`	Specifies the statistical criteria used in building the regression equation: (1) PIN = probability of F-to-enter, (2) POUT = probability of F-to-remove.
`  /DEPENDENT X9`	Specifies the dependant variable as X_9.
`  /METHOD=STEPWISE X1 X2 X3 X4 X5 X6 X7.`	Specifies the variable selection method as stepwise and the variables for analysis as X_1 to X_7.

CHAPTER 5: MULTIPLE DISCRIMINANT ANALYSIS AND LOGISTIC REGRESSION

2-Group Discriminant Analysis

`SET SEED 54321.`	Specifies a seed for the random number generator to generate a holdout sample.
`COMPUTE RANDZ=UNIFORM(1) > .65.`	Computes the variable RANDZ with a uniform distribution between 0 and 1.
`EXECUTE.`	
`DISCRIMINANT`	Initiates discriminant analysis and
`  /GROUPS X11 (0 1)`	specifies the grouping variable as X_{11} with a range of values of 0 and 1.
`  /VARIABLES X1 X2 X3 X4 X5 X6 X7`	Specifies the predictor variables used.
`  /SELECT=RANDZ(0)`	Selects cases with RANDZ equal to 0 for use in model estimation.
`  /ANALYSIS ALL`	
`  /METHOD MAHAL`	Specifies the method for selecting variables for inclusion.
`  /PIN=.05`	(a) PIN = probability of F-to-enter,
`  /POUT=.10`	(b) POUT = probability of F-to-remove.
`  /PRIORS SIZE`	Specifies the prior probabilities of group membership to be equal to the sample proportion of cases actually falling into each group.
`  /HISTORY STEP END`	Produces final summary report.
`  /STATISTICS ALL`	Prints all available statistics.
`  /PLOT=COMBINED SEPARATE MAP`	Plots combined- and separate-groups graphs and territorial map.
`  /CLASSIFY=NONMISSING SEPARATE`	Classifies cases that do not have missing values and uses the separate-groups co-variance matrix
`  /ROTATE=COEFF STRUCTURE.`	Specifies a rotated pattern matrix and a rotated structure matrix.

3-Group Discriminant Analysis

	There are two differences between a two-group and three-group discriminant analysis. One is the specification of the grouping variable and its range. The other is the CLASSIFY subcommand which is specified as POOLED The changes would appear as follows.
`DISCRIMINANT /GROUPS=X14(1,3)`	Now X_{14} is the grouping variable with values of 1 to 3.
`  /CLASSIFY=NONMISSING POOLED`	Case classification is based on the pooled within-group covariance matrices of the discriminant functions.

Logistic Regression Analysis

`SET SEED=123456.`	Specifies a seed for the random number generator to generate a holdout sample.
`COMPUTE RANDZ=UNIFORM(1)>.60.`	Computes the variable RANDZ with a uniform distribution between 0 and 1.
`LOGISTIC REGRESSION X11 WITH X1, X2, X3, X4, X5, X6, X7`	Initiates logistic regression with X_{11} as the dependent variable and X_1 through X_7 as independent variables.
`/METHOD=FSTEP`	Specifies a stepwise variable selection.
`/SELECT=RANDZ EQ 0`	Selects cases with RANDZ equal to 0 for use in model estimation.

/PRINT=ALL	Prints all available output.
/CRITERIA=ITERATE(50)	Specifies maximum iterations as 50.
/CASEWISE=PRED PGROUP RESID SRESID ZRESID LEVEL COOK DFBETA.	Specifies the diagnostic variables to be used in the casewise listing.

CHAPTER 6: MULTIVARIATE ANALYSIS OF VARIANCE

Multivariate Analysis of Variance (2 Group)

MANOVA X9 X10 BY X11(0 1)	Specifies the MANOVA procedure with X_9 and X_{10} as dependent variables and X_{11} as the independent variable with a range of 0 to 1.
/PRINT CELLINFO (MEANS CORR COV) DESIGN (COLLINEARITY)	Specifies the printed output: (1) cell information of means, correlation matrices and variance covariance matrices (2) Collinearity diagnostics for the design matrices.
HOMOGENITY(BARTLETT COCHRAN BOXM)	(3) homogeneity tests, Bartlett-Box F, and Box's M
SIGNIF(MULT UNIV STEPDOWN)	(4) significance tests of multivariate F tests for group differences and step-down tests
SIGNIF(EFSIZE)	(5) significance test of effect size for the univariate F- and t-tests
PARAMETERS (ESTIM)	(6) estimated parameters, including standard errors, t-tests, and confidence intervals
ERROR(CORR)	(7) significance tests for equality of covariances
/PLOT=ALL	Plots a normal and detrended normal plot, and a boxplot for the dependent variables.
/POWER T(.05) F(.05)	Requests that observed power values be calculated at .05 significance level.
/METHOD=UNIQUE	Requests the method of partitioning sums of squares corresponding to an unweighted combination of means (unique).
/ERROR WITHIN+RESIDUAL /DESIGN.	Specifies the structure of the model and must be the last subcommand of the model. Default (as shown) is the full factorial model.

Multivariate Analysis of Variance (3 Group)

	The only difference between a two-group MANOVA and a three-group MANOVA is the specification of the independent variable and its range of values, which is now X_{14} with a range of 1 to 3.
MANOVA X9 X10 BY X14(1,3)	Specifies the MANOVA procedure with X_9 and X_{10} as dependent variables and X_{14} as the independent variable with a range from 1 to 3.

Multivariate Analysis of Variance (2 Factor)

The only difference between a two-group MANOVA and a two-factor MANOVA is the specification of the independent variables and their range of values.

```
MANOVA X9 X10 BY X13 (0,1) X14(1,3)
```

Specifies the MANOVA procedure with X_9 and X_{10} as dependent variables and X_{13} and X_{14} as the independent variables.

CHAPTER 7: CONJOINT ANALYSIS

Designing the Stimuli: Generating an Orthogonal Fractional Factorial Design

Control cards used for the automatic generation of an orthogonal set of stimuli in conjoint analysis. Must set the SEED for exact replication.

```
ORTHOPLAN
  /FACTORS=
  MIXTURE 'Product Form' ('Premixed'
    'Concentrate' 'Powder')
  NUMAPP 'Num of Applic' ('50' '100' '200')
  GERMFREE 'Disinfectant' ('Yes' 'No')
  BIOPROT 'Biodegradable' ('No' 'Yes')
  PRICE 'Price/Applic' ('35 cents' '49
    cents' '79 cents')
  /HOLDOUT=4.
SAVE OUTFILE='CPLAN1.SAV'.
```

Uses ORTHOPLAN program to generate an orthogonal fractional factorial design for five factors, three 3-level factors and 2 two-level factors, labeling each level for each factor.

Four additional stimuli for holdout sample. Saves the generated plan for later use.

Designing the Stimuli: Specifying the Orthogonal Factional Factorial Design

The control cards necessary to replicate the plan used in the HATCO example. Also example of method to input specified design of stimuli rather than generation as in above method.

```
DATA LIST FREE
  /MIXTURE NUMAPP GERMFREE BIOPROT
  PRICE STATUS_CARD_.
```

Defines the factors and specifications
  STATUS_: 0—use for estimation
           1—holdout sample
           2—choice simulator data

```
BEGIN DATA.
    2.00    3.00    1.00    1.00    1.00    0    1
    3.00    3.00    1.00    1.00    1.00    0    2
    1.00    2.00    1.00    2.00    2.00    0    3
    3.00    3.00    1.00    2.00    2.00    0    4
    3.00    1.00    1.00    1.00    3.00    0    5
    2.00    3.00    2.00    2.00    3.00    0    6
    1.00    2.00    1.00    1.00    3.00    0    7
    1.00    3.00    1.00    1.00    2.00    0    8
    3.00    2.00    2.00    1.00    2.00    0    9
    2.00    1.00    1.00    1.00    2.00    0   10
    3.00    2.00    2.00    1.00    1.00    0   11
    2.00    2.00    1.00    1.00    3.00    0   12
    1.00    3.00    2.00    1.00    3.00    0   13
```

The levels of each factor that define each stimuli for use in estimation, validation, and the choice simulator.

```
     1.00      1.00      1.00      1.00      1.00      0      14
     2.00      2.00      1.00      2.00      1.00      0      15
     1.00      1.00      2.00      2.00      1.00      0      16
     2.00      1.00      2.00      1.00      2.00      0      17
     3.00      1.00      1.00      2.00      3.00      0      18
     2.00      2.00      1.00      1.00      2.00      1      19
     3.00      2.00      2.00      2.00      1.00      1      20
     3.00      3.00      1.00      2.00      3.00      1      21
     2.00      1.00      2.00      2.00      1.00      1      22
     1.00      1.00      1.00      2.00      3.00      2      23
     2.00      3.00      1.00      2.00      2.00      2      24
     3.00      3.00      1.00      2.00      1.00      2      25
END DATA.
```

```
SAVE OUTFILE='CPLAN1.SAV'.                Saves generated plan for later use.
```

Printing Plancards (Full-Profile Descriptions)

```
GET FILE='CPLAN1.SAV'.                    Recalls orthogonal plan.

PLANCARDS
  /FACTOR=MIXTURE NUMAPP GERMFREE BIOPROT  Specifies factors to use.
     PRICE
  /FORMAT BOTH                             Generates both cards and listing.
  /TITLE 'HYPOTHETICAL INDUSTRIAL
     CLEANSER )CARD'.                      Title appearing on each stimuli card. The
                                           )CARD places the card number on each
                                           stimuli.
```

Estimating the Conjoint Analysis Model

```
                                          The control cards necessary to (1) read in
                                          the preference data provided by
                                          respondents when evaluating the stimuli
                                          and (2) estimate the conjoint model.

DATA LIST FREE/ QN PROD1 to PROD22.
BEGIN DATA.
   104   4  6  5  4  4  6  4  4  4  4  4  4  4  5  5  4  4  4  4  6  6
   107   6  3  5  2  3  1  1  6  6  6  7  4  1  6  6  6  6  1  7  7  1  7
         The complete data set is listed at the end of the appendix
   417   5  5  2  5  1  1  2  2  2  2  1  1  1  4  6  3  2  1  4  4  2  3
   418   6  7  2  7  1  4  1  3  6  4  7  1  5  2  1  2  4  2  3  7  5  7
END DATA.

CONJOINT PLAN='CPLAN1.SAV'                 Retrieves the orthogonal plan.

  /FACTORS=                                Selects factors in conjoint estimation.
  MIXTURE 'Product Form' ('Premixed'
    'Concentrate' 'Powder')
  NUMAPP 'Number of Applications' ('50'
    '100' '200')
  GERMFREE 'Disinfectant' ('Yes' 'No')
  BIOPROT 'Biodegradable' ('No' 'Yes')
  PRICE 'Price per Application' ('35 cents'
    '49 cents' '79 cents')
  /SUBJECT=QN                              Selects variable QN as subject ID.
```

```
/SCORE=PROD1 PROD2 PROD3 PROD4 PROD5
  PROD6 PROD7 PROD8 PROD9 PROD10 PROD11
  PROD12 PROD13 PROD14 PROD15 PROD16
  PROD17 PROD18 PROD19 PROD20 PROD21
  PROD22
/UTILITY='UTIL.SAV'.
```
Specifies the preference response variables. They must be listed in the order of profiles in the orthogonal design.

Saves the part-worth estimates.

CHAPTER 8: CANONICAL CORRELATION

SPSS does not have a separate procedure for canonical analysis, but it can be performed through the MANOVA procedure with these commands.

```
MANOVA X9 X10 WITH X1 TO X7
  /PRINT=ERROR (SSCP COV COR) SIGNIF
    (HYPOTH EIGEN DIMENR)
```
Prints error matrices and significance tests.

```
  /DISCRIM=RAW STAN ESTIM COR ALPHA(1.0)
```
Produces the raw and standardized discriminant function coefficients (RAW and STAN), effect estimates (ESTIM) and the correlations between the dependent and the canonical variables (COR). All discriminant functions are reported (ALPHA(1.0)) for the canonical discriminant analysis.

```
  /RESIDUALS=CASEWISE PLOT
  /DESIGN.
```
Displays and plots casewise values and residuals.

CHAPTER 9: CLUSTER ANALYSIS

Hierarchical

```
PROXIMITIES
  X1 X2 X3 X4 X5 X6 X7
  /MATRIX OUT ('C:\WINDOWS\TEMP\SPSSCLUS.TMP')
  /VIEW=CASE
  /MEASURE=SEUCLID

  /PRINT=NONE
  /STANDARDIZE=NONE.
CLUSTER
  /MATRIX IN ('C:\WINDOWS\TEMP\SPSSCLUS.TMP')
  /METHOD=WARD(WCLUS)



  /PRINT=SCHEDUAL DISTANCE CLUSTER(2,5)



  /PLOT=DENDROGRAM VICICLE
  /SAVE=CLUSTER(2,5).




ERASE FILE=
  'C:\WINDOWS\TEMP\SPSSCLUS.TMP'.
```
Computes the proximity of observations to one another across the variables (X_1 through X_7) input in the cluster procedure.

Specifies the distance measure used, squared Euclidean distance (default).

Retrieves the proximities file.

Clustering method to be used is Ward's method, which requires squared Euclidean distances. WCLUS specifies a root name for saving clusters (see next command).

Prints the agglomeration schedule and distance matrix.

Prints cluster membership for each case, from 2 to 5 clusters.

Plots the dendrogram procedure.

Saves each case's cluster memberships for the two-, three-, four-, and five-cluster solution. The new variables are WCLUS5, WCLUS4, WCLUS3, and WCLUS2.

Erases the proximities file.

Nonhierarchical (Prespecified Cluster Seed Points)

QUICK CLUSTER X1 X2 X3 X4 X5 X6 X7
  /INITIAL=(4.46 1.576 8.9 4.926 2.992 2.51
  5.904 2.57 3.152 6.888 5.57 2.84 2.82
  8.038)
	Nonhierarchical clustering of X_1 to X_7.
	Supplies the initial seed points, reading the cluster centroids for group 1, then group 2, on variables X_1, X_2, . . . , X_7. In this example, the centroid for group 1 on X_1 is 4.46 whereas group 2 has a mean value on X_1 of 2.57.

/CRITERIA=CLUSTERS (2)	Specifies two clusters will be formed.
/PRINT=CLUSTER ANOVA	Prints ANOVA test for differences of each variable across clusters.
/SAVE=CLUSTER(NHCLUS).	Saves the cluster membership in NHCLUS.

Nonhierarchical (Random Selection of Cluster Seed Points)

SET SEED 345678.	Specifies a seed number for random number generator to ensure replication.
QUICK CLUSTER X1 X2 X3 X4 X5 X6 X7	Nonhierarchical clustering of X_1 to X_7.
/CRITERIA=CLUSTERS (2) NOINITIAL	Specifies two clusters will be formed from randomly selected initial cluster centers.
/PRINT=CLUSTER ANOVA INITIAL	Prints ANOVA test for differences of each variable across clusters.
/SAVE=CLUSTER(NHRCLUS).	Same as above, variable now NHRCLUS.

CHAPTER 10: MULTIDIMENSIONAL SCALING

Multidimensional Scaling (INDSCAL)

ALSCAL	Selects ALSCAL procedure for classical multidimensional scaling.
VARIABLES=var1 var2 var3 var4 var5 var6 var7 var8 var9 var10	Defines 10 similarity ratings.
/SHAPE=symmetric	Specifies that symmetric matrix is used so that upper portion of matrix does not need to be entered.
/INPUT ROWS(10)	Defines the number of rows.
/LEVEL=ratio (1)	Specifies metric level of analysis due to rating data used as input.
/CONDITION=MATRIX	Specifies that each respondent's data are unique.
/MODEL=INDSCAL	Selects INDSCAL (Individual Differences Scaling Model) procedure.
/CRITERIA=CONVERGE(.001) STRESSMIN(.001) ITER(50) CUTOFF(0) DIMENS(1,5)	Default estimation parameters.

```
/PLOT=DEFAULT ALL                                   Selects all possible plots.

/PRINT=HEADER.
```

Property Fitting (PROFIT) and Preference Mapping (PREFMAP)

See annotated control commands in PC-MDS section

Correspondence Analysis

```
DATA LIST FREE/ HATCO FIRM_A FIRM_B FIRM_C         Defines the attributes to the ratings of
  FIRM_D FIRM_E FIRM_F FIRM_G FIRM_H FIRM_I.          firms.

BEGIN DATA.
  4   3   1  13   9   6   3  18   2  10           The individual entries in the matrix are
 15  16  15  11  11  14  16  12  14  14             the number of times a firm is rated as
 15  14   6   4   4  15  14  13   7  13             possessing a specific attribute.
 16  13   8  13   9  17  15  16   6  12
 14  14  10  11  11  14  12  13  10  14
  7  18  13   4   9  16  14   5   4  16
  6   6  14  10  11   8   7   4  14   4
 15  18   9   2   3  15  16   7   8   8
END DATA.

ANACOR                                              Performs correspondence analysis based on
                                                      the cross-tabulation matrix.
  TABLE=ALL(8,10)                                   Specifies the number of rows and columns.
  /DIMENSION=2                                      Specifies the number of dimensions to be
                                                      computed.
  /NORMALIZATION CANONICAL                          Specifies method of normalizing the row
                                                      and column scores.
  /PRINT TABLE SCORES CONTRIBUTIONS PROFILES        Prints selected correspondence statistics.
    (PERMUTATION
  /VARIANCES ROWS COLUMNS SINGULAR
  /PLOT ROWS COLUMNS JOINT NDIM(ALL,MAX).           Plots separate (ROWS COLUMNS) and combined
                                                      (JOINT) row and column scores as well as
                                                      the two dimensions (NDIM).
```

ANNOTATED SAS CONTROL COMMANDS

CHAPTER 2: Examining Your Data

Creating the SAS Data File

```
DATA HATCO;                          Specifies a temporary data file name.

INPUT ID 4-6 X1 10-12 X2 16-18 X3 21-24
  X4 28-30 X5 34-36 X6 40-42 X7 45-48
  X8 51 X9 54-57 X10 61-63           Identifies variables and column location.
  X11 66 X12 69 X13 72 X14 75;

LABEL ID 'ID'
X1 'Delivery Speed'
X2 'Price Level'
X3 'Price Flexibility'
X4 'Manufacturer Image'
X5 'Overall Service'
X6 'Salesforce Image'
X7 'Product Quality'
X8 'Firm Size'                       Specifies a label for each variable.
X9 'Usage Level'
X10 'Satisfaction Level'
X11 'Specification Buying'
X12 'Structure of Procurement'
X13 'Industry Type'
X14 'Buying Situation';

CARDS:
  1   4.1    .6   6.9   4.7   2.4   2.3   5.2   0   32.0   4.2   1   0   1   1
  2   1.8   3.0   6.3   6.6   2.5   4.0   8.4   1   43.0   4.3   0   1   0   1
  .    .     .     .     .     .     .     .    .    .      .    .   .   .   .
  .    .     .     .     .     .     .     .    .    .      .    .   .   .   .
 A complete listing of the dataset is provided at the end of this Appendix
  .    .     .     .     .     .     .     .    .    .      .    .   .   .   .
  .    .     .     .     .     .     .     .    .    .      .    .   .   .   .
 99   3.1   2.2   6.7   6.8   2.6   2.9   8.4   1   42.0   4.3   0   1   0   1
100   2.5   1.8   9.0   5.0   2.2   3.0   6.0   0   33.0   4.4   1   0   0   1
RUN;
```

Descriptive Statistics

```
PROC UNIVARIATE DATA=HATCO NORMAL    Performs the tests necessary for testing
PLOT;                                the normality of the variables and
VAR X1 X2 X3 X4 X5 X6 X7 X9 X10;     identifies the number of missing values
RUN;                                 per variable.
```

CHAPTER 3: FACTOR ANALYSIS

Components Analysis

```
PROC FACTOR CORR MSA SCREE;          Initiates the factor procedure with a
ROTATE=VARIMAX;                      VARIMAX rotation producing correlations,
VAR X1-X4 X6 X7;                     MSAs and a scree plot.
RUN;
```

```
PROC FACTOR CORR MSA SCREE
ROTATE=PROMAX
VAR X1-X4 X6 X7;
RUN;
```
Initiates the factor analysis procedure with a PROMAX rotation producing correlations, MSAs and a scree plot.

Common Factor Analysis

```
PROC FACTOR METHOD=PRINIT CORR MSA
SCREE ROTATE=VARIMAX;
VAR X1-X4 X6 X7;
RUN;
```
Same as the principal component analysis except the method is specified as common factor analysis.

CHAPTER 4: MULTIPLE REGRESSION ANALYSIS

Multiple Regression

```
PROC REG;
```
Initiates the regression procedure.

```
MODEL X9=X1-X7
  /ALL SELECTION=STEPWISE PARTIAL;
```
Identifies the regression model to be used: X_9 as the dependent variable, X_1 to X_7 as independent variables. All statistics given with the stepwise entry procedure. Requests partial regression leverage plots for each independent variable.

```
PLOT R.*P.;
```
Specifies variables for plotting, residuals and predicted.

```
RUN;
```

CHAPTER 5: MULTIPLE DISCRIMINANT ANALYSIS

2-Group Discriminant Analysis

```
PROC DISCRIM METHOD=NORMAL;
POOL=YES LIST CROSSVALIDATE;
```
Initiates the discriminant analysis procedure with all statistics and validation.

```
CLASS X11;
```
Specifies X_{11} as dependent variable.

```
VAR X1-X7;
RUN;
```
Specifies predictor variables X_1 to X_7.

3-Group Discriminant Analysis

The only modification needed for a 3-group discriminant analysis is the identification of a new classification variable, X_{14}, a three-group variable.

```
CLASS X14;
```
Selects X_{14} as classification variable.

Logistic Regression Analysis

```
PROC CATMOD;
```
Initiates the categorical data modeling.

```
DIRECT X1-X7;
```
Specifies the independent variables containing design matrix values.

MODEL X11=X1-X7/ML CORRB FREQ ONEWAY PRED=PROB NOGLS XPX;	Specifies dependent variable as X_{11}, and independent variables, X_1 to X_7, plus optional additional output. For use as logistic regression, one must request ML and NOGLS. This uses maximum-likelihood estimates and suppresses computation of generalized (weighted) least-squares.
RUN;	

CHAPTER 6: MULTIVARIATE ANALYSIS OF VARIANCE

Multivariate Analysis of Variance (2 Group)

PROC GLM;	Initiates the general linear model (GLM).
CLASS X11;	Selects X_{11} as classification variable.
MODEL X9 X10 = X11;	Identifies the MANOVA model with X_{11} as the independent variable and X_9, X_{10} as the dependent variables.
MEANS X11 / BON SNK TUKEY;	Requests means for each level of X_{11} with Bonferroni t-tests, Student-Newman-Keuls multiple range tests, and Tukey's studentized range test on main effects.
MANOVA H=X11 / SUMMARY; RUN;	Specifies the effect employed as the hypothesis matrices and ANOVA tables for each dependent variable.

Multivariate Analysis of Variance (3 Group)

	The only modification needed for a 3-group MANOVA analysis is the classification variable, X_{14}, a 3-group variable.
CLASS X14;	Selects X_{14} as classification variable.

Multivariate Analysis of Variance (2 Factor)

	The only modification needed for a 2-factor MANOVA analysis is the classification variables, X_{13} and X_{14}.
CLASS X13 X14;	Identifies the classification variables as X_{13} and X_{14}.

CHAPTER 8: CANONICAL CORRELATION

PROC CANCORR ALL;	Initiates the canonical correlation procedure with all additional outputs.
VAR X9 X10;	Selects X_9 and X_{10} as dependent variables.
WITH X1-X7; RUN;	Selects X_1 to X_7 as predictor variables.

CHAPTER 9: CLUSTER ANALYSIS

Hierarchical

PROC CLUSTER M=WARD PSEUDO OUT SEED2;   Initiates the fast cluster procedure,
                                               specifies the maximum number of clusters
                                               to be 2.

VAR X1-X7;   Identifies variables for cluster analysis
                                               as X_1 to X_7.

PROC TREE N=5;   Initiates the tree procedure with the
                                               maximum number of clusters to be
                                               diagrammed as 5. Same as a dendogram.

PROC PRINT DATA=SEED2;
PROC FASTCLUS MAXC=2 MEAN=NEW OUT=TWOCLUS;   Specifies file (NEW) with cluster centers,
RUN;   identifies the new temporary file to be
                                               saved as TWOCLUS.

Nonhierarchical (Prespecified Cluster Seed Points)

PROC FASTCLUS SEED=NEW MAXC=2 OUT=TWOCLUS   Initiates the cluster procedure using the
                                               previously generated cluster means as
                                               seed points.

VAR X1-X7;   Identifies the variables to be X_1 to X_7.
RUN;

PROC ANOVA   Implements the ANOVA procedure with the
CLASS CLUSTER   two clusters and independent variables of
MODEL X1-X7=CLUSTER; X_1 to X_7. This tests for significant
                                               differences between the clusters on
                                               the variables used.

RUN;

Non-Hierarchical (Randomly Selected Cluster Seed Points)

PROC FASTCLUS MAXC=2 OUT=TWOCLUS   Same as above but with random selection of
  REPLACE=RANDOM   initial cluster centers.

VAR X1-X7;

PROC ANOVA;   Same as above.
CLASS CLUSTER;
MODEL X1-X7=CLUSTER;
RUN;

ANNOTATED LISREL VIII CONTROL COMMANDS

CHAPTER 11: STRUCTURAL EQUATION MODELING

Confirmatory Factor Analysis: Initial Model Specification

```
CONFIRMATORY FACTOR ANALYSIS          Title card.

DA NI=7 NO=100 MA=KM                  Specifies data file for number of
                                        variables (7), sample size (100), and
                                        data type (KM=correlation).

KM FU FI=C:\HATCO.COR FO=5            Reads data file from disk.
(7F9.0)

SELECT                                Selects correlations for analysis from
  1 2 3 4 6 7 /                         entire matrix. Note that variable 5 is
                                        omitted.

MO NX=6 NK=2 PH=ST TD=SY,FI           Model card defines number of exogenous
                                        indicators (6), number of exogenous
                                        constructs (2), and characteristics of
                                        associated matrices.

LA                                    Labels for variables in input matrix.
'DelvSpd' 'PriceLvl' 'PriceFlx'
'MfgImage' 'Service' 'SalesImg' 'Quality'

LK                                    Labels for exogenous constructs.
'Strategy' 'Image'

PA LX                                 Pattern matrix specifying loadings of
1 (1 0)                                 indicators on exogenous constructs. This
1 (1 0)                                 format is suggested as it corresponds
1 (1 0)                                 directly to the familiar format of factor
1 (0 1)                                 analysis and text discussions.
1 (0 1)
1 (1 0)
FR TD(1,1) TD (2,2) TD(3,3) TD(4,4) TD(5,5)   "Frees" the indicator error terms for
   TD(6,6)                            estimation.

OU SS TV RS MI                        Output card: requests standardized
                                        solution, t-values, residuals, and
                                        modification indices.
```

Confirmatory Factor Analysis: Model Respecification

```
CONFIRMATORY FACTOR ANALYSIS
DA NI=7 NO=100 MA=KM
KM FU FI=C:\HATCO.COR FO=5
(7F9.0)
SELECT
  1 2 3 4 6 7 /
MO NX=6 NK=2 PH=ST TD=SY,FI

LA
'DelvSpd' 'PriceLvl' 'PriceFlx' 'MfgImage'    Cards same as in the earlier model.
'Service' 'SalesImg' 'Quality'
```

```
LK
'Strategy'  'Image'
PA LX
1 (1  0)
1 (1  0)
1 (1  0)
1 (0  1)
1 (0  1)
1 (1  0)
FR TD(1,1)  TD  (2,2)  TD(3,3)  TD(5,5)  TD(6,6)
```

```
VA .005 TD(4,4)
```
Specifies error variance of variable 4 to be .005 as remedy for Heywood case.

```
OU  SS  TV  RS  MI
```
Same as in earlier model.

Confirmatory Factor Analysis: Estimation of Null Model

```
CONFIRMATORY FACTOR ANALYSIS — NULL MODEL
DA NI=7 NO=100 MA=KM
KM FU FI=C:\HATCO.COR FO=5
(7F9.0)
SELECT
 1 2 3 4 6 7 /
```
Cards same as earlier models.

```
MO NX=6 NK=1 PH=ST TD=SY,FI
```
Specifies single exogenous construct for null model.

```
LA

'DelvSpd'  'PriceLvl'  'PriceFlx'  'MgfImage'
'Service'  'SalesImg'  'Quality'
```
Same as earlier

```
LK
'Null Mod'
```
Labels single factor as null model.

```
PA LX
```
Specifies no loadings for indicators (see below).

```
6 (0)
VA 1.0 LX(1,1)  LX(2,1)  LX(3,1)  LX(4,1)
VA 1.0 LX(5,1)  LX(6,1)
FR TD(1,1)  TD  (2,2)  TD(3,3)  TD(4,4)
FR TD(5,5)  TD(6,6)
```
Specifies that construct loadings for all indicators equal 1.0 (no measurement error) and frees error terms for estimation.

```
OU  SE  TV  RS  SS  MI
```

Structural Equation Model (Path Model): Model Estimation

```
CAUSAL MODEL WITH MULTIPLE INDICATORS
DA NI=15 NO=136 MA=KM
KM FU FILE=C:\STRUC1.COR FO=5
(8F6.4/7F6.4)
```
Specifies a correlation file with 15 variables and sample size of 136 to be read from disk.

```
MO NX=13 NK=3 NY=2 NE=2 GA=FU,FI
  PS=SY,FI C BE=FU,FI TE=SY,FI PH=SY,FR
```
Model consists of 13 indicators for three exogenous constructs and two endogenous constructs with one indicator each (total = 15). Associated matrices also defined.

```
LA
'USAGE'  'SATISFAC'  'PRODQUAL'  'INVACCUR'
'TECHSUPT'  'NEWPROD'  'DELIVERY'
'MKTLEADR'  'PRDVALUE'  'LOWPRICE'  'NEGOTIAT'     Labels for variables in input matrix.
'MUTUALTY'  'INTEGRTY'  'FLEXBLTY'  'PROBRES'

LK                                                 Exogenous construct labels.
'FIRMPROD'  'PRICEFAC'  'RELATFAC'

LE                                                 Endogenous construct labels.
'USAGE'  'SATISFAC'

PA LX
1 (0 0 0)
1 (1 0 0)
1 (1 0 0)
1 (1 0 0)                                          Specification of measurement model for V_1,
1 (1 0 0)                                            exogenous indicators. Note that V_1, V_7,
1 (1 0 0)                                            and V_10 have no loading because each will
1 (0 0 0)                                            be set to 1.0 to control for scale
1 (0 1 0)                                            invariance (see below).
1 (0 1 0)
1 (0 0 0)
1 (0 0 1)
1 (0 0 1)
1 (0 0 1)
PA GA
1 (1 1 1)                                          Specifies exogenous coefficients for
1 (0 0 0)                                            structural equations.
PA BE
1 (0 0)                                            Specifies endogenous coefficients for
1 (1 0)                                              structural equations.
PA PHI
1                                                 Correlations among exogenous constructs.
1 1
1 1 1
PA PS
1                                                 No correlations among endogenous
0 1                                                 constructs.

VA 1 LX(1,1) LX(7,2) LX(10,3) LY(1,1)             Sets indicator loadings to 1.0 to control
  LY(2,2)                                           scale invariance.

VA 0.00 TE(2,2) TE(1,1)                           Sets measurement error to 0 for single
                                                    item indicators for endogenous
                                                    constructs.

OU SE TV RS SS MI AD=OFF                          Specifies output.
```

Structural Equation Model (Path Model): Null Model Estimation

```
CAUSAL MODEL WITH MULTIPLE INDICATORS—NULL MODEL
DA NI=15 NO=136 MA=KM
KM FU FILE=C:\STRUC1.COR FO=5
(8F6.4/7F6.4)
SELECT
  1 2 3 4 5 6 7 8 9 10 11 12 13 14 15/     Same as earlier except with new variables.
MO NX=15 NK=1 TD=DI,FR PH=SY,FR
LA
'USAGE' 'SATISFAC' 'PRODQUAL' 'INVACCUR' 'TECHSUPT'
'NEWPROD' 'DELIVERY' 'MKTLEADR' 'PRDVALUE' 'LOWPRICE'
```

```
'NEGOTIAT'  'MUTUALTY'  'INTEGRTY'  'FLEXBLTY'  'PROBRES'
PA LX
15 (1)
OU SE TV RS SS MI
```

Structural Equation Model (Path Model): Competing Model (COMPMOD1)

```
CAUSAL MODEL WITH MULTIPLE INDICATORS — COMPMOD1
DA NI=15 NO=136 MA=KM
KM FU FILE=C:\STRUC1.COR FO=5
(8F6.4/7F6.4)
MO NX=13 NK=3 NY=2 NE=2 GA=FU,FI C
  PS=SY,FI BE=FU,FI TE=SY,FI PH=SY,FR
LA
'USAGE'  'SATISFAC'  'PRODQUAL'  'INVACCUR'  'TECHSUPT'
'NEWPROD'  'DELIVERY'  'MKTLEADR'  'PRDVALUE'  'LOWPRICE'
'NEGOTIAT'  'MUTUALTY'  'INTEGRTY'  'FLEXBLTY'  'PROBRES'
LK
'FIRMPROD'  'PRICEFAC'  'RELATFAC'
LE
'USAGE'  'SATISFAC'
PA LX
1 (0 0 0)
1 (1 0 0)
1 (1 0 0)                              Same as earlier structural model.
1 (1 0 0)
1 (1 0 0)
1 (1 0 0)
1 (0 0 0)
1 (0 1 0)
1 (0 1 0)
1 (0 0 0)
1 (0 0 1)
1 (0 0 1)
1 (0 0 1)
PA GA
1 (1 1 1)                              Specifies that exogenous constructs are
1 (1 1 1)                              constructs now related to all endogenous
PA BE                                  constructs in the structural equations.
1 (0 0)
1 (1 0)
PA PHI                                 Same as earlier model.
1
1 1
1 1 1
PA PS
1
0 1
VA 1 LX(1,1) LX(7,2) LX(10,3) LY(1,1) LY(2,2)
VA 0.00 TE(2,2) TE(1,1)
OU SE TV RS SS MI AD=OFF
```

Correlation Matrix: Confirmatory Factor Analysis (HATCO.COR)

```
 1.000000
 −.349225   1.000000
  .509295   −.487213    1.000000
  .050414    .272187   −.116104    1.000000
  .611901    .512981    .066617    .298677    1.000000
  .077115    .186243   −.034316    .788225    .240808    1.000000
 −.482631    .469746   −.448112    .199981   −.055161    .177294    1.000000
```

Correlation Matrix: Strucutral Model Estimation (STRUC1.COR)

```
 1.0000   .4112   .2878   .3594   .2683   .2116   .2498   .3046
  .3284   .2679   .1423   .3289   .5187   .5100   .3411
  .4112  1.0000   .1676   .1585   .1413   .0810   .0603   .1270
  .1326   .0459   .1043   .0625   .0768   .0896   .1732
  .2878   .1676  1.0000   .7845   .6763   .5813   .6323   .6903
  .2931   .1844   .2890   .3266   .4130   .3297   .1540
  .3594   .1585   .7845  1.0000   .6370   .6222   .6436   .6667
  .2626   .1245   .2492   .2924   .3851   .2718   .1728
  .2683   .1413   .6763   .6370  1.0000   .6266   .5378   .5507
  .3365   .2296   .2993   .1119   .2647   .1919   .1818
  .2116   .0810   .5813   .6222   .6266  1.0000   .6986   .6251
  .2898   .2604   .2194   .2923   .3456   .1514   .1526
  .2498   .0603   .6323   .6436   .5378   .6986  1.0000   .6923
  .2065   .1096   .1550   .3258   .2593   .2098   .1070
  .3046   .1270   .6903   .6667   .5507   .6251   .6923  1.0000
  .1740   .2536   .2986   .2526   .2622   .2254   .1018
  .3284   .1326   .2931   .2626   .3365   .2898   .2065   .1740
 1.0000   .3009   .3071   .1790   .3472   .3479   .2718
  .2679   .0459   .1844   .1245   .2296   .2604   .1096   .2536
  .3009  1.0000   .6760   .1137   .2251   .2590   .0522
  .1423   .1043   .2890   .2492   .2993   .2194   .1550   .2986
  .3071   .6760  1.0000   .1727   .2048   .1616   .0933
  .3289   .0625   .3266   .2924   .1119   .2923   .3258   .2526
  .1790   .1137   .1727  1.0000   .4105   .5550   .2018
  .5187   .0768   .4130   .3851   .2647   .3456   .2593   .2622
  .3472   .2251   .2048   .4105  1.0000   .5318   .4112
  .5100   .0896   .3297   .2718   .1919   .1514   .2098   .2254
  .3479   .2590   .1616   .5550   .5318  1.0000   .4247
  .3411   .1732   .1540   .1728   .1818   .1526   .1070   .1018
  .2718   .0522   .0933   .2018   .4112   .4247  1.0000
```

ANNOTATED PC-MDS CONTROL COMMANDS

CHAPTER 9: MULTIDIMENSIONAL SCALING

Interpretation of the MDS Solution: PROFIT

```
1 10 2 8 0 0 3 0.0                        Programs control cards.
(2F9.0)
     .6077    1.2221
     .3501    1.3025
    −.6334     .9673                       Stimulus coordinates obtained from INDSCAL
   −1.1740    −.9958                          analysis.
   −1.4989    −.1782
     .5022   −1.3253
    1.4608    −.1062
     .3209   −1.6578
   −1.1906     .2491
    1.2552     .5224
(10F6.0)
Property 1
  6.94  7.17  7.67  3.22  4.78  5.11  6.56  1.61  8.78  3.17    Average evaluation of
Property 2                                                        each object on
  4.00  1.83  6.33  7.67  6.00  5.78  5.50  6.11  7.50  4.17    eight attributes.
Property 3
  6.94  5.67  3.39  3.67  3.67  6.94  6.44  7.22  4.94  6.11
Property 4
  4.00  3.39  7.33  6.11  7.50  4.22  7.17  4.33  8.22  5.56
Property 5
  5.16  3.47  6.41  5.88  6.06  4.94  5.29  4.82  8.35  4.65
Property 6
  5.11  1.22  5.78  7.89  6.56  3.83  4.28  6.94  8.67  4.72
Property 7
  5.33  3.72  6.33  5.56  6.39  4.72  5.28  5.22  7.33  5.11
Property 8
  4.17  1.56  6.06  8.22  7.72  4.28  3.89  6.33  7.72  5.06
HATCO
Firm B
Firm C
Firm D
Firm E
Firm F                                    Labels.
Firm G
Firm H
Firm I
Firm J
```

Incorporating Preferences into the MDS Solution: PREFMAP

```
10 2 6 0 1 0 2 4 0 1 1 25 0 0 1           Programs control cards.
(2F9.0)
     .6077    1.2221
     .3501    1.3025
    −.6334     .9673
   −1.1740    −.9958                       Stimulus coordinates obtained from INDSCAL
   −1.4989    −.1782                          analysis.
     .5022   −1.3253
    1.4609    −.1062
     .3209   −1.6578
   −1.1906     .2491
    1.2552     .5224
```

```
(2X,10F6.0)
  1  2  3   5  6  7  4  10  8   1  9
  2  5  2   7  6  9  3   4  1  10  8
  3  4  1   8  7  6  9   3  5  10  2
  4  4  3  10  2  7  8   6  1   9  5
  6  4  1   8  7  9  3   5  2  10  6
HATCO
Firm B
Firm C
Firm D
Firm E
Firm F
Firm G
Firm H
Firm I
Firm J
SUBJ 1
SUBJ 2
SUBJ 3
SUBJ 4
SUBJ 5
SUBJ 6
```
                                        Preference ratings of 10 firms by 5
                                          respondents.

                                        Labels.

Developing Perceptual Maps with Cross-tabulated Data by Correspondence Analysis: CORRESP

```
8 10                                    Programs control cards.
(10F3.0)
   4    3    1   13    9    6    3   18    2   10
  15   16   15   11   11   14   16   12   14   14
  15   14    6    4    4   15   14   13    7   13
  16   13    8   13    9   17   15   16    6   12
  14   14   10   11   11   14   12   13   10   14
   7   18   13    4    9   16   14    5    4   16
   6    6   14   10   11    8    7    4   14    4
  15   18    9    2    3   15   16    7    8    8
Product Quality
Strategic Orientation
Service
Delivery Speed
Price Level
Salesforce Image
Price Flexibility
Mfgr. Image
HATCO
Firm A
Firm B
Firm C
Firm D
Firm E
Firm F
Firm G
Firm H
Firm I
```
                                        Cross-tabulated data of attributes by
                                          objects (firms).

                                        Labels.

HATCO DATASETS

HATCO Database (X1 to X14)

	X1	X2	X3	X4	X5	X6	X7	X8	X9	X10	X11	X12	X13	X14
1	4.1	.6	6.9	4.7	2.4	2.3	5.2	0	32.0	4.2	1	0	1	1
2	1.8	3.0	6.3	6.6	2.5	4.0	8.4	1	43.0	4.3	0	1	0	1
3	3.4	5.2	5.7	6.0	4.3	2.7	8.2	1	48.0	5.2	0	1	1	2
4	2.7	1.0	7.1	5.9	1.8	2.3	7.8	1	32.0	3.9	0	1	1	1
5	6.0	.9	9.6	7.8	3.4	4.6	4.5	0	58.0	6.8	1	0	1	3
6	1.9	3.3	7.9	4.8	2.6	1.9	9.7	1	45.0	4.4	0	1	1	2
7	4.6	2.4	9.5	6.6	3.5	4.5	7.6	0	46.0	5.8	1	0	1	1
8	1.3	4.2	6.2	5.1	2.8	2.2	6.9	1	44.0	4.3	0	1	0	2
9	5.5	1.6	9.4	4.7	3.5	3.0	7.6	0	63.0	5.4	1	0	1	3
10	4.0	3.5	6.5	6.0	3.7	3.2	8.7	1	54.0	5.4	0	1	0	2
11	2.4	1.6	8.8	4.8	2.0	2.8	5.8	0	32.0	4.3	1	0	0	1
12	3.9	2.2	9.1	4.6	3.0	2.5	8.3	0	47.0	5.0	1	0	1	2
13	2.8	1.4	8.1	3.8	2.1	1.4	6.6	1	39.0	4.4	0	1	0	1
14	3.7	1.5	8.6	5.7	2.7	3.7	6.7	0	38.0	5.0	1	0	1	1
15	4.7	1.3	9.9	6.7	3.0	2.6	6.8	0	54.0	5.9	1	0	0	3
16	3.4	2.0	9.7	4.7	2.7	1.7	4.8	0	49.0	4.7	1	0	0	3
17	3.2	4.1	5.7	5.1	3.6	2.9	6.2	0	38.0	4.4	1	1	1	2
18	4.9	1.8	7.7	4.3	3.4	1.5	5.9	0	40.0	5.6	1	0	0	2
19	5.3	1.4	9.7	6.1	3.3	3.9	6.8	0	54.0	5.9	1	0	1	3
20	4.7	1.3	9.9	6.7	3.0	2.6	6.8	0	55.0	6.0	1	0	0	3
21	3.3	.9	8.6	4.0	2.1	1.8	6.3	0	41.0	4.5	1	0	0	2
22	3.4	.4	8.3	2.5	1.2	1.7	5.2	0	35.0	3.3	1	0	0	1
23	3.0	4.0	9.1	7.1	3.5	3.4	8.4	0	55.0	5.2	1	1	0	3
24	2.4	1.5	6.7	4.8	1.9	2.5	7.2	1	36.0	3.7	0	1	0	1
25	5.1	1.4	8.7	4.8	3.3	2.6	3.8	0	49.0	4.9	1	0	0	2
26	4.6	2.1	7.9	5.8	3.4	2.8	4.7	0	49.0	5.9	1	0	1	3
27	2.4	1.5	6.6	4.8	1.9	2.5	7.2	1	36.0	3.7	0	1	0	1
28	5.2	1.3	9.7	6.1	3.2	3.9	6.7	0	54.0	5.8	1	0	1	3
29	3.5	2.8	9.9	3.5	3.1	1.7	5.4	0	49.0	5.4	1	0	1	3
30	4.1	3.7	5.9	5.5	3.9	3.0	8.4	1	46.0	5.1	0	1	0	2
31	3.0	3.2	6.0	5.3	3.1	3.0	8.0	1	43.0	3.3	0	1	0	1
32	2.8	3.8	8.9	6.9	3.3	3.2	8.2	0	53.0	5.0	1	1	0	3
33	5.2	2.0	9.3	5.9	3.7	2.4	4.6	0	60.0	6.1	1	0	0	3
34	3.4	3.7	6.4	5.7	3.5	3.4	8.4	1	47.3	3.8	0	1	0	1
35	2.4	1.0	7.7	3.4	1.7	1.1	6.2	1	35.0	4.1	0	1	0	1
36	1.8	3.3	7.5	4.5	2.5	2.4	7.6	1	39.0	3.6	0	1	1	1
37	3.6	4.0	5.8	5.8	3.7	2.5	9.3	1	44.0	4.8	0	1	1	2
38	4.0	.9	9.1	5.4	2.4	2.6	7.3	0	46.0	5.1	1	0	1	3
39	.0	2.1	6.9	5.4	1.1	2.6	8.9	1	29.0	3.9	0	1	1	1
40	2.4	2.0	6.4	4.5	2.1	2.2	8.8	1	28.0	3.3	0	1	1	1
41	1.9	3.4	7.6	4.6	2.6	2.5	7.7	1	40.0	3.7	0	1	1	1
42	5.9	.9	9.6	7.8	3.4	4.6	4.5	0	58.0	6.7	1	0	1	3
43	4.9	2.3	9.3	4.5	3.6	1.3	6.2	0	53.0	5.9	1	0	0	3
44	5.0	1.3	8.6	4.7	3.1	2.5	3.7	0	48.0	4.8	1	0	0	2
45	2.0	2.6	6.5	3.7	2.4	1.7	8.5	1	38.0	3.2	0	1	1	1
46	5.0	2.5	9.4	4.6	3.7	1.4	6.3	0	54.0	6.0	1	0	0	3
47	3.1	1.9	10.0	4.5	2.6	3.2	3.8	0	55.0	4.9	1	0	1	3
48	3.4	3.9	5.6	5.6	3.6	2.3	9.1	1	43.0	4.7	0	1	1	2
49	5.8	.2	8.8	4.5	3.0	2.4	6.7	0	57.0	4.9	1	0	1	3
50	5.4	2.1	8.0	3.0	3.8	1.4	5.2	0	53.0	3.8	1	0	1	3
51	3.7	.7	8.2	6.0	2.1	2.5	5.2	0	41.0	5.0	1	0	0	2
52	2.6	4.8	8.2	5.0	3.6	2.5	9.0	1	53.0	5.2	0	1	1	2
53	4.5	4.1	6.3	5.9	4.3	3.4	8.8	1	50.0	5.5	0	1	0	2
54	2.8	2.4	6.7	4.9	2.5	2.6	9.2	1	32.0	3.7	0	1	1	1
55	3.8	.8	8.7	2.9	1.6	2.1	5.6	0	39.0	3.7	1	0	0	1

56	2.9	2.6	7.7	7.0	2.8	3.6	7.7	0	47.0	4.2	1	1	1	2
57	4.9	4.4	7.4	6.9	4.6	4.0	9.6	1	62.0	6.2	0	1	0	2
58	5.4	2.5	9.6	5.5	4.0	3.0	7.7	0	65.0	6.0	1	0	0	3
59	4.3	1.8	7.6	5.4	3.1	2.5	4.4	0	46.0	5.6	1	0	1	3
60	2.3	4.5	8.0	4.7	3.3	2.2	8.7	1	50.0	5.0	0	1	1	2
61	3.1	1.9	9.9	4.5	2.6	3.1	3.8	0	54.0	4.8	1	0	1	3
62	5.1	1.9	9.2	5.8	3.6	2.3	4.5	0	60.0	6.1	1	0	0	3
63	4.1	1.1	9.3	5.5	2.5	2.7	7.4	0	47.0	5.3	1	0	1	3
64	3.0	3.8	5.5	4.9	3.4	2.6	6.0	0	36.0	4.2	1	1	1	2
65	1.1	2.0	7.2	4.7	1.6	3.2	10.0	1	40.0	3.4	0	1	1	1
66	3.7	1.4	9.0	4.5	2.6	2.3	6.8	0	45.0	4.9	1	0	0	2
67	4.2	2.5	9.2	6.2	3.3	3.9	7.3	0	59.0	6.0	1	0	0	3
68	1.6	4.5	6.4	5.3	3.0	2.5	7.1	1	46.0	4.5	0	1	0	2
69	5.3	1.7	8.5	3.7	3.5	1.9	4.8	0	58.0	4.3	1	0	0	3
70	2.3	3.7	8.3	5.2	3.0	2.3	9.1	1	49.0	4.8	0	1	1	2
71	3.6	5.4	5.9	6.2	4.5	2.9	8.4	1	50.0	5.4	0	1	1	2
72	5.6	2.2	8.2	3.1	4.0	1.6	5.3	0	55.0	3.9	1	0	1	3
73	3.6	2.2	9.9	4.8	2.9	1.9	4.9	0	51.0	4.9	1	0	0	3
74	5.2	1.3	9.1	4.5	3.3	2.7	7.3	0	60.0	5.1	1	0	1	3
75	3.0	2.0	6.6	6.6	2.4	2.7	8.2	1	41.0	4.1	0	1	0	1
76	4.2	2.4	9.4	4.9	3.2	2.7	8.5	0	49.0	5.2	1	0	1	2
77	3.8	.8	8.3	6.1	2.2	2.6	5.3	0	42.0	5.1	1	0	0	2
78	3.3	2.6	9.7	3.3	2.9	1.5	5.2	0	47.0	5.1	1	0	1	3
79	1.0	1.9	7.1	4.5	1.5	3.1	9.9	1	39.0	3.3	0	1	1	1
80	4.5	1.6	8.7	4.6	3.1	2.1	6.8	0	56.0	5.1	1	0	0	3
81	5.5	1.8	8.7	3.8	3.6	2.1	4.9	0	59.0	4.5	1	0	0	3
82	3.4	4.6	5.5	8.2	4.0	4.4	6.3	0	47.3	5.6	1	1	1	2
83	1.6	2.8	6.1	6.4	2.3	3.8	8.2	1	41.0	4.1	0	1	0	1
84	2.3	3.7	7.6	5.0	3.0	2.5	7.4	0	37.0	4.4	1	1	0	1
85	2.6	3.0	8.5	6.0	2.8	2.8	6.8	1	53.0	5.6	0	1	0	2
86	2.5	3.1	7.0	4.2	2.8	2.2	9.0	1	43.0	3.7	0	1	1	1
87	2.4	2.9	8.4	5.9	2.7	2.7	6.7	1	51.0	5.5	0	1	0	2
88	2.1	3.5	7.4	4.8	2.8	2.3	7.2	0	36.0	4.3	1	1	0	1
89	2.9	1.2	7.3	6.1	2.0	2.5	8.0	1	34.0	4.0	0	1	1	1
90	4.3	2.5	9.3	6.3	3.4	4.0	7.4	0	60.0	6.1	1	0	0	3
91	3.0	2.8	7.8	7.1	3.0	3.8	7.9	0	49.0	4.4	1	1	1	2
92	4.8	1.7	7.6	4.2	3.3	1.4	5.8	0	39.0	5.5	1	0	0	2
93	3.1	4.2	5.1	7.8	3.6	4.0	5.9	0	43.0	5.2	1	1	1	2
94	1.9	2.7	5.0	4.9	2.2	2.5	8.2	1	36.0	3.6	0	1	0	1
95	4.0	.5	6.7	4.5	2.2	2.1	5.0	0	31.0	4.0	1	0	1	1
96	.6	1.6	6.4	5.0	.7	2.1	8.4	1	25.0	3.4	0	1	1	1
97	6.1	.5	9.2	4.8	3.3	2.8	7.1	0	60.0	5.2	1	0	1	3
98	2.0	2.8	5.2	5.0	2.4	2.7	8.4	1	38.0	3.7	0	1	0	1
99	3.1	2.2	6.7	6.8	2.6	2.9	8.4	1	42.0	4.3	0	1	0	1
100	2.5	1.8	9.0	5.0	2.2	3.0	6.0	0	33.0	4.4	1	0	0	1

Pretest of HATCO Database (X_1 to X_{14}) with Missing Data

(Denoted with a ., Read with same format at HATCO database)

201	3.3	0.9	8.6	4.0	2.1	1.8	6.3	0	41.0	4.5	1	0	0	2
202	.	0.4	.	2.5	1.2	1.7	5.2	0	35.0	3.3	1	0	0	1
203	3.0	.	9.1	7.1	3.5	3.4	.	0	55.0	5.2	1	1	0	3
204	.	1.5	.	4.8	1.9	2.5	7.2	1	36.0	.	0	1	0	1
205	5.1	1.4	.	4.8	3.3	2.6	3.8	0	49.0	4.9	1	0	0	2
206	4.6	2.1	7.9	5.8	3.4	2.8	4.7	0	49.0	5.9	1	0	1	3
207	.	1.5	.	4.8	1.9	2.5	7.2	1	36.0	.	0	1	0	1
208	5.2	1.3	9.7	6.1	3.2	3.9	6.7	0	54.3	5.8	1	0	1	3
209	3.5	2.8	9.9	3.5	3.1	1.7	5.4	0	49.0	5.4	1	0	1	3
210	4.1	3.7	5.9	.	.	.	.	.	.	.	0	1	0	2

211	3.0	2.8	7.8	7.1	3.0	3.8	7.9	0	49.0	4.4	1	1	1	2
212	4.8	1.7	7.6	4.2	3.3	1.4	5.8	0	39.0	5.5	1	0	0	2
213	3.1	.	.	7.8	3.6	4.0	5.9	0	43.0	5.2	1	1	1	2
214	.	2.7	5.0	.	2.2	.	.	1	.	3.6	.	1	.	1
215	4.0	0.5	6.7	4.5	2.2	2.1	5.0	.	31.0	4.0	1	0	1	1
216	.	1.6	6.4	5.0	.	2.1	8.4	1	25.0	3.4	0	1	1	1
217	6.1	0.5	9.2	4.8	3.3	2.8	7.1	0	60.0	5.2	1	0	1	3
218	.	2.8	5.2	5.0	.	2.7	8.4	1	38.0	3.7	0	1	0	1
219	3.1	2.2	6.7	6.8	2.6	2.9	.	1	.	4.3	0	1	0	1
220	6.5	.	9.0	7.0	3.2	3.7	8.0	0	33.0	5.4	1	0	0	1
221	.	1.6	.	4.8	2.0	2.8	.	0	32.0	4.3	1	0	0	1
222	3.9	2.2	.	4.6	.	2.5	8.3	0	47.0	5.0	1	0	1	2
223	2.8	1.4	8.1	3.8	2.1	1.4	6.6	1	39.0	4.4	0	1	0	1
224	.	.	8.6	5.7	2.7	3.7	6.7	0	.	5.0	1	0	1	1
225	4.7	1.3	.	.	3.0	2.6	6.8	0	54.0	5.9	1	0	0	1
226	3.4	2.0	9.7	4.7	2.7	1.7	4.8	0	49.0	4.7	1	0	0	3
227	3.2	.	5.7	5.1	3.6	2.9	6.2	0	.	4.4	1	1	1	2
228	.	1.8	7.7	.	3.4	1.5	5.9	0	40.0	5.6	1	0	0	2
229	5.3	1.4	9.7	6.1	.	3.9	6.8	0	54.0	5.9	1	0	1	3
230	4.7	1.3	9.9	6.7	3.0	2.6	6.8	0	55.0	6.0	1	0	0	3
231	3.7	0.7	8.2	6.0	2.1	2.5	.	0	41.0	5.0	1	0	0	2
232	.	.	8.2	5.0	3.6	2.5	9.0	1	53.0	5.2	0	1	1	2
233	4.5	.	.	5.9	.	.	8.8	1	50.0	.	0	.	0	.
234	2.8	2.4	6.7	4.9	2.5	2.6	9.2	1	32.0	3.7	0	1	1	1
235	3.8	0.8	8.7	2.9	1.6	.	5.6	0	39.0	.	1	0	0	1
236	2.9	2.6	7.7	7.0	2.8	3.6	7.7	0	47.0	4.2	1	1	1	2
237	4.9	.	7.4	6.9	4.6	4.0	9.6	1	62.0	6.2	0	1	0	2
238	.	2.5	9.6	5.5	4.0	3.0	7.7	0	65.0	6.0	1	0	0	3
239	4.3	1.8	7.6	5.4	3.1	2.5	4.4	0	46.0	5.6	1	0	1	3
240	.	1.5	9.9	2.7	1.3	1.2	1.7	1	50.0	5.0	0	1	1	2
241	3.1	1.9	.	4.5	.	3.1	3.8	0	54.0	4.8	1	0	1	3
242	5.1	1.9	9.2	5.8	3.6	2.3	4.5	0	60.0	6.1	1	0	0	3
243	4.1	1.1	9.3	5.5	2.5	2.7	7.4	0	47.0	5.3	1	0	1	3
244	3.0	3.8	5.5	4.9	3.4	2.6	6.0	0	.	4.2	1	1	1	2
245	.	2.0	.	4.7	.	3.2	.	1	.	3.4	0	.	1	.
246	3.7	1.4	9.0	.	2.6	2.3	6.8	0	45.0	4.9	1	0	0	2
247	4.2	2.5	9.2	6.2	3.3	3.9	7.3	0	59.0	6.0	1	0	0	3
248	.	.	6.4	5.3	3.0	2.5	7.1	1	46.0	4.5	0	1	0	2
249	5.3	.	8.5	3.7	3.5	1.9	4.8	0	58.0	4.3	1	0	0	3
250	.	3.7	.	5.2	3.0	2.3	9.1	1	49.0	4.8	0	1	1	2
251	3.0	3.2	6.0	5.3	3.1	3.0	8.0	1	43.0	3.3	0	1	0	1
252	2.8	3.8	8.9	6.9	3.3	3.2	8.2	0	53.0	5.0	1	1	0	3
253	.	2.0	9.3	5.9	3.7	2.4	4.6	0	60.0	6.1	1	0	0	3
254	3.4	3.7	6.4	5.7	3.5	3.4	8.4	1	47.0	3.8	0	1	0	1
255	.	1.0	.	3.4	1.7	1.1	6.2	1	35.0	4.1	0	1	0	1
256	.	3.3	7.5	4.5	2.5	2.4	7.6	1	39.0	3.6	0	1	1	1
257	3.6	.	.	5.8	3.7	2.5	9.3	1	44.0	4.8	0	1	1	2
258	4.0	0.9	9.1	5.4	2.4	2.6	7.3	0	46.0	5.1	1	0	1	3
259	.	2.1	6.9	5.4	1.1	2.6	8.9	1	29.0	3.9	0	1	1	1
260	.	2.0	6.4	4.5	2.1	2.2	8.8	1	28.0	3.3	0	1	1	1
261	3.6	.	.	6.2	4.5	.	.	1	.	.	.	1	1	2
262	5.6	2.20	8.2	3.1	4.0	1.6	5.3	0	55.0	3.9	1	0	1	3
263	3.6	.	9.9	.	.	.	.	.	.	4.9	1	0	0	3
264	5.2	1.3	9.1	4.5	3.3	2.7	7.3	0	60.0	5.1	1	0	1	3
265	3.0	2.0	6.6	6.6	2.4	2.7	8.2	1	41.0	4.1	0	1	0	1
266	4.2	2.4	9.4	4.9	3.2	2.7	8.5	0	49.0	5.2	1	0	1	2
267	3.8	0.8	.	.	2.2	2.6	5.3	0	42.0	5.1	1	0	0	2
268	3.3	2.6	9.70	3.30	2.9	1.5	5.2	0	47.0	.	1	0	1	3
269	.	1.9	.	4.5	1.5	3.1	9.9	1	39.0	3.3	0	1	1	1
270	4.5	1.6	8.7	4.6	3.1	2.1	6.8	0	56.0	5.1	1	0	0	3

Conjoint Data Set: Evaluations of 22 Stimuli by 100 Respondents

ID																						
104	4	6	5	4	4	6	4	4	4	4	4	4	4	5	5	4	4	4	4	4	6	6
107	6	3	5	2	3	1	1	6	6	6	7	4	1	6	6	6	6	1	7	7	1	7
109	7	7	6	7	3	3	3	6	3	5	2	3	2	2	6	3	3	3	6	3	3	3
110	5	5	7	7	1	5	1	1	4	5	5	5	1	5	5	2	6	1	3	5	1	5
120	7	7	1	6	1	1	1	1	1	5	5	1	1	1	7	1	1	1	5	6	1	1
123	7	7	5	5	1	1	1	2	2	1	7	1	1	2	7	7	7	1	7	7	1	7
129	7	7	7	7	7	7	7	7	6	7	6	7	6	7	7	6	6	7	7	6	7	6
133	5	7	2	7	2	2	2	3	3	2	5	2	1	6	5	2	2	2	2	6	2	7
135	7	7	2	5	1	1	1	2	2	2	7	1	1	7	4	5	2	1	2	5	2	5
144	6	7	3	6	6	2	3	3	5	5	7	7	2	2	7	3	6	5	5	5	5	3
150	7	7	5	5	3	6	3	4	7	7	5	2	5	7	6	7	4	3	6	7	3	6
155	7	7	3	6	2	1	1	5	1	5	1	3	2	2	7	1	1	1	5	1	1	1
156	7	6	5	3	5	5	2	6	5	2	7	2	5	6	7	5	5	6	6	7	2	6
161	6	6	5	5	1	1	1	3	2	5	2	1	1	5	7	3	1	1	5	2	1	2
162	7	6	3	6	3	5	3	6	3	2	6	6	2	5	7	4	2	2	4	7	6	5
167	4	5	3	3	2	3	3	2	6	5	6	3	3	3	4	2	3	5	3	6	5	5
168	7	7	3	2	2	2	1	1	2	3	7	1	1	7	7	7	3	2	3	7	2	7
170	6	6	4	2	1	3	2	1	2	5	7	3	3	1	7	2	6	2	4	7	2	5
173	6	6	4	4	3	3	3	5	2	4	6	3	3	5	5	5	4	3	4	6	5	6
174	6	5	6	6	6	1	3	6	1	6	1	5	1	6	7	1	1	5	6	1	3	1
178	7	7	2	5	1	1	1	1	1	5	1	1	1	1	6	1	1	1	5	3	2	1
180	7	7	2	6	6	1	2	2	2	5	2	7	1	7	7	2	1	5	7	2	7	2
181	6	6	3	5	4	2	2	3	3	5	6	4	2	3	6	4	6	2	6	5	4	6
187	6	7	6	7	2	2	2	5	5	6	6	3	1	4	7	5	2	3	6	7	3	6
190	6	6	5	5	1	1	1	3	5	5	7	1	1	3	6	3	5	1	5	5	1	6
192	7	7	5	6	2	1	2	2	2	2	2	2	1	4	5	1	1	2	5	3	3	3
193	7	7	6	6	2	1	6	6	1	5	4	5	2	3	7	1	2	2	7	2	7	2
194	2	6	6	2	6	2	2	4	2	4	2	4	3	4	6	1	4	6	4	2	6	2
195	5	5	5	5	5	4	5	5	4	5	4	5	4	5	5	4	4	4	4	4	5	4
197	6	7	5	6	3	3	4	6	4	4	6	3	7	4	4	4	5	3	3	7	6	6
198	7	7	5	7	4	6	5	7	5	4	2	7	2	4	7	1	1	2	5	2	6	2
200	5	4	2	2	3	5	5	4	5	3	4	2	2	2	3	1	2	3	5	4	3	1
209	6	6	1	2	2	2	1	2	5	2	5	1	1	2	6	1	1	1	2	7	1	4
211	5	7	5	7	2	2	1	5	5	2	5	1	2	2	7	5	2	2	1	7	7	7
222	7	7	3	3	2	2	5	5	3	6	5	2	2	6	7	3	4	2	5	5	2	5
225	7	7	4	6	2	2	2	6	2	4	3	3	1	4	5	2	2	3	5	3	4	2
229	7	7	7	7	1	4	5	3	6	2	6	3	1	2	2	7	1	4	1	1	5	1
231	5	5	2	2	1	1	1	5	1	2	6	2	1	3	5	2	2	1	6	4	1	3
233	6	5	2	1	7	5	6	2	1	7	5	3	3	3	7	6	2	1	4	4	1	4
234	5	5	4	5	4	4	5	5	4	4	4	4	4	5	4	4	5	6	4	5	4	4
235	5	7	6	6	2	3	6	3	3	6	7	6	3	7	7	6	6	3	5	5	3	5
236	4	6	2	4	1	1	1	2	2	2	4	1	1	5	7	2	1	1	2	2	1	2
240	2	2	5	7	2	6	2	2	4	2	2	2	2	2	7	6	1	5	4	6	6	4
246	7	3	6	4	2	6	5	6	5	5	5	4	3	3	2	2	3	3	1	1	4	1
249	4	4	6	7	2	1	2	5	5	4	3	2	1	4	6	4	2	2	6	4	2	3
250	7	4	3	3	4	5	3	3	3	4	7	3	5	5	4	5	4	4	3	4	5	7
251	4	4	1	4	1	1	1	2	2	2	2	1	1	4	3	2	1	1	1	3	5	2
254	5	7	2	1	1	1	1	3	1	3	3	1	1	2	4	1	1	1	3	1	1	1
258	7	1	1	3	1	2	1	1	3	1	7	1	3	1	7	1	1	3	4	4	1	4
260	5	5	5	5	5	2	4	5	2	5	2	6	2	5	5	1	2	2	5	2	2	2
261	6	6	4	6	2	2	2	4	6	6	6	2	2	4	6	4	4	2	4	4	2	4
266	6	7	7	5	5	6	4	2	3	3	3	2	2	1	4	3	4	2	1	2	1	1
271	7	5	4	3	3	2	3	4	5	6	6	3	4	4	2	2	2	1	2	1	1	1
272	7	7	6	5	6	6	3	3	4	5	5	6	2	2	3	4	1	4	1	3	3	1
277	6	6	5	5	4	3	4	3	3	3	5	4	1	2	3	1	2	2	2	1	1	1
285	5	4	2	3	5	5	4	3	4	2	4	4	3	3	1	1	5	2	2	1	2	1
287	3	5	2	2	1	1	1	1	1	1	3	1	1	2	2	3	1	1	1	5	1	2

ID																						
288	7	4	4	3	4	4	3	5	6	6	2	3	3	3	2	2	1	2	3	3	1	1
289	5	4	2	4	4	5	5	2	4	2	4	5	3	3	3	2	3	1	3	3	2	2
300	6	5	4	5	2	2	2	4	4	5	5	2	2	5	6	3	5	2	5	5	2	5
302	6	6	6	7	5	3	4	5	3	5	3	5	3	5	6	2	2	4	7	3	7	3
303	6	6	5	7	4	2	2	3	3	3	3	6	2	3	7	2	3	5	6	3	7	2
306	5	5	6	7	5	2	4	4	2	5	1	4	2	6	7	3	3	5	7	3	7	2
308	7	7	3	6	3	3	3	3	3	3	6	3	6	6	6	7	3	3	3	5	7	5
309	7	6	5	7	3	1	1	3	3	3	1	2	2	6	6	3	2	2	4	1	4	3
310	6	7	3	7	3	3	3	4	4	4	4	5	2	5	7	2	2	7	5	2	6	3
317	7	7	6	6	3	3	3	5	3	5	3	4	3	5	5	3	3	5	5	3	2	3
318	3	4	5	6	5	2	5	6	1	6	2	5	3	6	7	4	2	5	6	4	6	4
319	1	5	6	5	2	1	4	3	2	6	2	5	5	4	6	4	6	4	6	4	6	4
323	4	4	1	6	1	6	1	4	1	4	4	1	4	6	6	6	4	4	1	4	1	4
324	6	7	7	6	6	2	5	6	2	5	5	7	3	6	7	1	2	7	7	2	7	5
325	7	7	5	5	5	7	6	5	4	3	3	3	4	6	7	7	5	7	6	5	7	7
330	4	5	5	6	3	6	3	4	5	4	6	5	4	4	6	5	4	7	5	5	6	4
336	3	1	4	1	4	4	5	5	5	5	3	2	5	7	4	6	5	4	2	3	1	4
339	6	6	2	4	1	1	1	2	2	6	2	1	1	6	7	1	2	1	3	4	1	3
348	3	3	1	1	1	1	1	1	1	1	3	1	1	3	5	6	1	1	1	6	1	6
350	6	6	3	7	5	2	3	5	2	5	2	5	2	5	5	2	2	5	6	3	6	3
352	6	7	4	4	1	1	1	4	4	4	6	1	1	6	7	1	1	1	4	1	1	1
353	7	7	5	5	3	3	3	5	1	3	1	1	1	7	7	3	3	3	5	1	1	7
354	5	5	4	6	1	1	2	4	3	5	2	2	6	6	6	6	5	2	6	6	5	6
356	4	5	2	4	1	1	1	4	3	2	5	1	3	5	7	2	1	3	6	2	6	
363	7	6	4	6	1	1	1	5	3	6	7	2	6	5	7	6	5	2	6	5	5	6
366	5	5	6	3	3	3	3	4	4	3	4	2	2	5	6	4	2	1	6	4	3	3
368	7	7	6	6	3	1	2	4	2	5	3	2	1	3	6	3	2	1	3	2	2	2
370	6	3	5	5	3	2	4	6	5	5	2	4	4	6	7	5	4	4	6	3	2	3
372	7	6	7	5	5	3	5	2	2	5	4	3	3	6	7	2	3	4	7	2	3	2
381	1	3	3	2	2	1	7	2	2	4	4	3	3	2	2	1	2	2	3	3	4	2
382	3	3	3	2	2	1	2	4	7	5	2	2	1	2	3	3	3	4	4	6	1	1
385	4	7	7	5	7	5	7	5	6	4	7	1	4	3	4	4	4	3	1	2	1	1
394	6	6	4	5	1	2	3	7	6	7	1	1	1	6	2	3	6	4	5	4	5	5
396	7	7	5	5	1	1	1	5	5	5	7	1	1	7	7	7	5	1	5	7	1	7
399	4	5	2	1	1	1	1	1	2	2	4	1	1	4	5	2	1	1	1	4	1	2
401	6	6	3	3	1	1	1	3	1	3	5	1	1	5	7	6	1	1	4	5	1	5
403	5	5	6	5	1	1	1	3	2	5	2	2	1	0	5	4	2	1	5	4	5	2
412	7	7	6	6	1	1	1	5	1	6	1	1	1	7	6	1	1	1	5	1	1	1
413	1	1	3	1	1	1	1	1	1	1	1	1	1	1	1	1	1	1	3	1	1	1
414	6	6	5	7	3	4	3	5	2	5	2	2	3	4	6	2	1	2	5	4	3	2
416	5	7	2	6	2	2	1	4	3	1	6	3	2	1	7	2	1	1	4	5	2	2
417	5	5	2	5	1	1	2	2	2	2	1	1	1	4	6	3	2	1	4	4	2	3
418	6	7	2	7	1	4	1	3	6	4	7	1	5	2	1	2	4	2	3	7	5	7

Preference Data: Similarity Ratings for 18 Respondents

Respondents 1-5	Respondents 6-10	Respondents 11-15	Respondents 16-18
4	8	3	7
6 9	7 4	3 6	7 5
6 6 1	1 1 2	1 1 1	1 1 2
6 6 4 8	1 1 2 8	1 1 2 4	1 1 2 5
3 2 2 3 3	2 1 1 1 1	3 1 1 2 1	5 1 1 2 1
3 5 2 2 2 4	1 2 2 1 1 1	1 1 4 2 1 2	1 2 5 1 1 2
2 2 4 4 4 2 4	1 1 2 1 1 1 1	1 1 1 1 1 2 1	1 1 2 2 2 1 1
4 2 9 2 8 8 2 2	1 6 8 1 8 6 1 1	1 1 7 2 7 6 3 2	1 7 2 1 7 7 1 1
8 9 4 2 2 5 9 3 2	7 9 1 1 1 1 6 1 1	5 8 2 2 2 2 7 2 1	7 8 1 1 1 2 7 1 1
8	6	8	4

Respondents 1-5	Respondents 6-10	Respondents 11-15	Respondents 16-18
7 6	4 3	8 6	4 5
1 1 3	1 1 3	6 5 8	2 2 2
1 1 2 8	1 1 3 6	6 6 6 8	2 2 4 8
3 1 1 3 1	4 1 1 4 1	6 6 6 8 6	4 2 2 4 2
1 2 3 1 1 3	2 4 4 1 1 4	6 5 6 5 5 8	2 3 4 2 2 2
1 1 2 1 1 2 1	1 1 4 2 1 3 1	6 6 6 7 6 6 6	2 2 4 2 2 4 2
1 4 1 1 7 1 1 1	2 3 7 2 7 4 1 2	6 5 8 6 8 7 4 5	2 4 6 4 5 4 3 3
1 6 2 1 1 2 1 1 1	8 6 1 2 2 2 6 2 2	7 7 5 6 6 6 8 5 5	5 5 3 3 3 5 6 3 3
9	8	9	6
8 7	7 2	8 6	7 9
5 5 7	1 1 2	1 1 5	1 1 8
6 6 7 8	1 1 2 7	3 3 6 8	1 1 6 7
7 5 5 7 6	2 4 1 1 1	6 2 2 5 3	6 1 1 8 1
6 7 7 6 6 7	1 1 1 1 1 2	3 6 6 1 1 5	1 6 6 1 1 8
6 6 8 6 6 7 5	1 1 2 1 1 2 1	3 3 6 4 4 7 3	1 1 7 1 1 7 1
5 8 9 7 8 8 7 6	1 6 6 1 2 2 1 1	3 8 8 2 6 5 5 6	1 7 9 1 9 9 2 1
8 8 7 7 7 8 9 7 7	2 2 1 1 1 4 6 1 1	6 6 3 4 4 2 8 2 2	9 9 1 1 1 6 9 1 1
7	3	7	
6 4	3 4	4 4	
3 3 5	2 2 2	2 1 1	
2 2 3 8	2 2 2 6	1 1 3 5	
3 3 3 4 3	2 2 2 4 2	6 1 1 1 1	
3 5 4 3 3 4	2 3 3 3 3 3	1 2 5 1 1 1	
3 3 5 3 4 5 3	3 3 3 3 3 2 2	1 1 2 1 1 1 1	
3 4 8 4 8 4 3 3	2 5 7 2 8 7 4 2	1 3 3 1 7 4 1 1	
7 6 3 3 3 4 8 2 3	7 8 5 2 2 4 8 2 2	4 7 1 1 1 1 5 1 1	
5	9	8	
2 7	9 3	7 7	
2 2 1	1 7 2	5 6 7	
3 2 8 5	2 2 7 9	6 7 5 7	
3 3 2 2 2	2 2 2 7 2	6 5 5 7 6	
2 2 2 2 2 4	2 2 2 2 2 7	7 5 6 5 5 6	
2 2 2 2 2 1 2	2 2 2 2 2 7 3	5 6 5 5 5 6 5	
2 7 7 1 6 5 2 1	3 2 8 2 8 7 2 2	5 7 6 5 8 7 6 6	
5 5 2 1 1 1 6 2 2	8 8 2 2 2 2 8 2 2	7 8 6 5 5 6 8 6 6	

Index